MRE
Materials Research and Engineering
Edited by B. Ilschner and N. J. Grant

Lyle H. Schwartz Jerome B. Cohen

Diffraction from Materials

Second Edition

With 381 Figures and 30 Tables

Springer-Verlag
Berlin Heidelberg New York
London Paris Tokyo

Prof. LYLE H. SCHWARTZ

Director, Institute for Materials Science and Engineering,
U.S. Dept. of Commerce, National Bureau of Standards,
Gaithersburg, MD 20899/USA

Prof. JEROME B. COHEN

Dept. of Materials Science and Engineering,
Dean, The Technological Institute, Northwestern University,
Evanston, IL 60201/USA

Series Editors

Prof. BERNHARD ILSCHNER

Laboratoire de Métallurgie Mécanique,
Département des Matériaux, Ecole Polytechnique Fédérale,
CH-1007 Lausanne/Switzerland

Prof. NICHOLAS J. GRANT

Dept. of Materials Science and Engineering, MIT,
Cambridge, MA 02139/USA

Copyright for the 1st edition 1977: Academic Press, New York

ISBN 3-540-17114-2 Springer-Verlag Berlin Heidelberg NewYork
ISBN 0-387-17114-2 Springer-Verlag NewYork Heidelberg Berlin

Library of Congress Cataloging-in-Publication Data
Schwartz, Lyle H.
Diffraction from materials. (Materials research and engineering)
Includes bibliographies and index.
1. Diffraction. 2. Crystallography. 3. Materials–Optical properties.
I. Cohen, J. B. (Jerome Bernard), 1932–   . II. Title. III. Series.
QC415.S38   1987   620.1′1295   87-9799
ISBN 0-387-17114-2 (U.S.)

This work is subject to copyright. All rights are reserved, whether the whole or part of the material is concerned, specifically the rights of translation, reprinting, reuse of illustrations, recitation, broadcasting, reproduction on microfilms or in other ways, and storage in data banks. Duplication of this publication or parts thereof is only permitted under the provisions of the German Copyright Law of September 9, 1965, in its version of June 24, 1985, and a copyright fee must always be paid. Violations fall under the prosecution act of the German Copyright Law.

© Springer-Verlag Berlin, Heidelberg 1987
Printed in Germany

The use of registered names, trademarks, etc. in this publication does not imply, even in the absence of a specific statement, that such names are exempt from the relevant protective laws and regulations and therefore free for general use.

Typesetting: ASCO Trade Typesetting Ltd., Hong Kong
Offsetprinting: Ruksaldruck, Berlin. Bookbinding: Lüderitz & Bauer, Berlin
2161/3020-543210

Dedication

To us – our friendship for almost three decades, which has withstood the evolution from a student-teacher relationship, to that of two colleagues, to the preparation of two books, to a change of what interested us in our careers, and countless little things. "That's what life's about…". And to our present and former students who helped us learn (what always seems too little). For their sake, we hope we've got it right – they certainly gave their best to help us to do just that.

L. H. Schwartz · J. B. Cohen

Preface

The atomic arrangements in condensed matter play an ever increasing role in many areas of science and technology – Materials Science and Engineering, Chemistry, Physics, Geology, Biology and Electrical, Civil, Mechanical and Chemical Engineering. Exciting discoveries in these fields in this century often stemmed from studies of these arrangements using diffraction: the structure and functions of DNA and other biological molecules, the configuration of polymer chains, the crystalline nature of metals and their imperfections, semiconductors and insulators, and the links between their structures, their defects and material properties, and the interaction between materials and the environment.

The broad, interdisciplinary character of diffraction studies makes them particularly exciting. With new tools such as the high-resolution electron microscope, new detectors, new techniques (such as EXAFS and glancing angle diffraction) and the new sources, the horizons of this field greatly expanded in the 1950's and 60's. Pulsed neutron sources and high intensity storage rings that came on the scene in the late 70's have opened up possibilities for new study to such vast horizons that it is hard to sit here writing this – there's so much to be done!

Within the walls bounding each field of science or engineering, diffraction and structure is only one specialty. It is too easy for this topic to be developed in such a narrow way that sight is lost of the basic principles and broad possibilities. Chemists are trained only about structure determination (if at all) and materials scientists know how to take pictures with an electron microscope, but both often lack the basic knowledge about diffraction essential to use or advance the theory pertinent to their needs. And most importantly, everyone in materials related fields needs to know enough to understand the results of practitioners in this art.

Our intent is that this book help bridge this gap and the barriers between fields in this area. We wish it to be used to place diffraction methods in their proper perspective right from the first exposure to this field. It is intended for use in senior and graduate courses. The first four chapters include all the basics for a first course (for x-rays, electrons and neutrons). The last four chapters emphasize the three major topics of diffraction: structure determination, defects, and dynamical scattering. We introduce all these areas and the major mathematical topics associated with them.

This book evolved over twenty-five years of teaching, at Northwestern University, three (quarter-long) courses in this field, one at the senior level, two for graduate students. We cannot emphasize enough the importance of associated laboratory sessions. In our first course, after certain "set" sessions on crystallography, Laue and powder patterns, small groups of students do a short project on their own, such as measuring a stress or a lattice parameter, or the volume fraction of a phase. In the text, original films are used in some of the problems to help develop the connection to the experimental aspects of this science. In the graduate courses, we have offered open-ended mini-research projects, perhaps an application to a student's own thesis.

This text has evolved over the years, from a book by one of us (JBC) entitled *Diffraction Methods in Materials Science*, MacMillan, New York, 1966 to a joint effort, *Diffraction from Materials*, Academic Press, New York, 1977.

Have fun!

L. H. Schwartz · J. B. Cohen

Acknowledgments

Mrs. Caroline Harden and Ms. CaRole Lamb did much of the typing, while Mrs. Sally Dumas added considerably with her artwork. Drs. B. Batterman and H. Cole helped us with Chap. 8, which is an extensive expansion of a review they wrote on dynamical theory [Rev. Mod. Phys. 36, 681 (1964)]. Prof. J. Ibers reviewed Chap. 6, and Prof. C. N. J. Wagner offered his class notes on liquids. Prof. P. Georgopoulos drafted an initial version of the section on EXAFS.

The many students who took our courses helped us to see new ways of presenting this topic or that – ways from which they (and we) could benefit. Many of these were graduate students working with us. We have been fortunate that these have always been truly fine people and they prevailed on us, each in his own way, to do a little more to extend our thoughts on a topic or problem. They are really the teachers – and we are not always quick learners!

L. H. Schwartz · J. B. Cohen

Contents

1. Geometry of Crystal Structures 1

   1.1 Introduction 1
   1.2 Optical Crystallography 1
       1.2.1 The Law of Constancy of Angle 1
       1.2.2 Internal vs. External Structure 3
   1.3 Crystal Symmetry 4
       1.3.1 Symmetry Elements 4
       1.3.2 Form and Habit 6
   1.4 Lattices 7
       1.4.1 Periodic Arrays 7
       1.4.2 Bravais Lattices 9
   1.5 Naming Planes, Points, and Directions 10
   1.6 Atomic Size and Coordination 15
   1.7 Analytical Calculations in Crystallography 20
       1.7.1 Introduction to the Reciprocal Lattice 20
       1.7.2 Application of Reciprocal Lattice Vectors 23
   1.8 Graphical Methods in Crystallography 26
       1.8.1 Introduction to the Stereographic Projection 26
       1.8.2 The Standard Projection of a Cubic Crystal 28
       1.8.3 Using the Stereographic Projection – The Wulff Net . . . 29
   1.9 Practical Applications of Graphical Methods 34
References 39
Problems 39

2. The Nature of Diffraction 46

   2.1 Diffraction from a Grating 46
   2.2 Diffraction from Planes of Atoms 50
       2.2.1 Bragg's Law 50
       2.2.2 Interplanar Interference 51
   2.3 Diffraction from a One-Dimensional Crystal 54
       2.3.1 Diffraction from a Row of Like Atoms 54
       2.3.2 Diffraction Viewed Graphically – The Argard Diagram . . 56
       2.3.3 Conditions for Diffraction – The Ewald Sphere 60
       2.3.4 Diffraction from a Row of Unlike Atoms 61
   2.4 Diffraction from a Three-Dimensional Crystal 63
       2.4.1 Derivation of the Diffraction Conditions 63

		2.4.2	The Structure Factor and the Laue Conditions	66
		2.4.3	Calculation of the Structure Factor	71
	2.5	Summary		73

References . . . 74
Problems . . . 74

3. Properties of Radiation Useful for Studying the Structure of Materials . 77

3.1 Production of Radiation – X-rays . . . 77
    3.1.1 The Conventional X-ray Tube . . . 77
    3.1.2 Characteristic X-radiation . . . 82
    3.1.3 Non-conventional Sources – Synchrotron Radiation . . . 91
3.2 Production of Radiation – Neutrons . . . 97
3.3 Production of Radiation – Electrons . . . 101
3.4 The Interaction of X-rays with Matter . . . 104
    3.4.1 Polarization of X-rays . . . 104
    3.4.2 X-ray Scattering Factor . . . 107
    3.4.3 Resonant Absorption and Scattering of X-rays – Index of Refraction . . . 109
    3.4.4 Effects of Thermal and Static Atom Displacements . . . 116
    3.4.5 Compton Scattering of X-rays . . . 118
3.5 The Interaction of Electrons with Matter . . . 121
3.6 The Interaction of Neutrons with Matter . . . 122
    3.6.1 Nuclear Interactions . . . 122
    3.6.2 Magnetic Interactions . . . 126
3.7 The Absorption of X-rays in Matter . . . 131
    3.7.1 The Mass Absorption Coefficient . . . 131
    3.7.2 The Kramers-Kronig Dispersion Relation . . . 137
3.8 The Absorption of Neutrons in Matter . . . 138
3.9 Refraction of Radiation by Matter . . . 140
3.10 Detection of Radiation . . . 142
    3.10.1 Film Techniques . . . 142
    3.10.2 Counters and Associated Electronic Components . . . 143

References . . . 153
Problems . . . 153

4. Recording the Diffraction Pattern . . . 159

4.1 Introduction . . . 159
4.2 Using a Range of Wavelengths – Laue Patterns . . . 160
    4.2.1 Recording the Laue Pattern . . . 160
    4.2.2 Applications of Laue Patterns . . . 166
4.3 The Rotating Crystal Method . . . 169
4.4 The Weissenberg Method . . . 175
4.5 The Precession Method . . . 178
4.6 The Counter – Diffractometer . . . 182
    4.6.1 Orienting a Crystal for One Diffraction Spot . . . 182

Contents   XI

        4.6.2 Orienting a Crystal for an Arbitrary Point in Reciprocal Space 185
  4.7 The Powder Method 188
        4.7.1 Recording the Diffraction Pattern 188
        4.7.2 Indexing the Powder Pattern 195
        4.7.3 Applications of Powder Patterns 198
        4.7.4 Powder Diffraction Using Energy Analysis 206
  4.8 Patterns with Very Short Wavelength Radiation 209
  4.9 Measurement of Diffraction Intensity 212
        4.9.1 Integrated Intensities from Crystals 212
        4.9.2 The Effect of Absorption 217
        4.9.3 Measurement Techniques Using a Diffractometer 219
        4.9.4 The Effect of Crystal Perfection on Integrated Intensities . 222
        4.9.5 Integrated Intensities from Polycrystalline Specimens . . . 224
        4.9.6 Absolute Intensities and Intensity in Nondispersive
             Experiments 227
References 228
Problems 229

5. Crystal Symmetry and the Diffraction Pattern 238

  5.1 Introduction 238
  5.2 One-Dimensional Symmetry 238
  5.3 Two-Dimensional Symmetry – Point Groups 241
  5.4 Two-Dimensional Symmetry – Lattices 245
  5.5 Two-Dimensional Symmetry – Space Groups 247
  5.6 Three-Dimensional Symmetry 252
        5.6.1 Space Lattices 252
        5.6.2 Point Group Symmetry 254
        5.6.3 Space Group Symmetry 257
  5.7 Space Groups and Diffraction 264
        5.7.1 Diffraction and Symmetry 264
        5.7.2 Structure Factor Calculations 266
        5.7.3 Some Simple Crystal Structures 268
References 274
Problems 275

6. Determination of Crystal Structures 281

  6.1 Introduction 281
  6.2 Applications of Fourier Analysis to Diffraction 281
        6.2.1 Introduction to Fourier Series 281
        6.2.2 The Fourier Integral Theorem 284
        6.2.3 Fourier Series and Transforms in Vector Space 288
        6.2.4 Fourier Transforms of Periodic Functions –
             The Diffraction Pattern 290
  6.3 Electron Density Mapping 293
        6.3.1 Three-Dimensional Density Maps 293

		6.3.2 Sections and Projections	295
		6.3.3 Optical Diffraction	300
	6.4	The Patterson Function	303
		6.4.1 Definition of the Patterson Function	303
		6.4.2 Definition of the Convolution Function	304
		6.4.3 Use of the Patterson Function	306
		6.4.4 Patterson-Harker Sections	309
	6.5	Heavy Atom Techniques	312
	6.6	Intensity Statistics and Inequalities	317
		6.6.1 Detection of a Center of Symmetry, Mirrors and Rotation Axes	317
		6.6.2 Direct Methods – Inequalities	321
		6.6.3 Phase Determination Using Direct Methods	324
	6.7	Difference Techniques	327
		6.7.1 Difference Fourier Synthesis	327
		6.7.2 Difference Patterson Synthesis	330
	6.8	Helical Structures	335
	6.9	Refinement by Least Squares	339
		6.9.1 Single Crystal Data	339
		6.9.2 Powder Data	341
	6.10	The Use of Electron Diffraction	343
	6.11	Aperiodic Crystals	347
		6.11.1 Introduction	347
		6.11.2 Diffraction from Quasicrystals	351
References			354
Problems			355

7. What Else Can We Learn from a Diffraction Experiment Besides the Average Structure? . . . 360

	7.1	Introduction	360
	7.2	Thermal-Diffuse Scattering (TDS)	361
		7.2.1 The Equations for TDS	361
		7.2.2 Measuring TDS	365
		7.2.3 The Uses of TDS	369
	7.3	Distortion and Mosaic Size	372
		7.3.1 Introduction	372
		7.3.2 Imaging in the Transmission Electron Microscope	373
		7.3.3 Effects on Diffraction Peaks	377
		7.3.4 A More Quantitative Evaluation of the Effect of Mosaic Size and Distortion	378
		7.3.5 The Meaning of the Fourier Coefficients	380
	7.4	Slit Corrections	382
		7.4.1 Theoretical Considerations	382
		7.4.2 Evaluating the Slit Correction	387
		7.4.3 Errors in Fourier Coefficients	390
		7.4.4 Some Uses of Particle Size and Strain	391

7.5	Stacking Disorder		391
	7.5.1	The Geometry of Stacking Faults	391
	7.5.2	Fourier Analysis of Faulting	395
	7.5.3	Peak Shifts Due to Faults	396
	7.5.4	Peak Broadening Due to Faults	400
7.6	Local Ordering and Clustering		402
	7.6.1	Effects of Concentration Fluctuations on the Diffraction Pattern	402
	7.6.2	Including the Effects of Atomic Displacements	407
	7.6.3	Simplifying the Basic Equation	409
	7.6.4	Data Analysis	412
	7.6.5	The Minimum Volume for a Measurement	417
	7.6.6	The Significance of the Results	417
	7.6.7	Experimental Details	419
	7.6.8	Checking the Results	420
	7.6.9	More Complex Systems	422
7.7	Small-Angle Scattering (SAS)		423
	7.7.1	The Shape Transform	423
	7.7.2	The Guinier Region	428
	7.7.3	Small-Angle Scattering from Polymers	429
	7.7.4	The Porod Region	433
	7.7.5	Size Distributions	434
	7.7.6	Slit Corrections	435
7.8	Liquids and Amorphous Solids		436
	7.8.1	Single Component Substances	436
	7.8.2	Multicomponent Substances	439
	7.8.3	The Keating Approach	442
	7.8.4	The Anomalous Dispersion Approach	443
	7.8.5	The Bhatia-Thornton Equations	443
7.9	Extended X-ray Absorption Fine Structure (EXAFS)		446
	7.9.1	Introduction	446
	7.9.2	The Physics of the EXAFS Oscillations	448
	7.9.3	Quantitative Analysis of EXAFS	450
	7.9.4	Experimental Procedures for EXAFS	452
	7.9.5	Data Analysis	453
	7.9.6	Statistical Error Limits in EXAFS Data	455
References			458
Problems			460

8. The Dynamical Theory of Diffraction 469

8.1	Introduction		469
8.2	The Dielectric Constant in a Crystal		473
	8.2.1	Away from any Resonance	473
	8.2.2	Near an Absorption Edge	474
8.3	Waves that Satisfy Bragg's Law		475
	8.3.1	Maxwell's Equations Inside a Crystal	475

	8.3.2 Applications of the Equations for the Fields Inside a Crystal	479
	8.3.3 The Role of the Actual Index of Refraction	481
	8.3.4 The Branches of the Dispersion Surface and Tie Points	483
8.4	Boundary Conditions	485
	8.4.1 Fields at the Entrance Source	485
	8.4.2 The Wave Vectors at a Surface	486
	8.4.3 Selecting Tie Points	487
	8.4.4 The Center of a Bragg Reflection	489
	8.4.5 The Exit Conditions	489
8.5	Field Amplitudes	490
	8.5.1 The Laue Case	492
8.6	Poynting's Vector and Energy Flow in a Crystal	495
	8.6.1 What is the Poynting's Vector?	495
	8.6.2 The Use of Poynting's Vector in Dynamical Diffraction	497
	8.6.3 The Case of Spherical Waves	500
8.7	Absorption in More Detail	501
	8.7.1 Along the Normal to the Surface	501
	8.7.2 Along Poynting's Vector	503
	8.7.3 The Origin of the Borrmann Effect	505
8.8	Physical Interpretation of Anomalous Absorption	508
	8.8.1 For Energy Flow Parallel to the Diffracting Planes	508
	8.8.2 When the Energy Flow is not Parallel to the Diffracting Planes	509
8.9	Exit Beams Again – The Symmetric Laue Case	511
8.10	Intensity in Bragg Reflections	517
	8.10.1 Neglecting Absorption	517
	8.10.2 Including Absorption	519
	8.10.3 Shift in Position of the Standing Wave	523
	8.10.4 Remarks Concerning Integrated Intensities	524
8.11	X-ray Topography	526
	8.11.1 The Various Techniques	526
	8.11.2 Contrast Mechanisms	532
References		534
Problems		535

Appendix A: Location of Useful Information in International Tables for Crystallography . . . 539

Appendix B: Crystallographic Classification of the 230 Space Groups . . . 540

Appendix C: Determination of the Power of the Direct Beam in X-ray Diffraction . . . 542

    Method 1: Aluminum Powder . . . 544
    Method 2: Polystyrene . . . 547
    Method 3: Multiple Foils . . . 550
    Method 4: The Ionization Chamber . . . 551
    References . . . 552

Appendix D: Accuracy in Digital Counting 553
        D.1 Some Additional Information on Counting Electronics . 553
        D.2 Measurement of Dead Time 554

Answers to Selected Problems 558

Subject Index 583

1. Geometry of Crystal Structures

1.1 Introduction

The reader, if he or she has had even the slightest exposure to science or engineering, will certainly recall many examples of the relation of atomic arrangements to properties. Ball models of the atoms in crystals have become part of almost every laboratory in the last sixty years. Most of our knowledge concerning atomic arrangements comes from diffraction — simply because any diffraction grating expresses itself in its diffraction pattern; this pattern can tell us much about the nature of the grating.

Thus we can learn about atomic arrangements (in liquids, gases, or solids) and defects in these arrangements, relative amounts of phases in multiphase structures, phase transformations, size and composition of tiny precipitates, and so on. This information is vital for our understanding of the behavior of materials. Furthermore, in many of these areas, diffraction is the *only* tool for obtaining this information without gross assumptions. For this reason, it is important that a materials scientist be on intimate terms with the basic concepts of diffraction, regardless of whether or not he intends to use this tool extensively. Only then will he be able to assess the literature and to make full use of available information, or to decide if diffraction would be worth considering for his current problem. The purpose of this book is to provide the basic tools for this understanding.

Before describing the interaction of radiation with matter to produce a diffraction pattern, it is useful to first discuss the geometric description of atoms and molecules in crystals. Diffraction is by no means useless for studying non-crystalline materials, but most of its applications are with crystalline solids. We restrict the coverage in this chapter to elementary concepts and results of this geometrical crystallography, returning once again to this subject for a more detailed analysis in Chap. 5.

1.2 Optical Crystallography

1.2.1 The Law of Constancy of Angle

The history of geometric crystallography (see Schneer 1983) is intimately tied to the development of science and traces back to the ancient Greeks and the efforts of the Phythagoreans in the seventh century B.C. to relate the geometric form of the Universe to that of perfect geometric figures and harmonies of numbers. However, it was not until Kepler's model of the Universe (1595) that the effort reached its set goal. Kepler was able to correlate the ratios of the Copernican distances of the planets from the sun with the ratios of the inscribed spheres to those circumscribing the regular solids (cube, tetrahedron, dodecahedron, icosahedron and octahedron). Although these ideas were soon to be supplanted by more sophisticated descriptions of planetary motion, fascination with the geometry of perfect solid shapes was stimulated by Kepler's success. By the end of the sixteenth century a detailed knowledge of the geometry of regular solids had become a recognized part of the education of a gentleman. The science of crystallography was to emerge from this environment stimulated by a desire to find the same Pythagorean relationships on earth as Kepler had found in the heavens.

The reader has undoubtedly examined natural minerals, looked at particles of salt through a microscope or grown "rock-candy" (sugar) from solution at some time. These materials all have in common that they are solids bounded by plane faces. While cut gem-stones also appear to fit this description, their faces are man-made and do not reveal information about the internal atomic arragements. Thus we shall narrow our attention for the remainder of this chapter and define a *crystal as a homogeneous solid bounded by naturally-formed plane faces*. The observation of such naturally occurring mineral crystals for their aesthetic value must trace to the earliest periods of man's history. Their study from a scientific point of view is much more recent.

In one of the earliest recorded examples of such a study, we find Kepler again, as in 1611 he attempted to relate the structure of snow flakes to the regular packing of water droplets. In 1665 Robert Hooke explained the external shape of crystals of quartz and salt (angles between growth faces independent of size and overall shape) by a supposition of an internal structure (the packing of minute spheres in specific arrays). Hooke and his predecessors were tied in their studies to the Pythagorean ideas of perfect polygons, and consequently maintained 90° angles, hexagons, pentagons and equilateral triangles as the bases upon which to form these internal arrays. It was left to Nicolas Steno (1669) to take the decisive step, abandoning the Pythagorean assumptions and establishing the basis for the Law of Constancy of Angle. By cutting sections from real, distorted crystals and tracing their outlines on paper, Steno showed that analogous angles in the different sections, whatever the actual size and shape of the sections, were always the same. Thus sections cut at right angles to the vertical edges ab (Fig. 1.1) always had angles

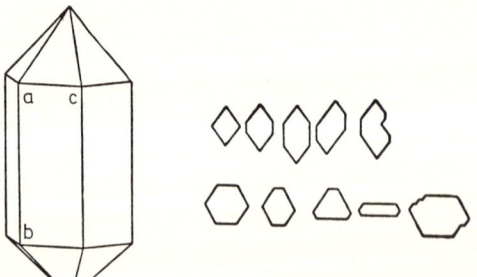

Fig. 1.1. Diagramatic reproduction of one of Steno's figures, showing sections of a quartz crystal. (After Philips, F.C., "An Introduction to Crystallography", 3rd ed. Longmans, Green, London, 1964)

1.2 Optical Crystallography

of 120°, though they were regular hexagons only in an undistorted crystal. Steno's observations on many types of naturally occurring mineral crystals were extended over the next hundred years by other workers and led to a basic conclusion: stated in modern language this Law of Constancy of Angle is *"In all crystals of the same substance, the angles between corresponding faces have a constant value"*.

1.2.2 Internal vs External Structure

In parallel with the development of the Law of Constancy of Angle during the eighteenth century several men were continuing Kepler's and Hooke's other theme of relating the observed external geometric forms of minerals to a limited number of simple forms. Most notable of these were de l'Isle and Hauy who may be credited with the molecular constructs which explained the Constancy of Angle. Rome de l'Isle formalized the Law of Constancy of Angle implicit in Steno's work. de l'Isle's assistant Carangeot devised the contact goniometer, a jackknife arrangement of two straight edges which made measurement of interfacial angles straightforward, eliminating the need for mechanical sectioning. Finally, it was de l'Isle who, more clearly than any of his predecessors saw that the Law defined the precise geometric form of each crystal and he who identified this form with the form of an integer building block — a tiny polyhedral block of the same shape as the crystal or as the polyhedral blocks into which a crystal characteristically cleaves when struck.

By the turn of the eighteenth century, Hauy was able to formulate the law relating the measured angles of crystal forms to an internal repetition of identical polyhedral blocks. A simple example of Hauy's construction is shown in Fig. 1.2. In Fig. 1.3, modern scanning electron microscopy offers us a revealing demonstration of Hauy's picture of internal structure by displaying the microstructure of a real opal crystal. However, it is important to emphasize that at the time of Hauy's work, chemistry was still in its infancy and the concepts of atoms and molecules were yet

Fig. 1.2. A crystal built up of very small cubic blocks can lead to faces that are flat to the eye or even to the optical microscope. [After Hauy, R., *Phil Mag.* 1, 35, 46, 287, 376 (1798)]

Fig. 1.3. Natural opal seen in high magnification on a scanning electron microscope. Photograph courtesy of P. Gilson

to be accepted. Nevertheless, Hauy had set the goal for crystallography — the definition of the molecules within the crystal and their structural relationship to these polyhedral building blocks.

1.3 Crystal Symmetry

1.3.1 Symmetry Elements

Lacking the tools now at our disposal for probing the internal atomic arrangements of naturally occurring mineral crystals, early students of these substances developed classification schemes based on the external appearance or form of these crystals. In this section we briefly introduce the vocabulary of this subject, while in Chap. 5 we will return to develop the classification scheme in full.

One of the first observations one may make upon examination of a crystal is to

1.3 Crystal Symmetry

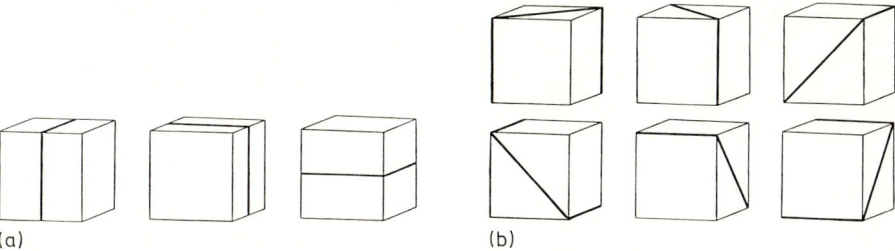

Fig. 1.4. a The three planes of symmetry parallel to cube faces. **b** The six diagonal planes of symmetry in a cube

note that the arrangement of faces is such that the edges formed by a number of them are parallel. Such a set of faces constitutes a *zone*, defined as a set of faces whose mutual intersections are parallel. The common direction of these parallel edges is known as the *zone-axis* of that particular zone.

Crystals often occur with similar faces in parallel pairs on opposite sides of the crystal. A crystal in which every face has a geometrically similar face parallel to it is said to possess a *center-of-symmetry* or *inversion center*. A cube, for example, has six sides, consisting of three pairs of parallel faces and exhibits a center-of-symmetry. By contrast, a regular tetrahedron has no such center since each face is opposite to a corner, rather than another face.

Additional symmetry may be present in crystals which can be divided by an imaginary plane in such a way that the half-crystal on one side of the plane is the mirror image of that on the other. Such a crystal exhibits a *plane-of-symmetry* or more simply a *mirror plane*. A cube exhibits not only a center of symmetry, but, as shown in Fig. 1.4, has three mirrors parallel to faces and six diagonal mirrors. The reader should note that a general rectangular parallelopiped has three mirrors parallel to faces as does the cube, but has no diagonal mirrors. A drawing of such a figure will reveal that a diagonal plane will divide the solid into geometrically similar wedges, but that these are not reflections of one another across the plane.

The third type of symmetry readily observable upon examination of a crystal is that many faces can be brought into coincidence with one another by rotation about a line termed an axis of symmetry. For example, if a cube is rotated about a line normal to one of its faces at its midpoint, it will turn into a congruent position every 90°, or four times during a complete revolution. Consequently, we say that this axis is one of fourfold rotational symmetry or a tetrad. A cube possesses three such axes, one normal to each of the three pairs of parallel faces. We define an *axis of symmetry* as a line such that after rotation about it through $360°/n$ the crystal assumes a congruent position; the value of the integer n determines the *degree* of the axis. In Table 1.1 we list the properties, names, and conventional geometric symbols for the five axes of rotational symmetry of importance in crystallography. In Chap. 5 we shall prove that these are indeed the only ones we must consider. The reader is left the task of demonstrating, using drawings or better by examination of a solid cube, that a cube possesses four triad axes through diagonal corners and six diad axes through diagonally opposite edges. These symmetry axes of a cube

Table 1.1. The crystallographic axes of rotational symmetry

n	Rotation angle	Name	Symbol
1	360°	Identity	
2	180°	Diad	▬
3	120°	Triad	▼
4	90°	Tetrad	■
6	60°	Hexad	⬢

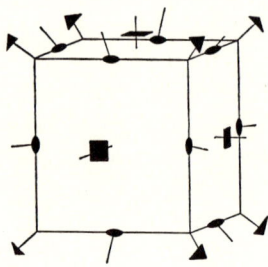

Fig. 1.5. The symmetry axes of a cube

Table 1.2. The six crystal systems

System	Axes
Triclinic	No axes
Monoclinic	1 diad
Orthorhombic	3 diads
Tetragonal	1 tetrad
Cubic (isometric)	4 triads
Hexagonal	1 hexad or 1 triad

are displayed in Fig. 1.5 and, with the inversion center and mirror planes define the full symmetry of a cube.

On the basis of the presence or absence of these symmetry elements, all crystals may be grouped into seven major divisions — the *Crystal Systems* — which we define in Table 1.2 in terms of the minimum axes of symmetry present.

1.3.2 Form and Habit

Crystals are rarely as simple in appearance as the perfect cube characteristic of table salt. In that instance all six faces of the cube are identical to one another. But consider the crystal depicted in Fig. 1.6a. We note that four triad axes are still present (through the centers of the triangular faces) so this must be classified as cubic among the seven crystal systems, yet it does not have the same shape as a perfect cube. This crystal, consisting of six octagonal faces and eight equilateral triangular faces is said to show faces of two different *forms*. If one octagonal face is present, the other five must also be there to maintain the cubic symmetry. Similarly if one corner of a cube

1.4 Lattices

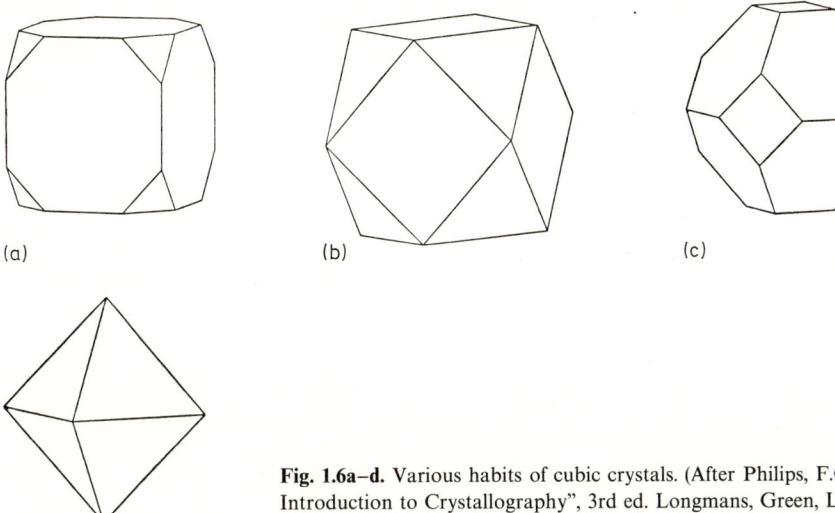

Fig. 1.6a–d. Various habits of cubic crystals. (After Philips, F.C., "An Introduction to Crystallography", 3rd ed. Longmans, Green, London, 1964)

is replaced by a triangular face all corners must be similarly replaced or the action of all of the triad axes would no longer produce congruence. We say that *form is the assemblage of all faces necessitated by the symmetry when one face is given.*

We may imagine Fig. 1.6a to have been produced by cutting away the corners of a cube; however, why stop with such a minimal material removal? If we continue to shave equal amounts off of all corners, we arrive eventually at the shape drawn in Fig. 1.6d after passing through various stages, two of which are depicted in Figs. 1.6b and c. Now the symmetry of these figures is identical, each possessing a center of inversion, nine mirror planes, three tetrads, four triads, and six diads. They differ only in the relative degree to which the faces of the cube or the corners appear. We call this relative development of faces the *habit of the crystal*: the general aspect conferred by the relative development of the different forms.

Natural crystals may exhibit different habits depending on the relative rates of crystal growth in different directions (see Problem 1.2), and are thus not readily recognized as representatives of the same crystal system (e.g. Figs. 1.6a and d), but measurement of interfacial angles would clearly reveal that fact. Thus we may now see the importance of the Law of Constancy of Angle in identifying the common symmetry exhibited by crystals of different habit.

1.4 Lattices

1.4.1 Periodic Arrays

We cannot go much further in the discussion of crystal symmetry without exploring the internal atomic arrangements. During the 19th century, before the discovery of x-radiation and its utility in probing the structure of matter, crystallographers were forced to speculate about all possible internal atomic arrangements. Guided by

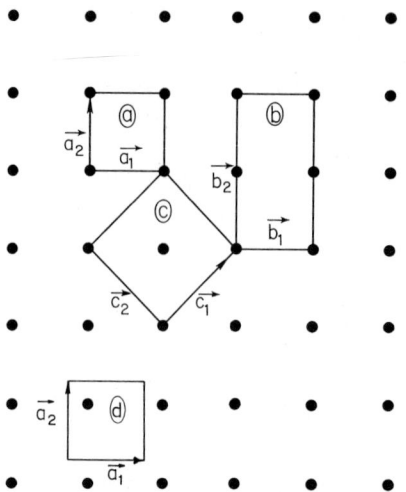

Fig. 1.7. A two-dimensional lattice with three possible unit cells. (a), (d) primitive; (b), (c) doubly primitive

ideas such as exemplified by the schematic of Haüy's polyhedral building blocks (Fig. 1.2), they developed the subject of geometric crystallography. One of the readily observable features of Fig. 1.2, and all similar building block models of crystals, is the three dimensional periodic repetition of the primary building block (polyhedron or *unit cell*). To complete our brief introduction of crystal symmetry, we will discuss the properties of such periodic repetition, once again deferring the full implications of our results to Chap. 5.

Let us begin with a two dimensional array of points as shown in Fig. 1.7. These points (imagined to extend indefinitely in all directions in the plane of the page) constitute a two-dimensional *net* or *lattice*. By placing an atom or molecule at every lattice point, we could build up a two-dimensional "crystal". The lattice then represents a convenient means for simplifying the representation of the crystal. A further simplification can be made by recognizing that the lattice itself can be constructed by following a simple recipe. Using the *unit translation vectors* $\mathbf{a}_1$ and $\mathbf{a}_2$ in Fig. 1.7(a), a *unit cell* is constructed. The lattice may then be reproduced by translation of the unit cell by vectors $\mathbf{L} = m\mathbf{a}_1 + n\mathbf{a}_2$ where m and n are positive or negative integers. Actually the unit cell need not be located as in Fig. 1.7(a) — the cell in Fig. 1.7(d) would do as well. These two cells have an important feature in common — they each contain a single lattice point (or cover an area equal to that per lattice point) and they are designated *primitive unit cells*. The cells in Figs. 1.7(b) and (c) are two of an infinite array of other cells which might be selected to represent the lattice. Since their areas are both equal to twice that of the primitive cell, they are designated "doubly primitive". For obvious reasons, cell (c) is also called "centered". Using these cells, the same lattice would be constructed by translations of the type $\mathbf{L}' = m\mathbf{b}_1 + n\mathbf{b}_2$ and $\mathbf{L}'' = m\mathbf{c}_1 + n\mathbf{c}_2$. The choice of cell is not entirely arbitrary, since the one usually chosen is selected to best represent the symmetry of the lattice — thus in this case the primitive cells (a) and (b) or the doubly primitive (c) would be likely choices since they exhibit the square symmetry of the lattice.

1.4.2 Bravais Lattices

When the concept of unit cells is extended to three dimensions, a third, out-of-plane translation vector is added to the basic vectorial relation defining periodicity, $\mathbf{L} = l\mathbf{a}_1 + m\mathbf{a}_2 + n\mathbf{a}_3$, where now l, m, and n are all positive or negative integers, and the vectors $\mathbf{a}_1$, $\mathbf{a}_2$, $\mathbf{a}_3$ define the unit cell. As we will show in Chap. 5, the requirement that three dimensional lattices are both periodic and consistent with the conditions imposed by axes of symmetry limits the number of three dimensional unit cells to fourteen. This was first correctly pointed out in 1848 by Auguste Bravais after whom the resultant space lattices or unit cells are known. The 14 Bravais lattices are depicted in Fig. 1.8 with spheres located at lattice points. Each cell

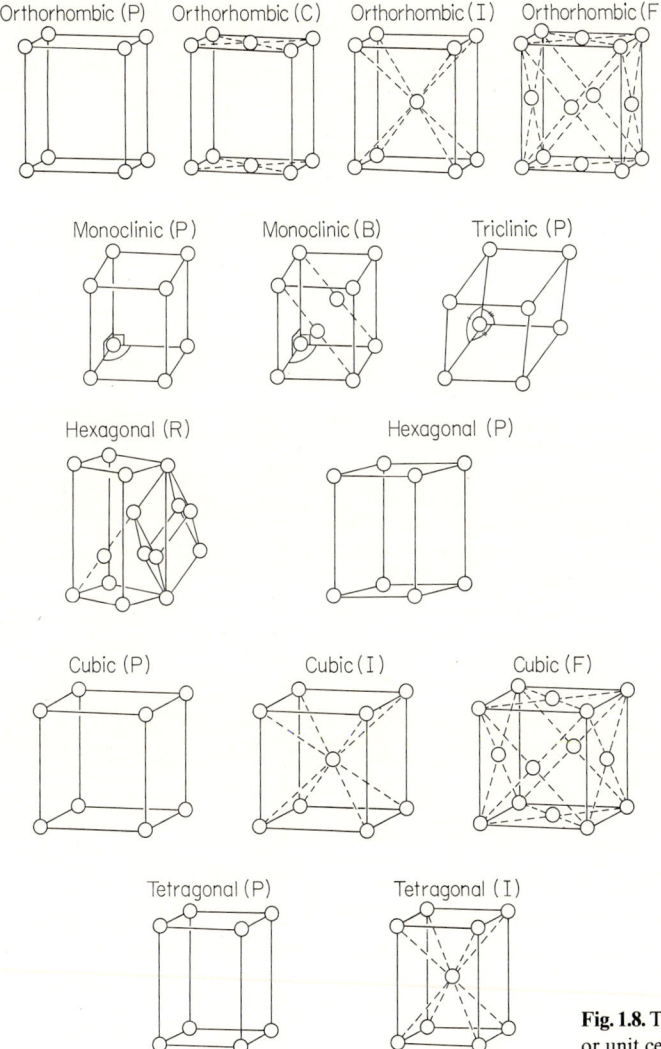

Fig. 1.8. The 14 Bravais space lattices or unit cells

Table 1.3. The three-dimensional systems

System	Axes	Minimum symmetry
Triclinic	$a_1 \neq a_2 \neq a_3$ $\alpha \neq \beta \neq \gamma$	None
Monoclinic	$a_1 \neq a_2 \neq a_3$ $90° = \alpha = \beta \neq \gamma$	One twofold axis (along $\mathbf{a}_3$)
Orthorhombic	$a_1 \neq a_2 \neq a_3$ $\alpha = \beta = \gamma = 90°$	Three twofold axes (along $\mathbf{a}_1, \mathbf{a}_2, \mathbf{a}_3$)
Tetragonal	$a_1 = a_2 \neq a_3$ $\alpha = \beta = \gamma = 90°$	One fourfold axis (along $\mathbf{a}_3$)
Hexagonal	$a_1 = a_2 \neq a_3$ $\alpha = \beta = 90°$ $\gamma = 120°$	One threefold axis (along $\mathbf{a}_3$) or One sixfold axis (along $\mathbf{a}_3$)
Cubic	$a_1 = a_2 = a_3$ $\alpha = \beta = \gamma = 90°$	Four threefold axes (along body diagonals of the cube of edge $\mathbf{a}_1$)

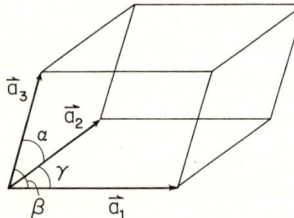

Fig. 1.9. A three-dimensional unit cell showing the definitions of the interaxial angles α, β, and γ. Note that α is opposite to $\mathbf{a}_1$, β is opposite to $\mathbf{a}_2$, and γ is opposite to $\mathbf{a}_3$

is identified with that crystal system with which its symmetry axes are consistent (see Table 1.2) and, in parentheses with a symbol indicating: primitive (*P*), base-centered (*B* or *C*), body-centered (*I*) or face-centered (*F*). In table 1.3 the Bravais unit cells are described and related to the minimum symmetry conditions defining the associated crystal systems. The notation adopted for this table is defined in Fig. 1.9.

With this brief introduction completed, we turn our attention in the remainder of this chapter to the development of convenient analytical and graphical tools necessary to describe crystal lattices and symmetry relationships.

1.5 Naming Planes, Points, and Directions

Planes on the surface of a crystal (as we saw in Fig. 1.2) arise from the presence of lattice points at the surface. For simplicity, let us first consider two dimensions, as in Fig. 1.10. The solid lines at angles represent the surface planes defined by a crystal in which matter is present in shaded unit cells, but not in the unshaded ones. Note that the dashed lines represent internal planes in the crystal parallel to the surface planes. Suppose we express the tangent of the angle in the figure in terms of the

1.5 Naming Planes, Points, and Directions

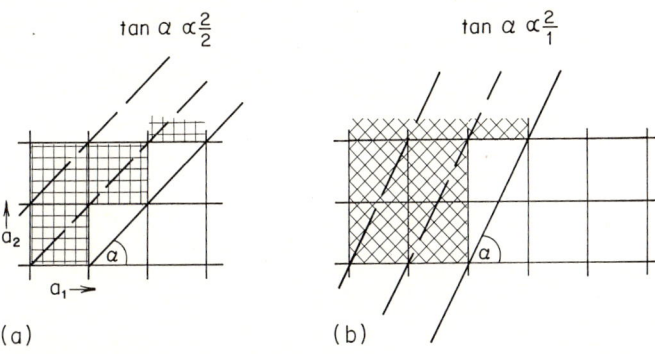

Fig. 1.10a, b. Planes in a crystal. (After Philips, F.C., "An Introduction to Crystallography," 3rd ed. Longmans, Green, London, 1964)

ratios of unit translations, (*not* in terms of the actual lengths of the unit translations which may be different along the two edges of the cell.) Then we can see that this tangent is the ratio of small whole numbers for planes passing through lattice points. These are known as *rational planes*. (There are of course certain "irrational" planes that never pass through such points, but they have little physical significance.) This idea that the tangents are simple ratios of integers for planes was recognized in the eighteenth century and led to a very simple shorthand for "naming" the planes, so that we do not need to have a lengthy discussion to indicate *what* plane we are talking about.

For a three-dimensional crystal, three noncoplanar axes are chosen, such as simple directions suggested by the habit of a macroscopic crystal. In Fig. 1.11a these three directions are labeled $\mathbf{a}_1$, $\mathbf{a}_2$, $\mathbf{a}_3$, and their lengths terminate at the dots. A *parametric plane* is chosen such as one of the exposed faces of the crystal or an imaginary plane inside the crystal. This plane cuts all three axes and fixes the positions of the dots, i.e., it defines the unit lengths along $\mathbf{a}_1$, $\mathbf{a}_2$, $\mathbf{a}_3$. The intercepts of any arbitrary plane on these three axes are then set equal to $|\mathbf{a}_1|/h'$, $|\mathbf{a}_2|/k'$, $|\mathbf{a}_3|/l'$. The set of values $h'k'l'$ define the plane. If there is a common factor n, we divide to form $h = h'/n$, $k = k'/n$, $l = l'/n$. The three numbers hkl are enclosed in parentheses (hkl) and are called the *Miller indices* of the plane after the man who popularized their use.

When intercepts are along negative axes, we put a bar over the corresponding index; thus $(h\bar{k}l)$, read "h, k-bar, l", implies intersection with the negative $\mathbf{a}_2$ axis. The parametric plane shown in Fig. 1.11a has intercepts $|\mathbf{a}_1|/1$, $|\mathbf{a}_2|/1$, $|\mathbf{a}_3|/1$, and indices $(hkl) = (111)$. Similarly in Fig. 1.11b, the intercepts are $|\mathbf{a}_1|/2$, $|\mathbf{a}_2|/3$, $|\mathbf{a}_3|/4$, $h' = 2$, $k' = 3$, $l' = 4$, so $(hkl) = (234)$. But for a plane with $h' = 2$, $k' = 4$, $l' = 4$, we divide by 2 to give $(hkl) = (122)$ (read as one-two-two, not one hundred twenty-two). Other examples are also given in the figure. Note particularly that in Fig. 1.11d, where the plane does not intersect the $\mathbf{a}_1$ and $\mathbf{a}_2$ axes, the values of h and k are zero because the intercepts are at infinity. It is obvious that in this procedure of naming planes by their Miller indices, it is necessary to clearly state the kind of unit cell and the lengths of the edges if everyone is to visualize the same plane.

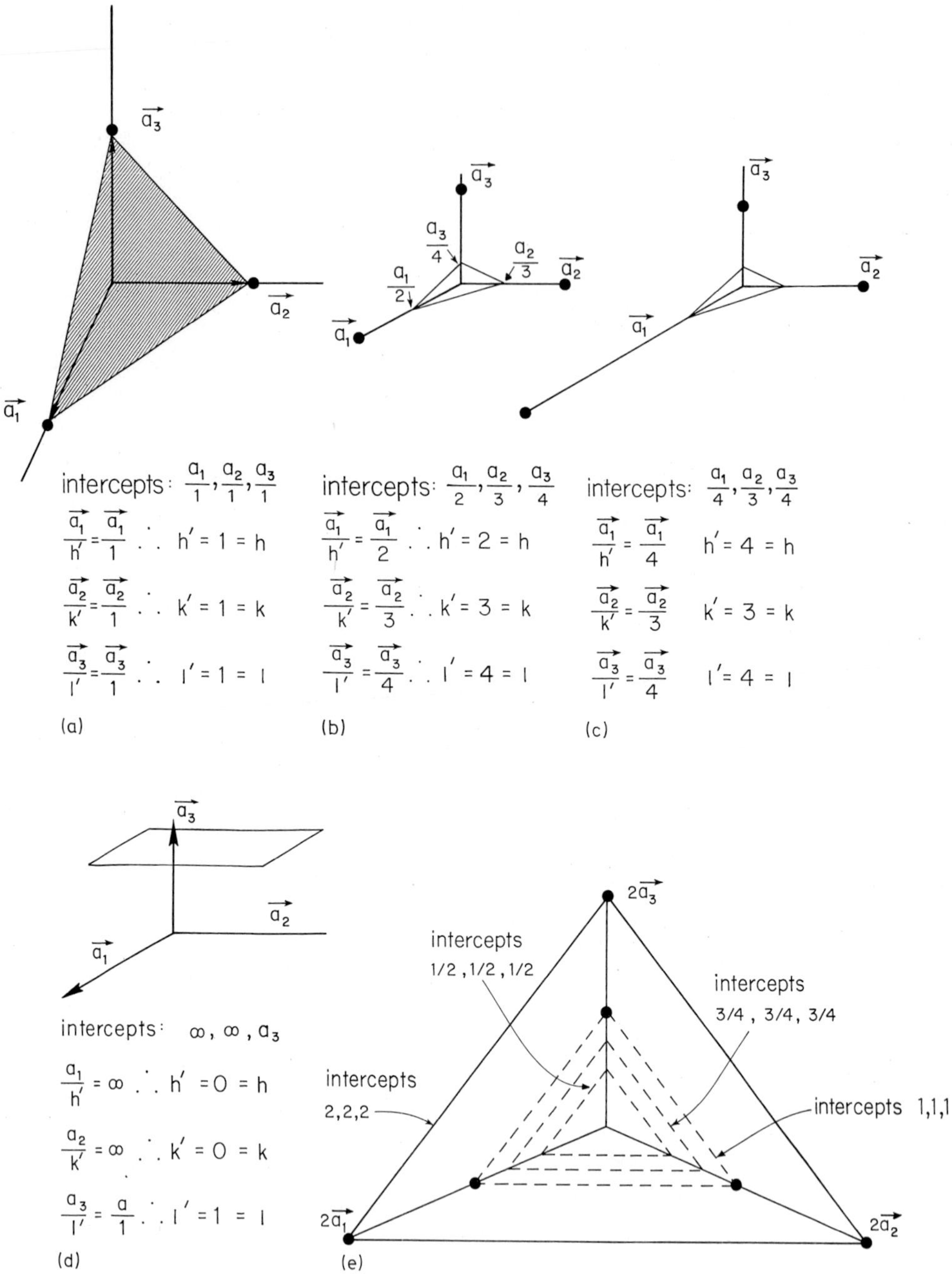

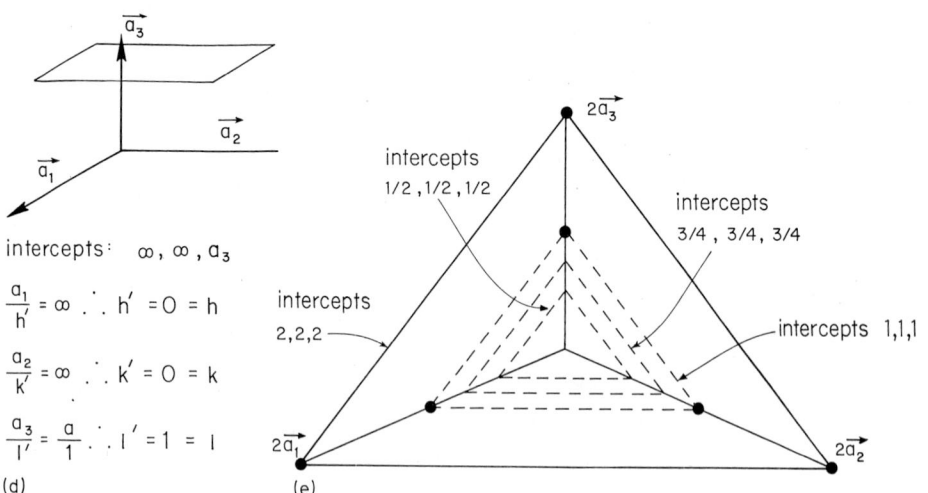

Fig. 1.11a–e. Examples of determination of Miller Indices. In **e**, all the planes shown have indices (111)

1.5 Naming Planes, Points, and Directions

By removing common factors in the intercepts, we lose the distinction between the plane we chose and all planes parallel to it in the crystal, i.e., the numbers (*hkl*) represent a whole set of parallel planes (note the parentheses used to indicate this set). This fact is illustrated in Fig. 1.11e, where all the illustrated planes have the same set of indices according to our procedures.

Because of the symmetry elements present in a crystal, any one kind of plane may be related by symmetry to other planes. As we have seen, the collection of planes generated in this way is called a form. We shall represent the planes of a form by the symbol {*hkl*}. For example, in a cubic crystal, {100} implies the planes (100), (010), (001), ($\bar{1}$00), (0$\bar{1}$0), and (00$\bar{1}$). There are six planes and this number is known as the multiplicity of the form. If the structure was, say, tetragonal, so that $\mathbf{a}_1 = \mathbf{a}_2 \neq \mathbf{a}_3$ then {100} would mean only (100), (010), ($\bar{1}$00), and (0$\bar{1}$0), but not (001) or (00$\bar{1}$). These last two planes are not generated from a (100) and the symmetry elements of a tetragonal space lattice.

In Fig. 1.6 it was shown that a cube and regular octahedron have identical symmetry elements. They are simply different figures for the same collection of symmetry elements; the only difference in the two crystals is the relative development of different forms. Other developments are also shown in Fig. 1.6 for the same system. This relative development of forms or habit was, and is, quite an annoying feature in examining the external features of a crystal. It soon became obvious, however, that it really was not important for defining a crystal in terms of its symmetries (although the growth conditions that cause different habits can be quite interesting). What *is* important in terms of the symmetry is the type of plane, its form, and its relation to other planes in the crystal, i.e., its angles with respect to other planes. In considering internal symmetry and structure we are not interested in the size of the plane at all, whether we look at it on a crystal or in the structure. What we are generally interested in is the *kind* of plane. In removing the common factor in the Miller indices, we are also eliminating this spurious information.

The *coordinates for points in a space lattice* (or coordinates of an atom or molecule) are written in terms of *x, y, z*, the fractional coordinates in a unit cell, with each edge as unit length. *Directions*, such as edges of a crystal where faces intersect, or rows of lattice points or atoms, are written by picking any point on a vector in the desired direction, writing its coordinates, clearing fractions and common factors, and enclosing in brackets, as [*uvw*]. An example is given in Fig. 1.12. As with planes, *symmetry equivalent directions comprise a form of directions, written as* [[*uvw*]].

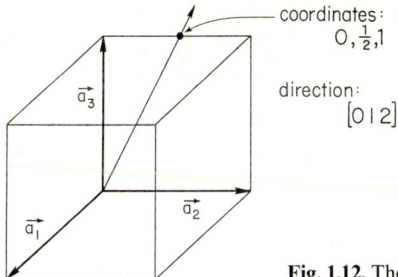

Fig. 1.12. The determination of the indices of a direction in a structure

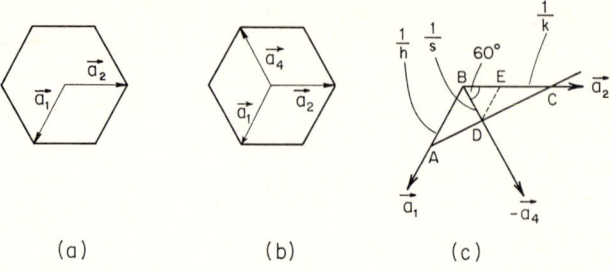

Fig. 1.13. a, b How the fourth axis arises in the hexagonal system. **c** The relationship between the fourth axis a_4 and a_1 and a_2.

These rules for choosing indices for planes and directions may be applied to all crystal systems; however, in the hexagonal system an alternate scheme is often used for naming planes. The unit cell for this system is defined by axes a_1 and a_2, as shown in Fig. 1.13a, and a third axis a_3 perpendicular to a_1 and a_2. Rotation by 120° (consistent with either a threefold or sixfold axis along a_3) produces the symmetry equivalent axis a_4 as shown in Fig. 1.13b. A set of "Miller-Bravais" indices for planes in this system is defined in terms of intercepts along the four axes a_1, a_2, a_4, and a_3, and written $(hkil)$. In Fig. 1.13c, the intersection of a plane perpendicular to the page is shown as the line ADC. The line DE, parallel to the a_1 axis, has been drawn from the plane's negative intercept on the new a_4 axis to the a_2 axis. Because triangles ABC and DEC are similar, and triangle BED is an equilateral triangle:

$$\frac{1/h}{1/k} = \frac{1/s}{1/k - 1/s} \quad \text{or} \quad h + k = s = -i.$$

Because of this relationship, the index along this third axis is not really necessary. Thus (1230) can also be written as (12·0), where the dot represents the i index. Everyone can calculate the extra coordinate from the relationship $i = -(h + k)$. However in certain cases, there is a definite advantage in using all the indices. Consider the form $\{11\bar{2}0\}$. All of the planes in this form are now written with both notations:

(11$\bar{2}$0) (11·0)   ($\bar{1}$2$\bar{1}$0) ($\bar{1}$2·0)

($\bar{1}$2$\bar{1}$0) ($\bar{1}$2·0)   ($\bar{1}\bar{1}$20) ($\bar{1}\bar{1}$·0)

(2$\bar{1}\bar{1}$0) (2$\bar{1}$·0)   ($\bar{2}$110) ($\bar{2}$1·0).

The three-index notation does not clearly indicate that all the planes are in the same form. With this four index Miller-Bravais notation, it is especially important to be careful and it is recommended that initially the student draw a grid as in Fig. 1.14 to help indexing.

For *directions in a hexagonal crystal*, the three axis scheme $[uvw]$ should be utilized. The indices of a direction based on four axes $[UVTW]$ are quite different as indicated in Fig. 1.14. For example, draw $[11\bar{2}1]$. What are the indices with the three-axis notation? They are [331]. The reader might try proving that in general

$$U = (2u - v)/3, \quad V = (2v - u)/3, \quad T = -(u + v), \quad W = w.$$

1.6 Atomic Size and Coordination

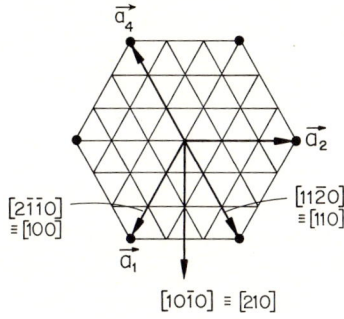

Fig. 1.14. A ruled net like this one is often helpful in indexing in the hexagonal system. (After Barrett, C.S., and Massalski, T.B., "Structure of Metals", 3rd ed. McGraw-Hill, New York, 1966)

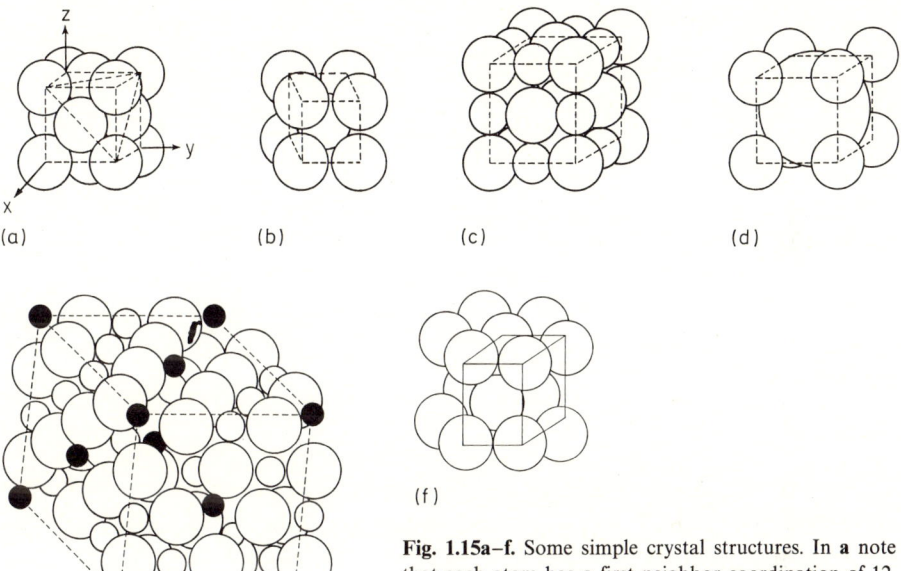

Fig. 1.15a–f. Some simple crystal structures. In **a** note that each atom has a first neighbor coordination of 12, that is, 12 nearest neighbors. In **b** this number is 8

1.6 Atomic Size and Coordination

The application of diffraction techniques allows the determination of atomic arrangement in materials. The results of such studies for crystalline solids may be conveniently displayed by a drawing showing the atomic arrangement relative to a unit cell of the Bravais lattice associated with the structure. The unit cells of several simple structures are given in Fig. 1.15. The structures shown in (a), (c), and (e) have *face centered cubic unit cells* (fcc), (b) is *body centered cubic* (bcc), and that shown in (f) is *hexagonal close packed* (hcp). In (a) there is only one *atom* at each lattice point, i.e., four per unit cell at coordinates 0, 0, 0; $\frac{1}{2}, \frac{1}{2}, 0$; $\frac{1}{2}, 0, \frac{1}{2}$; and 0, $\frac{1}{2}, \frac{1}{2}$. We may use the shorthand notation: four atoms at 0, 0, 0 + fct (face-centering translations).

In (c) there are four Na^+Cl^- ion pairs per cell: four Cl^- at 0, 0, 0 + fct and four Na^+ at $\frac{1}{2},\frac{1}{2},\frac{1}{2}$ + fct. In (e), a more complex structure is shown with eight "molecules" of AB_2O_4 per unit cell; the structure is still face-centered cubic, since for any atom in position x, y, z, there are three identical atoms located by the face-centering translations. The structure in (d) is primitive cubic with one ion pair of Cs^+Cl^- per cell, not bcc, since the Cs^+ ion at $\frac{1}{2},\frac{1}{2},\frac{1}{2}$ is not identical to the Cl^- ion at 0, 0, 0.

Atoms or ions in Fig. 1.15 are represented schematically as if they were spheres of well defined diameters. In the spirit of this model, we can assume that the atoms "touch" along a face diagonal in (a) and estimate the "hard-sphere diameter" for copper as one-half a face diagonal, $a/\sqrt{2}$. A more sophisticated interpretation of this estimate of atomic size is that it represents the diameter of the bonding electron orbitals. In this fashion, from measured dimensions of unit cells, we can obtain such information for the pure elements and then examine how sizes change in compounds due to electron transfer, etc.

The assumption of atoms touching along face diagonals for copper implies quite *closely packed planes of atoms*. One of these is outlined in Fig. 1.15a (dotted triangle) and in Fig. 1.16a. Why is there a threefold axis along a cube diagonal? To answer this, note the way close packed planes are stacked in the unit cell, in Figs. 1.15a and 1.16c, with each atom (shaded) over a space between atoms in the layer below.

The area occupied by one atom is defined by a circle inscribed in a hexagon, as in Fig. 1.16a. Calling the atomic radius r, the atom's area in the plane is πr^2. The hexagon can be constructed from 12 triangles of area

$$(r/\sqrt{3}) \times r/2 = r^2/2\sqrt{3} \,.$$

Therefore, the fraction of the total area occupied by the atom is

$$\pi r^2/[(12r^2)/(2\sqrt{3})] = 0.907 \,.$$

For the square array in Fig. 1.16b the fraction is only 0.785. The more "efficient" close packing is in fact quite logical for a metal such as copper. Because of the lack of directionality of the metallic bond, the atoms tend to pack themselves geometrically, with as many neighbors as possible. (This also occurs in crystals made up of molecules weakly bonded by van der Waal's forces. However, the lack of symmetry in the shape of the molecules often leads to very low symmetry unit cells.) There are six nearest neighbors to any atom in its close-packed plane (Fig. 1.16a)

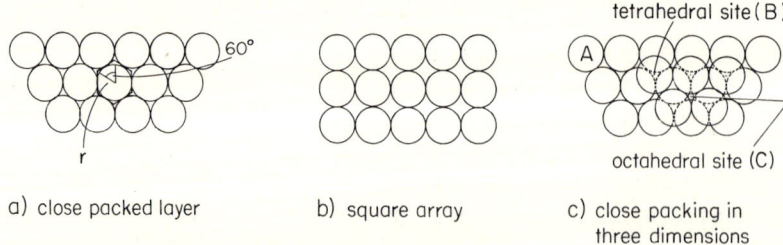

a) close packed layer   b) square array   c) close packing in three dimensions

Fig. 1.16a–c. Close-packed layers [in **c** two layers are stacked]

1.6 Atomic Size and Coordination

X Octahedral site 1/2, 1/2, 1/2

X Tetahedral site 3/4, 3/4, 3/4

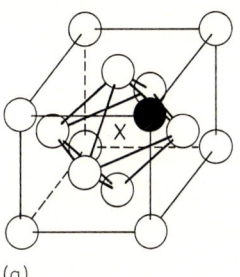

(a)

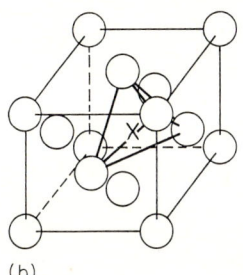

(b)

Fig. 1.17a, b. Interstitial sites in a close packed fcc cell

and three above and three below (Figs. 1.16c or 1.17a) for a total of 12. (Remember in looking at a unit cell that there are other cells in the structure surrounding this one.)

The volume fraction occupied by copper atoms in its unit cell can easily be found. There are four copper atoms per cell, the atom at the corner and the three atoms at faces nearest to the corner; the others "belong" to other cells. Alternatively, each corner atom is shared by eight cells, and each atom on a face is shared by two cells, again giving four atoms per cell. Thus, the total atomic volume in the cell is $4 \times (4/3)\pi r^3$. The cell edge can be expressed in terms of the atom's radius. Half of a face diagonal, $a/\sqrt{2}$, is $2r$, so that $a = 2\sqrt{2}r$ or $a^3 = 16\sqrt{2}r^3$. The volume fraction of atoms is then

$$(16 \pi r^3/3)/16\sqrt{2}r^3 = 0.74.$$

The rest of the volume is occupied by voids. Notice in Fig. 1.16c that there are two types of voids. One type has three atoms as nearest neighbors in a plane below and one atom above. This is referred to as a *tetrahedral site or hole*; as shown in Fig. 1.17b the neighbors to the hole form a regular tetrahedron. Another void type is midway between the planes, with three atoms above and three below. It is referred to as an *octahedral site (or hole)*; as shown in Fig. 17a its neighbors form an octahedron. These "voids" represent the empty space in the cell.

Another atom of a different size (radius R) might fit in one of these voids. If *all* the atoms touch, then for the smaller atom in the octahedral hole $2R + 2r = a$ (see Fig. 1.17a). But $a = 2\sqrt{2}r$ and therefore, $R/r = \sqrt{2} - 1 = 0.414$. For a smaller value of R/r, the smaller atom is not touching the surrounding copper atoms. If the size of the smaller atom is increased, the larger atoms no longer touch one another and there is no longer close packing. The ratio R/r may continue to increase until the atoms with radius R touch one another and the atoms with radius r are in octahedral voids. For this case, $R/r = 1/0.414$. Thus we find $0.414 \leq R/r \leq 1/0.414$ for this case of octahedral packing.

There are octahedral holes at the unit cell's center as shown in Fig. 1.17a, but also at the center of all 12 edges of the cell. Each edge is shared by four cells, so that there are three edge sites per cell, and with the one in the center of the cell there are four per cell, or one per atom. Examine the structure of NaCl in Fig. 1.15. The Na^+ ions are in octahedral sites, each surrounded by six Cl^- ions and vice versa. But the structure is not close packed; compare the ionic radii to the ratios calculated above. These ratios are not followed exactly [see O'Keefe (1977) for a discussion and further references].

The tetrahedral site in Fig. 1.17b has its center one quarter of the way along a body diagonal. There are eight such sites per cell, or two per atom. The radius ratio can be readily shown to be $R/r = 0.225$ for close packing.

Examine the bcc structure in Fig. 1.15b. Is it close packed? Locate the voids surrounded by six atoms (octahedral voids) and those surrounded by four atoms (tetrahedral voids). Note that in the bcc structure these voids are not perfect octahedra or tetrahedra as in the fcc, but are distorted. The radii of spheres which can fit into these voids are R (octahedral) $= 0.154r$ and R (tetrahedral) $= 0.291r$.

We have seen that the fcc structure is built up of close-packed planes of atoms, each one placed above the holes of the one below. Two such layers are shown in Fig. 1.16c. A third layer is on top of these two above the triangular spaces of the second layer whose apexes point downward. When a fourth layer is added in the same way, it is directly over the atoms in the first layer and the stacking pattern repeats. This sequence is often written ABCABC, with the letters indicating the planes of atoms shown in Fig. 1.16c. Note, however, that each layer also has triangular spaces between atoms with apexes pointing upward. If the third layer is placed above these spaces, it is directly over the atoms in the first layer and the stacking is ABABAB. If you build such a structure or sketch it in exploded view, you will readily see that the holes are the same kind and number as with copper, the density of packing is the same but the unit cell is hexagonal with two atoms in the cell at 0, 0, 0 and $\frac{2}{3}, \frac{1}{3}, \frac{1}{2}$. This is illustrated in Fig. 1.15f. The second atom is *not* at a lattice point. This structure is often referred to as hexagonal close-packed (hcp) even though the unit cell is primitive hexagonal. Many other stacking sequences occur in nature, with lesser frequency, such as $ABCABABCAB\ldots$, which is the stacking in the rare-earth samarium.

Many structures appear to be based on considerations of close packing or near-close packing. For example, in MgO (used for bricks for high temperature furnaces) O^{2-} anions are in a close-packed fcc array with the Mg^{2+} cations in octahedral sites; the structure is very much like NaCl. Barium titanate is another interesting oxide used as a transducer; it is shown in Fig. 1.18a. Titanium ions are

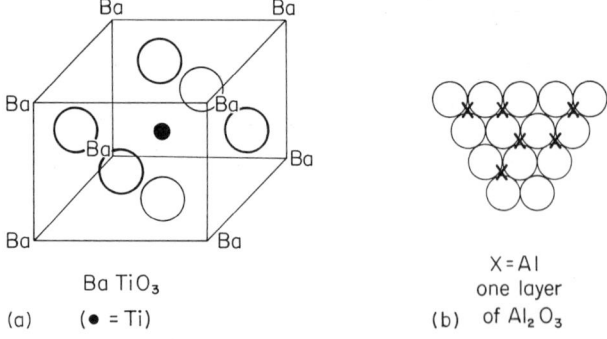

Fig. 1.18. Crystal structures of (**a**) $BaTiO_3$ and (**b**) Al_2O_3

1.6 Atomic Size and Coordination

octahedrally coordinated to the face-centered oxygen ions. In oxides such as Al_2O_3 (employed for spark plugs, phonograph needles, crucibles, etc.), the O^{2-} are close packed in the hexagonal arrangement, with Al ions in rows of octahedral sites, as shown in Fig. 1.18b. In these rows every third octahedral site is lacking an Al^{3+} ion and as there is one such site per oxygen this leads to the composition Al_2O_3. Spinels, AB_2O_4 (where A and B are metal ions of valences 2 and 3 respectively) are also interesting. As can be seen in Fig. 1.15e, the structure is essentially eight face-centered almost close packed unit cells of oxygen. In the cell of the spinel there are, therefore, 32 oxygen ions and 32 octahedral sites. There are also 64 tetrahedral sites. If there are 32 oxygens, from the formula there are 16 B atoms and 8 A atoms and these are distributed over these interstitial sites, depending on the specific metal ions involved. It is logical to suspect that the A's would be in tetrahedral sites and the B's in octahedral sites, because the B's have the larger valence and can therefore interact with more oxygen ions. This does occur in $ZnFe_2O_4$, but for $MgFe_2O_4$ and Fe_3O_4, tetrahedral sites are occupied by Fe^{3+} ions and the octahedral sites are shared by divalent and trivalent ions. [The structures can be written as $Fe^{3+}(Mg^{2+}Fe^{3+})O_4$ and $Fe^{3+}(Fe^{2+}Fe^{3+})O_4$.] These are called inverse spinels. From neutron diffraction, it is found that the magnetic moment of the iron ions is oppositely directed on the two sites. If the $MgFe_2O_4$ is perfectly inverse, the material is antiferromagnetic, but changes in heat treatment, composition, and additions of other elements can alter the occupation to some condition between inverse and normal making the material "ferrimagnetic". These materials also have magnetic hysteresis loops which make them ideally suited for switching; as a result they have been used as the memory elements in computers and "magnetic" tapes.

In all of the inorganic compounds we have discussed, one can write a simple molecular formula to indicate the numbers of atoms of each type, but when we look at the structure, there are no molecular units as there often are in organic structures. Each ion is bonded to many neighbors not just the ones indicated by the formula. For example, in NaCl it is the ionic forces between one Na^+ and shells of Cl^- that hold the structure together. A "molecule" of NaCl is not readily discernible. Even more distinct from "molecular" structures are solid solutions. In such structures, for example nickel dissolved in copper, the average unit cell is still face-centered cubic as for pure copper, but at any particular place in the crystal a unit cell might have anywhere from zero to four of the lattice points occupied by nickel rather than copper. The use of diffraction to describe such structures is the subject of Sect. 7.6 of this book.

It is easy to realize that much of the behavior of all materials (how the atoms move relative to one another when deformed or when subjected to fields) is associated with their structure. Despite the fact that symmetry limits the kinds of structures possible, it is not very easy to guess the structure for a given composition. In some cases now, with the numbers of structures that have been determined, we *can* make a good first guess, based on radius ratios, valence, bond type, and the structure of compounds of elements from similar positions in the periodic table. Much then remains, however, to obtain the actual atomic positions with accuracy. This is the subject of Chap. 6.

1.7 Analytical Calculations in Crystallography

1.7.1 Introduction to the Reciprocal Lattice

We now have a scheme for naming planes, directions, and points in all the crystal systems. The next items we shall examine are methods for determining relationships between planes and directions graphically and analytically. If we see a line on the face of a crystal, how can we determine in what plane or planes it lies? How do we calculate angles between planes and directions? The reader can undoubtedly answer all these questions for a cubic axial system for which edges are all equal and at right angles, because he is probably familiar with this orthogonal axial system from his early studies of vectors, but most crystals are *not* cubic! Furthermore, any graphical representation we choose must be one which emphasizes the angular relationships and deemphasizes the size and shape of planes. For highest accuracy and for deriving general relationships, the analytical approach is essential, and so we consider it first.

The location of any point in a lattice, A, can be specified by a vector sum of translations along the three axes of the unit cell:

$$\mathbf{A} = m_1 \mathbf{a}_1 + m_2 \mathbf{a}_2 + m_3 \mathbf{a}_3 \,. \tag{1.1}$$

It then seems logical to consider vector algebra as the foundation for our analytical representation. In Fig. 1.19, the basic operations of vector addition, subtraction, and scalar and vector products are presented. (The reader might quickly review the fundamentals of vector algebra at this point, although we will not require too much more than these operations.) We shall proceed by examining a number of questions about crystals with the aid of vectors. Some exercises with this approach can be found in the problems at the end of the chapter.

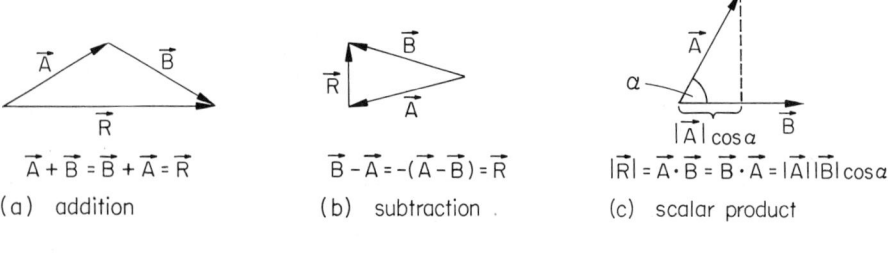

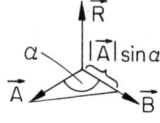

Fig. 1.19a–d. Illustrating simple vector algebraic operations

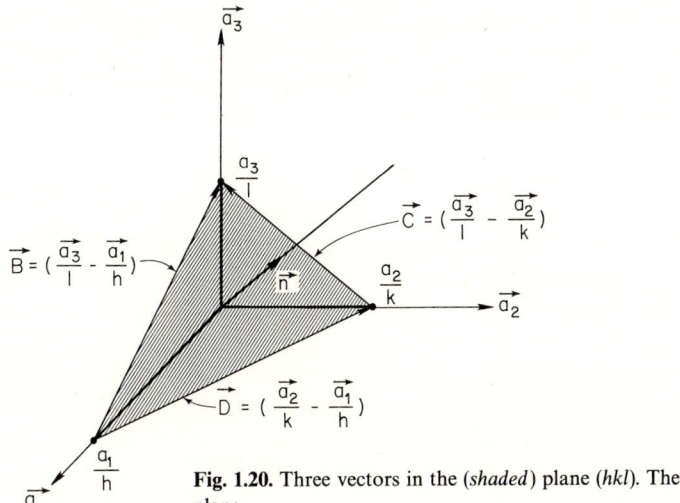

Fig. 1.20. Three vectors in the (*shaded*) plane (*hkl*). The vector **n** is normal to the plane

As we are not interested in the size or shape of a plane in a crystal or on its surface, but only its orientation, a useful representation is a vector normal to the plane. A plane with Miller indices (hkl) is shown in Fig. 1.20 with the intercepts described in terms of vectors. A vector normal to the plane must be normal to all vectors in the plane, such as $\mathbf{D} = (\mathbf{a}_2/k) - (\mathbf{a}_1/h)$ or $\mathbf{B} = (\mathbf{a}_3/l) - (\mathbf{a}_1/h)$, or $\mathbf{C} = (\mathbf{a}_3/l) - (\mathbf{a}_2/k)$. We may express the *normal to the plane* in terms of the unit cell vectors as

$$\mathbf{n} = u\mathbf{a}_1 + v\mathbf{a}_2 + w\mathbf{a}_3 , \tag{1.2}$$

where u, v, w are to be determined. Since perpendicular vectors have scalar products equal to zero,

$$\mathbf{n} \cdot \mathbf{D} = \mathbf{n} \cdot \mathbf{B} = \mathbf{n} \cdot \mathbf{C} = 0 . \tag{1.3a}$$

It is useful to expand one of these equations at this point. Let $\cos(a_i a_j)$ be the cosine of the angle between vectors $\mathbf{a}_i$ and $\mathbf{a}_j$. Then

$$\begin{aligned}\mathbf{n} \cdot \mathbf{D} = 0 &= u[(a_1 a_2/k)\cos(a_1 a_2) - (a_1^2/h)] \\ &+ v[(a_2^2/k) - (a_1 a_2/h)\cos(a_1 a_2)] \\ &+ w[(a_3 a_2/k)\cos(a_3 a_2) - (a_3 a_1/h)\cos(a_3 a_1)] . \end{aligned} \tag{1.3b}$$

It is important to remember that the angles between the vectors (or cell edges) are *not necessarily* right angles since our coordinate system is not always orthogonal. But as an example, consider a cubic crystal. Then all the interaxial angles in Eq. (1.3b) are 90° and $a_1 = a_2 = a_3$. Equation (1.3b) reduces to

$$0 = (-ua_1^2/h) + (va_2^2/k) , \quad \text{or} \quad u/v = h/k .$$

From the other equations (1.3a), it can similarly be shown that $u/w = h/l$ and $v/w = k/l$. In other words, except for a common factor, $u = h$, $v = k$, $w = l$. We conclude that *a normal to a plane in the cubic system always has the same indices as the plane itself*, but this is not true in other systems. In these equations we do not really need to solve for u, v, w; the ratio $u:v:w$ will suffice, for a [111] direction is the same as a [222]. Thus we really need to solve only two of the three equations (1.3a) to obtain these ratios.

There is a set of vectors, called *reciprocal lattice vectors*, that makes many of these kinds of derivations and calculations very simple. They are also quite useful in diffraction and in solid-state physics. These vectors are defined as

$$\mathbf{b}_1 = \frac{\mathbf{a}_2 \times \mathbf{a}_3}{\mathbf{a}_1 \cdot \mathbf{a}_2 \times \mathbf{a}_3}, \tag{1.4a}$$

$$\mathbf{b}_2 = \frac{\mathbf{a}_3 \times \mathbf{a}_1}{\mathbf{a}_1 \cdot \mathbf{a}_2 \times \mathbf{a}_3}, \tag{1.4b}$$

$$\mathbf{b}_3 = \frac{\mathbf{a}_1 \times \mathbf{a}_2}{\mathbf{a}_1 \cdot \mathbf{a}_2 \times \mathbf{a}_3}. \tag{1.4c}$$

The denominator of these expressions is a scalar quantity. As described in Fig. 1.19, $\mathbf{a}_2 \times \mathbf{a}_3$ is the vector normal to the plane of $\mathbf{a}_2$ and $\mathbf{a}_3$ with magnitude equal to the area of the parallelogram of sides $\mathbf{a}_2$ and $\mathbf{a}_3$. Therefore, $\mathbf{a}_1 \cdot \mathbf{a}_2 \times \mathbf{a}_3$ is the volume (V_a) of the cell defined by the $\mathbf{a}_i$. In Fig. 1.21, the axes $\mathbf{a}_i$ and $\mathbf{b}_i$ are shown superimposed for the case of the hexagonal crystal system. The $\mathbf{b}_i$ are located merely by employing the definitions of the $\mathbf{b}_i$ and the physical meaning of the cross product. These reciprocal lattice vectors have the useful *property of orthonormality*:

$$\mathbf{a}_i \cdot \mathbf{b}_j = \begin{cases} 0, & i \neq j, \\ 1, & i = j. \end{cases} \tag{1.5}$$

The reader should prove the statements of Eq. (1.5) using the definitions given in Eq. (1.4).

Consider the reciprocal lattice vector $\mathbf{r}_{hkl}^* = h\mathbf{b}_1 + k\mathbf{b}_2 + l\mathbf{b}_3$. If we take the dot product of this with vectors in the (hkl) plane, keeping in mind Eq. (1.5):

$$[\mathbf{a}_2/k - \mathbf{a}_1/h] \cdot \mathbf{r}_{hkl}^* = (k/k)\mathbf{a}_2 \cdot \mathbf{b}_2 - (h/h)\mathbf{a}_1 \cdot \mathbf{b}_1 = 0,$$

and

$$[(\mathbf{a}_3/l) - (\mathbf{a}_1/h)] \cdot \mathbf{r}_{hkl}^* = 0.$$

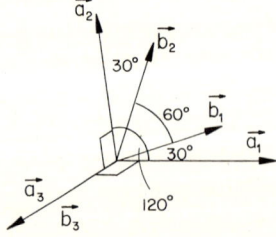

Fig. 1.21. An example of vectors in real space ($\mathbf{a}_i$) and reciprocal space ($\mathbf{b}_i$) in the hexagonal system

As $\mathbf{r}^*_{hkl}$ is normal to two vectors in the plane, it is normal to the plane. Hence we can immediately write the vector normal to an (hkl) plane in any axial system in terms of the reciprocal lattice vectors, $\mathbf{b}_i$. The magnitude of $\mathbf{r}^*_{hkl}$ is also interesting. If $\mathbf{N}$ is a unit normal to the plane (hkl), the projection of the vector $\mathbf{a}_1/h$ on $\mathbf{N}$ is

$$\mathbf{N} \cdot (\mathbf{a}_1/h) = d_{hkl},$$

where d_{hkl} is the spacing between (hkl) planes, i.e., the distance from the origin to the plane in Fig. 1.20 [there is another plane parallel to (hkl) through the origin]. A unit normal can also be written as

$$\mathbf{N} = \mathbf{r}^*_{hkl}/|\mathbf{r}^*_{hkl}|.$$

Thus

$$d_{hkl} = (\mathbf{r}^*_{hkl}/|\mathbf{r}^*_{hkl}|) \cdot (\mathbf{a}_1/h) = 1/|\mathbf{r}^*_{hkl}|$$

since $\mathbf{r}^*_{hkl} \cdot \mathbf{a}_1/h = 1$. Thus *the magnitude of $\mathbf{r}^*_{hkl}$ is equal to the reciprocal of the (hkl) interplanar spacing*. This fact will become quite important in our discussion of diffraction in Chap. 2.

Consider next, the *volume defined by the $\mathbf{b}_i$*, that is $V^* = \mathbf{b}_1 \cdot \mathbf{b}_2 \times \mathbf{b}_3$. Expanding, using the definitions of Eq. (1.4),

$$\mathbf{b}_2 \times \mathbf{b}_3 = [(\mathbf{a}_3 \times \mathbf{a}_1)/V_a] \times [(\mathbf{a}_1 \times \mathbf{a}_2)/V_a].$$

Let

$$\mathbf{a}_1 = \mathbf{b}, \qquad \mathbf{a}_2 = \mathbf{c}, \qquad \mathbf{a}_3 \times \mathbf{a}_1 = \mathbf{a}.$$

As

$$\mathbf{a} \times (\mathbf{b} \times \mathbf{c}) = \mathbf{b}(\mathbf{a} \cdot \mathbf{c}) - \mathbf{c}(\mathbf{a} \cdot \mathbf{b}),$$

$$\mathbf{b}_2 \times \mathbf{b}_3 = (1/V_a^2)\{\mathbf{a}_1[\mathbf{a}_2 \cdot (\mathbf{a}_3 \times \mathbf{a}_1)] - \mathbf{a}_2[\mathbf{a}_1 \cdot (\mathbf{a}_3 \times \mathbf{a}_1)]\}.$$

The first term in brackets is just V_a, the second is zero. Therefore,

$$\mathbf{b}_2 \times \mathbf{b}_3 = \mathbf{a}_1/V_a$$

and

$$V^* = \mathbf{b}_1 \cdot \mathbf{b}_2 \times \mathbf{b}_3 = \mathbf{b}_1 \cdot \mathbf{a}_1/V_a = 1/V_a. \tag{1.7}$$

Thus *the volumes defined by the two sets of vectors are also reciprocal*.

1.7.2 Application of Reciprocal Lattice Vectors

Let us now examine some common crystallography problems to see how useful these reciprocal lattice vectors can be. As we noted earlier, *a group of planes that have parallel intersections* — like (100), (010) and ($\bar{1}$00), (0$\bar{1}$0) in a cube — *is called a zone*. The direction parallel to the lines of intersection is called *the zone axis*. If we know the planes, what is *the zone axis*? As an extension of this, we could ask the question: What is the line or vector of intersection of one plane $(h_1k_1l_1)$ with another $(h_2k_2l_2)$? This is called the trace of $(h_1k_1l_1)$ on $(h_2k_2l_2)$. Conversely, if we see such a trace, what planes is it from? Such questions arise when examining, for example, a line produced by deformation of a polished crystal or a platelike precipitate appearing on a polished sample, when we want to know the plane of the matrix on which the precipitate has formed.

Let the zone axis be $\mathbf{r} = u\mathbf{a}_1 + v\mathbf{a}_2 + w\mathbf{a}_3$. This vector must be perpendicular to the normals of all the planes in the zone. Hence, for each (hkl) plane in the zone,

$$\mathbf{r} \cdot \mathbf{r}^*_{hkl} = 0 = hu + kv + lw \,. \tag{1.8a}$$

If we have a (100) plane, its zone axis can be [0vw], i.e., [002], [012], etc. Only u must be zero. For a (110) plane, $u + v = 0$. That is, directions of the type [1$\bar{1}$0], [1$\bar{1}$3], etc., are possible zone axes for a (110) plane. If we have *two* intersecting planes with indices $(h_1k_1l_1)$ and $(h_2k_2l_2)$, their zone axis or the trace of one in another can be found from

$$h_1u + k_1v + l_1w = 0, \tag{1.8b}$$

$$h_2u + k_2v + l_2w = 0 \,. \tag{1.8c}$$

Although there are only two equations for three unknowns, we only need the ratios u/v, v/w, u/w to define the direction $\mathbf{r}$. Dividing Eq. (1.8b) by l_1w and Eq. (1.8c) by l_2w, multiplying the former by h_2/l_2 and the latter by h_1/l_1, we obtain

$$(h_2h_1u/l_2l_1w) + (h_2k_1v/l_2l_1w) = -h_2/l_2 \tag{1.8d}$$

and

$$(h_1h_2u/l_1l_2w) + (h_1k_2v/l_1l_2w) = -h_1/l_1 \,. \tag{1.8e}$$

Subtracting Eq. (1.8e) from Eq. (1.8d), we obtain

$$(v/w)[(k_1h_2/l_1l_2) - (k_2h_1/l_1l_2)] = (h_1/l_1) - (h_2/l_2),$$

or

$$v/w = (h_1l_2 - h_2l_1)/(k_1h_2 - k_2h_1),$$

or

$$v = l_1h_2 - l_2h_1, \qquad w = h_1k_2 - h_2k_1 \,.$$

Similar relationships can be written for u/w, u/v and hence, u. It is then a simple matter to calculate the trace of one plane in another. Note how useful the reciprocal lattice vectors were in deriving the initial equations.

As a second practical example, suppose we are familiar with a structure and various indices of planes in one unit cell, but we now wish to reindex the planes employing another possible cell. This could be a laborious task with models of the structure, but vector algebra can help us. Consider a new set of axes $\mathbf{A}_i$ defined in terms of the old set $\mathbf{a}_i$ by the relations

1.7 Analytical Calculations in Crystallography

$$\mathbf{A}_1 = u_1\mathbf{a}_1 + v_1\mathbf{a}_2 + w_1\mathbf{a}_3,$$
$$\mathbf{A}_2 = u_2\mathbf{a}_1 + v_2\mathbf{a}_2 + w_2\mathbf{a}_3, \quad (1.9\text{a})$$
$$\mathbf{A}_3 = u_3\mathbf{a}_1 + v_3\mathbf{a}_2 + w_3\mathbf{a}_3.$$

Now, the distance between planes is the same, regardless of the coordinate system. Choosing the reciprocal lattice vectors $\mathbf{R}^*_{HKL}$ and $\mathbf{r}^*_{hkl}$ to represent this distance.

$$H\mathbf{B}_1 + K\mathbf{B}_2 + L\mathbf{B}_3 = h\mathbf{b}_1 + k\mathbf{b}_2 + l\mathbf{b}_3, \quad (1.9\text{b})$$

where the reciprocal lattice vectors $\mathbf{B}_i$ are defined in terms of the $\mathbf{A}_i$ by equations analogous to Eq. (1.4). Taking the dot product of both sides of Eq. (1.9b) with $\mathbf{A}_1$, we have

$$H = hu_1 + kv_1 + lw_1, \quad (1.9\text{c})$$

with $\mathbf{A}_2$,

$$K = hu_2 + kv_2 + lw_2 \quad (1.9\text{d})$$

and with $\mathbf{A}_3$,

$$L = hu_3 + kv_3 + lw_3. \quad (1.9\text{e})$$

These three equations can be arranged in matrix form as

$$\begin{pmatrix} H \\ K \\ L \end{pmatrix} = \begin{pmatrix} u_1 & v_1 & w_1 \\ u_2 & v_2 & w_2 \\ u_3 & v_3 & w_3 \end{pmatrix} \begin{pmatrix} h \\ k \\ l \end{pmatrix}. \quad (1.9\text{f})$$

Knowing the values of u_i, v_i, and w_i from Eq. (1.9a), it is possible to calculate (HKL) in the new cell from any (hkl) in the old cell. Or, if we know three (HKL)'s and their corresponding (hkl)'s, we can obtain the u_i, v_i and w_i.

In such a transformation the volume of the new unit cell is $\mathbf{A}_1 \cdot \mathbf{A}_2 \times \mathbf{A}_3$. Now

$$\mathbf{A}_2 \times \mathbf{A}_3 = u_2 v_3 (\mathbf{a}_1 \times \mathbf{a}_2) + u_2 w_3 (\mathbf{a}_1 \times \mathbf{a}_3) + v_2 u_3 (\mathbf{a}_2 \times \mathbf{a}_1)$$
$$+ v_2 w_3 (\mathbf{a}_2 \times \mathbf{a}_3) + w_2 u_3 (\mathbf{a}_3 \times \mathbf{a}_1) + w_2 v_3 (\mathbf{a}_3 \times \mathbf{a}_2)$$
$$= (u_2 v_3 - v_2 u_3)(\mathbf{a}_1 \times \mathbf{a}_2) + (w_2 u_3 - u_2 w_3)(\mathbf{a}_3 \times \mathbf{a}_1)$$
$$+ (v_2 w_3 - w_2 v_3)(\mathbf{a}_2 \times \mathbf{a}_3).$$

Recalling the definitions of $\mathbf{b}_i$ [Eq. (1.4)]:

$$\mathbf{A}_2 \times \mathbf{A}_3 = V_a[(u_2 v_3 - v_2 u_3)\mathbf{b}_3 + (w_2 u_3 - u_2 w_3)\mathbf{b}_2 + (v_2 w_3 - w_2 v_3)\mathbf{b}_1].$$

Then

$$V_A = \mathbf{A}_1 \cdot \mathbf{A}_2 \times \mathbf{A}_3 = V_a \begin{vmatrix} u_1 & v_1 & w_1 \\ u_2 & v_2 & w_2 \\ u_3 & v_3 & w_3 \end{vmatrix} \quad (1.10)$$

where the vertical lines denote the determinant of $u_i v_i w_i$.

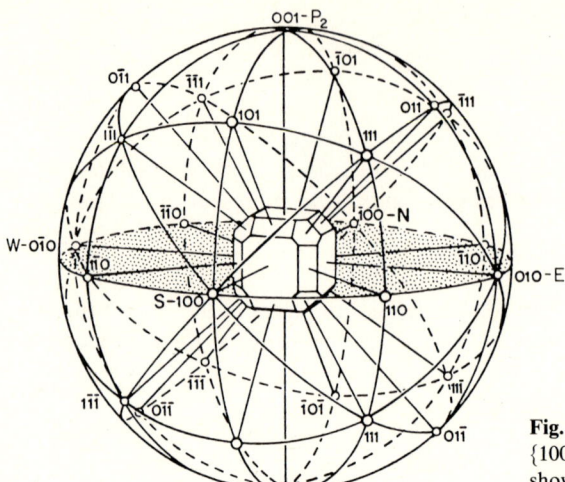

Fig. 1.22. Poles of the low index planes {100}, {110}, and {111} for a cubic system, shown as intersections of the normals to the planes with a reference sphere

1.8 Graphical Methods in Crystallography

1.8.1 Introduction to the Stereographic Projection

Let us turn now to graphical techniques. We have seen that a particularly simple way of representing a plane is by its normal (which we can write in terms of $\mathbf{r}^*_{hkl}$). These normals may be used to define angular relationships between the planes because the angle between two normals is equal to the angle between their planes. Imagine a very small crystal at the center of a large sphere with normals to the planes of the crystal extended to intersect the sphere. Such a situation is shown in Fig. 1.22. We can concentrate on the interrelation of the normals by examing the intersection of these normals with the sphere. These intersections are called *poles*. We could measure the angles between planes by measuring the angles between their poles on the surface of the sphere. We cannot carry such a sphere around with us, however, so the results are conveniently projected on to two dimensions. One of the most popular ways of doing this is the *stereographic projection* (which is often employed in map making). This projection is of most use to us, because it is "angle true", i.e., angular relationships on the surface of the sphere will be correctly represented in the projection. The poles are projected back to a point P_1 as illustrated in Fig. 1.23. The intersection of the projection with a great circle (called the primitive circle) is considered the representation of the pole. The index of a pole (the index of the corresponding plane) is *not* placed inside any kind of bracket, i.e., the notation *hkl* is used for the pole of (*hkl*) or its representation on the projection plane. We will often use the term pole to denote the projection as well.

It is clear from the figure that if P_1 (the $00\bar{1}$ pole in this case) is used as the projecting point, poles below the primitive circle will project *outside* of this circle. We can correct this by switching to projection point P_2 (the 001 pole) and by using open circles for these points and closed circles for poles projected to P_1.

1.8 Graphical Methods in Crystallography

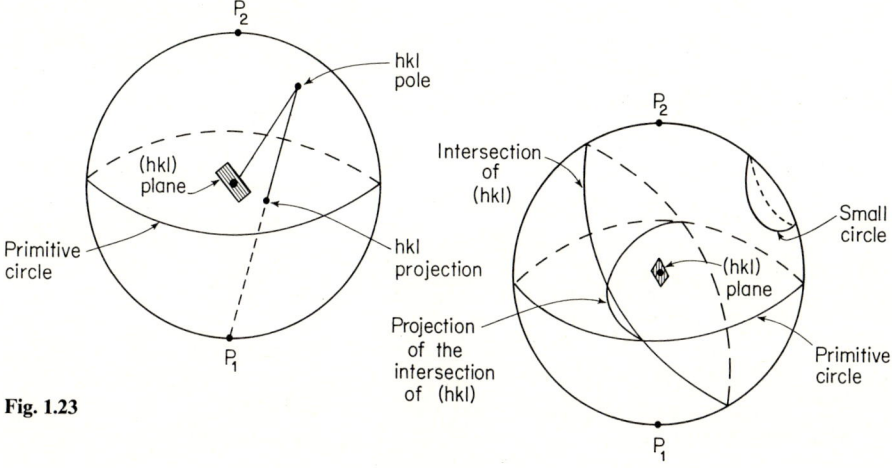

Fig. 1.23. The stereographic projection of the *hkl* pole using P_1 as the projection point. The primitive plane is taken coincident with the equatorial plane

Fig. 1.24. Representation of a plane may also be made by projecting the intersection of the plane with the reference sphere (trace of the plane on the sphere)

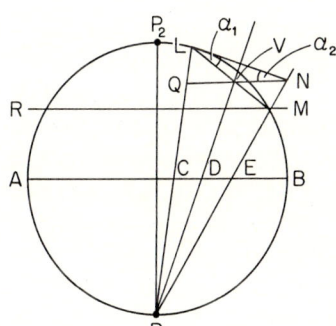

Fig. 1.25. A stereographic projection of the small circle whose trace is *LM* onto the plane of projection, here seen edge on as *AB*. (Adapted from Philips, F.C., "An Introduction to Crystallography", 3rd ed. Longmans, Green, London, 1964)

As alternative way to represent a plane is by its intersection with the projection sphere as shown in Fig. 1.24. We note for future reference that every point on this intersection of the plane (*hkl*) is 90° away from the pole of (*hkl*). We may wish to project this trace on to the primitive circle. We could do this point by point, but it will be far easier to show analytically that any circle on the sphere projects as a circle on the projection plane (then the projection of the intersection may be drawn with a compass). Once we show that this is true, we will automatically have proved our contention that *the stereographic projection is angle true*, since angles on the sphere are measured along circular arcs.

In Fig. 1.25 we have placed the sphere from Fig. 1.24 so that P_1 is on the bottom, and *AB* is the trace of the primitive circle on the sphere. The drawing shows the intersection of the sphere with a vertical plane. The chord LM represents the trace of a small circle on the sphere and V is the center of that circle. We wish to prove that the projection of the circle LM, which is CE, is also a circle. The right section (through LN) of the cone, of which LP_1N is a cross section, is therefore an ellipse. Hence LM is the trace of one of this cone's circular sections. QN is then the trace of a conjugate circular section as it is tilted so that $\alpha_1 = \alpha_2$. The plane MR is then drawn parallel to the primitive circle AB.

Hence

(a) angle $RMP_1 = P_1LM$ (they subtend equal angles on the entire circle),
(b) angle P_1LM = angle P_1NQ,
(c) therefore, angle RMP_1 = angle P_1NQ, and
(d) thus QN is parallel to RM and hence to AB. If the conical section QN is a circle, then the conical section CE (on a parallel plane) is also a circle. Note however that the center of the actual circle is *not* the center of the projected circle, i.e., CD does not equal DE.

Great circles at an angle to the axis P_1P_2 also project as circles, (see Fig. 1.24) but if they pass through the axis P_1P_2, they project as straight lines through the center of the projection. Projected great circles cut the primitive circle at diametrically opposite points since such a circle intersects the equator at diametrically opposite points.

1.8.2 The Standard Projection of a Cubic Crystal

As a start in examining the manner of using this projection, let us try to *locate the poles of the faces of the small cubic crystal* in Fig. 1.22 on the projection or primitive circle. We draw a large circle as in Fig. 1.26 and then label North, South, East, and

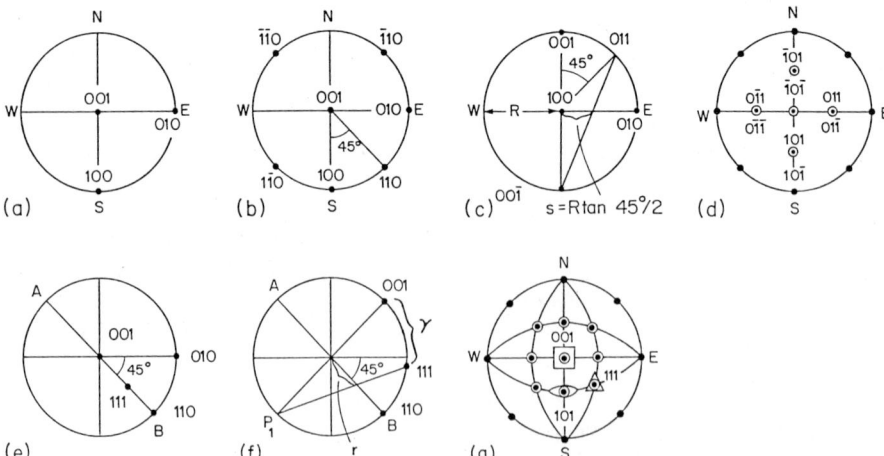

Fig. 1.26a–g. The location of poles of planes shown in Fig. 1.22. Poles above the projection plane are *filled circles*, those below are *open circles*

1.8 Graphical Methods in Crystallography

West on the projection plane, so that as we twist and turn it, we will not lose track of where we started. The (001) face has its pole at the center of the projection. All planes perpendicular to the primitive circle will have poles that lie on the perimeter of this circle. Thus the 100 pole is the "South pole" and the 010 is the "East pole" for this cubic crystal. This stage of our projection is illustrated in Fig. 1.26a, looking down on the projection plane from above. The poles of planes such as the (110), ($\bar{1}$10), (1$\bar{1}$0), and ($\bar{1}\bar{1}$0) also on the primitive circle are readily located if we know the angles between these planes and the {100} (45° in this case as the crystal is cubic).[1] This is illustrated in Fig. 1.26b. The normal to the (011) plane is 45 degrees from the 001 on the great circle through the 001 and the 010 poles. This pole will project somewhere along the East-West axis. Imagine the sphere to be rotated 90° back into the page about this axis so that the plane of projection is a horizontal line. This is the orientation shown in Fig. 1.26c. The pole of the (011) plane is located on the circumference at 45° from the 001 pole and projected back to the 00$\bar{1}$ pole, as shown. Note that the distance on the projection from the center is $s = R\tan(45°/2)$, where R is the radius we chose for the primitive circle. We can therefore locate the remaining {110} poles above and below the projection plane in our original primitive circle, as shown in Fig. 1.26d. The (111) plane makes an angle of $\gamma = 54°44'$ with the (001) plane, and in fact forms a zone with this plane and the (110) plane as can be seen in Fig. 1.22. Thus the normals to all three planes are in a plane which cuts the sphere as a great circle. This great circle containing 111, 110, and 001 poles is 45° from the East-West axis toward the 100 pole. In the projection its trace is the line AB in Fig. 1.26e. We can then locate the projection of the 111 pole exactly along this trace. To do this, we rotate the 001 around the AB axis to the edge of the circles as shown in Fig. 1.26f. The 111 pole would intersect the sphere at an angle of $\gamma = 54°44'$ from the 001 as shown. Its projection on AB is indicated at a distance $r = R\tan(\gamma/2)$ from the center. All of the poles in Fig. 1.22 above and below the primitive plane are shown in Fig. 1.26g. This is denoted a *standard* 001 *projection of cubic crystal*. Figure 1.26g also includes symbols for the rotation axes of symmetry located on the poles. The poles of other (hkl) planes may of course be added to such a standard projection.

1.8.3 Using the Stereographic Projection — The Wulff Net

Great circles contain all the poles to planes in some zone, as can be seen in Fig. 1.22. In Fig. 1.27 a great circle at an angle to the primitive circle is shown. On the plane of projection this circle projects as a longitude line on a map, denoted CD. The normal to the plane of this great circle (the zone axis) intersects the sphere at F' and projects on the primitive circle as shown at F. If we have a pole or zone axis in a

---

[1] The angles between planes (hkl) and $(h'k'l')$ can be calculated in any system from $\mathbf{r}^*_{hkl}$. In all systems but the cubic these angles are of course sensitive to the axial ratios. A table of interplanar angles for the cubic system can be found in the International Tables of Crystallography, Volume II, p. 120 (see Appendix A). Occasionally one can find such a table for some other structure in the literature, but with modern computers the calculations are simple.

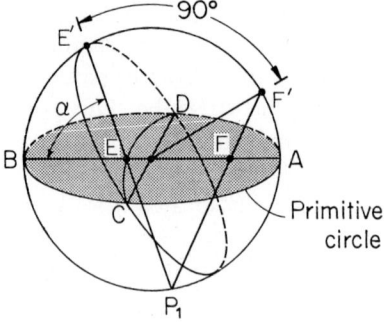

Fig. 1.27. Finding the normal to a plane in a stereographic projection. A plane extended from a tiny crystal in the center of the sphere, intersecting the sphere as a great circle, making an angle α with the primitive circle. The pole of this plane is at F, 90° from the plane

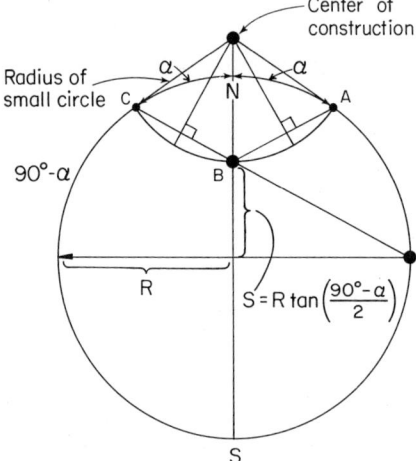

Fig. 1.28. How to draw a small circle of opening α around the North pole in a stereographic projection

projection we can always find the trace of its plane (the arc of longitude CD). We draw diameter AB from pole F through the center of projection, obtain the position of F' on the sphere by rotating about AB and projecting from P_1 to F'. Marking 90° from pole F' to E', we then find the projection of E' at E. We then draw CD perpendicular to AB. Points C, E and D are the construction points for the great circle in projection (three points are sufficient to define a circle). The intersection of the perpendicular bisectors of chords CE and DE is the center of construction of the great circle.

To eliminate repetition of these time-consuming steps each time a great circle is to be drawn, we may construct a net similar to graph paper. We begin by placing longitude lines every few degrees. This can easily be done, by displacing E' in Fig. 1.27 a few degrees at a time, and determining the trace of each great circle. Then we can also draw the projections of small circles or latitude lines by locating three construction points. The points A, B, and C in Fig. 1.28 are such points for a latitude of α from the North pole. Points A and C are located at angles α from the N pole. Point B is also at an angle α from N, but must be located in a manner analogous to that used to locate the 111 pole in Figs. 1.26e and f. Thus B is at a distance

1.8 Graphical Methods in Crystallography

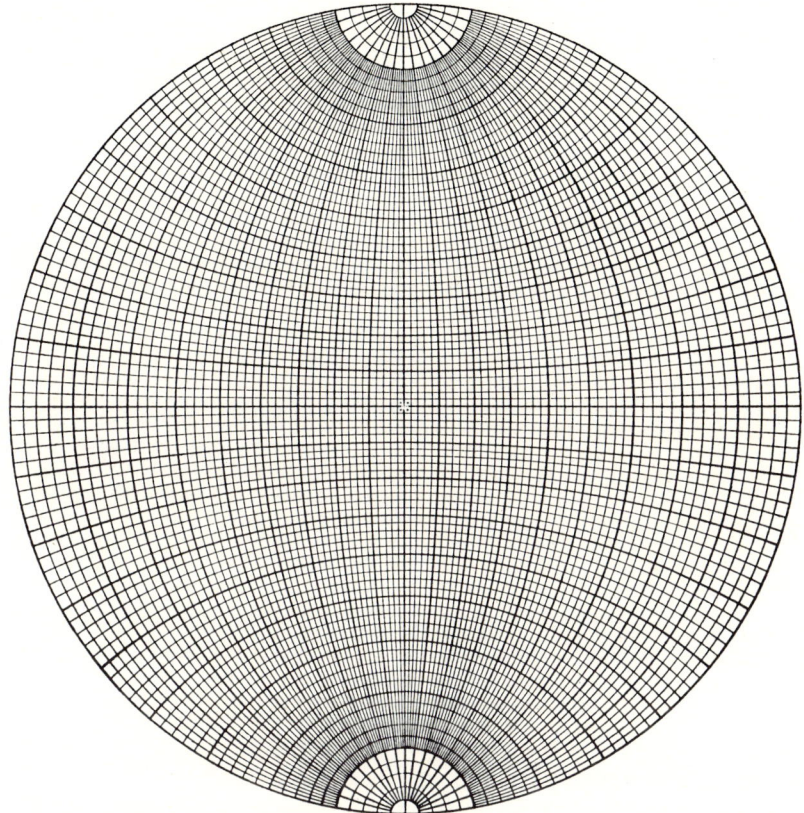

Fig. 1.29. A Wulff net. Intervals of longitude or latitude are 2° and are darker for every 10°. Similar nets can be purchased from Polycrystal Book Service, P.O. Box 27, Western Springs, IL 60558

$S = R \tan(90° - \alpha)/2$ from the center of the projection. After connecting A to B and B to C by chords, the perpendicular bisectors of these chords locate the center of construction. Changing the magnitude of α by several degrees at a time will give a series of small circles. We then have a map on which we can locate poles by their latitude and longitude. Such a *net of longitude and latitude lines is known as a Wulff net* in crystallography and is illustrated in Fig. 1.29 for angular intervals of 2°. Note that the equal degree markings are closer at the center than near the circle's perimeter. (We can see why this is so by differentiating the equation for S given above, with respect to angle.)

Using the Wulff net, many operations can be carried out quite simply. A tracing paper is placed over the net with a pin in the center to hold them together, and the primitive circle drawn, with points indicating North, South, East, and West directions. In what follows, it will be helpful to refer to Fig. 1.22 or to draw some latitude and longitude lines on an old tennis ball. It is often easier to visualize the operations as they would be performed on a sphere, and then make the appropriate adjustments on the projection, than to just work with the projection.

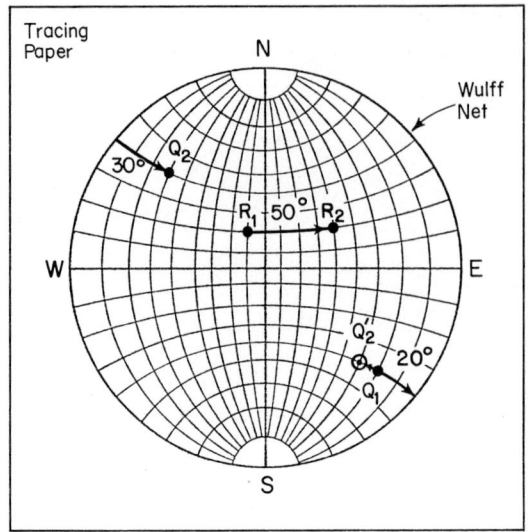

Fig. 1.30. Use of the Wulff net to rotate poles about the N–S axis. (After Cullity, B.D., "Elements of X-ray Diffraction". Addison-Wesley, Reading, Massachusetts, 1956)

(1) The pole at the intersection of two great circles (two planes) gives *the direction of the zone axis* of the two planes, and the trace (intersection direction) of one plane in the other.

(2) The normals or poles in a zone lie on a plane and therefore appear in projection as a great circle. It follows that *to measure the angle between two poles*, we should plot them on the tracing paper and then rotate the paper until both poles lie on the same arc of longitude, i.e., on the same great circle. Their angular separation is then their difference in latitude.

(3) If *a pole is rotated about any axis on a sphere, it will follow a small circle perpendicular to the rotation axis*. Thus rotations on the projection must be along projections of small circles. Hence rotation about the N–S axis moves poles along latitude lines of the Wulff net. For example consider Fig. 1.30. Suppose it is desired to rotate the poles R_1 and Q_1 by 50° from west toward east about the N–S axis. Rotation by 50° takes R_1 to R_2. Similarly rotation by 50° takes Q_1 to Q'_2 on the back of the sphere. As it is usually more convenient to work only with poles on the front of a sphere, we can replace the pole Q'_2 with its diametrical opposite Q_2. Alternately, we may change the center of projection as the pole reaches the perimeter of the Wulff net and continue the rotation along the same arc.

(4) *Rotation around the E–W axis* can be accomplished by placing the E–W axis of the tracing paper coincident with N–S of the Wulff net and following (3). After the rotation, the tracing paper is returned to its original position.

(5) *Rotation about a direction normal to the projection* is accomplished by rotating about the pin at the center, with the angle measured on the perimeter of the Wulff net.

(6) *Rotation around an axis inclined to the plane of the projection* can be accomplished by first rotating this axis to the center or N–S positions. In such an operation *all* poles in the projection *must* be moved in the same manner. After the desired rotation, the

1.8 Graphical Methods in Crystallography

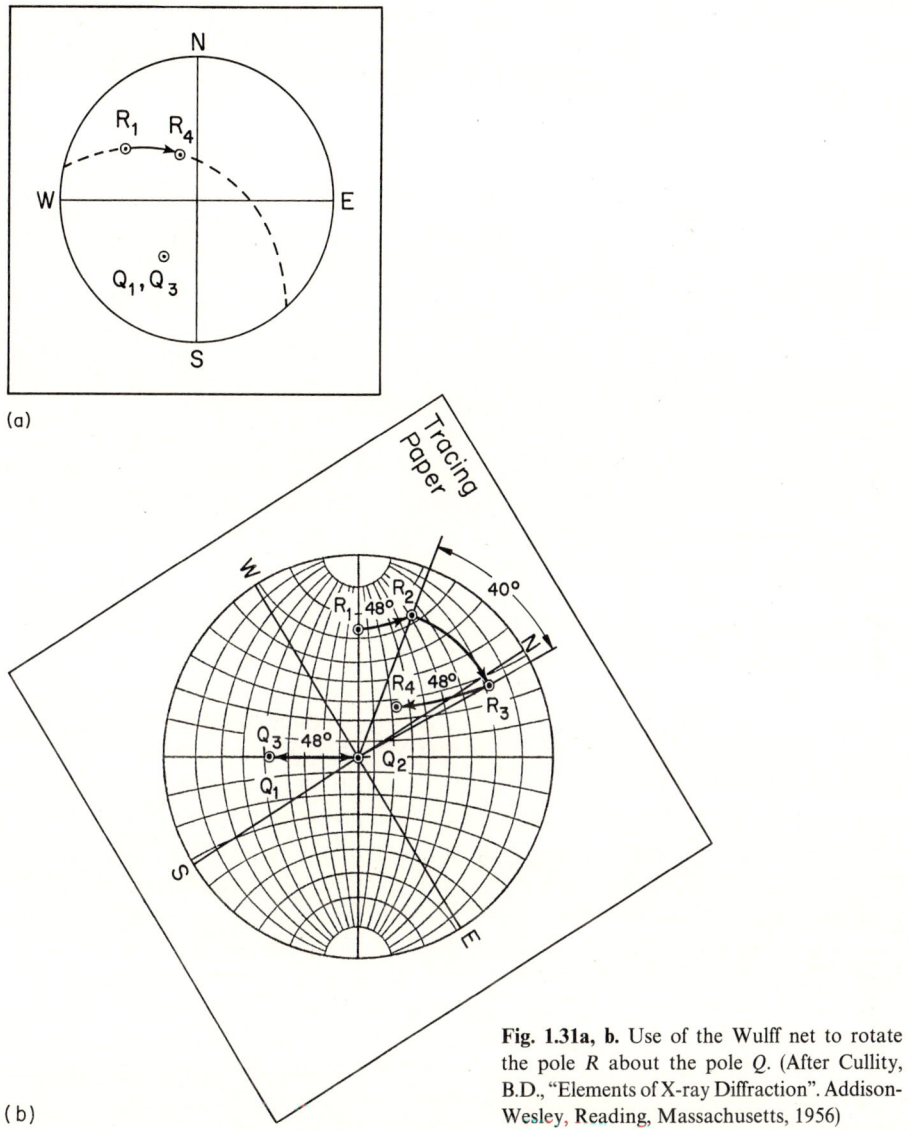

Fig. 1.31a, b. Use of the Wulff net to rotate the pole R about the pole Q. (After Cullity, B.D., "Elements of X-ray Diffraction". Addison-Wesley, Reading, Massachusetts, 1956)

pole under consideration and all others are rotated back until the pole of rotation is in its original position. This is done with the same rotations employed to place this pole in the center or N–S, but in the opposite sequence and sense. As an example, consider rotation of R_1 about Q_1 by an angle of 40° in a clockwise direction as shown in Fig. 1.31. The poles R_1 and Q_1 are shown in their original positions in Fig. 1.31a. The tracing paper is first rotated about the center to bring Q_1 to the equator of the Wulff net. Then a rotation of 48° about the N–S axis of the Wulff net brings Q_1 to the point Q_2 at the center of the net. This rotation takes R_1 to R_2 along a latitude line. Now the rotation about Q_2 is accomplished by marking off

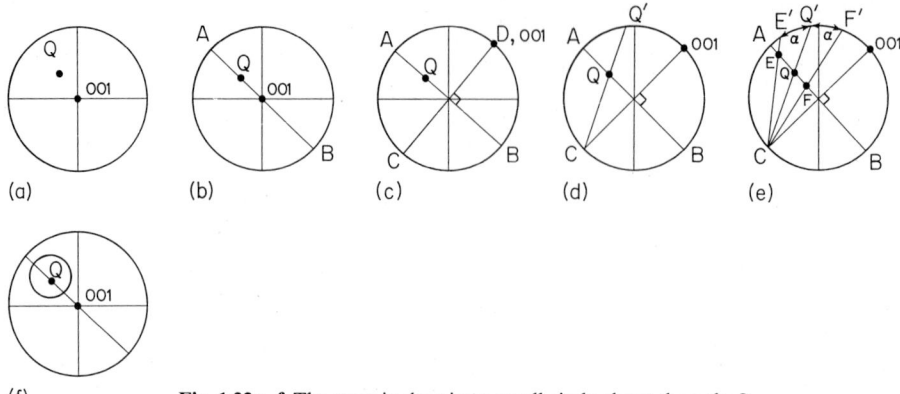

(f)   **Fig. 1.32a–f.** The steps in drawing a small circle about the pole Q

40° on the perimeter of the Wulff net, taking R_2 to R_3. This completes the desired rotation, but now we must return Q_2 to its original position by rotating back 48° about the N–S axis of the Wulff net, taking Q_2 to Q_3 and R_3 to R_4. Finally, we return the tracing paper to its original orientation to display Q_3 and R_4 as shown in Fig. 1.31. Rotation around Q_1 must be on a small circle perpendicular to Q_1, though in this example, the rotation angle is measured along the perimeter (a *small* circle which is a *great* circle as well.)

(7) The *pole (or normal) of any great circle* is readily found along a diameter that bisects the longitude line formed by the projection of its trace. Place the longitude on the tracing paper with its end on the N–S axis of the Wulff net. The E–W axis is then such a diameter. To find the pole, measure 90° along this diameter from the point where it intesects the longitude line.

(8) *A small circle around any pole Q*, the locus of all poles at a given angle α from Q, can be drawn as shown in Fig. 1.32. In (c) after the construction in (b), the projection is imagined to be rotated 90° about AB. The pole is projected to the circumference at Q' and the angles α marked around it. These marks are projected back on to AB, and by bisecting the distance between them, (EF), the construction center for the circle is located. (As mentioned above the construction center is not the pole itself.) Alternately, rotate the tracing paper until the pole Q is on a diameter of a Wulff net, measure the angle α on both sides of the pole along the diameter, bisect the distance for the construction center, draw the circle, and return the tracing paper to its original position.

1.9 Practical Applications of Graphical Methods

We now turn to some of the ways of using this stereographic projection in actual situations. We have so far dealt with a cubic crystal to simplify the presentation of the stereographic projection. Let us go to the opposite extreme and consider how we can plot the axes and poles and index them in *a projection of a triclinic crystal*.

1.9 Practical Applications of Graphical Methods

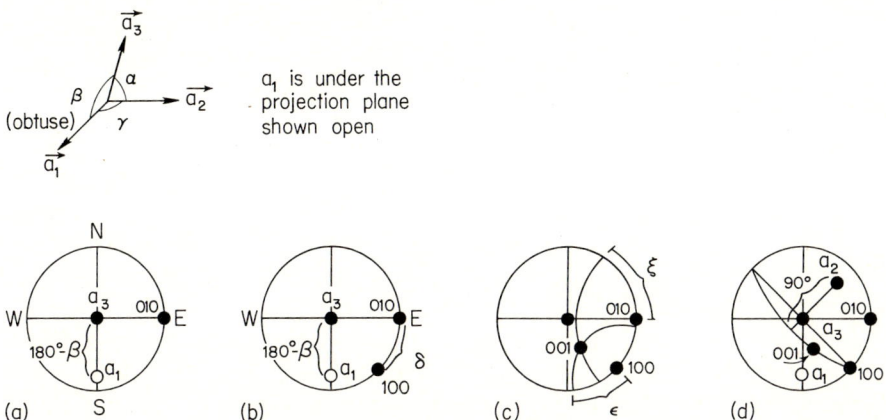

Fig. 1.33a–d. The stereographic projection of a triclinic crystal. (After Phillips, F.C., "An Introduction to Crystallography", 3rd ed. Longmans, Green, London, 1964)

We shall assume that the axial angle β (see Table 1.3) is found to be obtuse when measured on an actual crystal, and that the N–S axis is parallel to the $\mathbf{a}_1 - \mathbf{a}_3$ plane. Furthermore, $\mathbf{a}_3$ will be taken as normal to the projection so that this $\mathbf{a}_3$ axis is the center of our projection. The direction $\mathbf{a}_1$ is located as shown in Fig. 1.33 from knowledge of β. Now, the (010) plane is parallel to the $\mathbf{a}_1$ and $\mathbf{a}_3$ axes. Therefore, its normal or pole is perpendicular to $\mathbf{a}_3$ and must lie on the perimeter of the primitive circle, as shown in Fig. 1.33a. The 010 pole does *not* indicate the position of $\mathbf{a}_2$ in a triclinic crystal. As the (100) plane is parallel to $\mathbf{a}_3$ and $\mathbf{a}_2$ (*not* normal to the $\mathbf{a}_1$ axis), its pole is also on the perimeter. By calculating or measuring the angle δ between the (100) and (010) planes, the 100 pole is located as in Fig. 1.33b. The 001 pole is at the intersection of the circles around 100 and 010 poles of radii equal to the angles ε and ξ, where ε is the angle between (100) and (001), and ξ is the angle between (010) and (001) planes, as shown in Fig. 1.33c. The $\mathbf{a}_2$ axis is then the pole of the zone (that is, the normal to the great circle) containing 001 and 100 poles. This is readily located as in Fig. 1.33d.

Now that we have located the axial directions $\mathbf{a}_i$ and 100 type poles, we can locate any other pole in the projection by calculating angles between it and any two other planes such as the (001) and (010). But suppose instead we are dealing with a real crystal and could measure the required angles but did not know the new plane's indices. Recall that if $\mathbf{N}$ is a unit normal to a plane, $\mathbf{N} \cdot \mathbf{a}_1/h = Na_1/h \cos(Na_1) = d_{hkl}$. We can write two more such equations for $\mathbf{a}_2/k$ and $\mathbf{a}_3/l$. Thus,

$$(Na_1/h)\cos(Na_1) = d_{hkl} = (Na_2/k)\cos(Na_2),$$

or

$$h/k = (a_1/a_2)[\cos(Na_1)/\cos(Na_2)]. \tag{1.11}$$

A similar equation can be written for h/l. We do not need a third equation because $(h/l)/(h/k) = k/l$. We could solve these equations for h, k, l if we knew the $|\mathbf{a}_i|$, but suppose we do not. Recall that a (111) plane defines the axial lengths a_i. If we arbitrarily *choose* some pole to be the 111 pole, the angles between this pole and a_1 and a_2 directions determine the ratio a_1/a_2, and then we can index all other poles. *This arbitrarily chosen* plane is called a *parametric plane* as it defines a set of lattice parameters or unit cell edge lengths. In such indexing procedures, examining the indexed poles around the one being indexed helps considerable in establishing signs and the right sequence for *hkl*. Another useful thing to remember in assigning indices is that a pole at the intersection of two great circles must be the zone axis of the two planes represented by these circles. Of course, when a diffraction study is made, the true lattice parameters can be determined (as we shall see in Chap. 2). Then the Miller indices in terms of the parametric plane may be transformed to new indices using Eq. (1.9f).

If we wished to examine the symmetry of a crystal free from the vagaries of habit, we would proceed as described above to obtain the stereographic projection. This in turn could be examined for its symmetry elements. In order to measure angles between crystal faces for an actual crystal, we can shine a well-collimated light beam on the crystal oriented with a zone axis vertical and record the rotation necessary to obtain the next bright reflection by turning the crystal from one face to another. Then another zone can be examined and so on. A simple device for holding a crystal and tilting in a variety of different ways is the "goniostat" shown in Fig. 1.34. The crystal can be translated or rotated about two different axes that meet at a common point so that the crystal is not displaced during tilting. A

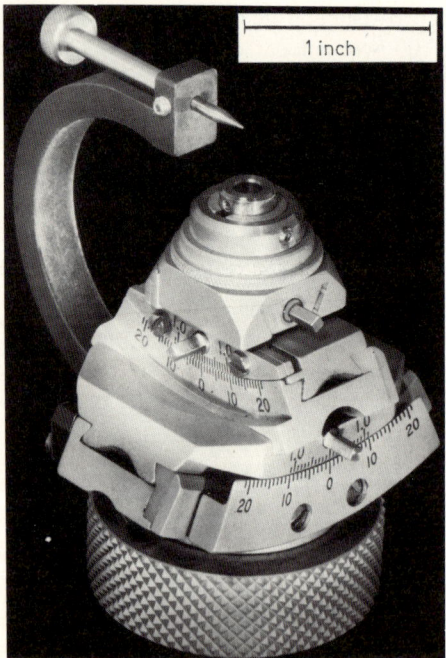

Fig. 1.34. A two-circle goniostat. The needle points to the intersections of the axes of the two arcs. (Courtesy of Electronics and Alloys, Inc., Ridgefield, New Jersey)

1.9 Practical Applications of Graphical Methods

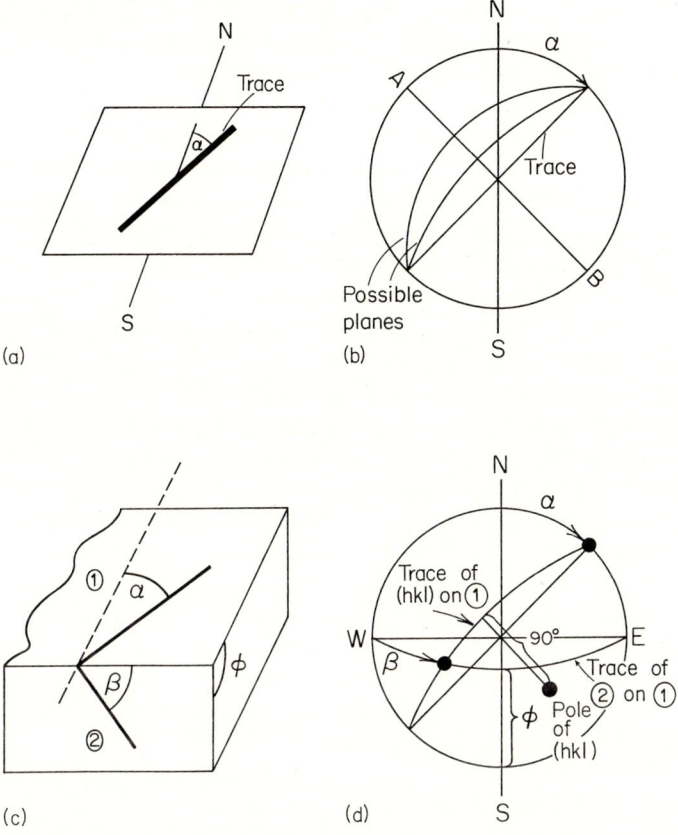

Fig. 1.35. Analysis of traces (**a**) on one surface (**b**) on two surfaces. (After Cullity, B.D., "Elements of X-ray Diffraction". Addison-Wesley, Reading, Massachusetts, 1956)

variety of forms of this device are available, depending on whether the crystal is big or small, whether the crystal is to be cut to expose a certain plane or just examined, etc.

Another useful application of the stereographic projection concerns learning something about *markings on the surface of a crystal*. These might be traces of shear planes due to deformation (slip lines), markings due to a transformation (a precipitate or local change in structure), etc. In Fig. 1.35 we show such a marking on a face of a crystal. It might be informative to know what is the slip plane, or what is the interface between the matrix and transformation product (known as the composition plane).

We shall take the crystal face containing the marking as the plane of projection. The trace of the desired plane must intersect the projection plane at an angle from the N–S axis as shown in Fig. 1.35b, but many planes will do this. From the trace in one plane we cannot tell how the intersecting plane slants back into the crystal, i.e., we do not know the angle shown in Fig. 1.35c. What we can be sure of is that the line *AB* is the locus of the poles of all possible planes. This is simply another statement of the fact that in a zone there are many planes [see Eq. (1.8)]. The trace of one plane on one face represents only the zone axis of the two planes. If there are several similar markings on a crystal face, then we can proceed further. First of all consider the number of possible different orientations of the lines. These cannot exceed the multiplicity (or different possible orientations) of the plane causing the trace. For example, in a cubic crystal there cannot be more than three orientations if a (100) plane is the cause. For each of these different orientations we can construct lines like *AB* as in Fig. 1.35b. Suppose we know the indices of the face. Then we can superimpose on this drawing a standard projection with the pole of the face at the center. We look for poles of one kind of plane (like 112, 211, etc.) that fall on all the lines like *AB* at the same time. This then tells us the indices of the planes causing the trace. But suppose we do not even know the indices of the face we are looking at. We can try many possible standard projections to see if one can be found where this superposition of a form of poles occurs. If we are lucky, this will identify both the face *and* the planes of the traces. This kind of procedure can also be employed in reverse. Suppose we know the planes causing the several orientations of markings (perhaps they are slip lines and we know the preferred slip planes in the crystal). We may then be able to identify the indices of the face.

By examining Eq. (1.8), we can see that to identify the unknown plane without trial and error what we need is its trace on two surfaces (i.e., two sets of *uvw*). If we know the orientation of the crystal so that we can index the traces, that is, determine [*uvw*] the line of intersection in each face, we can obtain the common (*hkl*) of the plane causing the trace from Eq. (1.8). We can also do this graphically, as illustrated in Fig. 1.35c and d. We know the orientation of face (1) and the angle φ between face (1) and face (2). We can therefore draw a stereographic projection with the normal of face (1) at the center, that is with face (1) as the plane of projection. Surface (2) is located as the great circle ϕ degrees from the south axis. As shown in the figure, we can fix the plane causing the traces on the two faces in this projection from the angles α and β measured on the crystal. That is, we can locate the great circle causing the observed traces from α and β. Then we can index its pole with Eq. (1.11). If we do not know the orientation of the crystal, we can still make the projection in Fig. 1.35d for several differently oriented markings. By superimposing different standard projections, we can try to identify the planes of the traces *and* the crystal orientation by trial and error.

In this first chapter we have presented information about the geometry and crystal structure of materials and learned how to deal with it quantitatively. We turn our attention now to the problem of how we can learn about this internal structure experimentally. However, real crystals are not perfectly periodic. The atoms are vibrating. There are imperfections, such as missing atoms (vacancies) and missing portions of planes (dislocations). The crystal is generally "broken up" into regions (subgrains or mosaic structures) tilted slightly with respect to each other. Although these play a key role in the properties, they are low in concentration compared to the atomic density. At first we shall ignore these imperfections, and concentrate on learning about the average periodic atomic arrangement.

References

Azaroff, L.V., "Introduction to Solids." McGraw-Hill, New York, 1960
Azaroff, L.V., "Elements of X-ray Crystallography." McGraw-Hill, New York, 1968
Barrett, C.S., and Massalski, T.B., "Structure of Metals," 3rd ed. McGraw-Hill, New York, 1966
Buerger, M.J., "Elementary Crystallography." Wiley, New York, 1956
Cullity, B.D., "Elements of X-ray Diffraction," 2nd ed. Addison-Wesley, Reading, Massachusetts, 1978
Dana, J.D., "Manual of Mineralogy" (revised by C.S. Hurlbut, Jr.) 18th ed. Wiley, New York, 1971
Holden, A., and Singer, P., "Crystals and Crystal Growing." Doubleday, New York, 1960
Koerber, G.G., "Properties of Solids." Prentice-Hall, New Jersey, 1962
Nye, J.F., "Physical Properties of Crystals." Clarendon, Oxford, 1957
O'Keefe, M., Acta Cryst. $A33$, 924 (1977)
Pearson, W.B., "A Handbook of Lattice Spacings and Structures of Metals and Alloys." Pergamon, New York, 1958
Phillips, F.C., "An Introduction to Crystallography," 4th ed. Wiley, New York, 1971
Sands, D.E., "Introduction to Crystallography." Benjamin, New York, 1969
Schneer, C.J. "The Renaissance Background to Crystallography" in American Scientist, 71, 254–263 (1983)
Wood, E.A., "Crystals and Light." Van Nostrand, Princeton, New Jersey 1964

For Wulff nets, write to: Polycrystal Book Service, P.O. Box 27, Western Springs, IL 60558.

Problems

1.1. Draw several different lines in a two-dimensional lattice. Which have the highest density of points? Which have the greatest interline spacing?

1.2. Consider a two-dimensional crystal. The rate of growth in the four x directions at 90° to each other is $\frac{6}{5}$ that of y directions which are 45° to the x directions. Draw the crystal at various stages of growth. Start with a seed with edges perpendicular to x and y and all of equal length. What conclusions can you reach about the relation between the observed habit of a crystal and the relative rate of growth of low-index (high-density) planes?

1.3. In the structure of diamond there are carbon atoms at the corners and face centers of a cubic unit cell, and also carbon atoms at $\frac{1}{4}, \frac{1}{4}, \frac{1}{4}; \frac{3}{4}, \frac{3}{4}, \frac{1}{4}; \frac{3}{4}, \frac{1}{4}, \frac{3}{4}; \frac{1}{4}, \frac{3}{4}, \frac{3}{4}$. (Note that these last four positions are the coordinates of corner and face atoms plus $\frac{1}{4}, \frac{1}{4}, \frac{1}{4}$.) What is the nearest neighbor distance in terms of the cell edge? How many near neighbors are there?

1.4. Certain intermetallic compounds have the fluorite structure, a cubic structure with four CaF_2 molecules per cell, where the cation has eightfold coordination. Others have the antifluorite structure with the cations and anions reversed; sketch the unit cell of the structure. Give ideal coordinates of the atoms. Calculate the radius ratios that are possible. Look up sizes of Ca^{2+}, F^- ions and compare to the calculated ratios.

1.5. The radius ratio for twelvefold coordination (i.e., 12 nearest neighbors) is unity. What prevents a simple ionic AB compound, having ions of equal size, from adopting a crystal structure in which each atom is coordinated by 12 ions? (*Hint*: What materials do you know that have twelvefold coordination?)

1.6. In a hexagonal close-packed structure, if all the atoms touch and are spherical, what is c/a?

1.7. GeSe is an orthorhombic covalently bonded crystal. Its density is 5.52 gm/cm^3. How many molecules are there in the unit cell? $a_1 = 4.40$, $a_2 = 10.82$ Å, $a_3 = 3.85$ Å.

1.8. (a) Consider a substitutional solution of Fe^{2+} in MgO. Can the Fe^{2+} ion be expected to fit in place of Mg^{2+}? Suppose Fe^{2+} is oxidized to Fe^3; will it fit in any other positions? (The Handbook of Physics and Chemistry contains tables of atomic and ionic radii.)
(b) The spinel $MgFe_2O_4$ precipitates from the solid solution $(Fe_xMg_{1-x})O$. Look up the phase diagram. Examine the crystal structures and the dimensions of the unit cell and matrix. Are there any planes that are similar? What do you expect the precipitates to look like in the microscope?

1.9. (a) Prepare a tracing of the plane pattern shown in Fig. P1.1. Select any convenient point on the tracing and mark all the corresponding points of the lattice.

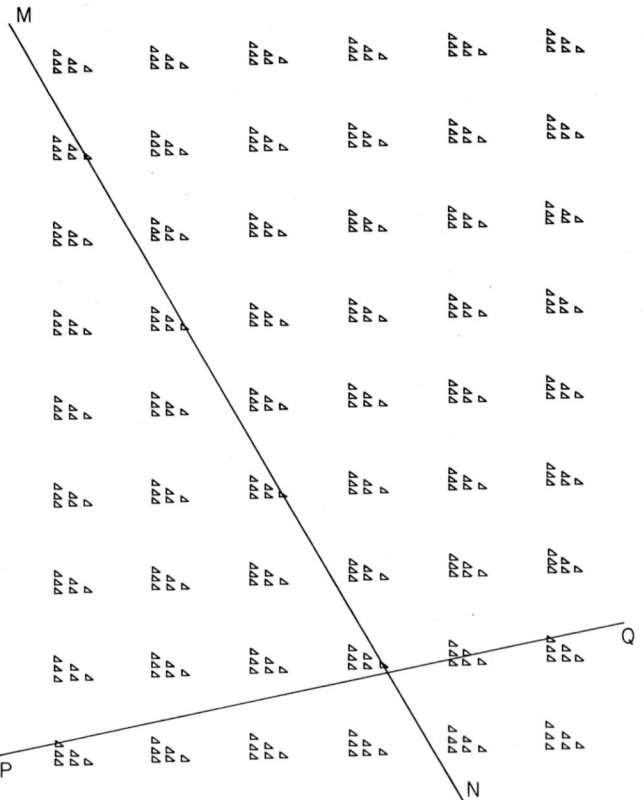

Fig. P1.1

(b) Outline the unit cell in several different ways and measure the cell dimensions a_1 and a_2 and the angle α in each case. (Note that there is *no* rectangular cell.)
(c) Draw a line parallel to MN through an adjacent lattice point and add all the lines of this set. Determine the Miller indices of this set for each of your different choices of unit cell. Treat similarly the set of planes corresponding to the line PQ.

1.10. Sketch the following planes and directions in each of the six crystal systems: $(\bar{1}00)$, $(2\bar{3}4)$, $[12\bar{1}]$.

1.11. Calculate the multiplicity of $\{h00\}$, $\{hk0\}$, and $\{hhh\}$ in the tetragonal and orthorhombic systems.

1.12. Using reciprocal lattice vectors find the angles between (111) and (100) planes in a cube.

1.13. Find the indices of a normal to a plane in terms of the actual crystal axes (i.e., not in terms of reciprocal vectors).

1.14. Show in which of the six crystal systems the normal to a plane always has the same indices as the plane. Are there any special cases?

1.15. What plane contains the zone axes for (111), $(\bar{1}11)$, and (100), $(\bar{1}00)$? Use: (a) reciprocal lattice vectors; (b) stereographic projection.

1.16. The Weiss zone rule says that all poles of possible faces lying in a zone between given faces $(h_1 k_1 l_1)$ and $(h_2 k_2 l_2)$ have indices of the type

$$mh_1 + nh_2, \quad mk_1 + nk_2, \quad ml_1 + nl_2,$$

where m, n are integers. Prove this using reciprocal lattice vectors.

1.17. Find a primitive unit cell for the fcc unit cell and write the equations to transform the Miller indices.

1.18. Using reciprocal lattice vectors, derive an equation for $1/d^2$ for a lattice based on a hexagonal unit cell, in terms of its Miller indices (hkl) and the lattice dimensions a_1 and a_3.

1.19. Using the stereographic projection without a Wulff net: (a) Find the opposite end of a pole. (b) Find all poles 30° from another pole. (c) Find the zone axis of a great circle. (d) Repeat (a)–(c) with a Wulff net.

1.20. Consider the rotation of pole R_1 about the inclined axis Q_1 as shown in Fig. 1.31. Locate a third point S on the small circle through R_1 and R_4, find C, the center of the circle, and draw the arc of the circle through R_1 and R_4.

1.21. Find the traces of the planes whose poles have the following coordinates:
(a) 0°N, 30°E; (b) 20°N, 0°E; (c) 40°S, 15°W.
These angular coordinates are measured from the center of the Wulff net.

1.22. (a) Find the angle between the plane A which is represented by the great circle passing through the N and S poles and the point 0°N, 70°W and the plane B represented by its pole at 30°N, 50°W.

(b) Find the angular coordinates of the intersection line of these two planes. Notice that this point represents a possible pole of a crystallographic plane. Draw in the trace of the plane corresponding to this pole.

1.23. A pole whose coordinates are 20°N, 50°E is to be rotated sequentially around the following axes. Find the coordinates of the final position of this pole and the path traced out during rotation. Be careful to rotate *only* on *small* circles.
(a) 100° rotation about the N–S axis, counterclockwise looking from N to S.
(b) 60° rotation about an axis normal to the plane of projection, clockwise to the observer.
(c) 60° rotation about the inclined axis B, whose coordinates are 10°S, 30°W, clockwise to the observe.

1.24. Construct a standard (001) projection of a cubic crystal showing the poles of {100}, {110}, and {111} planes. On the above projection, find the zone axis of the zone containing poles of ($\bar{1}$10), ($\bar{1}$01), (0$\bar{1}$1), and (1$\bar{1}$0) planes and identify its Miller indices.

1.25. If the indices of a pole are not known, they can be found from the stereographic projection as indicated in the text. This unknown pole represents a direction in space which is a normal to the plane of Miller indices (hkl) and makes angles of α, β, and γ with the coordinates axis a_1, a_2, and a_3. These angles are directly measurable on the stereographic projection. If the distance between the origin of all poles and the closest plane of the set (hkl) is designated as d, we have the following relations from the definition of Miller indices:

$$\cos\alpha = d/(a_1/h), \quad \cos\beta = d/(a_2/k), \quad \cos\gamma = d/(a_3/l)$$

or

$$h:k:l = a_1\cos\alpha : a_2\cos\beta : a_3\cos\gamma .$$

In the cubic system $a_1 = a_2 = a_3$, and therefore, $h:k:l = \cos\alpha : \cos\beta : \cos\gamma$. A further relationship also holds *only in the cubic system*. This is that the normal to a plane (i.e., the pole) and a lattice direction parallel to this normal have identical indices. For instance, the [111] direction is parallel to the 111 pole. On the projection constructed in Problem 1.24, find the indices of a plane in the [$\bar{1}$10] zone and at an angle of 25.2°W from (001). Find the zone axis of the zone containing ($\bar{1}$10), ($\bar{1}$11), and (1$\bar{1}$0). Determine the Miller indices of this zone axis.

1.26. From the (001) projection constructed in Problem 1.24, construct a standard (111) projection showing all poles of planes of the form {100}, {110}, and {111}. Perform this construction using a Wulff net to rotate all the poles from the (001) projection in such a way that the 111 pole is in the center of the new projection.

1.27. Sketch in a stereographic projection the positions of all poles of the {111} planes for a cubic system. Let the center of the projection be a pole normal to one of the cube axes. Index each of the poles. (You may use a Wulff net.) If the crystal suddenly expands normal to the plane of projection, changing from cubic to tetragonal with $a_1/a_3 > 1$, sketch on the projection what happens to these 111 poles.

Problems

1.28. Draw a (100) standard projection for the tetragonal structure of tin.

1.29. The stereographic projection may also be used in conjunction with well-formed crystals to identify the symmetry of the particular crystal as indicated in the text. The technique normally followed is to select a crystal with well-formed faces and place this on an optical goniometer. This instrument consists of a beam of light, a measuring telescope, and a goniometer stage for measuring angles. The crystal is mounted in such a manner that a large number of faces are parallel to the rotation axis. The crystal is then rotated so that the beam of light is reflected from the crystal face into the telescope and the angle of rotation is recorded. The crystal is then rotated until a second reflection is detected in the telescope and this angle is recorded. The difference between these readings is the angle between two crystal planes. The rotation axis which is parallel to the crystal faces is a zone axis. By repeating this procedure for other zone axes one can obtain the angles between all faces of the crystal. The poles of the faces can be plotted on a stereographic projection and the results examined for pertinent symmetry. Certain important directions may be chosen as the axes for the crystal and then one pole lying between the a_1, a_2, and a_3 axes can be selected as the pole of the parametric plane. This parametric plane should be parallel to a crystal face which intersects the axes a_1, a_2, and a_3. Note that the Miller indices of such a parametric plane are defined to be (111). From the geometry of the crystal and the angles on the stereographic projection the ratios a_1/a_2 and a_3/a_2 can be calculated allowing identification of all other poles (see Problem 1.25). Optical examination of a crystal gave the following measurements of interfacial angles along four zones. The faces have been lettered:

(1)	(2)	(3)	(4)
$b \brace m$ 50°53'	$c \brace p$ 71°40'	$b \brace p$ 53°13'	$b \brace n$ 27°43'
$m \brace m$ 78°14'	$p \brace m$ 18°20'	$p \brace l$ 36°47'	$n \brace c$ 62°17'
$m \brace b$ 50°53'		$l \brace p$ 36°47'	$c \brace n$ 62°17'
		$p \brace b$ 53°13'	$n \brace b$ 27°43'

(a) Draw a stereographic projection of the crystal and give its system. (The same letter repeated represents faces which appear similar on the crystal, not necessarily the same face.)
(b) Index the faces and find the axial ratios.
(c) If a (111) plane made traces on intersecting (100) and (001) faces, sketch the angles you would observe on the crystal due to these traces and give the indices of the trace directions.

1.30. Imagine that you are looking down on a (110) plane of a cubic material with [001] running North and South.
(a) With the aid of a stereographic projection sketch give the direction indices of lines made by the intersection (111), (001), (100), and ($\bar{1}$11) planes with the ($\bar{1}$10) plane.
(b) Project the face diagonals of the (100) on this (110) surfaces. What are the indices of these projections?
(c) Indicate how you would also solve (a) and (b) analytically.

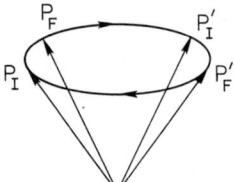

Fig. P1.31

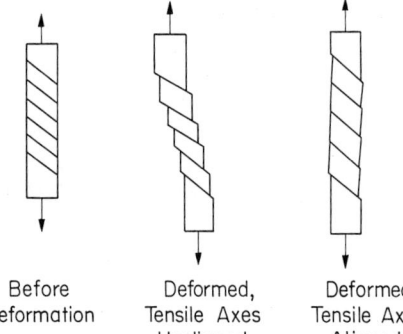

Before Deformation | Deformed, Tensile Axes Unaligned | Deformed, Tensile Axes Aligned

Fig. P1.32. (After Cullity, B.D., "Elements of X-ray Diffraction". Addison-Wesley, Reading, Massachusetts, 1956)

1.31. It sometimes happens that one part of a crystal is rotated about some axis of the rest of the crystal as a result of deformation. Suppose you wish to know what is the rotation axis and the amount of rotation. The situation is represented schematically in Fig. P.1.31. The subscript I represents the major portion of the crystal and F represents the rotated region. You know the location of several poles of both regions, such as P, P'. (These are *not* diametrically opposite to each other.)
(a) Where is $\mathbf{P}_I + \mathbf{P}_F$ with respect to the figure? How could you locate it on a stereographic projection? (The vectors are poles, i.e., unit normals or unit vectors to the planes.)
(b) Where is $\mathbf{P}_I - \mathbf{P}_F$ with respect to the figure and how could you locate it on a stereographic projection?
(c) How could you determine the axis of rotation?
(d) How could you determine the amount of rotation?

1.32. When a single crystal of a ductile material like a metal is deformed, it shears on specific planes. If the top and bottom are held in grips, the shear planes rotate toward being parallel to the stress axis, and so does the shear direction in this plane. See Fig. P.1.32. Make a (100) standard stereographic projection of a cubic crystal. Assume the stress axis is the North direction. In the projection show the slip plane and direction before and after some deformation. Assume shear is on the (111) plane in the [011] direction. [See Barrett and Massalski (1966) for further information about this problem.]

1.33. Fcc crystals often "twin" on {111} planes in [[112]] directions. In such a case, the "twin" is a part of a crystal which is a mirror image of the original crystal.

(a) Pick a plane normal to a {111} plane and containing a [[112]] direction and sketch the atomic arrangement. Include atoms above and below the plane, using different symbols for each level.

(b) Shear part of the crystal along the chosen (111) plane in a [[112]] type direction to obtain a twin. What is the exact index of the direction of shear you have chosen?

(c) On a stereographic projection place poles of the parent and twin. (*Hint*: Note that reflection is equivalent to twofold rotation in the lattice.) Are there any rotation axes or mirror planes? Why?

(d) Replace the face atoms with copper and the corners with gold (Cu_3Au). What is the unit cell? After twinning, what is the unit cell? Can you generalize this as to the effect of a twinning shear on atoms not at lattice points?

(e) For Cu_3Au indicate the motions required to restore the structure in the twin to that of the matrix.

2. The Nature of Diffraction

2.1 Diffraction from a Grating

If we observe a monochromatic beam of electromagnetic radiation sufficiently far from its source, the electric and magnetic fields associated with this beam travel as *transverse plane waves*.[1] By transverse waves we mean that the field oscillates perpendicular to its direction of propagation with maxima and minima, or crests, confined to planes perpendicular to the direction of propagation. Thus waves traveling in the x direction can be expressed as

$$A = A_0 \cos(2\pi/\lambda)(x - vt) . \tag{2.1}$$

The instantaneous magnitude or amplitude of the field A is determined by A_0, its maximum value, λ the wavelength, v the velocity of the wave, and t the time. If we graph this equation, first for time t_1 then for t_2, we note that the wave is moving to the right, as shown in Fig. 2.1. In the time $t = t_2 - t_1$, the wave moves a distance $x' = vt$. Two (or more) such waves from different sources can "interfere" when their paths overlap. They can *constructively interfere* or add if both are oscillating "in phase," that is both are passing through maxima and minima at the same point and time as in Fig. 2.2a; or *destructively interfere* if they are "out-of-phase," that is, if one is passing through a maximum when the other is at a minimum as in Fig. 2.2b. This destruction is complete only if both waves have the same wavelength *and* amplitude. Partial interference is also possible if the waves have different amplitudes or wavelengths. An example of this phenomenon is shown in Fig. 2.2c.

One of the earliest demonstrations of the fact that electromagnetic radiation (visible light) can produce interference effects was performed in 1800 by the English scientist Thomas Young. In his so-called *two-slit experiment*, Young allowed light to pass through an opaque screen containing two tiny slits. The transmitted light from the two slits produced an "interference pattern" of low and high intensity which could be understood by arguments such as led to Fig. 2.2. Young's two-slit experiment is discussed in every elementary physics text as it was crucial in establishing the wave nature of light.

---

[1] The wavefront emitted from a single scattering point is spherical, but may be considered as a plane wave when viewed by a small sample far from the source. When this plane wave approximation is sound, the resultant diffraction is known as Fraunhofer diffraction. On the other hand, when the spherical wavefront must be considered, the resultant is known as Fresnel diffraction. (See Sect. 2.4.1 for further discussion of this point.)

2.1 Diffraction from a Grating

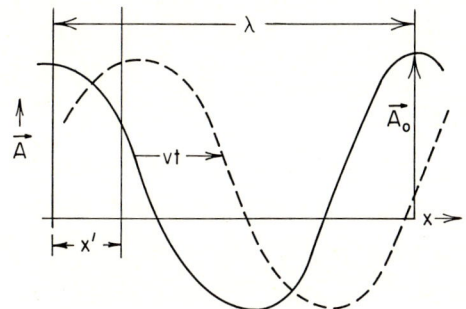

Fig. 2.1. A cosine wave moving to the right

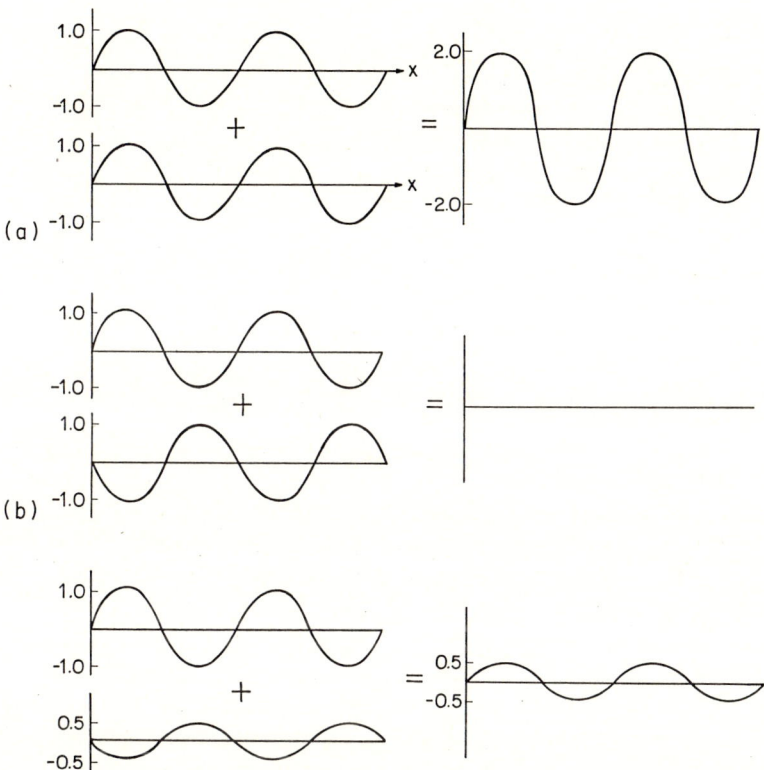

Fig. 2.2a–c. Interference between two waves. Both waves are assumed to have the same maximum amplitude (unity) except in c and the same wavelength in all cases. a Constructive interference; b complete destructive interference; c partial destructive interference

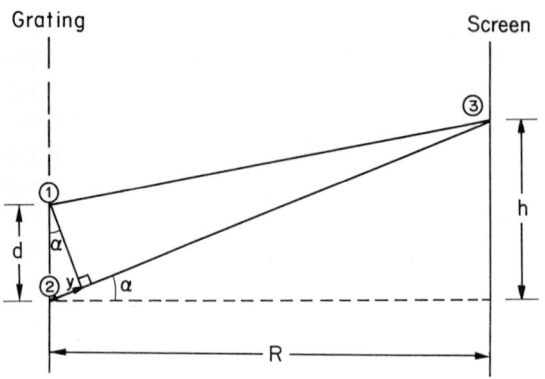

Fig. 2.3. Diffraction from a grating. The directions of the ray from two adjacent slits that intersect on the screen are shown

We shall consider here a generalization of Young's experiment, using a very large number of parallel slits, all the same width, and spaced at regular intervals d as shown in Fig. 2.3. Such an arrangement of slits is known as a diffraction grating. Radiation incident on the grating from the left is transmitted through all the slits, and these transmitted beams interfere, producing a pattern called a *diffraction pattern*. When this diffraction pattern is observed at a distance R very large compared to d, the contribution of each beam may be taken as a plane wave. The problem is to calculate the angles α at which these beams all constructively interfere to produce light on the screen. For two waves from the two adjacent elements (1) and (2) to arrive at (3) in phase, their difference in path y must be an integral number (n) of wavelengths. From the figure, the path difference for rays 1 and 2 is:

$$d \sin \alpha = y = n\lambda . \tag{2.2a}$$

The position of this interference maximum on the screen is at a height h:

$$h = R \tan \alpha , \tag{2.2b}$$

and therefore,

$$\lambda = (dh/nR) \cos \alpha . \tag{2.2c}$$

We can see from Eq. (2.2b) that if the diffraction pattern is to be seen, h/R cannot be too small for then α is so small that the scattered beams are too close to the direct beam. From Eq. (2.2a) we see that λ must be of the same order of magnitude as d; if it is very much smaller, again the first constructive interference ($n = 1$) will be too close to the direct beam to be seen. From the same equation it is also clear that d cannot be smaller than λ, for then $\sin \alpha$ is greater than unity. This requirement that $d \approx \lambda$ was well recognized in the study of diffraction of visible light at the turn of the century, but its implications for the study of the atomic arrangements in solids awaited one of those magic moments in science in which the right person is at the right place at the right time.

In 1912 there was a young German professor in Munich, Max von Laue, who was inspired by a discussion with a doctoral candidate of another professor, Sommerfeld. This student, Ewald, was studying refraction in periodic structures. Since many solids were thought to be made up of atoms (or groups of atoms) in periodic arrangements, Ewald was looking for the effects of this periodicity on refraction.

2.1 Diffraction from a Grating

Laue's primary interest was optics, and he had just finished a chapter for a book on diffraction from gratings. He wondered what would happen if radiation propagating in a periodic structure of atoms had a wavelength about that of the spacing of the atoms ... a diffraction pattern *should* occur. His colleagues discouraged him. Even if this was theoretically possible, say with x-rays (which were thought at the time to have a wavelength of the order of atomic spacings in crystals), thermal vibrations of the atoms would destroy any interference. Furthermore, it was not at all certain at that time that x-rays *were* waves. However, there was a strong crystallographic group present at the same university and Laue was aware that if this experiment could be carried out, it would be quite important for it could prove in one single experiment that crystals were periodic and that x-rays *were* waves. He persuaded Sommerfeld to allow one of his postdoctoral assistants, Friederich, to help him try this out. Friederich was about to study the nature of x-rays, anyway. Knipping, who had just finished his Ph.D. degree with Röntgen, the discoverer of x-rays, volunteered to assist for he knew how to use the equipment to produce x-rays, and this would minimize the time Friederich would be away from his work. Friederich's research equipment was soon ready, and they found a diffraction pattern from a crystal in a few days! In this one beautiful experiment they proved that *many solids were indeed periodic atomic arrays, and that x-rays were waves with a wavelength of the dimensions of the spacings in crystals*. This one beautiful piece of research has led to all our knowledge about the atomic arrangements in metals, alloys, ceramics, polymers, and biological molecules. It has also played a key role in the development of quantum theory and the modern theory of the solid state.

In this chapter, we are going to concentrate on learning about this diffraction process from materials, but we are going to postpone to a later chapter any detailed discussion of how atoms scatter, or what types of radiation are useful, and how they are produced.[2] We shall also make the following assumptions:

(1) Waves from the source can be treated as *traveling waves*, like ripples on water.
(2) The *path difference is a linear function of the spacing* of the scattering elements, as in Fig. 2.3. This implies we are far from the grating elements. (If the measuring position is close to the grating compared to the spacing of the grating elements, this is not the case.)
(3) There is *conservation of energy in the scattering process*. That is, there is no change in energy (and hence wavelength) of the scattered radiation compared to the incident radiation.
(4) *A once-scattered beam inside a material does not rescatter.*

These ideas form the basis of the *kinematic theory of diffraction*. We shall examine all of these assumptions in later chapters, but shall put that off to first look at the fundamentals of scattering of plane waves by a periodic grating. We do so because these assumptions are generally valid in the most widely used methods for studying diffraction from materials.

---

[2] We should keep in mind that the discussion applies not only to electromagnetic waves like light or x-rays, but also to the scattering of electrons and neutrons. As shown by de Broglie, when a particle of mass m is moving with velocity v, it behaves as if it were a wave of wavelength $\lambda = h/mv$, where h is Planck's constant.

2.2 Diffraction from Planes of Atoms

2.2.1 Bragg's Law

Consider a series of atom planes, struck by an incident beam of radiation with some wavelength of the order of the interplanar spacing d. In Fig. 2.4 a section through a series of planes is shown; the planes are seen from their edges. The atoms scatter the incident rays in some fashion to be explored later. We will assume the angle of incidence equals the angle of diffraction and justify this assumption at the end of this chapter. If the path difference GEH between rays ABC and DEF is an integral multiple (n) of λ, the scattered rays will reinforce and produce diffraction. Now,

$$GE = EH = d \sin \theta .$$

Therefore, for diffraction from planes,

$$n\lambda = 2d \sin \theta . \tag{2.3}$$

This *Bragg's law* was derived by Sir W.L. Bragg in 1912 just after he had completed his graduate training, while working with his father who was involved in studies of x-rays. They were most excited by Laue's findings, but Laue's formulation of the result appeared too complex.

Note that in Eq. (2.3) *if 2d is less than λ, no diffraction is possible*, for this would require $\sin \theta$ to be greater than unity. But if $2d$ is greater than λ, we can observe different *orders of diffraction*, that is $n = 1, 2, \ldots$, at different angles. One such higher-order reflection is shown in Fig. 2.4 as a dashed line. If we plot the magnitude of the resultant amplitude versus $n/d = (2 \sin \theta)/\lambda$, we would expect the pattern shown in Fig. 2.5. The peaks are sharp as there are not just two waves adding, but

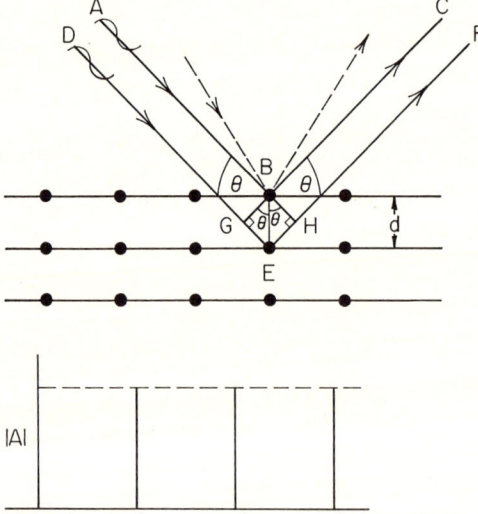

Fig. 2.4. Scattering from planes (seen "end-on") at an angle θ. The *dashed lines* represent a different incident and scattered direction but as with the *solid lines* they are at equal incidence and scattering angles

Fig. 2.5. Amplitude vs. $2 \sin \theta / \lambda$

2.2 Diffraction from Planes of Atoms

many from many planes, at some depth until the beam is absorbed (we shall shortly prove this).

2.2.2 Interplanar Interference

Suppose now that we interpose planes with the same types of atoms, at $d/4$, as in Fig. 2.6. For the first-order diffraction ($n = 1$) which results from a path difference λ between scattered beams from the planes A and B, the planes a and a' have path differences with A and B of $\lambda/4$, and scattering from these planes partially destroys the amplitude for $n = 1$. For $n = 2$ (planes like A and B scatter with a path difference of 2λ), the scattering from plane a (a') has a path difference with that from A (B) of $\lambda/2$. These waves are completely out-of-phase with those from A (B) and eliminate the peak that would occur if these extra atoms were not present. For $n = 3$ (third order) there is again only partial destructive interference. But at the fourth order, all planes are "in phase." (Those at $d/4$ have a scattering path exactly one wavelength different than that for scattering from A or B.) The resultant pattern is given in Fig. 2.7, showing the amplitude scattered versus n/d.

Now suppose that the atoms on the planes at $d/4$ are different from these on A, B, etc. The scattered waves, while still out-of-phase will not completely cancel since the scattered amplitudes are no longer the same for each plane. Then the total scattered amplitude will not completely vanish for the second order ($n = 2$); instead, there will be a small peak for which the amplitude depends on the difference in the scattering of the different types of atoms. Such intermediate planes of atoms may be required by the symmetry of the crystal we are investigating.

Let us now apply these thoughts to a real structure in a way analogous to the method used by Sirs W.H. and W.L. Bragg, father and son, in their pioneering studies of crystal structures. Sodium chloride crystals are known to be cubic from optical studies. (They have four threefold axes, see Table 1.3). We shall measure the magnitude of the amplitude of scattering of x-rays from (100), (110), and (111) planes with crystals tilted to allow these planes to make equal angles of incidence and

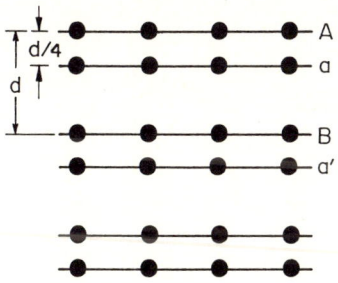

Fig. 2.6. Each plane (A, B, etc.) has another plane of atoms $d/4$ below it

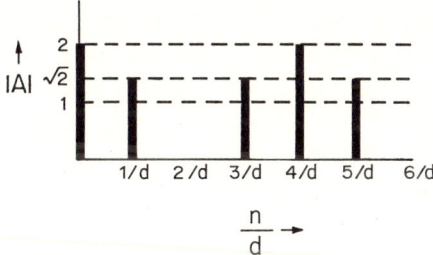

Fig. 2.7. Amplitude vs. $2\sin\theta/\lambda$ for the planes in Fig. 2.6. (Compare to Fig. 2.5)

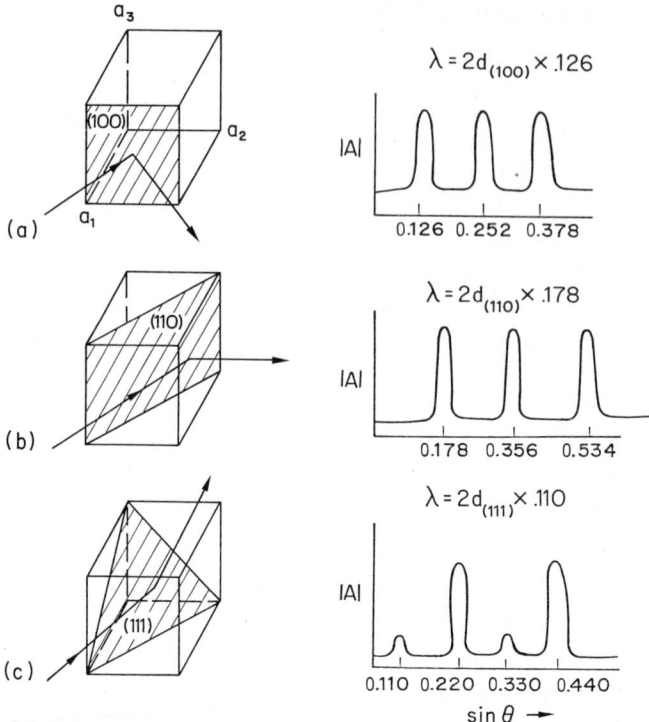

Fig. 2.8a–c. Diffraction from different faces of an NaCl crystal. [After Bragg, W.H., Proc. Roy. Soc. London *A89* 246 (1913)]

scattering, as in Fig. 2.8. We shall move a detector which records this scattering from the crystal[3] at the same time, to maintain the equality of angles. The device which accomplishes this motion is known as a diffractometer. Looking at the resultant patterns in Fig. 2.8, we might assume that the first diffraction peak in (a), (b), and (c) correspond to scattering from planes with interplanar spacings $d_{(100)}$, $d_{(110)}$, and $d_{(111)}$. We would like to use Bragg's law, Eq. (2.3) to obtain these d values from the measured $\sin\theta$, but at the time the Braggs did this work, there were no accurate values for the λ of x-rays. However, if the crystal is cubic, the ratios of interplanar spacings may be calculated. For example, $d_{(100)}/d_{(110)} = \sqrt{2}$. [The reader should be certain he can prove this either from simple geometry or from the relation for a cubic crystal $1/d^2 = (h^2 + k^2 + l^2)/a^2$ which can be derived from $r^*_{hkl} \cdot r^*_{hkl} = 1/d^2$.] Calculating this d ratio from the patterns in Fig. 2.8 and Bragg's law eliminates λ, and we can see that the results are in agreement with the expected ratio for the first peaks in Figs. 2.8a and b, i.e., $0.178/0.126 = \sqrt{2}$ and $d_{(100)}/d_{(110)} = \sqrt{2}$. Now $d_{(111)}/d_{(100)} = 1/\sqrt{3}$ for a cube, but from the measurements in Figs. 2.8a,

---

[3] Actually, a detector measures the scattered energy which is proportional to A^2. We will discuss the operation of such detectors in Chap. 3, but basically this energy exposes film or ionizes a media or produces electron-hole pairs.

2.2 Diffraction from Planes of Atoms

and 2.8c, $d_{(111)}/d_{(100)} = 0.126/0.110 = 2/\sqrt{3}$. Something is wrong... or is it? Notice that in the pattern from the (111) planes, the first and third peaks are weaker than the second and fourth suggesting partial cancellation of the diffracted waves for the former peaks. Let us then assume that in NaCl, the (111) planes are alternating planes of Na$^+$ and Cl$^-$ ions. If there is a (111) Cl$^-$ plane halfway between two Na$^+$ planes, the pattern in Fig. 2.8c is exactly what we would expect from our previous discussion of partial destructive interference. The Cl$^-$ is out of phase with respect to that from the Na$^+$ planes and partially cancels it. For this case, the correct expression for the first peak is not

$$\lambda = 2d_{(111)} \sin \theta,$$

but

$$\lambda/2 = 2d_{(111)} \sin \theta, \quad \text{or} \quad \lambda = 2(2d_{(111)} \sin \theta).$$

If this is employed with the patterns in Figs. 2.8a and c,

$$d_{(111)}/d_{(100)} = 1/\sqrt{3}$$

as it should. A structure for NaCl made up of alternating (111) Na$^+$ and Cl$^-$ planes has been shown in Fig. 1.15c. In this figure we can also see that there are planes between the {100} cell faces and also planes between the {110} planes compared to a simple cube. Furthermore, in both cases these planes and those in between them are the same composition, half Na$^+$ and half Cl$^-$. The scattering from these planes and the half way planes should be the same and the first peaks in Figs. 2.8a and b must therefore be for $n = 2$. The first order peak ($n = 1$) has zero amplitude due to destructive interference, but because there are "halfway" planes for (100), (110), and (111) types, it is still true that $d_{(100)}/d_{(111)} = 1\sqrt{3}$ and $d_{(100)}/d_{(110)} = \sqrt{2}$. We have therefore arrived at a structure which qualitatively satisfies the actual pattern. We want to know the actual atomic coordinates and we can get that from the scattering amplitudes, but we shall postpone the details at this point. To confirm the atomic arrangement, the Braggs examined KCl crystals having the same external crystal symmetry. It seemed reasonable, since these are similar salts, that the structure was the same. If the scattering depends on the number of electrons as it should if x-rays are electromagnetic radiation, K$^+$ and Cl$^-$ scatter x-rays identically, since their electron number is the same. There should be complete destructive interference for $n = 1, 3$, etc., in the pattern from (111) planes, and hence the small peaks in Fig. 2.8c should be missing; they are!

We can obtain even more information from this experiment now that the structure is known. We may write an expression for the density ρ of the crystal in terms of the cell edge a. From Fig. 1.15c there are four NaCl "molecules" per cell. Thus,

$$\rho = (\text{molecular weight} \times 4)/a^3 N_0,$$

where N_0 is Avogadro's number. *From experimental values of ρ and the molecular weight, we can calculate a and hence the d values for Bragg's law.* From the patterns in Fig. 2.8, λ can then be obtained and is henceforth a known quantity in diffraction experiments. This experiment was indeed a brilliant example of simple logical thinking.

Although this discussion reveals the essential beauty of the analysis of structure by diffraction methods, we can see many problems in this approach. It certainly would be difficult to handle complex structures. Secondly, in the derivation of Bragg's law, we have *assumed* that the angle of incidence equals the angle of diffraction. While that is true for reflection (and it was Sir W.L. Bragg's basic idea that scattering from each plane was like a reflection that led to his law), how do we know this is right? We do not get a pattern at all angles, as with a mirror, but just at certain angles. We need then to look more closely at the process of diffraction.

2.3 Diffraction from a One-Dimensional Crystal

2.3.1 Diffraction from a Row of Like Atoms

In Fig. 2.3 we saw that radiation transmitted through multiple slits formed maxima at certain angles due to constructive interference. Now let us extend these ideas to a scattering problem by considering a one-dimensional collection of N scattering points separated by identical distances a. Let the amplitude of radiation scattered by the j-th point be represented by A_j. We need not use a vector notation for amplitude as all the waves that will concern us are transverse. The situation is illustrated in Fig. 2.9. The incident beam arrives at the bottom of the row and plane waves are found (at some distance from the scattering points) propagating in a direction at an angle α to the incoming beam. A dashed reference plane is shown which is the maximum in the cosine wave. All other lines represent the plane waves at other than maximum amplitude, and are positions on the wave behind the maximum. At any one line we can add the waves from each source. Using Eq. (2.1), let $2\pi v t/\lambda = \phi$, and $2\pi x/\lambda = \beta$. If the phase difference β involves the path difference (x) between waves from points 0 and 1, then between 0 and 2 the phase difference is 2β, and so on. Thus at any one line or wave position in Fig. 2.9, the

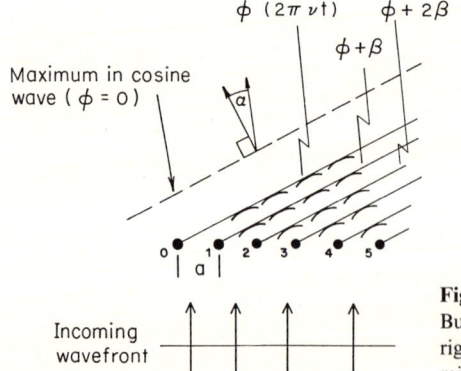

Fig. 2.9. Scattering from a row of atoms. [After Buerger, M.J., "Crystal Structure Analysis." Copyright 1960 by John Wiley & Sons. Reprinted by permission of John Wiley & Sons.]

2.3 Diffraction from a One-Dimensional Crystal

total amplitude A_T is[4]

$$A_T = A_0 \cos\phi + A_1 \cos(\phi + \beta) + A_2 \cos(\phi + 2\beta) + \ldots = \sum_{j=0}^{N-1} A_j \cos(\phi + j\beta)$$

$$= \cos\phi \sum_{j=0}^{N-1} A_j \cos j\beta - \sin\phi \sum_{j=0}^{N-1} A_j \sin j\beta .$$

We leave the A's in the sum, as in the general case the scattering points or atoms need not be all identical. Let

$$\sum_j A_j \cos j\beta = X, \qquad \sum_j A_j \sin j\beta = Y.$$

Then we can write

$$A_T = X \cos\phi - Y \sin\phi$$

$$= (X^2 + Y^2)^{1/2} \left(\frac{X}{(X^2 + Y^2)^{1/2}} \cos\phi - \frac{Y}{(X^2 + Y^2)^{1/2}} \sin\phi \right)$$

$$= (X^2 + Y^2)^{1/2} (\cos\gamma \cos\phi - \sin\gamma \sin\phi),$$

where $\tan\gamma = Y/X$. Thus

$$A_T = \left[\left(\sum_j A_j \cos j\beta \right)^2 + \left(\sum_j A_j \sin j\beta \right)^2 \right]^{1/2} \cos(\phi + \gamma)$$

$$= A'_T \cos(\phi + \gamma). \tag{2.3a}$$

The final form shows us that *the total amplitude is itself a single traveling wave with a new amplitude A'_T*. With ripples on a liquid we could measure the resultant amplitude, but with x-rays and other forms of short wavelength electromagnetic radiation we can only measure energy — the blackening of a film or ionization of a gas in a detector — not the amplitude. In fact, *we can only measure average intensity* (average energy per unit time, per unit area of the beam) because it takes a finite time for us to make a measurement. Consider a volume in the incoming or scattered beam which for x-rays are both oscillating electric fields. The energy per unit volume is proportional to the square of the field, so we must square both sides of Eq. (2.3a). To average the energy over time we integrate the squared amplitude over a period of oscillation (the period is the inverse of the frequency ν where $c = \nu\lambda$). In so doing we arrive at one-half times the square of the maximum amplitude $(A'_T)^2$ which is the term under the square root in Eq. (2.3a), i.e., the time average intensity $\langle A_T^2 \rangle$ is

$$\langle A_T^2 \rangle = \left(\sum_j A_j \cos j\beta \right)^2 + \left(\sum_j A_j \sin j\beta \right)^2 . \tag{2.3b}$$

The constant of one-half appears in the expressions for both incident and scattered

---

[4] The reader may wonder why the arguments are $\phi + \beta$ not $\beta - \phi$ as in Eq. (3.1). The incoming wave reaches all the points at the same time; so the wave scattered from the jth point is behind the one scattered by the point at $(j-1)$ at the line where we are summing.

intensities and hence cancels. That is, the constant appears in front of $\langle A_T^2 \rangle$, but also in front of the A_j terms on the right, as these depend on the incident intensities. We see that we can write intensities as amplitudes squared.

We can obtain the same result if we write waves in complex exponential notation. These are actually easier to manipulate than sines and cosines. We can express A_T as

$$A_T = \sum_j A_j e^{-i(\phi + j\beta)} = e^{-i\phi} \sum_j A_j e^{-ij\beta} \tag{2.4a}$$

or

$$A_T = e^{-i\phi} \left\{ \sum_j A_j \cos j\beta - i \sum_j A_j \sin j\beta \right\}. \tag{2.4b}$$

If we now multiply A_T by its *complex conjugate* A_T^* (the conjugate is formed by replacing all i's with $-i$), and recall that $i^2 = -1$:

$$A_T A_T^* = A_T^2 = \left(\sum_j A_j \cos j\beta \right)^2 + \left(\sum_j A_j \sin j\beta \right)^2. \tag{2.4c}$$

This is exactly the same result as that of Eq. (2.3b). It does not matter that the exponential form for a real amplitude contains imaginary terms; they are *not* present in the average intensity. As this is all we can measure and the complex form turns out to be a great computational aid, we shall employ it from now on.

2.3.2 Diffraction Viewed Graphically — The Argand Diagram

We can carry out the sum indicated in Eqs. (2.3) or (2.4) graphically in the complex plane in an *Argand diagram* as in Fig. 2.10 in which the axes are for real and imaginary components. Assume all the A_j's are identical and denoted f. We need not consider ϕ as it cancels when we form $A_T A_T^*$. Therefore, we put the first f along the horizontal axis, and then vectorially add another f to this one at an angle β with respect to the first, and so on. The vector sum has a component along the

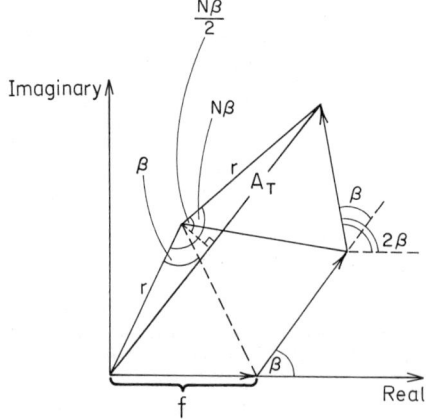

Fig. 2.10. Addition of waves of amplitude f

vertical axis which is the sine sum in Eqs. (2.3c) or (2.4c) because each vector has a component on the imaginary axis which is $f \sin j\beta$. Similarly, the vector sum has a horizontal component which is the cosine sum. We show just a few vectors and then notice that there is a circle of radius r that we can circumscribe around the partial polygon that forms. Now

$$\sin(\beta/2) = \frac{f/2}{r}, \quad \text{or} \quad r = \frac{f/2}{\sin(\beta/2)}.$$

Similarly,

$$\sin\left(\frac{N\beta}{2}\right) = \frac{1/2|A_T|}{r}, \quad \text{or} \quad r = \frac{1/2|A_T|}{\sin(N\beta/2)}.$$

Therefore,

$$|A_T| = f\frac{\sin(N\beta/2)}{\sin(\beta/2)}. \tag{2.5a}$$

Now A_T has a phase angle which is equal to the angle the middle vector makes with the real axis when N is odd, or the average of the angles of the two middle vectors when N is even, so that the phase factor can be written as $\exp[i(N-1)\beta/2]$. Then

$$A_T = f\frac{\sin(N\beta/2)}{\sin(\beta/2)}e^{i(N-1)\beta/2} \tag{2.5b}$$

and

$$|A_T|^2 = A_T A_T^* = \frac{f^2 \sin^2(N\beta/2)}{\sin^2(\beta/2)}. \tag{2.6}$$

[Again, we could multiply Eq. (2.5) by $e^{-i\phi}$ to include this term as well, but it vanishes in the intensity, Eq. (2.6), so we ignore this term.]

The reader might wonder what happens when the circle is complete or wraps partially around a second time. This will actually occur for any large N because $\beta = 2\pi x/\lambda$ is a large angle. (x is the path difference between two beams from two successive points in a row and x must be the order of magnitude of λ as we have already seen). But our equation is perfectly satisfactory even in this situation; as N increases, $\sin(N\beta/2)$ cycles through 2π and repeats as does the exponential term. For those who prefer a more mathematical derivation of Eq. (2.6), this is given in the footnote.[5]

---

[5] We start with Eq. (2.4b). The term $e^{-i\phi}$ can be ignored as it vanishes in forming the intensity (when multiplying A_T by A_T^*). We will assume all the A_j are identical and can therefore be factored from the sum. The remaining terms are a geometric series:

$$\frac{A_T}{A} = \sum_{j=0}^{N-1} e^{-ij\beta} = a + ar + ar^2 + \dots,$$

where $a = 1$, and $r = e^{-i\beta}$. Then,

$$\sum_{j=0}^{N-1} e^{-ij\beta} = \frac{a(r^N - 1)}{r - 1} = \frac{e^{-iN\beta} - 1}{e^{-i\beta} - 1}.$$

Multiplying by the complex conjugate to form the intensity, we obtain

$$\frac{A_T A_T^*}{A^2} = \frac{2 - e^{-iN\beta} - e^{+iN\beta}}{2 - e^{-i\beta} - e^{+i\beta}} = \frac{2 - 2\cos N\beta}{2 - 2\cos\beta} = \frac{\sin^2(N\beta/2)}{\sin^2(\beta/2)}.$$

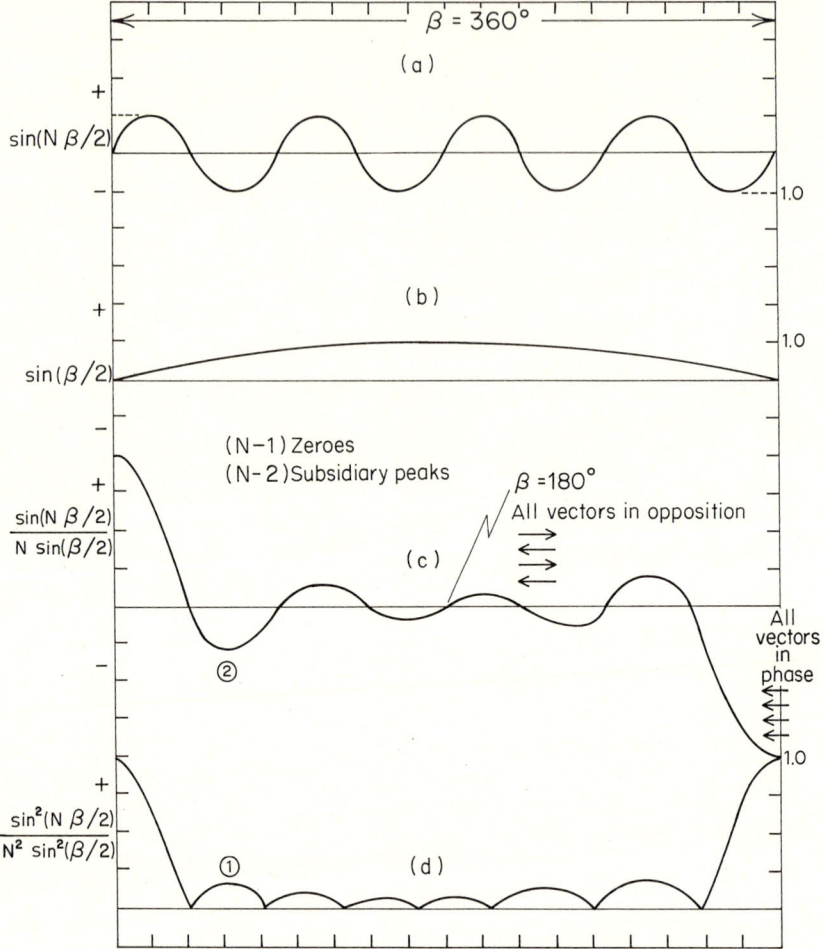

Fig. 2.11a–d. The interference function for $N = 8$. There are $N - 1$ zeros in **d** and $N - 2$ subsidiary maxima besides the two large ones for $\beta = 0$ and $360°$. Curves **c** and **d** have been normalized to unity. [After Buerger, M.J., "Crystal Structure Analysis." Copyright 1960 by John Wiley & Sons. Reprinted by permission of John Wiley & Sons]

Ignoring f, the rest of Eq. (2.5a) (for amplitude) or Eq. (2.6) (for intensity) are often referred to as *the interference functions*. In Figs. 2.11a and b, the numerator and denominator of the interference function are plotted versus β for $N = 8$, as well as the amplitude and intensity interference functions in (c) and (d). For large N, the numerator is varying much more rapidly than the denominator and we can locate the maxima and minima and other features by considering only the numerator. From this figure we can make the following statements:

(1) The largest value of the amplitude, N, occurs for $N\beta/2 = Nh\pi$, or $\beta = h2\pi$, where h is an integer. The maximum value of the intensity is N^2, also at $\beta = h2\pi$. The

2.3 Diffraction from a One-Dimensional Crystal

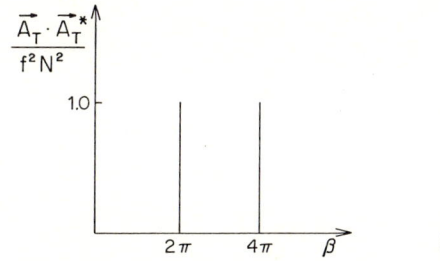

Fig. 2.12. Intensity vs. β when N is large

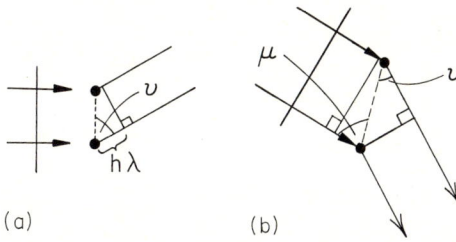

Fig. 2.13. Location of diffraction peaks for (a) normal incidence, and (b) inclined incidence

second maximum in intensity at (1) in Fig. 2.11d, which is the first minimum at (2) in Fig. 2.11c, occurs at about $N\beta/2 = 3\pi/2 + Nh\pi$. Putting this value for $N\beta/2$ in Eq. (2.6), the intensity of this second maximum is found to be only 5% of the first maximum, for $N = 8$. This second maximum and all other maxima get smaller rapidly as N increases. There are $(N - 2)$ of these subsidiary maxima in intensity.
(2) $(N - 1)$ zero values of amplitude and intensity occur when $N\beta/2 = Nh\pi + \pi$, i.e, for $\beta = 2h\pi + 2\pi/N$. Thus the width of the large peak in the interference function is of the order of $2(2\pi/N)$ in terms of β. If the maximum in intensity is N^2 and the width is proportional to $1/N$, the area is proportional to N. Therefore the largest peak sharpens and increases in area as N increases. For very large N, the pattern appears as shown in Fig. 2.12.
(3) We can see physically why the zeros in amplitude occur. Referring to Fig. 2.13a, the path difference between two adjacent scattering points is $a(\cos v)$. Then the phase difference is $2\pi a(\cos v)/\lambda$. As the first zero occurs for $\beta = 2\pi/N + 2h\pi$,

$$(2\pi a \cos v)/\lambda = (2\pi/N) + 2h\pi, \quad \text{or} \quad Na \cos v = (Nh + 1)\lambda.$$

Thus, for this scattering angle, the path difference between the first and Nth atom is only one wavelength and all the intermediate points destructively interfere. The first maximum occurs for $\beta = h2\pi$. For this scattering angle v',

$$(2\pi a \cos v')/\lambda = 2\pi h$$

or

$$a \cos v' = h\lambda. \tag{2.7a}$$

Therefore a large maximum in scattering amplitude occurs when the path difference between successive scattering points is λ (or a multiple of λ). That is, the maximum occurs when all the vectors representing the scattering points are parallel and have a phase angle of 360°.
(4) We can find the directions of the large maxima by searching for the angle for which the path difference between two successive points is an integral multiple of λ. If the incident beam is normal to the row, Fig. 2.13a,

$$\cos v = h\lambda/a.$$

More generally, if the beam is incident at an angle μ as in Fig. 2.13b,

$$\cos v - \cos \mu = h\lambda/a . \tag{2.7b}$$

Note that *the directions of the maxima depend only on the interpoint distance a for a given λ, while the intensity depends on f^2, the "scattering ability" of each point, and on N^2, the number of scattering points.*

2.3.3 Conditions for Diffraction — The Ewald Sphere

Actually, Eq. (2.7) describes cones of maxima in the radiation about the row of scattering points. To see this more clearly, we can write Eqs. (2.7a) and (2.7b) in vector notation. Let $\mathbf{S_0}$ be a unit vector in the direction of the incident beam and $\mathbf{S}$ be a unit vector in the direction of the diffracted beam. Then Eq. (2.7b) becomes (see Fig. 2.13)

$$(\mathbf{S} - \mathbf{S_0}) \cdot \mathbf{a} = h\lambda , \tag{2.8}$$

or, if $\mathbf{n}$ is a unit vector in the direction of $\mathbf{a}$,

$$[(\mathbf{S} - \mathbf{S_0})/\lambda] \cdot \mathbf{n} = h/a . \tag{2.9}$$

In Fig. 2.14 we have sketched the meaning of these equations. If we form the vector difference $(\mathbf{S} - \mathbf{S_0})/\lambda$, the vector $\mathbf{s}$ results. Equation (2.9) tells us that the component of $\mathbf{s}$ along $\mathbf{a}$ must be confined to planes $\pm 1/a$, $\pm 2/a$, etc., apart. This condition is satisfied for all vectors $\mathbf{S}/\lambda$ which lie on cones around the row of scattering points. The cone for $-1/a$ only is shown.

An alternative construction shown in Fig. 2.15 reveals more of the meaning of Eq. (2.9). The planes at separation $1/a$ represent the right-hand side of Eq. (2.9). The reader may generalize from our discussion of reciprocal lattice vectors in Chap. 1

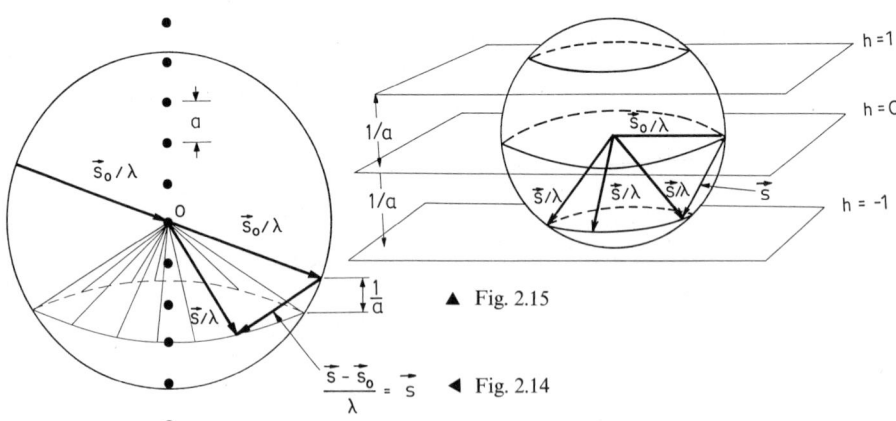

◀ Fig. 2.15

◀ Fig. 2.14

Fig. 2.14. One cone of scattering around a row of atoms. The incident beam is represented by unit vector $\mathbf{S_0}$, the diffracted beam by unit vector $\mathbf{S}$

Fig. 2.15. Ewald sphere representation of diffraction from a row of scattering points

2.3 Diffraction from a One-Dimensional Crystal

to note that *this set of planes is actually the reciprocal lattice for the one-dimensional row of points of separation a*. To find all possible solutions of Eq. (2.9) for a given incident beam direction $\mathbf{S}_0$, draw a sphere of radius $1/\lambda$ (recall that $|\mathbf{S}_0| = 1$) tangent to a reciprocal lattice plane. This sphere is known as *the Ewald sphere*. *The intersections of this Ewald sphere with the reciprocal lattice planes are the solutions to* Eq. (2.9). That is, the required left-hand side of Eq. (2.9) occurs for vectors $\mathbf{S}$ to the intersection circles, since for such vectors, the vector $\mathbf{s} = (\mathbf{S} - \mathbf{S}_0)/\lambda$ has projection on the unit vector $\mathbf{n} = \mathbf{a}/|\mathbf{a}|$ of $\pm h/a$. The figure is drawn to show the cone of $\mathbf{S}$ vectors which is allowed for $h = -1$.

Referring once again to the analysis of the Argand diagram, we note that for $\beta = 360°$, when N is even, the phase is such that the amplitude is negative; $\exp[i(N - 1)/\beta/2]$ is $e^{\pi i}$ or $e^{3\pi i}$, etc., which has the value (-1). When N is odd, the amplitude is positive. However, if we can only measure average intensity $A_T A_T^*$, we can never tell the difference.

Consider our wave addition as a polygon of vectors as in Fig. 2.10: (a) The radius r gets smaller and then bigger as β increases, with fixed N. (β varies with the spacing of the scatterers and the position of observation around them.) The resultant A_T is from the origin to the tip of the last vector, that is, to the end of the partial polygon. (b) For fixed β, as N increases, the circle has a fixed radius, but closes and opens.

If a is less than λ, no maxima can occur, as we can see from Eq. (2.7). As a becomes smaller and approaches zero, $\sin^2 \beta/2$ approaches $\beta^2/4$ and

$$|A_T|^2 = (4 \sin^2 N\beta/2)/\beta^2 .$$

This function has only one large maximum, for $N\beta/2 = 0$. Thus there is one maximum for $a \leq \lambda/2$ in the direction of the incoming beam. The beam is traveling right through the material. In this case, the Ewald sphere lies entirely between the planes in reciprocal space.

2.3.4 Diffraction from a Row of Unlike Atoms

It is a simple matter to extend this analysis to points of different scattering power, where a set of points or "molecules" is repeated at equal intervals, as in Fig. 2.16.

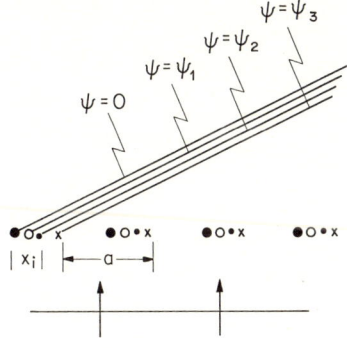

Fig. 2.16. Scattering from a periodic group of different scattering elements. [After Buerger, M.J., "Crystal Structure Analysis." Copyright 1960 by John Wiley & Sons. Reprinted by permission of John Wiley & Sons.]

We can consider each type of point as a row, spaced equal values of a apart. Thus each type represents the same problem we have just treated. But each is some distance (less than a) from another type. There is some phase difference between the rows of each type. Hence we can immediately write for the four types in the molecule (setting the common factor $e^{-i\phi} = 1$ since it drops out in the intensity):

$$A_T = A_0 \sum_{j=0}^{N-1} \exp(-ij\beta) + A_1 \exp(-i\psi_1) \sum_{j=0}^{N-1} \exp(-ij\beta)$$

$$+ A_2 \exp(-i\psi_2) \sum_{j=0}^{N-1} \exp(-ij\beta) + A_3 \exp(-i\psi_3) \sum_{j=0}^{N-1} \exp(-ij\beta) \quad (2.10a)$$

$$= [A_0 + A_1 \exp(-i\psi_1) + A_2 \exp(-i\psi_2) + A_3 \exp(-i\psi_3)] \sum_{j=0}^{N-1} \exp(-ij\beta) \quad (2.10b)$$

$$= A_m \sum_{j=0}^{N-1} \exp(-ij\beta). \quad (2.10c)$$

The terms in brackets in Eq. (2.10b) represent the combination of wavelets from points within one group. Note also that the same sum $\sum e^{-ij\beta}$ appears as with our simple row of scattering elements. Therefore, diffraction peaks will appear in the same place or angle, around the row; their *position is clearly dependent only on the spacing of the net, a*. However, *the intensity at any peak depends on the detailed structure of the collection of points being repeated at each net point*.

For the hth order diffraction, we have seen that identical points are $2\pi h$ out of phase. However, consider the relation between the different types of points in Fig. 2.16.

$$\psi_i / X_i = h2\pi/a, \quad \text{or} \quad \psi_i = (hX_i/a)2\pi.$$

Therefore, the *structure factor* A_m of Eq. (2.10c) may be written as:

$$A_m = A_0 + A_1 \exp[-2\pi i h X_1/a] + A_2 \exp[-2\pi i h X_2/a]$$

$$+ A_3 \exp[-2\pi i h X_3/a].$$

Letting x_n represent the fractional coordinate along the a axis ($x_n = X_n/a$), we obtain:

$$A_m = \sum_{n=0}^{3} A_n \exp(-2\pi i h x_n). \quad (2.11)$$

As an example of a structure factor calculation, let us consider a specific one-dimensional structure. We shall take the case of one "molecule" per cell consisting of C and D atoms located at coordinates $1/4a$, and $1/2a$, respectively. From Eq. (2.11), then

$$A_m = \sum_{n=0}^{1} A_n \exp(-2\pi i h x_n)$$

$$= A_C \exp[-2\pi i h(1/4)] + A_D \exp(-2\pi i h(1/2))$$

$$= \exp(-\pi i h/2)[A_C + A_D \exp(-\pi i h/2)].$$

2.4 Diffraction from a Three-Dimensional Crystal

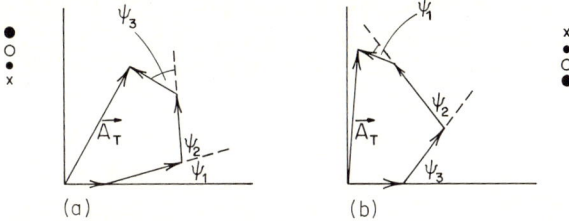

Fig. 2.17a, b. The magnitude of the total amplitude scattered from the molecule, $|A_T|$, is the same for both arrangements of scattering elements

The intensity or amplitude squared is

$$A_m^2 = \begin{cases} (A_C + A_D)^2, & h = 4n, \ n \text{ an integer}, \\ (A_C - A_D)^2, & h = 4n + 2, \\ (A_C^2 + A_D^2), & h \text{ odd}. \end{cases}$$

We see that interference between scattering from the two atoms in the unit cell produces partial cancellation for $h = 4n + 2$ and reinforcement for $h = 4n$, while for h odd the atoms scatter independently.

In this example, the structure amplitude A_m is real for h even, but is complex for h odd and must be described by its magnitude and phase. We can see this by simply representing the terms in Eq. (2.11) on a vector diagram in the complex plane, as we did for the simple row; this is done in Fig. 2.17. If we can only measure intensity, we lose the phase and can only obtain A_m^2. As shown in the figure, we lose information about the atomic arrangement (in this case the orientation) that we could have if the amplitude could be measured, not just its magnitude.

2.4 Diffraction from a Three-Dimensional Crystal

2.4.1 Derivation of the Diffraction Conditions

We are now ready to generalize the results we have just obtained to analyze diffraction from a three-dimensional body. Consider scattering originating from the two points labeled 0 and j in Fig. 2.18 and intercepted at a detector some distance away. The two scattered beam paths d and q differ by an amount obtainable from trigonometry:

$$q^2 = d^2 + r^2 - 2dr\cos\beta, \quad q = d\left[1 + \left(\frac{r^2}{d^2} - \frac{2r\cos\beta}{d}\right)\right]^{1/2}.$$

For $d/r \gg 1$, we may expand the square root and retain only the first two terms:

$$q \cong d\left[1 + \frac{1}{2}\left(\frac{r^2}{d^2} - \frac{2r\cos\beta}{d}\right)\right].$$

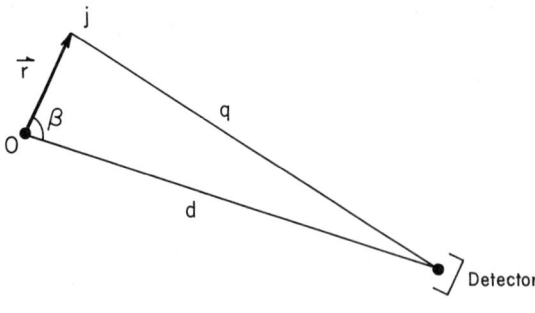

Fig. 2.18. The geometry of the path lengths to a point on the detector

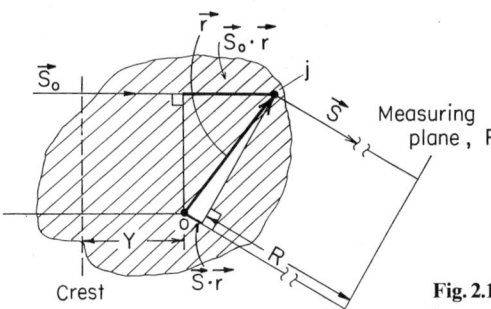

Fig. 2.19. Diffraction geometry for a volume of scattering material

For $d/r \gg 1$, we neglect the term in r^2/d^2 and the path difference $d - q \cong r\cos\beta$. If **S** is a unit vector in the scattered direction, $d - q \cong \mathbf{S}\cdot\mathbf{r}$. This approximation is equivalent to assuming that the scattered beams from points separated by r are parallel. In the usual diffraction experiment, $d \approx 0.1$ m, so the approximation is valid for $r \gtrsim 10^5$ Å. However, the more complete form for $(q - d)$ should be used whenever coherent contributions to the amplitude arise from portions of the sample more widely spaced than $\approx 10^5$ Å. Such a situation may arise in the scattering from perfect crystals discussed in Chap. 8. Usually imperfections in the material prevent this and we will assume parallel scattered beams in this chapter.

We now consider scattering from a chunk of matter, with a beam of radiation of wavelength λ, defined in direction by the unit vector $\mathbf{S}_0$ and incident at 0 in Fig. 2.19. This beam is scattered by all the scattering elements in the chunk (electrons, nuclei). We wish to find the resultant amplitude at a detector on some plane of observation P, at a large distance R from the chunk and in the direction defined by the unit vector **S**. Since the distances from source to sample and sample to detector are both large compared to the distance r, we use the above results and assume that all incident rays are parallel and all scattered rays are parallel. Then the incoming rays are all in phase when they reach the vertical line through 0. The difference in path of the two beams shown is $(Y + R + \mathbf{S}_0\cdot\mathbf{r}) - (Y + R + \mathbf{S}\cdot\mathbf{r}) = -(\mathbf{S} - \mathbf{S}_0)\cdot\mathbf{r}$. If the scattered amplitude per unit volume of material at position $\mathbf{r}$ is $(\rho_\mathbf{r} - \rho_0)$, where ρ_0 is the density of scattering ability of the surrounding matter, the amplitude scattered by a small volume $dV_\mathbf{r}$ is $(\rho_\mathbf{r} - \rho_0)\,dV_\mathbf{r}$. Then the instantaneous amplitude

2.4 Diffraction from a Three-Dimensional Crystal

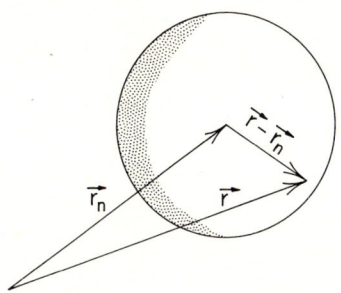

Fig. 2.20. Vector to the nucleus of the nth atom ($\mathbf{r}_n$) and the vector $\mathbf{r} - \mathbf{r}_n$ around the nucleus

of this scattered wave which reaches the measuring plane at distance R is given by

$$A(\mathbf{S}) = \varepsilon_I(\rho_\mathbf{r} - \rho_0)\,dV_\mathbf{r} \exp\left\{-\frac{2\pi i}{\lambda}[-(\mathbf{S} - \mathbf{S}_0)\cdot\mathbf{r} - vt]\right\},$$

where ε_I is the incident amplitude when the beam reaches 0, and $(\mathbf{S} - \mathbf{S}_0)\cdot\mathbf{r}$ is the additional distance the top beam must travel compared to that through the origin, 0. The total scattered amplitude, $A_T(\mathbf{S})$, is

$$A_T(\mathbf{S}) = \int \varepsilon_I(\rho_\mathbf{r} - \rho_0)\exp\left[\frac{2\pi i}{\lambda}(\mathbf{S} - \mathbf{S}_0)\cdot\mathbf{r} + vt\right]dV_\mathbf{r}. \tag{2.12}$$

Letting $(\mathbf{S} - \mathbf{S}_0)/\lambda = \mathbf{s}$, the scattering vector,

$$A_T(\mathbf{s}) = \exp[2\pi i/\lambda vt]\varepsilon_I \int (\rho_\mathbf{r} - \rho_0)\exp(2\pi i \mathbf{s}\cdot\mathbf{r})\,dV_\mathbf{r}. \tag{2.13}$$

Thus, the total scattered amplitude is proportional to an integral (known as the *Fourier transform*[6]) of a function, in this case the density of scattering ability per unit volume relative to that of the surrounding material. In a measurement of intensity $A_T(\mathbf{s})A_T^*(\mathbf{s})$, the exponential term before the integral drops out. Also considering ε_I as some plane wave (of the form Ae^{ix}), only the square of the magnitude of the incident amplitude will appear in the expression for average intensity.

We now have to specify the appropriate function for $(\rho_\mathbf{r} - \rho_0)$. We shall concentrate for the remainder of this discussion on scattering from a dense solid in a gaseous surrounding material like air, so we may neglect ρ_0. (In Sect. 7.7 we shall return to the use of $(\rho_\mathbf{r} - \rho_0)$ for the discussion of "small angle scattering.") We can consider $\rho_\mathbf{r}$ as the product of the scattering ability per unit scattering element times the density of scattering elements. Suppose the scattering is due to the electrons around the nucleus of an atom. Then the first part of the product (the scattering per electron) can be included as a constant k. We shall discuss this constant in Chap. 3. The second part is the electron density $\rho'(\mathbf{r})$.

Consider a group of N atoms. Let $\mathbf{r}_n$ be the vector to the nucleus of the nth atom. Then $\mathbf{r}$ is the vector to some point around the nucleus as shown in Fig. 2.20. For the most part, in this book we shall be interested in crystalline materials. Then we may write $\mathbf{r}_n$ in terms of the lattice vectors $\mathbf{a}_1$, $\mathbf{a}_2$, and $\mathbf{a}_3$ as

---

[6] The definition of the Fourier transform and its utility in diffraction analysis is discussed in Chap. 6.

$$\mathbf{r}_n = n_1 \mathbf{a}_1 + n_2 \mathbf{a}_2 + n_3 \mathbf{a}_3,$$

where n_1, n_2, and n_3 are integers or fractions. If the electron density of one atom is $\rho'(\mathbf{r})$, then the electron density of the crystal is that for each atom, repeated at each lattice point. We may write this as a sum in three dimensions:

$$\rho'(\mathbf{r}) = \sum_{n_1} \sum_{n_2} \sum_{n_3} \rho'(\mathbf{r} - \mathbf{r}_n) = \sum_n \rho'(\mathbf{r} - \mathbf{r}_n), \quad (2.14a)$$

where $\sum_n$ is a shorthand notation for the triple sum $\sum_{n_1}$, $\sum_{n_2}$, $\sum_{n_3}$, and n is interpreted as the triplet of numbers n_1, n_2, n_3. Substituting in Eq. (2.13):

$$A_T(\mathbf{s}) = \varepsilon_I k \sum_n \int \rho'(\mathbf{r} - \mathbf{r}_n) \exp(2\pi i \mathbf{s} \cdot \mathbf{r}) \, dV_\mathbf{r}. \quad (2.14b)$$

Multiplying by $\exp[(2\pi i \mathbf{s} \cdot \mathbf{r}_n) - (2\pi i \mathbf{s} \cdot \mathbf{r}_n)] \doteq 1$:

$$A_T(\mathbf{s}) = \varepsilon_I k \sum_n \left[\int \rho(\mathbf{r} - \mathbf{r}_n) \exp\{2\pi i \mathbf{s} \cdot (\mathbf{r} - \mathbf{r}_n)\} \, dV_\mathbf{r} \right] \exp(2\pi i \mathbf{s} \cdot \mathbf{r}_n). \quad (2.14c)$$

The term in brackets is the *scattering ability per atom relative to the unit scattering ability* (k), and we shall write this term as f_n. If, for example, the scattering is due to the electrons, then the scattering of an atom relative to the scattering of an electron is called *the atomic scattering factor* f_n. Equation (2.14c) can thus be simplified to

$$A_T(\mathbf{s}) = \varepsilon_I k \sum_n f_n \exp(2\pi i \mathbf{s} \cdot \mathbf{r}_n). \quad (2.15)$$

The atomic scattering factor *could* have an imaginary component [see Eq. (2.14c)] which we ignore at this point. We shall return to this in Chap. 3. If f_n is real, then in measuring $A_T(\mathbf{s}) A_T^*(\mathbf{s}) = |A(\mathbf{s})|^2$, we can only determine the modulus $|A_T(\mathbf{s})|$ as we saw for the one-dimensional case. We shall see below that this means that the structure will express itself in our measurement as if it has a center of symmetry even if it does not.

2.4.2 The Structure Factor and the Laue Conditions

Suppose we consider $A_T(\mathbf{s})$ as the scattering from one molecule or unit cell in a structure. Then Eq. (2.14b) (without the constants $k\varepsilon_I$) is the Fourier transform of the density of scatterers in the molecule or unit cell. We shall call *the amplitude per unit of structure* [Eq. (2.15) summed over the atoms in the unit cell] *the structure factor* $F(\mathbf{s})$. We drop the constant $k\varepsilon_I$ in what follows for simplicity. We shall then be writing the amplitude per unit cell or structure factor in terms of the scattering per unit scatterer, e.g., for x-rays, the scattering relative to the scattering by one electron; in this case our amplitude is in "electron units." Thus *for a unit cell containing M atoms, the structure factor is*

$$F(\mathbf{s}) = \sum_{n=0}^{M-1} f_n \exp(2\pi i \mathbf{s} \cdot \mathbf{r}_n). \quad (2.16)$$

Now suppose we displace the molecule or unit cell by the vector $\mathbf{a}$:

2.4 Diffraction from a Three-Dimensional Crystal

$$F_a(\mathbf{s}) = \sum_n f_n \exp[2\pi i \mathbf{s} \cdot (\mathbf{r}_n + \mathbf{a})] = \exp(2\pi i \mathbf{s} \cdot \mathbf{a}) \sum f_n \exp(2\pi i \mathbf{s} \cdot \mathbf{r}_n)$$

$$= \exp(2\pi i \mathbf{s} \cdot \mathbf{a}) F(\mathbf{s}).$$

There is an exponential term due to the displacement of the unit cell which depends on the size of the displacement $|\mathbf{a}|$. However, when we measure the *intensity*, this term disappears; although there will be regions of high and low intensity due to $F(\mathbf{s})$, we cannot get the *absolute* position of a group of atoms from such a measurement.

If we now consider *two* molecules or cells *separated* by $\mathbf{a}$:

$$A_T(\mathbf{s}) = F(\mathbf{s}) + F_a(\mathbf{s}) = F(\mathbf{s})(1 + e^{2\pi i \mathbf{s} \cdot \mathbf{a}})$$

$$= \begin{cases} 2F(\mathbf{s}), & \text{for } \mathbf{s} \cdot \mathbf{a} = 0, 1, 2, \ldots \\ 0, & \text{for } \mathbf{s} \cdot \mathbf{a} = \frac{1}{2}, \frac{3}{2}, \ldots, \end{cases}$$

and the total intensity I_T is

$$I_T = \begin{cases} 4|F(\mathbf{s})|^2, & \text{for } \mathbf{s} \cdot \mathbf{a} = 0, 1, 2, \ldots \\ 0, & \text{for } \mathbf{s} \cdot \mathbf{a} = \frac{1}{2}, \frac{3}{2}, \ldots. \end{cases}$$

The maximum intensity is four times that for one molecule at some places in $\mathbf{s}$ space while the intensity disappears on planes defined by $\mathbf{s} \cdot \mathbf{a} = \frac{1}{2}, \frac{3}{2}$, etc.

If we have a row of N unit cells or molecules spaced a apart,

$$A_T(\mathbf{s}) = F(\mathbf{s}) \sum_{n=0}^{N-1} e^{2\pi i \mathbf{s} \cdot n\mathbf{a}}. \tag{2.17a}$$

The reader may compare this result with that derived in Eqs. (2.4) and (2.5) for a row of scattering points. We have already seen that, for large N, such a sum is a set of continuous planes spaced $1/a$ apart. The sum can be written, by analogy with Eq. (2.5), as

$$A_T(\mathbf{s}) = F(\mathbf{s}) \frac{\sin \pi \mathbf{s} \cdot N\mathbf{a}}{\sin \pi \mathbf{s} \cdot \mathbf{a}} e^{\pi i \mathbf{s} \cdot (N-1)\mathbf{a}}. \tag{2.17b}$$

For large N the condition for a large amplitude is that $\pi \mathbf{s} \cdot \mathbf{a} = h\pi$ or $\mathbf{s} \cdot \mathbf{a} = h$. Large amplitude or intensity occurs in cones whose axes are in the direction of $\mathbf{a}$.

For a three-dimensional crystal, consisting of rows or groups of atoms in three dimensions, we can immediately write, by extension of Eq. (2.17a), that

$$A_T(\mathbf{s}) = F(\mathbf{s}) \sum_{n_1=0}^{N_1-1} \sum_{n_2=0}^{N_2-1} \sum_{n_3=0}^{N_3-1} \exp[2\pi i \mathbf{s} \cdot (n_1\mathbf{a}_1 + n_2\mathbf{a}_2 + n_3\mathbf{a}_3)]$$

$$= F(\mathbf{s}) \sum_{n_1=0}^{N_1-1} \exp(2\pi i \mathbf{s} \cdot n_1\mathbf{a}_1) \sum_{n_2=0}^{N_2-1} \exp(2\pi i \mathbf{s} \cdot n_2\mathbf{a}_2) \sum_{n_3=0}^{N_3-1} \exp(2\pi i \mathbf{s} \cdot n_3\mathbf{a}_3).$$
$$\tag{2.18a}$$

Each summation in Eq. (2.18a) is over one dimension and may be evaluated as in Eq. (2.17b) to produce

$$A_T(\mathbf{s}) = F(\mathbf{s}) \frac{\sin \pi \mathbf{s} \cdot N_1 \mathbf{a}_1}{\sin \pi \mathbf{s} \cdot \mathbf{a}_1} \frac{\sin \pi \mathbf{s} \cdot N_2 \mathbf{a}_2}{\sin \pi \mathbf{s} \cdot \mathbf{a}_2}$$

$$\times \frac{\sin \pi \mathbf{s} \cdot N_3 \mathbf{a}_3}{\sin \pi \mathbf{s} \cdot \mathbf{a}_3} \exp\{\pi i \mathbf{s} \cdot [(N_1 - 1)\mathbf{a}_1 + (N_2 - 1)\mathbf{a}_2 + (N_3 - 1)\mathbf{a}_3]\}, \quad (2.18\text{b})$$

and

$$A_T(\mathbf{s})A_T^*(\mathbf{s}) = I_T = |F(\mathbf{s})|^2 \frac{\sin^2 \pi \mathbf{s} \cdot N_1 \mathbf{a}_1}{\sin^2 \pi \mathbf{s} \cdot \mathbf{a}_1} \frac{\sin^2 \pi \mathbf{s} \cdot N_2 \mathbf{a}_2}{\sin^2 \pi \mathbf{s} \cdot \mathbf{a}_2} \frac{\sin^2 \pi \mathbf{s} \cdot N_3 \mathbf{a}_3}{\sin^2 \pi \mathbf{s} \cdot \mathbf{a}_3}. \quad (2.18\text{c})$$

For diffraction from such a three-dimensional crystal, we have three conditions that must be satisfied simultaneously to give constructive interference:

$$\mathbf{s} \cdot \mathbf{a}_1 = h, \quad \mathbf{s} \cdot \mathbf{a}_2 = k, \quad \mathbf{s} \cdot \mathbf{a}_3 = l. \quad (2.19)$$

These are the three conditions that Laue recognized must hold while he was thinking about the interaction of rays of short wavelengths in a three-dimensional solid. We have already encountered one such equation, Eq. (2.9), for the one-dimensional row of atoms. Each interference term in Eq. (2.19) represents (in **s** space) a set of planes perpendicular to $\mathbf{a}_i$ and with period $1/a_i$.

For a two-dimensional crystal, we have only the first two quotients in Eq. (2.18c), and I_T has significant magnitude only when $\mathbf{s} \cdot \mathbf{a}_1 = h$ and $\mathbf{s} \cdot \mathbf{a}_2 = k$, simultaneously. These conditions define lines perpendicular to the plane defined by $\mathbf{a}_1$ and $\mathbf{a}_2$, the intersections of two sets of planes perpendicular to a_1 and a_2, respectively, with periods $1/a_1$ and $1/a_2$. For a given incident ray the diffracted beams $\mathbf{S}/\lambda$ can take on any orientation in space on the Ewald sphere (the sphere of reflection) of radius $1/\lambda$ as we saw for one dimension. But actual measurable intensity occurs when we choose **S** so that the lines of intensity intersect this sphere, as shown in Fig. 2.21. Then as in Eq. (2.8), $\lambda \mathbf{s} \cdot \mathbf{a}_1 = (\mathbf{S} - \mathbf{S}_0) \cdot \mathbf{a}_1 = \lambda h$ and $\lambda \mathbf{s} \cdot \mathbf{a}_2 = (\mathbf{S} - \mathbf{S}_0) \cdot \mathbf{a}_2 = \lambda k$. On a film (as shown) there will be strong *spots* (not cones, as there were in one dimension) indexed with values h, k according to the particular intersection planes h/a_1, k/a_2. In effect, there are cones of scattering around each lattice row, but the scatter-

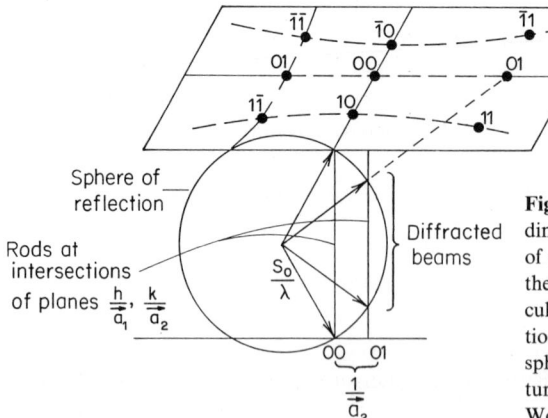

Fig. 2.21. Diffraction from a two-dimensional periodic structure. The sphere of reflection (Ewald sphere) is shown, and the lines of appreciable intensity perpendicular to the plane of the structure. Diffraction occurs when the lines intersect the sphere. A flat film placed above the structure will record diffracted spots. [After Wood, E.A., J. Appl. Phys. *35*, 1306 (1964)]

2.4 Diffraction from a Three-Dimensional Crystal

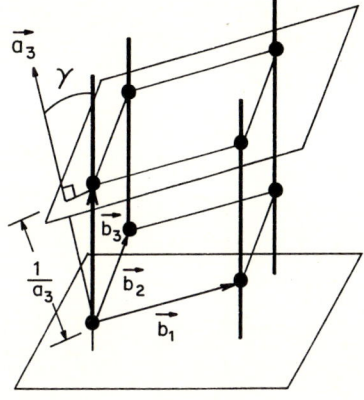

Fig. 2.22. The result of the Laue conditons in three dimensions. The vectors $\mathbf{b}_1$, $\mathbf{b}_2$, and $\mathbf{b}_3$ defined by these conditions are the reciprocal lattice vectors

ing has an appreciable value only at the intersection of these cones, where both interference functions have maxima. (The intersection of the cones with a flat film placed above the crystal are shown dotted in Fig. 2.21.)

In three dimensions the Laue conditions, Eq. (2.19) indicate that scattering is described by three sets of planes spaced h/a_1, k/a_2, and l/a_3 apart and perpendicular to $\mathbf{a}_1$, $\mathbf{a}_2$ and $\mathbf{a}_3$. All terms in Eq. (2.18c) must be large for the scattering to be appreciable, and this occurs at the intersections of these three sets of planes. The planes perpendicular to $\mathbf{a}_1$ and $\mathbf{a}_2$ intersect in the vertical lines shown in Fig. 2.22; these lines are in the direction of the reciprocal lattice vector $\mathbf{b}_3$ as they are perpendicular to $\mathbf{a}_1$ and $\mathbf{a}_2$. The planes perpendicular to $\mathbf{a}_3$ are also shown. Their separation along the vertical axis is $1/(a_3 \cos \gamma)$. The magnitude of the separation is the same as the magnitude of $\mathbf{b}_3$. We can proceed in the same manner with the planes perpendicular to $\mathbf{a}_2$ and $\mathbf{a}_1$. The vectors $\mathbf{b}_1$, $\mathbf{b}_2$, and $\mathbf{b}_3$ defined by the intersections of these three sets of planes are the reciprocal lattice vectors we defined in Chap. 1. Thus the interference function has significant values only at the points of the reciprocal lattice. The sphere of reflection must touch a reciprocal lattice point for diffraction to occur. As shown in Fig. 2.23a, this means that

$$(\mathbf{S} - \mathbf{S}_0)/\lambda = \mathbf{r}^*_{hkl} . \tag{2.20}$$

In effect there are three sets of cones of scattering with appreciable intensity at the common intersections as shown in Fig. 2.23b.

The intersections between the Ewald sphere and the reciprocal lattice points are not frequent as can be seen in Fig. 2.23a. To examine scattering for many spots, we can continuously change the incident beam $\mathbf{S}_0$ over all possible directions, so that the sphere describes a still bigger sphere, called the limiting sphere in Fig. 2.23a. But still the number of intersections is not very large. In many positions there are no such intersections. Instead of changing $\mathbf{S}_0$ we can move the crystal, and this of course is equivalent to moving $\mathbf{S}_0$. Or we can vary λ, thereby changing the radius of the sphere. Various techniques have been developed along these lines to enable exploration of the reciprocal space. These are described in Chap. 4.

Note that in one and two dimensions there are always intersections with the sphere but this is not the case in three dimensions. If, however, $2/\lambda < 1/a_i$, then

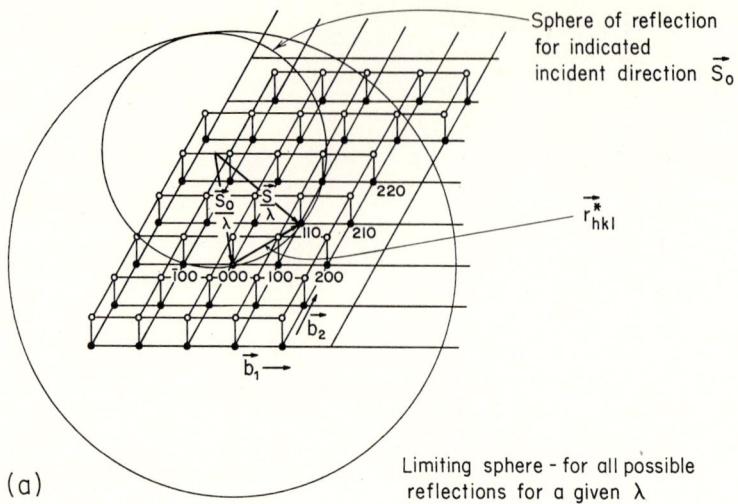

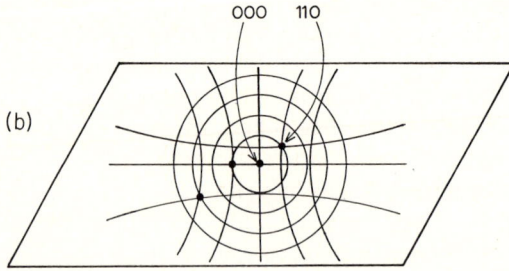

Fig. 2.23. a Diffraction from a three-dimensional structure occurs when the sphere of reflection (Ewald sphere) for a given direction of incidence S_0 touches a point in the reciprocal lattice. By changing the incident beam direction S_0, the sphere can be anywhere in the limiting sphere. (Note that a diffraction spot is given without parentheses; this convention will be used throughout the text). **b** On a flat film, appreciable intensity occurs at spots, the projection of **S** from where it touches the reciprocal lattice. In effect, these spots are the intersections of the three cones of scattering (shown) along the three directions in the crystal

regardless of the orientation of S_0 the sphere of reflection falls between diffraction planes (one-dimensional case) or lines (two-dimensional case) or spots (three-dimensional case) and there are never intersections, regardless of the direction of **s**.

In Eq. (2.18c) the total intensity is the product of $|F(\mathbf{s})|^2$ and the intensity interference functions. We can think of the value of $|F(\mathbf{s})|^2$ being sampled at the intersections of these interference functions with the sphere of reflection. The term $F(\mathbf{s})$ is really spread over a large region because it is a sum involving at best a small number of atoms in the repeating unit cell. But because of the sharpness of the interference functions (which leads to the reciprocal lattice), we are only sampling $|F(\mathbf{s})|^2$ at reciprocal lattice points. If $|F(\mathbf{s})|^2$ *has a zero value at a reciprocal lattice point, then even if the sphere of reflection touches that point of the reciprocal lattice, there would be no diffraction.*

We have seen in Eq. (2.20) that diffraction occurs when the sphere of reflection touches a point in the reciprocal lattice:

$$(\mathbf{S} - \mathbf{S}_0)/\lambda = \mathbf{r}^*_{hkl}.$$

2.4 Diffraction from a Three-Dimensional Crystal

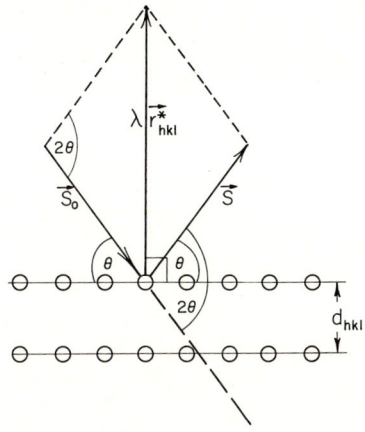

Fig. 2.24. Derivation of Bragg's law as a statement of the magnitude of the Laue conditions applied to a three-dimensional periodic structure

In Fig. 2.24, this condition has been drawn with the actual planes in a crystal seen end on. Vectors **S** and $\mathbf{S}_0$ are equal in magnitude as they are unit vectors. Because $\mathbf{r}^*_{hkl}$ is perpendicular to the *hkl* planes, the angle of incidence *must* equal the angle of diffraction if $\mathbf{S} - \mathbf{S}_0 = \lambda \mathbf{r}^*_{hkl}$ (recall that this was *assumed* true in the derivation of Bragg's law). Furthermore, the component of **S** or $\mathbf{S}_0$ along $\mathbf{r}^*_{hkl}$ can be seen to be $\sin \theta$. Hence the magnitude of Eq. (2.20) may be written as

$$2 \sin \theta = \lambda |\mathbf{r}^*_{hkl}|,$$

or since

$$|\mathbf{r}^*_{hkl}| = 1/d_{hkl},$$

$$\boxed{\lambda = 2 d_{hkl} \sin \theta.}$$

Bragg's law has been verified! Note though that it is not in the form $n\lambda = 2d \sin \theta$. We can see what this means by rewriting Bragg's equation:

$$\lambda = \frac{2}{n} d_{hkl} \sin \theta = \frac{2 \sin \theta}{n/d_{hkl}} = \frac{2 \sin \theta}{|\mathbf{r}^*_{nhnknl}|},$$

That is, the second-order ($n = 2$) term from, say, a (111) plane is represented in reciprocal space by the vector $\mathbf{r}^*_{222}$. Higher-order "reflections" are commonly indexed in reciprocal space by the notation *nh, nk, nl* — rather than writing "*n*th order of *hkl*."

2.4.3 Calculation of the Structure Factor

Finally, some comments on the calculation of the structure factor are appropriate:

$$F(\mathbf{s}) = \sum_n f_n \exp(2\pi i \mathbf{s} \cdot \mathbf{r}_n).$$

Now, if diffraction can only occur for $\mathbf{s} = \mathbf{r}^*_{hkl} = h\mathbf{b}_1 + k\mathbf{b}_2 + l\mathbf{b}_3$, and we write $\mathbf{r}_n$ as $x_n \mathbf{a}_1 + y_n \mathbf{a}_2 + z_n \mathbf{a}_3$, the structure factor may be written as

$$F_{hkl} = \sum f_n \exp[2\pi i(hx_n + ky_n + lz_n)]. \tag{2.21}$$

Let us consider two examples of the use of Eq. (2.21). Consider the face-centered cubic structure with identical atoms at $x, y, z = 000; \frac{1}{2}, \frac{1}{2}, 0; \frac{1}{2}, 0, \frac{1}{2};$ and $0, \frac{1}{2}, \frac{1}{2}$ having scattering factor f. Then

$$F_{hkl} = fe^{2\pi i(0)} + fe^{2\pi i(h/2+k/2)} + fe^{2\pi i(h/2+l/2)} + fe^{2\pi i(k/2+l/2)}$$

$$= f[1 + e^{\pi i(h+k)} + e^{\pi i(h+l)} + e^{\pi i(k+l)}]$$

$$= \begin{cases} 4f & \text{for } h,k,l \text{ all even or all odd}, \\ 0 & \text{for } h,k,l \text{ mixed even and odd}. \end{cases}$$

In this simple structure the scattering occurs only at those reciprocal lattice points for which h, k, and l are unmixed, and for others complete destructive interference occurs for waves scattered from the four atoms in the unit cell. An examination of the structure will show that there are many planes "in between" causing the destructive interference.

A second instructive example is the structure factor for NaCl, shown in Fig. 1.15c. Sodium ions are at $x, y, z = 000 +$ face-centering translations, and chlorine ions at $0, 0, \frac{1}{2} +$ face-centering translations. There are eight atoms per cell, i.e., eight terms from Eq. (2.21).

$$F_{hkl} = f_{\text{Na}^+}\{e^{2\pi i(0)} + e^{2\pi i(h/2+k/2)} + e^{2\pi i(h/2+l/2)} + e^{2\pi i(k/2+l/2)}\}$$

$$+ f_{\text{Cl}^-}\{e^{2\pi i(l/2)} + e^{2\pi i(h/2+k/2+l/2)} + e^{2\pi i(h/2+2l/2)}$$

$$+ e^{2\pi i(k/2+2l/2)}\}.$$

Factoring,

$$F_{hkl} = 4(f_{\text{Na}^+} + f_{\text{Cl}^-} e^{\pi il})\{1 + e^{\pi i(h+k)} + e^{\pi i(h+l)} + e^{\pi i(k+l)}\}.$$

We recognize the factor in braces as the same as above for the fcc structure. (In general we will be able to factor our terms due to the internal symmetry by which the atoms are distributed.) The restrictions on hkl for the fcc apply to this factor, so F_{hkl} is zero for h,k,l mixed and

$$F_{hkl} = 4(f_{\text{Na}^+} + f_{\text{Cl}^-} e^{\pi il}), \quad \text{for } h,k,l \text{ unmixed}$$

$$= \begin{cases} 4(f_{\text{Na}^+} + f_{\text{Cl}^-}), & l \text{ even}, \\ 4(f_{\text{Na}^+} - f_{\text{Cl}^-}), & l \text{ odd}. \end{cases}$$

The additional interference between the two types of atoms leads to partial cancellation for h,k,l all odd and reinforcement for h,k,l all even.

If the f_n's are real, the structure will *appear* in its diffraction pattern as if it has a center of symmetry, whether it does or not. To see this, expand Eq. (2.21) in sines and cosines:

$$|F_{hkl}| = [\{\sum f_n \cos 2\pi(hx_n + ky_n + lz_n)\}^2$$

$$+ \{\sum f_n \sin 2\pi(hx_n + ky_n + lz_n)\}^2]^{1/2}. \tag{2.22}$$

Thus $|F_{hkl}| = |F_{\bar{h}\bar{k}\bar{l}}|$ even if the structures do not have a center of symmetry, that is, even if there are not the same kinds of atoms in pairs at x,y,z and $\bar{x},\bar{y},\bar{z}$. This result is known as Friedel's law and will be discussed more fully in Chap. 6.

As we have seen, it is relatively easy to simplify the expressions of Eqs. (2.21) or (2.22) for structures that do not contain too many atoms. But if there are many atoms, it is a tedious job. In Chap. 5 we will return to this problem to demonstrate that a dramatic simplification may be made when we understand the internal symmetry of crystal structures.

2.5 Summary

We have seen that the diffracted intensity depends on the structure of the unit doing the scattering and that the positions of diffraction peaks are determined by the arrangement of these units. The mathematical expression of scattered amplitude may be written as a Fourier transform of the scattering density and represented in terms of the Fourier space variable $\mathbf{s} = (\mathbf{S} - \mathbf{S}_0)/\lambda$, the scattering vector of magnitude $(2 \sin \theta)/\lambda$. Diffraction from a periodic structure occurs only for the scattering vector $\mathbf{s}$ equal to a reciprocal lattice vector of the structure, $\mathbf{r}^*_{hkl}$. This reciprocal lattice is just the Fourier transform space for a triply periodic structure. Attached to each point in this reciprocal lattice is a term involving the scattering per repeating unit or unit cell, the structure factor. Diffraction occurs when a sphere of reflection of radius $1/\lambda$ intersects the reciprocal lattice provided the structure factor has a non-zero value at that point. The orientation of the direct beam relative to the crystal (or vice versa) and the wavelength determines what intersections, if any, will in fact occur. The structure factor of the unit cell extends over a large region of reciprocal space, but we are sampling it at the reciprocal lattice points.

Several problems need further attention. First, we have seen that we seem to lose certain information in measuring the intensity. Specifically, we do not know if the structure has a center of symmetry. Second, we have assumed that there is no change in wavelength in scattering, and third, we have assumed that radiation scattered by one row or plane does not rescatter from other rows or planes as it leaves the crystal.

In the next chapter we shall examine the production of radiation suitable for diffraction from materials. After this we shall consider the evaluation of the interaction of various radiations with matter and the scattering per unit scattering element and some other factors which affect the actual intensity. Then we shall examine more closely the required geometry for diffraction and the methods to obtain this geometry. Finally, we shall consider the determination of structure and some effects of imperfections. In each of these we shall develop appropriate techniques. The last assumption is left until the very end merely because it has been demonstrated experimentally that it is satisfactory for x-rays and neutrons, except in certain special cases and suffices qualitatively for electron diffraction. Because of imperfections in most crystals (small subgrains of a few hundred Angstroms with slight tilts between them, often called mosaic structure), multiple scattering does

References

Ewald, P.P., in "Fifty Years of X-ray Diffraction." N.V.A. Oosthock's Uitgeversmaats-chappij, Utrecht, The Netherlands, 1962

Guinier, A., "X-ray Diffraction." Freeman, San Francisco, 1963

James, R.W. "The Optical Principles of the Diffraction of X-rays." Bell, London, 1950

Problems

2.1. Derive the numerical values of the amplitudes in Fig. 2.7.

2.2. Calculate the height of the second and third maxima relative to the main maxima for $(\sin^2 N\beta/2)/(\sin^2 \beta/2)$ for $N = 4, 8, 16$.

2.3. For a row of scattering elements, with the incident beam normal to the row:
(a) What is the angular difference (v in Fig. 2.13) between a peak and its first zero? How will this vary with N? With order?
(b) If the diffracted beam is fixed, but not the incident beam, what is the angular difference (μ in Fig. 2.13) between a peak and its first minimum and how does this vary with N? With order?
(c) Repeat (a) with planes instead of rows and symmetric reflection.

2.4. By considering the interference function, evaluate the sums:

$$\sum_n n \cos nx, \quad \sum_n n \sin nx .$$

2.5. In the following sums for R solve for H and y in terms of f_i, x_i. Evaluate and sketch the result in Argand diagrams:
(a) $\exp(2\pi i x_1) + \exp(2\pi i x_2) = He^{2\pi i y} = R$
(b) $f_1 \exp(2\pi i x_1) + f_2 \exp(2\pi i x_2) = He^{2\pi i y} = R$
(c) $f_1 \exp(2\pi i x_1) + f_2 \exp(2\pi i x_2) + f_3 \exp(2\pi i x_3) = He^{2\pi i y} = R$.

2.6. Derive an expression for the structure factor F_{hkl} for a base centered unit cell. Sketch reciprocal space and compare to a primitive cell.

2.7. Assuming one atom per lattice point and all atoms in the unit cell identical, evaluate the structure factor F_{hkl} and $|F_{hkl}|^2$ for the following lattices:
(a) Simple cubic.
(b) Body-centered cubic.
Sketch the results in three-dimensional reciprocal space and indicate the unit cells in this reciprocal space. For what values of hkl does $|F_{hkl}|^2$ vanish?

Problems

2.8. Evaluate F_{hkl} and $|F_{hkl}|^2$ for a hexagonal close-packed unit cell. For what values of hkl does $|F_{hkl}|^2$ vanish? Sketch the results in three-dimensional reciprocal space.

2.9. In the structure of diamond there are carbon atoms at 000, $\frac{1}{4}, \frac{1}{4}, \frac{1}{4}$, and other positions given by face-centered translations added to these. Evaluate F_{hkl} and $|F_{hkl}|^2$. For what values of hkl is $|F_{hkl}|^2 = 0$?

2.10. In the diamond structure outlined in the previous example, suppose the 000 atom is S and the $\frac{1}{4}, \frac{1}{4}, \frac{1}{4}$ atom is Zn. (This is known as zinc blende.) Evaluate F_{hkl} and $|F_{hkl}|^2$. For what values of hkl does $|F_{hkl}|^2$ vanish?

2.11. Uranium is orthorhombic, with four atoms per unit cell in positions:

$u:$	0	0	$\frac{1}{2}$	$\frac{1}{2}$
$v:$	y	$-y$	$\frac{1}{2} + y$	$\frac{1}{2} - y$
$w:$	$\frac{1}{4}$	$\frac{3}{4}$	$\frac{1}{4}$	$\frac{3}{4}$

(a) To which Bravais lattice does this cell belong?
(b) For what reflections will F_{hkl} vanish regardless of y?

2.12. (a) Suppose that in a unit cell of a crystal there are six identical atoms in an octahedral array with a fourfold axis along $\mathbf{a}_3$. Choosing the origin at the center of the octahedron, derive an expression for $|F|^2$. Sketch the result for the $(h_1 h_2 0)$ plane in reciprocal space. Because this is a small group of atoms, $\mathbf{s}$ has a value other than $\mathbf{r}^*_{hkl}$; i.e. reciprocal space can be described by the vector $h_1 \mathbf{b}_1 + h_2 \mathbf{b}_2 + h_3 \mathbf{b}_3$ where the h_i are continuous variables.
(b) Suppose the octahedron moves around the $\mathbf{a}_3$ axis relative to the corners of the unit cell. Show how the sampling of the structure factor will change on a $(h_1 h_2 0)$ plane in reciprocal space.

2.13. Assume that in an orthorhombic crystal there are N electrons per unit cell distributed so that the electron density ρ (electrons per unit volume) is given by

$$\rho = \rho_1 \begin{cases} -a_1/4 < x < a_1/4, \\ -a_2/4 < y < a_2/4, \\ -a_3/4 < z < a_3/4, \end{cases}$$

$\rho = 0$ elsewhere.
(a) Derive an expression for the general structure factor F_{hkl} for this crystal.
(b) For what kind of reflections will F vanish?

2.14. In an orthorhombic crystal with axes $\mathbf{a}_1, \mathbf{a}_2,$ and $\mathbf{a}_3$, the electron density within one cell is given by

$$\rho(xyz) = (\alpha/\sqrt{\pi})^3 \{N_1 \exp[-\alpha^2(x^2 + y^2 + z^2)] \\ + N_2 \exp[-\alpha^2[(x-\tfrac{1}{2})^2 + (y-\tfrac{1}{2})^2 + (z-\tfrac{1}{2})^2]]\}.$$

Assume the electron densities are sharply concentrated (α^2 large) and that the appropriate integrals can be taken from $-\infty$ to $+\infty$.
(a) What is the Bravais lattice of this crystal?
(b) Derive an expression for F_{hkl}. [*Hint:* Use Eq. (2.13).]

2.15. (This problem is an introduction to a problem in Chap. 4 on twinning. If that problem is to be done, this one should be done first.) A fcc crystal often twins by shear on $\{111\}$ planes. Consider twins on one of the planes in this form and index "extra" spots in reciprocal space as a result of this twinning, in terms of the axes of the matrix (i.e. sketch the reciprocal space of the twin-matrix composite). How then can you detect twinning? [The twinning here makes the structure a mirror image across (111) plane. Rotation around a [111] direction will accomplish the same thing.]

2.16. At one stage in precipitation in some alloys we can imagine the solute distributed sinusoidally with wavelength λ (wave number $q = 2\pi/\lambda$). Consider a one-dimensional lattice of points p apart and average density of scattering material ρ_0. The electron density can then be written as $\sum_m \delta(x - mp)(\rho_0 + a\cos qx)$. The Dirac δ function has a value only when its argument is zero, so multiplying by this function causes the density to have a value only on the points of the lattice. Calculate the appearance of the diffraction pattern and sketch your result.

2.17. Consider diffraction from a rectangular parallelopiped of sides $N_1 a, N_2 a, N_3 a$ of primitive cubic material with lattice constant $a = 4$ Å. Sketch one unit cell of the reciprocal lattice and show (roughly to scale) the volumes in reciprocal space from which appreciable diffracted intensity could arises for the cases:
(a) $N_1 = N_2 = N_3 = (10)^4$,
(b) $N_1 = N_2 = (10)^4$, $N_3 = 5$ (a thin film!).
(*Hint*: Remember the interference function.)

3. Properties of Radiation Useful for Studying the Structure of Materials

We have seen in Chap. 2 that a diffraction pattern can occur when the wavelength of radiation incident on a periodic array is the same order of magnitude as the period. As we now know from diffraction studies, these spacings are of the order of 1 Å (10^{-10} m) for atoms in solids, thus we wish to study the properties of radiation in this wavelength range. We shall confine our attention to the most commonly used radiations: x-rays, neutrons, and electrons.

3.1 Production of Radiation — X-rays

3.1.1 The Conventional X-ray Tube

Röntgen discovered, in the summer of 1895, that in an evacuated tube in which a voltage was applied across two metal plates, unknown or "x"-rays emerged, penetrated matter, and exposed film. However, experiments to examine the nature of these rays, to determine whether they were particles or electromagnetic radiation, were difficult. Ruled grating experiments which should give diffraction patterns for waves gave poor results that were hard to resolve and therefore not definitive. We know now that very fine gratings were needed because the x-ray wavelengths are so small, but such gratings were not available at that time. The results of these studies did suggest that if the rays were waves, they were of quite small wavelength, and these experiments helped to stimulate Laue's thoughts that x-rays might have the correct wavelengths to diffract from crystals. As we shall see later, x-rays could be polarized like other electromagnetic waves, i.e., their amplitude of variation could be made to lie in only one direction perpendicular to their propagation direction. By contrast, x-rays could also eject electrons when they impinged on a material as in a particle collision. Experiments to reflect x-rays like waves also proved fruitless at first. Even if the results had been clear, all of these tests to identify x-radiation as wavelike or particlelike would have been superseded when the wave-particle duality of radiation and matter was established in the 1920's with the development of quantum mechanics. However, after some time, it became clear from their polarizability and the failure of the rays to be deflected by electric or magnetic fields that the rays were electromagnetic radiation. The emission of electrons when x-rays impinged on matter (the photoelectric effect) was explained by Einstein as being due to the fact that radiation was quantized into energy packets. Later experiments (such as Laue's) proved that the wavelength of x-rays was indeed of the same magnitude as the periodicity in crystals and that this was the order of angstroms (Å).

At the time, x-ray tubes were difficult to operate. Their operation depended on ionization of a gas at low pressure (0.01 mm) in a tube by a high voltage, followed by emission of electrons (cathode rays) at one metal plate by the bombardment of the ions and the acceleration of the electrons to the other (positive) plate, where x-rays were emitted. The control of gas pressure and purity, target cleanliness, etc., made operation of such tubes a tedious and nerve-racking affair. Current and voltage in the tube were dependent on each other. Today we use tubes [evacuated to 10^{-5} mm Hg] in which the electrons are emitted by heated tungsten filaments, allowing separate control of current and voltage, as in Fig. 3.1. The voltage and current to such a tube are usually supplied by a transformer after stabilizing the (ac) line voltage (variations in line voltage drastically affect emission from the hot cathode). Diodes are used with circuitry so that when ac voltage is applied, the tube

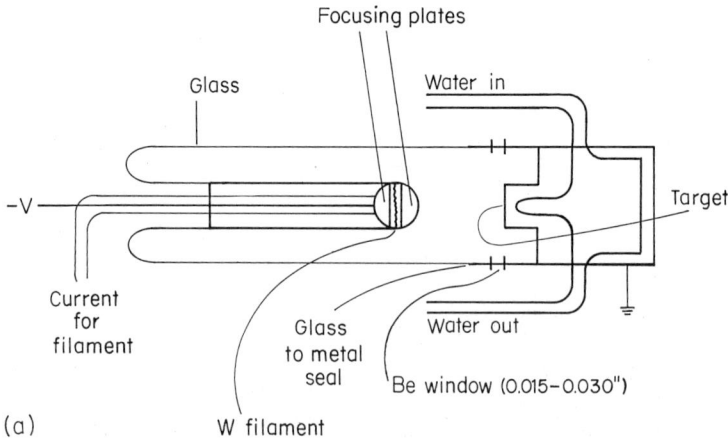

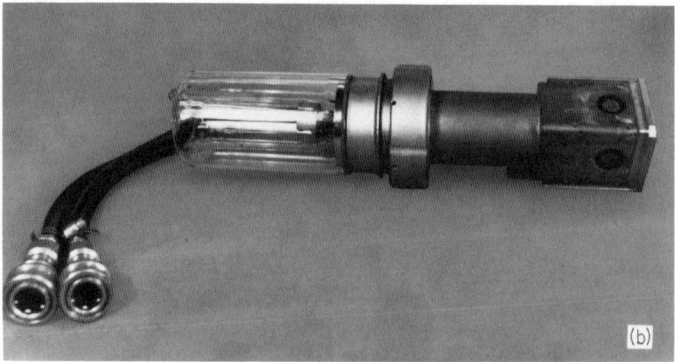

Fig. 3.1. a Schematic of an x-ray tube (the recess in the glass is to avoid contact of high voltage with the body). **b** An actual tube, about 0.3 m in length

conducts on both halves of the cycle. Condensers in the circuit provide voltage during the reverse cycle reducing the variations to a slight ripple or dc generation may be used. Safety circuits are included to shut off the unit if the voltage or current exceeds the ratings for the x-ray tube, or if the water flow (which cools the target) stops. Typically, stabilization will keep the x-ray output so constant that two intensity readings 24 h apart will differ by at most 0.3–3%. The intensity at the sample is of the order of 10^9 counts per second (cps). The sealed tube shown in Fig. 3.1 has a useful operating life of about 8000 hr when used at full current rating. These values are for general purpose x-ray diffraction tubes which are designed to operate in the range of 20–40 mA current depending on the target and construction. New tube designs allow operation up to 2000 mA producing x-ray intensities in the 10^{11} cps range. The target in these tubes is rotated so that one area of the target is not always under the electron beam. The tube is not permanently sealed, but can be opened to allow cleaning or changing the target and filament.

A common tool for examining diffraction patterns is the *diffractometer*, patterned after the original apparatus used by the Braggs in their study of NaCl. Fig. 3.2 shows a typical diffractometer with the x-ray tube at the far left. Radiation from the tube is collimated in a series of slits and directed at the sample. The scattered radiation

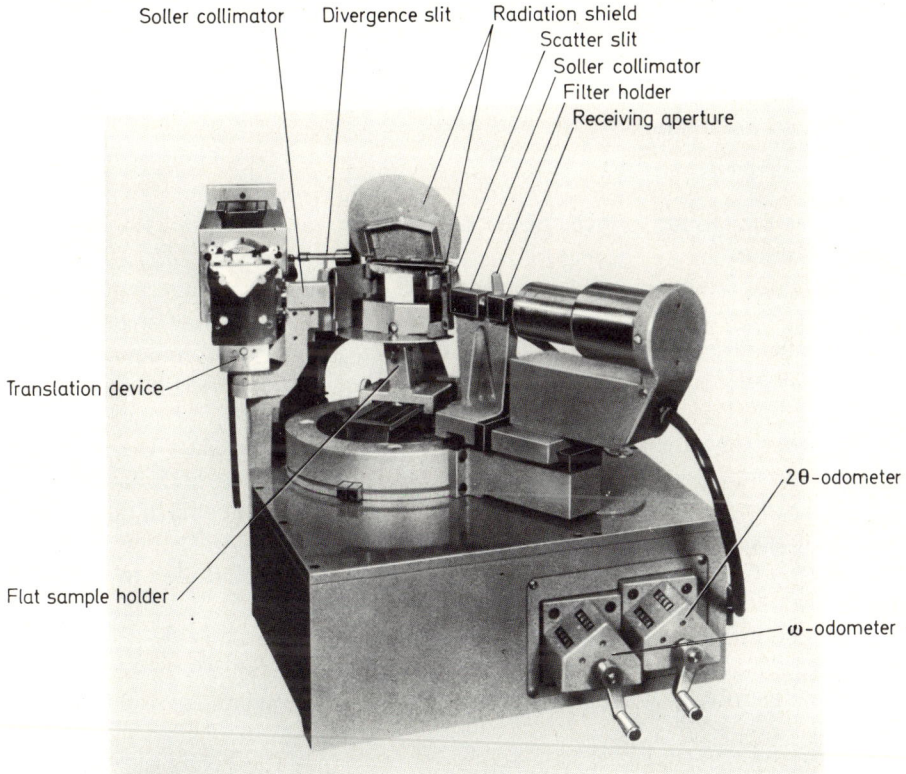

Fig. 3.2. X-ray diffractometer. One side of the unit is about 0.3 m long. (Courtesy of Picker Corp.)

is also collimated and then after passing through the receiving aperture, detected. We shall have much more to say about this apparatus later in this chapter and in Chap. 4, but we may now imagine using it in an experiment to examine the radiation coming from the tube. The detector moves around the specimen at a speed of $2\theta°$/min, while the sample moves $\theta°$/min; in this way the incident and diffracted beams always make equal angles with the specimen surface. (One reason for stabilizing x-rays is because the detector is moving; we want any variation in scattered intensity to be from the sample, not the source.)

Suppose we set up a NaCl crystal in the form of a cube with known d_{200} spacing,[1] apply voltage V_1 and current i to the tube, and examine intensity versus 2θ. The result is shown in Fig. 3.3. At each angle 2θ, Bragg's law is satisfied for some λ in the distribution of λ's coming from the tube. The diffracted intensity is proportional to the available intensity at that λ. We note that there is a range of λ's; diffraction starts at some minimum angle, corresponding to a minimum wavelength, λ_{min}, and occurs for a wide range of angles. If we increase the voltage to V_2 keeping the current constant, we get a similar curve with higher intensity. We find that λ_{min} *is characteristic of the tube voltage, and* λ_{max} *is the maximum wavelength that will reach the counter without being totally absorbed in the air.* (It is one of the fortunate facts of nature that x-rays of the right wavelength for diffraction from crystals are not completely absorbed in air! If vacuum paths were needed, as in electron diffraction, the fact that crystals are periodic arrangements of groups of atoms, and that these arrangements can be studied, might have waited many more years for verification!) It can be shown that the area under each curve of "white" or continuous radiation is

$$\text{area} \propto V^n Z^2 \,,$$

where Z is the atomic number of the target and n varies between 1.5 and 2 for different targets.

Now, if we return to our experiment, hold the voltage constant, but raise the current i as shown in Fig. 3.4, the curve is "lifted" vertically without any change in its fundamental features. *Altering the voltage changes the complete spectrum, but current changes only the number of events, not their type, and thus alters only the intensity.*

To understand the origin of this spectrum, we must consider the radiation scattered from an electron when it is decelerated upon collision with the target. Even though the complete treatment of this and other problems dealt with in this chapter is not contained in classical theory, the classical approach will allow us to see the important physical principles. As we proceed, modifications due to quantum mechanical theory will be point out.

From electromagnetic theory, an accelerating or decelerating charge radiates. In Fig. 3.5 a charge q is shown with acceleration **a**. At a distance **r**, the electric field which results from the acceleration is denoted by ε. The magnitude of this electric field is given by

---

[1] The reader will recall from Sect. 2.2 that the diffraction from planes with spacing d_{100} in NaCl is zero due to destructive interference.

3.1 Production of Radiation — X-rays

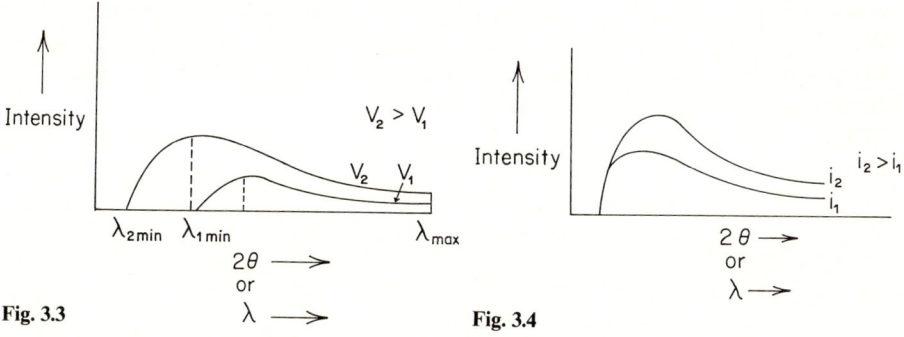

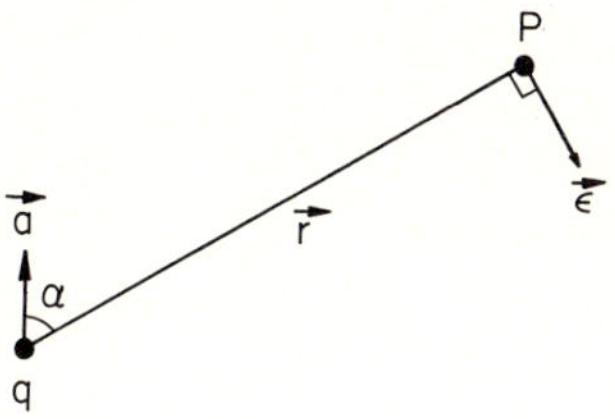

Fig. 3.3. The white spectrum from an x-ray tube, showing the effect of accelerating voltage

Fig. 3.4. The white spectrum from an x-ray tube, showing the effect of tube current

Fig. 3.5. A charge q with acceleration **a** produces a field ε at a distance **r** as indicated in Eq. (3.1)

$$\varepsilon = (\eta q a \sin \alpha)/c^2 r ,\tag{3.1}$$

where c is the velocity of light, $\eta = 1/(4\pi\varepsilon_0)$ and ε_0 is the permittivity of empty space.

The electric field is in the plane of **r** and **a** and its magnitude depends on the component $a \sin \alpha$. (The energy radiated from an accelerating or decelerating charged particle is electromagnetic radiation composed of perpendicular electric and magnetic fields. For clarity, only the electric field vectors are shown in Fig. 3.5.) Note also that since electromagnetic radiation is a transverse wave, the electric field vector ε must be perpendicular to the direction of propagation **r**. We shall discuss more implications of Eq. (3.1) later, but now we note that a decelerating electron converts its kinetic energy into electromagnetic radiation which is radiated away as x-radiation, if the voltage causing the acceleration is of the order of kilovolts.

In an x-ray tube, the electron is decelerated as it interacts with the coulomb fields of the atoms in the target. If we assume that all the potential energy of the applied voltage across an x-ray tube goes into kinetic energy of an electron, and this in turn is converted to radiation in one encounter with an atom in the target, conservation of energy gives

$$Ve = \tfrac{1}{2}mv^2 = hc/\lambda_{\min} ,$$

where V is the applied voltage, e the electron charge, m the electron mass, h Planck's constant, and c the velocity of light.[2] Expressing λ in Ångstroms (Å) and voltage in

---

[2] We could use, more exactly, the relativistic expression for the electron's energy

$$([m_0/(1 - v^2/c^2)^{1/2}] - m_0)c^2 .$$

However, at the voltage used, generally less than 50 kV, this represents a negligible correction.

kilovolts (kV), we have

$$\lambda_{min}(\text{Å}) \times 10^{-10} \times V(\text{kV}) \times 10^3$$

$$= \frac{6.625 \times 10^{-34} \text{ joule s} \times 2.998 \times 10^8 \text{ m/s}}{1.60 \times 10^{-19} \text{ coulomb}}$$

or

$$\lambda_{min}(\text{Å}) V(\text{kV}) = 12.40. \text{ (Å kV)}. \tag{3.2}$$

This tells us the short wavelength limit for a given voltage, but not all the electrons are decelerated in the same way. Some lose their energy in glancing collisions with atoms in the target, losing only some of their energy in each collision. The x-rays coming off from some depth in the target are absorbed on their way out. These effects account for the curve in Fig. 3.3, with a maximum at about $1.5\lambda_{min}$. This *radiation from the electron deceleration is called either Bremsstrahlung, slowing-down radiation, or by analogy with the visible spectrum, white radiation.*

3.1.2 Characteristic X-radiation

White radiation is not the only type of x-radiation observed. Let us continue our experiment, again holding current constant. If we raise the voltage above a certain level, sharp lines appear superimposed on the Bremsstrahlung, as shown in Fig. 3.6. In 1913, Moseley, following up some earlier work by Sir W.H. Bragg, showed that the square root of the frequency v of these sharp lines was characteristic of the atomic number of the target. Using many targets with different atomic numbers Z, he showed that a family of straight lines occurred when $\sqrt{v}$ was plotted versus Z. This is illustrated in Fig. 3.7. (He determined these frequencies by measuring the x-ray wavelengths with a crystal of known d spacing.) Moseley was aware of the atomic model that Bohr had just proposed, in which electrons were assumed to travel in

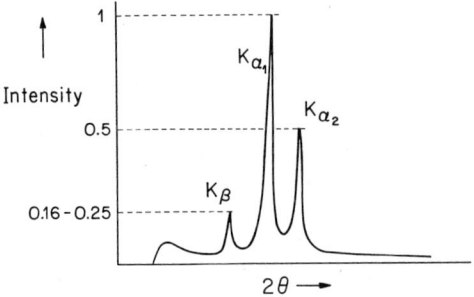

Fig. 3.6. The white and characteristic spectrum from an x-ray tube. The K_{α_1} transition has a higher energy than K_{α_2} and therefore a smaller wavelength. Hence the K_{α_1} peak occurs at a smaller 2θ since Bragg's law requires $\lambda = 2d \sin \theta$

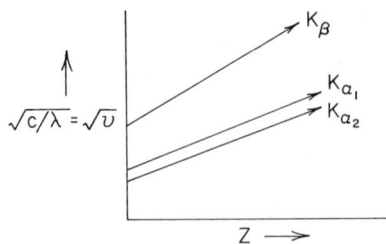

Fig. 3.7. Moseley's law. For each characteristic transition (K_{α_1}, K_{α_2}, etc.) the radiation frequency v and atomic number Z are related by $\sqrt{v} \propto Z$

3.1 Production of Radiation — X-rays

definite orbits around the nucleus. For the simple case of a circular orbit, balancing the force on an electron of mass m from the Z positive charges on the nucleus with the centripetal force, but assuming that the momentum in an orbit was "quantized", i.e., equal to nh (where n is an integer), Bohr had shown that the total energy of an orbit was

$$E = -2\pi\eta^2 mZ^2 e^4/n^2 h^2 .$$

Bohr assumed that when an electron moved from one level (1) to another (2), energy was absorbed or emitted in discrete packets or quanta of frequency v:

$$E_1 - E_2 = hv, \quad v = \frac{E_1 - E_2}{h} = \frac{2\pi^2\eta^2 mZ^2 e^4}{h^3}\left(\frac{1}{n_2^2} - \frac{1}{n_1^2}\right).$$

Note that since n_i is an integer, only certain energies are possible. Therefore:

$$\sqrt{v} = KZ\left(\frac{1}{n_2^2} - \frac{1}{n_1^2}\right)^{1/2} , \tag{3.3}$$

where K is a constant. Thus *each of Moseley's lines was for a transition from one shell to a lower one*. He had shown that the characteristic x-radiation was proof for Bohr's model.

As the theory of the atom and quantum mechanics developed, many corrections and more refined treatments of Bohr's revolutionary approach were evolved, involving fewer *a priori* assumptions. It was realized that when electrons moved from shell to shell, they were shielded somewhat from the nuclear charge Ze by electrons in inner shells [i.e., Z is to be replaced by $(Z - \delta)$]. Orbits could be elliptical and another integer or quantum number l was needed to describe the orbital angular momentum, or the minor axis, as well as n. Its values were shown to lie between 0 and $n - 1$. Momentum is a vector property and thus it was found that the orientation of the orbit relative to some axis in space was quantized with a quantum number m_l (with values $-l$ to $+l$). The electron spins around an axis and has a magnetic field. Thus m_s was found to be the quantum number describing this field, which could be only in two directions, opposed or aligned to an applied field; m_s was found to have a value $+\frac{1}{2}$ or $-\frac{1}{2}$. No two electrons could have the same set of quantum numbers (the Pauli exclusion principle), analogous to the fact that no two objects can occupy the same space at the same time. When these effects are all combined, the energy levels may depend on all four quantum numbers. In Fig. 3.8a, an energy-level diagram for an atom is presented. These levels are named or lettered after spectra that had been found and labeled prior to the knowledge of the level diagram. In this way the explanation of the emitted spectra was linked closely to the names then in use.

The interaction of spin and orbital angular momentum, or these with external electrostatic fields, splits levels of the same n and l into different energy levels $|m_l|$. The L shell is shown split in this fashion in Fig. 3.8a. (A magnetic field splits levels of the same n, l, m_l into different levels for the two values of m_s.) Transitions are possible for

$$\Delta l = \pm 1, \quad \Delta m_l = \pm 1, 0 .$$

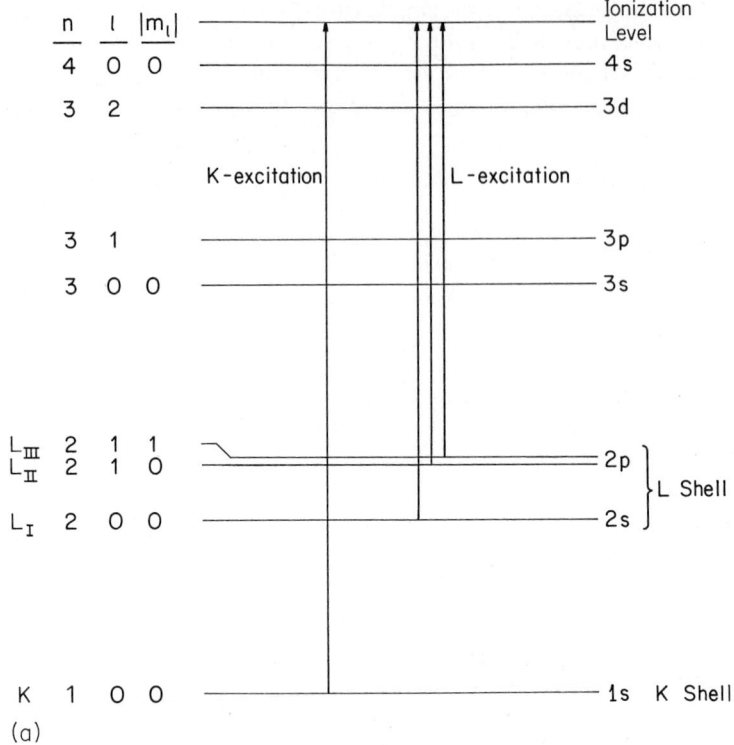

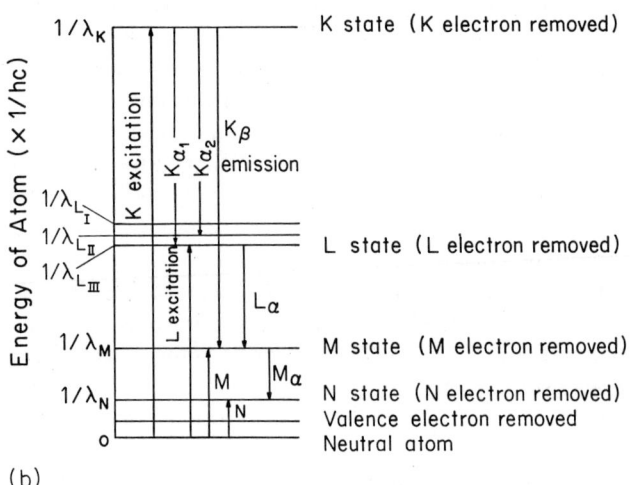

Fig. 3.8a, b. Schematic of the atomic energy levels with excitation and emission processes indicated by *arrows*. **a** Electron energy levels of copper metal with principal quantum numbers n, l, m_1, indicated. The *arrows* indicate electron transitions to form the K-state, L-state, etc., referred to in Fig. 3.8b. Each energy level is occupied by two electrons with $m_s = 1/2$ and $m_s = -1/2$. **b** Removal of a K electron (increasing the atomic energy) is followed by emission of characteristic x-radiation

3.1 Production of Radiation — X-rays

If electrons in an x-ray tube eject an inner electron from the K shell of a target atom, the energy of the atom is raised to the K state as shown in Fig. 3.8b. The energy is lowered to the L state, for example, by an electron from one of the allowed 2p states of the L shell falling to the K shell, emitting an x-ray of energy corresponding to the difference in energy between the K and L states. The final state is then the same as that which would be produced if an electron was ejected from the L shell. From Fig. 3.8b, we have for example,

$$h\nu_{K_{\alpha_1}} = hc/\lambda_{K_{\alpha_1}} = (hc/\lambda_{K_{ioniz}}) - (hc/\lambda_{L_{III ioniz}})$$

or

$$1/\lambda_{K_{\alpha_1}} = (1/\lambda_{K_{ioniz}}) - (1/\lambda_{L_{III ioniz}}). \tag{3.4}$$

Notice that the transition from L_I to K is missing from Fig. 3.8b. This and many other transitions are excluded by quantum mechanical selection rules. In this example $\Delta l = 0$ and violates the requirement that only $\Delta l = \pm 1$ is allowed.

We can measure the K or 1s ionization potential by lowering the voltage until a line disappears, i.e., by plotting $I_{K_{\alpha_1}}$ versus V and finding the voltage (V_{ioniz}) corresponding to $I_{K_{\alpha_1}} = 0$. X-rays with $V > V_{ioniz}$ have just enough energy to ionize the K (or 1s) electron. [We may then use Eq. (3.2) to obtain the λ_{ioniz} corresponding to V_{ioniz}: $\lambda_{ioniz}(\text{Å})V_{ioniz}(\text{KV}) = 12.4$.] We would note that all the lines K_{α_1}, K_{α_2}, K_β disappeared at once if we did this since all of these transitions require ionization of a K electron. The intensities of the different peaks in Fig. 3.6 are a measure of the probability for any type of transition in our mass of atoms, and a whole cascade of transitions occur when we exceed a tube voltage which will give the electrons from the filament enough energy to knock out a 1s (K) electron. The wavelength associated with this energy, written as $\lambda_{K_{ioniz}}$ or $\lambda_{K_{abs}}$ or the K absorption edge, is shorter than the wavelengths associated with the transitions. This wavelength is often referred to as the wavelength for the K absorption edge or $\lambda_{K_{abs}}$ because of the sudden rise in absorption of an x-ray beam when its λ has this value for the absorbing material. An incident x-ray of sufficient energy can ionize an inner electron and be absorbed in the process.

If the incident electron can ionize a K electron, it has sufficient energy to ionize L and M electrons with the resultant production of other x-rays. We do not always see all of these transitions because some of the rays (of long λ) have too low an energy and are absorbed in the air before they reach the counter or film. If the electrons have only enough energy to knock L electrons out, we get only spectra associated with transitions to the L level. We see that x-ray spectra can be quite useful in examining the inner energy levels in atoms.

Examination of a table of wavelengths of various spectra, as well as the ionization λ's (see International Tables, Volume IV, pp. 6–43) shows that *as the atomic number goes up* (i.e., as the positive charge on the nucleus increases) *the K ionization potential increases* (the nucleus has more positive charge to hold the inner electrons), *and the wavelengths associated with this ionization and the K spectra decrease*. With some tube targets such as tungsten, with the usual operating voltage of 50 kV, we see only the L spectrum from the tube. Because of the higher energy of this spectrum for targets of high atomic numbers, the L spectrum can reach the recording device.

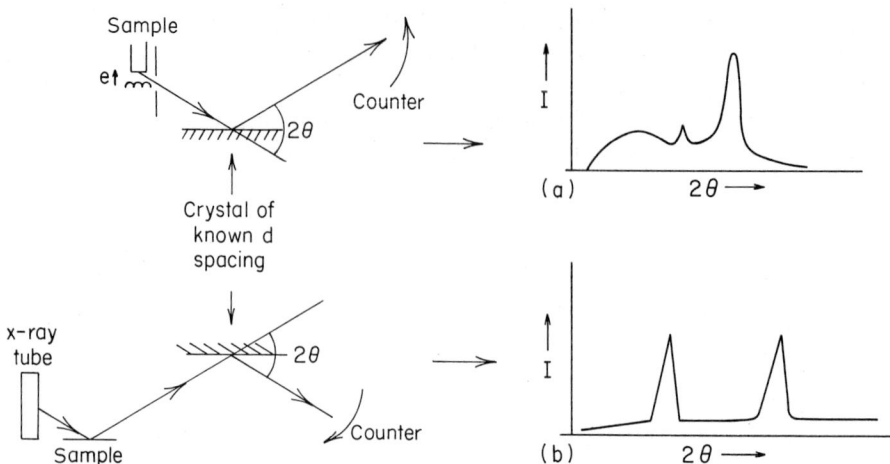

Fig. 3.9a, b. Two schemes for fluorescent analysis. **a** Sample fluoresced by electrons, and **b** sample fluoresced by x-rays

Because *the sharp peaks in the distribution are characteristic* of the types of atoms and not of their structure or state, *they can serve as a means for chemical analysis*. This has become a very powerful tool under the name of *fluorescent analysis*. Two schemes for this are illustrated in Fig. 3.9. In Fig. 3.9a the sample is bombarded as a target by electrons inside a tube which can be opened to change the target. In Fig. 3.9b, x-rays are used to cause the ionization so that the specimen can be outside the tube, allowing for more rapid sample exchange and a greater variety of sizes and shapes of specimens. [Note (and explain) the differences in the background in Figs. 3.9a and b.] Vacuum paths are often used to increase the range of wavelengths which can be detected and increase the number of elements that can be analyzed. Low power tubes with transmission targets at the end of a cavity are becoming increasingly important. The background is reduced and there is higher efficiency for this geometry as ejected and scattered electrons in the cavity also produce x-rays. Because a factor of 10 lower power is required, these tubes allow the construction of portable equipment. Radiation from radioactive isotopes is also employed in such equipment.

To analyze very small sample regions (~ 1 μm^2), focused electron beams are employed, in special instruments called electron microprobes, or in scanning electron microscopes. Although x-rays are the preferred source because they deposit 10^3–10^4 less energy, and thus minimize heating and sample damage (electrons interfere strongly with the outer electrons), x-rays cannot as yet be focused to such a small size.

In general, after a qualitative analysis reveals the elements, the method must be calibrated for quantitative work by comparing peak intensities to intensities from samples of known composition, not by comparing directly to the intensity from a pure piece of each element, although techniques for direct calibration are improving. There are two basic reasons for this:

1. The K_β or K_α from one element in the sample may excite radiation from another, diminishing radiation from the first and enhancing that from the second (if the latter is energetic enough to reach the counter).
2. Because of absorption in the matrix increasing with increasing atomic number, less radiation will emerge from, e.g., iron in silver than iron in aluminum.

For a more complete discussion of this topic, the reader is referred to Jenkins (1974) and "Advances in X-Ray Analysis," Volume 19 (1976).

The intensity of a characteristic peak is

$$I_{char} = ki(V - V_{K_{ioniz}})^n, \qquad (3.5)$$

where $n \approx \frac{3}{2}$. Now

$$i = \text{power to tube}/V = P/V.$$

Therefore,

$$I_{char} = kP(V - V_{K_{ioniz}})^{3/2}/V.$$

The ratio (R) of the intensity of a characteristic peak to the total white radiation, including the fact that P is proportional to V, is

$$R = \frac{[k'V(V - V_{K_{ioniz}})^{3/2}]/V}{Z^2 V^2} = \frac{k'(V - V_{K_{ioniz}})^{3/2}}{Z^2 V^2}. \qquad (3.6)$$

This ratio levels off at V about equal to three times $V_{K_{ioniz}}$. The peak-to-background ratio cannot be improved by raising the voltage above about three times the excitation voltage or by raising the current, and this is an important point to keep in mind in using x-ray equipment. (At normal operating voltages, the ratio of K_α to white is about 10 depending on how sensitive our measuring device is to various wavelengths.) But one should also be aware that even if the ratio does not change, the difference between peak and background increases with voltage, and higher voltages may be useful. Suppose the ratio is $\frac{4}{1}$. Doubling the voltage will raise this to $\frac{8}{2}$. Weak peaks are thus more readily detected from the background at higher voltage.[3]

In most commercial x-ray tubes for fluorescence or diffraction, the electrons are focussed on the target on an area 1×10 mm. As shown in Fig. 3.10 we can obtain a narrow beam ("line focus") or a square spot ("spot focus") depending on how we look at the target (the "take-off" angle). Using electromagnetic fields, the electrons can be focused to about a micron, and with a well-collimated beam from such a target analysis can be carried out over very tiny regions — grains, inclusions, etc. Scattered electrons and x-rays can be used to image the distribution of elements in the area of interest, and this is possible both in transmission and reflection in so-called scanning electron microscopes, in which the electron beam is made to scan across an area. There are also x-ray units with focussing to ~ 10–100 μm.

---

[3] This is true provided the wavelengths associated with the higher voltage do not produce fluorescent radiation from the specimen, which would increase the background.

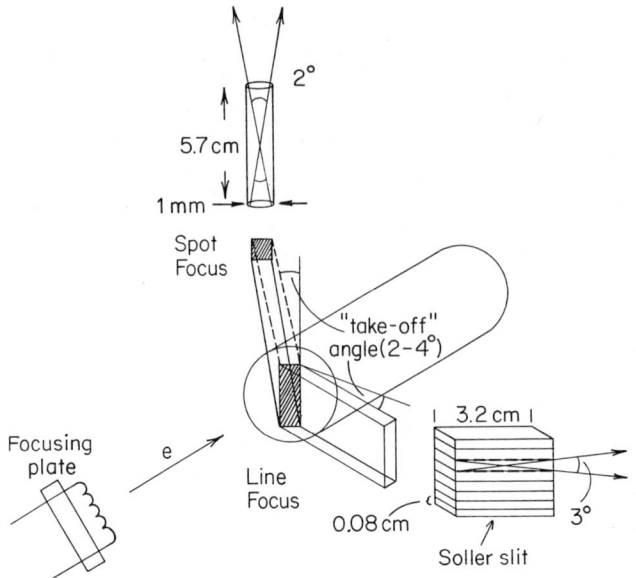

Fig. 3.10. Spot and line focus from an x-ray tube and typical slits for limiting divergence of the beam (divergence shown by *arrows*)

The slits in Fig. 3.10 reduce the horizontal and vertical angular divergences of a beam to the order of a degree so that all parts of the radiation used fulfill the Bragg condition very close to the same position on a film or at a counter, which keeps the peaks sharp. Slits from a tenth of a degree up to a few degrees are usually available, so that one can get more intensity, if necessary (with a loss in resolution). With a pinhole near the tube, one can get a picture of the distribution of x-rays across the target, i.e., how uniformly the electrons hit the target. This is important to know because a very nonuniform x-ray distribution can severely distort a peak shape, and it is best to be aware of possible distortions other than those from the diffracting material. Such distortions, if they are really from the material being studied, can tell us something about the internal structure as we shall discuss in Chap. 7.

If we pass x-radiation through a specimen, it will be absorbed as heat motion of the atoms, by scattering and by ionization. We shall discuss these interactions in detail later, but at this point, it is worth mentioning that because of ionization there is another method of chemical analysis. If we pass an x-ray beam from a tube through a specimen and then analyze it with a crystal, the result may look like that of Fig. 3.11. The sharp drop in transmitted intensity occurs for the K (or L) excitation of an element in the specimen. The absorbed energy is given off in *all* directions as the K (or L) spectrum of the element, which is why the *transmitted* beam has less intensity. The position of this drop in transmitted intensity is characteristic of the type of atom (not the state of aggregation), and the drop in intensity may be used for qualitative or quantitative analysis as we shall see soon. (It is also a way to measure $\lambda_{K_{abs}}$ or $\lambda_{L_{abs}}$.)

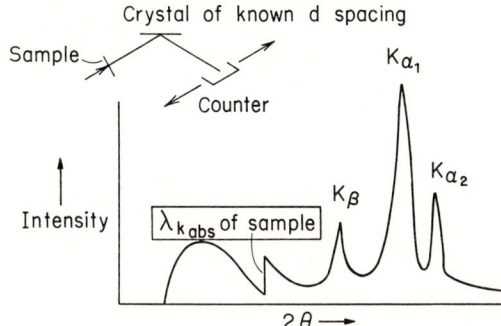

Fig. 3.11. Chemical analysis by absorption

For a target of atomic number Z the drop in intensity for element $Z - 1$ is often between the λ_{K_β} and λ_{K_α} from Z. *A foil of element $Z - 1$ of proper thickness can be used to clean up or filter the spectrum from a tube for diffraction*, by reducing the K_β intensity to say $\frac{1}{500}$ of its original intensity while only reducing the K_α by 50 to 60%. Tables for such filters with their effects on K_α, K_β can be found in the International Tables, Volume 3, pp. 75, 76.

Although it is usually not necessary, except for very weak scattering effects, *we can get a very clean spectrum from an x-ray tube by diffracting one wavelength from a crystal*. Even if there were not any sharp spectral lines in the beam from the tube, such a technique would allow us to choose one wavelength for diffraction from another crystal. Fortunately, the characteristic lines are strong and sharp so that this additional complication is not usually required. However, the cleaner spectrum from such a crystal *monochromator* can also be used in fluorescent analysis. Impurities present in only parts per million by weight can often be detected in this way. Without such a monochromator, the white radiation in the spectrum is diffracted from the sample, causing a background that limits detection of impurities to ≈ 0.1 wt. %.

X-ray quanta are wave packets or pulses of energy and these packets are the sum of many independent waves of slightly different wavelengths. Our treatment of diffraction in effect takes each one of these as an incident wave and adds the diffraction from them. It is because of the independence of the waves in the packet that we can do this. Actually, K_{α_1} has a wavelength spread of the order $\Delta\lambda/\lambda = 0.5 \times 10^{-3}$ due to the range of wavelengths present in the wave packet. To see how this affects the angular spread of a diffraction peak, we differentiate Bragg's law:

$$\Delta(\lambda) = d \cos\theta \, \Delta(2\theta)$$

(2θ is the variable we usually measure, the angle between the incident and diffracted rays, see Fig. 2.4) or,

$$\Delta(2\theta) = \Delta(\lambda)/d\cos\theta \qquad (3.7a)$$

and as $1/d = (2\sin\theta)/\lambda$,

$$\Delta(2\theta) = (2\Delta(\lambda)/\lambda)\tan\theta .$$

Multiplying by $360/2\pi$ to convert to degrees 2θ, we obtain

$$\Delta(2\theta)° = (720\Delta(\lambda)/2\pi\lambda)\tan\theta. \tag{3.7b}$$

For a low angle peak, say at $2\theta = 30°$, and $\Delta\lambda/\lambda = 0.5 \times 10^{-3}$,

$$\Delta(2\theta)° = 0.015° \, 2\theta,$$

and at $2\theta = 160°$,

$$\Delta(2\theta) \sim 0.3° \, 2\theta.$$

At large angles this effect is a very significant contribution to the breadth of a diffraction peak, which is actually about this magnitude for a large crystal. Other effects such as small crystal or mosaic size can also broaden the peaks, as we saw in Sect. 2.3.

Wavelengths of x-radiation have now been determined both by ruled grating experiments and by using density and crystal diffraction, as the Braggs did. At the time most of the measurements were made, there was an error in Avogadro's number, hence there was a difference in the wavelength for the two types of measurements. Calling the latter "kx" units, 1 kx unit $\cong 1.00202$ Å. There is a difference of 0.2%. At the present time, lengths, and hence d-spacings can be measured to six significant figures, but the accuracy in the last place of the conversion is not well established; it could be 1.00206, so that the accuracy can at best be only 1 part in 25,000. Precision can of course be higher, but it is always necessary to state the wavelength used as well as the precision in measurements if they are supposed to be of an accuracy higher than 0.2%, so the data can be adjusted, and the differences between reports of spacings from different investigators can be properly judged. This has not always been done, and care should be taken in examining the literature to ascertain the units, or wavelength. (The values in the International Tables are in angstroms.) One should also state how the position of a peak was measured! The positions of the center of gravity of a peak and its maximum can be quite different. Most tabulated values for the wavelength correspond to the maximum in the wavelength distribution. If the specimen whose d-spacings are being reported is an alloy of metals, or ceramic materials, do not forget to give the error in the chemical analysis also. This usually results in an error much higher than the error in the x-ray wavelengths! These points should not keep the reader from realizing that even with these uncertainties, the determination of d-spacings and lattice parameters is one of the most precise and accurate physical measurements that can be made. We shall discuss specific techniques for these measurements in Chap. 4.

The x-rays from a target are incoherent. We can see why by examining the result of adding two waves of the same amplitude emitted at different times so that their phases are different (see Problem 2.5a). The result is of the form

$$I \equiv A_T^2 = 2A_1^2 + 2A_1 \cos(\varphi_1 - \varphi_2).$$

It takes about 10^{-8} s for an atom to give off radiation. Thus at any one small region of the target during the seconds or minutes it takes us to make a measurement, there is a great range in phases and the cosine term averages to zero;

the intensity appears as if it is simply the sum of the intensities of independent sources. Suppose it happens that two atoms on the target emit at the same time, i.e., with same phase, and suppose they are 10^{-2} cm apart or 10^6 λ. Then even in the small angular range (1° or so) that we view the target, the difference in phase angle is large due to the path difference so that the cosine term goes through many cycles before the waves reach a detector or film. Again, the average intensity we measure over a time interval appears to be coming from independent sources.

3.1.3 Non-conventional Sources — Synchrotron Radiation

Recent years have brought a whole new excitement in the use of x-radiation for diffraction and spectroscopy. The source of this excitement is the enormous intensities of radiation from synchrotron or storage ring sources. In these devices, electrons or positrons are injected in discrete bunches into a circular vacuum chamber from some preaccelerator device. These charged particles are steered around the storage ring and focused by electro-magnets and maintained at a constant high energy in the 100 million electron volt (GeV) range using radiofrequency (rf) boosters. This rf energy input is required to compensate for the energy loss to emitted synchrotron radiation due to acceleration of the charged particles [Eq. (3.1)].

In the 1970's scientists began to make use of this radiation (which had been an unused by-product of many high-energy physics experiments) by constructing add-on facilities at some accelerators. These "parasitic" facilities are limited in capability since their design and scheduled operation are compromises dictated by the primary use of the accelerator for high-energy physics research. Consequently, dedicated sources of synchrotron radiation were proposed and are now in the 1980's becoming available throughout the world. These large facilities are specialized for emphasis on x-ray (for diffraction) or ultraviolet (for spectroscopy) and outfitted with many experimental stations or beam-lines (Fig. 3.12). Through such devices the use of x-radiation in materials research has entered the era of "big science" requiring travel to the experimental site just as is the case for neutron scattering experiments (Sect. 3.2). Such a trip must be justified by the requirements of the experiment, but not the least value of working at such synchrotron radiation sources is the exchange of information and techniques among users, many of whom are at the forefronts of the variety of scientific fields which make use of this radiation.

While each synchrotron source is uniquely constructed, there are some common features [see, for example, Bienenstock and Winick (1983)]. The charged particle bunches are spaced 2–2500 ns apart, and are 200–2000 ps wide, (depending on the ring and operating conditions) so that as they pass any location (say a slit) a burst of radiation appears. A typical pattern of "brilliance" (that is photon pulses per second, per mm^2 of source, per milliradian of divergence, per 0.1 pct energy band pass width) is shown in Fig. 3.13 for several such sources, and comparisons are given to a normal sealed x-ray tube in Fig. 3.14. To give some further comparison of intensities between a sealed x-ray tube and a synchrotron source, consider typical intensities on a specimen. With the characteristic radiation from an x-ray tube, this

92 3. Properties of Radiation Useful for Studying the Structure of Materials

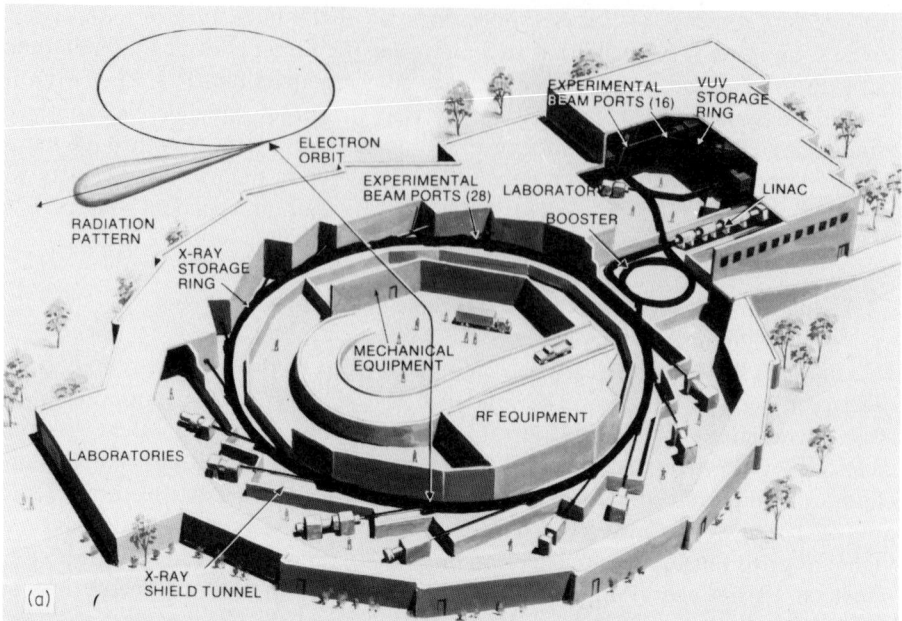

Fig. 3.12. a Beam line diagram. **b** Overview National Synchrotron Light Source, Brookhaven National Laboratory

3.1 Production of Radiation — X-rays

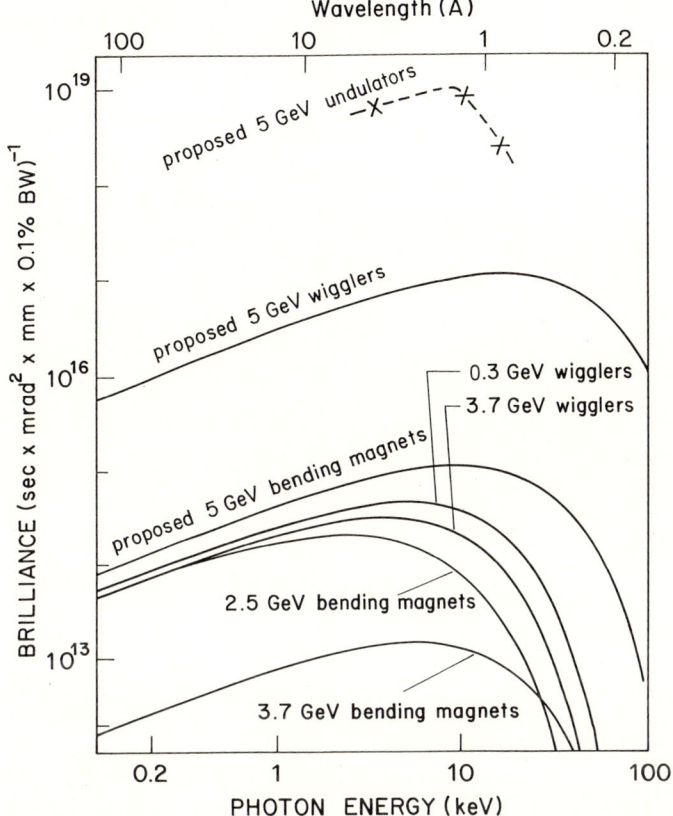

Fig. 3.13. Brilliance vs. energy at several synchrotron sources. (Courtesy of Dr. P. Eisenberger)

incident intensity is 10^7–10^9 c/s, (the lower value corresponds to the use of narrow slits and good crystal monochromators (see Sect. 4.7), whereas the larger is for normal filtered radiation). The corresponding white radiation is 10^3–10^5 c/s. At present (1985) synchrotron sources with high resolution the intensity is 10^{10}–10^{11} c/s over a wide range of wavelengths. Another important figure of merit of such sources is "emittance" — the product of source dimensions (~ 0.5 mm) and divergence angles (~ 0.05 mrad). From the definition of brilliance, it is inversely related to emittance. Thus emittance should be kept as small as possible, to maximize the intensity on a specimen.

The white spectrum in Fig. 3.13 can be characterized by a "critical" energy or wavelength, the wavelength that divides the curve in this figure into equal areas. This is given by:

$$\lambda_c(\text{Å}) = 5.59 R/E^3 \ . \tag{3.8}$$

Here R is the radius of the ring (in meters) and E its energy (in GeV). For example at the 2.5 GeV machine at Brookhaven National Laboratory $\lambda_c = 3$ Å, whereas at Cornell University's 8 GeV machine, $\lambda_c = 0.3$ Å. Thus, much higher intensities

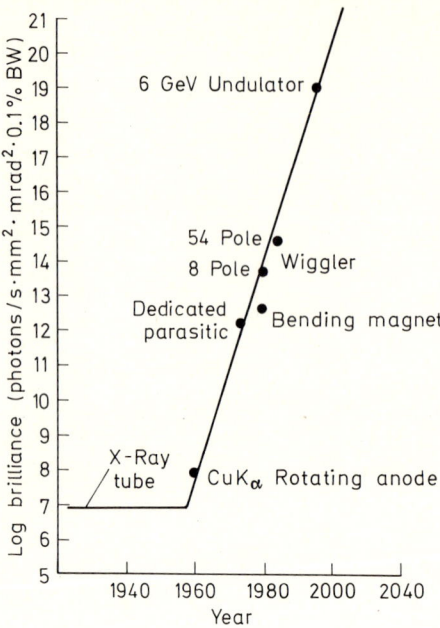

Fig. 3.14. The history of x-ray sources. (Courtesy of D.E. Moncton)

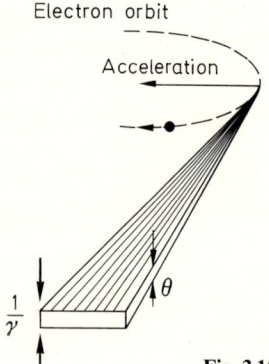

Fig. 3.15. Schematic of the x-ray fan from a synchrotron source

Fig. 3.16. a An insertion device — currently the magnets are made of a rare-earth cobalt material (SmCo$_5$). The beam oscillation is only 0.02 nm. Reprinted with permission from H. Winick, G. Brown, K. Halbach and J. Harris, Physics Today, May 1981, pg. 50. **b** In an undulator, the amplitudes of the radiation from the crests of the waves add constructively. The wavelength is adjusted by varying the magnet period. **c** Calculated spectrum from undulators. 3.0 GeV ring (from H. Winick, G. Brown, K. Halbach and J. Harris, Physics Today, May 1981, pg. 50)

3.1 Production of Radiation — X-rays

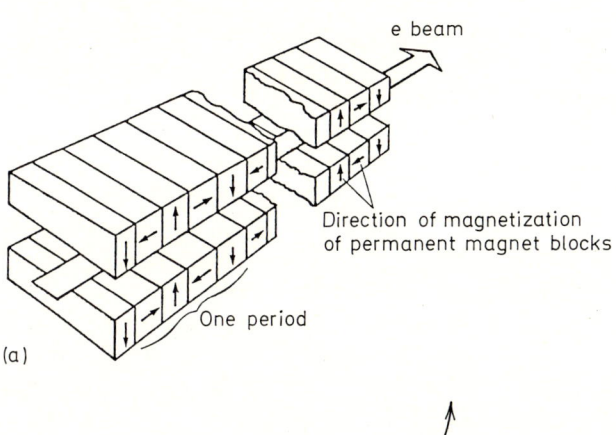

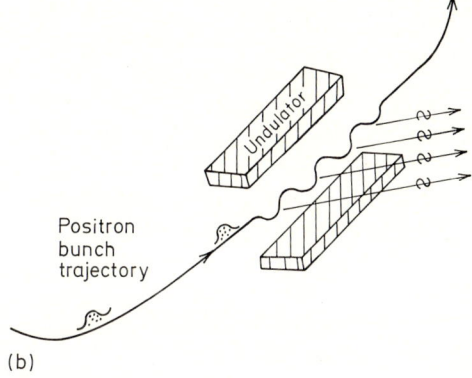

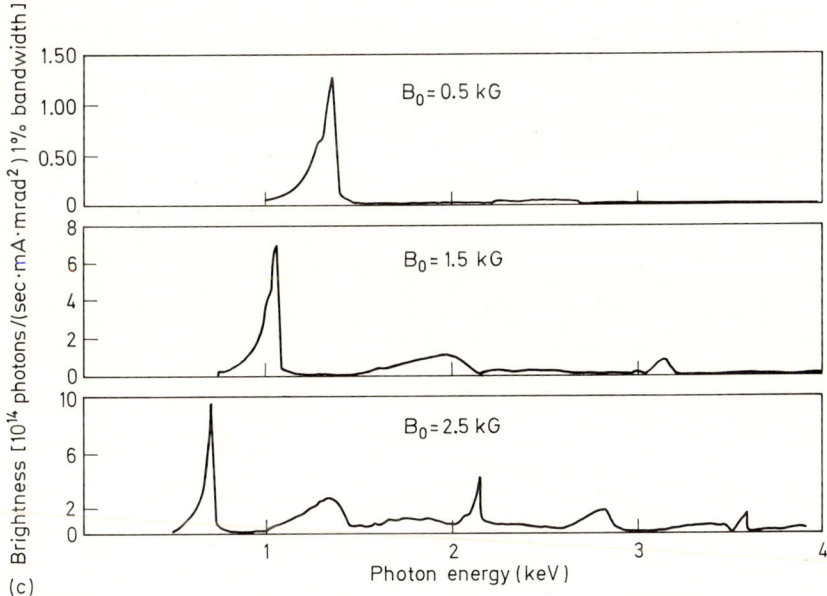

Fig. 3.16a–c

are available for very short wavelengths at Cornell University. For experiments requiring a single wavelength, a particular wavelength is chosen by interposing a crystal monochromating crystal.

When emitted, the radiation is already collimated very well in a direction normal to the ring (Fig. 3.15). In fact the divergence angle is $1/\gamma$, where γ is the ratio of the ring energy to the rest energy of the charge (for an electron or positron this rest energy is 0.5 MeV). Thus $\gamma = 5,000$ at Brookhaven and 16,000 at Cornell and the corresponding divergence angles are $\sim 0.012°$ at Brookhaven, $\sim 0.004°$ at Cornell. In the latter case, the beam spreads only ~ 1 mm in 15 m! In the horizontal direction, radiation with divergence of several milliradians is collected by a crystal and may be focused; we will discuss this technique in Sect. 4.7.

As suggested in Eq. (3.1), if we alter the course of the charges so that they bend around a smaller radius, the acceleration and consequently the intensity of the emitted radiation increases. This leads to the idea of "insertion devices", modified segments of the storage ring with rapidly varying local magnetic fields. There are three classes of these: 1) wigglers, 2) wavelength shifters and 3) undulators. The principle is illustrated in Fig. 3.16. A series of permanent magnets is arranged to "wiggle" the electron in its orbit, and, as indicated in Fig. 3.13, enormous gains in brilliance occur. With a wiggler the oscillations are relatively large, and the whole energy spectrum shifts upward, proportional to the number of periods in the magnet; the divergences remain the same. For an undulator, the magnetic field is kept low (~ 1 KG) so that the "wiggles" are slight (see Fig. 3.16b) the radiation from the crests is in phase; the brilliance rises proportional to nearly the square of the number (N) of periods and the divergence is $\sim 1/N\gamma$ in all directions. The energy is concentrated in one or a few wavelengths (Fig. 3.16c), which can be adjusted by varying the magnet spacing. A wavelength shifter is just a short undulator. Such devices are the order of a meter to several meters long, and must be inserted in straight sections of the ring. There are few of these straight sections on older storage rings which are essentially circles, but the newer ones will be polygons, to accommodate many more of these devices. The radiation intensity will increase at least 10^2 times with planned insertion devices on existing machines and 10^4 more in the next generation of machines with lower emittance and undulators. With these insertion devices, an x-ray microprobe with a beam size similar to an electron microprobe (or even smaller) will be possible, and the detection limit will be ~ 5000 times lower than with electrons because more of the energy is used for ionization. Another exciting feature of synchrotron sources is the promise of increased flexibility in the study of organic materials. Organic and biological samples are often damaged by exposure to x-radiation. A synchrotron source is helpful in such cases, because it appears that the damage is related to time in the beam, not the total dose, so that more data can be obtained with such a source before this damage becomes pronounced.

In addition to the higher intensities, the pulsed nature of the radiation allows study of transient behavior (either spectroscopic or structural) since a complete pattern may be obtained with a time resolution of a fraction of a nanosecond at some controlled time after initiating some change in the sample. Similar capability is also available from the x-rays produced by the interaction of powerful lasers with plasmas. These pulsed sources also permit diffraction patterns to be obtained in nanoseconds and are more convenient since they are laboratory size (Robinson 1979).

3.2 Production of Radiation — Neutrons

Many experiments have now verified that *all particles have wave properties* which may be expressed in terms of their momentum (mv) according to the relation proposed by de Broglie:

$$\lambda = h/mv . \tag{3.9}$$

The wave properties of neutrons were demonstrated in a diffraction experiment in 1936 by Halban and Preiswerk and by Mitchell and Powers. At that time the only source of neutrons was from the nuclear reaction of α particles with beryllium. With the development of peaceful uses of atomic energy after the second World War came the research reactor and high-intensity neutron beams. It is now becoming interesting to produce neutrons using the beams from nuclear reactions produced by high intensity-high voltage particle accelerators because the power of reactors has reached the current limits of technology. But most neutron scattering experiments are still done at research reactors. In these reactors the neutrons developed in the fission process [energies in the (MeV) range] are allowed to scatter from the light atoms in a moderator of heavy water or graphite. In these scattering collisions the neutrons lose energy, coming to thermal equilibrium with the moderator. The velocity distribution of these "moderated" or "thermalized" neutrons is approximately that of a monatomic gas, at the reactor temperature, T_m, with the largest number of neutrons occuring for a velocity v_m given in terms of T_m as

$$\tfrac{1}{2}mv_m^2 = \tfrac{3}{2}kT_m . \tag{3.10}$$

Since $T_m \approx 300°K$, the moderated neutron energies are in the millielectron volt (meV) range. Combining Eqs. (3.9) and (3.10), the wavelength of these neutrons is

$$\lambda_m = (h^2/3mkT_m)^{1/2} . \tag{3.11}$$

As the mass of a neutron is $\approx 10^{-27}$ kg, $\lambda_m \approx 1$ Å, just right for diffraction from crystals. The actual number of neutrons $N(\lambda)$ between λ and $\lambda + d\lambda$, per square centimeter per second, is of the form

$$N(\lambda) = (2N/\lambda)(E/kT)^2 e^{-E/kT} , \tag{3.12}$$

where N is the total number of neutrons of all wavelengths passing through an area of one square centimeter per second, i.e., a flux. (Typical fluxes at a reactor core are 10^{14} to 10^{16} neutrons/cm^2/s). E is the energy of any one neutron of wavelength λ. The distribution of the spectrum scattered from an analyzing crystal will look like that of Fig. 3.17a.

It is necessary in this case to use a monochromator crystal to choose a wavelength for diffraction because of the lack of any sharp intense spectrum such as the characteristic x-radiation. In general, if one wavelength is desired for diffraction, it is better

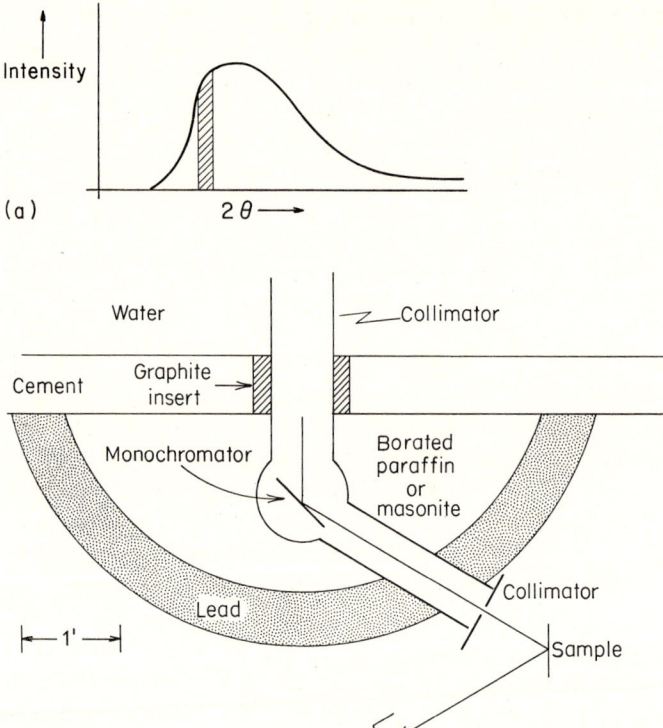

Fig. 3.17. a Spectrum from a thermal nuclear reactor. *Shaded region* is selected by monochromator. **b** A typical port for neutron diffraction. (Adapted from Bacon, G.E., "Neutron Diffraction", 2nd ed. Oxford Univ. Press, London and New York, 1962)

to choose a wavelength to the short wavelength side of the maximum in the distribution as shown in Fig. 3.17a. Then the intensity of $\lambda/2$, which will diffract in second order at the same position for first order from λ, is minimized. Alternatively, one may use a crystal whose structure factor is zero for the second order.[4] To see the effect of the monochromator on the shape of the peak, let $E = kT$. Then from Eq. (3.12), $N \approx N/\lambda$. If the angular range over which the monochromator diffracts strongly is $\Delta 2\theta$, the range of λ's reflected is [from Eq. (3.7a)]

$$\Delta \lambda = d \cos \theta \, \Delta 2\theta ,$$

and the number of neutrons in the beam from the monochromator, $N_T = N(\lambda) \Delta \lambda$ is

$$N_T = (N/\lambda) d \cos \theta \, \Delta 2\theta \, KF^2 = N[(\cot \theta)/2] KF^2 \, \Delta 2\theta , \tag{3.13}$$

---

[4] This effect occurs with x-rays also due to $\lambda/2$ in the white spectrum, and the same solution will help, although there are other solutions using electronic selection devices known as pulse height selectors. This will be discussed in Sect. 3.4.

where the diffracted intensity has been taken proportional to F^2, the structure factor for the monochromator. The more imperfect the crystal, i.e., the greater the value of $\Delta 2\theta$, the more divergent its beam will be, but also the higher the flux to the specimen. This intensity also is greater for a low-angle reflection from the monochromator not only because of the $\cot\theta$ factor in Eq. (3.13), but also because of geometrical factors we will discuss in Chap. 5. Hence, if possible, a monochromator should be chosen that has a large enough d-spacing to diffract the peak in the neutron distribution at low angles, and of course it should have a large structure factor for this diffraction; if resolution needs to be sacrificed for intensity, a crystal with a large degree of imperfection is required.

A typical arrangement for neutron diffraction is shown in Fig. 3.17b. The hydrogen in the paraffin or masonite shield slows down any fast neutrons which are then absorbed in the boron. The γ- or x-rays from the reactor are absorbed by the lead. This is for safety and to reduce the background. At the reactor face a typical neutron beam is the order of 2 to 8 cm^2 (to get adequate intensity). The collimator to the crystal monochromator is about 1 to 2 m long, and may be of wood with cadmium-plated steel channel (cadmium absorbs neutrons). There will be vertical plates spaced about 0.5–1 cm apart so that the horizontal divergence is of the order of 0.25°–0.5° and the vertical divergence is about 1°. A lead shutter and facilities for filling the collimator tube with water may be installed so that one can remove the shielding and get at the monochromator, or it can be removed from a hole at the top of the shielding which is normally plugged. External controls are available for tilting the monochromator to orient it. The entire monochromator assembly, shielding and exit collimator, may be mounted on a rotating plate to allow a choice of wavelength or there may be plugs in various places on the outside shield to obtain different wavelengths. The counter will be typically a 2-in. diameter, 1-ft-long cylinder filled with BF_3 enriched in the isotope ^{10}B or with ^3He. (We shall discuss counters more later; however, these isotopes emit α particles on absorption of neutrons, which temporarily ionize the gas, causing an electrical pulse between the casing and a central wire in the gas, as in a gas-filled x-ray counter.) When covered with shielding the unit weighs several hundred pounds! The goniometer to move the counter is usually much more massive than an x-ray goniometer, weighing several tons.

Knowing the dimensions of the collimator we can estimate what area of the neutron source is being seen, and hence the neutrons per second coming from the collimator; then when we know how to calculate the diffracted intensity in absolute units $[KF^2$ in Eq. (3.13)$]$, we shall find that the counting rate from the monochromator is 10^6-10^8 cpm, four orders of magnitude lower than with conventional x-ray sources. As we shall show later in this chapter, neutron absorption in matter is generally much less than absorption of x-rays. Consequently, large samples may be used to compensate for the low incident beam intensity, and recording the neutron diffraction pattern takes times comparable to that for a similar x-ray experiment. Because a reactor, unlike an x-ray generator, cannot be stabilized electronically, the beam from the monochromator is counted by a small low absorbing counter in the path of the beam, and counts from the specimen are measured for a fixed number of counts into this monitor counter.[5] Since the spectrum of neutrons is established by the thermalization temperature of the moderator [Eq. (3.11)], it is possible to obtain increased intensities of longer or shorter wavelength neutrons by shifting the moderator temperature down or up respectively. Such cold and hot sources are now becoming common in research reactors. Typically a hydrogen bearing material is inserted into the reactor core and maintained at a temperature different from that of the moderator. The neutrons within this source come to equilibrium at the higher or lower temperature by scattering from the hydrogen, losing or gaining energy in each collision. "Hot" neutrons are particularly useful in spectroscopic studies in which longer than millielectron energy transitions are involved. "Cold" neutrons are used for both smaller energy transfer spectroscopy, and for diffraction from structural features in the size range significantly longer than 1 Å. Such thermal sources can typically increase fluxes in the desired range by an order of magnitude or more.

The fluxes available from nuclear reactors are limited to the present levels of 10^{15} $n/cm^2/s$ (resulting in only $\sim 10^8$ cpm from a monochromator as mentioned above) by technical difficulty in removing the heat generated by the fission reaction. It is anticipated that no more than an order of magnitude improvement can be anticipated in new reactors of improved design. Consequently alternative sources of neutrons have been sought and found in accelerator driven pulsed sources.

In these pulsed neutron sources (Lander and Carpenter 1985) protons in the 10^3 MeV range from an accelerator are allowed to bombard a dense sample (e.g. tungsten or depleted uranium). The resultant nuclear reaction, known as spallation, is similar to the effect of the "break" in billiards, and results in pre-emission of copious high energy neutrons. These neutrons are thermalized in a hydrogenous moderator surrounding the spallation source and these thermalized neutrons may be used for spectroscopy and diffraction as are those from fission reactors.

Typical pulses of neutrons from such sources will be of 20–50 μs duration at frequencies of 10–50 s^{-1}. Thus the average neutron flux will be orders of magnitude lower than the peak flux in each pulse. With an anticipated peak flux of the order of 10^{16}–10^{18} $n/cm^2/s$, such sources would not appear particularly attractive at first glance; however, these are two features of pulsed sources which do justify their construction, and have led to prototype instruments at lower peak fluxes (10^{14}–10^{15} $n/cm^2/s$) in several countries.

These two features are high fluxes of "hot" neutrons and the inherent pulsed nature of the beam. Since the moderation of the spallation neutrons is typically incomplete (a thick moderator would be required for compete moderation, but that would worsen the time resolution of the pulses) a high energy (short wavelength) tail is present in the flux distribution. Thus spallation sources are inherently superior to reactors for large energy transfer spectroscopy. Use of the pulsed character of the beam is made in experiments in which the energy of the neutrons is determined by measuring neutron wavelength by the time-of-flight (TOF) of the

---

[5] This can also be done with x-rays to circumvent the instability of an x-ray generator or variations in the intensity due to absorption of x-rays by the air in the beam path, which varies with barometric pressure. The fluorescene from a thin foil in the x-ray beam to the sample can be measured by a second, monitor counter.

neutron from source to detector (Sect. 4.7). When carried out at conventional reactors, such experiments produce pulses of polychromatic neutrons by rotating absorbing material through the beam from the reactor so that a 20 μs pulse is produced 20 or so times per second. Such "chopped" beams throw away several orders of magnitude of neutrons which are absorbed in the absorber (chopper). Neutron pulses from a spallation source need not be chopped and so for TOF experiments their anticipated peak fluxes (10^{17} n/cm²/s) look very attractive compared to those likely to be available from the next generation of reactors (10^{16} n/cm²/s). As we shall see in Sect. 4.7, TOF experiments are particularly effective for studying polycrystalline samples of low symmetry materials. For such studies, spallation sources are likely to be superior to conventional reactors.

3.3 Production of Radiation — Electrons

The diffraction of electrons was first demonstrated in 1927 by Davisson and Germer (nine years before diffraction of neutrons was observed). Using the de Broglie relationship for electrons, Eq. (3.9), equating the applied field Ve to the kinetic energy of the electron as in deriving Eq. (3.2), and neglecting relativistic corrections:

$$\lambda(\text{Å}) = 12.3/\sqrt{V_{\text{volts}}}, \qquad (3.14\text{a})$$

where V is the accelerating voltage to which the electrons have been subjected. At very high voltages, 50 kV or more, relativistic corrections are required. In terms of the rest mass m_0, the mass and momenta are

$$m = \frac{m_0}{(1 - v^2/c^2)^{1/2}}, \qquad \text{or} \qquad (mv) = [(mc)^2 - (m_0 c)^2]^{1/2}.$$

Also, the kinetic energy is

$$(m - m_0)c^2 = Ve, \qquad \text{or} \qquad m = m_0 + (Ve/c^2).$$

Therefore,

$$mv = [2m_0 Ve + (e^2/c^2)V^2]^{1/2}$$

and

$$\lambda = 12.3/(V + 10^{-6} V^2)^{1/2}. \qquad (3.14\text{b})$$

The wavelength thus varies from about 1 Å at 150 V to about 0.03 Å at 100 kV, the highest voltage now in general use in electron microscopes. (Units with voltages in excess of 1 MeV are in use in some laboratories.) *At these high voltages (small wavelengths) the diffraction angles will be quite small*, several orders appearing from 0 to 2° 2θ. To a good approximation, $\sin\theta$ in Bragg's law can be replaced by θ.

As we shall see later in this chapter, electrons are strongly absorbed in matter. Consequently, electron diffraction must be carried out in transmission through thin foils, or by glancing incidence from thicker samples. A typical arrangement for an electron source uses an undoped fine W wire operating at 2400°K to produce a small flux of thermally emitted electrons. A typical electron "gun" is shown in Fig. 3.18. It is the same kind of electron gun that is employed in the electron microprobe

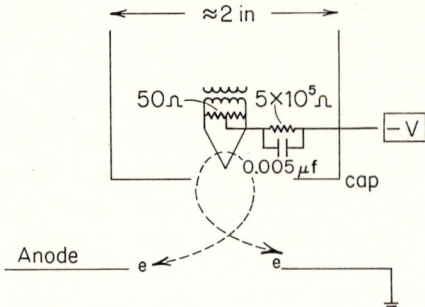

Fig. 3.18. Self-baising electron gun. (Adapted from Hall, C.E., "Introduction to Electron Microscopy", McGraw-Hill, New York, 1953)

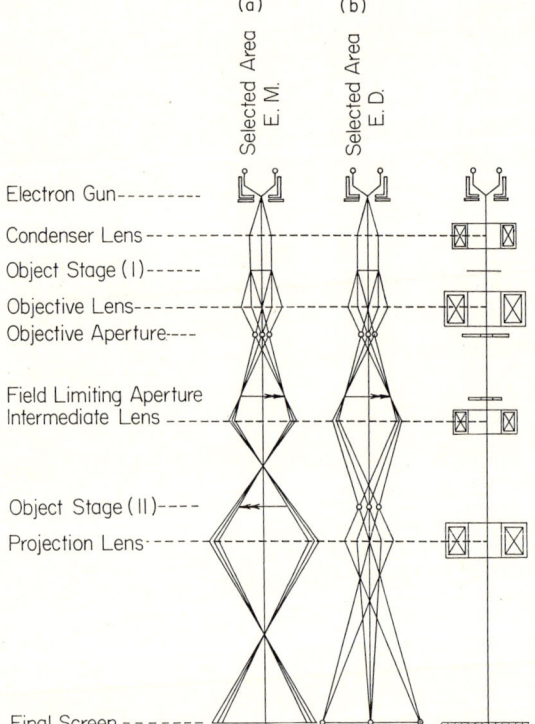

Fig. 3.19a, b

Fig. 3.19a–d. Schemes for electron microscopy and diffraction. **a–c** High voltage (approximately 50–100 kV) (**c** is the actual microscope). Where all diffracted rays from a point cross in **a** and **b**, an image is formed; where the same order of diffraction from different points in the sample cross, a diffraction peak occurs. The sample in **a** and **b** is on the first object stage; it could, however, also be placed on the second or "diffraction" stage, if an image is not required. In **a** and **b** the lens currents are altered to give either a diffraction pattern or image on the screen, i.e., to alter the location of the pattern or image relative to the final fluorescent screen. An area is "selected" with a field limiting aperture. The other apertures confine the beam to the center of the electromagnetic coils, where they act like optical lenses. (Courtesy of Hitachi, Ltd. and Perkin-Elmer Corp.) **d** Low voltage diffraction system (100 V). (Adapted from Germer, L., Phys. Today *17*, 1964 © American Institute of Physics)

3.3 Production of Radiation — Electrons

(for chemical analysis), the scanning electron microscope and the microfocus x-ray generator, as well as for the ordinary electron microscope.

The negative potential to the cap or "Wehnelt cylinder" around the filament increases as the current to the filament increases, so that at any filament current the beam current through the anode is stabilized, i.e., the heating of the filament and the negative bias are in balance. There is a "crossover" point of the electrons that is the effective source point. The beam intensity can be controlled by the depth of the filament inside the cylinder. A filament may last up to a week or two in continuous operation. Voltages are controlled very precisely electronically. Many modern microscopes use LaB_6 filaments for higher electron emission intensity. In some instruments in which ultra high vacuum is maintained in the microscope column, a field emission electron source is used for still higher intensity and sharper point source resolution. From this source point, the beam is controlled by a series of electromagnetic lenses. Typical arrangements are shown in Fig. 3.19. An aperture

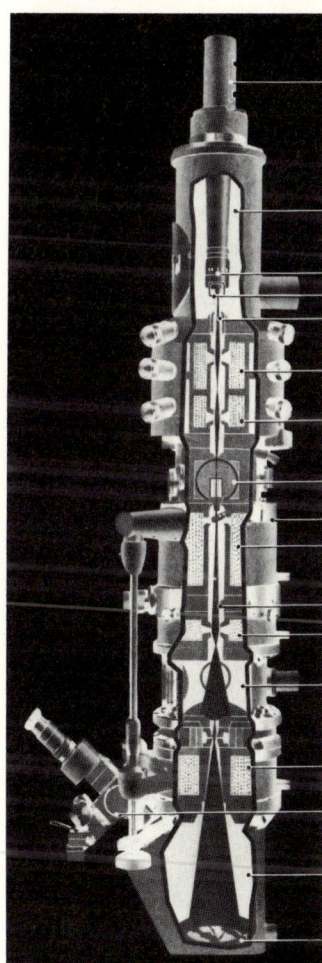

Cabledhead

Electron gun

Grid
Cathode
Anode

First condenser lens

Second condenser lens

Specimen chamber

Objective lens aperture
Objective lens

Field limiting aperture
Intermediate lens

Diffraction specimen chamber

Projection lens

Optical microscope

Viewing chamber

Fluorescent screen

Fig. 3.19c

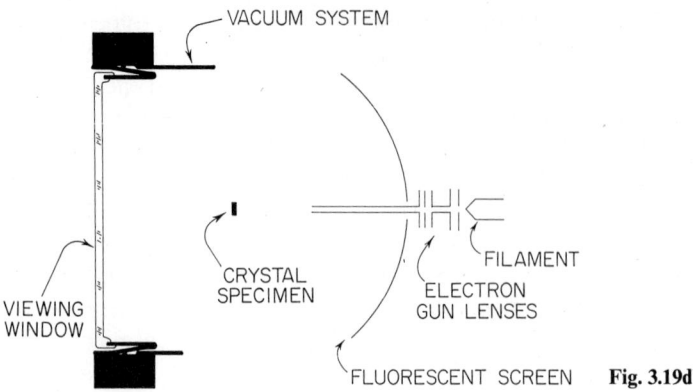

Fig. 3.19d

may be used to isolate a small fraction of the field for diffraction, as small as a few tens of angstroms. If the aperature blocks many of the orders of diffraction, the image will not be perfect. In Chap. 6, we shall show that the structure can be represented by a series, each term of which is related to a diffraction peak, and if terms are eliminated, the picture is not perfect.

The pattern is focused to be sharp, but must be calibrated for a given setting of the lenses or magnification with the diffraction pattern from a known specimen, if d-spacings are desired. In calibrating, it is necessary to measure distances to spots of known d-spacing in several directions, to correct for any eccentricity of the lenses.

3.4 The Interaction of X-rays with Matter

3.4.1 Polarization of X-rays

In Chap. 2, we discussed the physics of diffraction; our results involved a scattering factor per atom. We now turn our attention to this scattering factor and also to absorption and refraction of the useful kinds of radiation. The matters we are to deal with are associated with the interaction of the potential of the sum total of the atoms of the material and the radiation. This area can be dealt with in detail only with quantum theory and wave mechanics. We cannot hope to delve into these subjects in detail to provide the necessary background; as a result, we shall be concerned mainly with qualitative physical arguments to understand the nature of the phenomena involved, and to appreciate the quantitative relations presented.

When the electromagnetic field of an x-ray beam approaches an atom, it acts on the electrons causing them to scatter the radiation. *Consider the effect of a beam of x-radiation on a single free electron.* The time-dependent oscillating electric field of the radiation exerts a force on the electron which causes it to oscillate, accelerating with a time-dependent acceleration $\mathbf{a}(t)$. As we saw in Fig. 3.5 and Eq. (3.1), an

3.4 The Interaction of X-rays with Matter

accelerating electron gives off radiation with amplitude ε at distance r given by

$$\varepsilon = [e\eta |\mathbf{a}(t)| \sin \alpha]/c^2 r . \tag{3.15}$$

Note that the fields are proportional to $1/r$ not $1/r^2$; they are felt at a much larger distance than a stationary field. The reason this field is proportional to $1/r$ and not $1/r^2$ is easy to understand. As the radiation is emitted in all directions, the total energy of the propagating wave is distributed over a sphere. As the surface of the sphere increases proportional to r^2, the energy at any point of observation on the sphere must fall off as $1/r^2$. Since the energy at any point depends on the square of the field, the field at any point falls off as $1/r$.

The angle between the propagating field or observation point and the direction of oscillation is α (Fig. 3.5). Sin α *is a polarization term*. At the observation point we "see" only the component of vibration parallel to the observer, perpendicular to the direction of propagation. To verify this polarization factor, we can, as Barkla did in the nineteenth century, make a "polarizer" and an "analyzer" as in Fig. 3.20a using blocks of, say, amorphous carbon to scatter the x-rays without diffraction peaks.

From an x-ray tube, we can examine the polarization of the white and characteristic radiation as in Figs. 3.20b and c. The very short wavelengths in the white radiation are polarized (as we might expect), as this radiation comes from almost a complete deceleration in one step of the electrons incident along the y axis, and not by a series of collisions. The characteristic radiation is unpolarized, however, as shown in the figure.

As we deal mainly with characteristic radiation in diffraction experiments, we are then dealing with an unpolarized beam. Let us explore what happens when such a beam scatters from a single free electron. We shall consider the unpolarized instantaneous incident beam ε_0 in Fig. 3.21 resolved into two components $\varepsilon_{0_\parallel}$, $\varepsilon_{0_\perp}$ parallel and perpendicular, respectively, to the shaded plane. Both $\varepsilon_{0_\parallel}$ and $\varepsilon_{0_\perp}$ are perpendicular to the direction of propagation of the waves as these are transverse electromagnetic waves. *Even though the initial beam is unpolarized, after scattering to some point A, it is partially polarized*, because we are examining the scattering in a specific direction. We wish then to express the polarization as a function of diffraction angle, as the polarization must affect the measured intensity.

In Eq. (3.15), we can replace $\mathbf{a}(t)$ by the force on the electron, $\varepsilon_0(t)e$, divided by the mass m to give[6]

$$\varepsilon = [\varepsilon_0(t)e^2\eta^2/mrc^2] \sin \alpha .$$

Because of the mass term in the denominator, it is the electrons in an atom which are the primary producers of these fields, not the much heavier nucleus. We can measure only average scattered intensity I over some time which, as we have seen in Chap. 2, is related to the average of the square of the maximum amplitude of the field (represented by E_0). Therefore, for each component,

$$I = KE^2 = (KE_0^2 e^4 \eta^2/m^2 r^2 c^4) \sin^2 \alpha .$$

---

[6] $\varepsilon_0(t) = \mathbf{E}_0 \exp(2\pi i/\lambda)(x - vt)$.

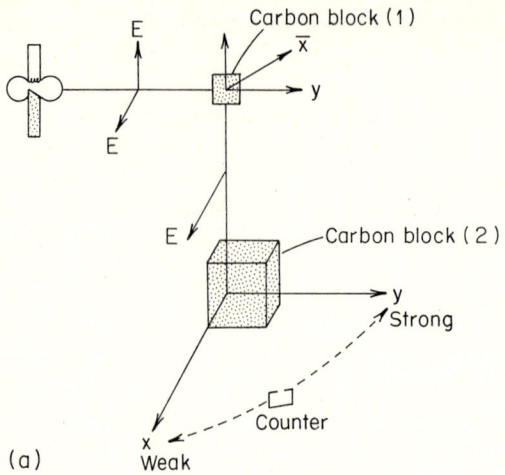

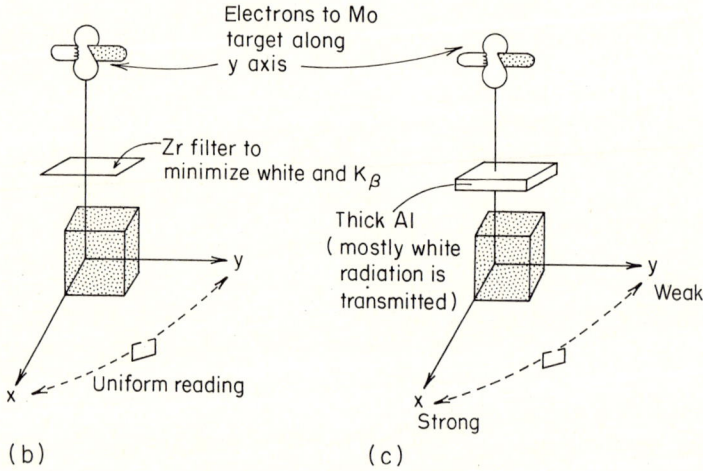

Fig. 3.20a–c. Examining the polarization of the spectrum from an x-ray tube. The scattering from carbon black is examined in the x-y plane. In **b**, Zr has an absorption edge just below the K_α wavelengths, so it acts as a filter, absorbing mostly the K_β and white radiation. In **c** Al absorbs more of the long wavelength radiation than the short; as we shall see later, this is the nature of x-ray absorption for all materials

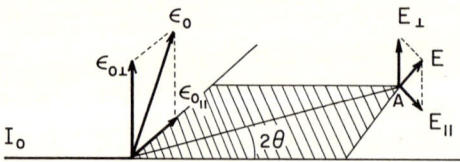

Fig. 3.21. Geometry for the calculation of the polarization factor in x-ray scattering

3.4 The Interaction of X-rays with Matter

At the measuring position A, the intensities of the two components are

$$I_\perp = I_{0_\perp}(e^4\eta^2/m^2r^2c^4), \quad \text{and} \quad I_\parallel = I_{0_\parallel}(e^4\eta^2/m^2r^2c^4)\cos^2 2\theta,$$

where $I_{0_\perp} = KE_{0_\perp}^2$ and $I_{0_\parallel} = KE_{0_\parallel}^2$. If the beam before diffraction is unpolarized, as for characteristic radiation,

$$I_{0_\parallel} = I_{0_\perp} = \tfrac{1}{2}I_0.$$

Thus, adding the two components,

$$I = \frac{I_0 e^4 \eta^2}{m^2 r^2 c^4}\left(\frac{1 + \cos^2 2\theta}{2}\right) = I_e. \tag{3.16}$$

I_e is known as *the Thomson scattering per electron* after J.J. Thomson who first derived the result given in Eq. (3.16). We note that $\eta^2 e^4/m^2 c^4 = 7.9 \times 10^{-30}$ m^2 per unit solid angle in mks units; $1/r^2$ is a measure of the solid angle over which the scattering is being observed. The scattering for one electron is quite small! The factor I_e will appear in all scattering expressions if the incident beam is unpolarized, and we often express the intensity as scattering per electron, dividing through by I_e. *The factor $(1 + \cos^2 2\theta)/2$ is often referred to as the polarization factor*. If the beam is scattered first by, say, a monochromator crystal and then by a sample, the total polarization will of course be different (see Problem 3.20).

The x-rays from a synchrotron ring are highly polarized in the plane of the ring. Therefore, diffractometers at such a source are mounted vertically, so that the diffraction plane define by **S** and **S**$_0$ is normal to the ring. In this way the polarization is largely parallel to the sample surface, and changes little during scattering.

3.4.2 X-ray Scattering Factor

We will find it useful later to know the total scattering per electron. The geometry of the problem is sketched in Fig. 3.22. The Thomson scattering derived in Eq. (3.16) gives the intensity scattered at angle 2θ into a unit solid angle. Multiplying by the angular element $(2\pi r \sin 2\theta)(r\, d(2\theta)) = 2\pi r^2 \sin 2\theta\, d(2\theta)$, and integrating, we obtain the scattered power P:

$$P = \int_0^\pi \frac{I_0 e^4 \eta^2}{m^2 c^4 r^2}\left(\frac{1 + \cos^2 2\theta}{2}\right) 2\pi r^2 \sin 2\theta\, d(2\theta) = \frac{8\pi\eta^2}{3} I_0 \frac{e^4}{m^2 c^4}. \tag{3.17}$$

The total scattering "cross section," scattering per unit incident intensity, is $P/I_0 = 6.7(10)^{-29}$ m^2/electron.

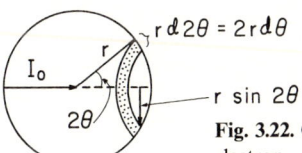

Fig. 3.22. Geometry for calculation of the total x-ray scattering from a single electron

We may now consider the scattering from all the electrons in an atom. *We define the quantity f as the scattering amplitude of an atom relative to that for a single electron.* The total scattering amplitude per atom at the angle 2θ may then be written as

$$I_e^{1/2} = \frac{I_0^{1/2} e^2 \eta}{mc^2 r} \left(\frac{1 + \cos^2 2\theta}{2}\right)^{1/2} f. \tag{3.18}$$

We know that the scattering factor f is [Eq. (2.14c)]

$$f(\mathbf{S} - \mathbf{S}_0) = \int \rho'(\mathbf{r} - \mathbf{r}_n) \exp\left[2\pi i \left(\frac{\mathbf{S} - \mathbf{S}_0}{\lambda}\right) \cdot (\mathbf{r} - \mathbf{r}_n)\right] dV_{(\mathbf{r} - \mathbf{r}_n)}, \tag{3.19a}$$

where $\rho'(\mathbf{r} - \mathbf{r}_n)$ is the electron density of the atom. When $\mathbf{S}$ and $\mathbf{S}_0$ are parallel, there is no phase factor between electrons in the atom, i.e., $f(0) = \int \rho'(\mathbf{r} - \mathbf{r}_n) dV$, and $f(0)$ is the atomic number Z. The phase factor is important at other angles because $\mathbf{S} - \mathbf{S}_0$ and $\mathbf{r} - \mathbf{r}_n$ are of the same magnitude, so that there is a significant path difference for scattering at different points around an atom.

To estimate the angular dependence of f, we shall assume that the electrons are distributed spherically around an atom. Then, as shown in Fig. 3.23,

$$2\pi i [(\mathbf{S} - \mathbf{S}_0)/\lambda] \cdot (\mathbf{r} - \mathbf{r}_n) = 2\pi i (2 \sin \theta) |\mathbf{r} - \mathbf{r}_n| \cos \varphi / \lambda.$$

Let $k = (4\pi \sin \theta)/\lambda$, and $|\mathbf{r} - \mathbf{r}_n| = r'$. Then

$$f = \int_0^\infty \int_0^\pi \rho'(r') 2\pi r'^2 \sin \varphi \exp[i k r' \cos \varphi] \, d\varphi \, dr'$$

or

$$f = \int_0^\infty 4\pi r'^2 \rho'(r') \frac{\sin kr'}{kr'} dr'. \tag{3.19b}$$

Thus, due to the factor $(\sin kr')/kr'$, f falls off with increasing k, i.e., as $\sin \theta / \lambda$ increases.

According to the quantum mechanical description of matter, the electron density ρ' at any one point is actually the product of the wave function ψ_e representing any electron times its complex conjugate, i.e., $\psi_e \psi_e^*$. The contribution to the scattering is calculated for electrons in all shells, and added, i.e., $f = \sum f_e$, $f^2 = (\sum f_e)^2$. The wavefunction can be evaluated in the following way, first devised by Hartree. Each electron is in the field of the others and the nucleus. A charge density is assumed for all the electrons and a potential is calculated from this and the field of the nucleus.

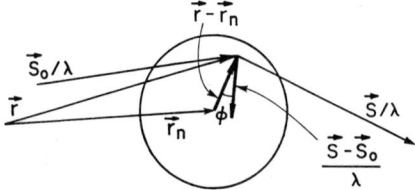

Fig. 3.23. Geometry for calculation of the x-ray scattering from electrons spherically distributed around the nucleus of an atom

3.4 The Interaction of X-rays with Matter

The motions of the electrons in this potential are calculated by wave mechanical methods. From the resulting wavefunctions, the charge density at any point is obtained and the problem is repeated until there is good agreement between the assumed and calculated densities. This works well for atoms with atomic numbers (Z) up to 25. For higher atomic numbers other methods are used. The calculations have also been refined to include any atomic asymmetry and interactions between charges that occur when atoms in a solid are close together. (See International Tables, Volume III, p. 201, and Volume IV, p. 71 in which values of f are tabulated versus $k/4\pi$, i.e., $\sin\theta/\lambda$.) In principle, measurements of the x-ray scattering intensities from known structures can be used to determine f and obtain information about the atomic bonding. Unfortunately, because of the $(\sin kr)/kr$ term in Eq. (3.19b), outer electrons (with large r) have only a very small effect on f, and it is very difficult to check the calculations of f for the effects of bonding. Attempts to check and refine calculations have been made, and although the accuracy of the calculations is quite good, it is still questionable whether the contribution from outer shells can be determined with any accuracy even in the most precise experiments, although in certain special cases it has been possible.

Once f is known for all atoms, the structure factor $F(\mathbf{s})$ developed in Chap. 2 may be calculated. *$F(\mathbf{s})$ is then the scattering amplitude of x-rays by a unit cell of atoms relative to the scattering by one electron.* With x-rays, we always use calculated values of the f's in $F(\mathbf{s})$ and these have been checked sufficiently to indicate there are no large discrepancies (i.e., of more than a few percent).

3.4.3 Resonant Absorption and Scattering of X-rays — Index of Refraction

In writing the scattering from an electron in an atom as in Eq. (3.18), we have used the expression for the scattering from a free electron, Eq. (3.16). This analysis neglects the fact that the electrons in atoms are bound to the nuclei. To consider this problem correctly requires a quantum mechanical treatment, however, we can learn a great deal about the connection between various aspects of the interaction of radiation with matter if we look at this bound scattering in a simple way, as if each electron is a mass on a spring attached to the nucleus. We shall adopt a treatment given by Feynman et al. (1963). The attraction of the nucleus to which the electron is bound is the analog to the restoring force of the spring. The applied force is proportional to the incident field and to be reasonably general, we shall also include a "frictional" term. We can see why we need this if we first neglect such a term. Writing Newton's second law for a one-electron system:

$$m_{\text{electron}}\, d^2 x/dt^2 = F - Kx.$$

Assume a general force of the form of a wave:

$$F = F_0 e^{i\omega t}$$

($\omega = 2\pi\nu$, where ν is the frequency). The force does not have imaginary terms. We are again interested only in the real part, but we shall use the exponential form for ease in handling the equation. The solution for this differential equation has the form $x = x_0 \exp(i\omega t)$, and substituting this solution, the maximum amplitude is

$$x_0 = \frac{F_0/m}{(K/m) - \omega^2}.$$

If we did not apply any force, the mass would oscillate with a natural frequency:

$$\omega = \sqrt{K/m} \equiv \omega_0.$$

so that

$$x_0 = \frac{F_0/m}{\omega_0^2 - \omega^2}.$$

As ω approaches ω_0 the displacement becomes unrealistically large — of course, $\omega = \omega_0$ corresponds to ionization but these should not be a large displacement $\omega < \omega_0$ even when ω is close to ω_0. Thus we shall include a damping term proportional to the velocity. We already have a term proportional to the displacement, so this is the natural thing to try. The equation becomes

$$m\, d^2x/dt^2 = -Kx - D(dx/dt) + F. \tag{3.20a}$$

Let $D = m\delta$, $K = m\omega_0^2$. Then,

$$F/m = (d^2x/dt^2) + \delta(dx/dt) + \omega_0^2 x, \tag{3.20b}$$

and with exponential forms for x, F:

$$x_0 = \frac{F_0/m}{(\omega_0^2 - \omega^2 + i\delta\omega)} = kF_0/m. \tag{3.20c}$$

There is no large displacement now when ω approaches ω_0. This is the displacement we need to know to calculate the acceleration, and hence the fields from an oscillating charge, Eq. (3.15). We may write the complex constant k as

$$k = 1/k_0 e^{i\alpha} = (1/k_0) e^{-i\alpha},$$

where

$$\tan \alpha = \delta\omega/(\omega_0^2 - \omega^2).$$

For x-ray scattering, $\omega \gg \omega_0$ for most of the electrons bound to atoms, and the displacement lags the applied field by as much as $\alpha \approx 180°$.

Suppose now that there are many atoms in a specimen being oscillated by the incoming field which has a different value at each electron in each atom. According to Eq. (3.20a), if we substitute $\sum_i x_i$ and $\sum_i F_i$, we get sums of differential equations, each just like Eq. (3.20a), each with its own independent solution. This is why we can assume the scattered waves from the electrons in atoms to be additive. If the motion satisfies a homogeneous differential equation (no powers of x or its derivatives), the solution for each electron is independent. We should then write the solution in Eq. (3.20c) as a sum over all electrons in atom, and then over all atoms irradiated by the x-rays.

The imaginary part of the oscillation (the phase lag) has several interesting consequences. First *the material has an index of refraction less than one for x-rays*.

3.4 The Interaction of X-rays with Matter

To see that this is so, let the incoming field for a wave traveling in the z direction be represented by $E_{inc} = E_0 \exp[i\omega(t - z/c)]$.[7] When it enters a material, it appears as if it is traveling with a speed c/n, where n is the index of refraction of the material. Thus it will take a smaller time increment to travel through a plate than through the same thickness outside the plate. If Δz is the thickness of a plate of the material, this time increment Δt is $(n-1)\Delta z/c$. Therefore, after the wave has passed through the plate, relative to the field incident on the material

$$E_{after} = E_0 \exp[i\omega(t - \Delta t - z/c)] = E_0 \exp[i\omega(t - (n-1)\Delta z/c - z/c)]$$
$$= \exp[-i\omega(n-1)\Delta z/c] E_{inc} . \tag{3.21a}$$

If we assume the index will be only slightly different from unity (we will check the validity of this assumption below), we can expand the exponent and retain only the first two terms

$$E_{after} = E_{inc}(1 - i\omega(n-1)\Delta z/c) = E_{inc} - i\omega(n-1)(\Delta z/c)E_{inc} . \tag{3.21b}$$

The first term is just the field of the source and thus the second must be the field produced by the oscillating charges in the plate (which of course depends on the size of the incoming field). The addition of the two can be made on the complex plane; the component due to the material is a small vector along the negative imaginary axis (it is tilted slightly counterclockwise from this axis, as can be seen by expanding the exponential term of E_{inc}, i.e., it has a real part).

An alternative expression for the field due to the accelerating electron in the plate may be derived. We already know the field due to a single accelerating electron [Eq. (3.1)]. All we need to do now is integrate this expression over the plate of material. Let b be the radius of a circle on the plane face of a plate of unit thickness and ρ be the electron density (per unit area). To get the acceleration of the electron, we shall describe its motion by $x = x_0 \exp[i\omega(t - r/c)]$ and differentiate twice with respect to time. Then the field due to the accelerating electrons in the plate $\sum \varepsilon$ is from Eq. (3.1):

$$\sum \varepsilon = \int_{b=0}^{\infty} \frac{-|e|\eta}{c^2 r} \omega^2 x_0 e^{i\omega(t-r/c)} \rho 2\pi b \, db .$$

We neglect the polarization term $\sin \varphi$ and any effects due to interference of the scattered waves (diffraction). Now $r^2 = z^2 + b^2$, where r is the distance from a point b on the plane to the measuring point and z is the distance normal to the plate ($z = 0$ is the origin of the coordinates).

Also $2r \, dr = 2b \, db$, as z is constant. Therefore,

$$\sum \varepsilon = \frac{-|e|\eta 2\pi\omega^2 x_0 e^{i\omega t}}{c^2} \rho \int_{r=z}^{\infty} e^{-i\omega(r/c)} \, dr .$$

Integrating, we obtain an expression for the field due to the oscillating electrons:

---

[7] Setting the velocity of light, $c = v\lambda = \omega\lambda/2\pi$, the expression for a plane wave becomes $\exp[2\pi i(ct - z)/\lambda] = \exp[i\omega(t - z/c)]$.

$$E_{osc} = \sum \varepsilon = \frac{-2\pi\eta\rho|e|}{c} i\omega x_0 e^{i\omega(t-z/c)},$$

Taking x_0 from Eq. (3.20), with $F_0 = eE_0$,

$$E_{osc} = \frac{-2\pi\rho\eta|e|^2}{c} \frac{i\omega E_0 \exp[i\omega(t-z/c)]}{m(\omega_0^2 - \omega^2 + i\delta\omega)}. \tag{3.22a}$$

For electrons in atoms, we let A_a be the number of atoms/unit area in an incident beam of amplitude $E_{inc} = E_0 \exp[i\omega(t-z/c)]$. The resonant frequency ω_{0q} and phase shift δ_{0q} are different for each electron shell. Taking this into account, and assuming all atoms identical,

$$E_{osc} = \sum \varepsilon = \frac{-2\pi A_a |e|^2 \eta}{c} \sum_q \frac{i\omega N_q E_{inc}}{m(\omega_{0q}^2 - \omega^2 + i\delta_q\omega)}, \tag{3.22b}$$

where $\sum_q$ is the sum over the electrons in one atom and N_q is the number of electrons of each type q. Comparing Eq. (3.22b) with the second term of Eq. (3.21b), we have

$$(n-1)\Delta z = \sum \frac{2\pi A_a |e|^2 N_q \eta}{m(\omega_{0q}^2 - \omega^2 + i\delta_q\omega)}.$$

If ρ_a is the number of atoms per unit volume, $\rho_a \Delta z = A_a$ and

$$n = 1 + \frac{\rho_a |e|^2 2\pi\eta}{m} \sum_q \frac{N_q}{(\omega_{0q}^2 - \omega^2 + i\delta_q\omega)}. \tag{3.22c}$$

When ω approaches ω_{0q}, the index rises and then falls as ω exceeds ω_{0q}. For the case of x-rays, $\omega \gg \omega_{0q}$ for most of the electrons in each atom, and n is smaller than unit. How much smaller? Let $\omega(= 2\pi c/\lambda)$ be $\gg \omega_0$, i.e., neglect ω_0 and also the imaginary term. If N_0 is Avogadro's number, A is the atomic weight, Z the atomic number, ρ the mass density, then assuming all electrons the same, so that

$$\rho_a \sum N_q = \frac{N_0 Z \rho}{A},$$

we have

$$n \cong 1 - \frac{e^2 \lambda \eta N_0 Z \rho}{m 2\pi c^2 A}. \tag{3.22d}$$

If we substitute numerical values, n is less than 1 by a few parts per million for all materials.

We know that the index of refraction is the velocity of light in free space divided by that in the medium, i.e., $n = c/v$. But how can a wave travel with more than the speed of light, which is what an index less than unity seems to mean? X-rays are quanta or pulses of energy made up of many waves. If we add two waves of the same (unit) amplitude together, we obtain

3.4 The Interaction of X-rays with Matter

$$\exp\{i[\omega_1 t - (2\pi x/\lambda_1)]\} + \exp\{i[\omega_2 t - (2\pi x/\lambda_2)]\}$$
$$= \exp\{\tfrac{1}{2}i[(\omega_1 + \omega_2)t - 2\pi x(1/\lambda_1 + 1/\lambda_2)]\}$$
$$\times (\exp\{\tfrac{1}{2}i[(\omega_1 - \omega_2)t - 2\pi x(1/\lambda_1 - 1/\lambda_2)]\}$$
$$+ \exp\{-\tfrac{1}{2}i[(\omega_1 - \omega_2)t - 2\pi x(1/\lambda_1 - 1/\lambda_2)]\})$$
$$= 2\exp\{\tfrac{1}{2}i[(\omega_1 + \omega_2)t - 2\pi x(1/\lambda_1 + 1/\lambda_2)]\}$$
$$\times \cos(\tfrac{1}{2}[(\omega_1 - \omega_2)t - 2\pi x(1/\lambda_1 - 1/\lambda_2)]).$$

We have a resultant wave that travels with an average frequency $(\omega_1 + \omega_2)/2$ and average wavenumber $(1/\lambda_1 + 1/\lambda_2)/2$, and this wave is modulated by the differences in these two quantities for the two waves. This modulation or *group* has a velocity V_m:

$$V_m = \frac{\omega_1 - \omega_2}{2\pi(1/\lambda_1 - 1/\lambda_2)} = \frac{\Delta\omega}{2\pi\Delta(1/\lambda)},$$

whereas each wave has a velocity $\omega_i/(2\pi/\lambda_i)$. It is the modulation that is of interest. We want the velocity of the group because it is the square of the amplitude of the group that is proportional to the energy of the group, and *it is the group velocity or packet velocity that is carrying the energy through the material*. According to our formula for the index of refraction, assuming $n \cong 1$, we can write from Eq. (3.22d)

$$1 \cong 1 - \frac{K}{\omega^2} = 1 - \frac{K}{(\omega 2\pi c/\lambda)}, \quad \text{or} \quad \frac{2\pi}{\lambda} \cong \frac{2\pi}{\lambda} - \frac{K}{\omega c} = \frac{\omega}{c} - \frac{K}{\omega c}.$$

Differentiating, we obtain

$$\frac{d(2\pi/\lambda)}{d\omega} = \frac{\omega^2 + K}{\omega^2 c}, \quad \text{or} \quad V_m = \frac{d\omega}{2\pi d(1/\lambda)} = \frac{c}{(1 + K/\omega^2)}.$$

Thus, there is no anomaly here since the group velocity of a packet of waves *is* in fact less than the speed of light.

Without carrying out the calculations any further, it is clear now that the scattered amplitude and intensity could be calculated for an atom [by using the field of an oscillating charge and differentiating the displacement, Eq. (3.20c), to get the acceleration in the equation for the field]; it is also clear that the scattered amplitude may be complex!

When the wavelength of the radiation used in diffraction is near that of one of the absorption edges (i.e., one of the natural vibrational frequencies) of a material, we can see from the above equations that the electrons associated with the absorption move out of phase from the others. For example, for incident radiation with $\lambda \approx \lambda_K$, the resultant interference reduces f by roughly the scattering equivalent of four electrons. If the two K-shell electrons are out of phase with two other electrons, ($\omega < \omega_{0K}$ for the K electrons, but $> \omega_0$ for the others) the reduction in f from the classical calculation is equivalent to the elimination of these four electrons. The results of a more detailed calculation are shown in Fig. 3.24. The "corrections" to the calculated f values are important. In materials science we are often dealing with radiation close to $\lambda_{K_{abs}}$ for our specimen (e.g., copper radiation and a zinc specimen).

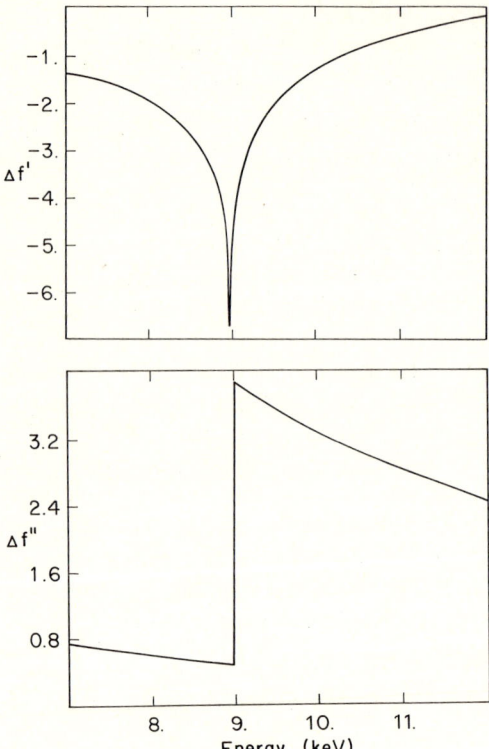

Fig. 3.24. Anomalous dispersion corrections for copper, calculated with the software supplied by D.T. Cromer (J. Appl. Cryst., 16, 437 (1983)

It also turns out, as we can see in our equation for E or x_0, that it is in the vicinity of this "dispersion" that the imaginary terms in the atomic scattering factor will be significant, as $\omega \approx \omega_0$. Thus *the true atomic scattering factor should be written*:

$$f = f_0 + \Delta f' + i\Delta f''. \tag{3.23}$$

(Tabulated values for these Δ-terms, often referred to as the Hönl corrections, can be found in the International Tables, Volume III, pp. 213–216, and Volume IV, p. 148.)[8]

Let us look at some of the consequences of these corrections to f. For example, consider β-brass, a CsCl structure with a copper atom at the corner of the cubic cell and a zinc at the center. From the lattice parameter of 2.45 Å we may calculate $\sin\theta/\lambda$ for the peaks of interest and then look up values of f in the tables. Certain peaks such as 200 have $F = f_{Cu} + f_{Zn}$, others such as the 100 have $F = f_{Cu} - f_{Zn}$.

---

[8] If $\Delta f'$ is negative, some of the scattering events, represented by f_0 do not take place. To conserve energy we must account for these photons associated with the negative amplitude $\Delta f'$. When an incident photon of energy $h\nu$ is less than, but near to the ionization energy of a K electron, $h\nu_K$, it may cause the simultaneous emission of an electron from the L shell and an x-ray of energy $h\nu - h\nu_L = h\nu'$. This emitted $h\nu'$ which gives rise to the $\Delta f'$ term in the scattering factor has been observed by Sparks (1974).

3.4 The Interaction of X-rays with Matter

Table 3.1

	Mo K$_\alpha$		Cu K$_\beta$	
	100	200	100	200
$f_{0(Cu)}$	23.5	16.3	23.5	16.3
$\Delta f'_{(Cu)}$	0.26	0.26	−5.2	−5.2
$f_{0(Zn)}$	24.1	17.2	24.1	17.2
$\Delta f'_{(Zn)}$	0.22	0.22	−2.8	−2.8

Table 3.1 is a table of f and $\Delta f'$ for the 100 and 200 peaks for Mo K$_\alpha$ and Cu K$_\beta$. In this example we may neglect $\Delta f''$ (if $\lambda/\lambda_k > 1$, $\Delta f'' = 0$ for the K electrons which is the largest contribution.) $F^2_{100}/F^2_{200} = 0.03\%$ with molybdenum radiation and the 100 peak is difficult to detect. However, if we use Cu K$_\beta$, $\lambda/\lambda_{K_{abs}}$ for copper is 1.0085, for zinc 1.08, and the corrections $\Delta f'_{(Cu)} = -5.2$, $\Delta f'_{(Zn)} = -2.8$ are large. Then, $F^2_{100}/F^2_{200} \cong 1.4\%$, and the 100 can be detected!

Now let us look at the imaginary terms. The atomic scattering factor is written as $f = f_0 + \Delta f' + i\Delta f''$. When we add the vectors from atoms in a unit cell for $F(\mathbf{s})$, those that have electrons with energies close to the wavelength being used have a small term with phase at 90° to the real terms. (Actually, all the f values have such a term, but it is large only for a wavelength close to λ_k.) The result of this effect for an hkl and $\overline{hkl}$ reflection from a crystal that does not have a center of symmetry in it is shown in Fig. 3.25a. The magnitudes of the F's are different! If the crystal has a center of symmetry, there are two identical atoms with imaginary terms on either side of the center, and these would cancel the effect as shown in Fig. 3.25b. It thus becomes possible, if the contribution to a given reflection by the atom involved is large enough, to know from diffraction if a center of symmetry is missing. As an example of this effect, consider the structure of ZnS which is fcc with 4 Zn at 0, 0, 0 + fct and 4 S at $\frac{1}{4}, \frac{1}{4}, \frac{1}{4}$ + fct. Using Mo K$_\alpha$ radiation ($\lambda = 0.711$ Å), the dispersion corrections for $f_{(S)}$ are small and will be ignored here, but for f_{Zn}, $\Delta f'_{(Zn)} = 0.3$ and $\Delta f''_{(Zn)} = 1.6$. The 111 and $\overline{111}$ reflections occur at $\sin\theta/\lambda = 0.16$ (the lattice constant, $a = 5.44$ Å), for which $f_{0(S)} = 12.4$ and $f_{0(Zn)} = 25.8$. The general expression for the structure factor becomes

$$F_{hkl} = [f_{0(Zn)} + \Delta f'_{(Zn)} + i\Delta f''_{(Zn)} + f_{0(S)}e^{\pi i(h+k+l)/2}]$$
$$\times [1 + e^{\pi i(h+l)} + e^{\pi i(h+k)} + e^{\pi i(k+l)}],$$

and

$$|F_{hkl}|^2 = \begin{cases} 0, & h, k, l \text{ mixed} \\ 16([f_{0(Zn)} + \Delta f'_{(Zn)}]^2 + [\Delta f''_{(Zn)} + f_{0(S)}]^2), & h+k+l = 4n+1 \\ 16([f_{0(Zn)} + \Delta f'_{(Zn)}]^2 + [\Delta f''_{(Zn)} - f_{0(S)}]^2), & h+k+l = 4n+3 \\ 16([f_{0(Zn)} + \Delta f'_{(Zn)} + f_0(S)]^2 + [\Delta f''_{(Zn)}]^2), & h+k+l = 4n \\ 16([f_{0(Zn)} + \Delta f'_{(Zn)} - f_0(S)]^2 + [\Delta f''_{(Zn)}]^2), & h+k+l = 4n+2. \end{cases}$$

The reflections affected the most by this dispersion effect are those for which $h + k + l$ are odd, and in particular, $|F_{111}|^2/|F_{\overline{111}}|^2 = 0.91$. Other techniques for

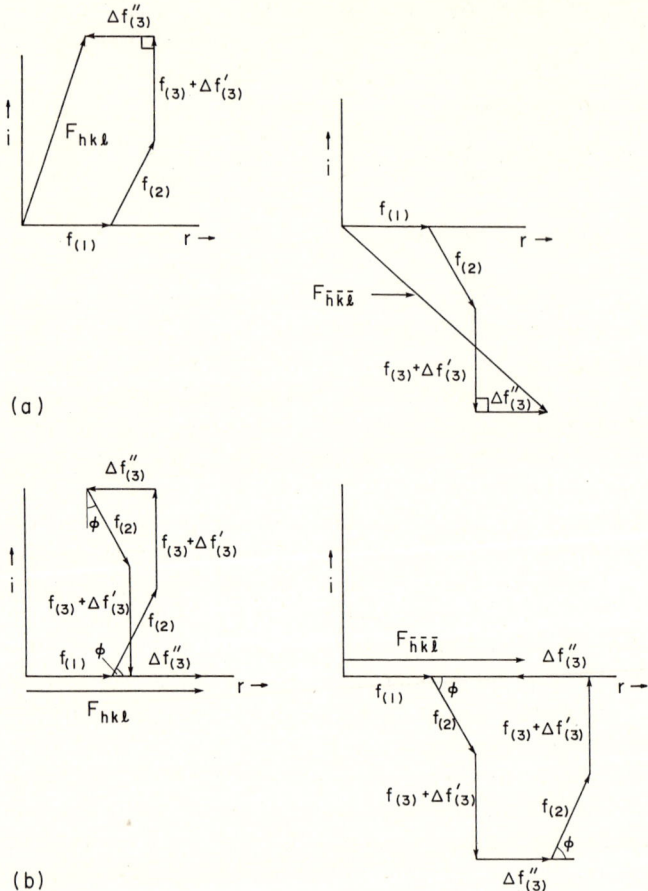

Fig. 3.25. a Non centrosymmetric molecule of three (numbered) atoms. The radiation is just higher than the K absorption edge of atom 3. The structure factors F_{hkl} and $F_{\bar{h}\bar{k}\bar{l}}$ show how the Hönl dispersion correction may make $F_{hkl} \neq F_{\bar{h}\bar{k}\bar{l}}$. The imaginary term for atom 3, $\Delta F''_{(3)}$, is generally small and is exaggerated in the drawing. **b** In a centrosymmetric structure the Hönl dispersion terms cancel, so $F_{hkl} = F_{\bar{h}\bar{k}\bar{l}}$. Note that atom 1 is located at the center of symmetry, so there is only one vector f_1

using the diffracted intensities to detect the lack of a center of symmetry are discussed in Sect. 6.6. As can be seen in Fig. 3.24, the changes $\Delta f'$ and $\Delta f''$ can be quite large in the vicinity of an absorption edge. When synchrotron radiation sources are used, the monochromator may be tuned to the appropriate wavelength to take maximum advantage of the effects described above.

3.4.4 Effects of Thermal and Static Atom Displacements

It is necessary in using f values to take into account the fact that *the atoms in a structure are vibrating with an rms displacement,* $(\langle u^2 \rangle^{1/2})$, 5–10% of the interatomic separation at room temperature. The tabulated f values assume no vibration. This

3.4 The Interaction of X-rays with Matter

introduces an additional phase factor. For any term $f\exp(2\pi i \mathbf{s}\cdot\mathbf{r})$, we can write at some instant in time $f\exp[2\pi i\mathbf{s}\cdot(\mathbf{r}_n + \Delta\mathbf{r}_n)] = f\exp(2\pi i\mathbf{s}\cdot\mathbf{r}_n)\exp(2\pi i\mathbf{s}\cdot\Delta\mathbf{r}_n)$, where Δr_n is the displacement due to thermal vibration of the atom. A proper treatment of this effect involves complete statistical mechanical treatment of the coupled vibrations in solids [see Chap. 7, James (1950) and Warren (1969)], but we can obtain an important and useful result here using simple classical arguments (Castellano and Main 1985).

We note that the time taken to record the diffraction pattern is long compared with the period of vibration of atoms ($\approx 10^{-13}$ s) so we may regard the x-rays as seeing each atom displaced by an amount $\Delta\mathbf{r}$. We describe the atomic vibrations as a system of classical harmonic oscillators with a Boltzmann distribution of energy. We further consider the component of momentum in the direction of the displacement ΔX which is parallel to the scattering vector $\mathbf{s}$. That is, for a crystal where the vibrational energy is independent of which unit cell the atom is in, the number of atoms per unit volume with momentum (p) in the interval $dp_x\,dp_y\,dp_z$ is

$$dN \propto \exp(-E/k_B T)\,dp_x\,dp_y\,dp_z, \tag{3.24}$$

where E is the oscillator energy and k_B is Boltzmann's constant. Since the energy of a classical oscillator is proportional to the square of the momentum and also to the square of the expected displacement, Eq. (3.24) yields the probability density of atomic displacements:

$$P(\Delta r) \propto \exp(-\alpha\Delta x^2), \tag{3.25}$$

where α is the proportionality constant between $(\Delta x)^2$ and $E/k_B T$. This Gaussian function can be normalized to give

$$P(\Delta x) = (2\pi\langle\Delta x^2\rangle)^{-1/2}\exp[-(\Delta x)^2/2\langle\Delta x^2\rangle], \tag{3.26}$$

where $\langle\Delta x^2\rangle$ is the average of $(\Delta x)^2$ for each atom over the whole range of energies.

For the term $\exp(2\pi i|\mathbf{s}|\Delta x_n)$, we may now calculate the average over all Δx_n:

$$\langle\exp(2\pi i|\mathbf{s}|\Delta x_n)\rangle$$

$$= (2\pi\langle\Delta x_n^2\rangle)^{-1/2}\int_{-\infty}^{\infty}\exp(2\pi i|\mathbf{s}|\Delta x_n)\exp[-(\Delta x)^2/2\langle\Delta x_n^2\rangle]\,d(\Delta x_n)$$

$$= \exp[-8\pi^2(\sin^2\theta/\lambda^2)\langle\Delta x_n^2\rangle]$$

$$= \exp[-B\sin^2\theta/\lambda^2] = \exp(-M). \tag{3.27}$$

The expression e^{-M} is known as the Debye-Waller factor and is seen to be the result of the space average over a large number of classical oscillators with a Boltzmann distribution of energy.

In a calculation of $F(\mathbf{s})$, we substitute for f, fe^{-M}. For a simple structure of one atom type, the intensity of a peak is proportional to $|F^2|$ or to $f^2 e^{-2M} = f^2\exp(-2B\sin^2\theta/\lambda^2)$. It is a simple matter to plot log intensity (after correcting for the polarization factor and f^2) versus $\sin^2\theta/\lambda^2$ to obtain the slope $2B$ and to correct all values of $|F|^2$ for this *effect of thermal vibration*. If the structure has peaks where for some indexes,

$$|F_1|^2 = [f_A e^{-M_A} + f_B e^{-M_B}]^2,$$

and for others

$$|F_2|^2 = [f_A e^{-M_A} - f_B e^{-M_B}]^2 \, ;$$

then

$$|F_1| + |F_2| = 2f_A e^{-M_A}, \qquad |F_1| - |F_2| = 2f_B e^{-M_B}.$$

It is therefore possible to separate out the individual terms to get the correction term for each atom. For more complex structures, the thermal parameter B_i for each atom is added to the list of unknown position parameters x_i, y_i, and z_i. All of these parameters can be determined simultaneously by fitting the observed intensities to those calculated from $F(s)$ including thermal parameters, by a least-squares analysis. This procedure is discussed in Sect. 6.9.

Thermal vibration reduces many peaks at room temperature to only a half or a third of their intensity at, say, liquid nitrogen. [This can easily be seen by taking $\langle \Delta X_n^2 \rangle^{1/2} = 5\%$ of an interatomic distance in Eq. (3.27).] If at all feasible, it would be good to make measurements at low temperatures, e.g., at liquid nitrogen temperature. It is not possible to completely eliminate the effect of vibrations as there *are* some even at absolute zero; this term accounts for about 20% of the total $\langle \Delta X_n^2 \rangle^{1/2}$ at room temperature for many solids. A useful and interesting set of data on this "temperature depression" can be found in the International Tables, Volume III, pp. 232–245. The dynamic displacement is related to the Debye temperatures and to elastic constants, and can be a useful tool in itself for measuring these. However this topic is beyond the scope of this text, and the reader is referred to the literature (see James 1950; Warren 1969). We shall return to this topic in Sect. 7.2.

The depression factor measured in the manner described may include a static displacement because in alloys with close packed structures, differences in atomic sizes of the species may force atoms off lattice points. This static displacement $\langle \Delta X_{\text{static}}^2 \rangle^{1/2}$ can be determined by evaluating $2M$ at two temperatures and separating the two parts, as $2M \simeq 2M_{\text{static}} + 2M_{\text{dynamic}} + 2M_{\text{zero point vibration}}$, and $2M_{\text{dynamic}}$ is proportional to the absolute temperature. Furthermore, the last term can be approximately calculated (see James 1950).

3.4.5 Compton Scattering of X-rays

We have so far been assuming that the interaction of x-rays with electrons is completely elastic, i.e., that the energy of the scattered x-ray photon is unchanged by the scattering process. This assumption is usually satisfactory. For example, we have neglected the fact that the thermal vibrations of the atom do give rise to inelastic scattering of the x-rays. That is, *x-ray photons can gain or lose energy to the vibrational modes (phonons) of the sample*. However, since the energies associated with these vibrational modes of atoms in a solid are of order 0.01–0.1 eV, they represent a negligibly small modification of the energy of the x-ray photon which is of order 10 keV. This inelastic scattering does contain valuable information about the vibrational properties of matter, and we will discuss it in more detail in

3.4 The Interaction of X-rays with Matter

Sect. 7.2. A second source of inelastic scattering is associated with the anomalous dispersion term $\Delta f'$, as mentioned earlier.

A third type of inelastic scattering process is called *Compton scattering*. Since the electrons have finite mass, they may "recoil" when the x-ray beam is incident upon them. A.H. Compton predicted this in 1923 and also realized that there should be a larger effect with elements of low atomic number where the binding energy for the electrons would be lower. To test his ideas he performed the experiment illustrated in Fig. 3.26. Radiation from a Mo anode, filtered through a Zr foil to produce essentially pure Mo K_α, was analyzed by diffraction from the planes of a crystal. The peak corresponding to 0.71 Å is due to the characteristic Mo K_α radiation. After scattering from a carbon block, the spectrum contains a second broad peak corresponding to a mean wavelength of about 0.73 Å. A classical model for this Compton scattering is illustrated in Fig. 3.27 for scattering of a photon of energy hv_0 from a completely free electron initially at rest. Conservation of energy requires

$$hv_0 = hv + \tfrac{1}{2}mv^2,$$

where v is the final velocity of the electron and v is the frequency of the photon after collision. Conservation of momentum in the x direction yields

$$hv_0/c = (hv/c)\cos 2\theta + mv \cos \alpha,$$

and in the y-direction yields

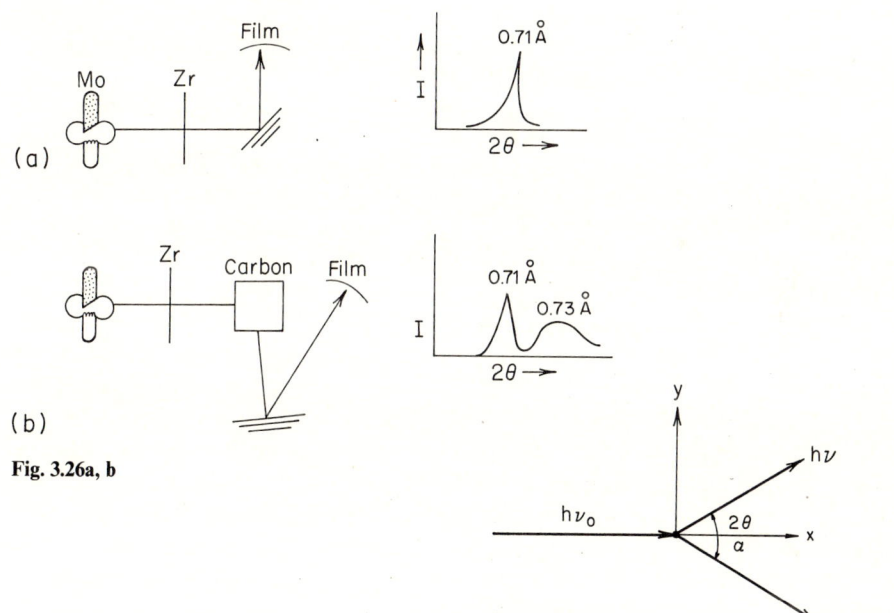

Fig. 3.26a, b

Fig. 3.27

Fig. 3.26. Scattering by loosely bound electrons in carbon showing the Compton modified scattering. The Zr filter is used to clean up the incident spectrum and make it more nearly monochromatic

Fig. 3.27. Geometry for the calculation of the wavelength change in Compton scattering. A photon of hv_0 is incident on an electron and when scattered, has energy hv

$$0 = (h\nu/c)\sin 2\theta + mv \sin\alpha \,.$$

Eliminating α and v, the result is

$$(\lambda - \lambda_0) = (h/mc)(1 - \cos 2\theta)$$

or

$$\Delta\lambda(\text{Å}) = 0.0243(1 - \cos 2\theta) \,. \tag{3.28}$$

The higher the 2θ, the greater the separation in wavelength; the separation $\Delta\lambda$ is independent of the initial wavelength. *This beam of changed wavelength* is incoherent as there is a random phase relation between Compton modified radiation scattered by different electrons. (The wavelength shift depends on the electron's position when it is struck by a photon.) Its intensity, as it is incoherent, depends on N, the number of scattering atoms and not on N^2 as diffracted beams do. This scattering does not contribute to diffraction but appears as background, which is difficult to avoid and it is sufficiently large in scattering from glasses or polymers (the atomic numbers of the elements are small) to be annoying. It can be eliminated experimentally though, as indicated in Problem 3.13.

It would appear that all the electrons should recoil so that there would be no elastic scattering. When analyzed with quantum mechanics, the total scattering for one bound electron is the sum of the incoherent scattering I_{inc} and the coherent scattering, $I_{\text{coh}} = I_e f^2$, and sum is given by the classical Thomson scattering, I_e:

$$I_{\text{coh}} + I_{\text{inc}} = I_e, \qquad \text{or} \qquad I_{\text{inc}} = I_e - I_e f^2 = I_e(1 - f^2) \,.$$

For many electrons, each with scattering factor f_j,

$$I_{\text{inc}} = I_e \sum_j^Z (1 - f_j^2) = I_e \left(Z - \sum_j f_j^2\right),$$

or

$$i_{\text{inc}} \text{ (in ``electron units'')} = \frac{I_{\text{inc}}}{I_e} = Z - \sum_j f_j^2 \,.$$

Consider the lithium atom which has two K electrons and one L electron. The various terms are shown schematically in Fig. 3.28. As $\sin\theta/\lambda$ increases, the incoherent scattering represents an increasing fraction of the total scattering. The heavier the element, the larger the ratio of coherent to Compton incoherent scattering. Values for incoherent scattering can be found in the International Tables, Volume III, pp. 247–253 and Volume IV, p. 48. There it will be seen that there is an additional term due to the Pauli exclusion principle which excludes two electrons from occupying the same state. Also there is a small correction due to the difference in wavelength of the incident radiation and the Compton scattering. The energy distribution of Compton intensity may be shown to be related to the distribution of momenta of the electrons in the sample and measurements of the Compton line profile are now made to learn more about the momentum distribution of electrons in solids.

The Compton scattering can be calculated and subtracted to obtain the coherent scattering in the most precise experiments when the incident beam intensity is accurately known, but this is not worthwhile unless a monochromator has been used and air scattering is measured (as in Fig. 3.29 by replacing the sample by a

3.5 The Interaction of Electrons with Matter

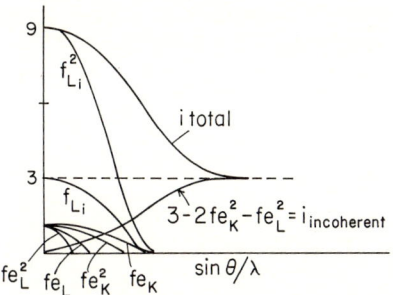
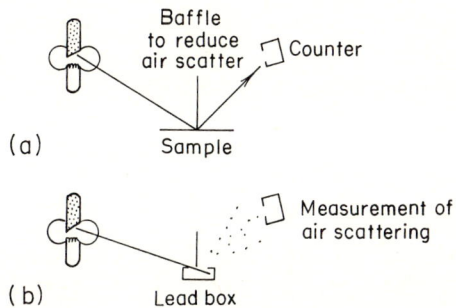

Fig. 3.28. Schematic of the contributions to the total scattering from a Li atom containing three electrons. At large values of $\sin\theta/\lambda$, the incoherent scattering dominates for this light atom

Fig. 3.29. a Using a baffle to minimize detection of radiation scattered by air in the beam path. **b** Measurement of residual air scattering not eliminated by the baffle

lead beam trap), or eliminated by evacuating the entire path from tube to counter. A discussion of measurements of the direct beam intensity and subtraction of the calculated Compton scattering is included in Appendix D.

3.5 The Interaction of Electrons with Matter

In considering the scattering of an electron by an atom, we have much the same theoretical problem as for x-rays, except that there is scattering by the nucleus as well as by the electron cloud, and the two have opposite signs. The scattering of a charged particle (the electron with $\lambda = h/mv$) by a point charge (the nucleus with charge $q = Ze$) may be analyzed by considering the electrostatic interaction between the two charges. This problem was solved in 1911 by Rutherford for the classic experiment in which α particles were scattered by atomic nuclei to establish the nuclear theory of the atom. In such a Rutherford scattering problem, the fraction of particles $d\sigma$ scattered at angle 2θ into solid angle $d\Omega$ is given by

$$|f(2\theta)| = |d\sigma/d\Omega|^{1/2} = qe/2mv^2 \sin^2\theta \,.$$

The derivation of this equation may be found in texts on modern physics. The dependence of the scattering on the product of the two charges and the scattering angle are immediate consequences of the Coulomb interaction between the two charged particles.

To account for the scattering from the electrons, we should consider the scattering from each electron and then add these effects for all electrons, including the appropriate phase factor to account for the distribution of electrons in space around the nucleus. We can factor this problem into two parts, the scattering by one electron, and a geometrical phase factor which depends only on the electron distribution. However, we have already considered the phase factor when we calculated the scattering of x-rays from this same electron distribution to obtain the scattering factor f [see Eq. (3.19)]. That is, we will obtain the desired result by replacing the electron cloud by a point charge at the nucleus with charge equal to $-f|e|$. Thus the total electron scattering factor is

$$f_e = \frac{(qe)_{\text{nuc}} + (qe)_{\text{elect}}}{2mv^2 \sin^2 \theta} = \frac{(Z-f)e^2}{2mv^2 \sin^2 \theta}.$$

or, using the de Broglie relation, $\lambda = h/mv$,

$$f_e \simeq \frac{me^2 \lambda^2}{2h^2 \theta^2}(Z-f), \tag{3.29}$$

where we approximate $\sin \theta$ by θ since the angles are small because λ is small.

This equation is not valid at very small angles for which $f \to Z$, and extrapolation from higher angles is used. Values of f_e are given in the International Tables, Volume III, pp. 217–227 and Volume IV, pp. 154–175.

There are several important features of Eq. (3.29). Substituting numerical values for the physical constants; we have

$$f_e(\text{Å}) \cong 0.0239(Z-f)\lambda^2/\theta^2.$$

Note that the constant is much larger than the scattering of one electron for x-rays. The term $\lambda^2(Z-f)/\theta^2$ is larger than f up to about $\sin \theta/\lambda = 0.3$–$0.4$, but then is lower than f. However, because of the constant, the electron scattering is always greater than that for x-rays. Consequently, weak scattering effects are more easily detected, e.g., from small irradiated volumes. This is the basis of selected area diffraction procedures in the electron microscope. A small adjustable aperture can be placed in the image before the second lens to define a specific region of the sample and examine its diffraction pattern.

The form of Eq. (3.29) has been tested for the range 15–80 kV with reasonable agreement. The scattering factors for low voltages are still questionable. Equation (3.29) must be multiplied by $I_0^{1/2}/r$ to obtain the total scattering. Also the temperature effect, e^{-2M}, is identical to that for x-ray scattering, i.e., one replaces f_e by $f_e e^{-M}$.

Electrons can exist in two spin states and therefore electrons can be "polarized." In practice, the internal fields of structures are usually not large enough to do this to the incoming electrons. However, polarization effects have been detected in scattering from magnetic materials. When an appropriate detector sensitive to the electron spin polarization is used in a scanning electron microscope, it becomes possible to observe details of the magnetic domain structures in ferro-, ferri-, or antiferromagnetic materials (Celotta 1985).

3.6 The Interaction of Neutrons with Matter

3.6.1 Nuclear Interactions

The interactions of a neutron with an atom include an interaction with the nucleus via short-range nuclear forces and an interaction of the magnetic moment associated

3.6 The Interaction of Neutrons with Matter

with the spin of the neutron and the magnetic moments of the atoms. This second effect occurs mainly for atoms with unfilled outer electron shells, such as the transition elements. We shall begin our discussion with the neutron-nucleus interaction.

When a neutron approaches the nucleus, scattering is essentially related to two phenomena. First there is a *nuclear force or potential scattering* for which the total intensity scattered to all angles is given by $\sigma = 4\pi R^2$, where R is the nuclear radius.[9] *Because the nucleus is so small ($R \approx 10^{-15}$ m) compared to the wavelength of thermal neutrons, there is no variation in the potential scattering with* $(\sin\theta)/\lambda$. Analyses of nuclear scattering experiments have shown that $R \cong 1.5(10)^{-13} A^{1/3}$ where A is the mass number ($A^{1/3}$ because nuclear matter has approximately constant density). The intensity scattered into a unit solid angle is then $\sigma/4\pi = R^2$ (dividing by the total solid angle in a sphere, 4π), and the potential scattering amplitude or potential scattering factor is R.

The second aspect of neutron-nucleus interaction is similar to absorption phenomena in x-ray scattering. When the neutron is close to the nucleus, its energy becomes distributed throughout the entire system — neutron plus nucleus. We may then think of a "compound" nucleus containing the neutron. A metastable state forms, then decays, reemitting the neutron. This is a physical phenomenon similar to the scattering of x-rays by a bound electron, with similar results. A resonance can occur, and for the appropriate energy, the neutron can be absorbed. As for x-rays, this results in a change in the scattering factor, and the total scattering factor (or amplitude) can be written as

$$b = R - (\gamma_n^{(R)}/2kE_R) = b_0 - \Delta b'. \tag{3.30}$$

Here $k = 2\pi/\lambda$, E_R is the resonance energy the neutron must have to form the compound nucleus and $\gamma_n^{(R)}$ is the resonance energy width for reemission of a neutron. For thermal neutrons, $\gamma_n^{(R)}$ is proportional to k, so this resonance term $\Delta b'$ is also independent of $\sin\theta/\lambda$. Since the resonance term may be larger than R, it is possible to have negative scattering amplitudes for certain nuclei (e.g., ^{1}H, ^{48}Ti, ^{62}Ni). The scattering for such nuclei is 180° out of phase with that from the nuclei with $b > 0$. For a very few nuclei (e.g., Cd, ^{10}B), thermal neutrons have energies very close to E_R, the absorption is very high, and the scattering amplitude also contains a small, imaginary wavelength dependent term $\Delta b''$. This imaginary term may be used to distinguish between centro- and noncentrosymmetric structures as discussed for x-rays in Sect. 3.4. The high absorption of Cd and its mechanical formability make it a useful shield material, as is lead for x-rays.

Experimental values of b are available, which is a big advantage over electrons or x-rays where we have only theoretical values. One is using values that have been measured, not calculated, and the measurements *may* be made to a high precision as b is independent of $(\sin\theta)/\lambda$. For comparison, we give in Table 3.2 scattering factors for x-rays, neutrons, and electrons for hydrogen, copper, and tungsten.

---

[9] The quantity $4\pi R^2$ is called the scattering cross section and represents the effective surface of the nucleus which is impenetrable to the oncoming neutron.

Table 3.2

	$b \times 10^{+14}$ m		$f_e \times 10^{+14}$ m[a]		$f_x \times 10^{+14}$ m	
$(\sin \theta)/\lambda =$	0.1	0.5	0.1	0.5	0.1	0.5
^{1}H	−0.378	−0.378	4530	890	0.23	0.02
^{63}Cu	0.67	0.67	51,100	14,700	7.65	3.85
W	0.466	0.466	118,000	29,900	19.4	12.0

[a] Multiply by $(1 - v^2/c^2)^{1/2}$, where v is electron velocity.

Neighboring atoms in the periodic table often have sufficient difference in their neutron scattering factor to aid in detecting weak peaks; e.g., the weak x-ray peaks in β brass (CuZn) discussed in Sect. 3.4 are easily detected. Light elements are more easily detected in electron diffraction than in x-ray diffraction, and even more readily with neutrons.

For atomic nuclei with even mass number and even charge, there is no interaction with the spin of the neutron because there is no net spin for the nucleus. But, if the angular momentum of the atom's nucleus is I, the neutron and nucleus may combine with spins parallel or antiparallel to give total spin of $I + \frac{1}{2}$ or $I - \frac{1}{2}$, with scattering amplitudes b_+ and b_-, respectively. Quantum mechanics restricts a spin with magnitude J to $2J + 1$ orientations in space, so the compound nucleus can have $[2(I + \frac{1}{2}) + 1] + [2(I - \frac{1}{2}) + 1] = 2(2I + 1)$ possible states. Of these, a fraction $w_+ = [2(I + \frac{1}{2}) + 1]/2(2I + 1) = (I + 1)/(2I + 1)$ corresponds to the parallel spin states and $w_- = I/(2I + 1) =$ corresponds to antiparallel spin states. A single isotope with $I > 0$ may then be described as if it were a random mixture of nuclei with atomic fractions w_+ and w_- and scattering amplitudes b_+ and b_-. Normally only this weighted average of b_+ and b_- can be measured. Independent determinations of b_+ and b_- can only be made when the nuclear spins can be easily polarized, as in ortho- and parahydrogen (proton spins in H_2, parallel and antiparallel, respectively), or at very low temperatures in the presence of very high fields.

In general, a naturally occurring element will be composed of several isotopes and we can reasonably expect that these isotopes will be randomly distributed on all sites occupied by the atoms of this element. Then we can write for the diffracted intensity from material with $\mathbf{r}_n$ the vector to an atom n [see Eq. (2.14c)]:

$$I(\mathbf{s}) = K \sum_n \sum_{n'} b_n b_{n'} \exp[2\pi i \mathbf{s} \cdot (\mathbf{r}_{n'} - \mathbf{r}_n)]$$

$$= \sum_0^{N-1} b_n^2 + \sum_n \sum_{\substack{n' \\ n \neq n'}} b_n b_{n'} \exp[2\pi i \mathbf{s} \cdot (\mathbf{r}_{n'} - \mathbf{r}_n)] \,,$$

where the subscript refers to the spin state, isotope, or element. We observe the average of $I(\mathbf{s})$ over possible atomic, isotopic, and spin arrangements:

$$\langle I(\mathbf{s}) \rangle = N \langle b_n^2 \rangle + \sum_n \sum_{\substack{n' \\ n \neq n'}} \langle b_n b_{n'} \rangle \exp[2\pi i \mathbf{s} \cdot (\mathbf{r}_{n'} - \mathbf{r}_n)] \,.$$

The occupation of the various lattice sites does not depend on the value of the

3.6 The Interaction of Neutrons with Matter

exponential term; the lattice is defined by one function, the occupancy by another, so that the average in the second term need not include the exponent. Assuming random occupation, the occupancy of one site does not affect any other, so $\langle b_n b_{n'} \rangle = \langle b_n \rangle \langle b_{n'} \rangle = \langle b_n \rangle^2$, and

$$\langle I(\mathbf{s}) \rangle = N \langle b_n^2 \rangle + \sum_{\substack{n \ n' \\ n \neq n'}} \langle b_n \rangle^2 \exp[2\pi i \mathbf{s} \cdot (\mathbf{r}_{n'} - \mathbf{r}_n)] .$$

Adding and subtracting the term for $n = n'$ missing in the sum:

$$\langle I(\mathbf{s}) \rangle = N(\langle b_n^2 \rangle - \langle b_n \rangle^2) + \sum_{n \ n'} \langle b_n \rangle^2 \exp[2\pi i \mathbf{s} \cdot (\mathbf{r}_{n'} - \mathbf{r}_n)] . \tag{3.31}$$

That is, *the scattering factor to be used for the coherent scattering* (in the sum) *is an average scattering factor based on the atomic fraction* (not on the weight fraction). The first two terms in Eq. (3.31) represent a continuous background.

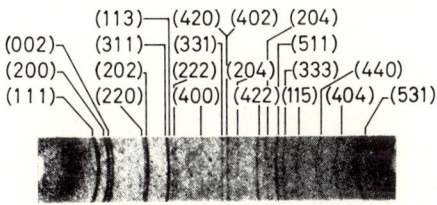

(a)

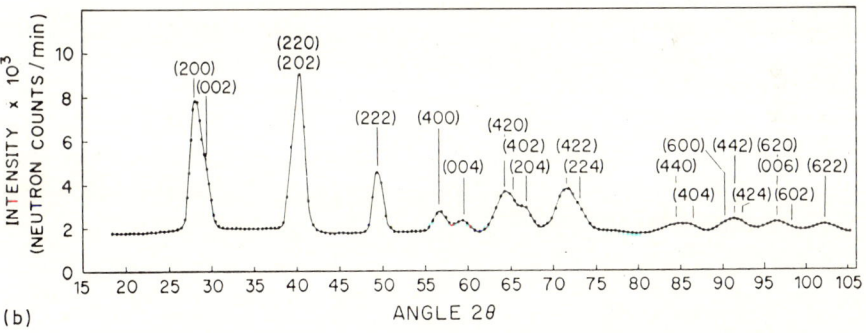

(b)

Fig. 3.30a, b. Diffraction patterns of $MD_{1.99}$ [$M = 62$ at. % Ti or 38% Zr, $D = {}^2$H]. **a** Filtered Cu; camera radius 114.6 mm. **b** Neutrons $\lambda = 1.09$ Å. Structure is similar to that for CaF_2 with M at the corner and fcc positions of a slightly distorted cubic cell and D at all tetrahedral sites, so that for hkl all even or all odd, $F_{hkl} = 4[\langle b_M \rangle + Cb_D \cos\{\pi/2(h+k+l)\}]$ and C is the mole fraction of D (1.99). For the Ti-Zr composition chosen, $\langle b_M \rangle$ is zero so that the neutron scattering is from the second term only, the structure formed by the interstitial deuterium which is a slightly distorted simple cubic cell. (Reprinted with permission from a paper by Sidhu, S.S., Heaton, LeRoy, and Mueller, M.H., J. Appl. Phys. *30*, 1323, (1959).) The indexing of the neutron diffraction pattern is the same as that for the x-ray pattern, i.e., 200 instead of 100, etc.

With x-rays or electrons, to calculate the structure factor for an alloy (not a compound), we must replace b_n with f_n in Eq. (3.31) and use a weighted f value at each atomic position, i.e., f_n is the sum of atomic fractions times the appropriate scattering factor. There is a continuous background, $N(\langle f_n^2 \rangle - \langle f_n \rangle^2)$, from an alloy with x-rays (called the Laue monotonic), or $N(\langle b_n^2 \rangle - \langle b_n \rangle^2)$ from a pure element for neutrons, even for a single isotope if the scattering nucleus has $I > 0$ (spin incoherent scattering). If the occupation of sites is not random, this continuous background will be modulated, i.e., it will show broad bumps indicating a tendency to ordering where peaks from an ordered cell might show up. We will discuss this in detail in Chap. 7.

Neglecting the structure, i.e., assuming independent atoms, as in a gas, the total scattering cross section per average atom for neutrons for a single isotope is

$$\sigma = 4\pi(w_+ b_+^2 + w_- b_-^2).$$

We can break this scattering into two parts:

$$\sigma = 4\pi(w_+ b_+ + w_- b_-)^2 + 4\pi(w_+ b_+^2 + w_- b_-^2) - 4\pi(w_+ b_+ + w_- b_-)^2,$$

or, since $w_+ = 1 - w_-$,

$$\sigma = 4\pi(w_+ b_+ + w_- b_-)^2 + 4\pi w_+ w_- (b_+ - b_-)^2 = l + P.$$

The P term in the scattering cross section is the spin incoherent scattering, while the l term is the coherent scattering, as it depends on the square of the average scattering amplitude as does the second term in Eq. (3.31).

For ^1H, $\sigma = 81.5 \times 10^{-28}$ m^2 (or 81.5 "barns") while l is 1.79 barns. For deuterium ^2H, $l = 5.4$ barns, $\sigma = 7.6$ barns. The background due to spin incoherent scattering is much lower in a diffraction pattern if ^2H rather than ^1H is involved. In studies of organic materials, it is often desirable to chemically replace the hydrogen atoms by deuterium to reduce this undesirable background.

Because of the mass of a neutron, the energy of a 1 Å neutron is the order of 0.1 eV, and there may be energy exchange with the thermal vibrational modes of the sample. This leads to an inelastic scattering, both coherent and incoherent. Since the energies of vibrational modes are of the same order as the incident neutron energy, these energy changes may be studied with great sensitivity. This contrasts with x-ray diffraction in which such energy changes cannot be detected. This subject is dealt with in greater detail in Sect. 7.2.

3.6.2 Magnetic Interactions

We now turn to a consideration of *the interaction of the neutron via its magnetic moment with the magnetic moments of materials*. We shall limit our attention to the simplest case, scattering from an ordered arrangement of spins on identical atoms, corresponding to monatomic ferromagnetic (all spins parallel) or antiferromagnetic (spins alternating parallel and antiparallel) materials. We further limit the discussion to the case where the magnetic moment on each atom is given by $\mu = 2S'\mu_B$, with S' the total of the unpaired spins in an atom, μ_B the Bohr magneton ($= eh/mc$), the

3.6 The Interaction of Neutrons with Matter

magnetic moment of a single electron, and m the mass of an electron. (The quantity S' is the spin quantum number and for a single unpaired electron is given by $S' = \frac{1}{2}$. Thus $\mu = 2 \cdot \frac{1}{2}\mu_B = \mu_B$ for a single unpaired electron.)

In 1939, Halpern and Johnson showed that the interaction of a neutron with magnetic moment $\mu_n = \gamma eh/m_n c$ (m_n is the neutron mass and γ is the ratio of magnetic moment to angular momentum, the so-called gyromagnetic ratio, $\simeq 1.9$ for neutrons) and an atom with spin S' in an ordered magnetic material will have a scattering amplitude:

$$p = (e^2/mc^2)\gamma S' f_{mag} . \qquad (3.32)$$

The form factor f_{mag} is exactly analogous to f in x-ray diffraction, taking account of the fact that the magnetic moment of the atom is distributed in space around the nucleus and interference effects due to scattering from unpaired spins in different volumes of space give rise to an angular dependence of the scattering amplitude p. As written in Eq. (3.32), f_{mag} is normalized, so that it is equal to unity at $\sin\theta/\lambda = 0$. Since f_{mag} is due only to the unpaired electrons, the measurement of f_{mag} in neutron scattering experiments provides a test of theoretical calculations of the distribution of unpaired spins in solids. Substituting numerical values, we find

$$p = (0.27)\mu f_{mag}(10)^{-14} \text{ m} .$$

Thus for $\mu = $ one Bohr magneton (one unpaired spin), p has the same magnitude as the nuclear scattering amplitude b for many elements. Consequently, *magnetic and nuclear scattering effects are of comparable intensities.* However, f_{mag} decreases rapidly with $(\sin\theta)/\lambda$, as it arises from the outer electrons.

The geometry of a typical experiment is shown in Fig. 3.31. Let $\mathbf{K}$ be a unit vector in the direction of the spin orientation in the sample (i.e., $\mathbf{K} = \mathbf{S}'/|\mathbf{S}'|$) and $\boldsymbol{\varepsilon}$ be a unit vector in the direction of the scattering vector $\mathbf{s} = \mathbf{S} - \mathbf{S}_0$ (i.e., $\boldsymbol{\varepsilon} = \mathbf{s}/|\mathbf{s}|$). Then a magnetic interaction vector $\mathbf{q}$ is defined as

$$\mathbf{q} = \boldsymbol{\varepsilon}(\boldsymbol{\varepsilon} \cdot \mathbf{K}) - \mathbf{K} . \qquad (3.33)$$

Thus, the vector $\mathbf{q}$ lies on the plane defined by $\boldsymbol{\varepsilon}$ and $\mathbf{K}$, and since $\mathbf{q} \cdot \boldsymbol{\varepsilon} = 0$, it is perpendicular to $\boldsymbol{\varepsilon}$, i.e., $\mathbf{q}$ lies on the reflecting plane. Since nuclear and magnetic scattering may occur from the same atom, the two processes can interfere and the intensity of neutrons scattered per unit solid angle (differential scattering cross section) has been shown by Halpern and Johnson to be:

$$I = d\sigma/d\Omega = b^2 + 2bp\mathbf{q} \cdot \boldsymbol{\eta} + p^2 q^2 . \qquad (3.34)$$

The second term in Eq. (3.34) is an interference term and includes a vector dot product showing that the scattering depends on the spatial orientation of the

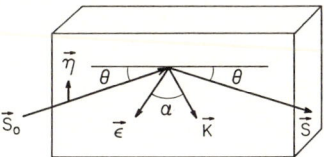

Fig. 3.31. Geometry for neutron scattering from a magnetic sample. The vectors describing the diffraction are: $\boldsymbol{\eta}$, the neutron polarization vector; $\boldsymbol{\varepsilon} = \mathbf{s}/|\mathbf{s}|$, the unit scattering vector; $\mathbf{K}$, the unit spin orientation vector

neutron spin polarization represented by the unit vector **η**, and on the magnetic spins in the sample **S'**, through the vector **q**.

In the simplest case of magnetic scattering, the incident beam of neutrons is unpolarized, i.e., the neutron polarization is equally likely to be up or down with respect to the vector **q**, so that averaged over all incident neutrons, $\mathbf{q} \cdot \mathbf{\eta} = 0$ in Eq. (3.34). Consequently, *the scattering cross section reduces to*

$$I = d\sigma/d\Omega = b^2 + p^2 q^2 \tag{3.35}$$

for unpolarized neutron beams. Since the interference term is absent, nuclear and magnetic intensity are strictly additive for an unpolarized beam and the resultant structure factor for a given *hkl* reflection becomes

$$F^2 = F_{\text{nuc}}^2 + q^2 F_{\text{mag}}^2, \tag{3.36}$$

where $q^2 = 1 - (\mathbf{\varepsilon} \cdot \mathbf{K})^2 = \sin^2 \alpha$, and α is the angle between **ε** and **K** shown in Fig. 3.31. Schematic drawings of the magnetic and nuclear neutron scattering from simple ferromagnetic and antiferromagnetic materials are shown in Fig. 3.32 (see also Problem 3.29). In Eq. (3.36), F_{nuc}^2 is the nuclear structure factor per unit cell with nuclear scattering amplitude b per atom, while F_{mag}^2 is the magnetic structure factor per unit cell with magnetic scattering amplitude p [(given by Eq. (3.32)] per atom. For a sample which contains a single magnetic domain, $\sin^2 \alpha$ is the same for all atoms, but in general for a multidomain sample, some average over spin orientation occurs. For ferro- and ferrimagnetic samples, an external magnetic field

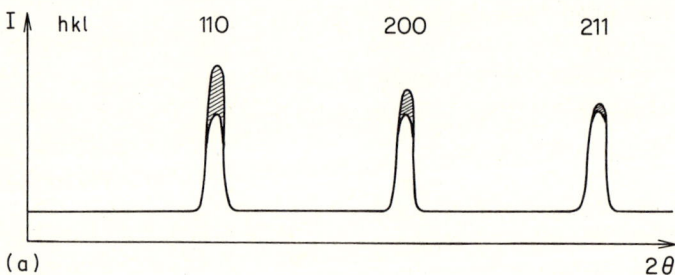

(a)

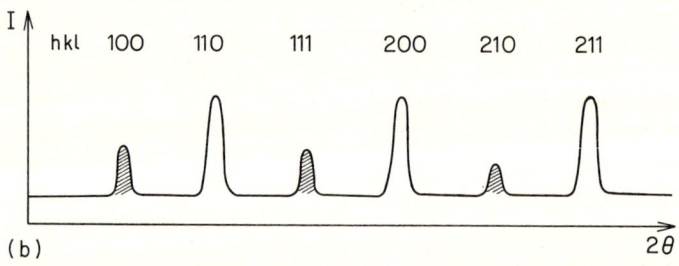

(b)

Fig. 3.32. Schematic neutron diffraction patterns from polycrystalline (**a**) bcc ferromagnet and (**b**) bcc antiferromagnet. The *shaded area* represents the contribution of the magnetic scattering which decreases with θ due to the decrease in magnetic scattering factor f_{mag}

3.6 The Interaction of Neutrons with Matter

may be applied to orient the spins, such that first $q^2 = 0$, then $q^2 = 1$. Measurements of F^2 for both situations allow the magnetic and nuclear scattering to be separated. Then the magnetic structure may be studied by examining F_{mag}^2 for various hkl reflections. In fact, almost everything we know about spin arrangements in materials has come from such neutron diffraction studies. The extinction conditions on F_{mag} allow us to determine the spin orientations, and the intensities of the observed reflections yield the magnitude of the spin per atom. In such studies, one usually uses calculated values for f_{mag}. See Gurevich and Tarasov (1968) for values of f_{mag}.

An extremely powerful tool for the study of magnetic scattering involves *the use of polarized neutrons*. How can we obtain a beam of polarized neutrons? Consider the scattering of an unpolarized beam from a ferromagnetic single crystal polarized by an external magnetic field **H**. If we take **H** (and thus **K**) in the reflecting plane and perpendicular to $\boldsymbol{\varepsilon}$, then the value of **q** in Eq. (3.34) is $(-\mathbf{K})$ and $\mathbf{q} \cdot \boldsymbol{\eta}$ in Eq. (3.33) is ± 1 depending on whether the incident neutron has spin antiparallel or parallel to the field **H**. For both cases, $q^2 = K^2 = 1$. Thus for the two polarization directions the scattering cross section is

$$I = \frac{d\sigma}{d\Omega} = \begin{cases} b^2 + 2bp + p^2 = (b+p)^2, & \text{antiparallel} \quad (3.37a) \\ b^2 - 2bp + p^2 = (b-p)^2, & \text{parallel.} \quad (3.37b) \end{cases}$$

The *total scattering from the unpolarized beam* is $\frac{1}{2}[(b+p)^2 + (b-p)^2] = b^2 + p^2$ as before, but for a given polarization, the nuclear and magnetic scattering amplitudes interfere either constructively or destructively. If we can find a material for which $b = p$, we can completely eliminate the parallel spin orientations by making $b - p \equiv 0$ in Eq. (3.37b). In fact, for an alloy of 92 at. % Co and 8 at. % Fe, the 111 reflection has $b \cong p$ and single crystal monochromators of this fcc alloy may be used to produce monochromatic beams of neutrons with as high as 99.7% of the neutrons having the same polarization.

To see how useful such a beam of polarized neutrons can be, consider the scattering from the 111 reflection of ferromagnetic Ni. For this reflection, $b = 1.03(10)^{-14}$ m and $p = 0.10(10)^{-14}$ m. Using unpolarized neutrons, the magnetic field on the specimen may be varied as shown in Fig. 3.33a and b to give scattered neutron intensities proportional to $b^2 + p^2$ and b^2, respectively. In Fig. 3.33b, **H**, and hence **K**, is parallel to $\boldsymbol{\varepsilon}$ so **q** in Eq. (3.34) is zero. From Eq. (3.35), $d\sigma/d\Omega = b^2$. The relative intensity in this case is $(b^2 + p^2)/b^2 = 1.009$, i.e., a 0.9% change in intensity, an effect easily missed when intensities are often measured to only 1% accuracy! On the other hand polarized neutrons may be used as shown in Fig. 3.33c and d. The direction of the neutron polarization may be "flipped" from up to down after leaving the Co(Fe) monochromator by the use of a radio frequency field. The two extremes in this case correspond to $(b+p)^2/(b-p)^2 = 1.48$ or a 48% change. We can see how the use of such a technique would allow for the detection of *very* weak magnetic scattering in the presence of strong nuclear scattering. This technique has proved particularly useful in the study of the variation of magnetic scattering with scattering vector **s**, i.e., in the measurement of f_{mag}. From measured values of f_{mag}, the magnetization or unpaired spin density on a single atom may be evaluated and compared with theory.

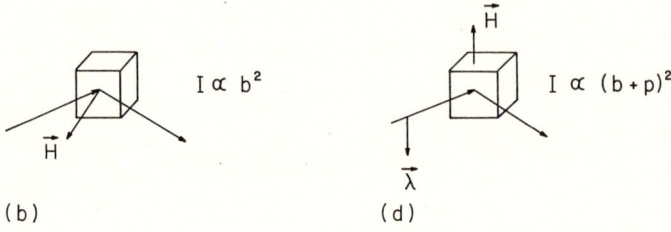

Fig. 3.33. Magnetic scattering using (**a, b**) unpolarized incident neutron beam, and (**c, d**) polarized incident neutron beam. (Adapted from Bacon, G.E., "Neutron Diffraction", 2nd ed. Oxford Univ. Press London and New York, 1962)

This brief introduction to magnetic neutron scattering has dealt with ordered arrays of magnetic spins. It should be noted that extensive studies of diffuse magnetic scattering (similar to the Laue monotonic scattering) have been made emphasizing the short-range order in spin arrays above magnetic ordering temperatures. Furthermore, the energy spectra of the inelastically scattered neutrons have been studied yielding information about time dependent fluctuations in magnetic spin arrangements. For further study, see Bacon (1962) and Gurevich and Tarasov (1968).

The interaction between photons and magnetic moments of electrons means that magnetic structures can also be studied using x-rays; however, the cross section for this scattering is five to seven orders of magnitude smaller than that for Bragg scattering from the atomic structure, and consequently quite difficult to observe. Experimental demonstration of the magnetic scattering from antiferromagnetic NiO was made by Bergevin and Brunel in 1972 using a sealed x-ray tube, but it is only with the advent of high intensity synchrotron sources that these experiments have now become practical. The intensity of a magnetic peak using a wiggler source at the Stanford Synchrotron Radiation Laboratory is one half that obtainable from high flux reactors. X-ray scattering studies of magnetic structures open up opportunities for improved spatial resolution, study of smaller samples than those required for neutron experiments and study of surface magnetic states. The exploration of the polarization of the scattering will allow for the distinction between orbital and spin magnetic scattering (which cannot be done with neutrons) and may also be useful in significantly reducing the charge scattering background (Blume 1985).

3.7 The Absorption of X-rays in Matter

3.7.1 The Mass Absorption Coefficient

In calculating the diffracted intensity, we shall have to know N, the number of atoms scattering, as was shown in Chap. 2. This will require knowing the depth to which the beam penetrates and how its intensity varies with depth. We then need to examine the sum total of interactions with material and the resultant effect, absorption. In addition, absorption processes are of considerable use in their own right.

As an x-ray beam passes through a material, radiation is removed from the direct beam by a variety of processes summarized in Fig. 3.34a. We have discussed

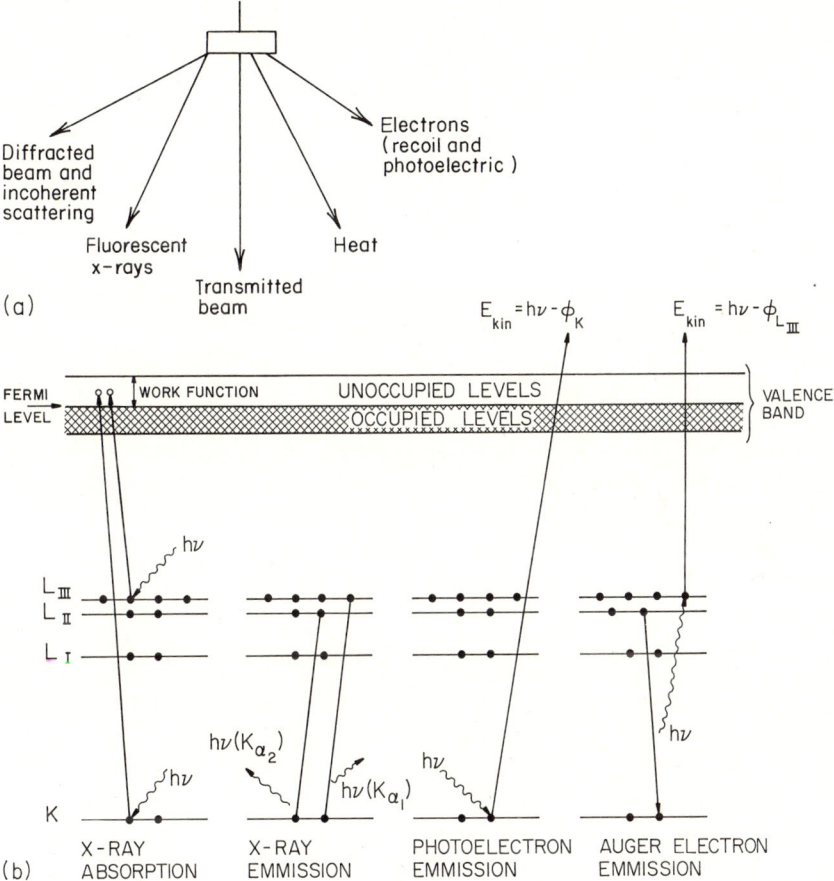

Fig. 3.34. a Contributions to x-ray absorption. (After Henery, N.F.M., Lipson, H., and Wooster, W.A., "The Interpretation of X-ray Diffraction Photographs". Macmillan, London, 1951.) **b** Various emission processes which give rise to spectroscopic analytical tools. Energy levels are shown schematically for atoms bound in a solid. (Adapted from Jenkins, R., "An Introduction to X-ray Spectroscopy". Heyden, New York, 1974)

coherent scattering and Compton (incoherent) scattering in some detail but have not dealt with such processes as photoelectron emission and Auger electron emission. Together with fluorescent x-ray emission, these processes have been used extensively in recent years for chemical analysis and as a means of studying the electron binding energies in solids. A schematic description of these important spectroscopic tools is summarized in Fig. 3.34b. The energies allowed for an electron are represented by several atomic levels including the K and L levels plus the valence band. The valence band contains both occupied levels (filled with electrons) and unoccupied levels, separated by the Fermi level. In the x-ray absorption process, quanta of energy $h\nu$ are absorbed, exciting the electrons to unoccupied levels in the valence band. The subsequent emission of x-ray quanta, as electrons fall back to the low energy states, and the original x-ray absorption reveal similar information about the differences in energies between atomic levels. When the incident photon has sufficient energy to supply the work function of the surface of the specimen, and the binding energy of a K electron, ϕ_K, the excited electron is emitted as a photoelectron with kinetic energy $E_{kin} = h\nu - \phi_K$.

In the Auger process also shown in Fig. 3.34b, an x-ray photon which arises from the transfer of an L_{II} electron to the K shell does not escape, but simultaneously ejects an electron from the L_{III} shell. These two-level Auger electrons have kinetic energy, $E_{kin} = h\nu - \phi_{L_{III}} = \phi_K - [\phi_{L_{II}} + \phi_{L_{III}}]$, and give us considerable information about the energy levels of atoms in solid. Furthermore, since these Auger electrons are not very energetic, they only escape from the sample if they originate near the sample surface. (Depths of a few angstroms are probed rather than the several microns of the other spectroscopies.) This makes the study of the emitted Auger electrons an excellent tool for the study of chemistry at the surface of materials. For a more complete discussion of these various types of spectroscopy, the reader is referred, for example, to Jenkins (1974).

The *attenuation of the incident beam by all the processes summarized in* Fig. 3.34 *is referred to as absorption.* We shall now present a quantitative estimate of this effect. Consider first the absorption due to the production of fluorescent radiation. Earlier in this chapter we treated this process and saw that it led to a complex index of refraction n. Writing $n = n_r - in_i$, we may substitute in Eq. (3.21a) for the electric field after passing through the material of thickness Δz,

$$E_{after} = \exp(-\omega n_i \Delta z/c)\exp[-i\omega(n_R - 1)\Delta z/c]E_{inc}. \tag{3.38a}$$

The first factor is an exponential term representing a decreasing amplitude as Δz increases, an absorption in the material. Thus the imaginary term of the index of refraction is associated with absorption. (This in turn, is connected to the imaginary part of the scattering factor, as we have seen. This is why $\Delta f_K''$ is zero for $\lambda/\lambda_K > 1$; the K-shell electrons cannot then be involved in absorption.)

When the amplitudes of the electric fields in Eq. (3.38a) are squared to give intensities, the result is

$$I_{after} = I_{inc} \exp(-2\omega n_i \Delta z/c), \tag{3.38b}$$

or in differential form for an infinitesimally thin slab of material,

$$dI = -2(\omega n_i/c)I\, dz. \tag{3.38c}$$

3.7 The Absorption of X-rays in Matter

It is convenient to lump all of the absorption processes of Fig. 3.34 into a single equation of the form of Eq. (3.38c), and write for a monochromatic beam:

$$dI = -\mu I\, dz. \tag{3.39a}$$

The proportionality constant μ *is known as the linear absorption coefficient* and has dimensions m^{-1} if z is measured in meters. For a sample of finite thickness z, we integrate Eq. (3.39a), taking $I = I_0$ at $z = 0$, to obtain

$$\ln(I/I_0) = -\mu z$$

or

$$I = I_0 e^{-\mu z}. \tag{3.39b}$$

We may multiply and divide μ by the density of the sample to give

$$I = I_0 e^{-(\mu/\rho)\rho z}, \tag{3.39c}$$

in which the quantity (μ/ρ) is the *mass absorption coefficient* (dimensions of m^2/kg and is independent of the state of the matter — gas, liquid, or solid.

By measuring intensity to a detector with and without an absorber, and knowing z and ρ, μ/ρ for the absorber may be experimentally determined. When we do this, we find that μ/ρ is of the order of 1 to $(10)^2$ m^2/kg and is strongly dependent on the wavelength of the incident x-rays. This latter fact suggests that *fluorescence is the dominant cause of absorption of x-rays*, but can we completely neglect the effect of scattering? The effect of coherent and incoherent scattering on absorption can be calculated as soon as we complete our calculation of the total intensity in a peak (not just the maximum value) in Sect. 4.9. It would be necessary to calculate the intensity for all the possible diffracted peaks for $\lambda/2d < 1$, i.e., we would have to calculate the total intensity in all the peaks on the limiting sphere of radius $2/\lambda$. We can take another approach and calculate the scattering per electron and sum for all the electrons in the material. This would be the situation for a gas, for example, or a single crystal not oriented for diffraction. As we have shown in Eq. (3.17), the total scattering per electron is $\sigma_e = P/I_0$ which has a value of 6.7×10^{-29} m^2. The transmitted beam will then be reduced by an amount σ_e per electron, i.e., σ_e is the "absorption" per electron due to scattering. We can write the absorption due to scattering by a sample containing many electrons with A the atomic weight, as:

$$\mu/\rho = \sigma_e \times \text{(no. of electrons/g)} \quad \text{or} \quad \mu/\rho = \sigma_e N_0 Z/A.$$

since $Z/A \approx \frac{1}{2}$, and $N_0 \approx 6(10)^{23}$, $\mu/\rho \cong 0.02$ m^2/kg. (Here we have also neglected Compton scattering, but this is generally small in comparison to this already small term.) Thus the contribution to the absorption from x-ray scattering is indeed negligible compared to the measured $1-10^2 m/kg$, at least away from Bragg peaks.

Barkla, the man who demonstrated polarization of x-rays, was also one of the pioneers in the study of absorption of x-rays. In the nineteenth century he showed the fundamentals of the variation of absorption with wavelength of the x-ray and atomic number of the absorber. His experimental arrangement is illustrated in

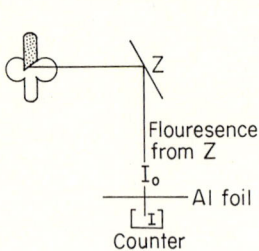

Fig. 3.35. Measuring the absorption of an Al foil for fluorescent x-rays from a specimen of atomic number Z

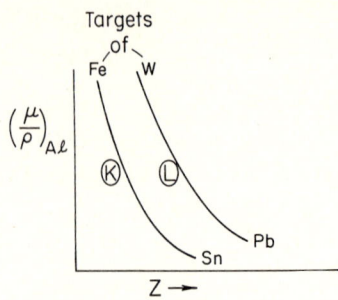

Fig. 3.36. Measured values of $(\mu/\rho)_{Al}$ (mass absorption coefficient for x-ray absorption by Al) versus atomic number Z of the target used in x-ray production. The two curves correspond to the characteristic K and L radiations

Fig. 3.35. (Fortunately, K_α was a strong component of the radiation from the tube, as monochromatic radiation was assumed in the work!) Some of his results are shown in Fig. 3.36 where the *absorption is seen to vary strongly with atomic number of the target Z*. (He arbitrarily called one curve the K curve, the other the L curve, for the nature of the spectrum was then not understood. These letters are still associated with the atomic shells from which electrons are ejected in producing fluorescence.) In each curve of Fig. 3.36, $(\mu/\rho)_{Al} \propto Z^{-6}$. Moseley later demonstrated that $\sqrt{v} \propto Z$, or $\lambda \propto Z^{-2}$. Hence $\mu/\rho \propto \lambda^3$ for a given absorber and a given electron shell absorption (K or L, etc.). Also, $\mu/\rho \propto Z_{abs}^4$, where Z_{abs} is the atomic number of the absorber. Thus,

$$\mu/\rho \cong (KZ_{abs}^4)\lambda^3 , \tag{3.40}$$

for each branch of the curve between absorption edges. For any given element, μ/ρ versus λ appears as in Fig. 3.37. The three edges of the L spectra arise because, as indicated earlier in this chapter, there are three energy levels in the L shell. (But recall that only two transitions can occur from this level to the K level, producing the K_{α_1} and K_{α_2} radiation.)

If there are several atom types present in a sample, because absorption depends only on the number of atoms present, we can imagine each atom type in one section of the sample, as in Fig. 3.38. Let w_i be the weight fraction of the *i*th element, t_i its thickness, and $\rho_i t_i = $ kg/m^2 of the *i*th element. Then, the intensity passing through the sample in Fig. 3.38 is absorbed by each layer:

$$I = I_0 \exp[-(\mu/\rho)_1 \rho_1 t_1] \exp[-(\mu/\rho)_2 \rho_2 t_2] \exp[-(\mu/\rho)_3 \rho_3 t_3] \cdots ,$$

$$I = I_0 \exp\left[-\sum_i (\mu/\rho)_i \rho_i t_i\right].$$

Substituting $\rho_i t_i = M w_i$, where M is the total mass per unit area,

$$I = I_0 \exp\left[-\sum_i [(\mu/\rho)_i] w_i M\right],$$

3.7 The Absorption of X-rays in Matter

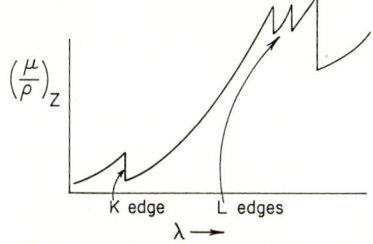

Fig. 3.37. Mass absorption coefficient $(\mu/\rho)_z$ for an element of atomic number Z as a function of incident x-ray wavelength λ

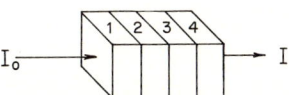

Fig. 3.38. Absorption from a material containing four elements. Since absorption is an atomic process, the calculation is simplified by imagining slabs of each of the four pure elements

but $M = \rho t$, where ρ and t are the density and thickness of the entire sample. Therefore,

$$I = I_0 \exp\left\{\left(-\sum_i [(\mu/\rho)_i w_i \rho t]\right)\right\} . \tag{3.41}$$

As an example of absorption calculations, let us calculate the absorption of Cr K_α radiation in 1 m of air. Absorption coefficients for x-rays are listed in the International Tables, Volume IV, pp. 47 ff. The average molecular weight of air is 29.

$(\mu/\rho)_{\text{oxygen}}$ for Cr $K_\alpha = 35(10)^{-1}$ m^2/kg,

$(\mu/\rho)_{\text{nitrogen}}$ for Cr $K_\alpha = 22(10)^{-1}$ m^2/kg

$$\frac{I}{I_0} = \exp\{-[1/5(35)(10)^{-1} + 4/5(22)(10)^{-1}]\frac{29(10)^{-3} \text{ kg}}{22.4(10)^{-3} \text{ m}^3} \times 1 \text{ m}\} \cong \frac{1}{10}.$$

Because of this absorption in the air, the use of radiation with wavelength very much greater than the 2.2 Å of Cr K_α requires the use of evacuated beam paths. A similar calculation will also show that $I/I_0 \approx 1/100$ in 20 μm of most solids samples so that x-ray diffraction is generally from a thin layer at the sample surface.

We can make use of the drop in absorption at the K edge (Fig. 3.37) to make the x-ray beam more nearly monochromatic. We take advantage of the fact that for transition metal targets with atomic number Z, the adjacent element with atomic number $Z - 1$ has its K absorption edge between the K_β and K_α of the target. This is illustrated in Fig. 3.39. Using a foil absorber of element $Z - 1$, we may reduce the white radiation and the K_β to negligible amounts while maintaining a useful amount of the K_α radiation. A discussion of filtering techniques may be found in Volume III of the International Tables, pp. 73 ff. Generally, a filter thickness is employed to reduce the K_β intensity to $\frac{1}{500}$ of its original value. The K_α intensity is then reduced by $\frac{1}{2}$–$\frac{1}{3}$ by this filter.

Absorption can be very useful in a practical way. In steel mills the roll separation is often automated to provide very uniform thickness in thin sheet by adjusting automatically to maintain a constant transmitted x-ray intensity. Platings on tin cans are checked by measuring the total fluorescence from tin or the diffraction

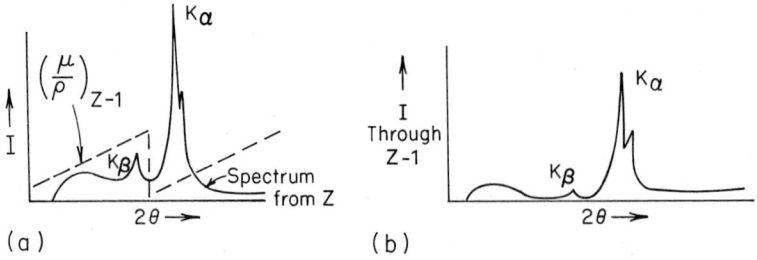

Fig. 3.39. a The x-ray spectrum from a target of the element with atomic number Z and the mass absorption coefficient $(\mu/\rho)_{Z-1}$ for the element with atomic number $(Z-1)$. **b** The x-ray spectrum of **a** after passing through a filter of element $(Z-1)$. Note the significant reduction in white radiation and K_β, while K_α is only slightly reduced

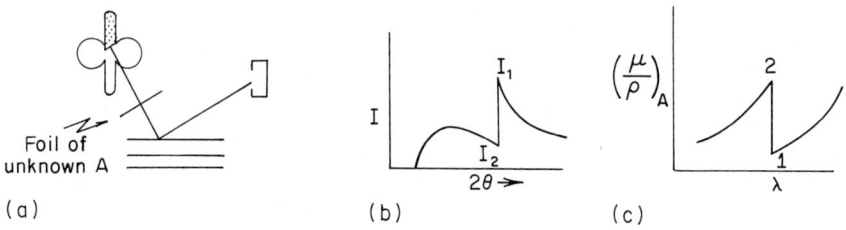

Fig. 3.40a–c. Chemical analysis by absorption. **a** X-rays pass through a foil containing an unknown concentration of element A and are diffracted by a crystal. The intensity appears as in **b** due to the absorption edge of A (as shown in **c**)

intensity of a peak from the iron under the tin which depends on the thickness of the tin plate. Also, in fluorescent analysis, secondary sources are sometimes used, rather than the total radiation from an x-ray tube. This reduces the background and hence, decreases the minimum level of an element that can be detected. A material whose fluorescent radiation (produced by an x-ray tube) is just on the short wavelength side of the absorption edge from the element of interest in the unknown is chosen to increase the efficiency of absorption by this element in the unknown compared to other exciting wavelengths.

We can also use absorption for quantitative analysis. As shown in Fig. 3.40, the sharp edge in a spectrum can be used to identify an element. For element A, μ/ρ varies discontinuously as shown in Fig. 3.40c from $(\mu/\rho)_{A_1}$ to $(\mu/\rho)_{A_2}$, so that:

$$I_1 = I_0 \exp\{-[(\mu/\rho)_{A_1} w_A + (\mu/\rho)_B w_B + \cdots]\rho t\},$$

$$I_2 = I_0 \exp\{-[(\mu/\rho)_{A_2} w_A + (\mu/\rho)_B w_B + \cdots]\rho t\},$$

$$\ln(I_1/I_2) = [(\mu/\rho)_{A_2} - (\mu/\rho)_{A_1}] w_A \rho t,$$

$$\frac{\ln(I_1/I_2)}{(\mu/\rho)_{A_1} - (\mu/\rho)_{A_2}} = w_A \rho t = \frac{\text{mass}}{\text{m}^2} \text{ of } A\,. \qquad (3.42)$$

This technique is not as general for chemical analysis as fluorescence because of the fact that one must transmit through the specimen and thus the specimen must be quite thin. Also the μ/ρ's for the element must be extrapolated from the tables to the absorption edge according to Eq. (3.40), which is only approximate.

3.7.2 The Kramers–Kronig Dispersion Relation

With increasing use of synchrotron radiation sources, materials scientists are more frequently selecting x-rays near an atomic absorption edge to enhance the difference in scattering from atoms close in the periodic table in the unknown structure. The resultant scattering depends on the magnitude of $\Delta f'$ and $\Delta f''$, both of which depend on electronic structure of a given element in the material under study. For precise results, measurements of $\Delta f'$ and $\Delta f''$ are preferred to calculated values. We will focus in this section on the relationship between $\Delta f'$ and $\Delta f''$. As we noted in Sect. 3.4.3, absorption arises because of the imaginary component of the index of refraction. In Sect. 8.2 we will show more formally that

$$\mu = (2\pi/\lambda)\Gamma(\Delta f''), \qquad (3.43\text{a})$$

where

$$\Gamma = 2.818 \cdot 10^{-15} \lambda^2 / \pi V_{\text{u.c.}}, \qquad (3.43\text{b})$$

and $V_{\text{u.c.}}$ is the unit cell volume. Thus, a direct measurement of absorption vs. λ will provide us with $\Delta f''$ vs. λ.

We can relate $\Delta f''$ to $\Delta f'$ and we begin by defining $(dg/d\omega)\,d\omega$ as the number of oscillators between ω and $\omega + d\omega$, that is, the oscillating charge density. Then, we can define an "oscillator strength" g, say for K electrons, as:

$$g(\text{K}) = \int_{\omega_k}^{\infty} (dg/d\omega)\,d\omega \qquad (3.44)$$

From Eq. (3.20c) for an oscillating charge we can write that:

$$x_0 = \frac{F_0/m}{\omega_0^2 - \omega^2 + i\delta\omega}.$$

From the second derivative with respect to time of this expression we obtain the acceleration of the oscillating charge to employ in Eq. (3.15) for the field from such a charge. Finally, the scattering factor, f, is obtained by dividing the amplitude of the scattered wave by the scattering for one electron, $(e^2\eta/mc^2 r)$ and including a minus sign to account for the fact that the scattered wave is opposite in phase to the incident beam (Sect. 3.4.3):

$$f = \frac{\omega^2}{\omega^2 - \omega_0^2 - i\delta\omega}. \qquad (3.45)$$

Therefore, the real part of the scattering factor is:

$$f' \equiv \frac{\omega^2(\omega^2 - \omega_0^2)}{(\omega^2 - \omega_0^2)^2 + \delta^2\omega^2}. \qquad (3.46\text{a})$$

Assuming δ is small and excluding the region very close to ω_0:

$$f' \cong \frac{\omega^2}{\omega^2 - \omega_0^2}, \qquad (3.46b)$$

or for all the charges in an atom:

$$f' = \sum_s \frac{g(s)\omega^2}{\omega^2 - \omega_s^2} = \sum g(s) - \sum \frac{g(s)\omega^2}{\omega_s^2 - \omega^2}. \qquad (3.47)$$

The first term is the sum of oscillator strengths over all electrons in the atom, or Z, the atomic number. We focus on an incident frequency near ω_K so that K electrons are most significant in the second term, and substituting from Eq. (3.44), obtain:

$$f' = Z + \sum_K \int_{\omega_K}^{\infty} \frac{\omega^2 (dg/d\omega)\, d\omega}{(\omega_K^2 - \omega^2)}. \qquad (3.48)$$

Therefore, for the K electrons, we have $\Delta f' = f' - Z$, or

$$\Delta f' = \sum_K \int_{\omega_K}^{\infty} \frac{\omega^2 (dg/d\omega)\, d\omega}{\omega_K^2 - \omega^2}. \qquad (3.49)$$

This is called the Kramers-Kronig dispersion relation.

Now absorption, μ, is the energy absorbed per atom, times the number of atoms per unit length (ρ_l), divided by the incident energy. It can be shown that the energy absorbed per oscillator is $2\pi^2 e^2 \eta/mc$ so that:

$$\mu = \frac{2\pi^2 e^2 \eta}{mc} E(\omega)(dg/d\omega)\, d\omega\, \rho_l / [E(\omega)\, d\omega], \qquad (3.50a)$$

or

$$dg/d\omega = \frac{mc}{2\pi^2 e^2 \eta}(\mu/\rho_l). \qquad (3.50b)$$

As μ is related to $\Delta f''$, with this equation for $dg/d\omega$ and Eq. (3.43a) we now have a relation between $\Delta f'$ and $\Delta f''$. Measuring absorption vs. wavelength will allow us to obtain both! For practical details, see J.J. Hoyt, D. deFontaine and W.K. Warburton, J. Appl. Cryst., 17, 344 (1984).

3.8 The Absorption of Neutrons in Matter

With neutrons, as with electrons, the scattering contribution to absorption is more important than for x-rays.[10] The scattering factors are smaller than those for x-rays;

---

[10] Little data are available on the absorption of electrons. However, because of the large scattering factors and the fact that diffraction angles are small so that a whole plane of reciprocal space is often scattering at once, the contribution due to the scattering is important and must be included. Multiple reflections are important in intensities of electron diffraction and we shall discuss these in more detail in Chap. 8. Intensity expressions neglecting multiple scattering will not be useful except for crystals less than 100 Å in thickness.

3.8 The Absorption of Neutrons in Matter

Table 3.3. True absorption coefficients for neutrons and total absorption coefficients for x-rays

	Neutrons, m²/kg, 1.08 Å	X-rays, m²/kg, 1.54 Å
Be	0.00003	0.150
Al	0.0003	4.86
Cu	0.0021	5.29
W	0.0036	1.72
B	0.24	0.24

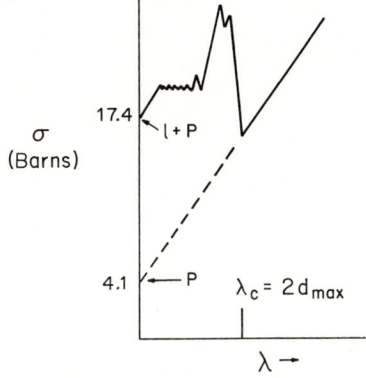

Fig. 3.41. Neutron cross section for polycrystalline Ni. (After Bacon, G.E., "Neutron Diffraction". Oxford Univ. Press, London and New York, 1962)

however, since true absorption is much smaller, there is more penetration, and hence scattering is more important. True absorption coefficients for neutrons are listed in Table 3.3 along with some data for the total absorption for x-rays. This true absorption, similar to fluorescence with x-rays, is due to neutron capture. It varies as $1/v$ or is proportional to λ (as $\lambda = h/mv$), compared to λ^3 for x-rays. For a very few elements such as boron, the absorption is very large at the usual wavelengths because we are near the resonant peak, and the absorption may be as large or larger than that for x-rays. (Values are tabulated in the International Tables, Volume III, pp. 197–199.)

To these true absorption factors for neutrons we must add the calculated scattering, coherent and incoherent, elastic and inelastic. We have already seen how to calculate the incoherent elastic term (Sect. 3.6). We shall see how to calculate the coherent elastic contribution in Chap. 6. As an example of the type of plot expected, we show results for powdered nickel in Fig. 3.41. At very long wavelengths diffraction is not occurring since $\lambda > 2d_{max}$ (and hence Bragg's law can not be satisfied since $\sin\theta$ cannot exceed unity), and at very short wavelengths the diffraction effects average out to a constant. As the wavelength increases, the discontinuities correspond to changes in the number of possible diffraction peaks from the powder. In practice, *the easiest way to determine the total absorption coefficient for neutrons is simply to measure it for the given specimen under the given experimental conditions* by determining the attenuation of the direct beam by the sample.

Since the absorption of neutrons is generally much less than that of x-rays, it is possible to use larger samples in neutron diffraction experiments. This results in an increase in scattering intensity which partly compensates for the lower incident intensities of neutrons. For example, a comparison has been made for the scattering of x-rays and neutrons from powdered CuCl. With comparable instrumental resolution, and optimum sample sizes, scattered x-ray intensities were typically only an order of magnitude larger than scattered neutron intensities, even though the incident x-ray beam (from a conventional sealed x-ray tube) was more than three orders of magnitude more intense. The neutron diffraction study for this comparison was done at a small research reactor, and one may expect an extra factor of 10 or so for the modern high-intensity research reactors available in the US and elsewhere. We see that the reduced absorption of neutrons allows us to use samples so large that measuring times are similar for comparable instrumental resolutions using x-rays from sealed tubes and neutrons. Of course x-ray studies at synchrotron sources gain four to five orders of magnitude in intensity.

3.9 Refraction of Radiation by Matter

We have already seen that for x-rays the index of refraction is less than unity. This means that, whereas with light there is total internal reflection, with x-rays there is total extrernal reflection, as shown in Fig. 3.42. This is why the initial experiments (mentioned in Sect. 3.1) to reflect or refract x-rays failed. In passing through a material, x-rays are bent away from the normal to the surface. The difference of n from unity is so small, however, that it is not practical to make lenses for x-rays. They would be absorbed by a lens thick enough to produce any focusing. More important, is the fact that the wavelength is different inside the material, and the observed diffraction angle may be slightly different than the true one. (In other words, Bragg's law applies for the wavelength inside the crystal.) The measured d spacings will be in error in the fourth or fifth decimal place unless a correction is applied. Examples of how to correct for this are given in Problems 3.22 and 3.23. In reflection, the apparent angle of a peak is too large.

Because there is total reflection at low glancing angles, it is possible to minimize the contribution of diffraction from the bulk, by scattering near or below the critical angle. The refracted beam traverses the surface, and serves as source for diffraction from the outermost layers (Marra, Eisenberger and Cho 1979). At slightly larger

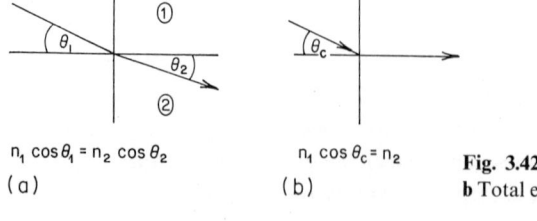

Fig. 3.42. a Refraction of radiation for $n_2 < n_1$. **b** Total external reflection occurs for θ less than θ_c

3.9 Refraction of Radiation by Matter

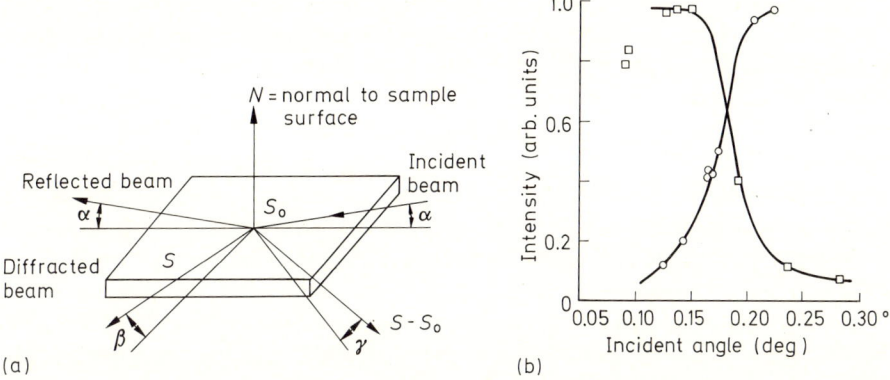

Fig. 3.43. a Schematic diagram of x-ray total-external reflection — Bragg diffraction. **b** Reflected intensity of MoK$_\alpha$, (□) and integrated intensity of 444 Bragg peak (○) as a function of incident angle. From R. Comstock, Ph.D. thesis, Northwestern University, June 1983

angles, beyond total reflection, there is still shallow penetration, and by adjusting this angle the depth sampled can be varied [see, for example, Nielson et al. (1981)]. With these techniques, it is possible to examine the rods in reciprocal space from the structure of the surface layers (Figs. 2.21 and 3.43). This is often done now by LEED (low energy electron diffraction) but this tool requires high vacuum to maintain the electron beam, and because of the high electron scattering factor, intensity calculations must include multiple scattering and are very complex. The x-ray technique (especially at a synchrotron where the intensity and collimation are high) can be carried out in air or under a fluid. It is even possible to sample the interface between two materials with different indices of diffraction. For recent examples of this type of study, see Robinson et al. (1984) and Marra et al. (1982). Minimizing the scattering from the support by total reflection is also useful in fluorescent analysis [H. Aiginger, and P. Wobrauschek, Adv. in X-ray Analysis *28*, 1 (1985)].

For electrons, the field inside the specimen due to the ions and electrons retards the electrons, and the index of refraction is given for an applied voltage V by:

$$n = (V + V_{\text{specimen}})^{1/2}/\sqrt{V} \ . \tag{3.51}$$

Thus n is greater than 1. The specimen diffracts at a lower apparent angle, as if the λ in the specimen was less than the value outside.

For neutrons, the index of refraction is

$$n = 1 - \left[\lambda^2 \sum_i N_i b_i / 2\pi \right], \tag{3.52}$$

where N_i is the number of i nuclei/m^3. The index n can be greater or less than unity because b_i can be positive or negative. The index differs from unity in any case by 1 ppm or less. Either total internal or external reflection will occur if b_i is negative or positive, and a value for b_i can be obtained by measuring the angle for which total reflection just starts. This technique for measuring b_i has been largely replaced by determinations of b_i from diffraction intensities from materials with known structures, because it is difficult to prepare clean surfaces for reflection. Total reflection has been employed (at one of the most powerful research reactors in the world, Institut Laue-Langevin, Grenoble, France) to construct neutron guide tubes. These guide tubes consist of nickel plated glass lined collimators. Total external reflection of neutrons from the thin nickel plating maintains the neutrons in the beam path even if their initial direction is slightly off the collimator centerline. Using such guide tubes, neutrons may be transported a distance of 100 m from the face of a reactor with less than 30% loss in intensity. At this great distance there is adequate floor space to accomodate many experimental stations rather than the few which could be crowded in next to the reactor face. As seen from Eq. (3.52) the deviation of n from unity (and hence the angular range over which total external reflection occurs) increases with increasing λ. Thus these guide tubes are most effective when coupled to a cold source (see Sect. 3.2). These tubes transport the neutrons to large distances from the reactor without appreciable loss, spreading experiments out like the spokes of a wheel and allowing for better shielding.

3.10 Detection of Radiation

3.10.1 Film Techniques

In electron, neutron, and x-ray diffraction, intensities can be measured with films or with counters. We shall discuss the techniques in some detail for x-rays, as this is the area which is most thoroughly developed, and then indicate any unique features with other sources of radiation.

Most *films for x-ray detection* (AgBr in gelatin) *have a linear range of density* $(D = \log_{10} I_0/I)$ *vs. exposure* (intensity times time), as shown in Fig. 3.44. This range extends over four orders of magnitude of I. The user should specify to the manufacturer that he wants film for x-ray work, in order to get one with a good linear range and steep slope. This linearity occurs because one quantum of x radiation is enough to create a nucleus that will form metallic silver on developing. At very high exposures, a sensitized region is being exposed over and over again, and the curve "bends over" and saturates. The curve depends on the film and developing procedure (time, developer). To measure I with a film it is only necessary

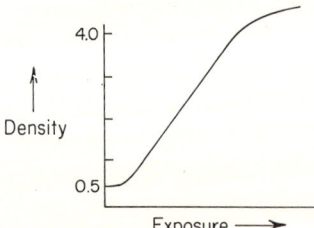

Fig. 3.44. Response of film to x-radiation

3.10 Detection of Radiation

then to prepare a calibration by causing a diffraction spot from a specimen to strike the film for different exposure times. If the development procedures are kept fixed, then by comparing spots on a film with this calibration set, a relative scale of "intensity" can be set up for all the spots in a pattern. It is best, for this comparison, to use a spot from the specimen being studied, so that the shape of the spot will be the same in the calibration and in the actual pattern; in this way the viewer is not "fooled" by the shape of the spot. With considerable practice, precisions of the order of $\pm 5-10\%$ can be achieved by eye. The linear range can be extended by using stacks of several thicknesses of films in the calibration, recording a medium strength spot to estimate the true intensity on, say, the second film with respect to that on the first. Then the use of multiple films in taking the actual pattern will allow accurate measurement of very strong peaks. More precise measurements of the spot "blackness" can be made with photometers; several types of these are available commercially and are known as film densitometers. To reduce the time for qualitative studies, intensifiers are sometimes employed, often in conjunction with Polaroid film. The film is held in close contact with the intensifying screen. X-rays pass through the film and cause light to be emitted from the intensifier. The light then increases the total exposure of the film. Accurate diffraction intensities cannot be measured this way since the degree of intensification is highly irreproducible, depending on film-intensifier contact pressure, etc.

With electrons, the linearity range of films is greatly reduced to one order of magnitude or less, but the principle of calibration is the same. With neutrons, as was mentioned before, the low intensities require broad slits and the angular resolution is not too good; films are generally used only for rough qualitative work, using a substance such as ^{10}B or ^{6}Li in a ZnS film over the film, so that the α particles emitted when these isotopes absorb neutrons cause light flashes in the ZnS, which then expose the film.

3.10.2 Counters and Associated Electronic Components

There are two types of counters commonly in use with x-rays. One of these is a gas-filled tube with a potential drop described schematically in Fig. 3.45a. In Fig. 3.45b some of the associated electronics are shown. Ionization of the atoms of the fill gas is produced by the x-rays. The electrons move quickly to the central wire and the positive ions move, more slowly, to the outer shell. During the motion, the ions collide with the gas atoms causing more ionization, and thus there is considerable amplification of 10^5-10^6. This discharge produces a current pulse of electrons which is picked up by the circuitry. As the slower positive ions move out they shield the positive central wire reducing the field from this wire, and hence the current is reduced; also quench gases are added, which are dissociated by the incoming beam and "pick-up" electrons, further reducing the time duration of the pulse. The pulse is also shaped somewhat by a capacitor circuit. The total amplification in the counter tube and in the subsequent, electronic amplifiers leads to pulses of the order of 1 to 10V. Shortly after a photon goes into the counter, subsequent

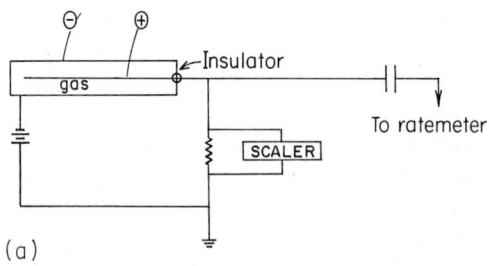

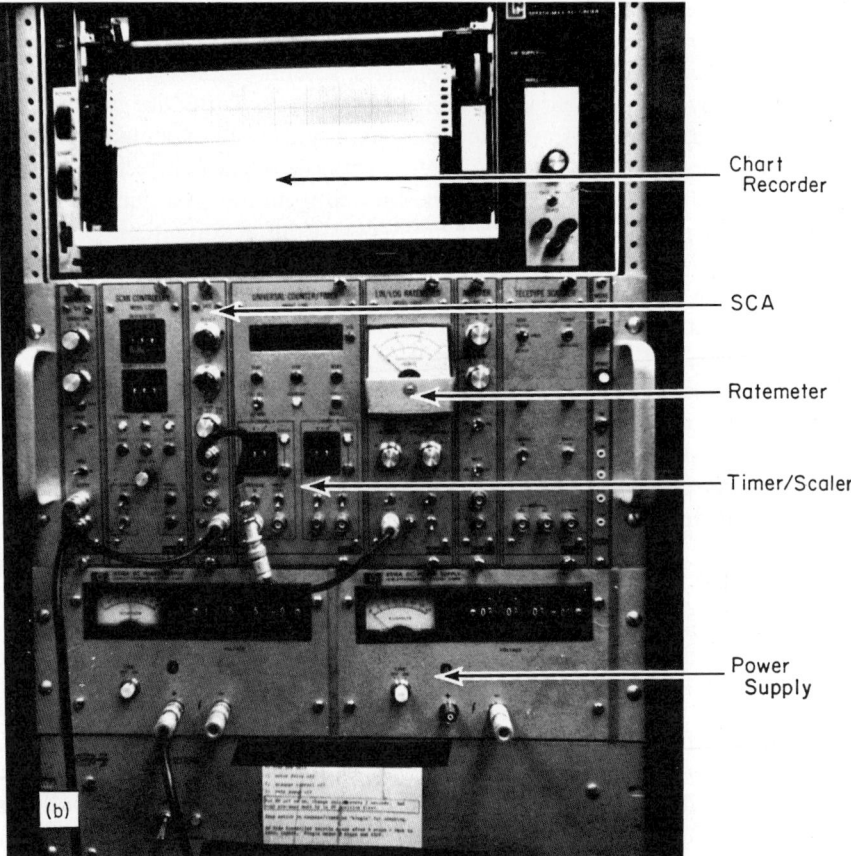

Fig. 3.45.a Gas-filled counter. This is one type of circuit in which the voltage drop across the detector is the detected pulse. In other circuits using field effect transistors (FETs), the charge from the pulse is injected into the middle of a transistor that has p and n ends to produce a current pulse. **b** Associated electronics

3.10 Detection of Radiation

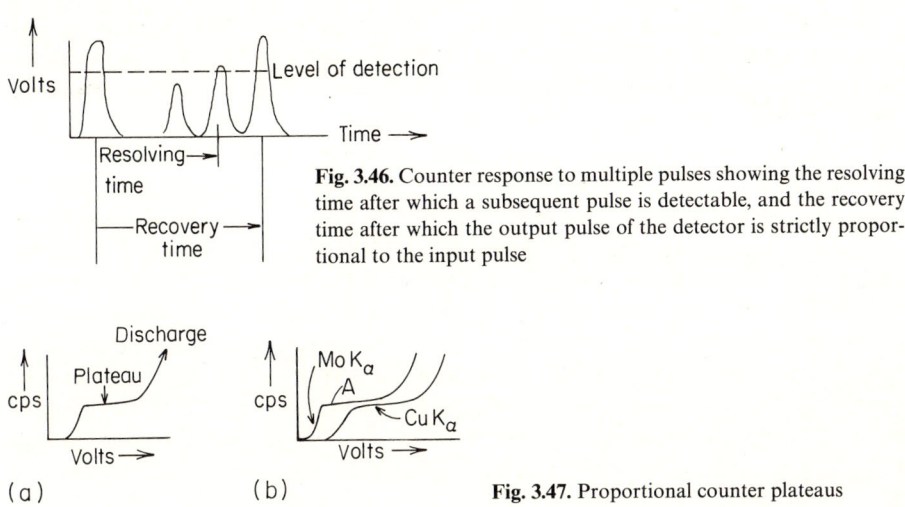

Fig. 3.46. Counter response to multiple pulses showing the resolving time after which a subsequent pulse is detectable, and the recovery time after which the output pulse of the detector is strictly proportional to the input pulse

Fig. 3.47. Proportional counter plateaus

pulses look like those shown in Fig. 3.46. Pulses cannot be "seen" for a time due to the positive ions shielding the wire, producing a volume that has no field across it. In modern proportional gas counters, this resolving time is about 0.2 μs. With a resolving time of 1 μs, a counter can distinguish 10^6 counts without "choking up," and there will be only a 1% loss in the measured counts when the intensity is 10,000 cps (because the counts are arriving in a random way, and some may be "bunched up" in time). A proportional counter may then be quite linear — up to 50,000 cps. Most scalers, which total the pulses by using the pulses in the counter to trigger lights, will have a resolving time of only 1 μs, so that the overall electronic system is linear, at best, to about 10,000 cps. Procedures for measuring dead time are given in Appendix D.

All counters have "plateaus," that is, there is a flat portion in a plot of counts per second versus voltage to the counter (see Fig. 3.47a), between the initial voltage sufficient to provide a measurable pulse and complete discharge. This plateau occurs because, with a certain gas pressure and gas in the tube, the amplification is limited in space (to a plane perpendicular to the wire) for a finite range of voltage (of the order of 150 to 250 V). The slope is the order of 0.05%/V, so that to keep the error in intensity to $\pm 0.2\%$, the voltage to the counter need only be stabilized to ± 4 V, when the plateau is in the vicinity of 1500 V (a typical value). Beyond this voltage plateau, the spatial extent of the ionization increases and the count rate increases again.

The *proportional counter* has another interesting feature, which led to its name. The voltage of the pulse is proportional to the incident photon energy so that, as shown in Fig. 3.47b, the plateau is shifted for different wavelengths. If you set the voltage at A for Mo K_α, the counter will not detect very much of Cu K_α. This property led to the development of the pulse height analyzer (PHA) mentioned earlier in this chapter. The PHA is an electronic device which examines the voltage V of each incoming pulse and compares it to two preset values E_1 and E_2. If the condition $E_1 \leq V \leq E_2$ is satisfied, the PHA transmits a signal causing an electronic scaler to increase its memory by one. The values of E_1 and E_2 may be set independently, or alternatively E_1, and a "window" $\Delta E = E_2 - E_1$ may be varied. Using this device with a proportional counter, all incident pulses outside of the range ΔE are rejected, selecting only a narrow band of wavelengths. Because only one scaler is involved, this system is often called a *single-channel analyzer* (SCA). The procedure for setting a SCA is discussed in Appendix D.

There are a variety of counters, filled with argon or krypton, using different pressures, shapes, etc., primarily because the sensitivity of the tube can be adjusted so that, e.g., the counter may be highly sensitive to Cu K_α but not $\lambda/2$. This too can be helpful in "cleaning up" the spectrum, and obviously such a counter will not yield high intensities with Mo K_α. With certain counters, if the x-ray tube's voltage is high enough, some of the white radiation entering the tube will excite inner shells of the gas atoms and pulses of E_λ and E_λ-$E_{\text{fluorescence}}$ will be produced. When using a SCA, these lower energy "escape" pulses will be removed and the counter may appear quite nonlinear. For this reason, krypton-filled counters should be used with the x-ray tube at 35 kV or less.[11]

Another type of counter currently in use is *the scintillation counter*. One version of this consists of a NaI crystal activated with thallium. Light pulses caused by the x-rays striking the crystal pass to a film of a CsSb compound behind the crystal, where they produce photoelectrons. The electrons enter a dynode, a tube with a series of photoelectric plates, in which the electrons are accelerated by a voltage. Several electrons are produced at each plate and after 10 such stages, an amplification of the order of 10^7 is achieved to produce pulses of several volts. Although the scintillation counter is also a proportional counter, the pulse-energy distribution is about twice as broad for a scintillation counter as for a gas proportional counter, so resolution with a SCA is poorer.

There are also a variety of detectors that permit recording a wide range of scattering angles all at one time, greatly reducing the measuring time, the so called position sensitive detectors. These are available for recording one or two dimensions of data, and involve a variety of detection modes, including TV tubes, scintillators and "channel plate" (tiny hollow glass tubes that are coated internally with compounds that emit electrons so that radiation causes electron emission which is multiplied along the tube). A comprehensive reference list of such detectors may be found in James (1984). To illustrate the principles we describe here a one dimensional gas-filled proportional counter. The method of detecting a photon in a position sensitive proportional counter (PSPC) is the same as in a classical proportional counter. The photons are absorbed in the gas by the photoelectric effect, and for x-rays of the energy of 1–20 keV, the photoelectrons are emitted mostly at right angles to the incident beam. The electrons are accelerated sufficiently through a potential field to produce gas multiplication of the ion pairs. A charge pulse is induced on the central anode wire in the counter which is amplified to give

---

[11] These escape peaks can also occur for the characteristic radiation from the x-ray tube which ionizes the inner shells of the gas atoms in the detector tube. Escape peaks of energy E_λ-$E_{\text{fluorescence}}$ can also occur with the other detectors discussed in this section whenever the x-radiation can fluoresce the detector materials.

3.10 Detection of Radiation

a signal corresponding to the localized multiplication of the primary ionization. The position of the avalanche may be determined by observing the amplitude, or rise time of the pulses occuring at the two ends of the anode.

A high resistance electrode is used in the amplitude or current ratio method. The charge induced on the electrode leaks off to both ends of the wire if they are effectively grounded. The sharing of the current, i_0, between the left (L) and right (R) flow directions is inversely proportional to the resistive paths, thus directly related to positions X_R and X_L:

$$X_L = \frac{i_R}{i_R + i_L} L ; \quad X_R = \frac{i_L}{i_R + i_L} L , \tag{3.53}$$

where

$$i_R + i_L = i_0 \tag{3.54a}$$

$$X_R + X_L = L, \text{ the wire length.} \tag{3.54b}$$

A ratio circuit is necessary to determine the position.

The rise time method determines the position by utilizing the time difference between rise times exhibited at each end of the anode. The rise time depends on the time constant, RC, where R is the resistance along the anode wire and C is the effective capacitance seen by the charge. If the load capacitance of the preamplifier at each end is matched, the time difference is directly related to the resistance and thus to the position. Since they are operated in the proportional range, these detectors may be coupled with appropriated energy analysis electronics and used as part of a SCA while simultaneously yielding pulse detector position. Spatial resolutions of 50–200 μm are typical, which translate to a few hundredths of a degree in 2θ for the usual sample to detector distances. Such a detector can be curved to fit the circle of a diffractometer.

At synchrotrons, the count rates are often too high for proportional detectors. Instead an "ionization chamber" is often employed. This is a modern form of the very first detectors employed in the early 1900's. Two plates at a voltage typically 500V (about half that for a proportional counter) are embedded in a gas-filled container. The small current produced by the discharge when x-ray quanta arrive is amplified by $\sim 10^7$ and fed to a voltage-to-frequency converter and the cycles are counted. Pulse height discrimination is no longer possible, but such a detector is useful over a wide counting range. However, it is *not* useful at low intensities because the background is troublesomely large. More details on this detector are given in Appendix D.

Solid state detectors have been used for many years in the study of nuclear and high energy physics, and are now being used in x-ray detection. A crystal of pure Ge or Li-doped Ge or Si is used as the primary detection device. To form these detectors, p-type Si or Ge is doped by diffusion of Li, an electron donor, to produce an intrinsic region which has high resistance and is hence not subject to conduction by thermal activation. The incident photon excites electrons from the valence band or an impurity level into the conduction band, producing electron-hole pairs. Under an applied voltage these carriers are "swept" from the intrinsic region of the crystal;

Table 3.4. Characteristics of x-ray detectors[a]

Detector	Resolution (keV)	Efficiency[b]
Gas proportional	1.7	0.5
Scintillation	4.8	1.0
Solid state (intrinsic Ge)	0.2	0.8

[a] Co K_α radiation, $E = 6.93$ keV.
[b] Intensity relative to scintillation detector.

the small current which results is proportional to the incident photon energy. The signal is amplified and then fed to the usual SCA circuitry. When Si(Li) or Ge(Li) detectors are used, they must be kept at liquid nitrogen temperatures to avoid diffusion of the interstitial Li ions. Warming to room temperature will destroy the properties of these detectors. This inconvenience has now been eliminated with the availability of pure (intrinsic) Ge detectors which do not have to be cooled in storage. They are cooled during use however, to eliminate the "noise" due to thermal excitation of electrons into the conduction band. However, the escape peaks from Ge are sometimes troublesome.

The principal advantage of the solid state detector is its very high resolution, typically better than $\Delta E = 200$ eV. Used with a SCA, this detector can exclude radiation outside a band of $\lambda \pm 0.02\lambda$ which far exceeds the 20% resolution typical of gas proportional counters. Solid state detectors are still five times as costly as gas proportional or scintillation detectors, and are only used for special purposes requiring such high energy resolution. The properties of these three detectors are summarized in Table 3.4.

The theoretical limit for energy resolution of detectors may be easily estimated. It takes on the average 27 eV of energy to ionize one gas atom. If a 10-keV x-ray photon loses its energy in the gas, it produces $10^4 \div 27$ or 370 ions. This average figure is subject to an uncertainty arising from the chance element in the process of collision and ionization. On the assumption of a bell-shaped probability curve, the uncertainty is measured by the square root of 379 or 19. In this example the uncertainty sets a theoretical limit of $19 \div 370$ or 5% on the energy resolution. A further uncertainty arises from the gas amplification process giving a net resolution for gas proportional detectors of $\approx 20\%$ in the 10 keV range. When solid state detectors are used, the ionization energy is reduced to ≈ 1 eV, 10^4 electrons per 10 keV photon are produced, and a theoretical resolution of 1% is predicted. Other factors account for the fact that the actual resolutions noted in Table 3.4 are three to four times larger than this simple analysis indicates.

High resolution solid state detectors have been coupled with multichannel analyzers to give new possibilities of equipment for x-ray diffraction and chemical analysis. As the name suggests, a *multichannel analyzer* (MCA) is a composite of many single channel analyzers. In circuitry called the analog-to-digital converter (ADC), the voltage of an input pulse is examined. This voltage charges a capacitor, and is converted to an appropriate digital number proportional to the voltage, by timing the capacitor discharge. A count is added to the channel corresponding to

3.10 Detection of Radiation

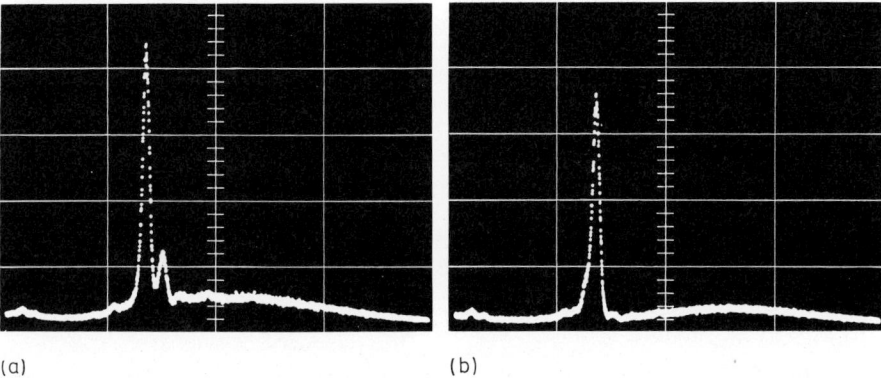

(a)   (b)

Fig. 3.48a, b. Oscilloscope traces of multichannel analysis of the direct beam from a Cu target x-ray tube. The tube was operated at 12.5 keV (just above the excitation voltage for Cu K radiation, so the ratio of K radiation to white radiation is smaller than is ordinarily seen) and 1 mA current. Intensity is recorded vs. channel number (energy), so the K_α with the lower energy (larger λ) is to the left of the K_β. **a** Unfiltered radiation. **b** Filtered with 0.9 mm Ni foil, showing the reduction of K_β

that digital number. Suppose an analyzer with 256 channels is set to receive pulses in the range 0–10 V. If the input pulse has pulse height of 4.5 V a count is added to channel 115 [$(4.5/10) \times 256 = 115$]. The oscilloscope display of intensity versus channel number (energy) will than contain the spectrum of the incident radiation. A typical display is shown in Fig. 3.48.

One application of the MCA is in nondispersive diffraction. Consider white radiation incident on a powder specimen with a solid state detector at angle 2θ from the incident beam. Crystals in the sample select the radiation of the appropriate wavelength λ to satisfy Bragg's law, $\lambda = 2d \sin \theta$, and an entire powder diffraction pattern is recorded when the data are fed to the MCA. This technique is particularly useful when the sample is confined in a highly absorbing container, such as is often the case in high pressure experiments. The requirement of only one entrance and one exit port for the radiation makes the design of the high-pressure cell much easier.

A second application of the MCA is in chemical analysis. The incident x-ray beam causes the sample to fluoresce, and the MCA may be used to identify the energies of the radiation, and hence the elements present and their concentrations. Using a crystal monochromated source of radiation, elements with concentrations as low as parts per billion may be detected, compared with the usual limits of parts per million for the usual x-ray fluorescence apparatus. When white radiation is used, chemical analysis and diffraction may be done simultaneously to give information about chemical composition and crystal structure. In this case, scattering is measured at two angles to separate the diffraction peaks from the fluorescence peaks (the former will appear at different energies at the two angles). In these applications, the uniform response of the solid state detector to x-radiation over a wide range of energies is a considerable advantage in interpreting intensities. This is in sharp contrast to gas proportional counters which depend on atomic

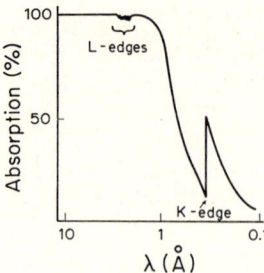

Fig. 3.49. Efficiency of a 1 atm. Xe-10% CH_4 gas-filled proportional counter vs. incident x-ray wavelength, showing the sharp change in efficiency at the Xe K- and L- absorption edges. The CH_4 gas is used to "quench" the tube, assuring rapid recovery after a photon is detected. The absorption curve for a Ge-Li solid-state detector would be less strongly dependent on x-ray wavelength but would show a Ge absorption edge at 1.1 Å, and decreasing efficiency for wavelengths less than 0.3 Å

absorption for their sensitivity, and hence have sharp variations in efficiency at the gas absorption edges (see Fig. 3.49).

An essential component used with any counter is an RC circuit (the rate meter in Fig. 3.45b) that smooths the pulses to a steady current, so that the average intensity can be read on the meter or fed to a chart recorder for continuous scanning of a pattern. The time constant, or rate of decay of this RC circuit can be adjusted so that we can be sure that the pen on a chart recorder is responding to the counts without an unnecessary time delay. A good rule of thumb is to be sure that the pen is oscillating as it passes through a sharp peak. The rate meter and chart should occasionally be calibrated against the scaler.

The counts arriving at the detector are governed by Poisson statistics so that for N counts, the variance $\sigma^2(N) = N$.[12] To obtain the intensity in the peak, however, one must subtract background due to other sources and this affects the accuracy of the results. A common procedure involves summing the counts for each of n angles traversing the peak to give a total count Q_T (including background). The background rates b_1 and b_2 on either side of the peak are then counted and subtracted to give the desired total number of counts in the peak, Q, as

$$Q = Q_T - n(b_1 + b_2)/2 .$$

The variance of Q, $\sigma^2(Q)$ is the sum of the variances of the individual measurements,

$$\sigma^2(Q) = \sigma^2(Q_T) + (n^2/4)[\sigma^2(b_1) + \sigma^2(b_2)] .$$

Since $\sigma^2(N) = N$,

$$\sigma^2(Q) = Q_T + (n^2/2)b ,$$

where $b = (b_1 + b_2)/2$. Thus, the fractional uncertainty in measuring Q, $\sigma(Q)/Q$, may be very large when the background is large even if the fractional uncertainty in Q_T, $\sigma(Q_T)/Q_T$ is made small by counting for long times.

If the slits on the counter are going to be quite wide, then it is necessary to be sure the counter has a uniform response across its opening. This can be checked by

---

[12] The square root of the variance, σ, is the standard deviation, a measure of the uncertainty in the measurements. For repetitive measurements, 67% of the repetitions will fall within $\pm \sigma$ of N. A convenient compilation of statistical principles may be found in Meyer (1975).

3.10 Detection of Radiation

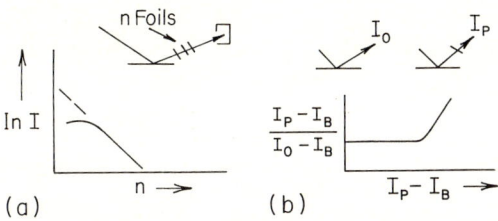

Fig. 3.50a, b. Methods of checking linearity

scanning a narrow direct beam with no receiving slit at the counter. (Occasionally, intense peaks will appear at the edges of such a scan, i.e., near the edges of the window. These are due to fluorescence from the metal counter wall caused by the direct beam.) If the central wire in a gas detector is too thick, it may cause a dead spot in the center of the counter window and the slits should be placed to avoid this or the counter should be tilted. To insure good precision, we want to be sure that we operate in the linear range. This should always be checked with a given counter and electronics. There are two simple ways of doing this, illustrated in Fig. 3.50. With a series (n) of foils of the same thickness (t_1), we can use the absorption law, $I = I_0 \exp(-\mu t_1 n)$. Using a single crystal such as Si cut to a 111 plane, we may obtain a beam of x-rays containing only one wavelength ($\lambda/2$ is absent since $F_{222} = 0$ for Si). A plot of $\ln I$ (corrected for background) versus n will be linear as long as the detection system is linear. The foils should be as similar in absorption as possible to avoid scatter in the plot. This can be arranged by placing the foils on rings in front of the counter and rotating each one until I/I_0 for each foil is the same within say $\pm 0.6\%$, which is the general stability of any unit.[13] Remember to take long counts in doing this, so that the statistical error is 0.1% or less. Another technique is to plot the ratio of the intensity from a crystal with and without one foil in the beam's path versus the intensity without the foil. As shown in Fig. 3.50b, this ratio should be constant until the counting rate, which can be adjusted by varying the current of the tube, exceeds the range of linearity. With this method, only one foil is used and its exact thickness is not too important, as long as it is reasonably uniform. Again, this should be checked with x-rays, not with a micrometer, by surveying the absorption of the foil with a small beam (a movable pinhole near the foil will help). In these ways we can choose operating conditions to achieve a conservative counting rate and be certain of linearity, at least within a few per cent. If a correction for nonlinearity is required, it will be necessary to measure the linearity more carefully. This subject is covered in more detail in Appendix D.

Electronic counting systems for neutron diffraction experiments are similar to those described above for x-rays; however, since neutrons do not carry charge, they cannot directly cause the ionization required in counters. Instead, the gas counters

---

[13] Thin metal foils for this purpose and for filters can be obtained from A.D. Mackay, Inc., 104 Old Kings Highway North, Darien. CN. The material chosen for the multiple-foil technique should absorb λ and harmonics such as $\lambda/2$ about the same amount. Otherwise, the plot $\ln I$ versus n will not be linear even when using a monochromator (if it diffracts $\lambda/2$ as well as λ). For example, Ni foils are suitable with Cu K_α.

are filled with either BF_3 gas enriched in ^{10}B or with ^3He gas. The ^{10}B and ^3He nuclei have high absorption probabilities for neutrons, and in each case this nuclear reaction results in an ionized α-particle (^4He nucleus) which produces a voltage pulse in the counter similar to that seen in x-ray counters. Position detectors are also available for neutrons (see Tompson et al. 1984).

Because of the low and variable intensities available from research reactors, a monitor counter and step scanning are commonly employed with long counts at each scattering position. The equipment is made automatic with a printer to record the data, and the positioning of a crystal, the recording of data, and its analysis can be tied to a computer, increasing the efficiency of the process. Many x-ray units are now automated in this same manner.

Solid state detectors are beginning to appear as neutron detectors. The α particles from a ^6LiF film on a semiconductor produce conducting charges — electrons and holes — by exciting them across the band gap. An electric field applied to the semiconductor then causes a current pulse before these charges return to the valence band. Such a device is shown in Fig. 3.51.

Hardly any work is now done in electron diffraction with counters. However, the "Faraday cage" is used to measure electron current in electron generators and some researchers are now using it for such measurements. It is essentially an insulated metal cup into which the electrons pass and, as shown in Fig. 3.52, the charge buildup causes pulses through a capacitor. Solid state counters may also be useful in this field.

Now that we have examined the general properties of radiation for diffraction, we turn to the geometry of recording diffraction patterns. The reader is to be warned that all three of these radiations are dangerous; just as they ionize gases in counters, so can they ionize atoms in the body, and the beams cannot be seen except with fluorescent screens. Caution is required to avoid placing parts of the body in the direct or diffracted beams. With reasonable care and shielding of equipment, one will not expose oneself to more than that from cosmic radiation. High-energy electrons can cause x-rays when they strike parts of equipment, and neutrons can make some substances radioactive, thus all parts of an experimental arrangement should be carefully thought out. It is a good idea to have a portable counter on

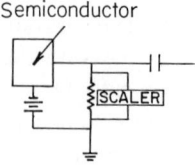

Fig. 3.51. Solid state detector circuitry

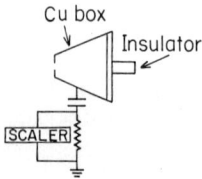

Fig. 3.52. Faraday cage detector circuitry

hand to look for stray radiation and to always keep away from the source. The National Bureau of Standards Handbook (1972) and Moore and McDonald (1968) contain recommendations for safety that should be examined by anyone involved in diffraction.

References

"Advances in X-ray Analysis", Vol. 19. Kendall Hunt, Dubuque (1976)
Aiginger, H., and Wobrauschek, P., Adv. in X-ray Analysis 28, 1 (1985)
Bacon, G.E., "Neutron Diffraction", 2nd ed. Oxford Univ. Press, London and New York, 1962
Bienenstock, A., and Winick, H., Physics Today, June 1983, pp. 48–58
Blume, M., J. Appl. Phys. 57, 3615 (1985)
Buerger, M.J., "Crystal Structure Analysis", Wiley, New York, 1960
Castellano, E.E., and Main, P., Acta Cryst. A41, 156 (1985)
Celotta, R.J. and Pierce, D.T., Science, 234, 333 (1986)
Feynman, R.P., Leighton, B., and Sands, M., "The Feynman Lectures on Physics," Vol. 1., Addison-Wesley, Reading, MA 1963
Gurevich, I.I., and Tarasov, L.V., "Low-Energy Neutron Physics", Wiley, New York, 1968
James, M., Ph.D. Thesis, Northwestern University, 1984
James, R.W., "The Optical Principles of the Diffraction of X-rays", Bell, London, 1950
Jenkins, R., "An Introduction to X-ray Spectrometry", Heyden, New York, 1974
Klug, H.P., and Alexander, L.E., "X-ray Diffraction Procedures", Wiley, New York, 1974
Lander, G.H., and Carpenter, J.M., in Conf. Proc. on "Neutron Scattering in the Nineties", International Atomic Energy Agency, Vienna (1985)
Marra, W.C., Eisenberger, P.E., and Cho, A.Y., J. Appl. Phys. 50, 6927 (1979)
Marra, W.C., Fuoss, P.H., and Eisenberger, P.E., Phys. Rev. Letters 49, 1169 (1982)
Meyer, S.L., "Data Analysis for Scientists and Engineers", Wiley, New York, 1975
Moore, T.M., and McDonald, D.J., "Radiation Safety Recommendations for X-ray Diffraction and Spectrographic Equipment," HEW, October 1968
National Bureau of Standards Handbook III, "Radiation Safety for X-ray Diffraction and Fluorescence Analysis Equipment", U.S. Dept. of Commerce, June 1972
Nielsen, M., Als-Nielsen, J., Bohr, J., and McTague, J.P., Phys. Rev. Letters 47, 582 (1981)
Pinsker, Z.G., "Electron Diffraction", Butterworth, London, 1953
Robinson, A.L., Science 295, 1239 (1979)
Robinson, I.K., Kuk, Y., and Feldman, L.C., Phys. Rev. B29, 4762 (1984)
Sparks, C.J., Phys. Rev. Letters 33, 262 (1974)
Thomas, G., "Transmission Electron Microscopy of Metals", Wiley, New York, 1962
Tompson, C.W., Mildner, D.F.R., Mehregany, M., Sudol, J., Berliner, R., and Yelon, W.B., J. Appl. Cryst. 17, 385 (1984)
Warren, B.E., "X-ray Diffraction", Addison-Wesley, Reading, MA, 1969

Problems

3.1. A copper target x-ray tube is run at 40 kV peak (full wave rectified) and a tube current of 20 mA so that the power input P is

$$P = \int_0^{T/2} \frac{Vi \sin 2\pi t/T \, dt}{T/2}.$$

154   3. Properties of Radiation Useful for Studying the Structure of Materials

The efficiency of an x-ray tube is so low that only on the order of 1 % of the energy is converted to x-rays.
(a) If there were no dissipation of heat by water cooling, conduction, radiation, etc., how long would it take a 100 g copper target to start to melt?
(b) If the target is cooled by a water supply of 2 l/min how much does the temperature of the water rise?
(c) What problems do you think might occur with an x-ray tube after long use?

3.2. A copper x-ray tube, with the filament 5 cm from the anode, is operating at 50 kV, 10 mA.
(a) Find the acceleration of the electrons to the anode.
(b) Find their maximum velocity if classical theory holds, and with relativistic corrections.
(c) Draw a sketch of the spectrum from the tube giving λ_{K_β}, $\lambda_{K_{\alpha_1}}$, $\lambda_{K_{\alpha_2}}$.
(d) What is the excitation voltage for the tube?

3.3. (a) To hold the output for the characteristic lines to $\pm 0.5\%$, how well must the voltage of an x-ray tube and its current be stabilized?
(b) With this stability how much will the maximum in the distribution of the white radiation vary? [*Hint*: Differentiate Eq. (3.5) to obtain dI/I as a function of di and dV.]

3.4. To hold the wavelength in electron diffraction to four significant figures, how well must the current and voltage be stabilized?

3.5. Calculate the angles (θ) for electron diffraction for the first four orders from the (200) planes from a crystal of copper at
(a) 50 kV,   (b) 100 V,   (c) 100 kV.
(d) If the distance from the sample to the film is about 2 ft, what will be the distance to these orders on the film (from the center)? Sketch the experimental arrangement with particular attention to the required orientation of the crystal with respect to incident and diffracted beams.

3.6. With Cu K_α x-radiation and a distance from the sample to a flat film of 5.73 cm, how far from the direct beam will the 200 and 400 from a crystal of iron appear? How far with Mo K_α? Sketch the arrangement with particular attention to the required orientation of the crystal to the incident and diffracted beam to permit these diffraction peaks.

3.7. The following information is obtained by studying the x-ray emission spectrum from Mo (Z = 42):

$V_{K_{ioniz}} = 20 \text{ kV}$, $\lambda_{K_{\alpha_2}} = 0.714 \text{ Å}$, $\lambda_{K_{\alpha_1}} = 0.709 \text{ Å}$, and $\lambda_{K_\beta} = 0.633 \text{ Å}$.

Transmission of monochromatic beams of these three radiations through thin sheets of Al give the following values for mass absorption coefficients, μ/ρ, in m²/kg: 0.53, 0.53, and 0.373 for the K_{α_2}, K_{α_1}, and K_β radiations, respectively. From these data and the physical relations discussed in the text, answer the following questions:
(a) What is the ionization voltage for electrons in the L_{III} state, $V_{L_{III_{ioniz}}}$?
(b) What is $V_{K_{ioniz}}$ for Cu (Z = 29)?

Problems

(c) What is the mass absorption coefficient for the transmission of Cu K_α radiation through Al metal, $(\mu/\rho)_{Al}$?

3.8. As part of an industrial research program, you have cast a Cu–Zn alloy, and you are fairly sure it is 50 wt-% Zn. In order to check this and get an accurate value, you send a sample to the laboratory for analysis. A few hours later someone calls you and says the alloy is 55.0% Cu. He says he has done the analysis with x-ray fluorescent techniques. You check his patterns and notice the Cu K_α, Cu K_β, Zn K_α, and Zn K_β. The Cu K_β is about $\frac{1}{5}$ the Cu K_α but the Zn K_β is very much weaker (see the International Tables, Volume III p. 71). The analyst says he has determined the composition by comparing the intensities of the K_α lines from the alloy with those from the pure metals. Is there anything wrong? What do you recommend he do?

3.9. An x-ray film is taken by scattering white radiation from an amorphous material. The film is made of AgBr. Look up the excitation wavelengths and sketch how the blackening of the film might appear. (Assume that the amorphous specimen scatters an intensity independent of 2θ.)

3.10. Calculate the neutron wavelength diffracted by
(a) the (111) planes of a copper monochromator set at a glancing angle of 15°,
(b) the (422) planes set at a glancing angle of 45°.
If the angular range of the neutron beam striking the monochromator is 1°, what is the width of the band of wavelengths diffracted in either case? ($a_0 = 3.615$ Å for Cu.)

3.11. A steel containing 1.5 wt.% C and 13 wt.% Mn can be retained in the austenitic (fcc) state on quenching. Consider the structure factor and show whether or not it is possible to tell whether the carbon atoms are randomly distributed in tetrahedral or octahedral sites, using (a) x-ray diffraction; (b) neutron diffraction.

3.12. 5 mole% of FeO is dissolved in MgO. MgO has the same structure as NaCl.
(a) Can the Fe^{2+} ions fit in the octahedral or tetrahedral positions? If the iron ions had the valence $+3$ would they fit any better in the tetrahedral positions?
(b) Consider the structure factors of the 111, 200, 220 peaks of (Fe, Mg) O.
Can you tell whether the iron is in octahedral or tetrahedral sites, or arranged in a spinel-like region (the spinel Fe_2MgO_4 precipitates after some time at high temperatures)? Consider (1) x-rays (Co K_α or Mo K_α), (2) neutrons.

3.13. Suppose you wish to eliminate Compton scattering from a measurement, experimentally, not by calculation.
(a) Consider placing a monochromator before the specimen or after it. Can either arrangement accomplish this? How?
(b) Is there any other way you can think of doing this? Give a detailed example.

3.14. A monochromatic beam of antimony K_{α_1} is scattered by a block of paraffin. On a concentric film (AgBr in gelatin) arranged to catch the scattered radiation, an absorption edge appears due to the change in sensitivity of the film at the wavelength $\lambda_{K_{abs}}$ of silver. At what scattering angle does this edge appear and which side is darker, the high-angle or low-angle side? Why?

3.15. Tungsten L_{α_1} is scattered by a block of paraffin. The intensity of scattered radiation (elastic and inelastic) is measured by a counter in front of which is a nickel foil. On varying 2θ (the angle the counter makes with the direct beam) a discontinuous jump occurs in the intensity.
(a) At what 2θ angle does this occur?
(b) On which side of this 2θ is the reading greater?
(c) The counter is set just on the high-angle side of this 2θ and the intensity (I) is read. With the nickel filter in the incident beam the intensity is reduced to $I/64$. With the filter in the scattered beam, the intensity is $33I/128$. What fraction of the intensity is due to inelastic scattering?

3.16. You are building a device to hold a sample under an atmosphere for use with Cu K_α. How thick can a beryllium window be to reduce the radiation by $\frac{1}{2}$? (Remember the beam goes through the window to the sample and then out again as scattered radiation.) How thick would Pyrex have to be? Mylar?

3.17. An x-ray beam from a copper target is monochromated by a LiF monochromator. This crystal, which has a NaCl structure, is being employed with the surface a (100) plane. The first reflection for λ (1.54 Å) is the 200 peak. But there is a component of $\lambda/2$; i.e., its reflection from 400 occurs at the same position. If the diffracted beam is sent through an aluminum sheet 0.1 mm thick, it is reduced to 57% of its value without the sheet.
(a) What fraction of the intensity from the LiF comes from $\lambda/2$?
(b) To what percentage of its value with one foil of Al would the beam be reduced if a second sheet was placed in the beam from the LiF?

3.18. Two fcc metals have exactly the same atomic size, but $f_B = 3f_A$. A powder sample contains $N_A = 3N/4$ of A and $N_B = N/4$ of B atoms. (N is the total number of atoms.) Consider three cases:
(a) A powder sample is a mechanical mixture of A and B.
(b) A powder sample is made up of fcc crystals of a completely random solid solution.
(c) A powder sample is made up of a completely ordered arrangement of A and B (A at faces, B at corners of the unit cell).
Compute the values of I for the 200 reflection for all three cases. In (b), compute the Laue monotonic scattering and compare to the 200 peak.

3.19. A powder sample of copper contains vacancies randomly distributed. Assume you are using unpolarized Cu K_α radiation with $I_0 = 10^9$ cps, and a diffractometer with a radius of 5.73 in. What concentration of vacancies is required to detect the diffuse scattering if 1 cps is the lowest level you can detect? Would you do any better by attempting to measure a decrease in peak intensity? Assume you could detect a 1% change.

3.20. Derive the polarization expression for double scattering, as shown in Fig. P3.20. To obtain this trigonometric term, express the intensity I_2 in terms of I_1. This trigonometric term is then the polarization expression for scattering from a monochromator used as a source.

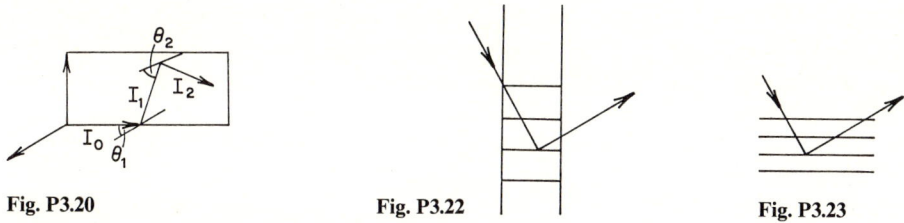

Fig. P3.20   **Fig. P3.22**   **Fig. P3.23**

3.21. Assume there is double scattering within the crystal, as in the figure for Problem 3.20. Express I_2 as a function of I_0 and explain how you might be able to tell if there was double scattering.

3.22. Derive the refraction correction to Bragg's law for the transmission case, shown in Fig. P3.22. (Remember $n = \lambda_{outside}/\lambda_{inside}$, and Snell's law of refraction.) Assume symmetric reflection and that the planes diffracting are perpendicular to the surface.

3.23. Derive the refraction correction to Bragg's law for reflection, as shown in Fig. P3.23. Consider that the sample is Ni. Calculate the correction for the 200 and 400 peaks, for neutron and x-ray diffraction.

3.24. Derive an equation for the critical angle for total external reflection for x-rays. Discuss its wavelength dependence. Calculate the angle for λ for Cu K_α and for $\lambda/2$. Would this be a useful way to eliminate $\lambda/2$ from a beam of x-radiation?

3.25. Derive an equation for determining the value of b using total reflection. Estimate the angle involved for one material with a negative b and one with a positive b.

3.26. Would vanadium be a good material for a sample holder or container for neutron diffraction? Why?

3.27. What fraction of a monochromatic neutron beam ($\lambda = 1$ Å) is lost due to air scattering and absorption in a path length of 4 m. The scattering cross sections for nitrogen and oxygen can be taken as 11.4 and 4.2 barns (1 barn = 10^{-28} m^2), and the "true" absorption cross sections as 0.7 and 0.0 barns, respectively.

3.28. You wish to build a counter for neutrons with $\lambda = 1$ Å using BF_3 gas with the isotope ^{10}B. What isotope enrichment is required to make the counter 90% efficient in a counter one foot long filled with the gas at 1 atm, and 20°C?

3.29. Calculate $F^2 = F_{nuc}^2 + \langle q^2 \rangle F_{mag}^2$ for neutron scattering of an unpolarized beam from polycrystalline samples of (a) a bcc ferromagnet, and (b) a bcc antiferromagnet (with spin at the body-centered atom equal and oppositely directed to that of atoms at corner sites). Assume in all cases that (i) the external magnetic field is zero, (ii) spin polarization is along a [[100]] direction in any given magnetic domain, and (iii) that a random distribution of magnetic domains exists in the polycrystalline sample. Compare your results to the schematic neutron diffraction pattern shown in Fig. 3.32.

3.30. (a) Derive an equation relating photon energy in kilo electron volts and $d \sin \theta$, starting with Bragg's law. Evaluate all constants.

(b) Make a plot for each of two d spacings of 2θ vs. energy. From these plots, is it better to use a high or low angle to separate different spacings with (i) an ordinary diffractometer (angle-dispersive), (ii) a MCA "energy-dispersive" system scattering at one angle.

4. Recording the Diffraction Pattern

4.1 Introduction

The choice of technique for recording the diffraction pattern depends on what specific type of information is desired, the nature of the sample material, and often what experimental facilities are available. If detailed structural information is desired, a single crystal is imperative in low symmetry systems, but in materials with cubic, hexagonal, tetragonal, and, in some cases, orthorhombic symmetry, a polycrystalline or powder sample may suffice. If the structure of a crystal is known, one may wish to use the diffraction pattern to determine the orientation of the crystal for use in measurements of orientation-dependent properties. Often only the precise values of unit cell parameters are desired and these may be most conveniently determined from powder or polycrystalline samples. In all of these records of the diffraction pattern, we must choose a beam of radiation with one wavelength or many wavelengths and a sample and detector orientation such that we satisfy the conditions for diffraction.

We have seen in Sect. 2.4 that diffraction can occur only if a reciprocal lattice point intersects the Ewald sphere, and then only if the structure factor F is nonzero for that reciprocal lattice point. For an arbitrary choice of crystal orientation and radiation wavelength, no intersections need occur, so we must explore techniques for varying both orientation and wavelength to produce a map of the reciprocal lattice and the diffracted intensities associated with the reciprocal lattice points.

The relative size of reciprocal lattice spacings and Ewald sphere affect the diffraction geometry. The geometrical conditions for diffraction are contained in the vector form of the Laue conditions:

$$(\mathbf{S} - \mathbf{S}_0)/\lambda = \mathbf{r}^*_{hkl} . \tag{2.20}$$

If interatomic spacings are of the order of 4 Å, the reciprocal lattice is 0.25 Å^{-1} on an edge, while the radius of the Ewald sphere for 1 Å radiation is 1 Å^{-1}. The curvature of the sphere is important relative to the lattice and some adjustment of orientation is required for the sphere to intersect the reciprocal lattice. With high energy electrons of $\lambda \simeq 0.03$ Å, the radius of the Ewald sphere is 34 Å^{-1} which is large compared with the reciprocal lattice. In this case, essentially a planar section of reciprocal space is seen all at once and some of the difficulty in obtaining and interpreting the diffraction pattern is removed. We shall defer a discussion of electron diffraction to the end of this chapter and begin with a discussion of methods

4.2 Using a Range of Wavelengths — Laue Patterns

4.2.1 Recording the Laue Pattern

Suppose we place a crystal in the path of a polychromatic beam of x-rays from the white radiation from an x-ray tube. The geometry for the incident and diffracted beams is shown in Fig. 4.1 for the case where diffracted beams *transmitted* through the crystal are recorded, and in Fig. 4.2 for the case where a record is made of the *back-reflected* diffracted beams. With such white radiation, we have a range of wavelengths, from the short wavelength limit λ_{min} to the largest wavelength from the tube that will reach the film or counter without nearly complete absorption in air, about 2 Å, λ_{max}. We, therefore, have a continuum of Ewald spheres of radii $1/\lambda$

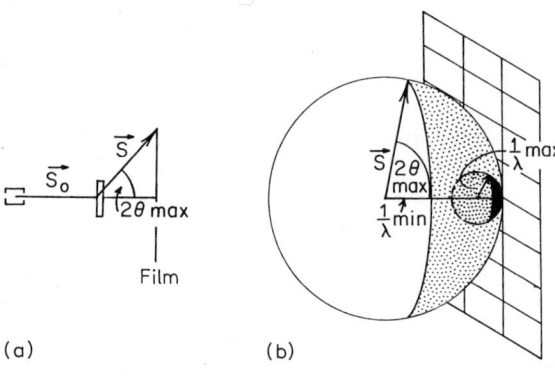

Fig. 4.1a,b. Geometry for Laue patterns in transmission geometry

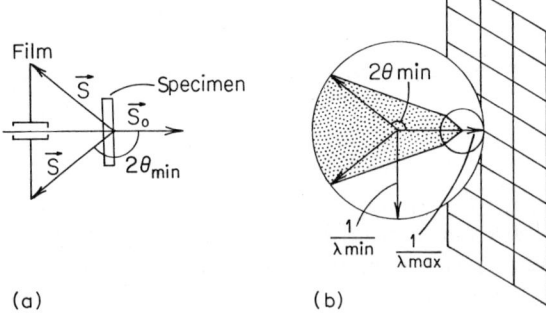

Fig. 4.2a,b. Geometry for Laue patterns in back-reflection geometry

4.2 Using a Range of Wavelengths — Laue Patterns

and *all reciprocal lattice spots lying between these two limiting spheres of radii $1/\lambda_{min}$ and $1/\lambda_{max}$ will satisfy* Eq. (2.20) *and diffract.* If a flat film is used to record the pattern as in Figs. 4.1a and 4.2a, only a certain fraction of the spots will be intercepted by the film as indicated by the shaded portions of Figs. 4.1b and 4.2b. *This technique is known as the Laue technique, and the diffraction spots produced are known as Laue spots.*

How will the spots appear on the film? Because of the profusion of spheres involved in our drawings with many wavelengths, it will be easier to see what is going on if we rewrite the Laue conditions for diffraction in the form

$$\mathbf{S} - \mathbf{S}_0 = \lambda \mathbf{r}^*_{hkl} \,. \tag{4.1}$$

Thus, we can draw an Ewald sphere of reflection of unit radius, and the reciprocal lattice points will be rods directed toward the origin, of lengths given by the spread of available wavelengths. This is shown in Fig. 4.3. The spots are the reciprocal lattice points for λ_{K_α}. The diffraction condition is satisfied for every intersection of a rod with this Ewald sphere.

We found in Sect. 1.8 that all the reciprocal lattice vectors that are normal to planes in a zone lie in a plane of the reciprocal lattice normal to the zone axis and passing through the origin of reciprocal space. These rods then are in such planes which intersect the sphere of reflection in small circles as can be seen in Fig. 4.3. In Fig. 4.4, with the crystal at the center of the reference sphere, the incident beam $\mathbf{S}_0$ enters at I and the transmitted beam exits at T. The point representing the zone axis lies on the circumference of the basic circle and the poles of planes belonging to the zone lie on the great circle shown (only two poles P_1 and P_2 are shown for clarity). Let the diffracted beam from plane P_1 intersect the sphere at some point D_1. The direction of the diffracted beam $\mathbf{S}$ for the plane P_1 can be found using Bragg's law and Fig. 4.4. The directions given by I, P_1, D_1, and T are coplanar, so D_1 must lie on the great circle through I, P_1, and T. The angle between I and P_1 is $90° - \theta$, and from Bragg's law, this must equal the angle between P_1 and D_1.

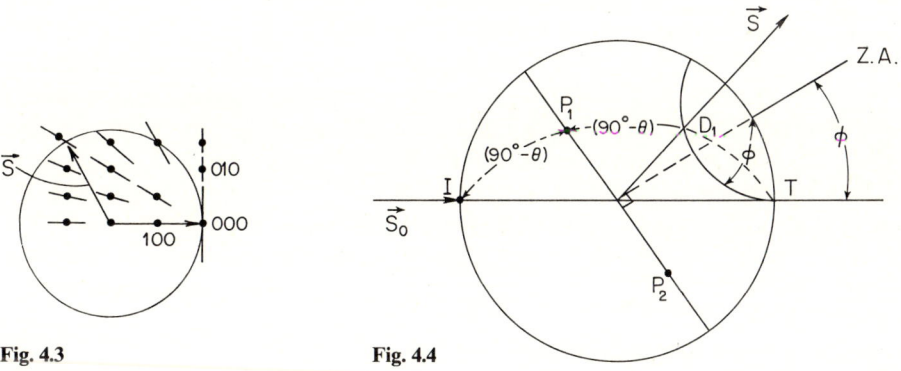

Fig. 4.3   **Fig. 4.4**

Fig. 4.3. The Ewald sphere of reflection drawn with unit radius. The lines through reciprocal lattice points represent the range of wavelengths available in taking a Laue pattern

Fig. 4.4. Demonstration that the Laue reflections from planes of a zone lie on the surface of a cone

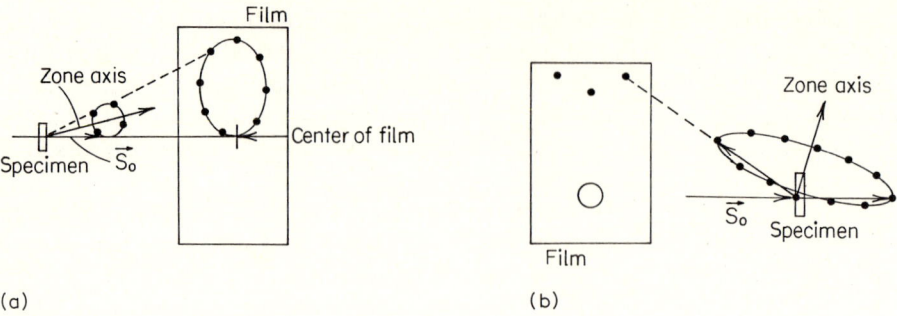

Fig. 4.5. a Transmission Laue pattern, showing the ellipse of spots from a zone. **b** Back-reflection Laue pattern, showing the hyperbola of spots from a zone

Continuing this analysis for every plane of the zone P_i, the diffracted beams D_i will lie on a small circle centered around the zone axis and intersecting the transmitted beam direction. As a result, *the diffraction spots from a zone will be in the form of a cone around the zone axis, with the incident beam as one edge of the cone.* This is illustrated in Figs. 4.5a and b. The intersection of a flat film with the cones of diffraction from a single zone produces conical sections: ellipses in the transmission Laue patterns and hyperbolae in the back-reflection Laue patterns.

Note from Eq. (4.1) that each spot in the cone will have a different wavelength at the intersection with the Ewald sphere. That is, *each plane selects the correct wavelength (if it is available) from the continuous radiation.*

Note also in Fig. 4.3 that a Laue spot can be at a lower or higher 2θ angle than a Bragg spot from the characteristic radiation. As the crystal is rotated, a Laue spot will move toward or away from the Bragg spot.

It is also possible to see, from Fig. 4.3 and Eq. (4.1) that at the same 2θ scattering angle, the 330 plane might diffract due to wavelength λ, and the 220 due to $3\lambda/2$; i.e., the third order from the 110 due to λ might superimpose on the second order from $3\lambda/2$. Each Laue spot may, in general, be made up of several harmonics. The presence of multiple reflections contributing to the same Laue spot minimizes the use of the Laue technique as a means for determining intensities, but, as we shall see, still leaves us with a very powerful tool for mapping the geometry of reciprocal space. (We have seen in Sect. 3.2 that the beam of neutrons available from a nuclear reactor is polychromatic and we might expect the Laue technique to be applicable with neutrons. In practice, however, the low beam intensities and limited sensitivity of film techniques have led to limited use of this technique with neutrons. An interesting variation is the use of a position sensitive detector to cover a small range of angles. Combined with the time-of-flight technique this neutron-Laue technique can distinguish between the multiple harmonics at a Laue spot and be useful for intensity analysis.)

Laue patterns are used mainly in two ways:

1. With a crystal whose structure is unknown, taking a Laue normal to a well-developed face will reveal certain aspects of the crystal symmetry. This application of Laue patterns will be discussed in Sect. 5.7.1.

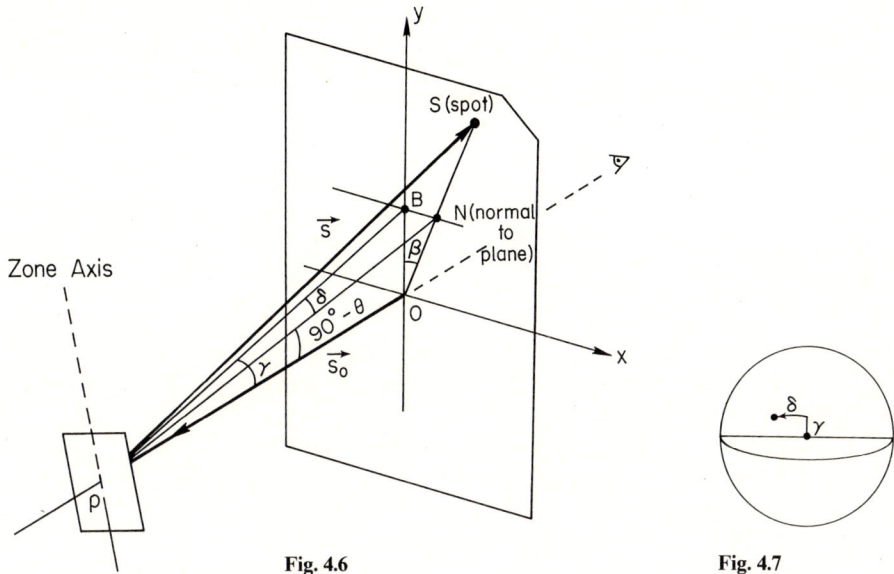

Fig. 4.6. Geometry for transforming a back-reflection Laue spot location to the stereographic projection. (After Cullity, B.D., "Elements of X-ray Diffraction". Addison-Wesley, Reading, MA 1956)

Fig. 4.7. Stereographic projection of the Laue spot of Fig. 4.6, looking in the direction of the incident beam

2. *Laue patterns are a valuable aid in determining the orientation of a crystal grown in a container.* For such crystals, faces do not arise due to the confines of the container. With this method we can determine the orientation before we test the crystal or cut it. We shall expand on this problem to see the techniques that have been developed to make interpretation easier.

Consider first *the back-reflection technique* (Fig. 4.6). We wish to figure out some scheme, such that from a diffraction spot we can locate the pole of the plane causing the spot, and identify the plane. Since our crystal might have any orientation, indexing the reciprocal space will be difficult. We can accomplish our goal if we can locate several normals and measure the angles between them. That is, if we can measure γ, δ (in Fig. 4.6), we can locate a pole (as shown in Fig. 4.7) in a stereographic projection and then measure angles between poles to identify them. (A good table of these angles for cubic crystals can be found in the International Tables, Volume II, pp. 120, 121.)

We shall adapt the excellent summary of the method given by Cullity (1956). From the drawing in Fig. 4.6, for the given diffraction spot S,

$$x = OS \sin \beta; \qquad y = OS \cos \beta; \qquad x/y = \tan \beta, \tag{4.2a}$$

$$OS = OP \tan(180° - 2\theta) = (x^2 + y^2)^{1/2} \tag{4.2b}$$

$$\tan \beta = \frac{BN}{BO} = \frac{PB \tan \delta}{PB \sin \gamma} = \frac{\tan \delta}{\sin \gamma} \tag{4.2c}$$

$$\tan(90° - \theta) = \frac{BN/\sin\beta}{PB\cos\gamma} = \frac{PB\tan\delta/\sin\beta}{PB\cos\gamma} = \frac{\tan\delta}{\sin\beta\cos\gamma} \quad (4.2d)$$

Here, OP is fixed experimentally. We can, therefore, determine γ, δ, the angles for the normal from x, y, the position of the spot (or vice versa) in the following way:

1. Measure x, y on the film.
2. Calculate β from Eq. (4.2a).
3. Calculate OS from Eq. (4.2a).
4. Calculate $(180° - 2\theta)$ and hence $(90° - \theta)$ from Eq. (4.2b).
5. Solve Eqs. (4.2c) and (4.2d) simultaneously for γ, δ.
6. Locate the pole on a stereographic projection as in Fig. 4.7.
7. Repeat for several poles and identify their Miller indices by measuring angles between them and comparing to a table for the system.
8. It is helpful to connect poles of the same zone with great circles. Common intersections of these circles must be prominent poles and could be easier to identify than individual poles.

This process can be made considerably more rapid by making a "net" of γ, δ for each x, y to place over the film. First, using a fixed OP, let us plot x and y as a function of δ (for various γ's), and then x and y as a function of γ for various fixed δ's. For example, with constant γ, $x/y = \tan\beta = \tan\delta/\sin\gamma$. For each δ (with constant γ), we get a value of β and x/y. Then OS can be calculated for a fixed OP, using Eqs. (4.2b) and (4.2d). Thus, with Eq. (4.2a) we can get x, y. We then plot x, y on a piece of transparent paper and label it with its values of γ, δ, and connect points of equal γ. Each resulting curve for constant γ will be a hyperbola, the locus of spots from planes on one zone. Our net is shown in Fig. 4.8 with darker lines every 10°

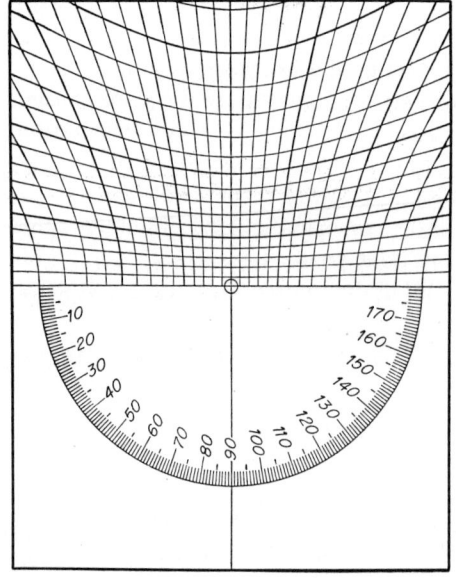

Fig. 4.8. Greninger net. (From Barrett, C.S., "Structure of Metals", 2nd ed. Copyright 1943, 1952 by McGraw-Hill, Inc. Used by permission of McGraw-Hill, New York)

4.2 Using a Range of Wavelengths — Laue Patterns

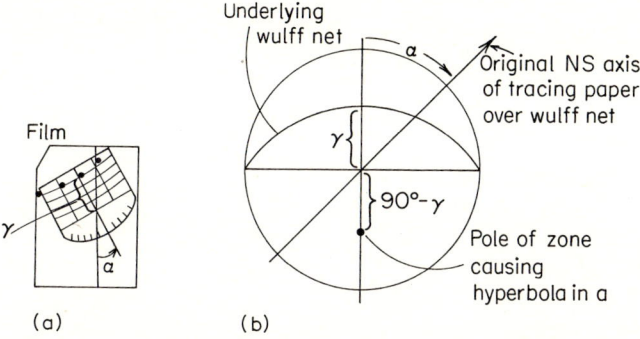

Fig. 4.9a,b. Locating a zone causing a hyperbola of spots in the Laue back-reflection pattern. (Compare position of cut corner on film in **a** with film in Fig. 4.6.) (After Cullity, B.D., "Elements of X-ray Diffraction". Addison-Wesley, Reading, MA, 1956)

in γ and δ. (The curves running vertically are for constant δ, varying γ.) *Such a net is named after its inventor, Greninger*, who devised it as part of his Ph.D. thesis. We can then read the angles of the normal to a plane right from the spot by placing this net over the film, on the side of the film away from the x-ray tube. (Cutting a corner of the film before exposure as shown in the figures will help us keep track of the film's orientation.) Now we can locate each pole. Two poles will fix the orientation if we know the system, a third or fourth are a useful check.

As all planes having poles of the same γ lie in a zone, they lie on a great circle in a stereographic projection; we can, therefore, plot the great circle or its pole, immediately. As shown in Fig. 4.9a, the net is placed over the film and rotated until one hyperbola of constant γ matches a hyperbola of spots. The direction and amount of rotation (indicated by the compass marked on the bottom of the net) are noted. As indicated in Fig. 4.9b, a tracing paper over a Wulff net is rotated in the opposite sense (why?) and the great circle located. This is much quicker and easier than reading each spot. The intersections of great circles and poles of these circles can be indexed by measuring the angles between them and comparing them to a table of angles between normals to planes. Angles between spots can be measured right on the film along a coincident hyperbola of the net, and the rotation angle between two hyperbolas is the angle between two planes or their great circles. With this in mind and some practice comparing the stereographic projection to the Laue pattern, the reader will soon be able to spot important symmetry axes and index spots right on the films without the projection. (Inking in the hyperbolas and/or spots on the film, or on a tracing, make them stand out more clearly.)

Now let us consider the same analysis for *the transmission method*, as in Fig. 4.10. Again we may seek equations to relate x, y to γ, δ, maintaining the orientation during the analysis so that we are looking in the direction in which the beam is going. *The net of contours of constant γ, δ for the transmission geometry is known as a Leonhardt chart.* Figure 4.11 shows the Leonhardt chart for the analysis of these transmission patterns and a stereographic plot of one pole.

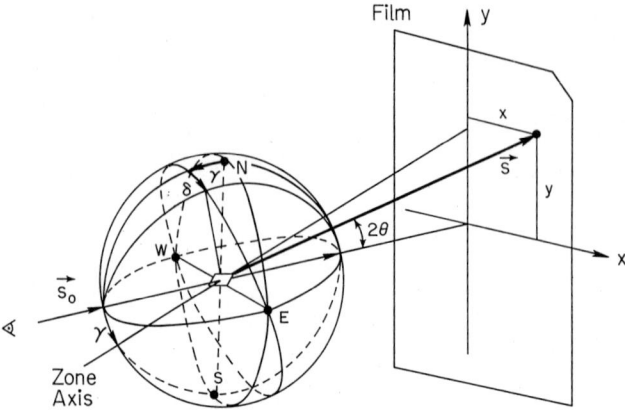

Fig. 4.10. Geometry for transforming a transmission Laue spot to the stereographic projection. (After Cullity, B.D., "Elements of X-ray Diffraction". Addison-Wesley, Reading, MA, 1956)

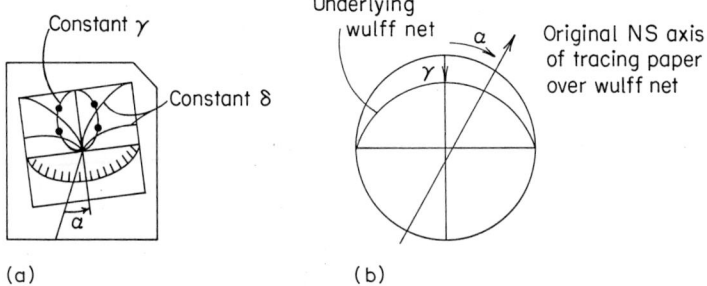

Fig. 4.11a,b. Locating an ellipse in a transmission Laue pattern. **a** Using the Leonhardt chart with the film. **b** Stereographic projection of the great circle of normals to planes giving rise to the ellipse of **a**. (Compare position of cut corner on film in **a** with film in Fig. 4.10.) (After Cullity, B.D., "Elements of X-ray Diffraction". Addison-Wesley, Reading, MA, 1956)

4.2.2 Applications of Laue Patterns

Suppose, after having determined the orientation, we wish to rotate the crystal to another orientation, e.g., to cut a face with a specific orientation. In order to examine the motions necessary to accomplish this, assume we mount a cubic crystal on a goniostat (a sturdy one if the crystal is to be held for a cut, and with the crystal cast in a plastic or resin if it is small). Prior to the first film, we locate the common pivot point of the arcs with a telescope mounted on a track with the camera and goniostat. We first focus on our specimen, tilt it, and locate the common point which will not defocus on tilting. Then by focusing on the collimator to be used we can ascertain if the beam will hit the specimen at this point. If not, the height of the goniostat can be adjusted. We then place the goniostat so that the uppermost arc (*A*) has its axis parallel to the beam, as in Fig. 4.12a. Suppose the first film gives us a stereographic projection, as in Fig. 4.12b, but we wish the orientation in Fig. 4.12c. Suppose that

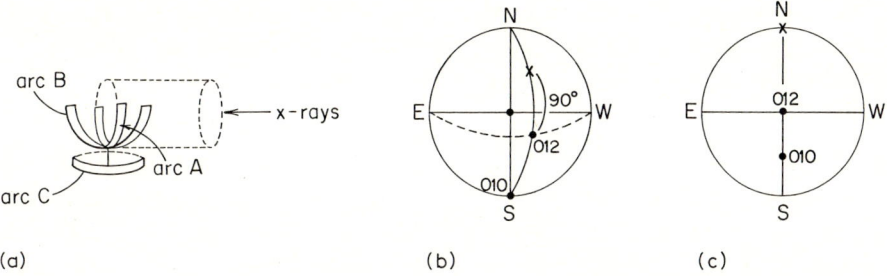

Fig. 4.12a–c. Orienting a cubic crystal to place the 012 pole along the beam direction

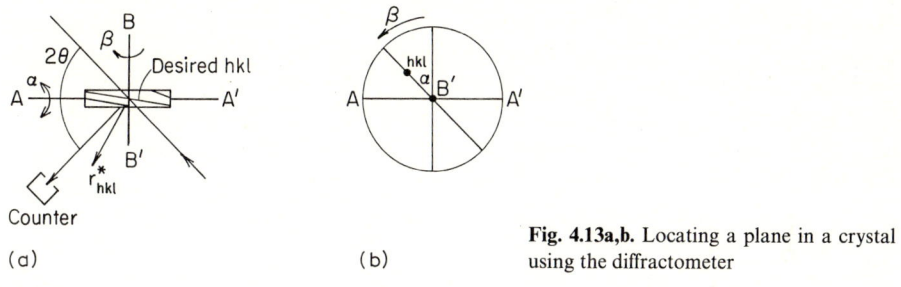

Fig. 4.13a,b. Locating a plane in a crystal using the diffractometer

our projection is made by back reflection using a Greninger net, i.e., the beam is going into the projection at its center, and the center is also the axis of arc A. We locate a point on the great circle connecting 012 and 010 poles, 90° from 012 (the point x in Fig. 4.12b). In the final orientation, this point is to be the North pole. Looking in the direction in which the beam is going, we rotate about the upper axis A around a small circle (around the center of the projection) to bring x to the N–S axis. Rotation about this upper axis does not affect the lower axes B, C. We then rotate about axis B, the E–W axis, to bring x to the North pole, and then about axis C (the N–S axis) to bring (012) into the center.

If we wished to cut a surface parallel to (012), it would only be necessary to rotate the 012 pole into the center of the projection. This could be done with only two of the circles; a rotation first about the B axis to put (012) on the E–W axis, and then around C to put (012) in the center. The orientation should be checked with another Laue and then the goniostat can be transferred to the bed of a milling machine or cutoff wheel where a track for the goniostat has been aligned accurately parallel to the cutting wheel. Dovetail tracks at the x-ray tube and on the bed of the cutoff wheel will provide for an accurate transfer.

This orientation could be done more accurately on a diffractometer than the $\pm\frac{1}{4} - \frac{1}{2}°$ possible with the usual Laue methods (due to the size of the spots and the interpolation of the nets). The counter is set at the 2θ for reflection from the plane desired, for the characteristic radiation of the tube, as in Fig. 4.13. The specimen is then rotated α around AA' and β around BB' until the reflection is located. The normal to the desired plane is now bisecting the incident and diffracted beams

$[(\mathbf{S} - \mathbf{S}_0)/\lambda = \mathbf{r}_{hkl}^*]$. Imagining a stereographic sphere around the specimen, with BB' through the center, the pole is located as in Fig. 4.13b. A narrow receiving slit in front of the counter can be added to refine the angles to a hundredth of a degree if necessary. At this point the specimen can be cut, or another pole located if orientation is desired. (Commercial counter units, in fact, came into being in the second world war to satisfy a need for a rapid device for cutting crystals for piezoelectric devices.) We shall have more to say about diffractometer techniques in Sect. 4.6.

There are many uses for such orientation determinations, of which we shall describe briefly only three.

1. For a single crystal being used, say for a tensile test, such an orientation determination perpendicular to the axis of the specimen locates the tensile axis as the N–S axis in the projection. After a small amount of deformation, slip markings on one or two surfaces will identify the slip plane. (See Sect. 1.9 and Problem 1.32) Laue patterns can be taken for increasing strains and the tensile axis plotted, say in a standard (100) projection. Because the slip direction rotates into the tensile axis, an arc connecting successive positions of the tensile axis must pass through the pole of the slip direction, on the great circle representing the slip plane. When other slip planes and directions become active, these can be noted.
2. Unusual striations, which might be twins, can be identified by taking Laue patterns, one away from the marking and one in or straddling the marking. [For an introductory treatment of twinning, see e.g., Reed-Hill (1964).]
3. Using a Laue pattern from a grain or crystal containing a precipitate, and two surface analysis, the precipitate orientation relative to the parent phase can be identified. Microfocus x-ray tubes with focal spots of the order of microns are especially helpful in taking patterns from grains in a polycrystalline specimen.

The spots on a Laue pattern are usually spread out due to the divergence of the beam. This means that in reciprocal space $\mathbf{S}_0$ does not have one direction but a range of directions; hence a variation of the intersection of the reciprocal lattice results from spheres of reflection tilted slightly. If the specimen is distorted, so that the planes are bent over a range of angles, say around one axis, the reciprocal lattice exists over the range of these angles. Again there is a volume of intersection. If this is in excess of the divergence due to the beam, it can be recognized, and in simple cases the axis of rotation can be identified, e.g., if the spots spread along a hyperbola, the axis of tilt is the zone axis of the planes contributing to the hyperbola. If there is a mosaic, with subgrains tilted relative to each other, we can visualize this as discrete reciprocal lattices tilted slightly with respect to each other. If the tilt is $1-2°$ or more, a Laue spot will be composed of small discrete spots. With microfocus tubes or fine collimation, tilts of only a few minutes of arc can be readily detected.

If the spots are quite broad due to mosaic, or distorted due to severe deformation, it may be difficult to gather enough spots with a flat film in any reasonable time. This situation can be readily overcome by using a cylindrical film. As shown in Fig. 4.14, focusing can be obtained at least in the horizontal plane by using a cylindrical film; the exposure will be greatly reduced as a result of the focusing. Charts for reading such films are described in the International Tables, Volume II.

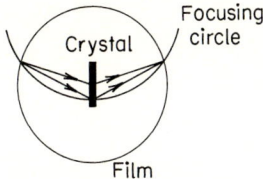

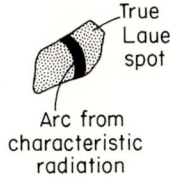

Fig. 4.14. Focusing action of a circular film

Fig. 4.15. Spread of a Laue spot due to lattice distortion

Some of the spots are formed partly with characteristic radiation. If the spots are spread out due to distortion, this more intense radiation will produce an arc as in Fig. 4.15 in a short time. A very long exposure (≈ 2 days) will usually bring out the full spot in a heavily deformed specimen.

Finally, it is important to remember that as white radiation is used for Laue patterns, and this is proportional to Z^2, it is best to use a tube with a high Z, unless the characteristic radiation, which is so much stronger than the white, will fluoresce the sample. Sometimes it can be helpful to cover the film or central portions of the film with a series of concentric aluminum foils of increasing diameter to reduce blackening from such effects.

For further information about Laue techniques, the reader is referred to Barrett (1952) and Amoros et al. (1974). We now turn our attention to methods of examining the diffraction pattern from a single crystal which can be used to determine the diffracted intensities. We shall examine several of these methods in which monochromatic radiation is used, and the crystal is oscillated in some manner, sweeping the Ewald sphere through reciprocal space and causing many Bragg reflections to occur.

Oscillating crystal methods are the most powerful of the diffraction techniques. Generally film techniques are employed with filtered radiation, followed by counter methods for precision intensity measurements. Although considerable attention has been given to automating the diffractometer to enable it to do the entire process to be described, film methods still provide the advantage of a convenient record of a great number of spots.

4.3 The Rotating Crystal Method

A film is wrapped around the crystal in the form of a cylinder and filtered radiation is employed. The crystal is placed, with some prominent axis along the axis of the cylindrical film and rotated, as shown in Fig. 4.16. The reflecting sphere cuts the reciprocal lattice (a) forming horizontal layers of spots (b) and (c) because an axis in real space is perpendicular to a plane in reciprocal space. In the full 360° rotation, the reflecting sphere generates a toroid, as shown in Fig. 4.17. All the points within the toroid can diffract and are observed, depending on the limits set by the film height and camera radius.

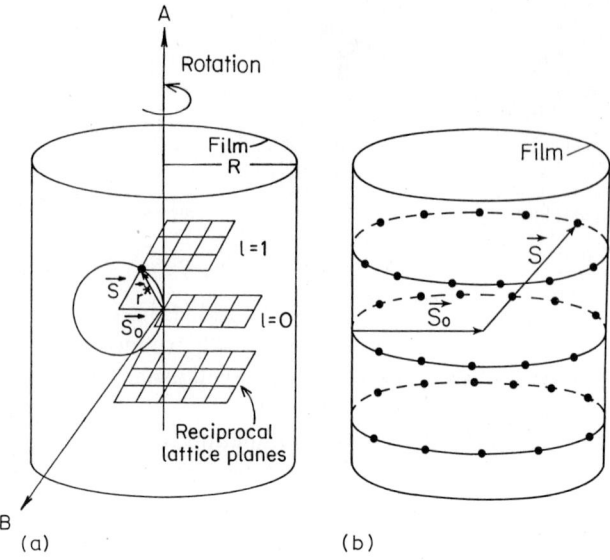

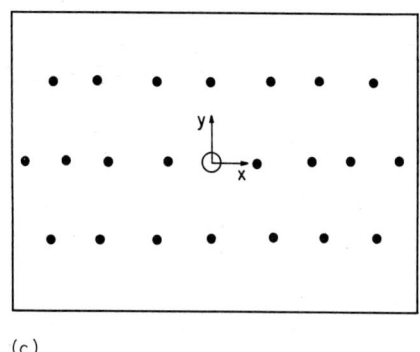

Fig. 4.16a–c. Geometry of formation of the rotation pattern. In c, the film has been unrolled and laid flat

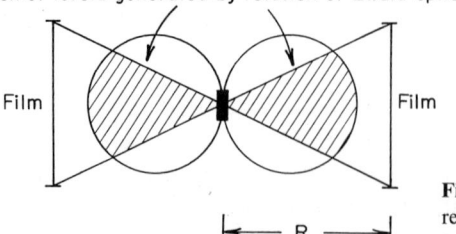

Fig. 4.17. The *shaded area* is the region of reciprocal space examined by the rotation pattern (looking along *B* in Fig. 4.16)

4.3 The Rotating Crystal Method

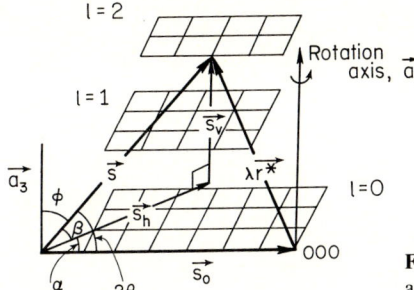

Fig. 4.18. Geometry for location of a diffraction spot on a rotation pattern

Consider the rotation axis as the $\mathbf{a}_3$ axis of the crystal as shown in Fig. 4.18. From the Laue conditions:

$$(\mathbf{S} - \mathbf{S}_0) \cdot \mathbf{a}_3 = l\lambda, \tag{4.3a}$$

$\mathbf{S}_0$ is perpendicular to $\mathbf{a}_3$; therefore,

$$\mathbf{S} \cdot \mathbf{a}_3 = l\lambda, \tag{4.3b}$$

or

$$|\mathbf{a}_3| \cos \varphi = l\lambda, \tag{4.3c}$$

or

$$\sin \beta = l\lambda/|\mathbf{a}_3|$$

Thus all diffracting spots with the same l, regardless of h, k have the same β and lie on the same layer, i.e., *each cone of spots is for one l value*, as can also be seen in Figs. 4.16 and 4.18. By measuring the separation between the film center and a layer (the distance y in Fig. 4.16c) or better yet, half the distance between two layers, and dividing by the radius of the camera, $\tan \beta$ and then $|\mathbf{a}_3|$ can be calculated. Thus, if you have a natural crystal with prominent faces, it is a simple matter to align it along the axes of the crystal and determine the true lengths of the unit cell. The accuracy of this method is about 1%. (With Laue patterns, the important axes can be located even if the crystal does not have well-developed faces.) Other axial lengths can be determined by tilting to another prominent direction, or indexing the spots on the layers.

If rotation is around the axis represented by $u\mathbf{a}_1 + v\mathbf{a}_2 + w\mathbf{a}_3$, this vector has a length that is the inverse of the distance between the planes (uvw) in reciprocal space. Let

$$\mathbf{d}_{uvw} = \frac{u\mathbf{a}_1 + v\mathbf{a}_2 + w\mathbf{a}_3}{1/|\mathbf{d}_{uvw}|}.$$

Then, $\mathbf{d}_{uvw}$ is a unit vector perpendicular to the reciprocal lattice layers, i.e., along the direction $[uvw]$. If we take the scalar product of $\mathbf{d}_{uvw}$ with an $\mathbf{r}^*_{hkl}$ (to a point in reciprocal space on a given layer):

$$\mathbf{r}^*_{hkl} \cdot \mathbf{d}_{uvw} = \frac{hu + kv + lw}{1/|\mathbf{d}_{uvw}|}. \tag{4.4a}$$

But $\mathbf{r}^*_{hkl} \cdot \mathbf{d}_{uvw} = n(|\mathbf{d}_{uvw}|)$, i.e., the length of the projection of a vector to the nth layer line onto the direction $\mathbf{d}_{uvw}$. Thus,

$$n = hu + kv + lw. \tag{4.4b}$$

If we have a layer n (say $n = 1$) and know the axis of rotation (say $[110]$), then on this layer of spots all indices must be of the form

$$h, \quad (\bar{h} + 1), \quad l.$$

Conversely, if we know the layer and the indices of two spots, we can determine the axis of rotation. This is simply the cross product of the two reciprocal lattice vectors.

Next, knowing the unit cell dimensions, we can plot the reciprocal lattice and lay out circles representing sections through the reflecting sphere at various levels. This is done in Fig. 4.19, where α is the angle between $\mathbf{S}_0$ and the component of $\mathbf{S}$ on a horizontal plane, i.e., it is the angle determined by the distance x on the film (Fig. 4.16c) and the radius (R) of the camera ($\alpha = x/R$). Thus, by calculating α for a spot on a given layer, the indices can be found. However, if the crystal is rotated continuously, the 360° rotation leads to an uncertainty; several spots may have the same α. For such uncertain spots, we can replace the gears on the camera by cams and oscillate over a finite angular range to see which of the possible spots is the correct one. This oscillation is indicated in Fig. 4.20. In (a) the oscillation is illustrated for spots on the equator, and in (b) for spots on some l layer. The shaded area is the region of possible intersection. It is necessary to check all possible indices for any given spot, not just one or two, as there may be several overlapping at one point in a full rotation.

The indexing of spots can be greatly simplified with the aid of a chart known as the *Bernal chart*. Consider Fig. 4.18. If we can measure the horizontal and vertical components of $\lambda \mathbf{r}^*$, $\lambda \mathbf{r}^*_h$, and $\lambda \mathbf{r}^*_v$, and φ directly from the film, we can quickly index a spot. Now,

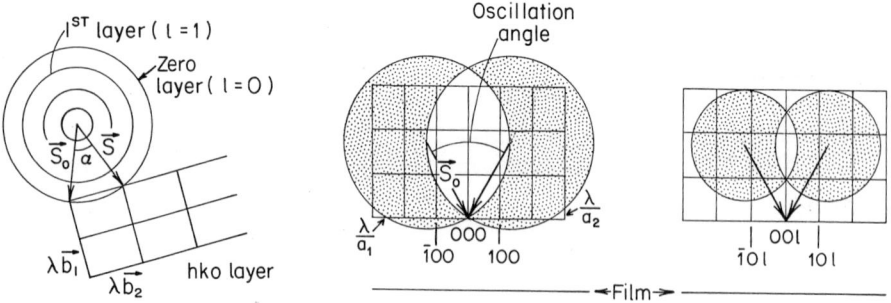

Fig. 4.19

Fig. 4.20

Fig. 4.19. Indexing the rotation pattern. *Circles* represent the intersection of the Ewald sphere with reciprocal space planes at the layer levels indicated (looking along A in Fig. 4.16)

Fig. 4.20. Oscillation pattern geometry showing the region of reciprocal space sampled on the zero and lth layers (looking along A in Fig. 4.16)

4.3 The Rotating Crystal Method

$$\mathbf{S} - \mathbf{S}_0 = \lambda \mathbf{r}^*.$$

Breaking the incoming and outgoing beams and $\mathbf{r}^*$ into horizontal and vertical components:

$$\mathbf{S}_v - \mathbf{S}_{0_v} = \lambda \mathbf{r}_v^*$$

But, since $\mathbf{S}_0$ is horizontal,

$$|\mathbf{S}_{0_v}| = 0.$$

Therefore,

$$\mathbf{S}_v = \lambda \mathbf{r}_v^*. \tag{4.5a}$$

Referring to Fig. 4.18:

$$\sin \beta = \frac{|\mathbf{S}_v|}{|\mathbf{S}|} = \lambda |\mathbf{r}_v^*|, \tag{4.5b}$$

where $\tan \beta = y/R$ (see Fig. 4.16c). From Bragg's law,

$$2 \sin \theta = \lambda |\mathbf{r}^*|.$$

Hence,

$$4 \sin^2 \theta = \lambda^2 (|\mathbf{r}_h^*|^2 + |\mathbf{r}_v^*|^2). \tag{4.5c}$$

The geometric situation in Fig. 4.18 is that of a spherical right triangle so that

$$\cos 2\theta = \cos \alpha \cos \beta.$$

Therefore,

$$1 - \cos 2\theta = 1 - \cos \alpha \cos \beta, \quad \text{or} \quad 2 \sin^2 \theta = 1 - \cos \alpha \cos \beta.$$

Substituting from Eq. (4.5c),

$$\lambda^2 (|\mathbf{r}_v^*|^2 + |\mathbf{r}_h^*|^2) = 2(1 - \cos \alpha \cos \beta).$$

But from Eq. (4.5b),

$$\lambda^2 |\mathbf{r}_v^*|^2 = \sin^2 \beta.$$

Therefore,

$$\sin^2 \beta + \lambda^2 |\mathbf{r}_h^*|^2 = 2(1 - \cos \alpha \cos \beta). \tag{4.5d}$$

In addition,

$$\alpha = x/R, \tag{4.5e}$$

where as before, x is the horizontal distance from the center line of the film to a spot (i.e., along a layer), and R is the camera radius. We can plot this equation as follows. For various y positions on the film in a camera of fixed radius R, calculate $\sin \beta$, and hence $\lambda |\mathbf{r}_v^*|$ from Eq. (4.5b). This yields a series of horizontal lines, which we label as $\lambda |\mathbf{r}_v^*|$. Then for a given value of $\sin \beta$, that is one horizontal line, pick values of x and for each value, this determines one α from Eq. (4.5e). With this α,

Fig. 4.21. Bernal chart. (Reprinted from Henry, N.F.M., Lipson, H., and Wooster, W.A., "The Interpretation of X-ray Diffraction Photographs". Macmillan, London, 1951)

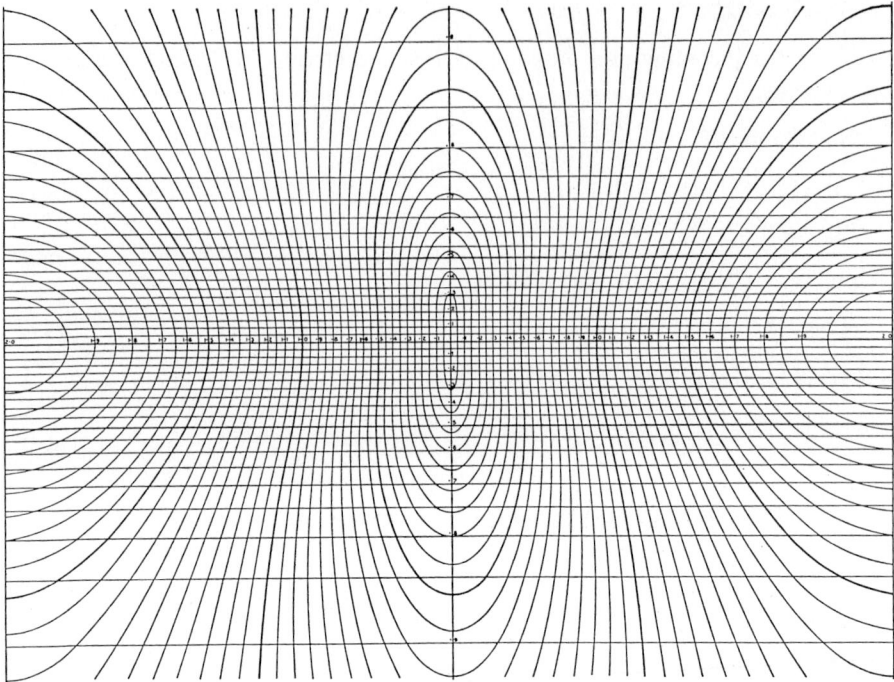

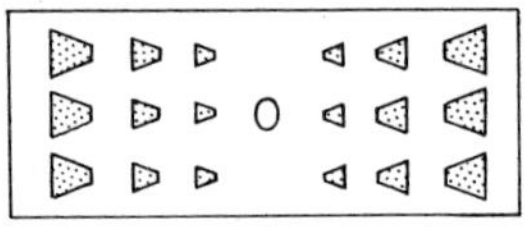

Fig. 4.22. Rotation pattern from a poorly aligned crystal. The sharp spots are spread into broad segments

solve Eq. (4.5d) for $\lambda|\mathbf{r}_h^*|$. We plot these x's on the film and label them $\lambda|\mathbf{r}_h^*|$. The completed chart is shown in Fig. 4.21. To use this chart to determine a cell dimension or axial length, the $\lambda|\mathbf{r}_v^*|$ distance between two corresponding layers above and below the center is read, divided by 2 and λ, and then inverted. To index the layers themselves, a scaled plot of $\lambda\mathbf{b}_1$ and $\lambda\mathbf{b}_2$ is made. The $|\mathbf{r}_h^*|$ values are read and, with a divider set at each of these values, the possible diffracting points can then be ascertained by swinging the dividers around 000.

In a small-angle oscillation we may come close to a point that should not diffract; based on our assumption of the initial orientation, it is just outside the sphere of reflection. If it appears, it will help us to refine the initial orientation by making us realize we were 1–2° off the original choice, or that the lattice parameters need adjustment. The former situation will cause the spots out near the edges of the film to be wide, as shown (exaggerated) in Fig. 4.22.

Examples of the use of rotation films are given in the problems at the end of this chapter. Additional examples may be found in Henry et al. (1961).

4.4 The Weissenberg Method

Oscillation photographs are one method of eliminating uncertainties in the rotation technique; however, several films may be necessary to explore even a small region of reciprocal space. A simple method devised by Weissenberg eliminates this complication. *The film is made to translate along the rotation axis while the crystal rotates so that spots at different rotation angles but with the same $\lambda |\mathbf{r}_h^*|$ do not overlap.* This device is sketched schematically in Fig. 4.23. A screen is placed between the crystal and the film such that diffraction from all but one reciprocal lattice layer is intercepted by the screen. The geometry of this selection is illustrated in Fig. 4.24. The Ewald sphere is of unit radius so distances are not $\mathbf{b}_i$, but $\lambda \mathbf{b}_i$. Let u be the complement of the angle between the direct beam and the rotation axis A and β is

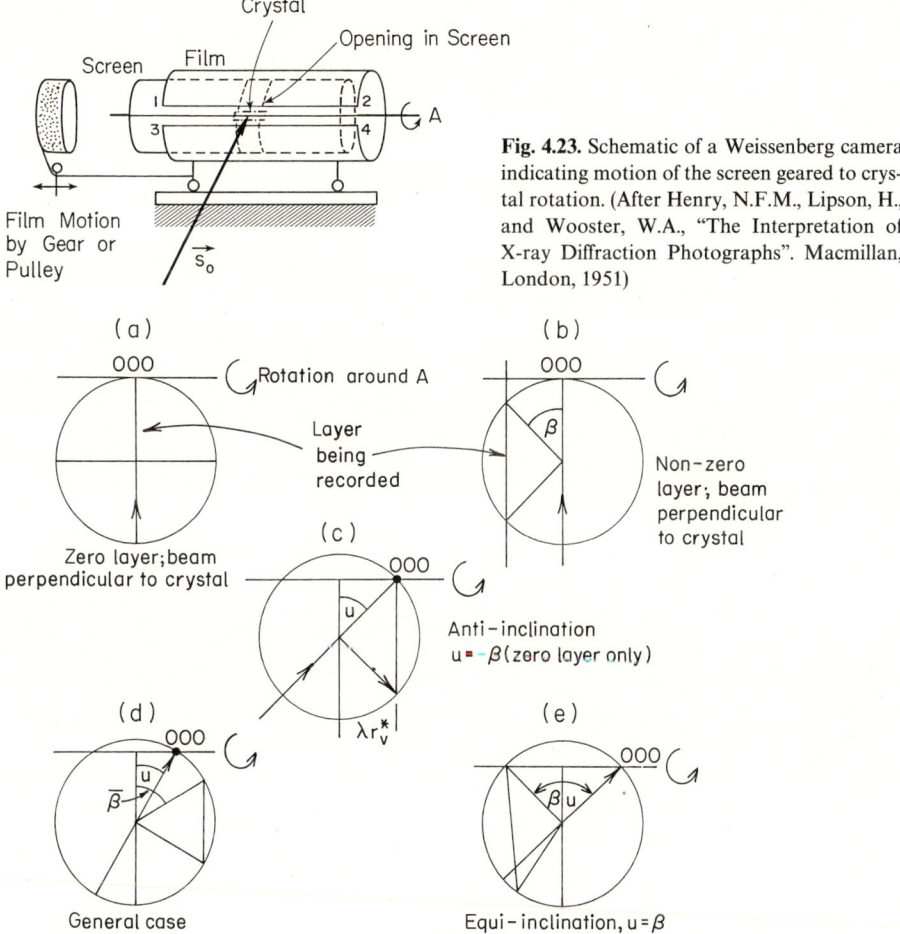

Fig. 4.23. Schematic of a Weissenberg camera indicating motion of the screen geared to crystal rotation. (After Henry, N.F.M., Lipson, H., and Wooster, W.A., "The Interpretation of X-ray Diffraction Photographs". Macmillan, London, 1951)

Fig. 4.24a–e. Some ways of using a Weissenberg camera. (After Henry, N.F.M., Lipson, H., and Wooster, W.A., "The Interpretation of X-ray Diffraction Photographs". Macmillan, London, 1951)

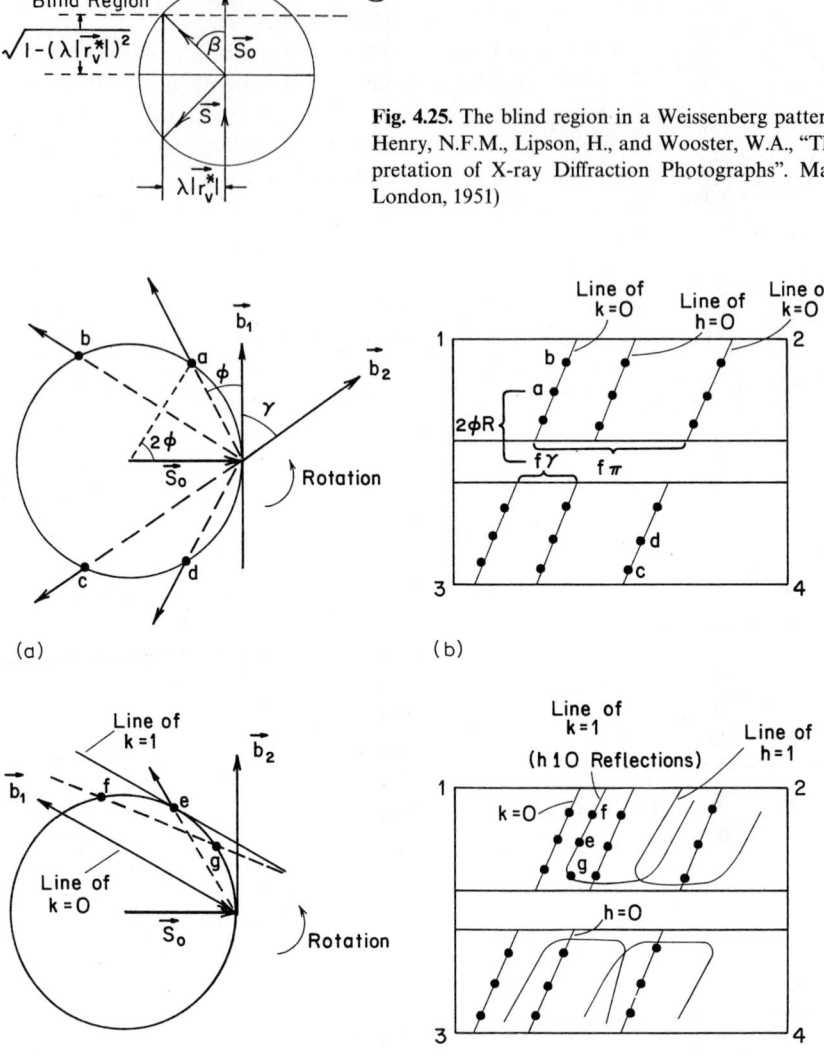

Fig. 4.25. The blind region in a Weissenberg pattern. (After Henry, N.F.M., Lipson, H., and Wooster, W.A., "The Interpretation of X-ray Diffraction Photographs". Macmillan, London, 1951)

Fig. 4.26a–d. How the Weissenberg pattern forms. The film is moving to the left. Note that the *numbered corners* in the film correspond to corners indicated in Fig. 4.23. In the reciprocal space drawings, one is looking along A in Fig. 4.23. *Arrows* indicate rotation and *dotted lines* are subsequent positions of a reciprocal lattice row as the crystal is rotated

4.4 The Weissenberg Method

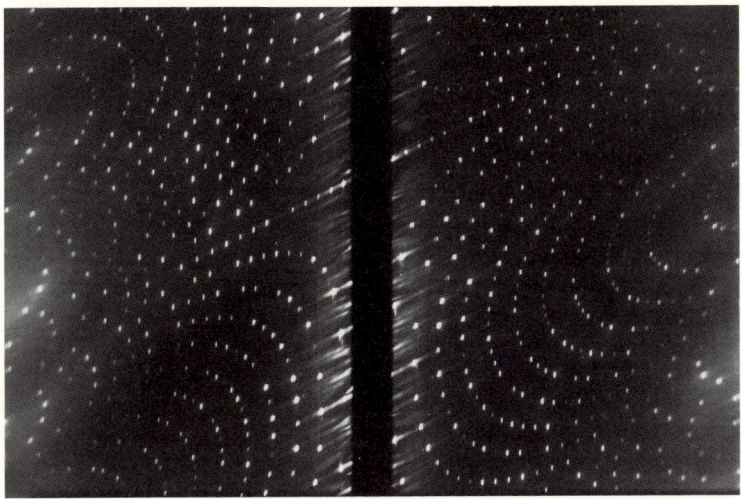

Fig. 4.27. A typical zero-level Weissenberg film, reproduced by permission of Prof. J. Ibers of Northwestern University

$\sin^{-1} \lambda |\mathbf{r}_v^*|$ as defined in Sect. 4.3. From such drawings and the radius of the camera, we can calculate the location of the screen relative to the direct beam. If the radius of the camera is R, the displacement for Fig. 4.24b is $R \tan \beta$. Figures 4.24c–e show geometries in which the direct beam is not normal to the rotation axis. One reason for using such "inclination" methods is illustrated in Fig. 4.25. We can see that there is a "blind spot" for $u = 0$. Points closer than $1 - (1 - |\lambda \mathbf{r}_v^*|^2)^{1/2}$ and further than $1 + (1 - |\lambda \mathbf{r}_v^*|^2)^{1/2}$ cannot be seen.

The geometry of the technique can be followed by examining reciprocal space and the moving film. This is done in Fig. 4.26 looking down along the rotation axis. In (a) and (b) the reciprocal lattice row passes through the origin. A crystal rotation by ϕ causes the point a to intersect the Ewald sphere resulting in diffraction. During this rotation the film has translated a distance f_ϕ (where f is a proportionality factor, commonly 1 mm for 2° in ϕ). The corresponding spot on the film appears at a horizontal distance f_ϕ from the 000 spot and displaced upwards by $R_{2\phi}$, where R is the camera radius. If the lattice row rotating is the $\mathbf{b}_1$ axis, all points lying on it have $k = 0$ and appear on the film along a straight line whose trace is shown in Fig. 4.26c. Rotation by 180° again brings $\mathbf{b}_1$ normal to the direct beam and the $k = 0$ line intersects the direct beam trace once again. If $\mathbf{b}_2$ is not 90° from $\mathbf{b}_1$, this can be readily ascertained and the angle γ measured as shown in Figs. 4.26a and b.

A reciprocal lattice row not passing through the origin ($k = 1$) is also shown in Fig. 4.26c, along with its trace on the film. In Fig. 4.26d, the $\mathbf{b}_2$ axis is shown. Can you prove that for an orthogonal crystal the trace of the $h = 0$ row on the film is parallel to that for $k = 0$ and displaced by a distance $f \times 90°$? A Weissenberg pattern is shown in Fig. 4.27. This analysis demonstrates how the Weissenberg method produces a distorted picture of the reciprocal lattice. Charts for reading such films may be constructed and are available from Polycrystal Book Service.

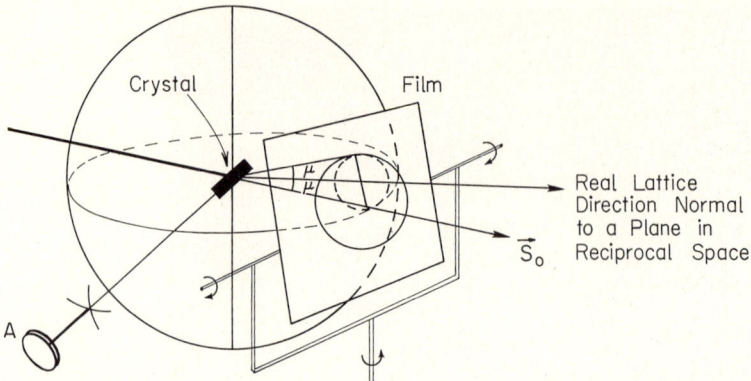

Fig. 4.28. The precession camera. The *small circle* around the lattice direction precesses around S_0. (After Henry, N.F.M., Lipson, H., and Wooster, W.A., "The Interpretation of X-ray Diffraction Photographs". Macmillan, London, 1951)

The reader is referred to Henry et al. (1961), Buerger (1942), and Azaroff (1968) for more detailed accounts of this technique.

4.5 The Precession Method

Buerger and de Jong and Bouman pointed out that interpretation of the diffraction pattern would be considerably easier if one could really see a picture of the reciprocal lattice undistorted rather than getting involved in the geometry of the crystal motion and the film shape. They accomplished this with a flat film! We shall discuss here the precession method of Buerger. Imagine the crystal at the center of the reflecting sphere of unit radius, rather than at the tip of S_0.[1] From the crystal we imagine some zone axis in real space that is, therefore, perpendicular to a layer in the reciprocal lattice of the crystal, as pointed out in Sect. 1.7. We shall *consider first a zero layer*, i.e., a layer passing through the origin of the sphere of reflection, and we shall imagine the film attached to this layer but displaced from it. This situation is shown in Fig. 4.28.

The real lattice direction is made to precess about the direct beam, and thus a small circle is cut by each of the reciprocal lattice planes which is perpendicular to this real lattice direction as it swings around the origin of the reciprocal lattice. The film and crystal move together so that an undistorted picture of reciprocal space is obtained. If a precession camera is available in your laboratory, put a piece of paper in the film holder with a layer of a lattice ruled on it and watch the motion. A series of pictures showing the camera and this motion is presented in Fig. 4.29.

---

[1] As S_0 is a vector, we can translate it relative to the crystal as we wish. A vector has direction and magnitude, but no fixed origin.

4.5 The Precession Method

Fig. 4.29a–c. A commercial version of a Buerger precession camera with a grid of reciprocal lattice points located at the film position and showing through the screen at various positions of the camera

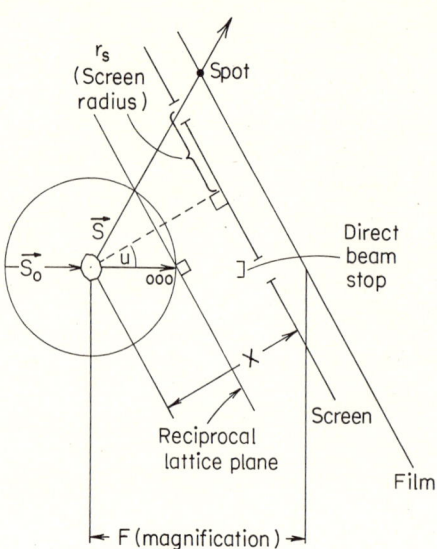

Fig. 4.30. Geometry of formation of a precession photograph for the zero-layer, looking along A in Fig. 4.28. (After Buerger, M.J., ASXRED Monogr. No. 1, "The Photography of the Reciprocal Lattice", 1944)

Each reciprocal lattice plane normal to the beam produces a cone of diffracting spots as the crystal precesses around the beam. In order to isolate the cone associated with one reciprocal lattice plane, we place a screen with an annular opening r_s, fixed to the crystal's motion at a distance x as in Fig. 4.30. The screen rotates normal to the zone axis and around the direct beam. (The direct beam can pass through the opening of the screen and is, therefore, stopped by a lead cup before it reaches the film.) From the figure of a known precession angle u and screens of definite annular openings, we can decide where to place each one (the distance x); this precession angle u is generally of the order of 10–15°. F is the magnification, i.e., as the reciprocal sphere is drawn as a sphere of unit radius, the actual value of F in millimeters is equivalent to the length of one unit of λr^*. (Note that F is not the perpendicular distance from the crystal to the film, but this distance over $\cos u$.) A typical film and the reciprocal lattice obtained from it are shown in Fig. 4.31. (The streaks are due to white radiation in the filtered direct beam, and appear since the film was overexposed.)

As with the rotation or Weissenberg cameras, in the precession technique the crystal can be initially orientated on a goniostat under an optical microscope, or with Laue patterns. However, alignment is also readily achieved by removing the screen and taking a precession photograph with unfiltered radiation. Sharp circles will appear marking the intersection of each layer of the lattice with the reciprocal sphere, i.e., the cutoff of possible reflection. If these circles are not centered on the direct beam, then the crystal must be tilted slightly to correct for this. The amount of tilt can readily be determined graphically with the aid of a rough scale drawing as in Fig. 4.30 or with formulas found in the manual supplied with the camera. Note that several zero layers can be quickly examined by choosing different normals, a choice which merely involves rotating the crystal around its axis. This is a simple

4.5 The Precession Method

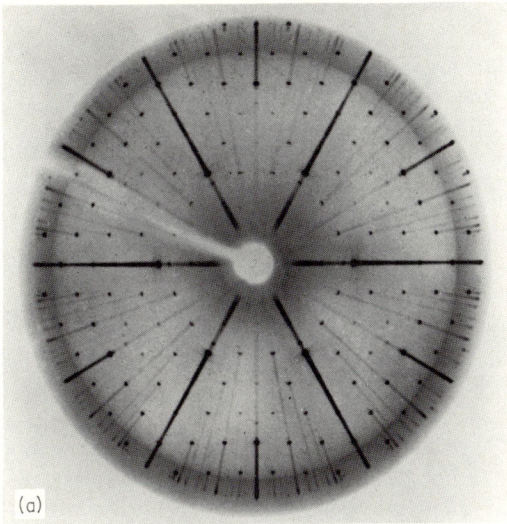

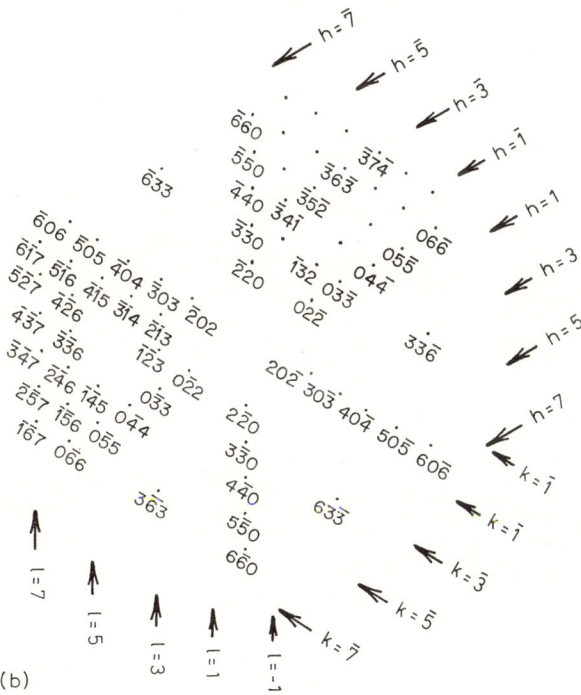

Fig. 4.31a,b. A typical precession film, taken by C. Fairhurst. The film is the zero-layer obtained by precessing the [111] axis of a crystal of cubic Ag-Hg; Mo Kα, Zr filter, $\mu = 20°$, $r_s = 15$ mm, crystal to film, 42 mm (perpendicular to film). This pattern was over-exposed to show streaking from white radiation. The *white region* is the shadow of the beam stop

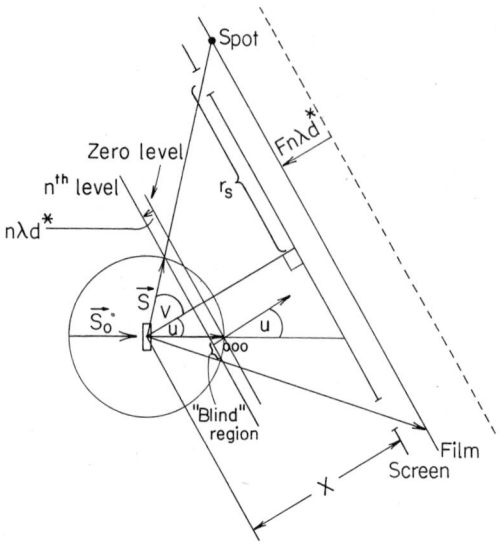

Fig. 4.32. Geometry of formation of a precession photograph for the nth layer, looking along A in Fig. 4.28. (After Buerger, M.J., ASXRED Monogr. No. 1, "The Photography of the Reciprocal Lattice", 1944)

manipulation on the actual camera, as can be seen in Fig. 4.29a; it involves turning the large wheel on the left.

For *nonzero layers*, the situation is shown in Fig. 4.32. For the nth level,

$$n\lambda d^* = (\cos u - \cos v) \, .$$

The angle v can be calculated for the nth layer if d^* is known. The film is moved closer to the crystal by $Fn\lambda d^*$ and the distance of the screen relative to the crystal readjusted so that $\tan v = r_s/x$. That is, for a screen of fixed r_s and u, x is chosen according to the value of v. Combining these two equations:

$$n\lambda d^* = (\cos u - \cos \tan^{-1}(r_s/x)) \, . \tag{4.6}$$

4.6 The Counter–Diffractometer

4.6.1 Orienting a Crystal for One Diffraction Spot

Using the film techniques described above, a large number of diffraction spots may be recorded and identified. If we are interested in the internal atomic arrangement in the unit cell, the diffracted intensities from these reciprocal lattice points must be determined and analyzed as discussed in Chap. 6. One may use a microdensitometer to measure film density, and hence scattered x-ray intensity (as discussed in Sect. 3.10), however the most precise method for measuring these intensities involves the use of diffractometers and counters. Monochromatic radiation is used as we must know the wavelength in order to analyze the data. With x-rays the filtered beam is

4.6 The Counter–Diffractometer

often sufficiently monochromatic due to the high intensity characteristic radiation, while with synchrotron sources or when using neutrons a monochromator must be used as only a white radiation spectrum is available. The approximate orientation of the crystal is assumed known, either by examining the habit with an optical goniometer or by the prior use of the Laue or precession techniques described above.

While many mechanical arrangements have been used to move crystal and counter so as to explore reciprocal space, the most common is the four-circle diffractometer illustrated in Fig. 4.33. The schematic drawing of Fig. 4.33a shows the incident and diffracted beam directions in the x, y diffraction plane defined by the three points: source (S), crystal (C), and detector position (D). We call the axis passing through the crystal and normal to the diffraction plane the main — or 2θ — axis. Coincident with this axis is the θ–axis. On most modern diffractometers the θ– and 2θ–angles may be set independently or coupled so that the counter rotates at twice the angular rate of the crystal, always maintaining the Bragg conditions in a so-called θ–2θ motion. The crystal is mounted at the rotation center of a goniostat which generally offers three rotational degress-of-freedom as shown in Fig. 4.33a: the ω-axis, coincident with the 2θ-axis; the ϕ axis passing through the crystal and making an angle χ with the main axis; and the χ axis parallel to the diffractometer plane. When χ is changed, any point on the ϕ axis travels in a circle, the χ–circle, perpendicular to the χ axis. In a three-circle diffractometer and in what follows, the ω angle is zero so that the χ circle lies in the perpendicular plane which includes the diffraction vector (bisecting the incident and diffracted beams) as shown in Fig. 4.33a. In Fig. 4.33b one example of a four circle diffractometer is shown with the rotation axes indicated.

We assume that preliminary alignment has been made using one of the film techniques described in earlier sections of this chapter. In Fig. 4.34a a crystal is aligned in a diffractometer for an $hk0$ peak. We would do this by aligning $\mathbf{a}_3$ vertical and setting $\mathbf{r}^*_{hk0}$ to bisect the angle between $\mathbf{S}$ and $\mathbf{S}_0$, as shown in Fig. 4.34b. The counter is set at 2θ calculated from Bragg's law, and ϕ and χ are set at zero. To examine another peak of type $h'k'0$, only the angle φ in Fig. 4.34a is needed (and, of course, the angle 2θ is changed). The angle φ is obtained from $\mathbf{r}^*_{hk0} \cdot \mathbf{r}^*_{h'k'0}$. This new orientation is shown schematically in reciprocal space in Fig. 4.34c. Thus, if we can identify a plane (say $hk0$) on a crystal and find its diffraction peak, then by simply tilting the crystal to φ and scanning θ–2θ, we find another reflection $h'k'0$.

If it is an $h'k'l'$ reflection we seek after $hk0$, we must tilt through ϕ, the angle between the component of $\mathbf{r}^*_{h'k'l'}$ on the $hk0$ plane and $\mathbf{r}^*_{hk0}$ and then by χ to bring the vector $\mathbf{r}^*_{h'k'l'}$ into the diffraction plane. These geometric relationships are shown in Fig. 4.35.

For the angle χ in the general case when the axes are not perpendicular:

$$\sin \chi = x/|\mathbf{r}^*_{hkl}|, \tag{4.7a}$$

where

$$x = (\mathbf{a}_3/a_3) \cdot \mathbf{r}^*_{hkl}. \tag{4.7b}$$

A vector in the direction of x is $\mathbf{x} = x\mathbf{a}_3/a_3$, so the vector in the horizontal plane is $\mathbf{y} = \mathbf{r}^* - \mathbf{x}$. The φ rotation may now be found from the dot product between this

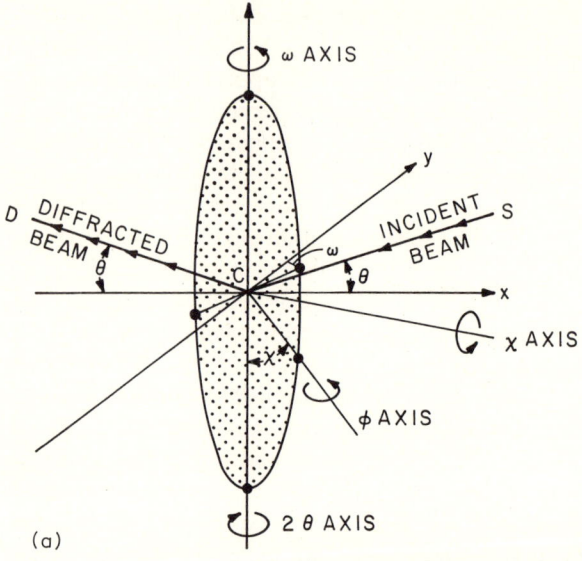

Fig. 4.33. a A four-circle diffractometer. **b** An attachment for locating reflections on a diffractometer, the General Electric eucentric goniometer

4.6 The Counter–Diffractometer

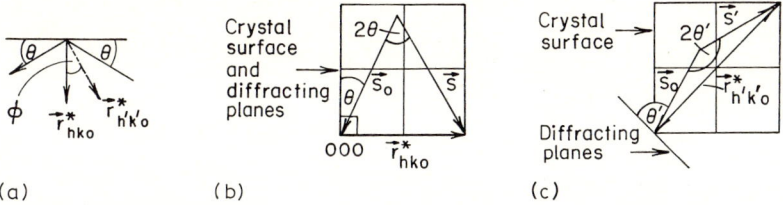

Fig. 4.34a–c. Changing crystal orientation for $hk0$ reflections to the $h'k'0$ reflection

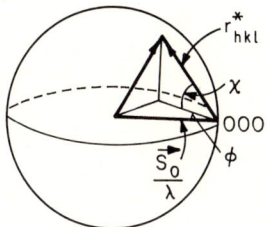

Fig. 4.35. Diffraction geometry for χ-ϕ rotation for a reflection $h'k'l'$

vector **y** and $\mathbf{r}^*_{hk0}$ (assuming as before that we have begun by locating some initial $hk0$ reflection). Each desired hkl reflection is then specified by a pair of angles χ, Φ calculated relative to the initial orientation, coupled with the value of 2θ obtained from Bragg's law.

These calculations are ideally suited for a computer. In many laboratories where work on crystal structures is the primary effort, a computer has been "attached" to the diffractometer along with motor drives for the angles θ, 2θ, φ, and χ. The computer evaluates the angular settings for a given peak, sets the crystal, collects intensity data, and then moves on to the next peak. Such automatic data collection systems are essential in the study of low symmetry, large unit cell crystals for which many thousands of individual diffraction peaks must be measured to determine the crystal structure. In many cases of interest, the crystal being studied can be damaged by the radiation used to determine the diffraction pattern. This is particularly true of the effects of x-radiation on organic crystals. Automatic techniques are particularly important in such cases to minimize the irradiation time and also to periodically return to selected reflections to demonstrate that they are unchanged during the measurements. With such a computer-controlled diffractometer, it is also possible to continuously refine the lattice parameter determinations as the measurements proceed and more reflections are observed, and to periodically check crystal alignment to detect possible movement of the crystal. With such automatic equipment, of course, large numbers of measurements with *any* specimen can be carried out much more rapidly, as discussed more fully below.

4.6.2 Orienting a Crystal for an Arbitrary Point in Reciprocal Space

Occasionally we wish to work with a crystal that does not have its principal axes clearly indicated from its shape. It may be a crystal of arbitrary orientation, cut

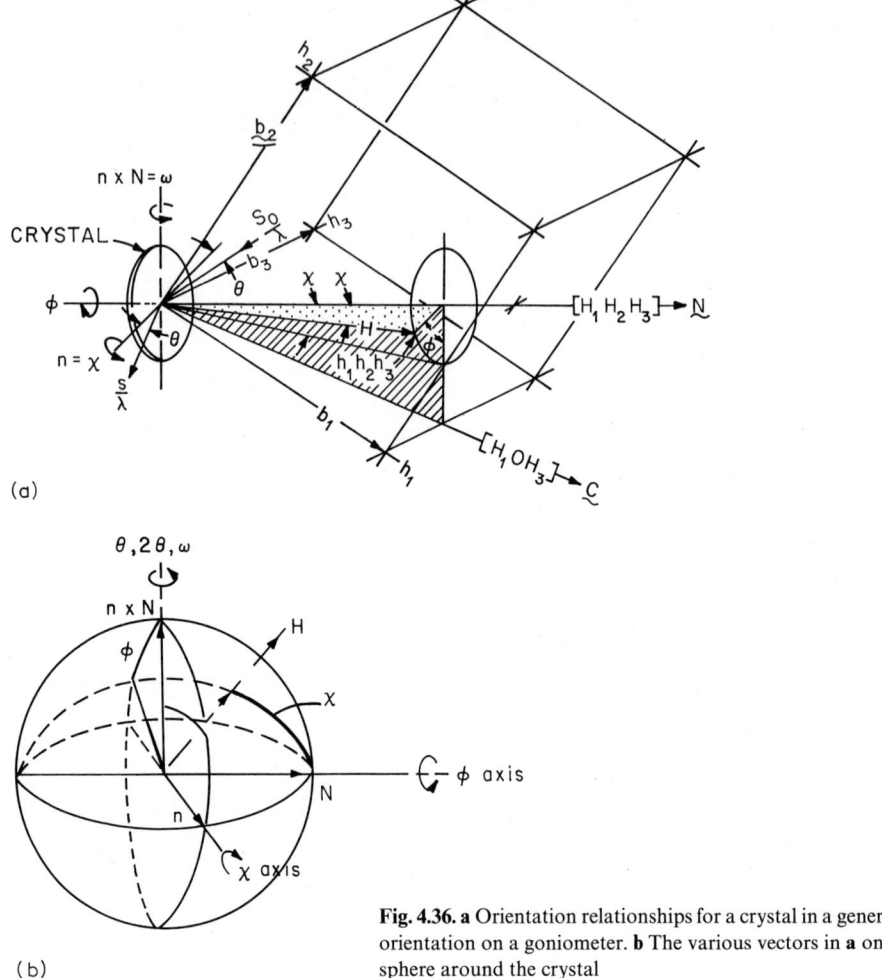

Fig. 4.36. a Orientation relationships for a crystal in a general orientation on a goniometer. **b** The various vectors in **a** on a sphere around the crystal

from a boule we have grown. We have two tasks in this case: 1) to determine the orientation, 2) to orient a given spot for diffraction. For simplicity in what follows we will consider a cubic crystal, although the procedures are quite general. Also, the procedure is suitable for locating *any* point in reciprocal space, not just a Bragg reflection.

Figures 4.36a and b illustrate the angular relationships. To establish the orientation, we need to know the vectors **n** in the surface and **N**, the surface normal. (These are unit vectors). Then we can put **H** in the diffraction position by placing it where **N** was originally. Suppose that the crystal is in the form of a flat disk. We can mount it on the goniometer so that it is parallel to the incident beam (by moving it to cut half the beam at 0°2θ, and being sure that on rocking it, the omega intensity decreases for both $+\omega$ and $-\omega$. [Before doing this, be sure that

4.6 The Counter–Diffractometer

the surface is at the intersection point of the χ, Φ, ω circles. A low-powered laser is helpful. The light's reflection from the sample surface is allowed to fall on a piece of paper, say on a wall some distance from the goniometer, and the position of the crystal on its mounting (a small goniometer as in Fig. 4.12a, with tilts and displacement motions x, y, z) is adjusted until the reflected spot does not precess with Φ. This also assures that the surface normal is along the Φ axis to start with.]

Then the χ tilt is obtained from:

$$\cos \chi = \frac{\mathbf{N} \cdot \mathbf{H}}{\mathbf{H}} = \frac{N_1 h_1 + N_2 h_2 + N_3 h_3}{(h_1^2 + h_2^2 + h_3^2)}. \tag{4.8}$$

With more than three reflections found by trial and error (an initial Laue pattern helps) the values of N_i can be obtained by least squares.

Now Φ is the angle between the plane defined by $\mathbf{N}$ and $\mathbf{H}$ and the plane defined by $\mathbf{N}$ and $\mathbf{n} \times \mathbf{N}$. Therefore,

$$\cos \Phi = \frac{(\mathbf{N} \times \mathbf{H}) \cdot \mathbf{n}}{|\mathbf{N} \times \mathbf{H}|}$$

$$= \begin{vmatrix} n_1 & n_2 & n_3 \\ N_1 & N_2 & N_3 \\ h_1 & h_2 & h_3 \end{vmatrix} / \{(N_2 h_3 - N_3 h_2)^2 + (N_3 h_1 - N_1 h_3)^2$$

$$+ (N_1 h_2 - N_2 h_1)^2\}^{1/2}. \tag{4.9}$$

Again, with more than three reflections the n_i are obtained. [A treatment of this process in terms of matrix algebra can be found in the International Tables Vol. IV, p. 276.] Once n and N are known, these same equations enable us to orient *any* position in reciprocal space for diffraction.

Conversely, with the aid of Fig. 4.36b we can obtain the vector $\mathbf{H}$ from the angles χ and Φ.

$$\mathbf{H} = \frac{2 \sin \theta}{\lambda} \{-\mathbf{n} \sin \chi \sin \Phi + \mathbf{N} \cos \chi + (\mathbf{n} \times \mathbf{N}) \sin \chi \cos \Phi\} \tag{4.10a}$$

or for individual components:

$$h_1 = \frac{2a \sin \theta}{\lambda} \{-n_1 \sin \chi \sin \Phi + N_1 \cos \chi + (n_2 N_3 - n_3 N_2) \sin \chi \cos \Phi\} \text{ etc.}, \tag{4.10b}$$

where θ is obtained from Bragg's law for the appropriate $h_1 h_2 h_3$ reflection. Thus by knowing the angles, we can find our coordinates in reciprocal space. Again this is perfectly general and we can explore regions between Bragg peaks as well as the peaks. By calculating $\partial h_i/\partial \Phi$, $\partial h_i/\partial \chi$ we can estimate the uncertainty in positions in reciprocal space due to uncertainties in Φ, χ.

4.7 The Powder Method

4.7.1 Recording the Diffraction Pattern

Single crystals are not always available for study and we must make do with powders or polycrystalline samples. As we shall see later, the techniques developed for studying such samples are so powerful that in some cases, for example, in the high precision determination of lattice constants, powders are generally the preferred sample material.

In this method, we have many small crystals or grains arranged randomly, and we use one wavelength, generally filtered radiation. We can think of this random orientation of crystals as one reciprocal lattice, with the Ewald sphere taking on all possible orientations around (000), as shown in Fig. 4.37a. The "limiting" sphere of radius $2/\lambda$ results. The number of possible points [neglecting points that do not diffract because $F(s) = 0$] is

$$\frac{(4/3)\pi(2/\lambda)^3}{\text{vol. primitive unit cell in rec. space}} = \frac{32\pi V_{u,c}(\text{real space})}{3\lambda^3}. \quad (4.11)$$

It is perhaps easier to understand the geometry of the diffraction pattern if we visualize this collection of crystal orientations by holding the Ewald sphere fixed and allowing each and every r^*_{hkl} to take on all possible orientations (Fig. 4.37b). The reciprocal space for such a polycrystalline aggregate is a series of concentric spheres of radii r^*_{hkl}. Each sphere intersects the Ewald sphere in a small circle leading

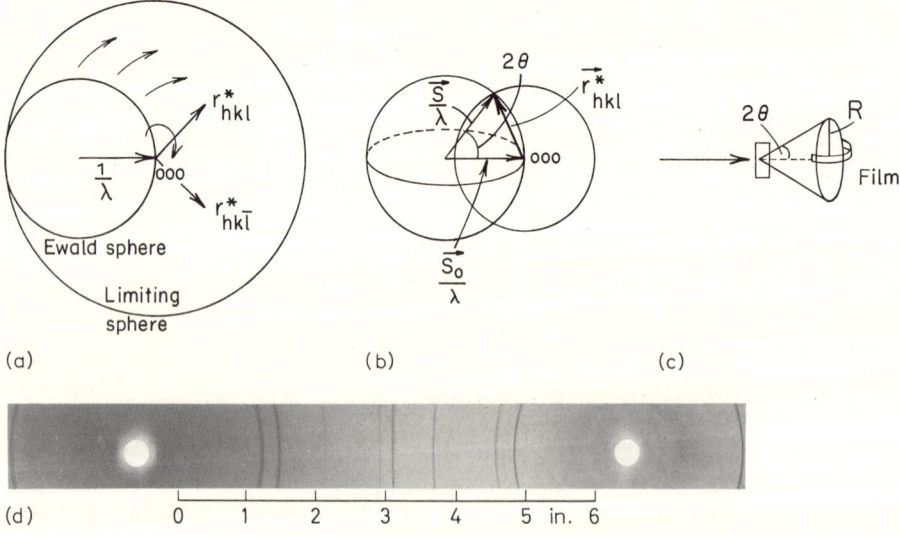

Fig. 4.37a–d. The powder pattern and reciprocal space. **a** The limiting sphere obtained by rotating the Ewald sphere through all orientations. **b** Intersection of the Ewald sphere with r^*_{hkl} sphere generated by random powder orientations. **c** A strip of film is wrapped around the powder sample, intersecting the diffracted cones of semi-apex angle, 2θ. **d** A typical powder pattern

to a cone of diffracted radiation with the direct beam as its axis and with 2θ as its semi-apex angle. Thus since the pattern depends only on the magnitude of $\mathbf{r}^*_{hkl}$, or on $1/d$, all planes of a form, such as $\{100\}$ [i.e., for a cube (001), (100), (010), ($\bar{1}$00), etc.], will diffract at the same angle and superimpose on the same cone. The multiplicity will, therefore, reduce the number of cones. For example, if we have a triclinic crystal, as diffraction introduces a symmetry center, every form has a multiplicity of 2. If the crystal has a unit cell volume of 50 Å^3 and we use Cu K$_\alpha$ radiation ($\lambda = 1.54$ Å), using Eq. (4.11) we find that there will be 230 such cones. With a diamond cubic structure of the same volume per cell there will be only 5.

As $\lambda = 2d \sin \theta$, we can write

$$\lambda^2/4d^2 = \sin^2 \theta \tag{4.12a}$$

or

$$1/d^2 = (4/\lambda^2) \sin^2 \theta = |\mathbf{r}^*_{hkl}|^2 . \tag{4.12b}$$

As the powder pattern depends only on the magnitude of $\mathbf{r}^*_{hkl}$, for a given λ it depends only on θ. *Planes with the same d-spacing, even if they have different indices, will also superimpose, as well as all those in one form.* For example, for a cubic crystal,

$$1/d^2 = (h^2 + k^2 + l^2)/a^2 = |\mathbf{r}^*_{hkl}|^2 . \tag{4.12c}$$

The $\{600\}$ and $\{442\}$ forms superimpose. This lack of "directionality" seriously reduces the information readily obtained from a powder pattern, although it is much easier to obtain a powder specimen. It is also easier to explore the information available in the pattern from a powder because the intensity is the same around the cone if the powder particles are randomly arranged. As shown in Fig. 4.37c, it is only necessary to wrap a narrow film around a powder rod or to move a counter around in one plane to record all the available information.[2] If a camera has a radius of 57.3 mm (so that around the circumference 1 mm = 1° 2θ), then the circular section of the cone at the film has a radius R equal to $57.3 \sin 2\theta$ mm. At $2\theta = 60°$, the radius (R) is about 50 mm and the circumference $2\pi R \cong 300$ mm. The film is normally about 25 mm wide, and therefore intercepts only about 1/12 of the cone on either side of the film. With a diffractometer, usually of radius 5.73 in. (so that 0.001 in. = 0.01° 2θ) and slit height of 0.5 in., only about 2% of the cone is intercepted. Weak peaks can be completely missed and this can cause additional trouble in ferreting out the structure.

It is important to be sure that the powder is randomly oriented, and to this end, brittle flat particles should never be pressed too hard in forming the sample. The powder should pass through at least a 250-mesh screen to assure a continuous cone. A hard mortar and pestle or the device a dentist uses to prepare amalgams can be used to make the powder. (This is merely a small tube with a pestle which is vibrated rapidly.) The specimen, a rod for a camera (or a flat specimen for the diffractometer),

---

[2] Occasionally a flat film perpendicular to the incident beam is used to get a series of circles; this is often referred to as a pinhole pattern.

is often rotated around the rod axis (or in the plane of the flat face) to bring other particles into position for diffraction and to reduce spottiness in the pattern (or lack of reproducibility with a counter), a result of too coarse a grain or particle size.

While the variation in intensity due to coarse grain size with sample position is bothersome in many applications of powder diffraction, it can be used to measure powder particle size, or grain size in a polycrystalline material. With modern equipment this can be done on-line during processing (say as a sheet of steel moves past an x-ray beam during rolling) and completely eliminates the time-consuming polishing and etching required for examination with an optical or electron microscope (Rinik et al. 1981.)

It is always desirable to choose a radiation that does not fluoresce the specimen as this raises the background. If a nickel specimen is used with copper radiation filtered at the x-ray tube, the fluorescence (caused by copper K_β getting through the nickel filter) can be quite high. A nickel tube is better. If enough filter material can be obtained to place over the film rather than just at the tube, a cobalt tube would be the best choice, because its filter Fe, would absorb most of the fluorescence from the nickel sample. (Only a small piece of filter will be needed in front of a counter in diffractometer work.)

Particular care should be taken in designing the equipment and slits to avoid having the beam strike other materials that might give a confounding pattern. The sample holder, if one is used and if it will be in the beam, should not give any pattern (i.e., it should be an amorphous material or one that gives no peaks at all, which is possible as we have seen with neutron scattering). In camera designs for film work, the slit is brought as close as possible to the specimen to avoid air scattering by the incident beam, which would fog the film. A beam trap is provided to prevent the direct beam from striking the film. The main features of the slits of a good camera are shown in Fig. 4.38 and most of the important points are available with commercial cameras. Note that both slits are tapered to minimize the lost angular region and to reduce scattering from inside the slits.

Air scattering can also be eliminated by evacuating the camera; this will also reduce the exposure time, as the absorption in the air of the direct beam and the diffracted peaks is reduced, but this is only necessary when the scattering factors are quite low. Another factor contributing to the background when using filtered radiation is the overlapping Laue patterns of all the grains due to the remnant white radiation which is always present even with properly selected filters. This can be eliminated by using a monochromator. If this is done, as proposed by Guinier (see

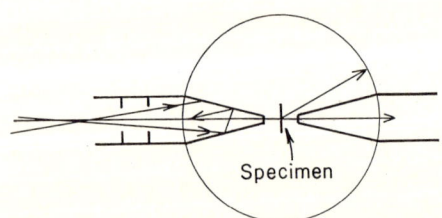

Fig. 4.38. Tapered slits for a powder camera

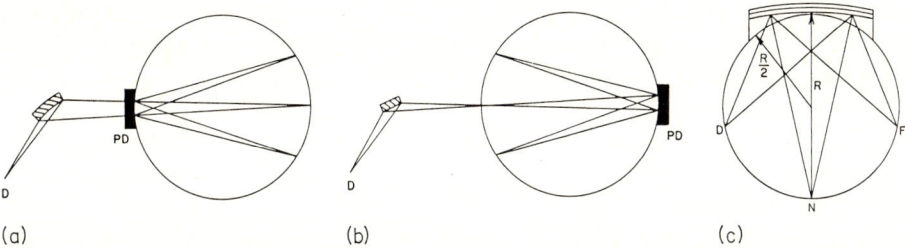

Fig. 4.39a–c. Focusing powder cameras equipped with monochromators. D is the radiation source, PD the powder specimen. **a** Forward reflection; **b** back reflection; **c** geometry for bending and polishing the monochromator. The crystal is ground to radius R, then bent so the planes have radius R and then the cut surface has radius $R/2$. (After Guinier, A., "La Radio Crystallographie," 3rd ed. Dunod, Paris, 1963. Published in English under the title, "X-ray Diffraction: In Crystals, Imperfect Crystals, and Amorphous Bodies". W.H. Freeman, 1963)

Fig. 4.39), exposure times are not increased too much by the double diffraction because the focusing allows us to use a broad region on the monochromator and a large powder specimen. As shown in Fig. 4.39c, the bend radius is R. If the crystal is ground to a radius R and then bent, the planes will be curved to a radius R while the ground surface will fit the focusing circle of radius $R/2$. Then normals to various parts of the crystal intersect at the point N, and a divergent beam from D is focused at F by the diffraction. The focusing action of such a monochromator produces sharp diffraction lines from the powder sample which can refocus the beam as shown in Figs. 4.39a and b. Also such a monochromator will help to reduce fluorescence from the sample caused by the white radiation. To further reduce the exposure time, the crystal may be bent in the plane perpendicular to the drawing, minimizing losses due to vertical divergence. With a doubly bent crystal, exposures are just about the same as for filtered radiation. [See Schwartz et al. (1963) for details of how to build and use such a monochromator.] In general, elastically bent quartz is used as a monochromator crystal when high resolution is important. Plastically bent LiF or pyrolytic graphite are used for maximum intensity, but resolution is poorer for these monochromators.

As we have mentioned previously, cameras are generally of a radius so that it is easy to read degrees 2θ in millimeters (57.3 mm, or for higher accuracy, 114.6 mm). However, this is only for rapid work. There are even transparent scales for a given camera radius and wavelength, which when placed over the film give a direct reading of d. Film shrinkage, however, must occur, and this must be corrected for in accurate work. Two marks placed a definite distance apart on the film during exposure, perhaps from the shadow of wires at fixed points around the circumference of the camera, will allow a correction for uniform film shrinkage. In some cameras there are many such marks around the film to allow for nonuniform shrinkage, or a scale can be printed on the film. (This shrinkage problem should be reduced as much as possible by slowly drilling and cutting holes in the film rather than punching it and wiping off any water patches on the film before drying.) In reading the position of a peak on a film with a viewer or light source behind the film, let the film sit on the

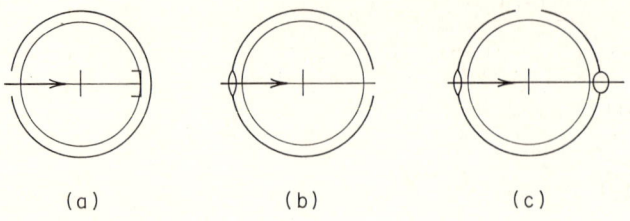

Fig. 4.40a–c. Three ways of placing film for a powder pattern. **a** Hole punched at exit slit; **b** hole punched at entrance slit; **c** holes punched at entrance and exit slits; this is known as the Straumanis mounting

viewer for 10 min or so before starting to read so that it will reach equilibrium with the warm viewer.

Film can be loaded in a powder camera in essentially three ways (Fig. 4.40). Which mounting in Fig. 4.40 is best? The most interesting is (c), as fiducial reference points (whose spacing must be measured) are not necessary to correct for shrinkage. One merely splits the distance between corresponding peaks on both sides of the holes at the entrance and at the exit of the beam, and the distance, center to center, must be 180° 2θ. If we differentiate Bragg's law we can see serious problems with the mounting in (a):

$$0 = 2d\cos\theta\Delta\theta + 2\Delta d \sin\theta,$$

$$\Delta\theta = -(\Delta d/d)\tan\theta; \quad \Delta d/d = -\Delta\theta/\tan\theta = \text{resolution}. \tag{4.13}$$

Thus the largest angular change for a small change in d is at high angles, and for some error $\Delta\theta$ in measurement, the error in d is smallest at large angles. But in the mounting in Fig. 4.40a this high-angle region is where the film shrinkage error would be most severe!

There are also different camera designs (Fig. 4.41). With both (b) and (c) large specimens can be used and even if they do not conform perfectly to the circle, there will be some focusing. Thus polished and etched metallographic sections may be used. Note that if Fig. 4.41a (with the Straumanis film mounting) or Fig. 4.41c is used, no reference marks are needed. Using the general features of each camera, $\Delta\theta$ is given for each camera design so that from Eq. (4.13) the resolution can be calculated. Notice that the focusing arrangements have twice the resolution of the Debye-Scherrer method, but cover only the high-angle regions.

This is a good place to examine *the features of a commercial diffractometer* when used to study diffraction from a polycrystalline specimen. One is shown in Fig. 4.33b. The tube is used in line focus with Soller slits (see Sect. 3.1) to minimize vertical divergence. Thus the geometry is cylindrical and a cross section of this geometry is the same for all heights of the beam. When used in the standard θ–2θ coupled mode, the flat powder specimen is made to turn at half the speed of the counter to maintain focusing, as shown in Fig. 4.42, because in order to maintain this focusing and get good resolution the normal to the specimen's surface must bisect the incoming and diffracted beams (as in Fig. 4.39c). If the specimen moves $\Delta\theta$, the angle of incidence changes by the same amount, thus the counter must move $2\Delta\theta$ to make the same angle as the incident beam does with the surface normal. Because of this motion,

4.7 The Powder Method

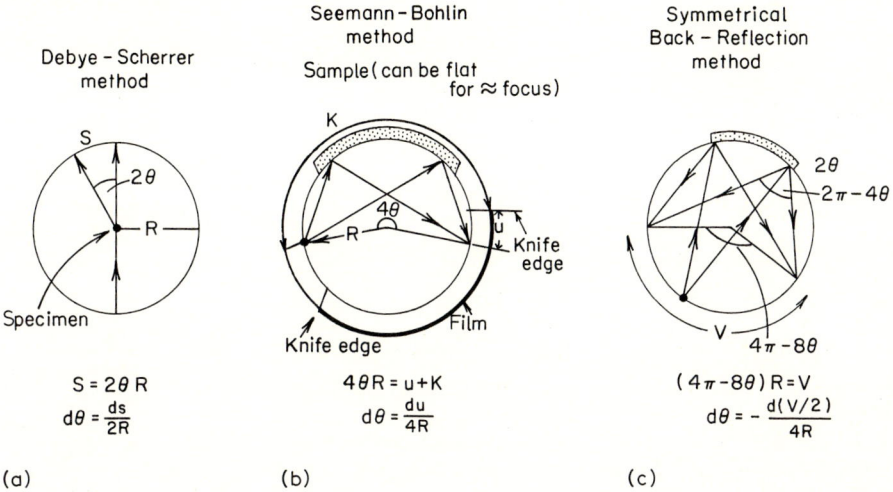

Fig. 4.41a–c. Various powder cameras and their resolution. a Debye-Scherrer camera; b, c focusing cameras. For b and c incident and diffracted beams cut equal areas on the circle so that if the incident beam is a line, the diffracted beam is a line

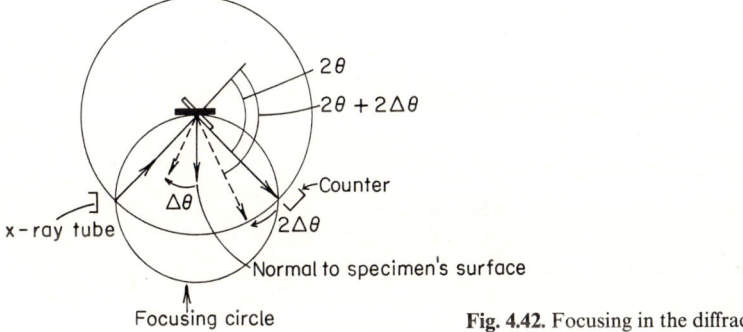

Fig. 4.42. Focusing in the diffractometer geometry

the radius of the focusing circle decreases with increasing θ. In Fig. 4.43 various ways are shown using a monochromator with a counter and maintaining the focusing. Commercial units have mechanical features to allow the operator to make sure that at 0° 2θ, the specimen surface is parallel to the beam's vertical direction and that the slits are parallel to the beam. The reader should realize that with a powder in a diffractometer, moving to maintain focusing, at the position for a 100 peak, one is measuring intensity from those grains with {100} parallel to the surface of the flat powder compact; at the 110 peak, only those grains with {110} parallel to the surface are being examined.

There are special advantages of doing powder diffraction at a synchrotron. Of course the high intensity means that much smaller samples can be employed. Furthermore, the high collimation and simple line shape (recall that a single wavelength is chosen) greatly improve resolution. A monochromator scheme often

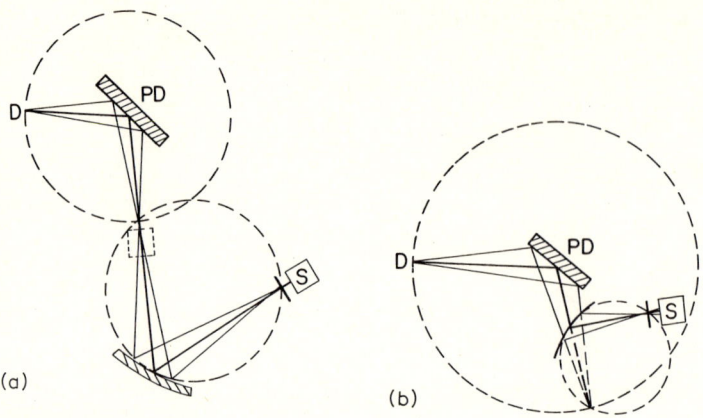

Fig. 4.43a,b. Monochromators with a diffractometer. The scattering geometry is independent of the beam direction, so one may interpret these diagrams with *D* as source of radiation, *PD* the powder specimen, and *S* the detector (diffracted beam monochromator), or with *S* as source and *D* as detector (incident beam monochromator). [After Lang, A.R., Rev. Sci. Instrum., 27, 17 (1956)]

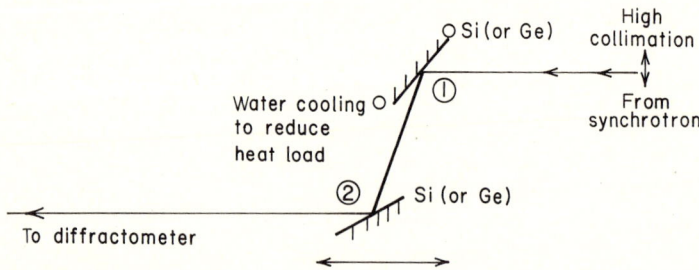

Fig. 4.44. Monochromator for a synchrotron. When θ (and hence λ) changes on the first crystal, the second crystal is translated laterally (as well as changing its θ) to diffract the beam from the first. The beam from the synchrotron spreads several milliradians normal to the page. The first crystal is water cooled as it takes the full heat load (of all λ's) from the synchrotron. Harmonics in the beam after the monochromator can be removed with a mirror set to reflect only the desired λ, or by using a solid state detector

employed for choosing any range of wavelengths, and yet maintaining the exit beam in a fixed position, is shown in Fig. 4.44. Because the crystals used are highly perfect, reflections are very narrow, of the order of 3–4 eV (see Chap. 8). [The divergence of the synchrotron ($1/\gamma$) is low but is still larger than that of the crystal reflection. With slits the divergence can be reduced further.] Recall that at a synchrotron the beam is highly collimated in the direction shown but several milliradians normal to the drawing, so a line focus is obtained. If necessary, the second crystal can be bent about a horizontal axis (an axis parallel to the page) to focus the beam in this long direction. (This bending degrades the energy resolution to about 5 eV.)

The tunable wavelength feature has many desirable advantages for powder diffraction. For example, consider taking the difference between two powder patterns, one with a wavelength close (1–4eV) to an absorption edge of an element

4.7 The Powder Method

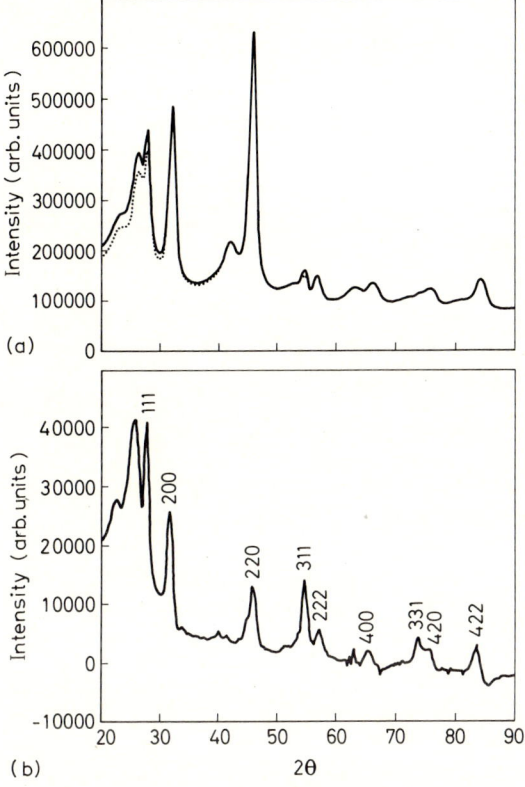

Fig. 4.45. a Diffraction patterns from a catalyst of 3.5 weight percent Pt deposited on Al_2O_3 powder. The *solid line* is for an incident beam of 11,508 eV (away from the Pt L_{III} edge at 11,568 eV) and the *dotted line* is at 11,558 eV, close to this edge. The peaks are all from the Al_2O_3 support; the Pt peaks are below these and are invisible. b The difference clearly reveals the fcc Pt peaks obscured in a. Note that both wavelengths employed were *below* the L_{III} edge, to eliminate fluorescence from the Pt. [From P. Georgopoulos and J.B. Cohen, J. of Catalysts 92, 211 (1985)]

present in one phase in a multiphase material, and one away (only 50–100 eV). The difference cancels everything but the pattern of this phase, since only its scattering factors change appreciably close to the absorption edge. An example of this technique is given in Fig. 4.45.

4.7.2 Indexing the Powder Pattern

Suppose we wish to determine if a sample is cubic, and if so whether it is simple cubic (all values of h, k, l possible), *body centered* ($h + k + l = 2n$), *or face centered* (h, k, l all even or all odd). The peaks will appear at increasing values of 2θ as hkl increases [Eqs. (4.12b) and (4.12c)]. For the three possible unit cells, the indices of the peaks are listed in Table 4.1. Note that even in the simple cubic, certain integral values of $h^2 + k^2 + l^2$ are missing. In both the simple and body-centered cubic cases, peaks are equally spaced in $\sin 2\theta$ range. Thus we can not readily discern whether the pattern is simple or body-centered cubic. The bcc structure identifies itself by the presence of a value $(r_x^{*2})/(r_1^{*2})$ of 7, which is not possible for the simple cubic. It is, therefore, important to use a radiation such that at least seven peaks appear on the pattern! The fcc structure also yields a characteristic set of ratios. Systematic

Table 4.1

Simple cubic			bcc			fcc		
hkl	$h^2 + k^2 + l^2$	$\sin^2\theta_x / \sin^2\theta_1$	hkl	$h^2 + k^2 + l^2$	$\sin^2\theta_x / \sin^2\theta_1$	hkl	$h^2 + k^2 + l^2$	$\sin^2\theta_x / \sin^2\theta_1$
100	1	1	110	2	1	111	3	1
110	2	2	200	4	2	200	4	4/3
111	3	3	211	6	3	220	8	8/3
200	4	4	220	8	4	311	11	11/3
210	5	5	310	10	5	222	12	4
211	6	6	222	12	6	400	16	16/3
220	8	8	321	14	7	331	19	19/3
300	9	9	400	16	8	420	20	20/3

absences of reflections ($F^2 = 0$) or very weak peaks which are not detected will make this analysis more-complex for cubic crystals with more than one atom per unit cell. One example is the diamond cubic structure (identical atoms at 000 and $\frac{1}{4}\frac{1}{4}\frac{1}{4}$ plus face-centering translation); in addition to the conditions for fcc given above, $F^2 = 0$ for $h + k + l = 4n + 2$ where n is an integer. Of those listed in Table 4.1, the 200, 222, and 420 would not be observed, and the ratios of $\sin^2\theta$ values (from Table 4.1) become $1 : \frac{8}{3} : \frac{11}{3} : \frac{16}{3} : \frac{19}{3}$, etc.

If the line positions cannot be indexed as cubic, we must look for a lower symmetry structure. For tetragonal or hexagonal (including rhombohedral indexed as hexagonal), one may use either graphical or analytical techniques. For orthorhombic and lower symmetries, analytical techniques are the only practical method of indexing powder patterns. With the common availability of digital computation, graphical techniques have been largely supplanted by analytical means. We will illustrate these with procedures applicable for tetragonal, hexagonal and orthorhombic systems based on the quadratic form of Bragg's law.

For tetragonal patterns, this form is:

$$\sin^2\theta = A_1(h^2 + k^2) + A_3 l^2, \tag{4.14a}$$

where $A_1 = (\lambda/2a_1)^2$ and $A_3 = (\lambda/2a_3)^2$. For hexagonal,

$$\sin^2\theta = A_1(h^2 + hk + k^2) + A_3 l^2, \tag{4.14b}$$

where $A_1 = \frac{1}{3}(\lambda/a_1)^2$ and $A_3 = (\lambda/2a_3)^2$. For orthorhombic,

$$\sin^2\theta = A_1 h^2 + A_2 k^2 + A_3 l^2, \tag{4.14c}$$

where $A_1 = (\lambda/2a_1)^2$, $A_2 = (\lambda/2a_2)^2$, and $A_3 = (\lambda/2a_3)^2$.

The analytical procedure depends on consideration of differences in $\sin^2\theta$ values for the various lines. Letting $q_i = \sin^2\theta_i$, we tabulate all possible values of $\Delta q = q_i - q_j$ for all pairs of lines i and j. We then search the table of Δq's for frequently occurring values, as these must represent frequently occurring linear combinations of the unknowns A_1, A_2, and A_3. For example, in the tetragonal system, from Eq. (4.14a),

Table 4.2. Indexing the powder pattern of KH_2PO_4

Line	$(\sin^2 \theta)_{obs}$	$(\sin^2 \theta)_{calc}$	hkl
1	0.0236	0.0229	101
2	0.0437	0.0428	200
3	0.0668	0.0657	211
4	0.0712	0.0700	112
5	0.0867	0.0856	220
6	0.0927	0.0915	202
7	0.1093	{0.1070, 0.1086}	{310, 301}
8	0.1212	0.1202	103
9	0.1512	0.1513	321
10	0.1559	0.1566	312
11	0.1633	0.1630	123
12	0.1708	0.1712	400
13	0.1945	{0.1941, 0.1946}	{411, 004}
14	0.2064	0.2057	303
15	0.2139	0.2141	420

$$\Delta q = A_1 \Delta(h^2 + k^2) + A_3(\Delta l^2).$$

For pairs of lines of common h, k, $\Delta q = A_3(\Delta l^2)$, where $\Delta l^2 = 1, 3, 4, 5$, etc. A pair of frequently occurring Δq's with ratio 3 : 1 is a good suspect for the determination of A_3.

As an illustrative example, consider the observed values, $\sin^2 \theta_{obs}$, listed in the second column of Table 4.2 (obtained using Cu K_α radiation, $\lambda = 1.542$ Å with powders of KH_2PO_4). All values of Δq are then tabulated and the three most frequently occurring values are found to be $\Delta q' = 0.043$ (appears eight times), $\Delta q'' = 0.049$ (five times), and $\Delta q''' = 0.086$ (six times) with no other value occurring more than four times. We note that $\Delta q''' = 2\Delta q'$ which makes this a good candidate for determining some unknown A_i. Further search of the table of Δq's reveals $\Delta q'''' = 0.127$ (four times) which is approximately 3 $\Delta q'$. If we assume $\Delta q' = A_1$ (corresponding to differences between $10l$ and $11l$ reflections), however, the value of a_1 computed is 3.72 Å which is physically too small for a large molecule such as KH_2PO_4. Thus $\Delta q'$ may be nA_1 where n is an integer. We try various values of n and assuming there are reflections of the type $hk0$, we compare calculated values of $\sin^2 \theta$ with the observed values. When we try $\Delta q' = 4A_1$ (corresponding to differences between $10l$ and $21l$ reflections), we are successful. Using this value for A_1, $\Delta q'/4 = 0.0107$, and trying various combinations of $hk0$, we compute directly the values of $\sin^2 \theta_{calc}$ listed in Table 4.2 for lines 2, 5, 7, 12, and 15. Proceeding further, we try $\Delta q'' = \Delta l^2 A_3$, and find that with $A_3 = \Delta q''/4 = 0.0122$, all the remaining lines can be indexed as shown in Table 4.2. There is ambiguity in indexing lines 7 and 13, and with the given data, either one or both of the hkl values given may be assumed present. These results correspond to $a_1 = 7.43$ Å and $a_3 = 6.97$ Å.

Note that permissible values of $h^2 + k^2$ for tetragonal are 1, 2, 4, 5, 8, etc. The analysis may sometimes be simplified by tabulating q_i, $q_i/2$, $q_i/4$, etc., and searching for quotients which are equal to one another or to one of the q_i, which will occur for $hk0$ lines. In this way A_1 is readily identified and A_3 may be determined as described above. If no fit is found, one tries the same procedure for the hexagonal system remembering that in this case $h^2 + hk + k^2$ takes on values of 1, 3, 4, 7, 9, etc.

If no fit to tetragonal is obtained from the frequently observed Δq's, one tries hexagonal [Eq. (4.14b)] or orthorhombic [Eq. (4.14c)]. The systematic computation of Δq's, evaluation of frequency of observation, and checking of various possible solutions lend themselves easily to programming for high-speed digital computers and several procedures for doing this have been developed. The need for such computer techniques is clearly imperative for structures of still lower symmetry (four unknowns in monoclinic and six in triclinic). A comprehensive discussion of such computer methods and their application to low-symmetry structures may be found in Klug and Alexander (1974) and a computer program for that purpose is discussed by Evans et al. (1963). It should be apparent at this point, however, that all reasonable efforts should be made to obtain a single crystal and use one of the procedures outlined in earlier sections of this chapter before resorting to the complexity of determining low-symmetry Bravais lattices from powder patterns.

There are several *important experimental details to note.* In a case where there are some lines which do not seem to fit, they may be from K_β radiation as filtered radiation is generally used and a weak K_β intensity is always present. From Bragg's law we can see that for a fixed value of d:

$$\sin^2 \theta_\beta / \sin^2 \theta_\alpha = \lambda^2_{K_\beta} / \lambda^2_{K_\alpha}. \tag{4.15}$$

That is, if we multiply the $\sin^2 \theta$ value for the suspicious line by $\lambda^2_{K_\alpha}/\lambda^2_{K_\beta}$, if it is a K_β this will bring the $\sin^2 \theta$ value into coincidence with another line. A final check can be made by reexposing the film (or rerecording with a counter) without the filter. The intensity of the suspected K_β line will increase considerably more than K_α lines. A similar procedure can be utilized for detecting lines from impurities on the x-ray tube target. These are generally W_L lines due to tungsten evaporated from the tube filament. Also, if the collimating slits of the camera are damaged, there may also be extra peaks due to the beam striking these slits. It is good practice to run a known sample occasionally to be aware of such spurious sources of radiation from the tube and camera.

4.7.3 Applications of Powder Patterns

There are a great many uses for powder patterns, despite the fact that unless the structure is reasonably simple it will be difficult to determine with this technique. Several applications depend on accurate determination of lattice parameters:

1. Determination of coefficient of thermal expansion.
2. As the lattice parameter of a solid solution generally varies with solute additions,

4.7 The Powder Method

chemical analysis can be carried out from a master curve. For example, with a diffusion couple, layers may be taken off and the parameter determined to obtain the concentration profile.

3. In contrast to a solid solution, in a region of a phase diagram involving two solid phases the composition of both phases does not vary, only their amount. Thus by determining the parameter of one phase in a series of alloys of varying composition, it is easy to locate the solvus line. This method is in fact the most useful for phase boundaries involving solid state reactions because they are sluggish, and heat evolution or length changes are not often satisfactory for accurately determining the alloy composition with the first small amount of a second phase. Quantitative metallography is also often not sufficiently accurate.

4. Defect structures (vacant sites, interstitials, etc.) can be detected by comparing the macroscopic density and that which is calculated from the lattice parameters and chemical composition.

5. Accurate measurements of unit cell parameters can be made for any crystal structure, once the lattice is known. What can we expect in terms of precision lattice constant determinations? The following systematic errors are possible:

a) If film is used, uniform film shrinkage (nonuniform shrinkage would be a random error).
b) Effective center of specimen is not the mechanical center due to absorption of the beam in the specimen and the limit of accuracy in alignment.
c) Radius of camera or location of reference marks are not known accurately.
d) Incorrect location of the peak (the systematic error, not the random portion).
e) Temperature fluctuations.

We have already discussed (a). Both (a) and (c) can be eliminated by using the Straumanis film mounting. As shown in Fig. 4.46, (b) causes a peak displacement:

$$\frac{\sin \Delta 2\theta}{\delta/\sin \theta} = \frac{\sin(180° - 2\theta)}{R},$$

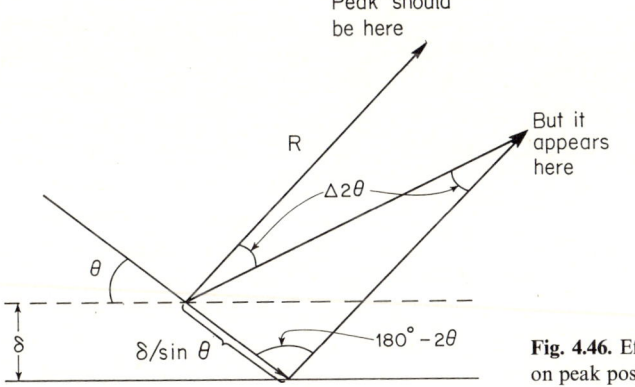

Fig. 4.46. Effect of sample displacement δ on peak position $\Delta 2\theta$ for a diffractometer

where R is specimen to film distance. Therefore, the angular error

$$\Delta 2\theta = -2\delta \cos \theta / R,$$

produces an error in lattice constant of

$$\frac{\Delta a}{a} = \frac{\Delta d}{d} = -\frac{1}{2\tan\theta}\Delta 2\theta = \frac{\cos\theta}{\sin\theta}\frac{2\delta\cos\theta}{2R} = \frac{K\cos^2\theta}{\sin\theta}. \tag{4.16}$$

Thus a lattice parameter calculated from a low-angle line will have a greater error than one from a high-angle line. By extrapolating parameters from lines at various angles versus $\cos^2\theta/\sin\theta$, this error can be eliminated. In general, this is the largest error with cameras and a diffractometer and the same extrapolation function can be used for both. A similar extrapolation function, known as the Nelson-Riley extrapolation function, includes the effects of vertical divergence. This function is described in the International Tables, Volume II, p. 216.

Of particular difficulty is error (d). One of the principal advantages of the diffractometer over a film is that a trace of the intensity distribution of a peak is given, whereas with a film one must estimate the center of a dark band. The intensity varies with angle across a peak because of geometric factors in the total intensity expression, as we shall see, and because the intrinsic wavelength dispersion of the emitted x-radiation is not symmetric. More important, the intensity distribution from the x-ray tube may not be uniform, and as a different section of the beam is seen at different angles with any of the methods, this will affect the shape. Furthermore, Soller slits, used on a diffractometer to reduce vertical beam divergence, cause asymmetry. These effects can be corrected for by using the center of gravity of a peak which can be determined from the trace on a diffractometer. The effects of various factors on the center of gravity can be treated analytically. However, the tails of a peak carry a heavy weight in such a procedure; they are difficult to determine and it is not at all certain that the peak maximum would not be the easiest and most useful quantity. It is hardly affected by dispersion and geometric factors, and one can choose a tube with a uniform target distribution.

Temperature is not much of a problem these days. With film methods all peaks are being recorded at the same time. They will be slightly broadened by any temperature fluctuations but it is only necessary if one is reading the center of a line to control the mean temperature during an exposure. With a diffractometer, the room temperature must be carefully controlled or read at each peak. If there is only a 3° drift in room temperature during a recording of a material with a coefficient of expansion of say $10^{-5}/°C$, the error in lattice parameter for two peaks, one at high angles and one at low angles, will be 3×10^{-5} or $\approx 0.003\%$. Modern air conditioning, however, can maintain much closer temperatures than this for the few hours needed.

With the aid of a digital computer, the powder diffraction pattern can be analyzed even when there is significant overlap of adjacent lines. This is a common problem for large unit cells of low symmetry. The data from the diffractometer are fed into the computer as intensity versus 2θ and a theoretical curve is fit, establishing the line positions and intensities. In this analysis one may use either analytically or

4.7 The Powder Method

experimentally determined descriptions of the peak shapes to determine the theoretical curve. This analysis is described in more detail in Sect. 6.9.

With care and extrapolation, it is possible to have a precision in a unit cell parameter of $\pm 10^{-3}\%$. Routinely, with little care, $10^{-2}\%$ is possible. There are few experimental measurements capable of this precision with as little effort! Absolute accuracy is another question. A comparison of parameters determined on samples of a given material in laboratories throughout the world using all the same constants (wavelength, coefficient of expansion to correct data to one temperature, etc.) showed that although the precision was high in each laboratory, the agreement in parameter was only within $\pm 10^{-2}\%$. (Parrish 1960). The wavelength used should always be reported, since they are not yet known to an absolute accuracy of better than $4 \times 10^{-3}\%$ as indicated in Sect. 4.1.

We close this subsection with a brief reference to three other common applications of powder patterns: 6) identificaton of unknown phases present in a multiphase specimen; 7) characterization of preferred orientation or "texture" in polycrystalline samples; 8) residual stresses.

6. One of the most important uses of powder patterns is phase identification in conjunction with the card systems kept up by the Joint Committee on Powder Diffraction Standards (JCPDS, 1601 Park Lane, Swarthmore, Pennsylvania). This is a collection of data on the d-spacings and relative intensities of lines on powder patterns from a great variety of substances indexed in various ways.

First, there is a book (or deck) of cards, one for each substance with the d-spacings and relative intensities of many peaks, and whatever is known about the space group, density, etc. Then there is an index to the cards called the Davey index, which lists all the substances alphabetically, and for each gives the location of the card and the three strongest (most intense) lines. The Hanawalt index is listed according to the d-spacings of the three strongest peaks from which one can also locate the card. If two substances have the strongest line at the same d-spacing, the two are placed in order of the d value of their second strongest peak, etc. Each substance is listed three times; if the three strongest peaks have d values d_1, d_2, d_3 (with d_1 the strongest), the substance can be found under listing d_2, d_1, d_3, or d_3, d_1, d_2. This helps if the powder is not random. Then there is the Fink index, which uses the eight strongest lines and the permutations of the six strongest. This is needed because in electron diffraction patterns the intensity variation is different and not as strongly varying as in x-ray patterns. The latest addition to these systems is a computer tape search program. Over 40,000 substances are covered. If the unknown contains more than one substance (looking at the powder in the microscope will help to establish this), then one of the strongest lines may be an overlap of two lines from each of two substances, or two of the lines may be from one and the third from the other. In the latter case, we might look for one card based on two of the lines and one based on the third plus two of the other lines in the pattern. In the former case we might look for a card based on two of the three stronger lines and another card which contains two of the measured peaks other than the first three, but which also has a smaller peak at the position of the supposed overlap. The various possible permutations involved in such a procedure must be tried, and chemical analysis can be helpful at this point in eliminating certain cases. (If there are three phases, without other information, the job is almost impossible.)

7. The mechanical or magnetic properties of solid materials that are polycrystalline often depend on the texture or relative alignment of the grains. For example, the good magnetic properties of the iron-silicon sheet used in transformers depend on a high degree of alignment of grains, with [[100]] in the sheet. This is the direction of easy magnetization in a single crystal of iron-silicon, and hysteresis losses are reduced if the alignment is good. If there is texture in a specimen, then there will not be uniform intensity around the sphere of radius r^*_{hkl}. Consider the 110 sphere in reciprocal space for a cubic material in wire form, with [[100]] directions aligned nearly parallel to the axis of the wire in most of the grains. As shown in Fig. 4.47a, the $r^*_{\bar{1}10}$ values are confined to being perpendicular to this [[100]] direction, or at 45° to it, and the intersections result in the cone in Fig. 4.47b. This whole cone can be recorded in a pinhole pattern using a flat film and monochromatic radiation. A specimen may also be placed in a diffractometer with the counter placed at the 2θ position, say, for a 110 peak. The specimen is then tilted, say, χ in Fig. 4.17c, and then the intensity is recorded as Φ varies. These are the same angles we discussed in Sect. 4.6 for the orientation of a single crystal, and an automated device similar to the goniometers employed with crystals may be used to obtain this data.

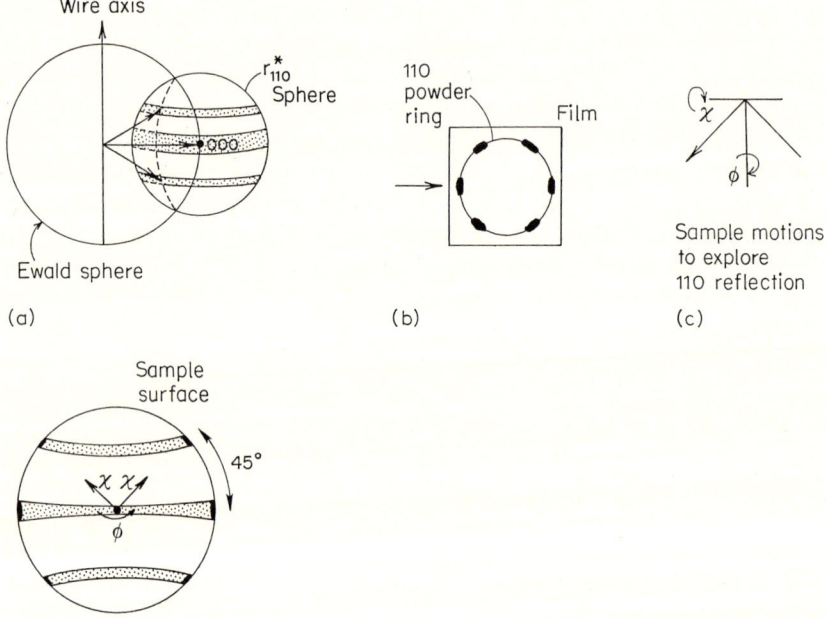

Fig. 4.47a–d. Examining texture in a polycrystalline sample. The example shown is for predominantly [[100]] directions along a wire axis of a cubic material and its influence on the 110 cone of reflection. [After Azaroff, L.V., Norelco Reporter 8, No. 2,33 (1961)]

4.7 The Powder Method

Contour lines can then be drawn at values of equal intensity in a stereographic projection as in Fig. 4.47d. (Transmission through the specimen rather than reflection is necessary to record the outer edges of the projection.) The projection represents the distribution of $\mathbf{r}^*_{\bar{1}10}$'s around the $\mathbf{r}^*_{\bar{1}10}$ sphere.

Such figures are useful representations, but they are incomplete. After specifying the likelihood of the orientation of $\mathbf{r}^*$ at some position in space, there is still the uncertainty of the orientation around this direction. Complete "orientation distribution functions" can be obtained from data on the intensity of one or several peaks as a function of χ, Φ and ω. By fitting such data in terms of surface spherical harmonics, various sections through the distribution can be calculated, and also this mathematical representation permits the calculation of the proper averages of various properties of a material with texture or preferred orientation (see Bunge 1982).

8. In many materials — metals, alloys, ceramics and polymers — there can be "locked-in" or residual elastic stresses, as a result of processing. For example, these stresses can arise during a heat treatment due to differential cooling of one part of a body with respect to another. (During a quench from high temperature, the surface cools more rapidly than the interior.) When a weld is made, the liquid pool contracts as it solidifies but is resisted by the colder solid metal around it. In both cases the two regions apply stresses on each other. The process may be visualized as in Fig. 4.48. Such stresses are important in material behavior. If a part will fail at a certain tensile stress, a tensile residual stress will reduce the load that is possible in service. Conversely, a compressive residual stress may prolong life. It is not surprising that engineers wish to measure these stresses after manufacturing, or during use. The only nondestructive method for doing this involves diffraction, and portable units have been built for rapid measurements in the field.

The principle of residual stress measurement is quite simple, and easily understood. In Fig. 4.49 a polycrystalline specimen with a compressive stress parallel to the surface is placed in a diffractometer, and the detector positioned near a particular reflection. Only those grains diffract that have the planes for this reflection nearly parallel to the surface. With a compressive stress parallel to the surface, these planes have a larger interplanar spacing than without stress, and (through Bragg's

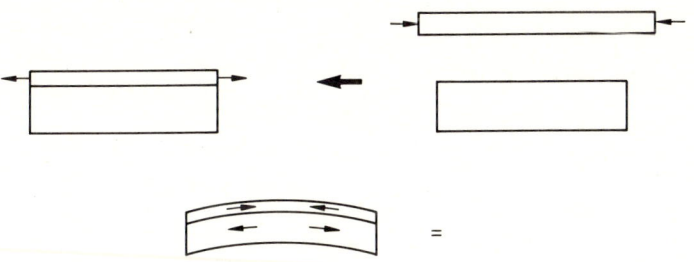

Fig. 4.48. The origin of residual stresses. Two parts have different dimensions. Both are deformed to fit each other, and joined, and the stress is released. The originally longer part tries to elongate, putting a tensile load on the other part. The thicker section holds the thinner one in place, applying compressive residual stress to it

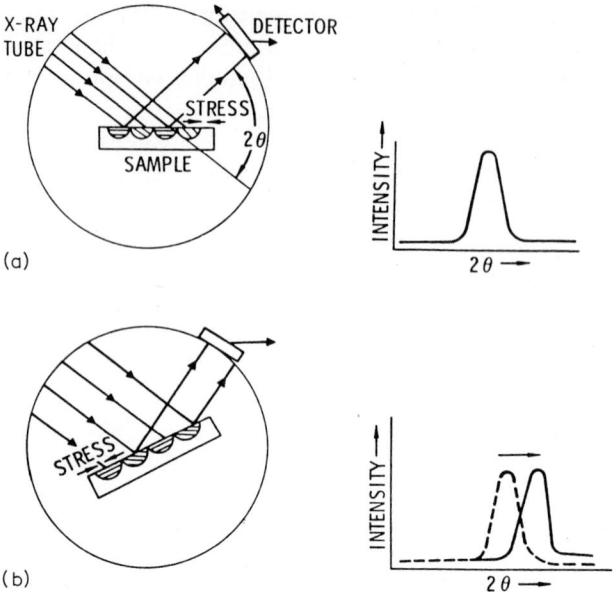

Fig. 4.49. a Schematic of a diffractometer. The incident beam diffracts x-rays of wavelength λ from planes that satisfy Bragg's law in crystals with these planes parallel to the sample's surface. If the surface is in compression, because of Poisson's ratio these planes are further apart than in the stress-free state. The d spacing is obtained from the peak in intensity vs. scattering angle 2θ and Bragg's law. **b** After the specimen is tilted, diffraction occurs from other grains, but from the same planes, and these are more nearly perpendicular to the stress. These planes are less separated than in **a**. Therefore, the peak occurs at higher angles of 2θ

law) shift to a lower 2θ angle. If the beam is tilted to the specimen, (or vice versa) diffraction occurs for different grains which are more nearly perpendicular to the stress. Their spacing is less than in the first case, and the peak is at higher 2θ. In effect, the d spacing is an internal strain guage. Through these peak shifts, strains ($\Delta d/d$) are sampled. As seen from differentiation of Bragg's law,

$$\Delta 2\theta = -2\left(\frac{\Delta d}{d}\right)\tan\theta. \tag{4.17}$$

Once obtained, the residual strains may be converted to residual stresses using the elastic constants of the material. Note that a high angle diffraction peak is preferred; due to the tangent function, a given $\Delta d/d$ produces a large peak shift ($\Delta 2\theta$) at large 2θ. Typically $\Delta d/d$ is 10^{-4}.

To be more precise consider the axial system in Fig. 4.50. The P_i axes describe the sample, the L_i the stress measuring system. We measure the spacing of planes perpendicular to L_3, at some tilt, ψ, of the specimen from the normal parafocussing arrangement. Strains in the L_i coordinate system will be primed, those in P_i, unprimed. Then a simple change in orientation of the strains from P_i to L_i, and inserting the tensor relationship between stress (σ_{ij}) and strain (ε_{ij}) leads to:

4.7 The Powder Method

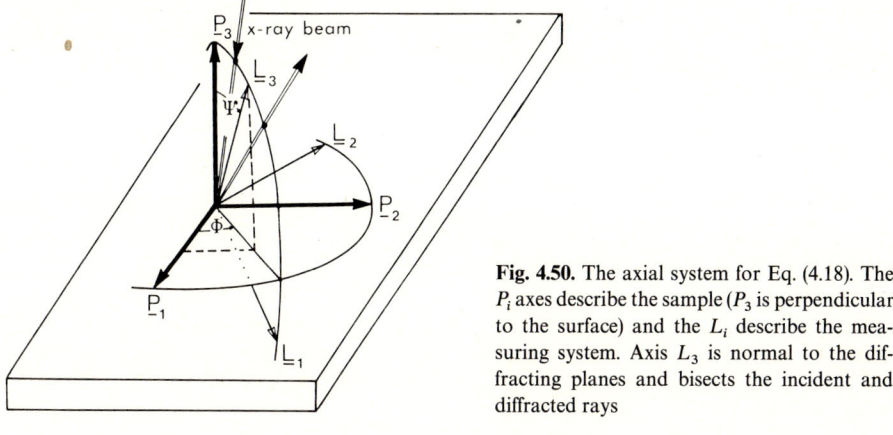

Fig. 4.50. The axial system for Eq. (4.18). The P_i axes describe the sample (P_3 is perpendicular to the surface) and the L_i describe the measuring system. Axis L_3 is normal to the diffracting planes and bisects the incident and diffracted rays

$$\varepsilon'_{33} \equiv \frac{d_{\Phi,\psi} - d_0}{d_0} = (S_2/2)\sigma_\Phi \sin^2\psi + (S_2/2)\sigma_{33} + S_1[\sigma_{11} + \sigma_{22} + \sigma_{33}]$$

$$+ (S_2/2)[\sigma_{13}\cos\Phi + \sigma_{23}\sin\Phi]\sin 2\psi . \tag{4.18}$$

Here the S_i are functions of the elastic constants ($S_2/2 = (1+v)/E$, $S_1 = -v/E$ for an isotropic solid, where v is Poisson's Ratio and E is Young's Modulus) and it is to be understood that the stresses and strains are averaged over the penetration depth of the beam.

In many cases the stresses in the third direction σ_{i3} are zero, or small as in differential cooling. Although the general case can be solved, for the sake of examining the principles, we assume this to be the case. Then:

$$\varepsilon'_{33} = (S_2/2)\sigma_\Phi \sin^2\psi - S_1[\sigma_{11} + \sigma_{22}] . \tag{4.19}$$

Furthermore,

$$\frac{d_{\Phi,\psi=0} - d_0}{d_0} = +S_1[\sigma_{11} + \sigma_{22}] . \tag{4.20}$$

Thus,

$$\frac{d_{\Phi,\psi} - d_{\Phi,\psi=0}}{d_0} = (S_2/2)\sigma_\Phi \sin^2\psi . \tag{4.21}$$

The term d_0 in the denominator of the left hand side can be replaced by $d_{\Phi,\psi=0}$, as the changes in d are only in the fourth decimal place. Therefore we do not need to know d_0, and the slope of a plot of $d_{\Phi,\psi}$ vs. $\sin^2\psi$ is $(S_2/2)\sigma_\Phi$ as shown in Fig. 4.51. Knowing S_2, one obtains the stress σ_Φ in the Φ direction on the sample surface. In fact, through Eq. (4.17), $\Delta d/d$ in Eq. (4.21) can be replaced by $-\Delta 2\theta/(2\tan\theta)$ to yield:

$$\Delta 2\theta = S_2 \sigma_\Phi \tan\theta \sin^2\psi . \tag{4.22}$$

A plot of $\Delta 2\theta$ vs. $\sin^2\psi$ also gives σ_Φ. At $\approx 156°$ 2θ for steel, a stress of 700 MPa gives a peak shift of $2°$ 2θ at $\psi = 60°$, compared to $\psi = 0°$. [For further details, see Noyan and Cohen (1987).]

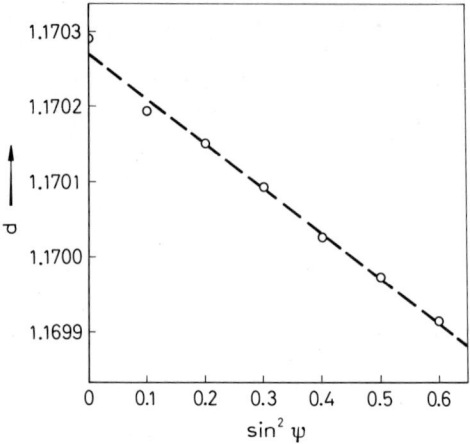

Fig. 4.51. Interplanar spacing, "d" vs. $\sin^2 \psi$ for the bcc (ferrite) phase in AISI 1010 steel, deformed in tension. 211 peak, CrK_α radiation. From the slope the stress was determined to be compressive and 107 MPa. (R.H. Marion, Ph. D. Thesis, Northwestern University, Jun, 1973)

4.7.4 Powder Diffraction Using Energy Analysis

Recently a new powder diffraction technique has been developed and used with neutron and x-radiation sources. The technique is based on the ability to determine the energy of the scattered radiation and will be termed *energy analysis*. We shall describe it in some detail for the neutron experiment and comment briefly on the x-ray technique.

We have seen in Sect. 3.2 that the neutrons available from a nuclear reactor consist of a white spectrum with wavelengths ranging from less than 0.5 Å to greater than 5 Å. One usually selects only a small fraction ($\approx 1\%$) of these with a monochromator and carries out standard diffractometer type experiments. If the apparatus described schematically in Fig. 4.52 is used, all the available wavelengths may be utilized. A chopper emits a burst of neutrons at regularly spaced intervals. These neutrons strike the powder sample and are scattered. At any Bragg angle $2\theta_0$, a detector "sees" a pattern consisting of Bragg peaks (elastic scattering from powder crystals correctly oriented to satisfy Bragg's law for that angle θ_0) and a white radiation background (due primarily to incoherent scattering from all the crystals in the sample). These neutrons arrive at the counter at different times t_i, as their time-of-flight over the fixed chopper-sample-detector distance L depends inversely on their velocity v_i, and from Eq. (3.9), $t_i/L = v_i = h/m\lambda_i$. The pulses from the detector are fed to a multichannel analyzer restarted as each burst of neutrons leaves the chopper, so that each channel corresponds to neutrons with a fixed time-of-flight (or a fixed wavelength). A typical pattern of number of neutrons counted vs. wavelength is shown in Fig. 4.53 for Ni powder. Since the entire diffraction pattern is viewed at a fixed scattering angle, this technique is also referred to as "non-dispersive". This technique is ideally suited for use at the new pulsed neutron sources (Sect. 3.2), since the neutrons are already arriving in periodically timed bursts.

This *energy analysis* (*time-of-flight* or TOF) *technique* has higher resolution than neutron diffractometers for large d-spacing reflections ($d \gtrsim 1.0$ Å) and is thus of

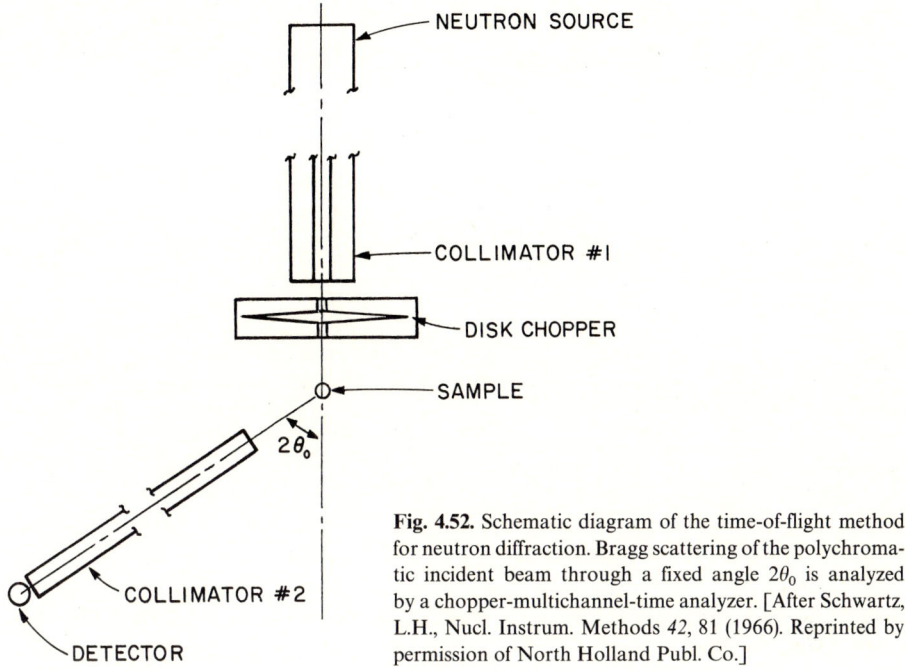

Fig. 4.52. Schematic diagram of the time-of-flight method for neutron diffraction. Bragg scattering of the polychromatic incident beam through a fixed angle $2\theta_0$ is analyzed by a chopper-multichannel-time analyzer. [After Schwartz, L.H., Nucl. Instrum. Methods *42*, 81 (1966). Reprinted by permission of North Holland Publ. Co.]

some interest in obtaining data from complex, large unit cell materials. To see that this is so, define the resolving power D for a particular reflection as

$$D_{hkl} = \Delta\lambda/\lambda ,\qquad(4.23a)$$

where $\Delta\lambda$ is the full width at half maximum of the reflection from the hkl planes at angle θ_0. Since θ is fixed, it follows that

$$D_{hkl} = \Delta d/d_{hkl} ,\qquad(4.23b)$$

where $d_{hkl} = \lambda_0/2\sin\theta_0$. In the TOF method the uncertainty in d-spacing arises from uncertainties in the measurement of neutron flight time D^t_{hkl}, and in the measurement of angle D^θ_{hkl}. Now $\lambda = Ct$, where t is the time-of-flight of the neutron of wavelength λ and $C = h/mL$ is a constant of the instrument; h is Planck's constant, m the neutron mass, and L the neutron flight path. Then

$$D^t_{hkl} = (\Delta\lambda/\lambda_0) = (C\Delta t/2\sin\theta_0)(1/d_{hkl}) .\qquad(4.24a)$$

The angular uncertainty can be obtained by differentiating Bragg's law:

$$|D^\theta_{hkl}| = |(\Delta d/d)_\theta| = \cot\theta_0\, \Delta\theta \qquad(4.24b)$$

Assuming Gaussian forms and independence in the time and angular uncertainties, the resolving power for the time-of-flight method is given by

$$D^{\text{TOF}}_{hkl} = \left[\left(\frac{h\Delta t}{2mL\sin\theta_0}\frac{1}{d_{hkl}}\right)^2 + (\cot\theta_0\,\Delta\theta)^2\right]^{1/2} .\qquad(4.24c)$$

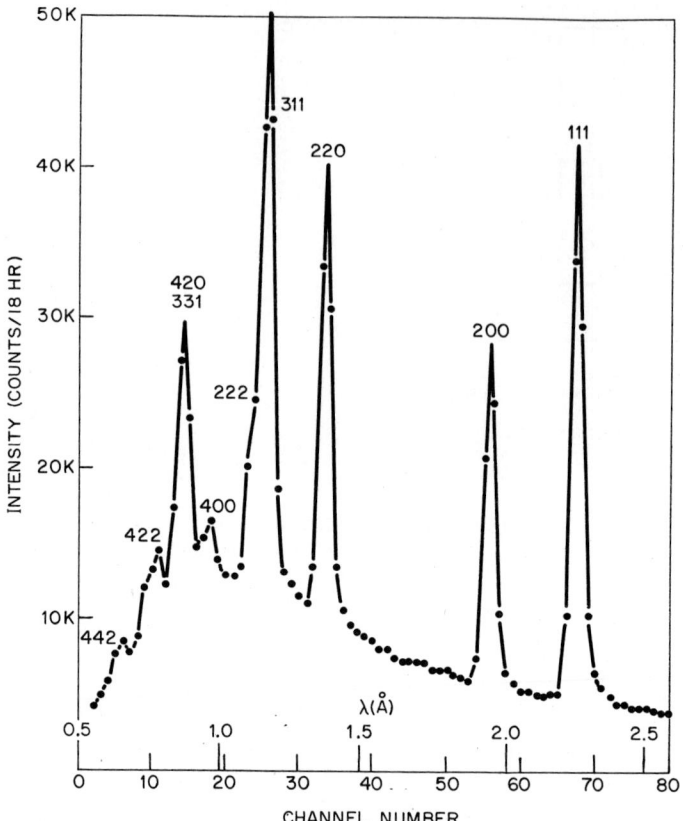

Fig. 4.53. Bragg scattering from Ni powder by the time-of-flight method. Intensity in counts/18h exposure vs. channel number (or wavelength). The broad peak in the background is due to incoherent scattering from the sample. [After Schwartz, L.H., Nucl. Instrum. Methods *42*, 81 (1966). Reprinted by permission of North Holland Publ. Co.]

It is significant to compare the resolving power of the TOF method with that of the conventional diffractometer (D). For the latter the uncertainty in angle is again given by Eq. (4.24b) where now $\theta = \theta_{hkl}$ is a variable and $\Delta\theta = \Delta\theta_s$ is the horizontal divergence of the spectrometer. The uncertainty in wavelength, due to finite divergence of the beam to the monochromator crystal is given by Eq. (4.24b) where $\theta = \theta_M$ is the monochromator Bragg angle and $\Delta\theta = \Delta\theta_M$ is the combined horizontal divergence of the primary collimator and mosaic spread of the monochromator. Combining these expressions gives

$$D_{hkl}^D = [(\cot\theta_{hkl}\,\Delta\theta_s)^2 + (\cot\theta_M\,\Delta\theta_M)^2]^{1/2} . \tag{4.25}$$

In practice L is made so large (several meters) and Δt so small (≈ 10 μs) that D_{hkl}^t is negligible compared to D_{hkl}^θ in Eq. (4.24c) for d-spacings greater than 1 Å. By contrast, for such large d-spacings the $\cot\theta_{hkl}$ term dominates Eq. (4.25) and makes resolution increasingly poorer for larger d-spacings (smaller θ_{hkl}). In typical experimental systems, $D^D_{hkl} > D^{TOF}_{hkl}$ for d-spacings greater than 0.5–1 Å.

The fixed angle detector of the TOF technique also means that radiation need enter the sample at the direct beam slit and exit at only one, or at most a few narrow slits. This has proved particularly useful in studies in which the sample is subjected to high pressures. In these experiments, sample containment vessels may be designed with a few narrow exit slits, but often not with a wide angular slit.

An analogous x-ray experiment has become possible with the advent of high-resolution solid state detectors of the Si(Li), Ge(Li), or intrinsic Ge type as discussed in Sect. 3.11. Again a white spectrum is used (e.g., a tungsten tube operating as for Laue patterns), a powder specimen, and a fixed angle detector are coupled to a multichannel analyzer used in pulse height mode. The energy of the pulse from the detector, proportional to the energy of the scattered radiation, determines into which channel of the multichannel analyzer a count is added. Thus one obtains a pattern of intensity vs. energy, or $1/\lambda$, which looks quite similar to that shown in Fig. 4.53 (see Fig. 3.48). As the entire pattern is recorded at the same time as in a film method, variations in sample environment from one diffraction peak to the next are eliminated. In both the x-ray and neutron energy analysis technique, one may couple the multichannel analyzer to a minicomputer and analyze the spectra as the counts are accumulated, allowing for repetitive measurements with changing sample environment (which might also be controlled by the minicomputer). A discussion of errors in this x-ray technique can be found in the paper by Sparks and Gedcke (1975). These authors demonstrate that standard deviations in d-spacings of 0.001 Å may be readily obtained.

4.8 Patterns with Very Short Wavelength Radiation

In electron diffraction, powder patterns are analyzed in much the same way as those just discussed, taking into account that the whole cone is intersected and that, as previously mentioned, it is necessary to correct for magnification and eccentricity with a known test specimen. The lines will be much closer together because of the small λ generally used.

As was pointed out in the introduction to this chapter, the sphere of reflection is quite large relative to the spacing of reciprocal space for electron diffraction. When using single crystal specimens, or looking at a single grain in an aperture selected area, because of mosaic or subgrains (regions of slight tilt) in the specimen, it is quite possible to record a whole plane of reciprocal space (the 0-layer, or plane passing through the origin of reciprocal space). For example, for a beam incident along an [001] in a cubic material with $a = 4.8$ Å we can ask what mosaic tilt will be required to record the 080 spot. Using Bragg's law in the form $\theta = \lambda/2d$, for $\lambda = 0.05$ Å^{-1}, the 080 will occur for $\theta = 2.6°$. Consider Fig. 4.54, a drawing of the 0kl plane with the incident beam along [001]. The mosaic spread of the spot must be 0.3a^* for this

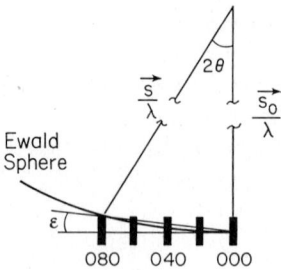

Fig. 4.54. Electron diffraction from the 0kl plane, incident beam along the [001] direction

080 spot to diffract. To see this, calculate ε from

$$\varepsilon/|r^*_{080}| = 2.6° \, 2\pi/360 \,.$$

From the usual mosaic spread, with a single pattern from within a grain in the sample, we are generally seeing a whole layer of the reciprocal lattice, especially with the short wavelengths at 50 kV and above. Because of the low angles of diffraction with these voltages in a transmission experiment, this layer is essentially perpendicular to the beam so that if the spots can be indexed, the cross product of the reciprocal lattice vectors for any two spots in the pattern gives the reciprocal lattice vector along the beam (and hence the $\{hkl\}$ of the surface of the specimen if it is not tilted). This orientation is only approximate, of course, due to the mosaic spread.

When using electron diffraction, we generally know the material's structure and want only to know the orientation. Indexing of the spots can be done by comparing the ratio of the distances from the center of the film (direct beam spot) to the diffraction spots with a table of such distances. Table 4.3 is such a table, prepared for the cubic system.

Recall (Fig. 3.20) that by adjusting the lenses, we can either see an image or the diffraction pattern.[3] If the intersection of two markings from known planes (such as slip traces) can be found, the orientation can be determined without the diffraction pattern. It is only necessary to ask what two lines on each of the planes can make the observed angle of intersection. The cross product of these lines gives the normal to the foil's surface. This procedure reduces the error in the method involving the diffraction pattern due to broad spots, and is useful if such markings can be found. If both edges of some known marking, such as a slip plane, can be seen in a transmission specimen, the thickness of the foil can be easily calculated once the orientation is known.

When the crystal is small, we can estimate its size in directions in the observed plane of the reciprocal lattice from the length of the spot in these various directions, because the breadth is proportional to $1/N$, where N is the number of unit cells in

---

[3] Because the lenses are electromagnetic, the image is rotated an amount which depends on the lenses and currents. This rotation is calibrated with a test specimen of known orientation or with a small crystal whose edges are of known indices. This is important in comparing the image and the diffraction pattern, and has been corrected for in the problems at the end of the chapter.

4.8 Patterns with Very Short Wavelength Radiation

Table 4.3. Ratios of reciprocal lattice vectors in the cubic system[a,b]

	100	110	111	200	210	211	220	221 300	310	311	222	320	321	400	322 410	330 411	331	420
100	1	1.414	1.732	2.00	2.24	2.45	2.83	3.00	3.16	3.32	3.46	3.61	3.74	4.00	4.12	4.24	4.36	4.47
110	0.707	1	1.23	1.41	1.58	1.73	2.00	2.12	2.24	2.34	2.46	2.55	2.64	2.83	2.92	3.00	3.09	3.16
111	0.577	0.817	1	**1.15**	1.29	1.41	**1.64**	1.73	1.83	**1.92**	**2.00**	2.08	2.16	**2.31**	2.38	2.45	**2.52**	**2.58**
200	0.500	0.707	**0.866**	1	1.12	1.23	**1.41**	1.50	1.58	**1.66**	**1.73**	1.80	1.87	**2.00**	2.06	2.12	2.18	2.24
210	0.448	0.634	0.776	0.898	1	1.10	1.27	1.34	1.41	1.49	1.55	1.62	1.68	1.79	1.85	1.90	1.96	2.00
211	0.408	0.577	0.707	0.816	0.912	1	1.15	1.22	1.29	1.35	1.41	1.47	1.53	1.63	1.68	1.73	1.78	1.83
220	0.345	0.500	**0.612**	**0.707**	0.789	0.868	**1**	1.06	1.12	**1.17**	**1.22**	1.28	1.32	**1.41**	1.46	1.50	**1.54**	**1.58**
221 300	0.333	0.472	0.578	0.667	0.745	0.817	0.914	1	1.05	1.11	1.16	1.20	1.25	1.33	1.37	1.41	1.46	1.49
310	0.316	0.448	0.548	0.634	0.707	0.775	0.895	0.949	1	1.05	1.10	1.14	1.19	1.27	1.30	1.34	1.38	1.41
311	0.302	0.426	**0.522**	**0.604**	0.674	0.739	**0.854**	0.905	0.955	**1**	**1.05**	1.09	1.13	**1.21**	1.24	1.28	**1.31**	**1.35**
222	0.289	0.408	**0.500**	**0.578**	0.645	0.707	**0.817**	0.866	0.913	**0.957**	**1**	1.04	1.08	**1.15**	1.19	1.22	**1.26**	**1.29**
320	0.277	0.392	0.480	0.554	0.619	0.679	0.784	0.831	0.877	0.918	0.960	1	1.04	1.11	1.14	1.18	1.21	1.24
321	0.267	0.378	0.463	0.535	0.597	0.655	0.757	0.802	0.845	0.887	0.927	0.965	1	1.07	1.10	1.13	1.17	1.20
400	0.250	0.354	**0.433**	**0.500**	0.559	0.612	**0.707**	0.750	0.791	**0.830**	**0.866**	0.902	0.936	**1**	1.03	1.06	**1.09**	**1.12**
322 410	0.242	0.343	0.420	0.485	0.542	0.594	0.686	0.727	0.767	0.803	0.840	0.875	0.907	0.970	1	1.02	1.06	1.08
330 411	0.236	0.334	0.409	0.472	0.527	0.578	0.667	0.707	0.747	0.782	0.817	0.852	0.883	0.944	0.973	1	1.03	1.06
331	0.229	0.324	**0.397**	**0.458**	0.512	0.561	**0.649**	0.687	0.725	**0.760**	**0.795**	0.827	0.858	**0.017**	0.945	0.971	**1**	**1.03**
420	0.224	0.316	**0.387**	**0.446**	0.500	0.548	**0.632**	0.670	0.707	**0.740**	**0.774**	0.807	0.837	**0.891**	0.922	0.947	**0.975**	**1**

[a] Prepared by O. Kimball.
[b] Bold face ratios are for an fcc lattice.

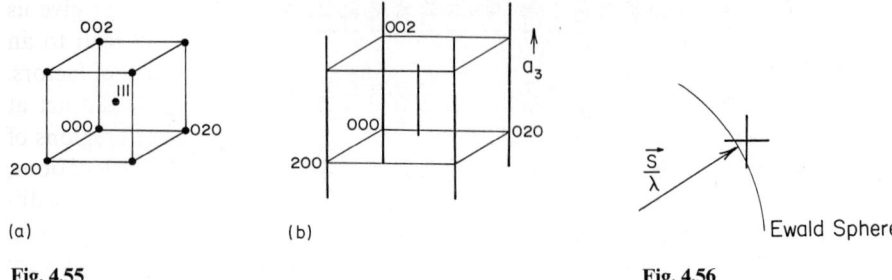

Fig. 4.55

Fig. 4.56

Fig. 4.55. a Reciprocal lattice of fcc material. **b** Same as in **a**, but for a sample one unit cell thick in the $\mathbf{a}_3$ direction

Fig. 4.56. Small crystals of regular dimensions sometimes produce multiple reflections in electron diffraction patterns

that direction. The crystal may also be quite thin in the direction of the beam, of the order of one unit cell. Although this is not generally true for metal crystals, it can be true for organic materials with large spacings. Then we have effectively two-dimensional diffraction, and the spots in reciprocal space spread in a direction perpendicular to the specimen or parallel to the beam. Extra reflections may appear on the 0-layer as in Fig. 4.55 (We can think of the size of a crystal as applying "pressure" to the spot due to the terms

$$(\sin^2 \pi \mathbf{S} \cdot N_i \mathbf{a}_i)/(\sin^2 \pi \mathbf{S} \cdot \mathbf{a}_i)$$

in the interference function. If the size goes down in any direction, the spot "oozes out" in that direction.) Finally, because of this spreading, small crystals of regular dimensions can produce multiple intersections with the Ewald sphere. An example of this is given in Fig. 4.56 for a small cubic crystal.

4.9 Measurement of Diffraction Intensity

4.9.1 Integrated Intensities from Crystals

When we considered diffraction from a periodic array of atoms in Sects. 2.3 and 2.4, we found that the information in the diffraction pattern could be divided into two types. The location of the Bragg peaks in reciprocal space may be used to identify the Bravais lattice, the unit cell size, and as we shall explore in detail in Chap. 5, some of the internal symmetry of the unit cell. To obtain detailed information about the atomic positions within the unit cell, however, one must examine the square of the structure factor, FF^*_{hkl}. To do this it is necessary to measure the intensity of the diffraction peak. The earlier sections of this chapter dealt with techniques for locating the positions of Bragg peaks. We now turn our attention to details of measuring their intensity and accurately determining FF^*_{hkl}.

4.9 Measurement of Diffraction Intensity

Although in principle the maximum intensity in a Bragg peak should give us the intensity information we desire to obtain, experimental problems lead to an alternative analysis. The breadth of the peak is dependent on a number of factors. If the incident beam is not collimated perfectly, different regions will diffract at slightly different angles. In nonperfect crystals (i.e., most crystals), small regions of the crystal are misoriented from one another by the distortions due to dislocations, subgrain boundaries, etc. These regions, called mosaic blocks, will diffract at different angles even if the beam is well collimated. The maximum intensity would correspond to the overlapping patterns from all regions. We would also need a very narrow slit to truly measure intensity. Instead, it is easier to work with a divergent beam and the total (integrated) intensity under a diffraction peak. We shall follow a treatment developed by Warren (1969).

Consider first *a small single crystal*, diffracting a roughly collimated monochromatic beam from the *hkl* planes, and rotating through a diffracting position with an angular velocity ω about an axis which is parallel to the (*hkl*) planes and perpendicular to $\mathbf{S}_0$ and $\mathbf{S}$. This situation is shown in Fig. 4.57.

At some receiving plane described by coordinates z', y', a film or counter is placed with a slit opened to get all the diffracted radiation during the rotation. During this rotation, all points of the crystal under the beam form the Bragg angle with each and every direction of the primary beam. (This rotation produces a spot on the equator of a rotation pattern.) We shall assume our crystal is a parallelepiped with N_1 repeating units along $\mathbf{a}_1$, N_2 along $\mathbf{a}_2$, and N_3 along $\mathbf{a}_3$.

Now the intensity (I) is the energy per unit area of film (A), per unit time (t), but since we collect all the scattered radiation, what we measure is energy E:

$$E = \int I \, dt \, dA . \tag{4.26a}$$

For computational convenience, instead of the crystal rotating, we shall consider $\mathbf{S}_0$ rotating. The time $\mathbf{S}_0$ spends between θ and $\theta + d\alpha$ is

$$dt = d\alpha/\omega .$$

The area on the measuring plane is

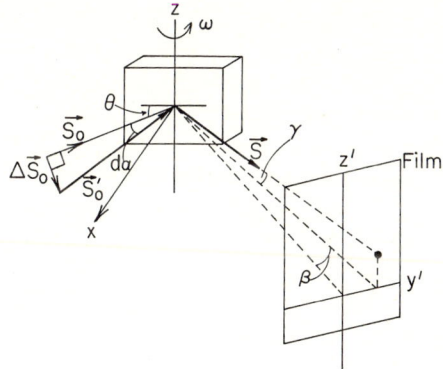

Fig. 4.57. Geometry for diffraction from a small rotating single crystal

$$dA = dy'\,dz' = R^2\,d\beta\,d\gamma\,,$$

where R is the distance from the crystal to the measuring plane. Thus,

$$E = \iiint I \frac{R^2}{\omega} d\alpha\,d\beta\,d\gamma\,. \tag{4.26b}$$

Our expression for I has a term depending on the scattering per scattering unit I_u. For example, for filtered x rays [from Eq. (3.15)],

$$I_u \equiv I_e = (I_0 e^4 \eta^2/m^2 c^4 R^2)(1 + \cos^2 2\theta)/2\,, \tag{4.27}$$

for one electron. I also includes FF^* and three terms like

$$\sin^2(\pi/\lambda)(\mathbf{S} - \mathbf{S}_0)\cdot N_1 \mathbf{a}_1 / \sin^2(\pi/\lambda)(\mathbf{S} - \mathbf{S}_0)\cdot \mathbf{a}_1 \tag{4.28a}$$

[see Eq. (2.18c)]. Now at any instant of time during the rotation, since we are considering directions of incident and diffracted beams which differ from Bragg's law directions, the terms of the type Eq. (4.28a) are evaluated for the vector $\mathbf{S}' - \mathbf{S}_0'$. Let $(\mathbf{S}' - \mathbf{S}_0') = \lambda \mathbf{r}_{hkl}^* + \Delta \mathbf{S}$, as shown in Fig. 4.58. Then

$$\frac{\sin^2(\pi/\lambda)(\mathbf{S}' - \mathbf{S}_0')\cdot N_1 \mathbf{a}_1}{\sin^2(\pi/\lambda)(\mathbf{S}' - \mathbf{S}_0')\cdot \mathbf{a}_1} = \frac{\sin^2(\pi/\lambda)\Delta \mathbf{S}\cdot N_1 \mathbf{a}_1}{\sin^2(\pi/\lambda)\Delta \mathbf{S}\cdot \mathbf{a}_1}\,. \tag{4.28b}$$

This is so because when $(\mathbf{S} - \mathbf{S}_0) = \lambda \mathbf{r}^*$, the phase angle between scattering elements changes by 2π and the magnitude of the $\sin^2 Nx/\sin^2 x$ term does not change. Thus the expression for energy becomes

$$E = I_e \frac{R^2}{\omega} FF^* \iiint \frac{\sin^2(\pi/\lambda)\Delta \mathbf{S}\cdot N_1 \mathbf{a}_1}{\sin^2(\pi/\lambda)\Delta \mathbf{S}\cdot \mathbf{a}_1} \frac{\sin^2(\pi/\lambda)\Delta \mathbf{S}\cdot N_2 \mathbf{a}_2}{\sin^2(\pi/\lambda)\Delta \mathbf{S}\cdot \mathbf{a}_2}$$

$$\times \frac{\sin^2(\pi/\lambda)\Delta \mathbf{S}\cdot N_3 \mathbf{a}_3}{\sin^2(\pi/\lambda)\Delta \mathbf{S}\cdot \mathbf{a}_3} d\alpha\,d\beta\,d\gamma\,. \tag{4.29a}$$

We can remove FF^* and I_e from the integral because they change very slowly compared to the interference terms. To remove the vectors from the equation, we represent $\Delta \mathbf{S}$ in terms of the unit vectors of reciprocal space, $\mathbf{b}_1$, $\mathbf{b}_2$, and $\mathbf{b}_3$, i.e., $\Delta \mathbf{S} = \lambda(h_1 \mathbf{b}_1 + h_2 \mathbf{b}_2 + h_3 \mathbf{b}_3)$. Then,

$$E = I_e \frac{R^2}{\omega} FF^* \iiint \frac{\sin^2 \pi N_1 h_1}{\sin^2 \pi h_1} \frac{\sin^2 \pi N_2 h_2}{\sin^2 \pi h_2} \frac{\sin^2 \pi N_3 h_3}{\sin^2 \pi h_3} d\alpha\,d\beta\,d\gamma\,. \tag{4.29b}$$

To procede further, we must relate the integration variables α, β, and γ to reciprocal lattice vectors. Now changes in the angles α, β, and γ are equivalent to increments in the vectors $\mathbf{S}_0'$ and $\mathbf{S}'$, the changes being normal to $\mathbf{S}_0$ and $\mathbf{S}$ as seen in Fig. 4.57. These vector increments correspond to changes in the vector $\Delta \mathbf{S}$ with components oriented to define a small volume as shown in Fig. 4.59. The volume of reciprocal space swept out by the terminal point of $\Delta \mathbf{S}$ is given by

$$\Delta V = d\boldsymbol{\alpha} \cdot (d\boldsymbol{\beta} \times d\boldsymbol{\gamma}) = d\alpha\,d\beta\,d\gamma \sin 2\theta\,. \tag{4.30a}$$

An alternative expression for the volume swept out by $\Delta \mathbf{S}$ is given from the reciprocal space definition,

4.9 Measurement of Diffraction Intensity

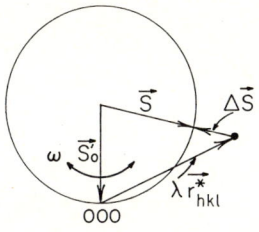

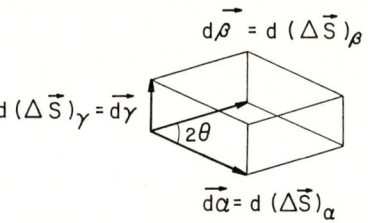

Fig. 4.58

Fig. 4.59

Fig. 4.58. For directions of incident and diffracted beams S' and S'_0 which differ from the Bragg's law directions S and S_0, $S' - S'_0 = \lambda r^* + \Delta S$

Fig. 4.59. Geometry of the volume of reciprocal space swept out by the vector ΔS. $d\alpha$ is perpendicular to S_0 and $d\beta$ is perpendicular to S; therefore, the angle between them is 2θ. $d\gamma$ is perpendicular to $d\alpha$ and $d\beta$

$$\Delta V = \lambda \, dh_1 \mathbf{b}_1 \cdot (\lambda \, dh_2 \mathbf{b}_2 \times \lambda \, dh_3 \mathbf{b}_3) = (\lambda^3/V_c) \, dh_1 \, dh_2 \, dh_3 , \tag{4.30b}$$

where V_c is the unit cell volume in real space. Equating these expressions for ΔV, we have

$$d\alpha \, d\beta \, d\gamma = (\lambda^3/V_c \sin 2\theta) \, dh_1 \, dh_2 \, dh_3 .$$

We have now obtained the right variables for the integration and may substitute in Eq. (4.29b):

$$E = \frac{I_e R^2}{\omega} \frac{FF^* \lambda^3}{V_c \sin 2\theta} \int_{-\infty}^{+\infty} \frac{\sin^2 \pi N_1 h_1}{\sin^2 \pi h_1} \, dh_1$$

$$\times \int_{-\infty}^{+\infty} \frac{\sin^2 \pi N_2 h_2}{\sin^2 \pi h_2} \, dh_2 \int_{-\infty}^{+\infty} \frac{\sin^2 \pi N_3 h_3}{\sin^2 \pi h_3} \, dh_3 . \tag{4.31}$$

These terms cannot be integrated from $-\infty$ to $+\infty$. There are an infinite number of peaks in this range. However, for any significant value for the N's, the terms have appreciable values only for very small values of πh, and for this range, $\sin^2 \pi h$ can be replaced by $(\pi h)^2$, i.e.,

$$\frac{\sin^2 \pi N_1 h_1}{\sin^2 \pi h_1} \cong \frac{\sin^2 \pi N_1 h_1}{(\pi h_1)^2} .$$

This new function has only one maximum, rather than the infinite number of the original function, but then we are only interested at the moment in one of the peaks. Now,

$$\int_{-\infty}^{+\infty} \frac{\sin^2 \pi N_1 h_1}{(\pi h_1)^2} \, dh_1 = N_1 .$$

Therefore,

$$E = (I_e R^2 \lambda^3 FF^*/\omega V_c \sin 2\theta) N_1 N_2 N_3 . \tag{4.32}$$

Now $N_1 N_2 N_3 = N$ is the total number of unit cells in the crystal, and $N(V_c)$ is the volume of the small crystal (ΔV). Therefore, multiplying the top and bottom of the above expression for total energy by V_c,

$$E = I_e R^2 \lambda^3 FF^* \Delta V / \omega V_c^2 \sin 2\theta . \tag{4.33}$$

For unpolarized x-radiation, taking I_e from Eq. (4.27),

$$E = \frac{I_0}{\omega} \left(\frac{e^4 \eta^2}{m^2 c^4}\right) \left(\frac{\lambda^3 FF^* \Delta V}{V_c^2 \sin 2\theta}\right) \left(\frac{1 + \cos^2 2\theta}{2}\right), \tag{4.34a}$$

or

$$E\omega/I_0 = Q\Delta V . \tag{4.34b}$$

We no longer need to consider the crystal as a parallelepiped, for our equation involves only the volume. The total energy is independent of R because in I_e there was a factor $1/R^2$. Q is often referred to as the scattering power.

If we are continually rotating the crystal, as in a rotation pattern, after n rotations in time t,

$$\omega = 2\pi n/t .$$

The total energy after n rotations is obtained from Eq. (4.33) by substituting for ω. If we are oscillating the crystal φ radians per oscillation (i.e., rocking the crystal),

$$\omega/\varphi = \text{oscillations/s}.$$

The total energy will be Eq. (4.33) multiplied by $(\omega/\varphi)t$. The quantity $E\omega/I_0$ given in Eq. (4.34b) *is often reported as the value for the integrated reflection as it is independent of the incident intensity and* ω. Known as the power per unit incident intensity, its units are

$$\frac{[\text{energy}][\text{rad}][\text{time}][\text{area}]}{[\text{time}][\text{energy}]} = [\text{rad}][\text{area}] .$$

Usually I_0 times the cross-sectional area of the incident beam is what is measured, rather than I_0, so that $E\omega/P_0$ (rad) is often given, where $P_0 = I_0 A$. (See Appendix C for techniques for measuring P_0.)

The term $1/\sin 2\theta$ that arose in the integration of Eq. (4.26) is known as *the Lorentz factor*—the polarization term in I_e and this term are often referred to as the *Lorentz-polarization (LP) factor*. The physical meaning of the Lorentz factor (L) can be made clearer, following the treatment by Buerger (1960). Consider Fig. 4.60. Since $|\mathbf{S} - \mathbf{S}_0| = 2 \sin \theta$,

$$\sin 2\theta = 2 \sin \theta \cos \theta = |\mathbf{S} - \mathbf{S}_0| \cos \theta .$$

Then

$$L = 1/\sin 2\theta = 1/|\mathbf{S} - \mathbf{S}_0| \cos \theta .$$

If the crystal is rotating at angular velocity ω about 000, the velocity of a diffracting point is $\omega|\mathbf{S} - \mathbf{S}_0|$, and the component of velocity normal to the sphere ($v_\perp$), is $\omega|\mathbf{S} - \mathbf{S}_0| \cos \theta$. The time ($t$) for a point to pass through the reflecting position is proportional to $1/v_\perp$,

4.9 Measurement of Diffraction Intensity

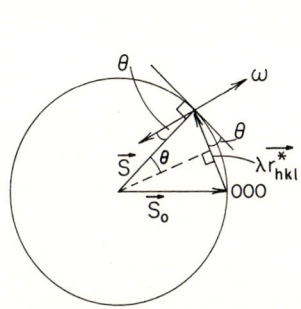

Fig. 4.60. Geometry for determining the Lorentz factor for diffraction from a zero-layer reciprocal lattice point

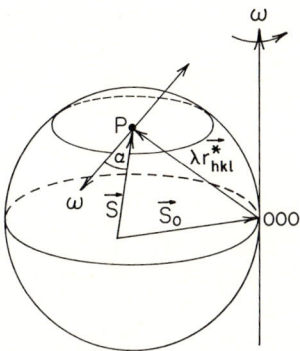

Fig. 4.61. Geometry for determining the Lorentz factor for diffraction form a general reciprocal lattice point

$$t = \frac{K}{v_\perp} = \frac{K}{\omega |\mathbf{S} - \mathbf{S}_0| \cos \theta} = \frac{KL}{\omega}. \tag{4.35}$$

Thus the Lorentz factor is simply related to the time it takes a point in the reciprocal lattice to move through the Ewald sphere. For the more general case of a point P moving in the direction shown in Fig. 4.61

$$v_\perp = \omega \cos \alpha, \quad L = 1/K' \cos \alpha.$$

Equations for the Lorentz factor for specific observation geometries can be found in the International Tables, Volume II, p. 266 and Volume III, p. 136.

4.9.2 The Effect of Absorption

In interpreting Eq. (4.34), we *have to evaluate ΔV which depends on absorption*. It is the effective volume of the crystal that diffracts, not the total volume, because the incident beam is reduced in intensity as it penetrates. As an example of this evaluation, *consider diffraction from a specimen in the form of a flat plate* whose dimensions are much larger than the incident beam. This situation is illustrated in Fig. 4.62 for a condition where the diffraction planes are not parallel to the surface of the plate. The volume irradiated in an element dZ thick at a depth Z is $(A_0/\sin \varphi) dZ$, where A_0 is the area of the incident beam. The incident intensity at this depth (attenuated by absorption in the path of length $Z/\sin \varphi$) is $I_0 \exp[-(\mu/\rho)\rho(Z/\sin \varphi)]$. The diffracted beam is also absorbed as it leaves the sample, and for this path length of $Z/\sin \xi$, the attenuation factor is $\exp[-(\mu/\rho)\rho(Z/\sin \xi)]$. Let b equal the fraction of the incident beam which is diffracted. Then for a crystal of thickness t, the diffracted beam emerging from the sample is

$$I_D = I_0 b \Delta V = I_0 b \int_0^t \frac{A_0}{\sin \varphi} \exp\left[-\left(\frac{\mu}{\rho}\right)\rho Z \left(\frac{1}{\sin \varphi} + \frac{1}{\sin \xi}\right)\right] dZ. \tag{4.36a}$$

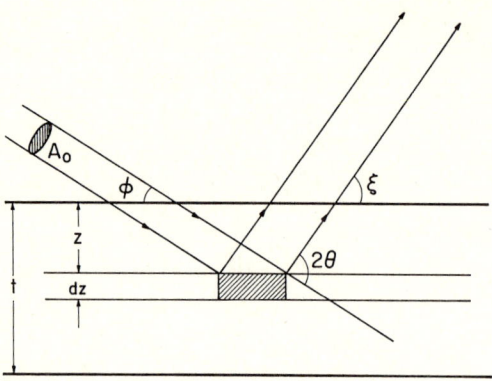

Fig. 4.62. Diffraction from a large flat plate. The incident and diffracted beam are attenuated by absorption in the crystal

For the particular specimen geometry often used in a diffractometer, $\varphi = \xi = \theta$ and the diffracting planes are parallel to the surface. Then

$$\Delta V = \int_0^t \frac{A_0}{\sin \theta} e^{-(2\mu/\rho)\rho(Z/\sin \theta)} dZ . \tag{4.36b}$$

For a thick crystal, we can let $t \to \infty$, and the integral has the simple *result for such an "infinitely thick" crystal*:

$$\Delta V = A_0/2(\mu/\rho)\rho . \tag{4.36c}$$

In this case the absorption is independent of θ, but only because the diffracting planes are parallel to the surface, and the crystal is thick enough so that the upper limit of integration may be taken as infinity. At small angles the path is large, but so is $A_0/\sin \theta$. Including this term for the diffraction volume, Eq. (4.34) becomes

$$\frac{E\omega}{I_0} = \frac{I_e}{I_0} \frac{R^2 \lambda^3}{V_c^2} FF^* \frac{1}{\sin 2\theta} \frac{A_0}{2(\mu/\rho)\rho} . \tag{4.37}$$

The criterion for an infinitely thick crystal is somewhat arbitrary and depends on our definition as to what constitutes a negligible contribution to the total intensity. Suppose we define infinite thickness as that for which the contribution to the intensity from the back side is $\frac{1}{1000}$ that of the intensity contributed from the front side. Then using the differential form of Eq. (4.36b) for $\varphi = \xi = 0$,

$$\frac{dI_D \text{ (at } Z = 0)}{dI_D \text{ (at } Z = t)} = \exp\left[2\left(\frac{\mu}{\rho}\right)\rho \frac{t}{\sin \theta}\right] = 1000$$

or

$$t = 3.45 \sin \theta/\mu . \tag{4.38a}$$

For typical values [$\mu \approx 2.5 \, (10)^4 \, \text{m}^{-1}$ and $\sin \theta \simeq 0.5$], $t \simeq 0.6 \, (10)^{-4}$ m or about 2.5 thousandths of an inch, a very thin "infinity," indeed.

We may rephrase this problem by asking *what contribution to the diffracted intensity is made by material at each depth in the sample*. We may express the diffracted intensity due to material at some depth as a fraction of the intensity that

4.9 Measurement of Diffraction Intensity

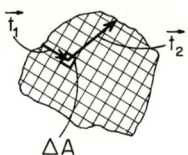

Fig. 4.63. Graphical procedure for determining the absorption factor for an odd-shaped crystal. $dV = $ beam height $\times \{\sum \exp[-\mu(t_1 + t_2)]\} \Delta A/n$ ΔA, where n is the number of parallelograms of area ΔA. [After Buerger, M.J., "Crystal Structure Analysis". John Wiley & Sons, (1960). Reprinted by permission of John Wiley & Sons, Inc.]

would be diffracted by an infinitely thick sample. Calling this fraction R_z, we have

$$R_z = \int_0^z dI_D \bigg/ \int_0^\infty dI_D = 1 - \exp\left[-2\left(\frac{\mu}{\rho}\right)\rho \frac{z}{\sin\theta}\right] \quad (4.38b)$$

for the symmetric geometry $\varphi = \xi = \theta$. Using the same values of $\mu \simeq 2.5\,(10)^4$ m^{-1}, $\sin\theta \simeq 0.5$, and $t \simeq 0.6\,(10)^{-4}$ m as above, we have $R_z = 0.9975$; i.e., 99.75% of the diffracted intensity comes from an outer layer of thickness $0.6\,(10)^{-4}$ m. Equations (4.38a) and (4.38b) emphasize the fact that the effective thickness of a sample depends on the scattering angle as well as the absorption.

The effective volumes ΔV are also easy to calculate for a cylindrical or a spherical crystal (these results can be found in the International Tables, Volume II, p. 291). For an odd-shaped crystal, the effective volume can be evaluated on a computer as shown in Fig. 4.63. This can also be done by ruling two pieces of transparent paper and laying them over a scaled drawing of the crystal at the correct angles of incidence and 2θ. By keeping the dimensions of the crystal (represented roughly by its average diameter d), small enough so that $\mu d < 0.5$, this absorption correction may be kept small and the shape of the crystal will be less important. This is why for other than the flat plate geometry of Fig. 4.62 it is better to use a crystal of the order of a few thousandths of a centimeter in size for x-ray diffraction. For many inorganic materials, $\mu \approx 5(10)^4$ m^{-1} and the required $d < 10^{-5}$ m is too small to be practical. Consequently, absorption corrections are one of the major sources of errors in intensity measurements from heavy inorganic materials. As we saw in Sect. 3.8, the absorption of neutrons by matter is many orders of magnitude smaller than for x-rays, so larger crystals may be used with neutron diffraction and still maintain $\mu d < 0.5$.

4.9.3 Measurement Techniques Using a Diffractometer

For the highest accuracy in measuring intensities, care must be taken in the integration to use a monochromator and to eliminate multiple reflections from the monochromator, i.e., effects due to $\lambda/2$. (A simple energy scan of a diffraction peak with a PHA will reveal such effects.) Usually, however, work is done with filtered radiation. Let us look more closely at this practice. [This discussion is based on the paper by Alexander and Smith (1962).]

An ω scan or rock with a stationary counter is not really quite good enough with filtered radiation. The crystal has finite dimensions (Fig. 4.64a), a mosaic spread may broaden the peak, the incident beam is not perfectly collimated, and there is a wavelength spread. (The angular terms due to these can be calculated in detail.) Thus r^*, in effect, exists over a range of values.

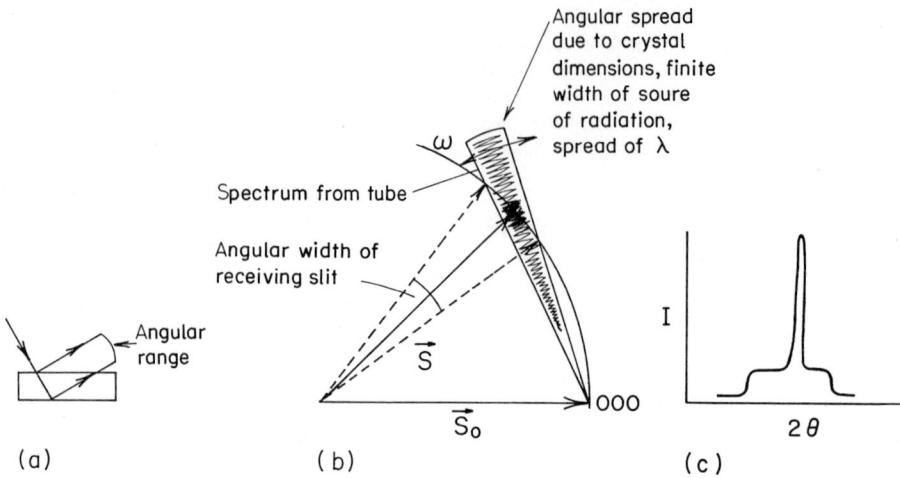

Fig. 4.64a–c. Collecting diffracted intensity in a ω-scan with a stationary detector. [After Alexander, L.E., and Smith, G.S., Acta Crystgallogr., *15*, 983 (1962)]

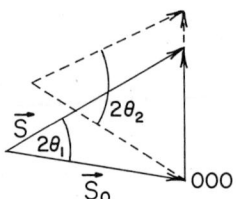

Fig. 4.65. Geometry for a θ–2θ scan in reciprocal space

Consider Fig. 4.64b. The shading represents the λ distribution from an x-ray tube with filtered radiation; the darkest region corresponds to K_α. As a result of the spread in r^*_{hkl} due to all factors the spot will intersect the sphere over a wide range, as shown in the figure. The stationary receiving slit must be open to receive the entire range for K_α. The spread due to mosaic is covered in the ω scan itself and is not included as part of the angular spread of λr^*_{hkl}. Note that during an ω scan the crystal literally falls off the wavelength distribution depicted in Fig. 4.64b. During the scan, the receiving slit sees nothing, then some white radiation, then K_α, white, and nothing again as shown in Fig. 4.64c. It is not possible to really decide where the true background is around the K_α due to this sharp drop.

In a θ–2θ *scan*, in principle, we are moving along a specific direction in reciprocal space, as shown in Fig. 4.65. This is because in this scan, the normal to the diffracting planes always remains as the bisector of **S** and **S$_0$**. In this case, crystal mosaic affects the peak shape. When the detector moves to a new angle, the crystal rotates bringing new regions of the crystal into the appropriate orientation to diffract. The amount of such scattering which is detected depends on the slit width of the counter, and wide slits are required to avoid missing some portion of the peak. The situation in reciprocal space is as shown in Fig. 4.66a. In this $\theta - 2\theta$ scan, the pattern is an examination of reciprocal space along the distribution of λ's (that is, along λr^*), and

4.9 Measurement of Diffraction Intensity

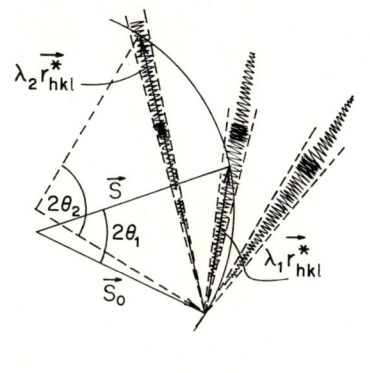

(a)

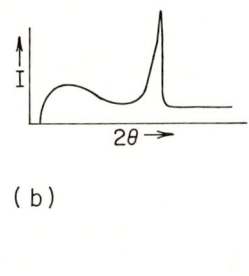

(b)

Fig. 4.66a,b. Collecting intensity in a θ–2θ scan with moving crystal and detector. [After Alexander, L.E., and Smith, G.S., Acta Crystallogr., *15*, 983 (1962)]

with filtered radiation, peaks will look like those shown in Fig. 4.66b. It is much easier to find the true background, and multiple wavelengths need not be eliminated, they contribute very similarly in the background on either side of the peak and under the peak. They are effectively removed by drawing a base line from one side of the peak to the other, or by measuring the background on either side of the peak, averaging, multiplying by the angular range of the peak, and subtracting.

As was mentioned before, in the ω scan the range of the crystal mosaic is covered in the rocking action, and the slit need only be wide enough to account for the other three factors, but in a θ–2θ scan it must be wide enough to include also the range due to any mosaic. A good rule is to keep opening the slits until, in a scan, the integrated intensity (above background) does not change. In general this need only be done for the broadest, high-angle peak. As the windows in commercial x-ray detectors have widths corresponding to 5–6° in 2θ, in certain situations with a highly mosaic crystal it may be necessary to use a monochromator to eliminate the white radiation background and take an ω scan.

It should be kept in mind that we sample some volume in reciprocal space at the tip of the diffraction vector, because of the divergences in the incident beam (due to slits and possibly a monochromator whose reflection shape depends on its perfection) and in the diffracted beam, due to the size of the receiving slits. This is illustrated in Fig. 4.67. We can sample this shape and adjust it, by placing a highly perfect crystal such as S_i in the sample position, and then performing θ–2θ scans at various ω tilts through the S_i reflection. Note that the shape of the volume sampled varies with 2θ. The drawing is two-dimensional and there is also a vertical dimension, so we need to repeat such scans for various χ tilts. The total vertical divergence β is obtained from measurement of $\chi_{\max} - \chi_{\min} = (\beta/\lambda)(2\sin\theta/\lambda)$. By opening up the receiving slits very wide, it can be seen from Fig. 4.67 that α is then very large and we sample α_0 only.

The shape of the sampled volume need not be a rhomboid. Suppose for example that we consider that the incident divergence is very narrow, as is the case at a synchrotron. The divergence is $1/\gamma$ in one direction, and a few milliradians in the other. Therefore the receiving slits really control the divergence, and with equal

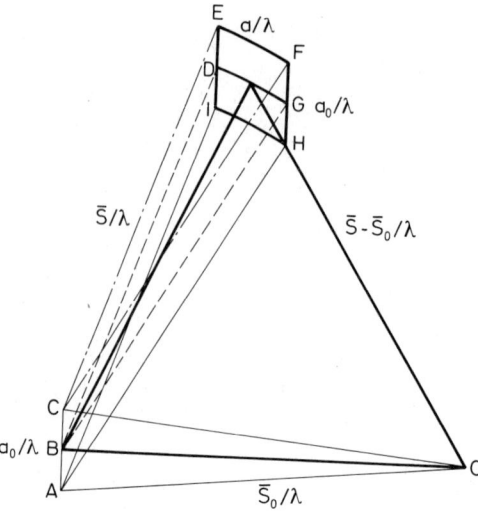

Fig. 4.67. Horizontal section of reciprocal space of a single crystal specimen, showing the section *EFHI* of the volume sampled. For simplicity the diagram is drawn for incident divergence α_0 equal to the diffracted beam's divergence α

horizontal and vertical receiving slit spacings, a small plate-like region normal to Fig. 4.67 is the sampling region.

4.9.4 The Effect of Crystal Perfection on Integrated Intensities

The analysis of this section has been based on diffraction from an *ideally imperfect crystal*, i.e. one composed of many small, slightly misoriented mosaic blocks. The diffraction from each mosaic block has been calculated as given in Eq. (4.34), and the total diffracted intensity taken as the sum of that diffracted from all blocks, as in the derivation of Eq. (4.37). Deviations from this idealization are classified as primary and secondary extinctions.

Primary extinction is a consequence of the perfection of a mosaic block. When such a block is oriented for diffraction, the once-diffracted beam may be rescattered into the incident beam direction as shown in Fig. 4.68. Since each scattering causes a phase lag (see Sect. 3.4), the twice-diffracted beam is out of phase with the incident beam and interferes with it destructively. Thus, it is not accurate to assume that the mosaic block is irradiated uniformly with an intensity I_0. Corrections for this effect have been developed, for example, by Zachariasen (1967). The general result of Zachariasen's analysis is that the measured diffracted power is reduced from that calculated in Eq. (4.34b), by a factor $\phi(\sigma)$, where $\sigma = QA$ is the diffracted power per unit length of beam path in the crystal, and A is the average cross-sectional area of the mosaic block normal to the beam. Expressions for $\phi(\sigma)$ depend on experimental geometry; but, for example, for a mosaic block shaped as an infinite plane parallel plate with incident and scattered beams making equal angles with the plate,

$$\phi(\sigma) = 1/(1 + \sigma \bar{t}), \qquad (4.39a)$$

where $\bar{t}$ is the mean path length of the beam in the mosaic block. When the crystal

4.9 Measurement of Diffraction Intensity

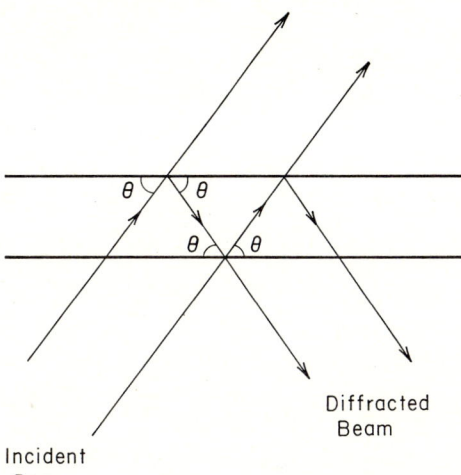

Fig. 4.68. In a perfect crystal the diffracted beam can be diffracted again, interfering destructively with the incident beam and causing primary extinction

has a mosaic structure with block size small compared to the crystal, the attenuation is

$$\phi(\sigma) = 1/(1 + \bar{\sigma}\bar{T}), \tag{4.39b}$$

where $\bar{T}$ is the mean path length through the whole crystal and $\bar{\sigma}$ is the average of σ for the distribution of blocks. Now $\bar{\sigma}$ is smaller than σ for a perfect crystal since not all blocks are oriented for diffraction, so reducing the size of mosaic blocks in a crystal will reduce the primary extinction. Crystallographers often attempt to decrease the perfection of the crystal to be studied by dipping it into liquid nitrogen. The resultant thermal shock will sometimes deform the crystal into smaller mosaic blocks. Small amounts of plastic deformation are also effective in reducing the mosaic size.

Even when a crystal is ideally imperfect, a second source of extinction exists. In addition to the normal absorption processes that were included in the derivation of Eqs. (4.36), when a crystal is oriented for diffraction additional energy is removed from the direct beam by scattering from blocks above it. This effect, known as *secondary extinction*, is equivalent to increasing the linear absorption coefficient from μ to μ' given by

$$\mu' = \mu + Q^*, \tag{4.40}$$

where Q^* is the scattering power per unit length of beam path.

Note that the effects of both primary and secondary extinction are most important for reflections that have the strongest intensities. Usually such strong reflections can be eliminated from consideration without interfering with the process of crystal structure determination. An extensive discussion of extinction corrections may be found in Dawson (1975).

A further source of error in intensity measurements is the possibility of scattering from many reciprocal lattice points simultaneously if several are on the Ewald sphere. Reflections can increase or decrease in intensity as a consequence of this

4.9.5 Integrated Intensities from Polycrystalline Specimens

We now turn our attention to *polycrystalline or powder specimens in a flat plate geometry* such as would be used in a diffractometer. For any single grain whose diffraction plane normal is in the plane of diffraction and whose plane is in the plane of the specimen's surface, the diffracted energy is given by Eq. (4.34a). For x-radiation, letting P be the expression for the polarization factor, this becomes

$$E = \frac{I_0}{\omega}\left(\frac{e^4\eta^2}{m^2c^4}\right)\frac{P\lambda^3 FF^* \Delta V}{V_c^2 \sin 2\theta} . \tag{4.41}$$

In a diffractometer a receiving slit of width w and height h accepts diffraction from all grains whose normals lie in the solid angle Ω indicated in Fig. 4.69. Division of w by $\sin\theta$ is necessary because it is the solid angle projected parallel to the slit which is of interest; division by the factor of 2 is necessary since angles measured at the slit position are in units of 2θ, and we have specified sample rotation as $\omega = d\theta/dt$. Thus from Fig. 4.69, the solid angle sampled is

$$\Omega = wh/4R^2 \sin\theta .$$

In a diffractometer Ω is sufficiently small that we may neglect variation in ω or polarization with grain orientation. The receiving slit however must be broad enough to see the entire reflection from one grain during the scan.

For a random powder in which n grains are irradiated and m_{hkl} is the multiplicity of the hkl reflection under consideration, the total number of hkl planes that could diffract is nm_{hkl}. Since the receiving slit accepts diffraction from only $4\pi\Omega$ of these, the total number of diffracting planes is $\Omega nm_{hkl}/4\pi$. Then the total diffracted energy accepted by the receiving slit is

$$E = \frac{I_0}{\omega}\left(\frac{e^4\eta^2}{m^2c^4}\right)\frac{P\lambda^3 FF^*}{V_c^2 \sin 2\theta}\frac{m_{hkl}wh}{4R^2 \sin\theta}\frac{n\Delta V}{4\pi} . \tag{4.42a}$$

For a flat plate sample, we found in the derivation of Eq. (4.36) that the effective

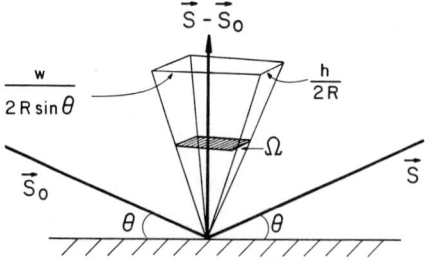

Fig. 4.69. Diffraction from a polycrystalline specimen. The solid angle $wh/4R^2 \sin\theta$ includes normals to all grains whose diffracted intensity will be collected in the detector

4.9 Measurement of Diffraction Intensity

volume irradiated ($n\Delta V$) is $A_0/2\mu$,[4] so

$$E = \left(\frac{I_0 A_0}{\omega}\right)\left(\frac{e^4\eta^2}{m^2 c^4}\right)\frac{\lambda^3 F F^*}{V_c^2}\frac{m_{hkl}wh}{64\pi R^2 \mu}\frac{P}{\sin^2\theta\cos\theta}. \qquad (4.42b)$$

The factor $1/\sin^2\theta\cos\theta$ is the Lorentz factor (L_p) for a powder. For the case of unpolarized incident x-radiation, tabulations of the combined Lorentz-polarization factor for single crystals and powders may be found in the International Tables, Volume II, pp. 268–273.

For accurate measurements of energy (integrated intensity), it is essential that the slits be set wide enough to obtain the entire reflection from a grain. If the slits are increased until the intensity above background is constant for a large mosaic single crystal, this will be more than adequate, as such a crystal generally has more mosaic spread than a small grain. It is also essential that the assumption of random grain orientation be satisfied. This means that powders should be fine mesh (<325 mesh) and that care in preparing samples of noncubic powders should be taken to avoid preferentially aligning powders. Never press a powder into a holder, sprinkle it. Also note that because of porosity and surface roughness, there may be an angularly dependent absorption factor. In some cases, this can produce errors of 30% or more. Suppose there is an element in the specimen of about the same atomic number as the radiation. It can be caused to fluoresce with some other radiation; if there was no surface roughness, this fluorescence would not vary with 2θ. The variation in the background is then a measure of this correction term. It vanishes at $\theta = 90°$ (with a pulse-height analyzer one can pick out the fluorescence from the normal diffraction with this other radiation; see Sect. 3.11).

To check for preferential orientation, set the diffractometer at a peak maximum and rotate the sample in the plane of the sample surface and tilt it from side to side. Variations of more than 10% in peak intensity with changing sample orientation suggests an unacceptable deviation from random grain orientation. Such problems are most severe in solid polycrystalline specimens which may show preferred grain orientation due to growth or deformation and recrystallization texture.

An alternate method of recording the diffraction from a powder sample is *the Debye-Scherrer technique* discussed in Sect. 4.7. Consider a monochromatic beam irradiating a sample containing n small crystals. For any set of *hkl* planes with spacing d_{hkl}, there will be a few crystals which are correctly oriented with the primary beam to allow a Bragg reflection. The diffracted beams make a fixed angle $2\theta_{hkl}$ with the primary beam forming a cone of semiapex angle $2\theta_{hkl}$. To calculate the scattered intensity, we need to know dn, the number of crystals having a set of planes *hkl* making an angle between θ and $\theta + d\alpha$ with the primary beam (the angles α, β, and γ in this discussion are those identified in Fig. (4.57). Assuming the small crystals are randomly arranged, with their normals on a sphere of radius R and m as the

---

[4] For a powder compact, the effective volume scattering is more complicated and depends on powder particle size and shape, and packing density ρ'. A first-order correction is obtained by multiplying the expression in Eq. (4.42) by ρ'/ρ. For a more complete discussion of this topic see Harrison, R. J., and Paskin, A., The Effects of Granularity on the Diffracted Intensity in Powders, U.S. Army Materials Research Agency, AMRA TR 63-02.

multiplicity,

$$dn = nm_{hkl} 2\pi R^2 \sin(90° - \theta)\, d\alpha / 4\pi R^2 = \tfrac{1}{2} nm_{hkl} \cos\theta\, d\alpha\,,$$

where R is also the sample to film distance. The intensity scattered from each crystal is given by Eq. (4.27). An area on the film surface is $R^2\, d\beta\, d\gamma$. Proceeding as before and integrating over all orientations and over the whole area of the receiving surface,

$$E = \int\int\int I \frac{nm_{hkl}}{2} \cos\theta\, d\alpha\, R^2\, d\beta\, d\gamma\,. \tag{4.43}$$

The triple integration is carried out as in the derivation of Eq. (4.32), and we find after some simplification that

$$E = I_0 \left(\frac{e^4 \eta^2}{m^2 c^4}\right) \frac{\Delta V \lambda^3 m_{hkl} F F^* P}{4 V_c^2 \sin\theta}\,, \tag{4.44}$$

where the total sample volume $\Delta V = n N V_c$, and we have neglected absorption in the polycrystalline sample. But this is the total power in the hkl reflection, which is spread over the Debye-Scherrer cone. On a cylindrical film which surrounds the sample (see Fig. 4.37), it is the power per unit length of the diffraction circle on the film which is of interest. The circumference of this circle is $2\pi R \sin 2\theta$, so the power per unit length of diffraction circle is

$$E' = E/2\pi R \sin 2\theta \tag{4.45a}$$

or

$$E' = I_0 \left(\frac{e^4 \eta^2}{m^2 c^4}\right) \frac{\Delta V \lambda^3 m_{hkl} F F^* P}{8 \pi R V_c^2 \sin\theta \sin 2\theta}\,. \tag{4.45b}$$

One important application of integrated intensities from powder specimens lies in their use for *determining the relative volumes of phases* simultaneously present in the sample. For example, if there are two phases a and b present and a pattern is taken, say of a flat powder sample in a diffractometer, the effective volume of a which scatters is

$$V_a = c_a \frac{A_0}{2(\mu/\rho)_m \rho_m}\,,$$

where c_a is the volume fraction of phase a and $(\mu/\rho)_m \rho_m$ is the absorption for the mixture. Then the relative integrated intensities for reflections hkl of the a phase and $h'k'l'$ of the b phase are

$$\frac{E'_{hkl-a}}{E'_{h'k'l'-b}} = \frac{F F_a^*(m_{hkl})_a c_a (V_c^2)_b [(1 + \cos^2 2\theta)/\sin^2\theta \cos\theta]_{hkl-a}}{F F_b^*(m_{h'k'l'})_b c_b (V_c^2)_a [(1 + \cos^2 2\theta)/\sin^2\theta \cos\theta]_{h'k'l'-b}} \tag{4.46}$$

All other terms such as R^2, λ^3, I_e, and $(\mu/\rho)_m \rho_m$ cancel, provided the same slits are used or both peaks. As $c_a + c_b = 1$, the expression can be solved for c_a or c_b. Note that the expression is most sensitive when $(V_c)_a$, $(V_c)_b$, and FF_a^* and FF_b^* are similar. Thus, for example, as little as 0.2 % retained austenite can be detected in a martensitic steel, if a monochromator is used to minimize the background (but the limit is about 3 % with filtered radiation). The technique can of course be extended to many phases by using a peak from each phase. If necessary, the volume fraction of one or more of these phases could be estimated with a microscope.

4.9.6 Absolute Intensities and Intensity in Nondispersive Experiments

It should be noted that absolute intensity measurements are only occasionally calculated in crystal structure analysis. To compare calculated intensities to measured values would then require a separate measurement of I_0 which is difficult due to its large magnitude (typically $\approx 10^9$ cps with conventional tube x-radiation). Techniques for such measurements are described in Appendix C. (Absolute intensities are used, however, in cases where the scattering is weak and it is necessary to calculate and subtract such contributions to the scattering as Compton scattering with x-rays or spin incoherent scattering with neutrons. Examples of such applications are discussed in Chap. 7.) Instead of using absolute intensities, all angular independent factors are lumped into a single constant K, and the measured energy for an hkl reflection from a single crystal is written as

$$E_{hkl} = KFF^*_{hkl}LP\Delta V, \qquad (4.47\text{a})$$

and similarly for a powder specimen,

$$E_{hkl} = K'FF^*_{hkl}m_{hkl}L_p P\Delta V. \qquad (4.47\text{b})$$

The constant K (or K') is simply considered as an additional parameter to be determined when fitting a calculated set of FF^*_{hkl} for an assumed atomic arrangement to the measured set of KFF^*_{hkl}.

As we noted in Sect. 4.7, *the technique of energy analysis* is being used increasingly in both neutron and x-ray scattering experiments. In this technique the incident beam for each Bragg reflection has a different wavelength, and hence the intensity I_0, wavelength λ, and absorption coefficient μ are different for each reflection. This requires measurements of I_0 as a function of wavelength. However, a more fundamental difference between these two techniques is due to the fact that in the standard angle-dispersive techniques, the intensity is integrated over angle while in the energy analysis techniques it is integrated over a range of wavelengths. To see how this affects the formulas for integrated intensities, let us look, for example, at the formula for a powder sample. Lumping all but the indicated factors into the constant K'', we may rewrite Eq. (4.42b) as

$$E = K'' I_0 \lambda^3 FF^* P(\theta) \Delta V / \sin^2\theta \cos\theta. \qquad (4.48)$$

In the energy-dispersive technique, I_0 is not a constant, but changes with wavelength, and θ is fixed at some angle θ_0. Let i_0 be the incident intensity per unit interval of wavelength. Then the intensity for a beam of wavelength interval from λ to $\lambda + d\lambda$ is $i_0 d\lambda$, and the integrated intensity for this beam is

$$dE = K''(i_0 d\lambda)\lambda^3(FF^*)P(\theta_0)\Delta V/\sin^2\theta_0 \cos\theta_0. \qquad (4.49)$$

The integrated intensity is then obtained by integrating Eq. (4.49) in the range of λ to $\lambda + \Delta\lambda$, where $\Delta\lambda$ is determined by Bragg's law and the angular divergence of the incident beam $\Delta\theta_0$. Hence,

$$\Delta\lambda = \lambda \cot\theta_0 \Delta\theta_0.$$

Assuming $\Delta\lambda \ll \lambda$ and that i_0 changes slowly with λ, the integration gives

$$E = K''i_0\lambda^4(FF^*)P(\theta_0)\Delta V \Delta\theta_0/\sin^3\theta \tag{4.50}$$

Techniques of measuring i_0 and applications of energy-dispersive diffraction are still in a developing stage and the interested reader is referred to the literature. See, for example, Buras (1975).

References

Alexander, L.E., and Smith, G.S., Acta Crystallogr. *15*, 983 (1962)
Amoros, J.L., Buerger, M.J., and Canut De Amoros, M.L., "The Laue Method." Academic Press, New York, 1974
Azaroff, L.V., "Elements of X-Ray Crystallography." McGraw-Hill, New York, 1968
Azaroff, L.V., and Buerger, M.J., "The Powder Method." McGraw-Hill, New York, 1958
Barrett, C.S., "Structure of Metals." McGraw-Hill, New York, 1952; Barrett, C.S., and Massalski, op. cit., 1966
Buerger, M.J., "X-Ray Crystallography." Wiley, New York, 1942
Buerger, M.J., "Crystal Structure Analysis." Wiley, New York (1960)
Bunge, H.J., "Texture Analysis in Materials Science", Butterworths, London 1982
Buras, B., Olsen, J.S., Gerward, L., Selsmark, B., and Anderson, A.L., Acta Crystallogr. *A31*, 327 (1975)
Cullity, B.D., "Elements of X-Ray Diffraction." Addison-Wesley, Reading, Massachusetts, 1956
Dawson, B., *in* "Advances in Structure Research by Diffraction Methods" (W. Hoppe and R. Mason, eds.), Volume 6. Pergamon, New York (1975)
Evans, H.T. Jr., Appleman, D.E., and Handwerker, D.S., Report No. PB216188, U.S. Dept. of Commerce, National Technical Information Service (1963)
Guinier, A., "La Radio Crystallographie," 3rd ed., pp. 178–1790, Dunod, Paris, 1956
Henry, N.F.M., Lipson, H., and Wooster, W.A., "The Interpretation of X-Ray Diffraction Photographs." Macmillan, London, 1961
James, R.W., "The Optical Principles of the Diffraction of X-Rays," Chap. 7. London (1954)
Klug, H.P., and Alexander, L.E., "X-Ray Diffraction Procedures." Wiley, New York, 1974
Noyan, I.C., and Cohen, J.B., "Residual Stress", Springer-Verlag New York (1987)
Parrish, W., Acta Crystallogr. *13*, 838 (1960)
Pearson, W.E., "A Handbook of Lattice Spacings and Structures of Metals and Alloys," Vols. 1 and 2. Pergamon, New York, 1958, 1967
Peiser, H.S., Rooksby, H.P., and Wilson, A.J.C., "X-Ray Diffraction by Polycrystalline Materials." Chapman and Hall, London, 1960
Reed-Hill, R.E., "Physical Metallurgy Principles", Van Nostrand, Princeton, NJ, 1964
Rinik, J., Hilliard, J.E., and Cohen, J.B., J. of Nondestructive Evaluation *2*, 133 (1981)
Schwartz, L.H., Morrison, L.A., and Cohen, J.B., in "Advances in X-Ray Analysis," Vol. 7. Plenum, New York, 1963
Sparks, C.J., and Gedcke, D.A., "Advances in X-Ray Analysis," Vol. 20. Plenum, New York, 1975
"Tables for converting 2θ to d spacings," PB13176K. Office of Technical Services, US Department of Commerce, Washington, D.C.
"The International Tables for X-Ray Crystallography," Vol. II, 3rd ed. Kynoch Press, Birmingham, England, 1969
"The International Tables for X-Ray Crystallography," Vol. III, 3rd ed. Kynoch Press, Birmingham, England, 1969

Warren, B.E., "X-Ray Diffraction." Addison-Wesley, Reading, Massachusetts (1969)
Wood, E.A., "Crystal Orientation Manual." Columbia University Press, New York, 1963
Wykoff, R.W.G., "Crystal Structures." Wiley (Interscience), New York, 1960
Zachariasen, W.H., Acta Crystallogr. *23*, 558 (1967)
For charts, graphs, etc.: (a) Polycrystal Book Service, P. O. Box 11567, Pittsburgh, 15238. (b) N. P. Nies, 969 Skyline Drive, Laguna Beach, California

Problems

4.1. A striation was found in a germanium crystal as indicated in Fig. P4.1. Laue patterns were taken at (a) away from the striation and at (b) in the marking, and are given below. What can you say about the marking?

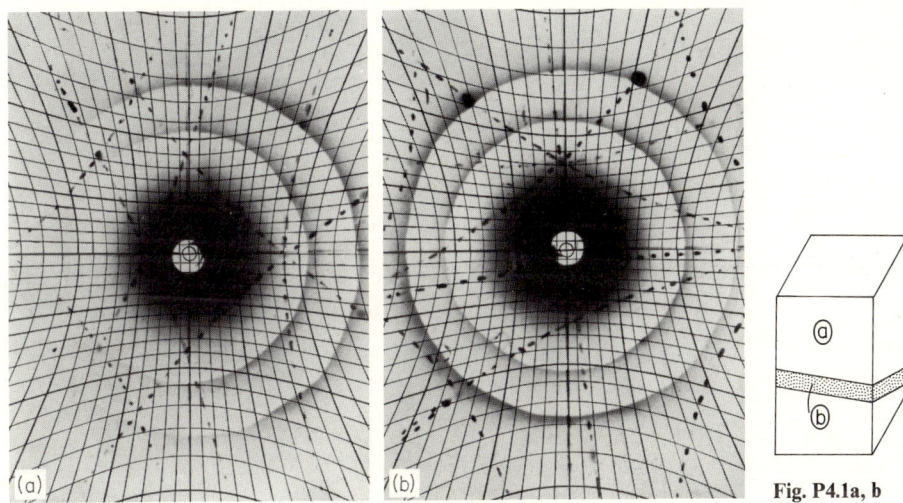

Fig. P4.1a, b

4.2. The two Laue back-reflection patterns shown in Fig. P4.2 are from an aluminum – 4 wt % copper single crystal (a) before deformation and (b) after an elongation of 47.4%. (They were taken by Professor J.G. Byrne of the University of Utah.) The tensile axis was normal to the beam and vertical. The specimen cross section was rectangular and the long edge was horizontal and perpendicular to the beam. The slip planes are known to be {111}.
(a) Plot the pole figures before and after deformation.
(b) Determine the slip direction. [*Hint*: Plot the tensile axis before and after deformation in one standard (001) projection. Find the slip plane initially with the maximum resolved shear stress (see Barrett, 1952, p. 345) and the direction in this plane to which the tensile axis rotates.]
(c) Just as slip started, what angles did the slip plane make with the faces of the crystal?
(d) What is the elongation you determine from the Laue patterns. (See Schmid. E., and Boas, W., "Kristallplastizität," p. 67. Springer, Berlin, 1935.)

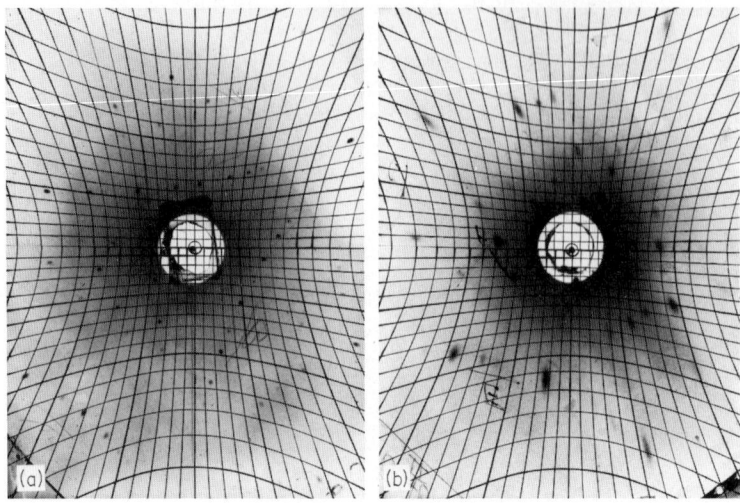

Fig. P4.2a, b

4.3. Silicon has a diamond cubic structure. An equator Weissenberg pattern is made using Mo K_α and oscillating the crystal π about a cube edge (a_3).
(a) What fraction of the $hk0$ points in reciprocal space have non-zero structure factors?
(b) Making reasonable approximations, how many diffraction spots should be observed on either half of the pattern?

4.4. Show with the aid of sketches of reciprocal space how to distinguish between the primitive, face-centered, body-centered, or c-centered orthorhombic lattice with a rotation camera.

4.5. Figure P4.5 is an oscillating crystal photograph of a hexagonal crystal (taken by Professor E.J. Freise of Northwestern University). The oscillation axis was the $\mathbf{a}_1$ axis, and the $\mathbf{a}_3$ axis was initially along the direct beam. The oscillation angle was $\pm 15°$, $\mathbf{a}_3 = 6.71$ Å. Index the intense reflections on the zero and first layers. Filtered copper radiation was used.

4.6. A single crystal (fcc) with identical atoms at each lattice point has a lattice parameter of 4 Å. A small portion of the crystal appears to be misoriented by a 90° rotation around [110], and the two pieces meet on {110}. With a 5-cm camera, an oscillation pattern ($\pm 15°$) is taken with the [001] as the oscillation axis and filtered copper radiation. Initially the beam is along the [110]. The beam hits the entire crystal. Draw the pattern to scale.

4.7. A new crystal AB has been grown. From chemical analysis and density it is known that there are four molecules in the unit cell. The crystal is a long plate as shown in the diagram of Fig. P4.7. A Laue back-reflection pattern perpendicular to the large flat face (1) showed a twofold axis of symmetry. Laue patterns from the other faces were inconclusive, but all faces are at right angles to each other.
(a) What can you say now about the possible crystal systems?
(b) A rotation pattern was taken with the sample oriented with the long dimension horizontal and the large flat face vertical. The layer line spacing (call it a_3) was

Problems

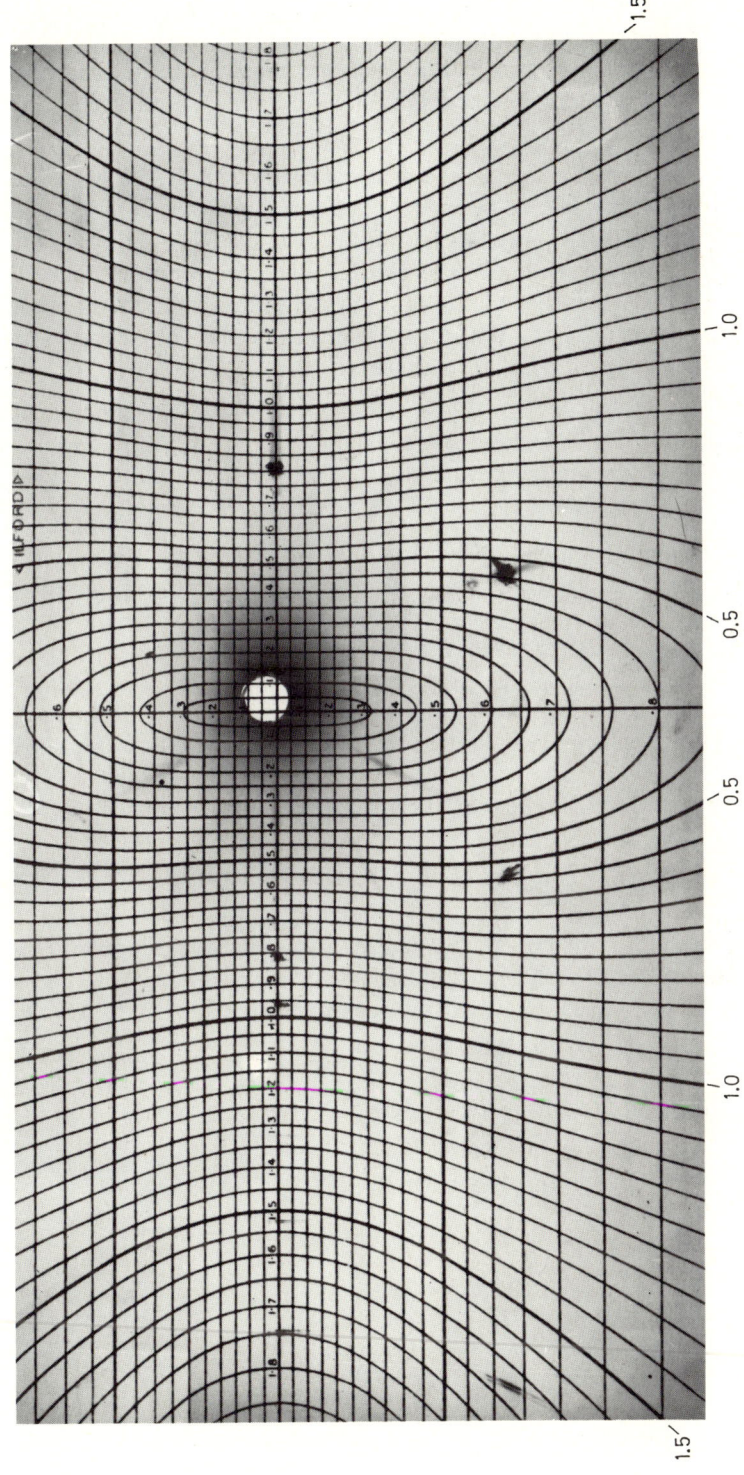

Fig. P4.5

4.40 Å. A pattern with the long dimension and the flat face horizontal gives $a_1 = 10.82$ Å. What is the crystal system?

(c) The rotation pattern shown in Fig. P4.7 was taken with the long direction vertical with filtered copper radiation (4 h, 40 kV, 15 mA). Determine a_2. What is the crystal system?

4.8. The rotation patterns in Fig. P4.8 were taken by Prof. R. De Angelis of the University of Kentucky, of a single crystal of copper shock-loaded to 435 kilobars. The rotation axis was [001]. In the following, (a) is a full 360° rotation, and (b) is a 15° oscillation to one side within 2–3° of [100]. Filtered copper radiation was used.
(a) Index the spots (Average readings from the film to the left and right of center.)
(b) Are there any extra spots? With the help of the oscillation patterns, what are the indices of these spots? What are they due to?

4.9. The powder pattern in Fig. 4.37d is of a copper-gold alloy, taken with filtered copper radiation by Dr. P. Gehlen, Battelle Laboratories.
(a) What is the unit cell and its parameter?
(b) What is the composition?

4.10. A powder-diffraction pattern made on a diffractometer shows lines at $\theta = 70°$ and 85°. The radiation used was Cu K_α and the $\alpha_1 \alpha_2$ doublet is resolved on the pattern.
(a) Calculate the angular separation in degrees of the two components of the doublet for the two lines.
(b) If the same specimen was run in a Debye-Scherrer camera of 57.3 mm radius, how far apart in millimeters would the components of the two lines appear on the film?
(c) How far apart would they be with a symmetrical back reflection focusing camera of the same radius?
(d) Qualitatively, how would the results in (a), (b), and (c) be affected by changing to chromium radiation?

4.11. Metal A and B form a body-centered cubic intermetallic compound γ. Specimens were prepared and heat-treated by quenching powders from the temperatures indicated, and the lattice parameters of the γ phase were measured with the following results:

Atomic %B	450°C	400°C	350°C	300°C	250°C	200°C
30	3.2215 Å	3.2268	3.2305	3.2338	3.2365	3.2386
35	3.2229					
40	3.2284					
45	3.2334					
50	3.2386					
55	3.2432					
60	3.2479					
65	3.2519					
70	3.2536	3.2528	3.2515	3.2506	3.2500	3.2496

Determine the boundaries of γ phase for the temperature range studied (i.e., plot this region of the phase diagram).

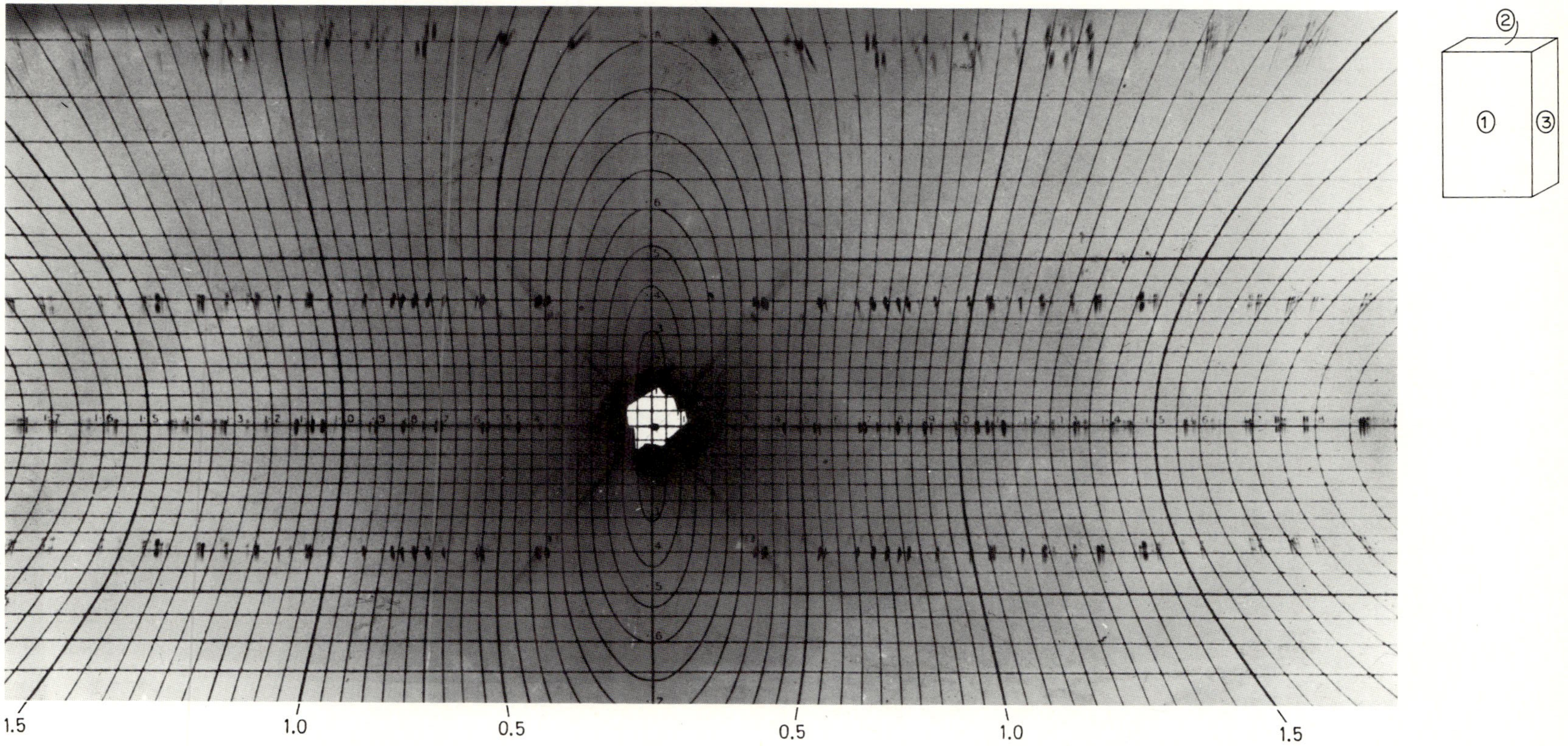

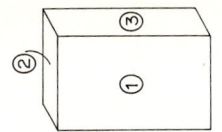

Fig. P4.7

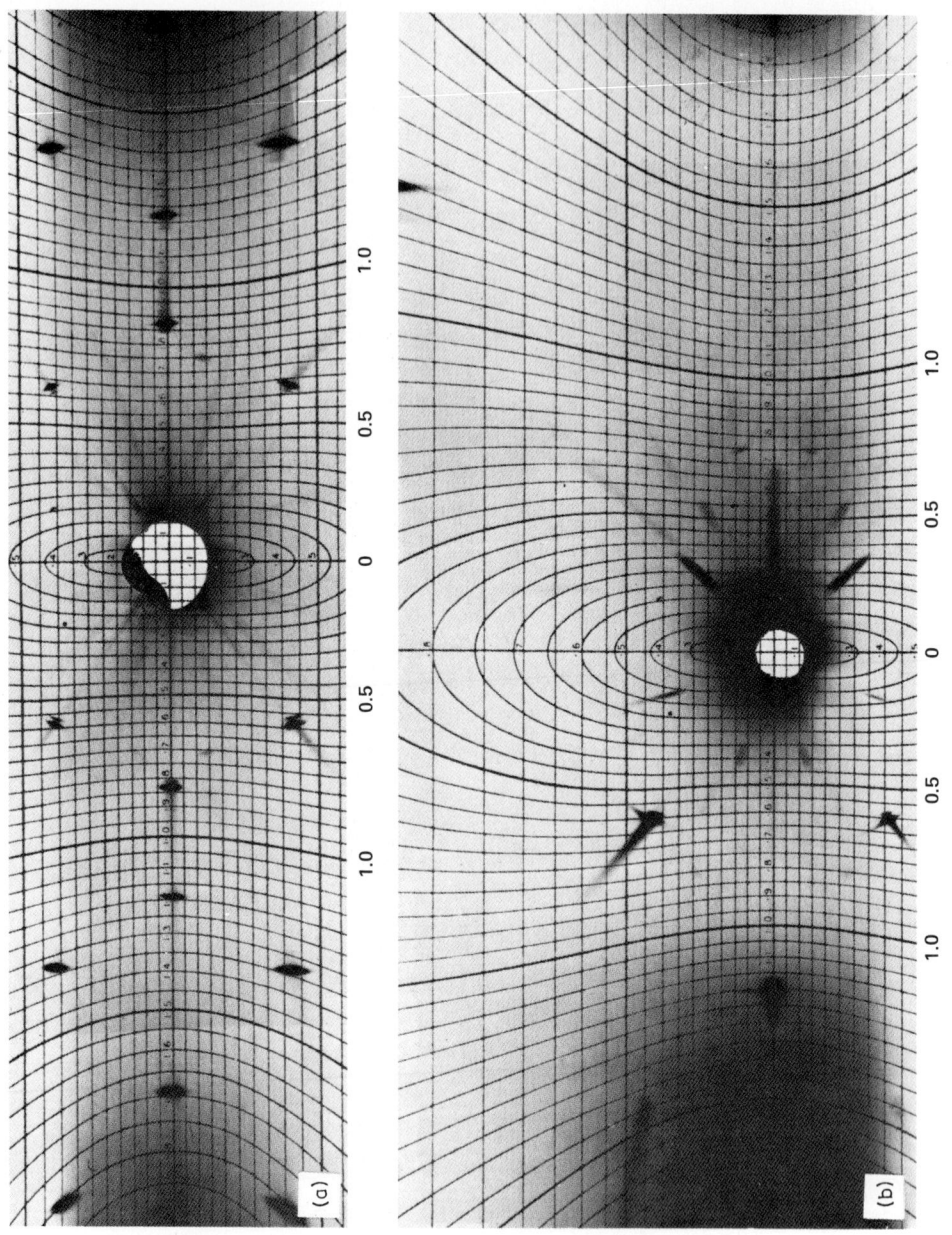

Fig. P4.8a, b

Problems

4.12. An alloy nominally of 50% nickel–50 at.% cobalt is a single-phase solid solution. A Debye–Scherrer film of this alloy made with Cu K_α radiation gave the following $\sin^2 \theta$ values:

0.1420   0.1894   0.3787   0.5207   0.5681   0.7574   0.8994

The measured density is 8.9 gm/cm³. Is this a defect structure? Assume one atom per lattice point.

4.13. The following data were obtained on an unknown white powder, using filtered copper radiation.

$2\theta°$	$I_{\text{above background}}/I_{\text{max}}$	$2\theta°$	$I_{\text{above background}}/I_{\text{max}}$	$2\theta°$	$I_{\text{above background}}/I_{\text{max}}$
28.2	1.00	55.6	0.193	75.8	0.046
35.6	0.078	60.0	0.036	87.3	0.118
46.9	0.672	68.5	0.054	94.2	0.029

Identify this material. Calculate its lattice parameter(s) and compare with the published data.

4.14. Using a Debye–Scherrer camera with fiducial markings every 10°-2θ and filtered Cu K_α radiation, the following data were obtained on an unknown specimen — a gray single phase powder. The sample was a cylinder 1 mm in diameter maintained at 25°C.

(a) Index all the lines and obtain the best possible lattice parameters by suitable extrapolation. (Do not use the card file at this point.)

Position of marking (cm)	Peak position (cm)	2θ	Intensity
20.415		20°	
21.710		30°	
	22.630		MS
	22.985		MS
23.010		40°	
	23.520		S
24.310		50°	
	24.955		M
25.615		60°	
26.915		70°	
	27.025		M
	27.905		VW
28.210		80°	
	28.555		M
	28.765		W
	29.135		MW
29.480		90°	
	29.550		VW
	30.175		W

(continued)

S, strong; M, medium; W, weak; MS, medium strong; MW, medium weak; VW, very weak

(*continued*)

Position of marking (cm)	Peak position (cm)	2θ	Intensity
30.755		100°	
	31.990		W
32.030		110°	
	32.885		M
33.285		120°	
	33.850		W
	34.285		W
	34.375		W
34.550		130°	

(b) Identify the material from both the *d*-spacings and the parameter(s).
(c) Calculate the expected relative intensities for this material to compare with the experimental data.

4.15. Two electron diffraction patterns are presented in Fig. P4.15, taken by Prof. M. Meshii of Northwestern University. The voltage was 75 kV.
(a) Calculate the magnification factors from the known material (b), which is chromium.
(b) Identify the unknown pure element in (a).

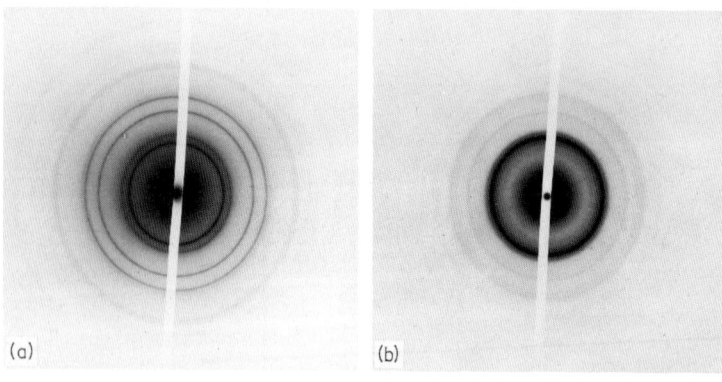

Fig. P4.15a,b

4.16. The single crystal electron diffraction patterns in Fig. P4.16 were taken by Prof. M. Meshii of Northwestern University from a grain of a polycrystalline gold specimen by transmission through a foil, (c) is in the matrix, (b) straddles the matrix and a marking shown in (a). Index the spots in (b) and determine the approximate orientation of the foil. State what you know about the marking. How thick is the foil? The magnification in (a) is 15,400 ×. (The marking runs north–south in (a) and in it there is a trace due to a dislocation which moved and broke the oxide coating on the specimen's surface.)

4.17. Derive the effect of sample displacement in a diffractometer on peak position, if the specimen is used in transmission. How does this compare with the case described in the text for reflection? Of what use might this difference be?

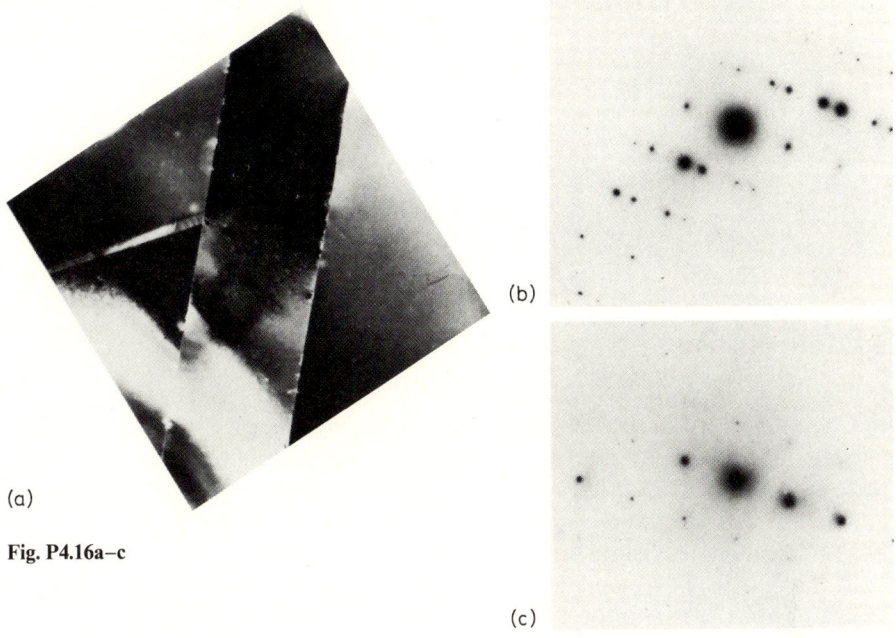

Fig. P4.16a–c

4.18. With a diffractometer and a cubic single crystal how would you set up to: (a) measure along the $h_1 00$ line; (b) explore the $h_1 h_1 0$ plane.

4.19. The dislocations, shown in Fig. P4.19, in a foil of a (fcc) nickel-titanium alloy lie on [111] planes. Estimate the foil's thickness, assuming the dislocations end on the two surfaces of the foil. (Films taken by Prof. S. Sass, Cornell University; magnification: 37,500 ×.)

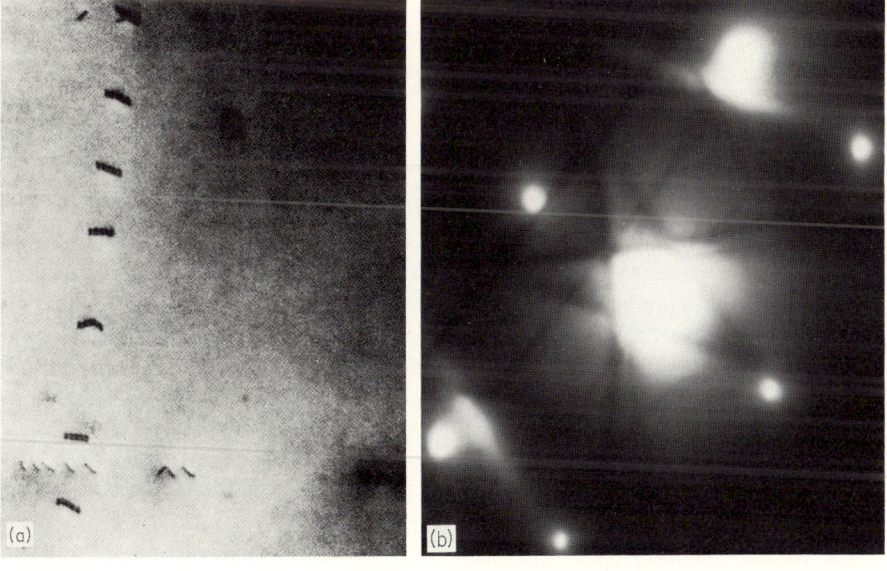

Fig. P4.19a,b

5. Crystal Symmetry and the Diffraction Pattern

5.1 Introduction

In Chap. 1 we introduced the concepts of external symmetry of crystals, form and habit, mirrors, rotation, and center of symmetry. We alluded to the relation between these external manifestations of regularity in solids and the internal regularity of atomic arrangements, but we deferred any detailed discussion of that topic until now. While we noted that in principle all n-fold rotational axes could occur, we said that only 1-, 2-, 3-, 4-, and 6-fold axes were compatible with the translational repetition of a crystal lattice. In this chapter we will prove that statement. We introduced the 14 Bravais lattices — we shall now demonstrate how these lattices may be identified as the only ones necessary to describe all crystalline solids. The reader might imagine that when we begin to discuss the arrangement of atoms in the unit cells of the Bravais lattices, an infinite number of arrangements would be possible. Instead we shall show that the limitations on atomic arrangements imposed by symmetry constrain the number of patterns to only two for one-dimensional crystals, seventeen for two-dimensional crystals, and two hundred thirty for three-dimensional crystals.

5.2 One-Dimensional Symmetry

We can see this effect of symmetry by starting in one dimension as in Fig. 5.1a. A linear, periodically spaced array of construction points is shown. In our discussion, we will assume the array is infinite. This periodic arrangement of imaginary points is referred to as a *space lattice* in three dimensions, or a *net* in one or two dimensions. This array of points in one dimension can be obtained by marking one point, choosing a vector **a** (called the unit translation) and putting its tail at this point. At its head, a new point is marked. The tail of the vector is now placed at this new point, another point is generated at the head, and so on. The periodic repetition could also be obtained by translating the *unit cell* outlined in Fig. 5.1a by **a**. This cell contains only one of the lattice points and it is called a *primitive* cell for this reason; it is therefore given the symbol p. If we choose the cell boundaries at the centers of two neighboring net points, then half of each point belongs to the outlined

5.2 One-Dimensional Symmetry   239

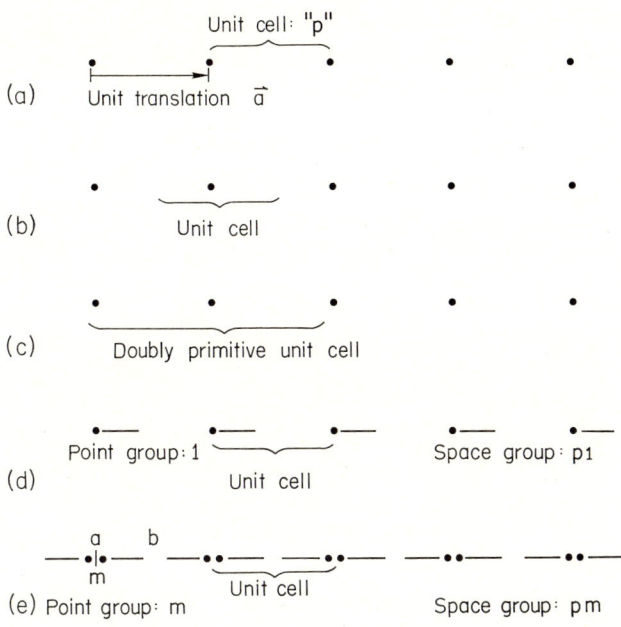

Fig. 5.1a–e. Periodic crystals in one dimension. **a** net; **b** alternate picture of the primitive unit cell in **a**; **c** a doubly primitive cell (one with two lattice points per cell instead of one as for a primitive cell); **d, e** possible arrangements of the molecule

cell and half to the cell to the left (or right) of the one indicated, as in Fig. 5.1a. Alternately, if we start the outline of the primitive cell just to the left of one point as in Fig. 5.1b, it ends just to the left of the next point to the right, and clearly the cell contains only one net point. We could also choose a cell that is not primitive, i.e., one containing two (or more) net points with the understanding that we carry the included point with us when we translate the cell to form any net. Such a cell is outlined in Fig. 5.1c. Again, we can take the origin of the cell just to the left of one net point, to see clearly that there *are* two points in the cell.

In Figs. 5.1d and e, we have placed one-dimensional asymmetric "molecules" (•—) at each of these points, in two different ways. Note that these are the *only* two possible ways; we could change the spacing of the points or the size of the molecule, but these changes do not alter the *kind* of arrangement. The reader should try other arrangements and convince himself that they are all equivalent to these two cases.

In Fig. 5.1d, our molecule is arranged at the point in a certain way. A 360° rotation about the net point brings the molecule into coincidence with itself. In Fig. 5.1e, the molecules are related at a net point by *reflection in a plane* perpendicular to the translation vector (labeled *m*), or by 180° rotation around the net point. Repeated application of either of these two *symmetry operations* will bring the molecule back on itself again, but since only one of the two operations is needed to describe the arrangement, the reflection operation will be employed. There are also *centers of inversion* at and between lattice points. That is, each molecule can be produced from another by drawing lines from points on it through these centers an equal distance on the other side of the center. Notice that these *operations at a net point* or *point symmetries* not only bring one molecule into coincidence with another but do the same thing with *all* the molecules in the structure. That is, all the molecules to the left of a net point are related to those to the right by reflection in the net point. Studying Figs. 5.1d and e, several facts are obvious:

(1) *All net points have identical surroundings* — this *must* be so to have a periodic arrangement.
(2) The unit cell in Figs. 5.1d and e is primitive; it contains only one net point even though in Fig. 5.1e there are two molecules in the cell. The repeating arrangement can be represented by a net point between them (or some place else), and the net will be the same as that in Fig. 5.1a. In Fig. 5.1e, if *a* is a net point, *b* is not because the surroundings are *not* the same at the two locations. Either position is a possible net point, but the next one is **a** away from it. In fact, the choice is arbitrary and *any* starting point for the net would do as well. However, one choice might be more convenient than another because it more clearly represents the chemical nature of the molecule, or it makes it easier to do calculations.
(3) The arrangement of molecules around a point such as a net point exhibits certain symmetry to the eye, which can be represented by imaginary *symmetry elements*. These cause the pattern to be brought into coincidence with itself. *The collection of such symmetry elements at a point is called a point group.* In Fig. 5.1d, there is only the redundant 360° rotation axis. This axis is a *onefold axis*, and this point group can be written simply as "1." In Fig. 5.1e, the point group can be described by a mirror for the reflection symmetry element, (with a symbol *m*) or an *inversion center* (*i*).
(4) In one dimension only two types of periodic molecular arrangements are possible, denoted *p*1 and *pm*. These are the two *space groups* in one dimension.
(5) Other symmetry elements can arise, such as the inversion centers and *twofold rotation axes* at and halfway *between* two net points, and the mirrors halfway between the points in Fig. 5.1e (the reader should locate these symmetry elements), even though we only need a mirror plane *at* the points and translation to construct the figure. Conversely, we can construct the figure with translation and the twofold axes, and mirrors will then also be present at and between lattice points as well as a twofold axis between the points and the inversions. Or, we could employ only translation and a mirror *between* lattice points and the other elements would appear. All of these combinations describe identical arrangements of molecules, and hence are equivalent.

The entire spatial arrangement, *the unit cell (or translation) combined with the symmetries at a point (point group), define what is known as a space group.* The symbols for one dimension are, as shown in Fig. 5.1d and e, *p*1 and *pm*. These symbols define the lattice and the symmetry at the lattice points required to construct the one-dimensional crystal. The additional symmetry elements that arise in pm are not needed for the construction and are therefore not indicated in the symbol. *Clearly, periodicity limits the possible atomic arrangements.*

5.3 Two-Dimensional Symmetry — Point Groups

We could relax our one-dimensional restriction a bit, perhaps by allowing our molecule to be finite in shape in a direction perpendicular to the single dimension employed above and to have freedom in the second dimension on the page. Rotation around the normal to the page and reflection would not then be equivalent, and even if the net was a one-dimensional row of points, new space groups would arise. (The reader might try to examine these himself at this point. Such one-dimensional schemes might arise in making long-chain molecules common in the fields of biology and polymer science.) We shall, however, go directly to two dimensions, now requiring that our configurations be strictly two-dimensional and that all operations on our molecule be in two dimensions. As with one dimension, we start with one point and put vector $\mathbf{a}_1$ in one direction, vector $\mathbf{a}_2$ in the other direction. (Of course we must define the angle between $\mathbf{a}_1$ and $\mathbf{a}_2$, and we will soon consider if there are limitations on possible values of this angle.) Then at the end of each vector, we put new points. At each of these two new points, we put our vectors $\mathbf{a}_1$ and $\mathbf{a}_2$ again, parallel to the directions they were pointing at the first point and so on to form a two-dimensional net such as that shown earlier in Fig. 1.7. Alternately we can form the net by translating a unit cell parallel to its two edges, stacking these so they just touch but do not overlap. Two such cells which are primitive are labeled (a) in Fig. 1.7. The reader can see clearly in one of these that the cell contains but one lattice point. In the other, each of the four net points at the corners of the cell belong $\frac{1}{4}$ to the drawn cell, and $\frac{3}{4}$ to cells around it. Cells (b) and (c) are doubly primitive. (b) is "edge-centered," while (c) is "centered." Not only do these cells contain two lattice points, but the area is twice that of the primitive cell.

A good two-dimensional object to represent a molecule is a "stick-hand" shown in Fig. 5.2. Consider first symmetry operations about a point. Figures 5.2a and c [or (b) and (d)] are related by two successive reflections through mirrors m_1 and then m_2, by twofold rotation, or by inversion through the point at the intersections

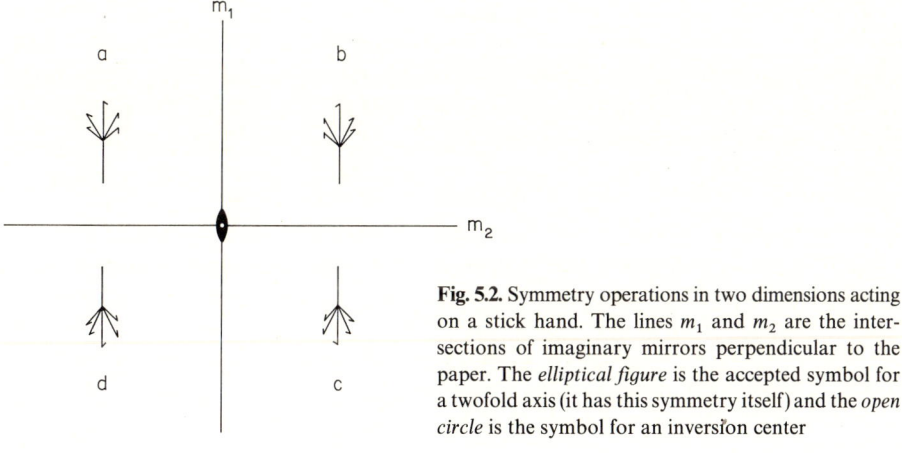

Fig. 5.2. Symmetry operations in two dimensions acting on a stick hand. The lines m_1 and m_2 are the intersections of imaginary mirrors perpendicular to the paper. The *elliptical figure* is the accepted symbol for a twofold axis (it has this symmetry itself) and the *open circle* is the symbol for an inversion center

of the mirrors. In these operations called *congruent* a left hand remains a left hand and (a) and (c) [or (b) and (d)] are congruent figures. But (a) and (b) are related by reflection only; a left hand is turned into a right hand. The single reflection is called an *enantiomorphous operation* and figures (a) and (b) are enantiomorphous. Thus in two dimensions a single rotation and a single reflection produce different results; this was not so in one dimension.

[If we relaxed our two-dimensional conditions and allowed rotation out of the paper, could (a) and (b) be related by a twofold (180°) rotation axis parallel to m_1? Yes! But with your own hands this is not possible, because of the difference between the front and back of a hand! Your hands are three-dimensional and reflection becomes a more unique operation as the dimensionality increases.]

We see in this figure that we can have reflection, rotation, and inversion as symmetry operations, as in one dimension. Inversion is equivalent to twofold rotation in two dimensions and therefore we will not need to consider it further. But *reflection and rotation are no longer equivalent as they were in one dimension.* Furthermore, some combinations of operations are equivalent to others such as two reflections being equivalent to an inversion or a twofold rotation.

There is no reason why we have to restrict our consideration to onefold or twofold axes at a point. There could be eightfold rotations (360°/8 = 45°) or fivefold axes (360°/5 = 72°) or *n*-fold axes (360°/*n*). However, the requirement of periodicity (that each net point have identical surroundings) places restrictions on the amount of rotation allowed at each point. These rotations act on the surroundings of a point, that is, on the entire net, and must bring it into coincidence with itself. This is illustrated in Fig. 5.3. If there is an *n*-fold axis at any one lattice point, periodicity requires that an *n*-fold axis appear at every lattice point. When this rotation axis operates on the translation vector $\mathbf{a}_1$, it rotates it by an angle $\alpha = 360°/n$ clockwise (or counterclockwise) to end on another lattice point, since each rotation must bring the lattice into coincidence with itself. Two such rotations, as indicated in the figure, produce a vector $\mathbf{b}$ parallel to the $\mathbf{a}_1$ axis. As the vector $\mathbf{b}$ extends from one lattice point to another, it must have length $|\mathbf{b}| = l|\mathbf{a}_1|$, where l is an integer. But from the figure we see

$$la_1 = b = a_1 + 2a_1 \cos \alpha , \tag{5.1a}$$

or

$$\cos \alpha = (l - 1)/2 = L/2 , \tag{5.1b}$$

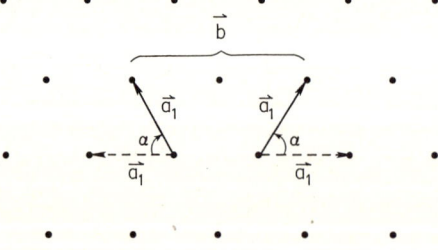

Fig. 5.3. Why only certain rotations are possible in a net. (After Buerger, M.J., "Elementary Crystallography." Copyright 1956, John Wiley & Sons. Reprinted by permission of John Wiley & Sons, Inc.) Note that in their original positions (*dotted*) the $\mathbf{a}_1$ vectors terminate on lattice points

5.3 Two-Dimensional Symmetry — Point Groups

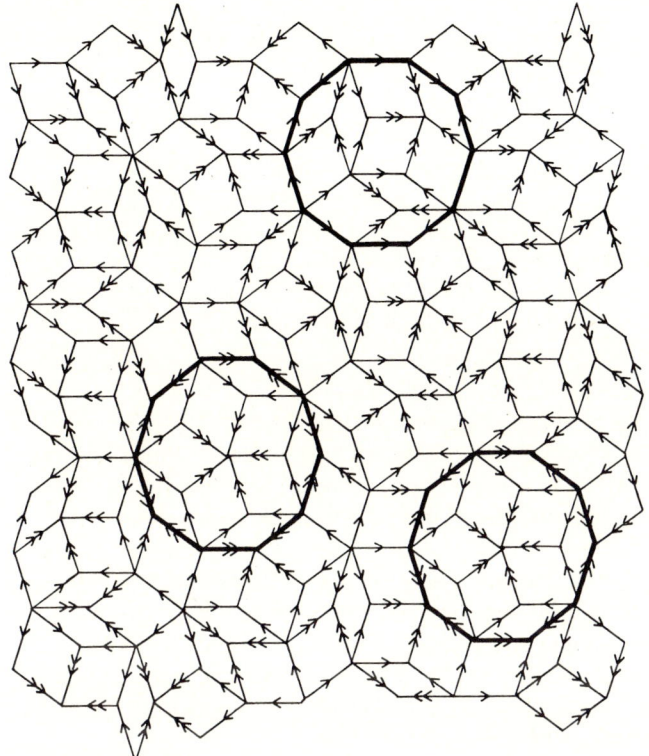

Fig. 5.4. A two-dimensional array composed of two unit cells, rhombuses of acute angles of 72° and 36°, fit together to fill space and yet exhibit local five-fold symmetry (note the outlined decagons). The associated diffraction pattern is discussed by Nelson and Halperin (1985)

where L is an integer. The possible values of L are 0, ± 1, ± 2 since $|\cos \alpha| \leq 1$. Thus α can be 90°, 360° or 180°, 60° or 120°. *Only one-, two-, three-, four-, or sixfold rotational axes are possible in a two-dimensional net*, although a value of rotation of $360°/n$ where n is any integer is possible at a point that is not part of a net.[1]

---

[1] Any rotation of $360°/n$ will cause coincidence about a single point; however periodic repetition of a single unit cell containing a five-fold rotational axis is not possible. Real materials need not be periodic, and may, under certain restricted conditions exhibit five-fold symmetry. It is possible, for example, to fill space with an admixture of two distinct unit cells. Each type of unit cell can be imagined to be occupied by a fixed arrangement of atoms. The existence of these two unit cells, occurring in an irrational proportion allows deviations from the rules derived above and includes the possibility of five-fold rotational symmetry. The requirement that space be filled places rigid constraints on the allowable repetition patterns of these two unit cells, and equally rigid constraints on the resultant diffraction patterns. A two-dimensional example of such an admixture of two cells is shown in Fig. 5.4 (see Nelson and Halperin 1985). Materials exhibiting these unusual symmetries have been dubbed "quasi-crystals" and have recently been observed in some rapidly solidified metal alloys [see Shechtman et al. (1984) and Sect. 6.11].

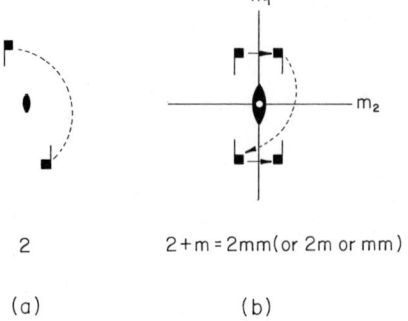

Fig. 5.5a, b. The point groups 2, 2mm. Note in **b** that one mirror (m_1) and a twofold axis result in a second mirror (m_2)

Just as in the case of one dimension, we now wish to list all combinations of symmetry elements which may exist at a point (point groups), all the unique choices of vectors $\mathbf{a}_1$ and $\mathbf{a}_2$ (nets, or two-dimensional lattices), and finally all the unique combinations of point groups and nets to form the two-dimensional space groups. *Combining the possible rotations with mirror planes, 10 point groups consistent with two-dimensional periodicity are possible*: 1, 2, 3, 4, 6, 1*m*, 2*mm*, (or *mm*), 3*m*, 4*mm* (or 4*m*), and 6*mm* (or 6*m*).[2] Remember when trying to draw these, that to maintain a two-dimensional figure, the mirrors and axes are normal to the plane of the drawing. The development of two of these is shown in Fig. 5.5. Several points are worth noting. First the type of axis is represented by drawing a simple filled geometric figure, which is brought into coincidence with itself by the rotation (an ellipse for a twofold axis, an equilateral triangle for a threefold axis, a square for a fourfold, a hexagon for a sixfold axis).

A combination of symmetry elements may create others as we saw in the one-dimensional examples. For example, in Fig. 5.5b, m_2 is created from the combined effects of m_1 and the twofold axis. The configuration of molecules at a point must exhibit symmetry due to m_1 and the twofold axis if these two elements are present, but then it also exhibits additional symmetry. There is also an inversion center in Fig. 5.5b represented by the open circle in the ellipse; we did not need it to construct the figure, but it is there when we do so! The reader should draw figures for the remaining eight point groups. Use a figure such as the flag symbol shown in Fig. 5.5, and let the symmetry elements of the point group act on this flag, reproducing it in positions required by the point symmetry. Search for other symmetry elements in your figures, symmetries which are not required to draw the figures, but appear after they are drawn.

---

[2] Certain abbreviations are common. 1*m* is often written as *m*. 2*mm* is often written as *mm* because a twofold axis always occurs at the intersection of two perpendicular mirrors. The second mirror arises anyway from the combination of a twofold axis and one mirror and it relates part of the pattern in a different orientation than the first mirror so both mirrors are used in the notation. Similarly 4*mm* and 6*mm* are often abbreviated as 4*m* and 6*m*.

5.4 Two-Dimensional Symmetry — Lattices

To determine the possible lattices, we must find combinations of noncollinear vectors $\mathbf{a}_1$ and $\mathbf{a}_2$ which satisfy Eq. (5.1). Let us return to Eq. (5.1) and calculate $|\mathbf{b}|$ for each possible α. We find the following:

(1) *For a one- or twofold axis, there are no restrictions on the net*, i.e., $\cos \alpha = 1$ or -1 and $|\mathbf{b}| = 3a_1$, a_1, respectively, and a_2 may have any magnitude. *Every* net of any form has 180° and 360° rotational symmetry, because points in a net have no shape. The unit cell can be a general parallelogram (p) or a rectangle, (r). To see that this is so, draw some two-dimensional lattices.

(2) *For three- or sixfold axes, a rhombus is the basic shape* since $|\mathbf{b}| = 0$ or $2a_1$, $\alpha = 60°$ or 120°, and $|\mathbf{a}_2| = |\mathbf{a}_1|$. Some readers may wonder why an equilateral triangle is not used for the sixfold axis ($\alpha = 60°$). It can fill all space as do the other figures we are using, but it will not fill space just by translation parallel to its sides as shown in Fig. 5.6a. By contrast, the rhombus does fill all space by translation as shown in Fig. 5.6b, and contains the allowed angles of $\alpha = 60°$ and 120°.

(3) *For a fourfold axis the figure is a square* since $\alpha = 90°$, $|\mathbf{b}| = a_1$, and $|\mathbf{a}_2| = |\mathbf{a}_1|$.

As in the one-dimensional case, alternate nonprimitive cells may be chosen to define each of the nets; however, there will *always* be a square cell when only a fourfold axis is present, or a rhombus when a sixfold axis is present. These cells have the desirable feature that they clearly exhibit the symmetry of the net they represent.

A *net or two-dimensional lattice* can then be defined by stating whether the cell edges are equal or not, and the angle between these edges; there are only four such *axial systems* in two dimensions. The four systems have primitive cells defined by these edges and angles: parallelogram, rectangle, rhombus, and square. The nets resulting from the translation of these cells have at least onefold and twofold symmetry (and obviously more if three-, four-, or six-fold axes are involved) and all

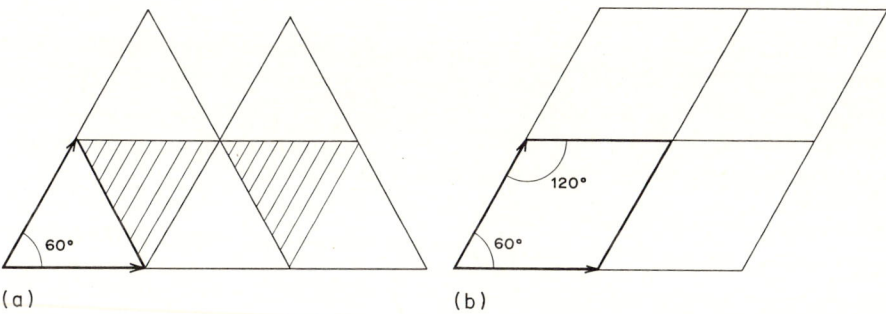

Fig. 5.6. A possible choice for a unit cell with $\alpha = 60°$ is the equilateral triangle, but as shown in **a**, translation parallel to lattice vectors does not fill the *shaded spaces*. The rhombus shown in **b** satisfies the angular requirements and does fill space by translation alone

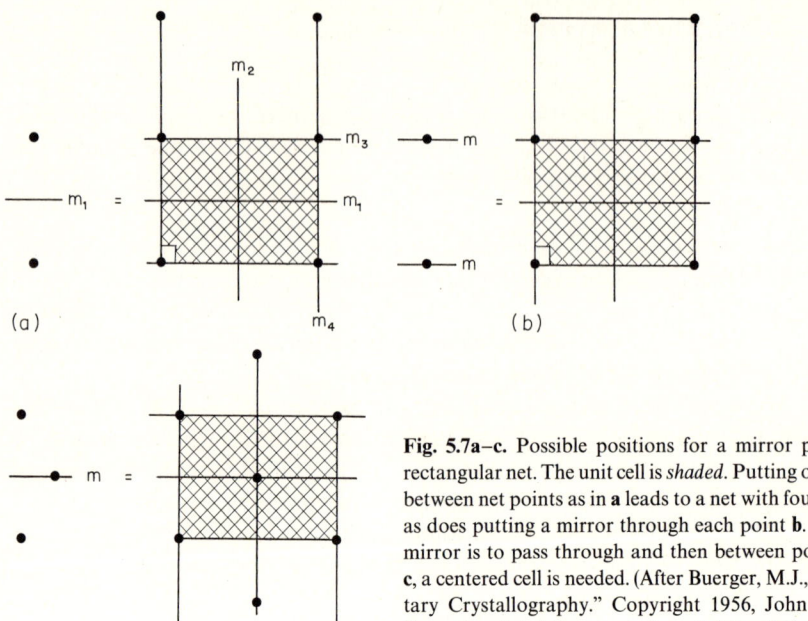

Fig. 5.7a–c. Possible positions for a mirror plane in a rectangular net. The unit cell is *shaded*. Putting one mirror between net points as in **a** leads to a net with four mirrors, as does putting a mirror through each point **b**. But if the mirror is to pass through and then between points as in **c**, a centered cell is needed. (After Buerger, M.J., "Elementary Crystallography." Copyright 1956, John Wiley & Sons. Reprinted by permission of John Wiley & Sons, Inc.)

but the parallelogram have mirror planes. The reader should draw a net based on a rhombus or a square and try to find these mirrors.

A periodic structure based on a general parallelogram can not host mirror planes. With a rectangular axial system, there *can* be mirror planes and so point groups *m* and *mm* can only appear in such a cell. If we draw a net based on a rectangular system, there are mirrors halfway between net points and mirrors through lattice points, as shown in Fig. 5.7a. In fact if we start with two net points and require only one mirror plane, say, between lattice points as in Fig. 5.7a, we can immediately see why a rectangle is required because the two points must lie on a line perpendicular to the mirror. Let us explore this procedure further. If we begin again with two lattice points as in Fig. 5.7b and pass the mirror through the lattice points, we arrive at the same result as that for the mirror between points as shown in Fig. 5.7a. Both have four mirror planes, though we started with only one.

Now consider staggered columns of lattice points represented by the three points indicated in Fig. 5.7c. When mirrors are passed through these lattice points, so that a given mirror passes between points in one column and through points in the next, a new kind of cell arises. This cell has a lattice point in the center of the cell and is denoted *c* for centered. This is a nonprimitive cell for this lattice, containing two lattice points per cell, but it clearly shows the presence of the mirror symmetry operations. Of course a smaller primitive cell could be drawn by connecting the centered point to corners; however, this primitive cell would not be rectangular and would conceal the presence of the mirror symmetry. Searching for additional combinations, we might try this procedure with the mirror between two columns of points, then through two columns of points etc. We would find that the *pattern is the same* as that when alternating every column. In fact, no further possibilities exist. For example, if we try this alternation with a square, a cell with a centered point also occurs but a square primitive cell can be drawn in this case by connecting this centered point to corners. *This* primitive cell does exhibit the mirror and fourfold symmetry, so it is not a new pattern.

5.5 Two-Dimensional Symmetry — Space Groups

We thus have four axial systems but five possible unit cells. In Fig. 5.8 these are shown, with the 10 point groups we have found appropriate to each cell, listed on the left of the figure, and the possible space groups on the right. Additional symmetry elements in the net are also shown. These extra symmetry elements are present in

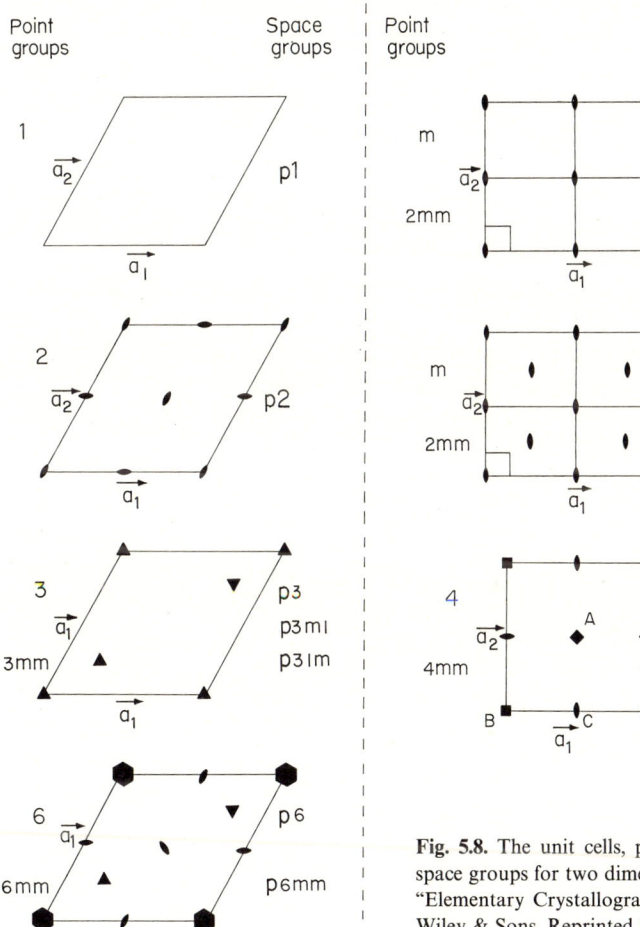

Fig. 5.8. The unit cells, point groups and associated space groups for two dimensions. (After Buerger, M.J., "Elementary Crystallography." Copyright 1956, John Wiley & Sons. Reprinted by permission of John Wiley & Sons, Inc.)

the *net* (but not necessarily in the structure obtained by placing a "molecule" at each lattice point). For example in the drawing of the square unit cell, the fourfold axes at the net points that led to the square cell cause twofold axes along the edges such as at C and a fourfold axis in the center at A. The symbol for the axis at A is shown tilted by 45° compared to the symbol at the corners, because the arrangement of lattice points is tilted by 45° around A, compared to the arrangement around corners. This extra symmetry is the result of combining rotation and translation. For example, consider a point between B and C in the pattern for the square in Fig. 5.8. Rotate it 90° counterclockwise around B to form a second point and translate it to the right by $\mathbf{a}_1$ to form a third. The first and third positions are related by the fourfold axis at A.

Point group $3m$ can be placed in its cell in two orientations, one with the mirror along the short diagonal of the rhombus ($p31m$) and one with it rotated by 90° to lie along the long diagonal ($p3m1$) so two patterns or space groups arise from this point group.

The total number of space groups described in Fig. 5.8 *is* 13, *but this does not exhaust all possible two-dimensional arrangements.* To identify *all* the possible space groups which describe different or unique patterns, we need to insert molecules and examine the pattern required by the symmetry elements. For this purpose we will employ a flag to represent a general asymmetric shape. We will take the upper left corner as origin and let the x axis be down, the y axis to the right, and thus if we wish, we can give coordinates to our molecular positions. Let x be a fractional coordinate measured in units of a_1 and y measured in units of a_2. We insert a flag at some general position (x, y) near the origin and then let all the symmetry elements operate on it (including point operations and translations), remembering that there are neighboring cells as well. One example for the point group m in the c cell, space group cm, is shown in Fig. 5.9a with the mirrors perpendicular to the y axis. If we start with molecule A, the mirror plane at the lattice point produces B and, as every lattice point is identical, we can place such figures at all the points, corners, and

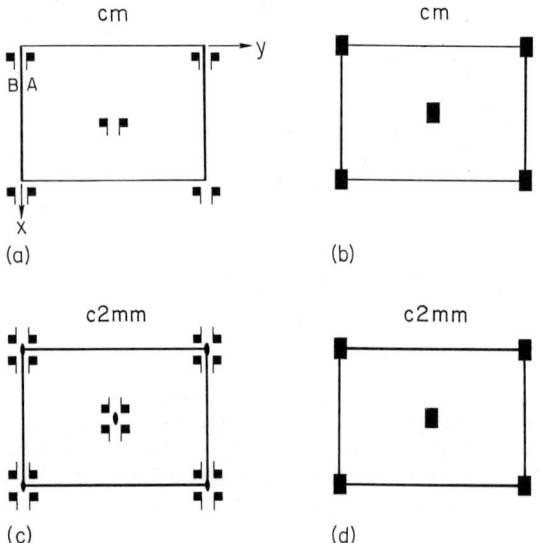

Fig. 5.9a–d. Space groups cm and $c2mm$. Molecules represented by *flags* are shown at positions corresponding to the general equipoint $[(x, y)$ and symmetry related points] in **a** and **c**, and at the special equipoint $[(0, 0)$ and $(\frac{1}{2}, \frac{1}{2})]$ in **b** and **d**

5.5 Two-Dimensional Symmetry — Space Groups

center. It is immediately apparent that many of the symmetry elements of the *c unit cell* (Fig. 5.8) are *not* present in this arrangement of flags. The horizontal mirrors and twofold axes are all missing. If we wish to place a molecule at a lattice point in *cm*, i.e., at coordinates $(0, 0)$, this molecule must itself have at least mirror symmetry, as shown in Fig. 5.9b. In Fig. 5.9c, the pattern for a molecule at (x, y) in the *c2mm* space group is shown. All of the symmetry elements of the *c* net *are* present in this higher symmetry pattern. Note that when a twofold molecule is placed at $(0, 0)$ in *c2mm*, as shown in Fig. 5.9d, the resultant pattern is the same as that obtained by putting that molecule at $(0, 0)$ in *cm*. In this case the higher symmetry space group is the appropriate description. The reader should similarly examine the patterns produced by the other space groups shown in Fig. 5.8.

The space groups listed in Fig. 5.8 do not yet exhaust all the possibilities for two dimensions. To obtain all possible spatial arrangements of periodic structures, we need to consider not only placing the point groups consistent with periodicity in the various nets at each net point, but also *combined operations*. We have aleady considered one such combination, *rotation followed by translation*, when we considered the arrangement at a point followed by translation. This led to the discovery of additional symmetry elements but not new patterns. But if we consider *combining reflection and translation* at a general angle as in Fig. 5.10a, the combined operation produces a displaced right-handed figure from a left-handed one, and we shall see that this leads to new patterns. As indicated in Fig. 5.10b, the result can be considered to be a single operation called a *glide plane* (g) displaced from the mirror. Now the resultant translational component is parallel to the glide plane. The combined operation is represented by a broken line. As a "left" is produced from a "right", the glide operation must occur at even multiples between lattice points to reproduce the same arrangement (the right again) at the next lattice point. Thus, the translational component must be half a cell edge (or one-fourth, etc.). The reader might consider why a glide translation of, say, $a/4$ or $a/6$ really need not be considered — new patterns are not produced.

In Fig. 5.11a the space group *p2mm* is shown. In Fig. 5.11b one of the mirrors has been replaced by a glide and a new pattern results. A flag located at (x, y) is repeated at $(x + \frac{1}{2}, -y)$ by the glide operation, at $(-x, -y)$ by the twofold rotation axis at the origin, and at $(-x + \frac{1}{2}, y)$ by the twofold rotation axis at the $(\frac{1}{2}, 0)$ position. Note that this pattern reveals horizontal mirrors which were not present in *p2mm*. The pattern in Fig. 5.11b is clearly a new space group and is denoted *p2mg*.

In Fig. 5.11c, both mirrors are replaced by glides, and a third structure occurs, denoted *p2gg*. Note that the group of molecules at the center is oriented differently than the ones at corners, so this is not a centered net. Also the glide planes are

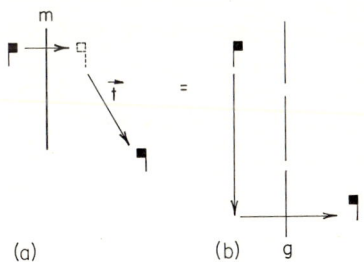

Fig. 5.10a,b. Reflection and translation *at an angle* to the reflection lead to a new operation called glide. The glide plane is given the symbol g in two dimensions, and is represented by a *dashed line*. Note that the glide plane is displaced from the original mirror. (After Buerger, M.J., "Elementary Crystallography." Copyright 1956, John Wiley & Sons. Reprinted by permission of John Wiley & Sons, Inc.)

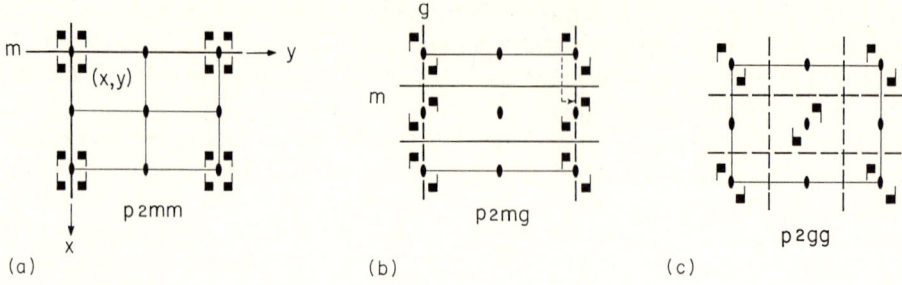

Fig. 5.11. How glide planes lead to new space groups

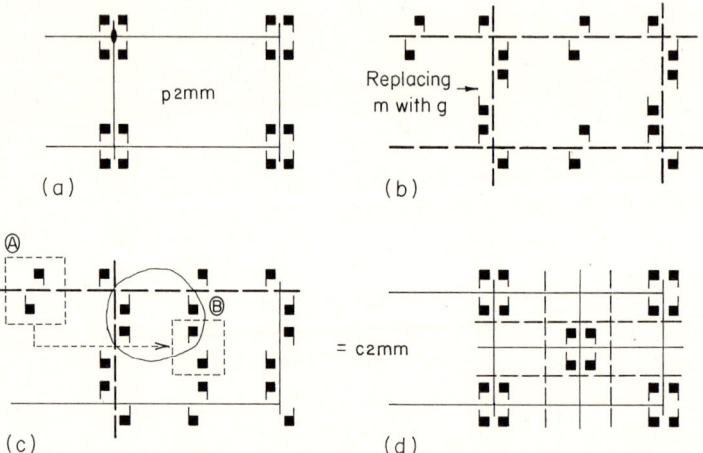

Fig. 5.12a–d. Putting two glide planes at the same positions as mirror planes does not produce a new space group. Starting in **a**, *p2mm*, the mirrors are replaced by glides resulting in **b** when only molecules in one cell are considered. In **c** the effect of glide on one group of molecules in another cell is considered. The resultant *circled groups* are the same *pattern* as in **d**, *c2mm*

displaced from the mirrors. If we try a figure with the two mirrors replaced by glides without this displacement, *c2mm* results, as shown in Fig. 5.12. (We could perhaps anticipate this because in Fig. 5.9c of *c2mm*, there are already glide planes present, but not indicated, halfway between all the mirror planes.) Two new space groups have therefore been derived by changing mirrors to glides in *p2mm*, these are *p2mg* and *p2gg*. Similar considerations with the other space groups will lead to the addition of only *pg* derived from *pm*, and *p4gm* derived from *p4mm*, for *a total of 17 space groups in two dimensions.*

The reader will find these patterns in common use in wall paper designs, men's ties, floor tiling, etc. Manufacturers of these items are well aware of these 17 basic structures. They can only obtain a new design by changing the shape of the molecule or by eliminating the periodicity!

We have placed our flag molecule in a "general position" (x, y) in the space group's unit cell. *The set of molecules produced by the symmetry operations in the unit cell is called an equipoint, and the number of these molecules in an equipoint is*

5.5 Two-Dimensional Symmetry — Space Groups

called the rank of the equipoint. The molecule (or some part of a molecule) is often called the *asymmetric unit.* If we were to place the molecule at certain *special* positions, such as *on* symmetry elements, then the molecule must have a shape consistent with the symmetry of these elements and the rank of the equipoint would change. This is illustrated in Fig. 5.13, a drawing similar to those in the International Tables for X-ray Crystallography. The drawing on the right of Fig. 5.13 shows the symmetry elements in the unit cell. The drawing on the left shows the positions of molecules which result when the symmetry of the group operates on a molecule placed in the general position (x, y). The positions of this equipoint of rank 4 are indicated below the drawing, where the bar over a coordinate means "minus."

Suppose we put our first molecule not at (x, y) but on the mirror plane parallel to the a_2 axis; that is, we put our first molecule at $(\frac{1}{4}, y)$. To allow this, the molecule must have horizontal mirror symmetry. The glide operation requires a left-handed molecule at the position $(\frac{3}{4}, \bar{y})$. But this molecule is related to a molecule at $(1 + \frac{1}{4}, y)$ by the twofold axis at $(1, 0)$. The latter molecule has the same configuration as the molecule at $(\frac{1}{4}, y)$ because it is located by translation a_1 from $(\frac{1}{4}, y)$. An alternate way to come to this result is obtained by setting $x = \frac{1}{4}$ in the listed coordinates of the general equipoint. When this is done, the left-handed molecule at (x, y) and the right-handed molecule at $(\frac{1}{2} - x, y)$ "coalesce" into a single molecule. The rank of this equipoint is two, and the point symmetry at each of the two points $(\frac{1}{4}, y)$ and $(\frac{3}{4}, y)$ is m. We urge the reader to try to obtain the number and positions of the equipoints, for both general and special positions, for several of these two-dimensional space groups. The solutions may be found in Volume I of the International Tables (1st edition).[3]

We have found in previous chapters that diffraction experiments on crystals allow us to determine the Bravais lattice, the lengths of the edges of the unit cell and the hkl indices of the reflecting planes. From the lengths of the cell edges we could calculate the area (A) of a unit cell (in square meters). If we could measure the density of our crystal ρ in kilograms per square meter (we are dealing with two dimensions), we could calculate the mass in one cell as ρA. A chemical analysis would tell us the mass of each element in our crystal, and with the known atomic weights of these elements, the number of atoms in each cell could then be determined. Suppose that for the space group shown in Fig. 5.13, it turns out that within the usual analysis error of a few tenths of a percent, the composition for one cell is $D_4 B_2$ (or two molecules of $D_2 B$). Then since all D atoms are identical, the D atom must be in the rank 4 equipoint, or in 2 rank 2 equipoints with different y parameters, and B must be in one of the rank 2 equipoints. Note that the magnitude of the y coordinate for the D atom in the general equipoint (x, y) is not required to be the same as the y coordinate for the B atom. We are assuming we have a compound. If we have a solid solution, D and B can be intermixed on one rank 4 and one rank 2 equipoint, or we might have all atoms in rank 2 positions with different y coordinates. This latter possibility might be prevented by the atomic sizes of the atoms involved. From the atomic dimensions and the size of the cell, it might turn

---

[3] A convenient list of information to be found in the International Tables may be found in Appendix A of this volume.

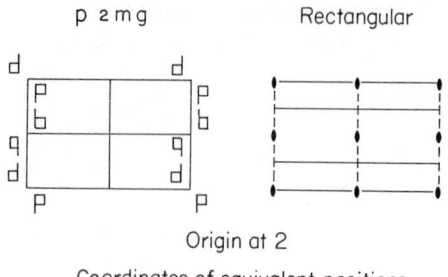

Fig. 5.13. An example of a two-dimensional space group. (Adapted from "International Tables for X-ray Crystallography," Vol. I, 3rd ed. Kynoch Press, Birmingham, England, 1969.) *Open flags* represent left-handed figures, *closed flags* are right-handed figures

Rank of equipoint and symmetry at each of its positions		Coordinates of equipoints
4	1	$x, y; \bar{x}, \bar{y}; \frac{1}{2}+x, \bar{y}; \frac{1}{2}-x, y$
2	m	$\frac{1}{4}, y; \frac{3}{4}, \bar{y}$
2	2	$0, \frac{1}{2}; \frac{1}{2}, \frac{1}{2}$
2	2	$0, 0; \frac{1}{2}, 0$

out that not all of the rank 2 equipoints are possible, without placing the atoms closer together than the sum of the radii. Thus one equipoint may be favored. Unfortunately, this is not usually the case, and in any event we usually want to know precisely the x, y coordinates. In Chap. 6 we will see how this information about atomic coordinates can be determined with the aid of diffraction. (In fact, most of our experimental information on atomic sizes and how they change with structure comes from such studies!)

Before proceeding, the reader should fully understand the terms net (or lattice), system, unit cell, symmetry element, point group, space group, and equipoint. We could now relax our requirements for two-dimensional nets to allow the atoms or molecules to have slight displacements above and below the plane (so that the two sides of the plane are distinct). This is a real situation; surface films must be arranged in this way and the reader might try to see how this relaxation adds to the space groups so far determined. Instead, we shall continue on to three dimensions. We shall give less detail concerning the derivations in this case, as most of the principles have been well illustrated in our considerations of one and two dimensions. The results of this analysis are tabulated in Volume I of the International Tables, so we shall focus our attention on the new features which arise in three dimensions, allowing the reader to understand the notation and results in the Tables. A systematic development of all three-dimensional space groups can be found in the monograph by Buerger (1956) listed in the references at the end of this chapter.

5.6 Three-Dimensional Symmetry

5.6.1 Space Lattices

Our search for space lattices in three dimensions is greatly simplified by the fact that a periodic three-dimensional array of points contains periodic planes of points

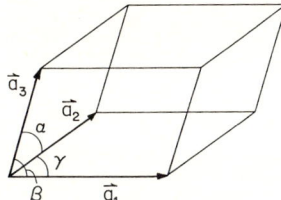

Fig. 5.14. A three-dimensional unit cell showing the definitions of the interaxial angles α, β, and γ. Note that α is opposite to $\mathbf{a}_1$, β is opposite to $\mathbf{a}_2$, and γ is opposite to $\mathbf{a}_3$. This cell is appropriate for the triclinic system

which in turn must have the symmetry of one of the two-dimensional nets, and we have already listed all of these. We can then build up the three-dimensional lattices by stacking two-dimensional nets. Let us start with a net consistent with the point group 1. The unit cell is a parallelogram defined by translation vectors $\mathbf{a}_1$ and $\mathbf{a}_2$. Translation of this net by an arbitrary vector $\mathbf{a}_3$ produces a three-dimensional lattice when $\mathbf{a}_3$ is not in the plane defined by $\mathbf{a}_1$ and $\mathbf{a}_2$. The unit cell for this lattice is defined by the vectors $\mathbf{a}_1$, $\mathbf{a}_2$, and $\mathbf{a}_3$, with all edges and interaxial angles unequal as shown in Fig. 5.14. This is called the *triclinic system*, since all three angles must be specified to define the cell. (If we specialize the angles or axial lengths in any way consistent with periodicity, we will arrive at other cells, the same ones we are about to consider by stacking other two-dimensional nets.)

Next suppose we start with the two-dimensional net consistent with a twofold axis. Again this is a parallelogram but there are four twofold axes per cell (see Fig. 5.8). To maintain the twofold symmetry in three dimensions, a twofold axis in the second layer must lie above one in the first layer. However, the second layer can be stacked with its corner twofold axis over any twofold in the first layer. If we put a corner directly over the same corner in the plane below, the cell in Fig. 5.15a, arises. Note that the cell edge connecting the two layers, $\mathbf{a}_3$, must be perpendicular to the layers as the twofold axis is perpendicular to the layers, i.e., two of the three interaxial angles are right angles but all three axes are unequal. This system is called *monoclinic* as only one angle need be specified.

A second possible stack is shown in Fig. 5.15b with the corner of the second layer above the cell center of the first layer. If we connect the corner points in these two layers, we get a parallelopiped for a cell with third axis $\mathbf{a}'_3$. However, if we continue the stacking, i.e., put a third on top of the second at the same translation $\mathbf{a}'_3$ as the second relative to the first, we get a cell which has the third axis $\mathbf{a}_3$ perpendicular to the first two. This cell is doubly primitive, with a lattice point in the cell center, but its shape reveals the twofold axial symmetry around $\mathbf{a}_3$, so it is the one usually chosen.

The third possibility is stacking the second layer with a corner over one of the twofold axes in the first layer which is along an edge. This arrangement is shown in Fig. 5.15c in which we view the two-dimensional nets from above. When a third layer is added, a cell with lattice points on two faces arises (called "B centered" as the face is defined by the $\mathbf{a}_1$ and $\mathbf{a}_3$ axes). But, as shown in this figure, this cell may be redrawn to give a centered cell. Thus the monoclinic system has only two distinct unit cells.

If we start with nets consistent with a fourfold axis, the corner of the second plane can be above that of the first, or above the fourfold axis in the center of the

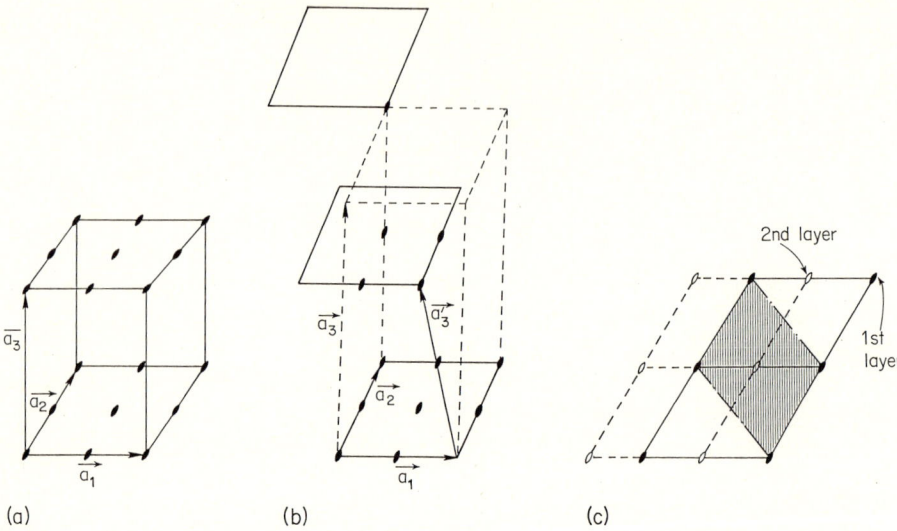

Fig. 5.15a–c. The stacking of two-dimensional nets to make three-dimensional space groups for the *P*2 net. In **a**, stacking a second layer so that its corner twofold axis is over the corner of the cell below leads to a primitive cell. In **b** if the corner twofold of the upper layer is over the twofold in the center of the net below, and a third layer is added in the same way, a centered cell results. In **c** only the twofold axes at lattice points are shown. Viewed from above, the second net is placed over a twofold axis along a cell edge, but an alternate centered cell is shown *shaded*. Thus the lattices in **b** and **c** are identical. Note that in all three cases the third direction is normal to the two-dimensional net, but that none of the three translation distances are equal. (Adapted from Buerger, M.J., "Elementary Crystallography." Copyright 1956, John Wiley & Sons. Reprinted by permission of John Wiley & Sons, Inc.)

unit cell in the first plane, since it has a fourfold axis as well. In the first case, a primitive unit cell arises with all edges at right angles, two axes (in the net) equal, the third, which is different, is perpendicular to the net. In the second manner of stacking, a body-centered cell arises but again all axes are perpendicular, and two are equal. These are cells in the tetragonal (axial) system. The third edge *could* be set equal to the other two, in which case a cubic axial system results.

Proceding in this manner the 14 *Bravais space lattices or unit cells in six axial systems* result. (Bravais was the French physicist who first showed in 1848 that there were only 14 space lattices.) The reader is referred to Chap. 8 of Buerger (1956) for the derivation of all cases. The unit cells have already been shown in Fig. 1.8 and their characteristics are given in Table 1.3 using the notation illustrated in Fig. 1.9.

5.6.2 Point Group Symmetry

Since three-dimensional space contains two-dimensional space as a subset, *all of the point groups for two dimensions are also valid in three dimensions.* However, the additional dimension permits many more combinations of mirrors and rotation axes through a point. For example, we might imagine two twofold axes intersecting

5.6 Three-Dimensional Symmetry

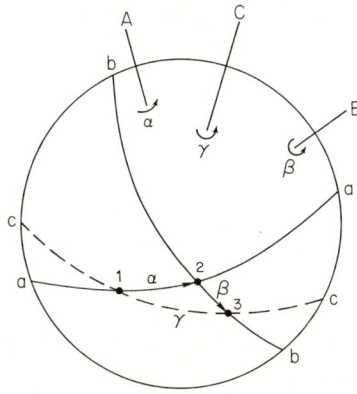

Fig. 5.16. If there are three axes A, B, and C at angles to each other at a point, rotation about A, then about B must be equivalent to rotation around axis C. (From Buerger, M.J., "Elementary Crystallography." Copyright 1956, John Wiley & Sons. Reprinted by permission of John Wiley & Sons, Inc.)

at a point. Our familiarity with the geometrical symmetries of two-dimensional space leads us to ask two questions: (1) Is there some restriction on the angle between these two axes; and (2) Do these two axes together imply the presence of other symmetry elements? The answers to both of these questions are yes. To see this, consider the intersection of the axes A and B at the center of the sphere shown in Fig. 5.16. Let A be an n-fold axis with equator aa, where $n = 360°/\alpha$, and B be an l-fold axis with equator bb, where $l = 360°/\beta$. Choose point 1 on aa at an angle α from the intersection of aa and bb. Then the effect of axis A is to rotate from 1 to 2 by angle α. The effect of axis B is to rotate from 2 to 3 by angle β. As shown in the figure, this is equivalent to rotation from 1 to 3 by some angle γ around an axis C whose equator cc passes through 1 and 3. The fact that we have found only a few allowable n-fold axes for A, B, and C consistent with translation in each two-dimensional lattice plane and the geometry of the spherical triangle limit the number of such three-axis combinations to the *six polyaxial point groups* shown in Fig. 5.17. (The figures drawn in Fig. 5.17 are incomplete. For example, the fourfold axis in 224 operates on the two twofold axes, repeating them every 90°.)

In three dimensions it is possible to have *mirrors perpendicular to rotation axes* as well as through them. We shall designate such a configuration by n/m. Thus $4/m$ refers to a fourfold axis with a perpendicular mirror. The reader should convince himself that when $4/m$ operates on a molecule above the plane of the mirror, it produces eight symmetry-related molecules (four above and four below the plane). We *may add perpendicular mirrors to all the rotation axes and the polyaxial point groups*. We may also include the groups m, $2mm$, $3m$, $4mm$, and $6mm$ discovered earlier. There are still more three-dimensional point groups, but to find these we shall have to consider the inversion operator once again.

Let us look at *inversion through a point*. If you hold your hands parallel to each other, palms inward with fingers of the left hand up and those of the right down, then your two hands are related by an inversion center. Whereas in two dimensions we found that 180° rotation and inversion were the same, inversion appears to be quite unique in three dimensions. We may designate this inversion as i, but also note that it is equivalent to a onefold rotation followed by inversion (denoted $\bar{1}$). We then need to examine n-fold rotation-inversion or roto-inversion axes (denoted

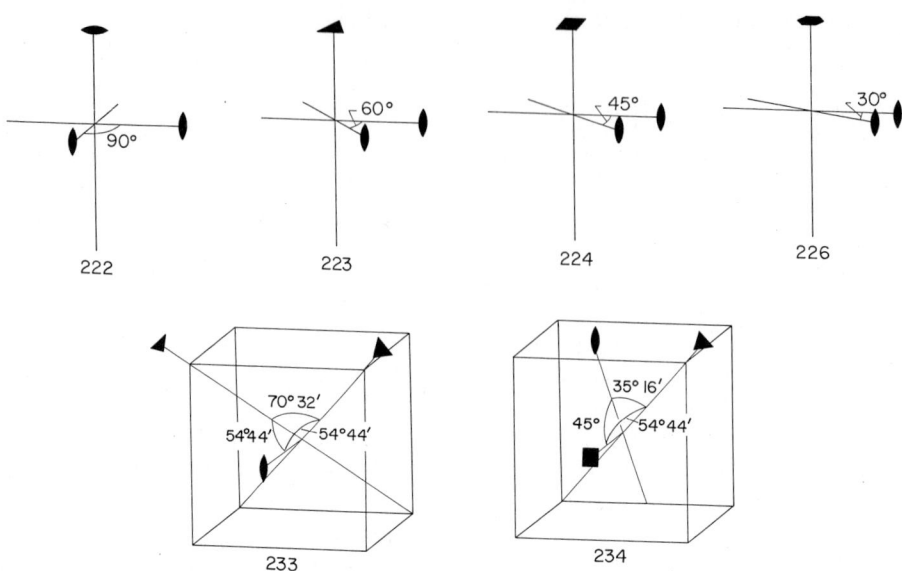

Fig. 5.17. The six permissible polyaxial point groups. (After Buerger, M.J., "Elementary Crystallography." Copyright 1956, John Wiley & Sons. Reprinted by permission of John Wiley & Sons, Inc.)

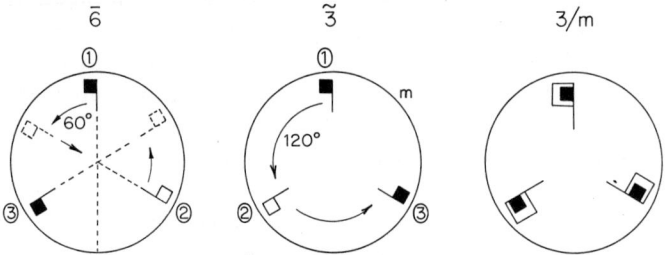

Fig. 5.18. The combined operations of roto-inversion ($\bar{n}$) and roto-reflection ($\tilde{n}$). Only parts of the operations are shown. The complete figure is obtained by continuing the operation until the original position is obtained. *Open figures* are below the plane of the drawing, *closed ones* are above it. The *dotted figures* are the positions of the molecule after the first part of the combined operation, e.g., the rotation prior to inversion in roto-inversion

$\bar{n}$). The designation $\bar{n}$ implies the following: Consider a starting molecule, rotate by $360°/n$ and then invert through the origin to locate a second symmetry-related molecule, repeat until coincidence with the starting molecule is realized. When this is done with a $\bar{2}$ axis a figure identical to that obtained from the point group $1/m$ is obtained. Similarly, $\bar{6}$ is equivalent to $3/m$ as shown in Fig. 5.18. However, $\bar{3}$ and $\bar{4}$ produce new patterns and along with $\bar{1}$ add new point groups to our list. Similarly, when roto-inversion axes are added to the polyaxial groups, most of the groups are equivalent to those already listed, but new ones are also discovered.

We note here that the relationship between your opposed hands could also be described by reflection across a mirror between them, followed by 180° rotation

5.6 Three-Dimensional Symmetry

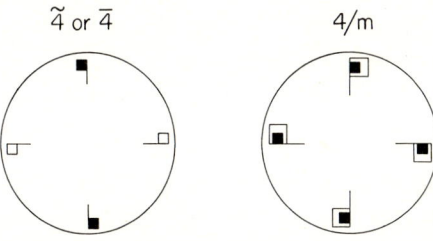

Fig. 5.19. The combined operations $\bar{4}$ and $\tilde{4}$ are equivalent, but differ from 4 plus a perpendicular mirror

around an axis normal to the mirror producing a *roto-reflection axis* (designated $\tilde{n}$ in general, or $\tilde{2}$ in this case). That is, twofold roto-reflection $\tilde{2}$ is equivalent to onefold roto-inversion ($\bar{1}$). Similar equivalences exist between $\tilde{3}$ and $\bar{6}$, $\tilde{4}$ and $\bar{4}$, etc. We could choose one description or the other, but roto-inversion has become the accepted choice. In Fig. 5.18, the operations $\bar{6}$, $\bar{3}$, and $3/m$ are illustrated with open symbols below the plane and closed ones above. The figures for $\bar{6}$ and $\bar{3}$ are not complete, but are intended only to illustrate the sequence of operations; however, if these operations are continued until the starting point is reached again, it will be seen that the operations $\bar{6}$, $\bar{3}$, and $3/m$ produce identical arrangements and hence are equivalent. In Fig. 5.19, the equivalence of $\bar{4}$ and $\tilde{4}$ is illustrated, as well as the unique $4/m$. Note, finally that a *roto-inversion axis is not the same as adding the inversion operator to a rotation axis*. Thus the fourfold roto-inversion axis $\bar{4}$ produces the figure to the left in Fig. 5.19, while the presence of a fourfold axis *plus* inversion is equivalent to $4/m$ shown at the right of Fig. 5.19.

Allowing for rotation axes, polyaxial combinations, mirrors through and perpendicular to axes, and roto-inversion axes, we can derive 32 unique three-dimensional crystallographic point groups. These groups are illustrated in Fig. 5.20 with their conventional designations indicated. On the right of each pair of circles, the symmetry elements are given, while on the left the molecular arrangement consistent with these symmetry elements is shown. In this case, for each drawing the point group has been represented as it would appear in a stereographic projection. Note how roto-inversion axes are indicated; look for example at the $\bar{6}$ axis. Note also that the point groups have been arranged in sets to indicate the axial systems with which they are consistent.

5.6.3 Space Group Symmetry

Proceeding as we did in the development of the two-dimensional space groups, we wish to combine the 32 three-dimensional point groups with the appropriate Bravais lattices to form *three-dimensional space groups*. In two dimensions, the combination of translational periodicity and reflection caused the appearance of a new symmetry element, glide. New kinds of important symmetry also arise in three dimensions when the combination of translation and other elements is considered.

(1) *Glide planes* can be parallel to three planes defined by the crystallographic axes or edges, $\mathbf{a}_1, \mathbf{a}_2, \mathbf{a}_3$ — the $\mathbf{a}_1\mathbf{a}_2$, $\mathbf{a}_2\mathbf{a}_3$, or $\mathbf{a}_1\mathbf{a}_3$ planes. The glide translation in any

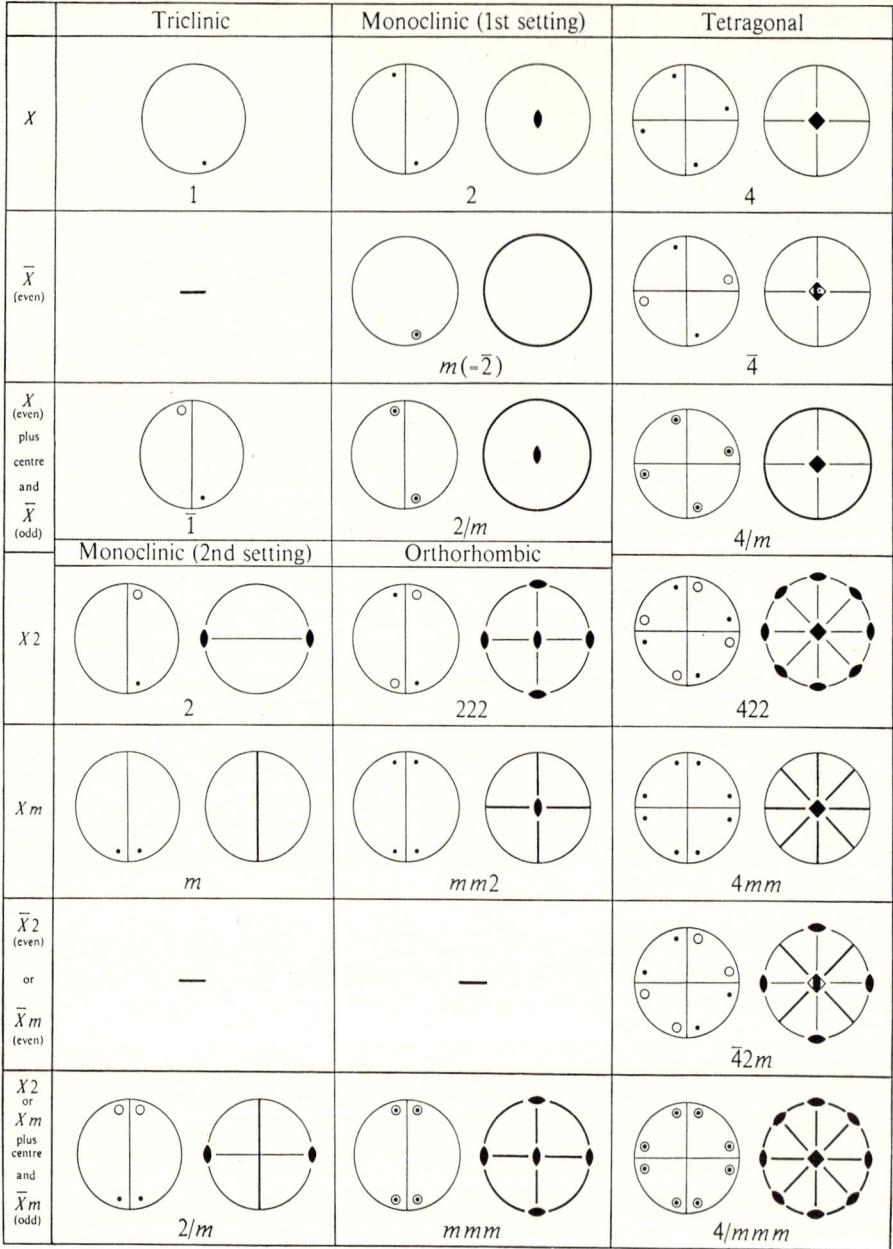

(a)

Fig. 5.20a,b. The 32 point groups or crystal classes. On the right of each pair of figures are given the symmetry elements, on the left the molecular arrangements. (From "International Tables for X-ray Crystallography." Vol. I, 3rd ed. Kynoch Press, Birmingham, England, 1969.)

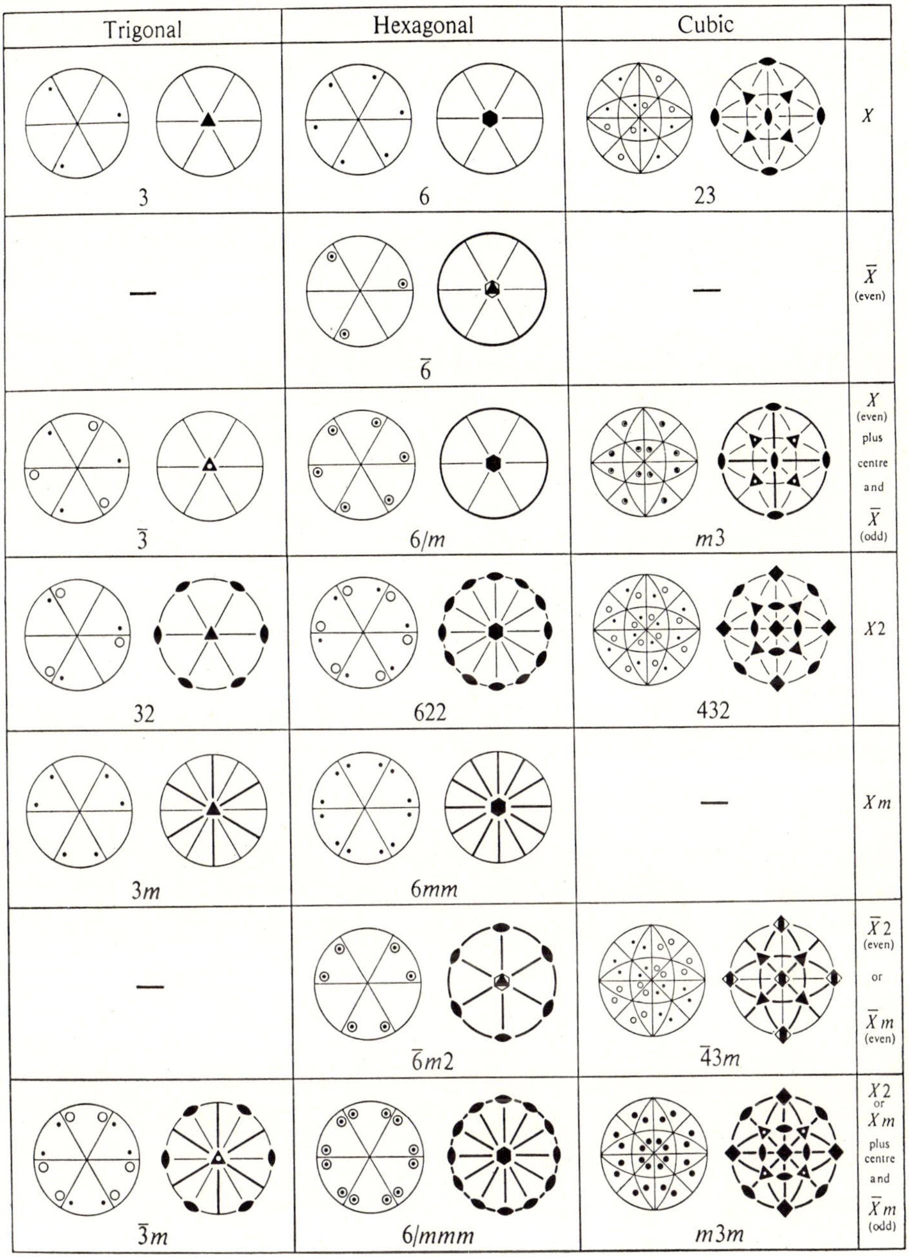

5.6 Three-Dimensional Symmetry

Trigonal	Hexagonal	Cubic	
3	6	23	X
—	$\bar{6}$	—	$\bar{X}$ (even)
$\bar{3}$	$6/m$	$m3$	X (even) plus centre and $\bar{X}$ (odd)
32	622	432	$X2$
$3m$	$6mm$	—	Xm
—	$\bar{6}m2$	$\bar{4}3m$	$\bar{X}2$ (even) or $\bar{X}m$ (even)
$\bar{3}m$	$6/mmm$	$m3m$	$X2$ or Xm plus centre and $\bar{X}m$ (odd)

(b)

Fig. 5.20b

plane can be in either of two directions, (along $\mathbf{a}_1$ or $\mathbf{a}_2$ for example in the $\mathbf{a}_1\mathbf{a}_2$ plane, called "a" or "b" rather than g), or there can be diagonal glides of $(\mathbf{a}_1 + \mathbf{a}_2)/2$ (referred to as "n"), or "diamond" glide $(\mathbf{a}_1 + \mathbf{a}_2)/4$ (referred to as "d"). In drawings of the space groups, glide in a plane parallel to the plane of the drawing is indicated by the corner of a rectangle with an arrow in the direction of glide. Glide normal to the drawing is indicated by broken lines — a dotted line if the translation is normal to the page, dashed if it is parallel, and dot-dash if it is diagonal.

(2) When two-dimensional periodicity was discussed, we examined rotation followed by translation perpendicular to the rotation axis. This only led to additional symmetry elements. But in three dimensions we can also have rotation combined with translation *parallel* to the axis. This leads to a new screw-like relationship between molecules, which is illustrated in Fig. 5.21. If we introduce such a motion

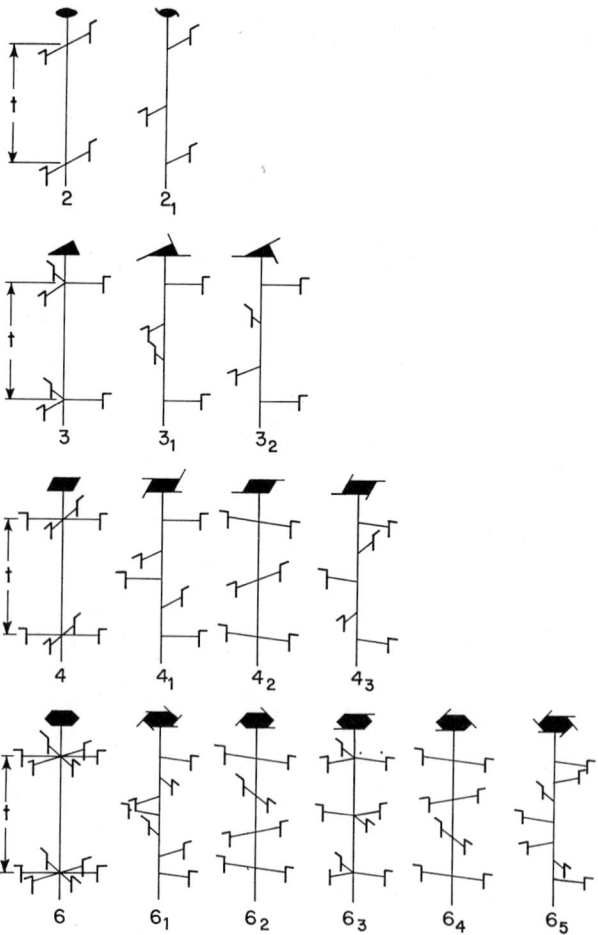

Fig. 5.21. Possible screw axes. (After Azaroff, L.V., "Introduction to Solids." McGraw-Hill, New York, 1960)

5.6 Three-Dimensional Symmetry

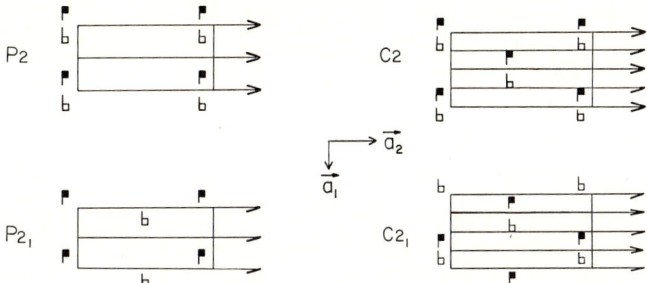

Fig. 5.22. Possible space groups based on the monoclinic point group 2. The twofold axis is parallel to a_2. The inclined axis is a_3. Screw axes are represented by *half arrows* when, as in this case, they lie parallel to the plane of the drawing. (Adapted from Phillips, F.C. "An Introduction to Crystallography," 3rd ed. Longmans, Green, and Co., London, 1961)

and repeat it n times, the total translation t must be a multiple m of the cell edge (say a_1) in the direction of translation if the structure is to be periodic. Therefore $t = (m/n)a_1$. Starting with a threefold axis, there would be a translation of $a_1/3$, $120°$ rotation, another $a_1/3$ plus another $120°$ rotation; finally, the third $a_1/3$ translation (and $120°$ rotation) returns a molecule to its original orientation at the next lattice point. This axis is designated as a 3_1 screw axis, the translational component $(m/n)a_1$ being $a_1/3$. Another possibility is $2a_1/3$ before each $120°$ rotation, producing a 3_2 screw axis. In this case the original molecule is brought into coincidence at the second lattice point from the starting point. (Coincidence at the first lattice point from the starting point arises from a molecule in the cell below the starting point.) These two axes appear as right-handed and left-handed screws, respectively, in the second line of Fig. 5.21. The symbols for the rotation axes (squares, triangles, etc.) have "tails" added to them when they represent screw axes, not pure rotations.

The combinations of all these symmetries lead to 230 *unique space groups*. Thus, a three-dimensional crystal cannot have *any* atomic pattern, it must be one of the 230. Of course, in different crystals the atomic coordinates or the occupied equipoints are different. These 230 space groups are listed in Volume I of the International Tables (1st and 2nd editions) and discussed in detail in Buerger's monograph (1956). While we cannot develop all of these possibilities in detail here, it will be instructive to consider two simple examples. First consider the monoclinic system, and one of its point groups, 2. The space lattice can be P or C so we can have $P2$, $P2_1$, $C2$, or $C2_1$. These are drawn in Fig. 5.22 with arrows indicating twofold axes in the plane of the drawing, and half-arrows indicating screw axes. (A summary of all the symbols used for symmetry elements can be found on pp. 49 and 50 of the International Tables[4], Volume I). As was the case with rotation axes, the presence of screw axes in $P2_1$ generates other screw axes. The presence of the centered-lattice point in the C-cell, produces screw axes. Therefore $C2$ and $C2_1$ are the same. Only the origin is different in the figure and as we have seen this is arbitrary. Thus only $P2$, $P2_1$, and $C2$ are unique.

---

[4] Pages refer to the old International Tables unless otherwise indicated.

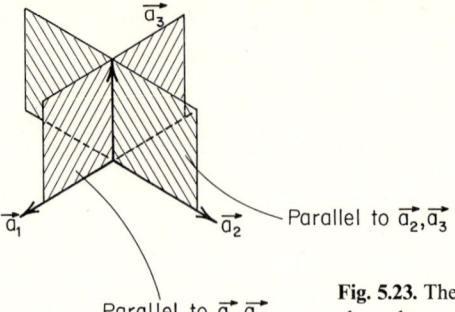

Fig. 5.23. The two mirror planes in $mm2$ (i.e., with twofold axis along the $\mathbf{a}_3$ axis)

As our second example, consider the orthorhombic point group $2mm$, and the primitive lattice P. Let the a_3 axis be the twofold axis (and of course it is also the line of intersection of the two mirror planes), as shown in Fig. 5.23. Then we must align the point group with orientation $mm2$.[5] Consider first, that the plane parallel to the axes $\mathbf{a}_2\mathbf{a}_3$ is a mirror plane, but that the $\mathbf{a}_1\mathbf{a}_3$ plane may be a mirror or an a-glide (glide in the $\mathbf{a}_1$ direction), a c-glide (glide in the $\mathbf{a}_3$ direction) or an n-glide (diagonal glide). The possible space groups are $Pmm2$, $Pma2$, $Pmc2_1$, or $Pmm2_1$. Note that in two cases, the presence of glide changes the twofold rotation axis to a twofold screw, even though we are considering only the mirror planes in deriving these groups. If the plane parallel to the $\mathbf{a}_2$ and $\mathbf{a}_3$ axes is a c-glide, $Pca2_1$, $Pcc2$, and $Pcn2$ result. The fourth possibility, $Pcm2_1$, is the same pattern as $Pmc2_1$ as only a change in axial notation is involved (i.e., the $\mathbf{a}_1$ and $\mathbf{a}_2$ axes are interchanged). The pattern of the space group is the same since only the $\mathbf{a}_3$ axis is unique because the twofold axis is along this axis. In the International Tables (Volume I, p. 545), there are tables of the equivalent notations of the space groups as a result of such relabeling. Although the axial lengths are not equal in the orthorhombic cell, which one is called a_1 or a_2 or a_3 is not important. Two workers may not choose the same axial schemes. It is not the dimensions of the cell that are important, but the symmetry of the molecular arrangement.

Other possible space groups are $Pba2$ and $Pbn2_1$. $Pbm2$ is possible but it is equivalent to $Pma2$, and $Pbc2_1$ is equivalent to $Pca2_1$. Finally, letting the plane parallel to $\mathbf{a}_2$ and $\mathbf{a}_3$ be a diagonal glide results in only one new arrangement, $Pnn2$. A total of 10 space groups occur for the point group $mm2$ in the primitive cell of the orthorhombic system.

All of these ten possible space groups are drawn in Volume I of the International Tables, on pp. 111–120 (1st edition), along with the equipoints. A typical example is given in Fig. 5.24 for $Pmc2_1$. The symbols for all symmetry elements are tabulated on pp. 28, 49, and 50 of this volume of the International Tables, and the reader is urged to consult these pages, and understand the symbols in this figure. The reader

---

[5] The first symbol in the orthorhombic space groups is the plane perpendicular to the $\mathbf{a}_1$ axis, or the rotation axis along $\mathbf{a}_1$, the second is the plane perpendicular to $\mathbf{a}_2$, or the rotation axis along $\mathbf{a}_2$, the third is the plane perpendicular to $\mathbf{a}_3$, or the rotation axis along $\mathbf{a}_3$. The notation for other axial systems is summarized on pp. 28 and 29 of Volume I of the International Tables.

5.6 Three-Dimensional Symmetry

$Pmc2_1$
C_{2v}^2 No. 26 $Pmc2_1$ $mm2$ Orthorhombic

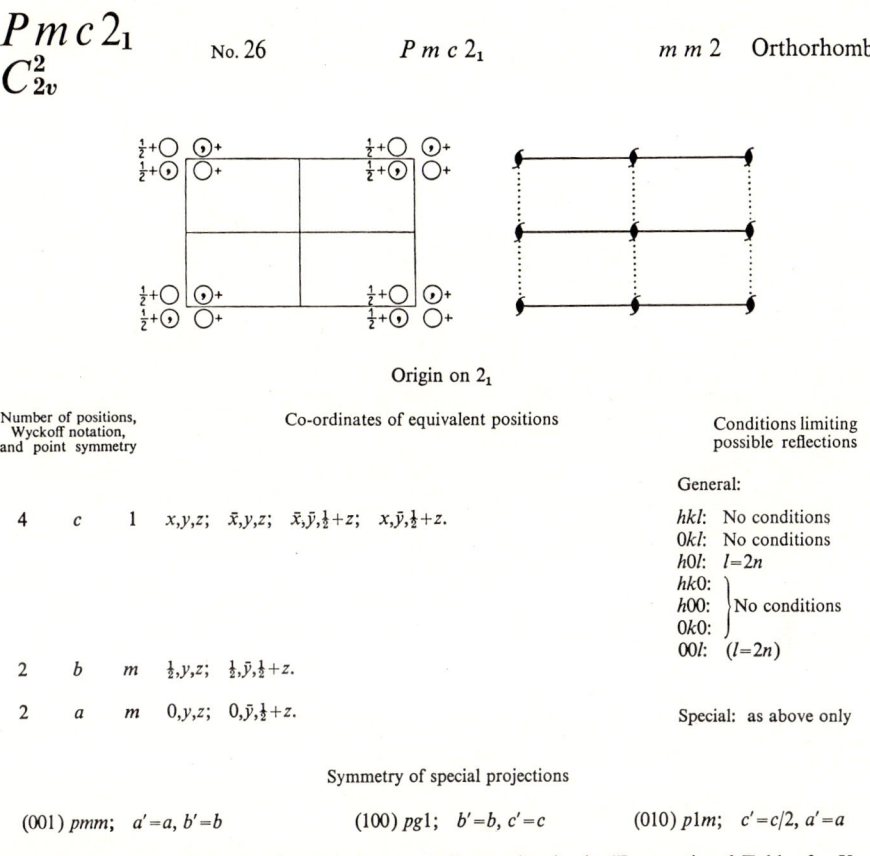

Fig. 5.24. A typical summary of a space group similar to that in the "International Tables for X-ray Crystallography" (Vol. I, 3rd. ed. p. 112. Kynoch Press, Birmingham, England. 1969)

should be sure that he or she can generate the general and special equipoints starting with a single molecule. (Other useful information in the Tables is listed in Appendix A of this volume.) For further details, in addition to the International Tables, the reader should consult Phillips (1964), Buerger (1956), and Sands (1969). A table prepared by S.K. Dickinson, Jr. summarizing all the information on geometrical symmetry in three dimensions is given in Appendix B.

The fact that there are only 230 space groups or structural patterns was discovered in the late nineteenth century simultaneously by a Russian mineralogist, Federov, a German mathematician, Schoenflies, and an English businessman, Barlow. Without any consideration as to the types of atomic bonding, the requirements of three-dimensional periodicity indicate that there are severe limits as to the possible patterns and that for molecules in certain structural locations (special equipoints), the molecule itself must have certain symmetry. (Of course, which structure occurs *will* depend on these interatomic bonding forces.) At the time of the discovery there was really no way of seeing if this amazing result of periodicity applied to real crystals. There simply was no way then of examining the internal atomic arrangements. By examining the external appearance of natural crystals, it was sometimes possible to determine the system or point group, but little else. (Occasionally physical properties may reveal the absence of inversion centers or presence of screw axes — see Sect. 5.7.1.) This is why the point groups are sometimes referred to as the *crystal classes*. It remained for diffraction to provide the necessary tool.

We have shown that all *geometrical* patterns of molecules periodic in three dimensions may be described as one of only 230 possibilities. This by no means exhausts all possible symmetries in three-dimensional space, because the atoms can have properties often referred to as "color." Consider, for example, that many crystals contain atoms with a net magnetic spin. The direction of the spin may depend on the location of the atom in the crystal. If we allow only those possibilities where the spin is "up" or "down" with respect to a given direction in the crystal, (denoting these as "black" and "white" atoms), the number of space groups is increased to 1651 as shown by the Russian mathematician Shubnikov. More general spin orientations increase the number of space groups still further. This area is still under study since it is possible to determine these spin orientations using neutron diffraction (see Sect. 3.6).

5.7 Space Groups and Diffraction

5.7.1 Diffraction and Symmetry

When a single crystal of the unknown material is available, taking a Laue pattern normal to a well-developed face will reveal certain aspects of the point group symmetry as viewed along an axis normal to the face. That is, *the symmetry of the pattern should be the symmetry of the point group, plus a center of symmetry* (as we have seen, in an intensity measurement a center is generally added unless there are imaginary terms in the scattering factors).

We can easily see that the point group symmetry is the same in reciprocal space as it is in real space. Any symmetry operation that takes a plane into another plane, takes the normal into the normal of this second plane, and therefore the reciprocal lattice, made up of normals, will show the same point group symmetry as the crystal. As the Laue pattern is a "picture" of reciprocal space around the origin, we see point symmetry. However, a center of symmetry is added, so that *of the* 32 *point groups, only* 11 *can be found by examining Laue patterns* taken along important axes of the crystal. These eleven *Laue Groups* can be generated as an exercise by the student by adding the symmetry $\bar{1}$ to all of the 32 point groups. For example $1 + \bar{1} = \bar{1}$, so the two point groups 1 and $\bar{1}$ are indistinguishable in a Laue pattern and are represented by the single Laue Group, $\bar{1}$. Similarly, 2, *m*, and 2/*m* become identical when the $\bar{1}$ is added, and all degenerate to the single Laue Group 2/*m*. (See Problem 5.9 and its solution in the International Tables, Vol. I, p. 30).

5.7 Space Groups and Diffraction

Fig. 5.25. How a screw axis changes the polarization of light. (After Feynman, R.P., Leighton, R.B., and Sands, M., "The Feynman Lectures on Physics," Volume I. Addison-Wesley, Reading, Massachusetts, 1963)

We may note that for given symmetries of the atomic arrangement, both the real and imaginary parts of the structure factor are identically zero for certain reflections and hence the intensity for those reflections is identically zero. These systematic absences are valuable clues in identifying the proper space group of an unknown, and are consequently tabulated in the International Tables. For example, referring again to Fig. 5.24 for the group $Pmc2_1$, we see in the column on the right that for the 4c general equipoint, $h0l$ reflections are present only if $l = 2n$ (revealing the effects of the 2_1 symmetry along the z-axis)[6], while all other combinations of hkl have no conditions imposed on them (except for the trivial 001 which is a special case of $h0l$ and hence enclosed in parentheses). Note — if the atoms are in certain special equipoints in some space groups — additional restrictions on reflections may arise, but if some atoms are in a general equipoint and others in a special one, the general conditions apply, *not* the special ones.

Nevertheless, the combination of the systematic absences and the Laue symmetry taken together suffice to uniquely determine only about 60 of the 230 space groups. For the remaining groups other evidence must be obtained. If the crystal is sufficiently well developed that the external symmetry is revealed, the point group can be established, and that combined with systematic absence is sufficient to enable identification of nearly all 230 space groups.

Generally, the unknown crystal is of sufficient irregularity that the point group remains in doubt. In this event, other physical measurements may sometimes be of value. Screw axes may sometimes be detected if a crystal is transparent. The screwlike arrangement influences the phase relationships in the scattered light; we can see why in Fig. 5.25. Polarized light, with its electric field in the **y-z** plane will

---

[6] To see this, calculate the structure factor for the general equipoint with coordinates of equivalent points: $x, y, z; \bar{x}, y, z; x, \bar{y}, \frac{1}{2} + z; \bar{x}, \bar{y}, \frac{1}{2} + z$. Using Eq. (2.21), we may combine real and imaginary terms and simplify. For example, letting A be the real terms,

$A = [\cos 2\pi(hx + ky + lz) + \cos 2\pi(-hx - ky + lz + l/2)]$

$\quad + [\cos 2\pi(-hx + ky + lz) + \cos 2\pi(hx - ky + lz + l/2)]$

$\quad = 2\cos 2\pi(hx + ky - l/4)\cos 2\pi(lz + l/4)$

$\quad + 2\cos 2\pi(-hx + ky - l/4)\cos 2\pi(lz - l/4)$

$\quad = 4\cos 2\pi hx \cos 2\pi(ky - l/4)\cos 2\pi(lz + l/4)$

We see directly that for $l = 2n + 1$, $A = 0$ if $k = 0$. A similar result is found for the imaginary terms. Thus for an $h0l$ reflection to have a non-zero structure factor we must have $l = 2n$.

move ions up and down. If there is a screw axis present, these ions will be forced to move in the **x** direction as well as along the **z** axis. As the ions scatter the light, this **x** component of motion leads to a rotation of the direction of the field of the incoming beam, the amount of rotation depending on the thickness of the crystal. We can sometimes detect the lack of a center of symmetry by attacking the crystal briefly with acid. Myriads of small etch pits develop and with the right etchant these will have smooth crystallographic sides; if there is no center they will appear differently on opposite faces. Also if there is no center, the crystal will be piezoelectric. It will develop a voltage across opposite faces when a stress is applied. These effects may, however, be so weak as to defy unambiguous detection.

A useful optical method for detecting the lack of a center of symmetry is known as second-order harmonic generation. The lack of a center of symmetry allows atoms to vibrate anharmonically since the restoring force on the atom is less for displacements in some directions than for others. The result of this is that when a beam of light of frequency v passes into such a crystal, harmonics of frequency $2v$, $3v$, etc., are generated. These effects are small and only observable when the source of light is one of very high intensity, e.g., from a pulsed laser. While not generally used, this technique promises to be a very sensitive tool for identifying noncentrosymmetric structures, at least those that are transparent to light.[7]

We have seen in Chap. 4 how diffraction experiments may be used to identify the Bravais lattice and measure unit cell axial lengths and interaxial angles. Coupled with determination of the Laue group and systematic absences in the diffraction pattern, this information goes a long way towards identification of the structure of the unknown. However, as we noted as early as Chap. 2, the atomic coordinates within the unit cell are revealed not in the positions of the Bragg reflections, but rather in their intensities — or more precisely in the structure factor which determines these intensities. We shall now show that the systematic compilation of the space groups leads to a great simplification in calculations of the structure factor.

5.7.2 Structure Factor Calculations

As we saw in Chap. 2, it is relatively easy to simplify the expression for the structure factor contained in Eq. (2.21) for structures that do not contain too many atoms. But if there are many atoms, it is a tedious job. This can be simplified in the following way. In any space group, the general equipoint will have certain symmetry. If there is one kind of atom in one coordinate set x, y, z of this equipoint, the same atom will be in all the sets. The atomic scattering factor can then be factored out of the sum for all the sets for this equipoint and the trigonometric terms simplified to terms involving one set of x, y, z. Then one only has to insert a set of actual values for the atom at x, y, z, multiply by the scattering amplitude per atom, f, and sum for all equipoints in the cell. Only the first coordinate in a set for an equipoint needs

---

[7] Commercial instrumentation for second harmonic analysis of powders or single crystals is described by Dougherty, J.P., and Kurtz, S.K., J., Appl. Crystallogr. 9, 145 (1976).

5.7 Space Groups and Diffraction

to be used since all other terms are included in the simplification. To see how these expressions arise, consider the monoclinic space group B2 (International Tables, Volume I, No. 5, p. 80). The coordinates of the general equipoint are x, y, z and $\bar{x}, \bar{y}, z$ plus the base-centering translation, $\frac{1}{2}, 0, \frac{1}{2}$. The structure factor for atoms on this equipoint may be obtained from Eq. (2.21) as

$$F_{hkl} = f\{e^{2\pi i(hx+ky+lz)} + e^{2\pi i(-hx-ky+lz)} + e^{2\pi i(h[x+1/2]+ky+l[z+1/2])}$$
$$+ e^{2\pi i(h[-x+1/2]-ky+l[z+1/2])}\}$$
$$= f(e^{2\pi i(hx+ky)} + e^{-2\pi i(hx+ky)})e^{2\pi ilz}(1 + e^{2\pi i(h+l)/2})$$
$$= 4f\cos^2(2\pi/4)(h+l)\cos 2\pi(hx+ky)[\cos 2\pi lz + i\sin 2\pi lz].\ ^8$$

Writing A and B for real and imaginary parts,

$$F_{hkl} = fA + ifB, \quad \text{and} \quad |F_{hkl}| = [(fA)^2 + (fB)^2]^{1/2}.$$

For more than one occupied equipoint per unit cell, the general expression is

$$|F_{hkl}| = \left[\left(\sum_e f_e A_e\right)^2 + \left(\sum_e f_e B_e\right)^2\right]^{1/2}, \tag{5.2}$$

where A_e and B_e are the simplified real and imaginary terms called *symmetry factors*, and the summation is taken over all occupied equipoints in the cell. Because there are only a small number of space groups, a tabulation of these simplifications could be quite useful. Such a tabulation can be found at the back of Volume I of the International Tables. Two precautions in their use need mentioning:

(1) Be careful to choose the origin of the unit cell the same as in the Tables when actually inserting values for x, y, z.
(2) The equations are developed for the most general equipoint. If special positions are occupied and the number of equivalent positions is reduced, the symmetry factors must be multiplied by (rank of special equipoint)/(rank of general equipoint).

As an example, consider the orthorhombic space group *Pmmm* (International Tables, Volume I, No. 47, p. 133). The general equipoint has a rank of 8, and if we take the origin at the intersection of the three mirrors, there is also an inversion center there. Thus the symmetry factor is centrosymmetric. On p. 400 of Volume I, we find that

$$B = 0, \quad A = 8\cos 2\pi hx \cos 2\pi ky \cos 2\pi lz.$$

Suppose that for the structure we wish to consider, there is one atom type with scattering amplitude f_1 in the t equipoint and another atom type with amplitude f_2 in the h equipoint. According to the Tables, the first has a rank of 2, the second a rank of 1. The actual coordinates are on p. 133 of Volume I: $\frac{1}{2}, \frac{1}{2}, z$ for the first, $\frac{1}{2}, \frac{1}{2}, \frac{1}{2}$ for the second. Therefore,

---

[8] The substitution $2\cos^2 2\pi(h+l)/4 = 1 + \exp[2\pi i(h/2 + l/2)]$ is justified only for h and l integers.

$$|F(s)| = (f_1)(\tfrac{2}{8})(8)\cos \pi h \cos \pi k \cos 2\pi lz$$
$$+ (f_2)(\tfrac{1}{8})(8)\cos \pi h \cos \pi k \cos \pi l.$$

There is an unknown z that can only be determined by comparison of the calculated $|F(s)|^2$ with experimental intensities.

5.7.3 Some Simple Crystal Structures

Once the crystal system is identified and all reflections indexed, systematic absences of certain types of reflections will help to reduce the number of possible space groups to only a few. Your choice of unit cell may not correspond to that used in the International Tables, and it is often necessary to try interchanging the labeling of axes and then reexamining the extinctions and the Tables to find possible matches. The tables on p. 545 of Volume I of the International Tables will help in this operation. The combined point group, Bravais lattice, and screw axes and glide planes define the *diffraction symbol*. On pp. 347–352 of Volume I of the Tables, the possible space groups for each diffraction symbol are listed.

The next step in the analysis involves selecting possible equipoints for all the atoms in the unit cell. It is assumed that the composition of the crystal is known from chemical analysis, and that the density has been measured. From these data and the unit cell size, the number of atoms of each kind per unit cell may be calculated. For very simple structures with only a few atoms per cell, this information may be sufficient to reduce the number of possible equipoints to a few. The structure determination in this case involves calculating the possible FF^* for the various combinations of atomic arrangements (eliminating those in which significant atomic overlap is required, and relying when possible on knowledge of the nature of the chemical bonding which may suggest preferential atomic groupings). When the coordinates of only one or a few atoms remain unknown, calculations of certain FF^*_{hkl} versus assumed coordinates may be helpful. For example, if a z coordinate of one atom is unknown, calculations of FF^*_{001} versus z for several l-values may be compared with the measured FF^*_{001} to quickly arrive at an estimate of the best value for z. Then the entire set of FF^*_{hkl} are calculated and variable parameters, thermal parameters, and scale factor varied to give the best agreement with experimental values of FF^*. The best fit is identified as the solution to the problem. In this section we consider three examples of such determinations, delaying until Chap. 6 the complexities associated with solving more difficult structures with many atoms per cell.

A simple example comes out of our salt shaker. Examination of table salt under an optical microscope shows cubic symmetry. Powder x-ray patterns confirm this, and identify the Bravais lattice as face-centered cubic since hkl's are all even or all odd. The lattice parameter combined with density indicates four "molecules" of NaCl per cell. Nothing else in the way of restrictions appears when we examine a single crystal pattern, no indication of screw axes or glide planes. Consulting p. 352 of the International Tables with the diffraction symbol F gives the following possibilities: $F23$, $Fm3$, $F432$, $F\bar{4}3m$, $Fm3m$. In all of these the rank 4 equipoints have

5.7 Space Groups and Diffraction

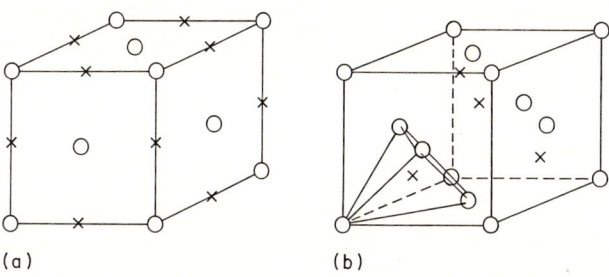

Fig. 5.26a,b. Two possible structures for NaCl. **a** Point symmetry is $m3m$; **b** point symmetry is $\bar{4}3m$

fixed coordinates. Trying all possibilities we find only two basic structures both of which are interpenetrating *fcc* cells as shown in Fig. 5.26. The point symmetry at Na$^+$ and Cl$^-$ ions in Fig. 5.26a, is $m3m$, while for Fig. 5.26b, it is $\bar{4}3m$. This further narrows the choice of space groups to $Fm3m$ and $F\bar{4}3m$. Hand calculations of FF^* for the 200 and 220 reflections for these two space group assignments give the following:

hkl:	200	220
$FF^*(Fm3m)$:	20.8	17.7
$FF^*(F\bar{4}3m)$:	3.6	17.7

After correcting the measured intensities for absorption and the *LP* factor, FF^* is found to be nearly identical for the 200 and 220 reflections, and the structure is confirmed as $Fm3m$, halite. The reader may recall from Sect. 2.2 that this structure was the first ever determined. How much easier it is to use the present results of space group symmetry than to go through the deductive reasoning that had to be used by the Braggs!

As a somewhat more complex example which summarizes our present knowledge, consider the material Cu_2O. By examining the symmetry of the diffraction pattern we find that Cu_2O is cubic, and from Bragg's law we determine the interplanar spacings and find that the lattice parameter (cell edge) $a_1 = 4.2696$ Å. The measured density ρ is found to be 6.1 gm/cm^3. From this information we can obtain the number of molecules per unit cell, Z, from the relation

$$Z = N_0 \rho V_c / \sum_j A_j, \qquad (5.3)$$

where N_0 is Avogadro's number, and A_j is the atomic weight of the *j*th atom. Substituting, we obtain

$$Z = N_0 \rho a_1^3 / (2A_{Cu} + A_O) = 1.99 \cong 2.$$

That is, there are two molecules of Cu_2O or $4Cu^+$ and $2O^{2-}$ per unit cell.

The diffraction pattern from Cu_2O shows no extinctions or systematic absences of hkl reflections, so the unit cell is primitive. However, for reflections of the type $0kl$, the only reflections observed have $k + l = 2n$, where n is an integer. Turning

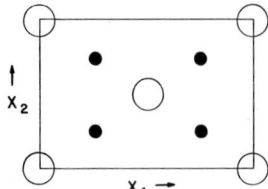

Fig. 5.27. The unit cell of Cu_2O, $Pn3m$

to p. 54 of the International Tables, Volume I, we find that this condition implies a diagonal glide (n) on a {100} plane. On p. 28 of the Tables, we find that symmetry along [[100]] is specified first in the space group notation for cubic space groups, so we can write a *diffraction symbol* for Cu_2O as Pn .... (The second position in the notation would refer to symmetry along [[111]], while the third position would specify symmetry along [[110]] but the diffraction pattern reveals no information about restrictions on hkl which affect these directions in Cu_2O.) These diffraction symbols would include any information on the point group symmetry if it were available, preceding the information on the space group.

The space groups that can have each of these diffraction symbols are listed in the International Tables, Volume I. From the list of such symbols on p. 352 of the tables for cubics, we find that the only two possible space groups are $Pn3$ and $Pn3m$. In $Pn3m$ (origin at $\bar{4}3m$ on p. 335 of the Tables), the two O^{2-} ions would have to be in the $2a$ equipoint, while the four Cu^+ could be in the $4b$ or $4c$ equipoints. Similarly, in the $Pn3$ space group (origin at 23 on p. 309 of the Tables), the O^{2-} is at the $2a$ equipoint and the Cu^+ at the $4b$ or $4c$ equipoints. The atomic coordinates are the same for the two space groups, and the same atomic arrangements occur in each case (see Fig. 5.27). However, in each arrangement, the O^{2-} ion is surrounded by four Cu^+ ions arranged in a regular tetrahedron. Thus the symmetry of the O^{2-} position must be that of a tetrahedron, $\bar{4}3m$. This is the symmetry of the rank $2a$ equipoint in $Pn3m$, but not in $Pn3$. Thus the correct space group is $Pn3m$. We have now determined the space group and located the atoms. If there were any unknown atomic coordinates, they could be determined by comparing measured and calculated scattering intensities as discussed in Chap. 6.

Let us now consider a slightly more complex structure, that of Ag_3Sn. This is the compound a dentist mixes with Hg to fill your teeth and has such unique features as a very good match with the thermal expansion of the tooth (better than gold); it is easy to shape in the amalgam and the proportion of Hg controls the expansion during the setting reactions to form Ag_2Hg_3 and Hg-Sn compounds. It is also a difficult x-ray problem as Ag and Sn are close in atomic number, and hence the scattering factors are similar. Also it has a very high absorption coefficient requiring careful corrections for absorption.

A portion of the Ag-Sn phase diagram is shown in Fig. 5.28. Cooling as indicated by the arrow produced small single crystals of Ag_3Sn embedded in the solidified Sn-rich liquid. When dissolved in concentrated HCl, crystals of Ag_3Sn were freed from the ingot. There was some dispute in the literature over the indexing of powder patterns of Ag_3Sn, as to whether the structure is orthorhombic or hexagonal. The material is plastic, and filing to make powders broadens the peaks thus making it

5.7 Space Groups and Diffraction

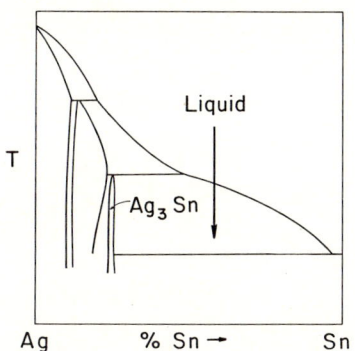

Fig. 5.28. A schematic drawing of the Ag–Sn phase diagram

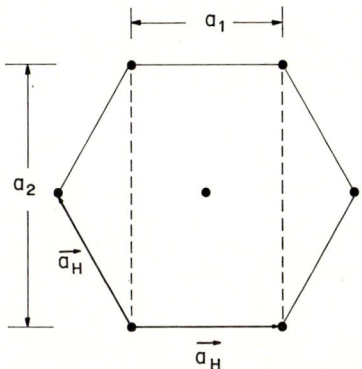

Fig. 5.29. Alternative hexagonal and orthorhombic cells for a hexagonal space lattice

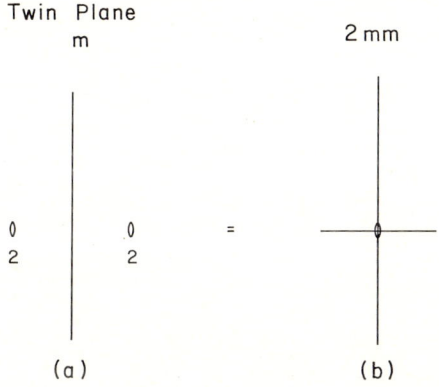

Fig. 5.30. The effect of using a twinned crystal to obtain the diffraction pattern. In this example the apparent symmetry 2mm is due to the combination of the true twofold symmetry with the mirror symmetry of the twin. (After Buerger, M.J., "Crystal Structure Analysis." Copyright 1960, John Wiley & Sons. Reprinted by permission of John Wiley & Sons, Inc.)

difficult to resolve neighboring peaks. This indexing problem is not unique, as one can always choose an orthorhombic cell for a hexagonal space lattice as shown by broken lines in Fig. 5.29. For this *relabeling* of axes, $a_1 = a_H$, $a_2 = \sqrt{3}a_H$, so $a_2/a_1 = 1.732$. After examination of single crystals with a precession camera, and indexing as orthorhombic, the lattice constants were found to be $a_1 = 2.984$ Å, $a_2 = 5.184$ Å, and $a_3 = 4.780$ Å. The measured ratio of $a_2/a_1 = 1.722$ suggested that the hexagonal cell is not a proper alternative and Ag_3Sn is orthorhombic. Furthermore, powder patterns from *well-annealed* filings showed the required splitting into orthorhombic lines. For example, what would be a 201 peak from an *hcp* structure was found to be four peaks, the 330, 412, and the 014 K_{α_1} and K_{α_2} doublet.

It turns out that this material's plasticity is associated with twinning. This deformation mode relates one part of a crystal to another by rotation about some axis, or by reflection. For example, consider a crystal with point group 2 which is twinned by a mirror plane. On each side of the twin plane the crystal exhibits only the twofold symmetry as shown in Fig. 5.30a, but when the crystal as a whole is examined, it has the common symmetry of the two individual crystals in twin orientation plus the symmetry of the twin operation (*m*), and thus appears as 2*mm* as shown in Fig. 5.30b. Other examples can be found in Buerger (1960, Chap. 5). It is critical to examine crystals carefully under a microscope to avoid the time and expense of studying anything but a single crystal, although such optical examination by no means eliminates all twinned crystals.

The crystal structure analysis proceeded by detailed analysis of extinctions observed in precession photographs. To distinguish the scattering from Ag ($Z = 47$) and Sn ($Z = 50$), Ag K_α radiation was chosen. This wavelength is close to the absorption edge of Sn and leads to a large correction to f_{Sn} which increases the intensity of peaks which depend on the difference in scattering factors of Ag and Sn. Films exposed for 48 h then showed that the $\mathbf{a}_1$ axis should be doubled[9]; there were weak peaks at half the distance of the strong peaks along $\mathbf{b}_1$. The density plus these cell edges lead to the fact that there are two Ag_3Sn units per cell. The indexed reflections revealed these conditions:

hkl: no restrictions
*h*00: $h = 2n$ (2_1 screw axis possible)
0*k*0: no conditions
00*l*: $l = 2n$ (2_1 screw also?)
*h*0*l*: $h + l = 2n$ — now we see that the previous extinctions were just special cases of this, which is a diagonal glide $(\mathbf{a}_1 + \mathbf{a}_3)/2$ on a plane perpendicular to the $\mathbf{a}_2$ axis.
0*kl*, *hk*0: no restrictions.

The Laue pattern for all orthorhombic structures shows *mmm*. The diffraction symbol is thus *mmm Pn* which means space groups $Pnm2_1$ (International Tables, p. 117), or *Pmmn* (pp. 147, 148). The first has no center of symmetry; the position of *n* in the second group indicates this diagonal glide plane is perpendicular to the $\mathbf{a}_3$ axis, so we have to interchange what we called a_2 and a_3 to conform to the International Tables. Thus, $a_1 = 5.968$ Å, $a_2 = 4.780$ Å, $a_3 = 5.184$ Å.

One way of checking whether the crystal is centrosymmetric or not would be to examine measured $|F_{hkl}|$ and $|\overline{F_{hkl}}|$. With a radiation close to an absorption edge, these might differ (see Sect. 3.4.3, Fig. 3.25). However, the absorption corrections are too extensive for this material to trust this procedure in this case.

If we consider the size of the atoms in the unit cell, they occupy 72% of it, compared with 74% for a close-packed structure (like fcc Cu). The 200 reflection [we are now using the *hkl* values for the interchanged axes] is weak; the 400 reflection, strong. This suggests close-packed layers along the $\mathbf{a}_1$ axis with planes at $a_1/4$ and $3a_1/4$ all Ag and with alternate planes half Ag, half Sn to arrive at the composition Ag_3Sn. A centrosymmetric structure corresponding to this arrangement is shown in Fig. 5.31. Sn atoms in the second layer are above Ag atoms, and it takes three layers in the $\mathbf{a}_2$ direction to repeat the cell. The reader should locate

---

[9] At a synchrotron radiation source, a wavelength closer to the Sn edge could be chosen, and with the high available intensity, this same pattern could be taken in minutes.

5.7 Space Groups and Diffraction

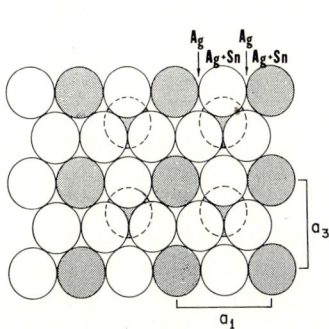

Fig. 5.31. The centrosymmetric structure proposed for Ag$_3$Sn. (After Fairhurst, C.W., Ph.D. Thesis, Northwestern University, 1966)

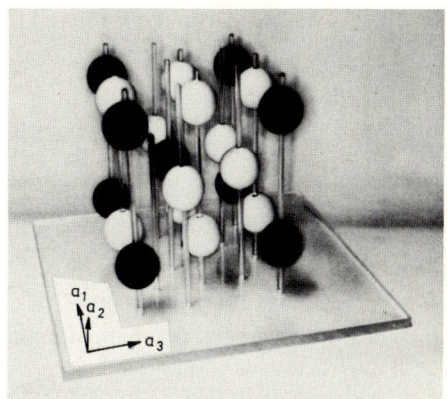

Fig. 5.32. A hard sphere model of the structure shown in Fig. 5.30. (After Fairhurst, C.W., Ph.D. Thesis, Northwestern University, 1966)

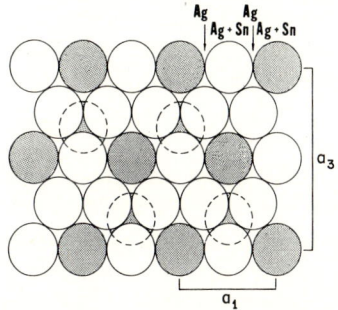

Fig. 5.33. The noncentrosymmetric structure proposed for Ag$_3$Sn. (After Fairhurst, C.W., Ph.D Thesis, Northwestern University, 1966.) *Open circles*: Ag; *shaded circles*: Sn

the diagonal glide and the inversion center. They are not in the plane of the drawing, and can be seen more clearly in Fig. 5.32. Building a model will help to clarify this point.

A second (noncentrosymmetric) structure is shown in Fig. 5.33. Here the a_3 axis is twice the measured value and hence the structure is unacceptable. Other possible structures might be formed by varying the atomic locations slightly to distort the *Pmmm* structure of Fig. 5.30 to fit the *Pm2$_1$n* symmetry. However, if these two space groups are examined, we note that they differ in a simple way. In *Pmmn*, atoms are related by inversion centers, while in *Pm2$_1$n* they are related by two-fold screw axes. If the atoms are spherical, there is no difference between these symmetries.

Hence we conclude from arguments of packing density, symmetry, and limited use of measured peak intensities that the space group is likely to be *Pmmn*. We use the setting given in the International Tables, p. 148. We have six atoms of Ag, two of Sn. There is no rank 6 equipoint and so the Ag must be on ranks 4 + 2. The rank 4d, 4c equipoints involving atoms at inversion centers cannot be right, because in our model the Ag atoms are not at the inversion center. Furthermore, Ag atoms at these coordinates in a unit cell of these dimensions would overlap (close-packed metallic radii for Ag and Sn are 1.44 and 1.58 Å, respectively). The remaining possibilities are:

Equipoints			
1	2	3	4
Ag: $4e + 2a$	$4e + 2b$	$4f + 2a$	$4f + 2b$
Sn: $2b$	$2a$	$2b$	$2a$

Furthermore, a shift of the origin of the unit cell by $0, \frac{1}{2}, 0$ takes $2a \to 2b$ and $2b \to 2a$, but does not shift $4f$ to $4e$, so only possibilities 1 and 3 are unique. Reflections with odd h are very weak. It is easy to show (see Problem 5.22) that this is not possible for the six atoms in $Pmn2_1$, but is for the atoms in $Pmmn$, so this is further evidence that we have made the correct choice. The atomic arrangement corresponding to possibility 3 is similar to the close-packed drawing of Fig. 5.30, so we can use this model to make some initial guesses about the undetermined coordinates in these equipoints. Our problem is now reduced to one of finding the best combination of undetermined atomic coordinates (x, y, z), thermal vibration parameters (B_{Ag}, B_{Sn}) and proportionality constant [K in Eq. (4.47a)] to fit simultaneously calculated FF^* to the measured integrated intensities. This procedure called structure refinement is discussed in Sect. 6.9.

We can readily judge from the preceeding examples that the complexity of structural determination increases rapidly as the number of atoms per unit cell increases. Even with the aid of the space group tables, the ad-hoc methods which were sufficient for structural determinations of NaCl, Cu_2O, and Ag_3Sn would quickly be found to be inadequate in the study of many inorganic compounds with tens to hundreds of atoms per cell, not to mention biological structures with many thousands of atoms per unit cell. Clearly systematic procedures are necessary. These are introduced in Chap. 6 which is devoted to crystal structure determination.

References

Azaroff, L.V., "Introduction to Solids." McGraw-Hill, New York, 1960
Buerger, M.J., "Elementary Crystallography." Wiley, New York, 1956
Buerger, M.J., "Crystal Structure Analysis." Wiley, New York, 1960
Dana, J.D., "Manual of Mineralogy." (revised by C.S. Hurlbut, Jr.) 18th ed. Wiley, New York, 1971
Holden, A. and Singer, P., "Crystals and Crystal Growing." Doubleday, New York, 1960
Koerber, G.G., "Properties of Solids." Prentice-Hall, New Jersey, 1962
Nelson, D.R. and Halperin, B.I., Science 229, 233 (1985)
Nye, J.F., "Physical Properties of Crystals." Clarendon, Oxford, 1957
Pearson, W.B., "A Handbook of Lattice Spacings and Structures of Metals and Alloys." Pergamon, New York, 1958
Phillips, F.C., "An Introduction to Crystallography." 4th ed. Wiley, New York, 1971
Sands, D.E., "Introduction to Crystallography." Benjamin, New York, 1969
Shechtman, D.S., Blech, I., Gratias, D., and Cahn, J.W., Phys. Rev. Lett. 53, 1951 (1984)
Wood, E.A., "Crystals and Light." Van Nostrand, Princeton, New Jersey, 1964

Problems

5.1. What symmetry elements can you find in your own external appearance, i.e., in the human body?

5.2. In deriving the possible unit cells in two dimensions, one cell was derived by placing a mirror *between* points of one column of lattice points, then *on* the points of a second column. Suppose we put the mirror between points in two successive columns, then on points in two columns as in Fig. P5.2. Is this array a lattice?

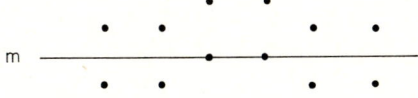

Fig. P5.2

5.3. Figure P5.3 shows diagrams of the arrangements produced by the operation of the 10 possible two-dimensional point groups. For each arrangement, insert the symmetry elements present and write down the conventional point group symbol.

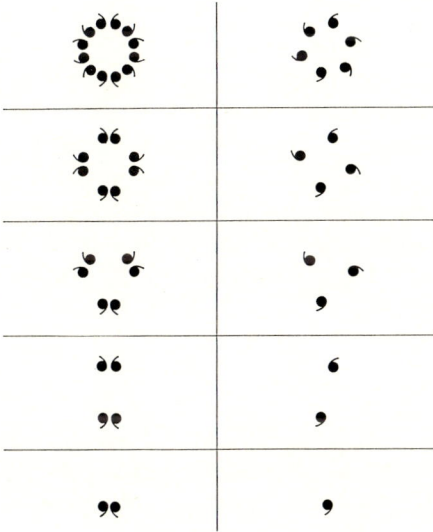

Fig. P5.3

5.4. Figure P5.4 shows drawings of the two-dimensional lattices with some symmetry elements inserted. Treat the corners of the unit cells as lattice points. Select an initial general position and complete several drawings by operating the symmetry elements so as to produce all the equivalent general positions. Then insert all the additional symmetry elements which arise. You will find it best to select the initial general position inside the unit cell and fairly close to a lattice point. The lines drawn inside the unit cell are to help you in placing your equivalent points. When you

The 17 Two Dimensional Plane Groups

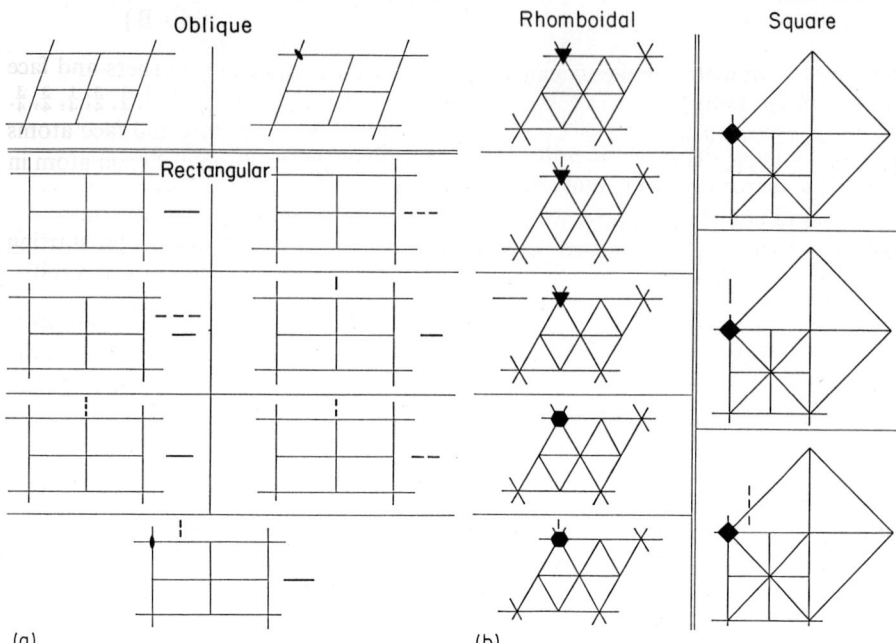

Fig. P5.4a, b. The 17 two-dimensional point groups

have completed your drawings, determine the space group symbol appropriate to each.

5.5. Give the equipoints and their rank for the two-dimensional space group *Pm*.

5.6. In discussing the glide operation in two dimensions, the glide translation has been taken as half a cell edge. Show what results if $\frac{1}{4}$ the translation is employed.

5.7. Consider *p2mm* of the two-dimensional groups. Allow for atomic motion above and below the plane and show some of the new space groups.

5.8. Show that replacing mirror planes by glide planes introduces nothing new for the two-dimensional space groups *p31m*, *p3m1*, and *p6m*, i.e., no new space groups are produced.

5.9. Add a center of symmetry to those three-dimensional point groups which do not have one. Group the 11 that are now indistinguishable to form the Laue Groups.

5.10. Show what happens to a 4_1 axis if (a) a center of symmetry is added, and (b) a mirror is added parallel to and intersecting the axis.

5.11. Consider a tetragonal lattice centered on the faces of the unit cell parallel to the fourfold axis. Does this produce a true space lattice, other than the 14 described in the text? Explain.

5.12. What point groups in the tetragonal system can have their rotation axes replaced by screw axes, when placed in a lattice? (Consult Appendix B.)

5.13. In the structure of diamond there are carbon atoms at the corners and face centers of a cubic unit cell, and also carbon atoms at $\frac{1}{4}, \frac{1}{4}, \frac{1}{4}; \frac{3}{4}, \frac{3}{4}, \frac{1}{4}; \frac{3}{4}, \frac{1}{4}, \frac{3}{4}; \frac{1}{4}, \frac{3}{4}, \frac{3}{4}$. (Note that these last four positions are the coordinates of corner and face atoms plus $\frac{1}{4}, \frac{1}{4}, \frac{1}{4}$.) What is the point group of the interatomic forces on a carbon atom in diamond?

5.14. Show, in a series of drawings, how the space group $Pnma$ develops, starting with a single molecule in a general position and the symmetry elements in a cell.

5.15. (a) GeSe is an orthorhombic covalently bonded crystal; its space group is $Pbnm$. Find this in the International Tables. (*Hint*: You may have to change the axial system. Look on p. 548 in the International Tables, Volume I.) From Problem 1.7 we found that there are four molecules in the unit cell.
(b) What are the possible atom distributions among the equipoints?

5.16. By considering its effect on one atom, show how the structure factor F_{hkl} and $|F_{hkl}|^2$ are changed by 4, 4_1, 4_2, axes along the a_3 direction. For example, for what hkl value is $|F_{hkl}| = 0$? Sketch the results in three-dimensional reciprocal space. [*Hint*: Put an atom in the general position x, y, z, work out the coordinates of other atoms required by the fourfold axis, and then use Eq. (2.21).]

5.17. Show the effect on the structure factor F_{hkl} and $|F_{hkl}|^2$ of a $c/2$ glide on a (100) plane and a (100) mirror. For what hkl values is $|F_{hkl}|^2 = 0$? Sketch the results in three-dimensional reciprocal space. [*Hint*: Put an atom in the general position x, y, z, work out the coordinates of other atoms required by the symmetry operation, and then use Eq. (2.21).]

5.18. Examine space group $C\ 2/m\ 2/m\ 2/m$. Note the conditions for reflection on the right-hand side of the page.
(a) Find out from the International Tables what the parentheses on certain of the conditions mean.
(b) Indicate what symmetry elements give rise to each condition, including the special conditions.

5.19. Evaluate in as simple a form as possible the structure factor for the rank 4 equipoint in the two-dimensional space group Pmg. (See coordinates in the International Tables, Volume I. Remember that there is a table of structure factors at the end of that volume so that you can check your answer.)

5.20. Evaluate in as simple form as possible the structure factor for space group $Pmm2$. Check your answer with the International Tables.

5.21. By using the Structure Factor Tables in the International Tables, Volume I, write an expression for the intensity of the 222 peak from normal spinel, AB_2O_4, in terms of f_A, f_B, f_O. (See Fig. 1.15e.)

5.22. Consider the Ag_3Sn described in Sect. 5.7.3. The diffraction symbol $mmmPn$ leads to two possible space groups $Pm2_1n$ and $Pmmn$. The (a_3, a_1) projection is close to the atomic arrangement shown in Fig. 5.31 for the *Pmmn* space group and *hkl* reflections are weak for all reflections with *h* odd. Examine the structure factor and show that the reflections with *h* odd could *not* be weak for $Pmn2_1$ but could be low for *Pmmn*. (*Hint*: Use the structure factors at the back of Volume I of the International Tables for each space group.)

5.23. Ruby is a dilute solution of Cr_2O_3 in Al_2O_3 (space group $R\bar{3}c$) and is a good laser. In order to understand the fundamental behavior, it is desirable to know the positions of the ions to see the effect of the field of aluminum and oxygen ions on the chromium energy levels. You have obtained intensity data with MoK_α from an Al_2O_3 and a ruby crystal (4 mole % Cr_2O_3). After correcting for absorption and the *LP* factor, the intensities are:

hkl	Al_2O_3	Ruby
00.6	4410	2476
00.12	37048	42562
00.18	33936	34582
00.24	558	722
00.30	15086	17136

(a) Evaluate B_0, B_{Al}, z_{Al} from the data on Al_2O_3 [$f = f_0 \exp(-B \sin^2 \theta / \lambda^2)$ and z is the atomic coordinate along the a_3 axis]. It is known that for Al_2O_3 and Cr_2O_3, the oxygen atoms are at $\pm(0.36, 0, \frac{1}{4}; 0, -0.36, \frac{1}{4})$, and therefore these positions are the same for all solid solutions. There are six formula weights per unit cell with oxygen in the 18e equipoint and the metal atoms in the 12c positions. The structure is almost close-packed hexagonal oxygen with metal atoms in octahedral holes. The lattice parameters are $a = 4.7686$ Å and $c = 13.018$ Å.
(b) Evaluate z_{Cr}, B_{Cr} from the data for the ruby crystal.

5.24. As you have seen in Problems 1.7 and 5.15, GeSe is orthorhombic, its space group is *Pnma*, and there are four "molecules" per cell with the four atoms of germanium and four of selenium on the 4c equipoint. You have measured the integrated intensity with CuK_α, corrected it for absorption and the *LP* factor, and taken its square root. Determine the atomic coordinates. Ignore any effects of temperature and use the data in Problems 1.17 and 5.15, if needed.

hkl	(Relative intensity)$^{1/2}$
202	0.17
200	0.14
400	1.00
600	0.28
004	0.09
113	0.20

Problems

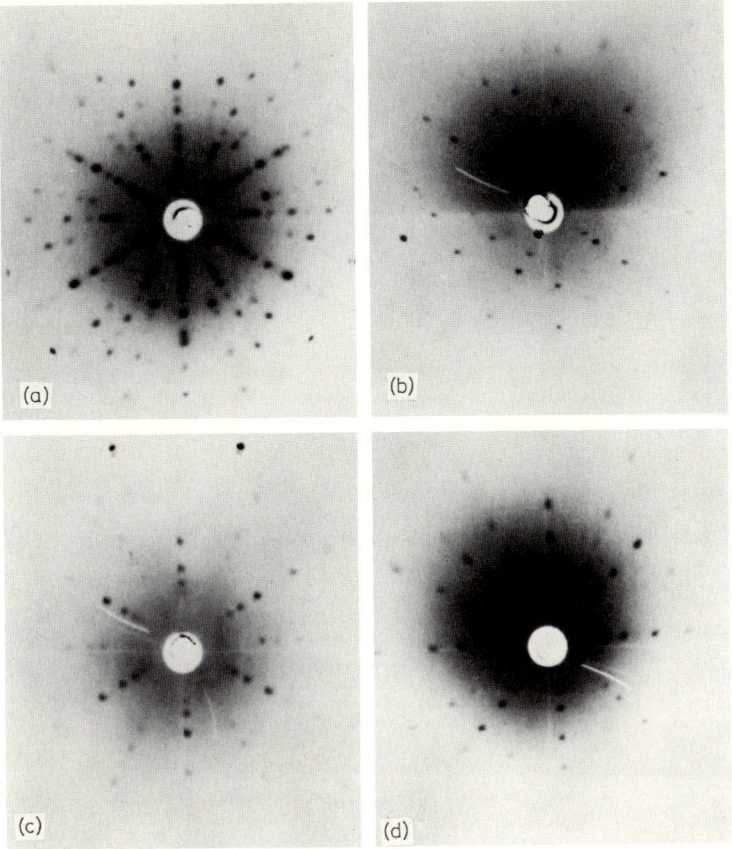

Fig. P5.25a–d. Four back-reflection Laue patterns (3-cm distance)

5.25. Four back-reflection Laue patterns (3-cm distance) are given in Fig. P5.25. They have been taken along prominent axes of a crystal. Indicate the point group symmetry in each.

5.26. Figure P5.26 shows two-dimensional, low-energy electron diffraction patterns (LEED) (taken by Dr. J. J. Lander, Head, Chemical Electronics Research Dept., Bell Telephone Laboratory, Murray Hill, New Jersey). They are of two surfaces of a single crystal of germanium after iodine has been absorbed. Determine the two-dimensional space group for each and compare to the group expected with a clean surface.

5.27. Refer to Problem 4.7 and carry out the following operations:
(a) Make a reciprocal lattice plot of λ/a_1, λ/a_3, and index as many spots as you can. Note any uncertainties and any systematic absences. (This can best be done by placing the indexed spots in separate columns for $h00$, $hk0$, hkl, hhl, etc.) You are given that from rotations around other axes:

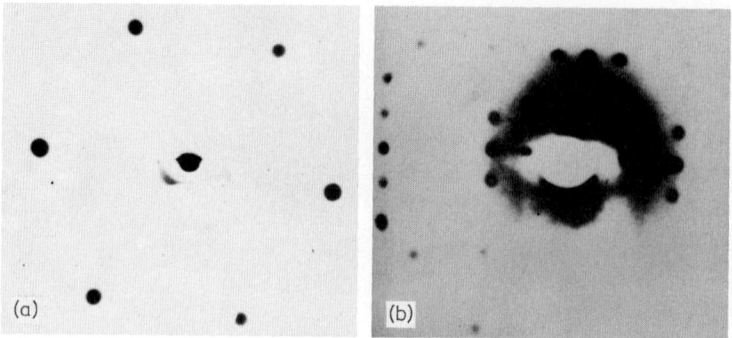

Fig. P5.26. Two-dimensional, low-energy electron diffraction patterns (LEED) (taken by Dr. J.J. Lander, Head, Chemical Electronics Research Dept., Bell Telephone Laboratory, Murray Hill, New Jersey)

$hk0 \quad h = 2n, \qquad h00 \quad h = 2n, \qquad 0k0 \quad k = 2n$

What is the diffraction symbol? (In reading data from the film, average values to the left and right of center.)

(b) Certain of the conditions you have set up in (a) from the film are uncertain. Pick one of these "problem areas" and sketch how you would check out some of the uncertainties with a diffractometer. Calculate the expected angles for the reflection(s) you are going to check.

6. Determination of Crystal Structures

6.1 Introduction

In the early chapters of this book we examined techniques for describing and studying the periodic arrays of atoms in matter which we call crystals. We showed that using radiation with wavelengths of the order of magnitude of the interatomic spacing, we could produce a diffraction pattern from the crystal, a pattern which contained the detailed structural information we need to describe that atomic arrangement. We showed in Chap. 5 that only a limited number of symmetries were possible for periodic atomic arrangements, and indicated how the diffraction pattern could be used to narrow the possible choices. It became obvious, however, that more sophisticated techniques would be required to extract that information with increasing complexity of the crystal — and the corresponding increase in complexity of the reciprocal space diffraction pattern. In this chapter we will survey many of the modern techniques used by crystallographers to determine atomic arrangements in periodic solids. It must be emphasized that the resultant structural determination usually refers to the average atomic arrangement. Deviations from that average, the defects such as strain, stacking faults, local order, etc. which make up such an important part of the field of materials science, are discussed in the next chapter.

6.2 Applications of Fourier Analysis to Diffraction

6.2.1 Introduction to Fourier Series

In our development of the diffraction of plane waves from an array of scattering centers in Chap. 2, we obtained results which included sums and integrals of trigonometric functions. In the remarks following Eq. (2.13), we noted that the amplitude scattered from such an array in the kinematical theory is proportional to the Fourier transform of the electron denisty and that reciprocal space is really Fourier transform space. Fourier series and transforms are, therefore, so common in all diffraction studies that we shall take a short detour from the main topic of this chapter to develop their properties and examine their usefulness.

If a function $f(x)$ is periodic with period (a), has no more than a finite number of discontinuities in a finite interval of x, and the following integral is finite (these restrictions are satisfied by most functions encountered in natural phenomena):

$$\int_{-a/2}^{a/2} |f(x)|\, dx,$$

then $f(x)$ can be decomposed into *an infinite Fourier series of sine and cosine terms*:

$$f(x) = A_0 + 2 \sum_{n=1}^{\infty} A_n \cos \frac{2\pi n x}{a} + 2 \sum_{n=1}^{\infty} B_n \sin \frac{2\pi n x}{a}, \tag{6.1}$$

where A_0, and all the coefficients A_n and B_n must be determined.

One of the important features of this Fourier series is the orthogonality of the individual terms, i.e., the fact that different terms do not interact. We can see this from the following integrals:

$$\int_{-a/2}^{+a/2} \cos \frac{2\pi n x}{a} \cos \frac{2\pi m x}{a}\, dx = \begin{cases} 0 & \text{if } n \neq m, \\ a/2 & \text{if } n = m, \end{cases} \tag{6.2a}$$

$$\int_{-a/2}^{+a/2} \cos \frac{2\pi n x}{a} \sin \frac{2\pi n x}{a}\, dx = 0, \tag{6.2b}$$

$$\int_{-a/2}^{+a/2} \sin \frac{2\pi n x}{a} \sin \frac{2\pi m x}{a}\, dx = \begin{cases} 0 & \text{if } n \neq m, \\ a/2 & \text{if } n = m, \end{cases} \tag{6.2c}$$

This orthogonality of the trigonometric functions suggests the way to determine the coefficients of Eq. (6.1). For the given function $f(x)$, substituting for $f(x)$ from Eq. (6.1) and using Eqs. (6.2), we have

$$\int_{-a/2}^{+a/2} f(x) \cos \frac{2\pi m x}{a}\, dx = A_m a, \tag{6.3a}$$

$$\int_{-a/2}^{+a/2} f(x) \sin \frac{2\pi m x}{a}\, dx = B_m a, \tag{6.3b}$$

$$\int_{-a/2}^{+a/2} f(x)\, dx = A_0 a, \tag{6.3c}$$

The last relation tells us that the area of the function in one period is proportional to the first coefficient, A_0. Furthermore, if the function is even (i.e., symmetrical about the $x = 0$ axis), then from Eq. (6.3b), all $B_n = 0$ and a cosine series results. If $f(x)$ is odd, or inverted through the origin (i.e., antisymmetric), then all the $A_n = 0$ from Eq. (6.3a) and a sine series results. For a general function $f(x)$, all terms in Eq. (6.1) are necessary.

We can write our Fourier series in a slightly different way, if we note from Eqs. (6.3a) and (6.3b) that $A_{(-n)} = A_n$ and $B_{(-n)} = -B_n$. Then,

$$f(x) = \sum_{-\infty}^{+\infty} A_n \cos \frac{2\pi n x}{a} + \sum_{-\infty}^{+\infty} B_n \sin \frac{2\pi n x}{a}. \tag{6.4}$$

The restrictions we have placed on the coefficients maintain a general series for $f(x)$.

For example, when $n = -4$, the sine term is negative, and if B_n did not change sign in Eq. (6.4), this sine term would cancel the term for $n = +4$ and, similarly, for all n, $-n$; only the cosine series would remain. Note that $B_0 \equiv 0$, so it may be included in this compact notation.

We can write the series in a still more compact way, if we define a complex coefficient:

$$C_n = A_n + iB_n .\tag{6.5}$$

Then

$$f(x) = \sum_{-\infty}^{+\infty} C_n e^{-2\pi i n x/a} .\tag{6.6a}$$

Negative values of n correspond to clockwise rotations on the complex plane, whereas positive values are counterclockwise rotations. Replacing C_n by Eq. (6.5) and recalling that $e^{i\theta} = \cos\theta + i\sin\theta$.

$$f(x) = \sum_{-\infty}^{+\infty} \left(A_n \cos\frac{2\pi n x}{a} + B_n \sin\frac{2\pi n x}{a} \right)$$

$$- i \sum_{-\infty}^{+\infty} \left(A_n \sin\frac{2\pi n x}{a} - B_n \cos\frac{2\pi n x}{a} \right).\tag{6.6b}$$

The second sum must be zero to correspond to Eq. (6.4), which is for a real function. This condition is satisfied since $A_{(-n)} = A_n$ and $B_{(-n)} = -B_n$ (i.e., $C_{(-n)} = C_n^*$ where the asterisk denotes "complex conjugate"). Using the orthogonality relations, Eqs. (6.2), the reader may show that the complex Fourier coefficients C_m can be obtained from

$$aC_m = \int_{-a/2}^{+a/2} f(x) e^{(+2\pi i m x/a)} dx .\tag{6.7}$$

Consider two functions $f_1(X)$ and $f_2(X)$ each with period a, multiplied together and averaged over this period,

$$\frac{1}{a}\int_{-a/2}^{+a/2} f_1(x) f_2(x) \, dx = \frac{1}{a}\int_{-a/2}^{+a/2} \left\{ \sum_{n=-\infty}^{+\infty} (C_1)_n e^{(-2\pi i n x/a)} \right\} f_2(x) \, dx$$

$$= \frac{1}{a} \sum_{n=-\infty}^{+\infty} (C_1)_n \int_{-a/2}^{+a/2} f_2(x) e^{(-2\pi i n x/a)} dx$$

$$= \sum_{n=-\infty}^{+\infty} (C_1)_n (C_2)_n^* .\tag{6.8}$$

Using the Fourier series representations of these functions, we have shown that *the average value of the product of the functions is the sum of the products of the coefficients.* Suppose the two functions are identical. Then we are taking the average value of $[f(x)]^2$ over the period. (In the diffraction problem, this might be the intensity averaged over the unit cell in reciprocal space.) This result may be written as

$$\frac{1}{a}\int_{-a/2}^{+a/2}[f(x)]^2\,dx = \sum_{n=-\infty}^{+\infty}|C_n|^2 = \sum_{n=-\infty}^{+\infty}(A_n^2+B_n^2). \tag{6.9}$$

According to Eq. (6.9), which is known as *Parseval's theorem*, the sum of the squares of the Fourier coefficients will give us the average "intensity" of the function. We can see this result directly from the series itself; look at Eq. (6.4), and consider the average value of the square of one term, say, $A_2 \cos 2\pi 2x/a$. The average value of the square is $\frac{1}{2}A_2^2$. But there are $+n$ and $-n$ values, so the average of the pair of these is A_2^2. The same is true for all terms. The sum of the squares of the coefficients is merely the sum of the average "intensity" in the unit cell of each wave in the series, and this is the same average we obtained by squaring the total series to arrive at Parseval's theorem. This is true because the terms do not interact.

If we plot the coefficients A_n and B_n as a function of n/a, such a plot will consist of a series of discrete lines spaced $1/a$ apart. As the period a increases, these lines move closer together and as the period approaches the limit of infinity, a continuous curve of coefficients develops. This limiting case of a Fourier series will allow us to represent nonperiodic functions as combinations of orthogonal plane waves.

6.2.2 The Fourier Integral Theorem

Rewriting the series, Eq. (6.1), and using the definitions of the coefficients,

$$f(x) = \frac{1}{a}\int_{-a/2}^{+a/2} f(\mu)\,d\mu + 2\sum_{n=1}^{\infty}\frac{1}{a}\cos\frac{2\pi n x}{a}\int_{-a/2}^{+a/2} f(\mu)\cos\frac{2\pi n\mu}{a}\,d\mu$$

$$+ 2\sum_{n=1}^{\infty}\frac{1}{a}\sin\frac{2\pi n x}{a}\int_{-a/2}^{+a/2} f(\mu)\sin\frac{2\pi n\mu}{a}\,d\mu$$

$$= \frac{1}{a}\int_{-a/2}^{+a/2} f(\mu)\,d\mu + \frac{2}{a}\sum_{n=1}^{\infty}\int_{-a/2}^{+a/2} f(\mu)\cos 2\pi n\frac{(x-\mu)}{a}\,d\mu, \tag{6.10}$$

where μ is a dummy integration variable. Let the period a approach infinity. The first term becomes zero (if the function still satisfies the conditions for a Fourier series, the integral is finite). If we further state that when $1/a$ approaches zero, as n approaches infinity n/a approaches zero, then n/a can be considered as a continuous variable, which we shall call s. With these conditions, Eq. (6.10) becomes

$$f(x) = 2\int_0^{\infty} ds \int_{-\infty}^{+\infty} f(\mu)\cos 2\pi s(x-\mu)\,d\mu. \tag{6.11}$$

This important result is known as *Fourier's integral theorem*.

Expanding the cosine of Eq. (6.11),

$$f(x) = 2\int_0^{\infty} ds \int_{-\infty}^{\infty} f(\mu)\{\cos 2\pi s\mu \cos 2\pi s x + \sin 2\pi s\mu \sin 2\pi s x\}\,d\mu$$

$$= 2\int_0^{\infty} \cos 2\pi s x\,ds \int_{-\infty}^{\infty} f(\mu)\cos 2\pi s\mu\,d\mu$$

$$+ 2\int_0^{\infty} \sin 2\pi s x\,ds \int_{-\infty}^{\infty} f(\mu)\sin 2\pi s\mu\,d\mu. \tag{6.12}$$

6.2 Applications of Fourier Analysis to Diffraction

Representing the cosine and sine intergrals of $f(\mu)$ by $a(s)$ and $b(s)$,

$$a(s) = \int_{-\infty}^{\infty} f(\mu) \cos 2\pi s\mu \, d\mu ,\tag{6.13a}$$

$$b(s) = \int_{-\infty}^{\infty} f(\mu) \sin 2\pi s\mu \, d\mu ,\tag{6.13b}$$

Eq. (6.12) becomes

$$f(x) = 2 \int_{0}^{\infty} \{a(s) \cos 2\pi sx + b(s) \sin 2\pi sx\} \, ds .\tag{6.14}$$

Multiplying and dividing by $[a^2(s) + b^2(s)]^{1/2}$,

$$f(x) = 2 \int_{0}^{\infty} [a^2(s) + b^2(s)]^{1/2} \left\{ \frac{a(s)}{[a^2(s) + b^2(s)]^{1/2}} \cos 2\pi sx \right.$$

$$\left. + \frac{b(s)}{[a^2(s) + b^2(s)]^{1/2}} \sin 2\pi sx \right\} ds .$$

Now, letting $R_s = [a^2(s) + b^2(s)]^{1/2}$ and $\tan \varphi_s = -b(s)/a(s)$,

$$f(x) = 2 \int_{0}^{\infty} R_s \cos\{2\pi sx + \varphi_s\} \, ds$$

$$= \int_{0}^{\infty} R_s \{\exp[i(2\pi sx + \varphi_s)] + \exp[-i(2\pi sx + \varphi_s)]\} \, ds .$$

From the definitions,

$$a(-s) = a(s), \qquad \varphi_{-s} = -\varphi_s$$
$$b(-s) = -b(s), \qquad R_s = R_{-s},$$

so we can rewrite the integral for $f(x)$ in two parts:

$$f(x) = \int_{0}^{\infty} R_s \exp(-2\pi isx) \exp(-i\varphi_s) \, ds$$

$$+ \int_{0}^{\infty} R_s \exp(-2\pi isx) \exp(-i\varphi_s) \, d(-s) .$$

Thus,

$$f(x) = \int_{0}^{\infty} R_s \exp(-2\pi isx) \exp(-i\varphi_s) \, ds$$

$$+ \int_{-\infty}^{0} R_s \exp(-2\pi isx) \exp(-i\varphi_s) \, ds ,$$

$$= \int_{-\infty}^{\infty} R_s \exp(-2\pi isx) \exp(-i\varphi_s) \, ds .$$

Finally, letting

$$F(s) = R_s \exp(-i\varphi_s) = a(s) + ib(s) ,\tag{6.15}$$

$$\boxed{f(x) = \int_{-\infty}^{+\infty} F(s) e^{-2\pi i s x} \, ds \,.} \tag{6.16}$$

This is the more common expression of Fourier's integral theorem. Note from Eqs. (6.13), when $f(x)$ is real and even, $b(s) = 0$ for all s, while for $f(x)$ real and odd, $a(s) = 0$ for all s. Substituting Eq. (6.13) in Eq. (6.15), we obtain

$$F(s) = \int_{-\infty}^{+\infty} f(x) \cos 2\pi s x \, dx + i \int_{-\infty}^{+\infty} f(x) \sin 2\pi s x \, dx \,,$$

or

$$\boxed{F(s) = \int_{-\infty}^{+\infty} f(x) e^{+2\pi i s x} \, dx \,.} \tag{6.17}$$

The function $F(s)$ is known as the Fourier transform of $f(x)$. Note the similarity between Eqs. (6.16) and (6.17). The two functions $F(s)$ and $f(x)$ related by these equations are known as a Fourier pair, and the variables s and x are known as a conjugate pair. We can abbreviate these two equations as

$$Tf(x) = F(s) \tag{6.18a}$$

and

$$T^{-1} F(s) = f(x) \,, \tag{6.18b}$$

where T and T^{-1} are read as "transform of" and "inverse transform of", respectively. We can think of Eqs. (6.16) and (6.17) in terms of two spaces: a Fourier transform space s for $F(s)$, and a "real" space x where our function $f(x)$ is plotted. From the nature of the Fourier transform, we can see certain basic properties:

$$F(0) = \int_{-\infty}^{+\infty} f(x) \, dx \quad \text{or} \quad f(0) = \int_{-\infty}^{+\infty} F(s) \, ds \,. \tag{6.19}$$

The integral of a function in one space equals its transform at the origin of the other space. Other important properties to note, which the reader can readily prove, are (see Problem 6.4).

$$Tcf(x) = cF(s) \tag{6.20a}$$

$$T \sum_i c_i f_i(x) = \sum_i c_i F_i(s) \,, \tag{6.20b}$$

$$Tf(cx) = \frac{1}{|c|} F\left(\frac{s}{c}\right), \tag{6.20c}$$

where c is a constant. An important example occurs for $c = -1$ in (6.20c):

$$Tf(-x) = F^*(s) \,. \tag{6.20d}$$

Finally, the transforms of nth derivatives of $f(x)$ may be expressed in terms of $F(s)$ by

6.2 Applications of Fourier Analysis to Diffraction

$$T(d^n/dx^n)f(x) = (-2\pi is)^n F(s) . \tag{6.20e}$$

We are often interested in a function repeated at many positions in real space; for example, the electron density of an atom, repeated at all the coordinates in a unit cell occupied by that atom. Shifting the origin of $f(x)$ to x_0, the transform of $f(x - x_0)$ is

$$Tf(x - x_0) = \int_{-\infty}^{+\infty} f(x - x_0) e^{+2\pi isx} dx .$$

Changing variables to $x_1 = x - x_0$,

$$Tf(x - x_0) = \int_{-\infty}^{+\infty} f(x_1) \exp[+2\pi is(x_1 + x_0)] dx_1$$

$$= \exp(+2\pi ix_0 s) \int_{-\infty}^{+\infty} f(x_1) \exp(+2\pi isx_1) dx_1$$

$$= \exp(+2\pi ix_0 s) F(s) . \tag{6.21a}$$

Combining this with Eq. (6.20b), we have

$$T \sum_{n=-\infty}^{+\infty} f(x - nx_0) = \sum_{n=-\infty}^{+\infty} F(s) \exp(+2\pi insx_0) . \tag{6.21b}$$

It is also easy to find *Parseval's theorem for transforms*:

$$\int_{-\infty}^{+\infty} f_1(x) f_2(x) dx = \int_{-\infty}^{+\infty} F_1(s) F_2^*(s) ds ,$$

$$\int_{-\infty}^{+\infty} [f(x)]^2 dx = \int_{-\infty}^{+\infty} F(s) F^*(s) ds = \int_{-\infty}^{+\infty} |F(s)|^2 ds . \tag{6.22}$$

As an example of a transform, consider the function plotted in Fig. 6.1a,

$$f(x) = \begin{cases} 1 & \text{for } |x| \leq 1/2, \\ 0 & \text{for } |x| > 1/2 . \end{cases}$$

Its transform is then

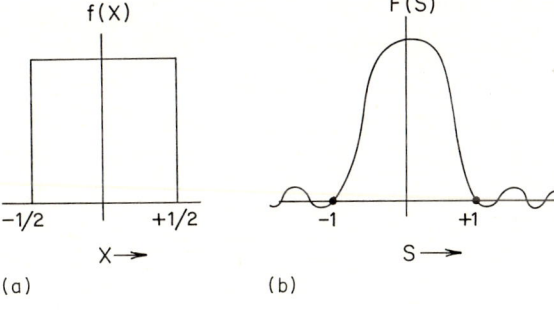

Fig. 6.1. A schematic representation of (a) a rectangular function $f(x)$ and (b) its Fourier transform, $F(s)$

$$F(s) = \int_{-\infty}^{+\infty} f(x) e^{+2\pi i s x} \, dx = \int_{-1/2}^{+1/2} e^{+2\pi i s x} \, dx,$$
$$= (e^{\pi i s} - e^{-\pi i s})/2\pi i s = \sin \pi s/\pi s, \tag{6.23a}$$

and

$$F(0) = \int_{-1/2}^{+1/2} f(x) \, dx = 1.$$

The transform is plotted in Fig. 6.1b. A similar function is $f(x) = 1$ for $|x| \le t/2$, $f(x) = 0$ for $|x| > t/2$. The transform of this function is

$$F(s) = \sin \pi t s/\pi s. \tag{6.23b}$$

Now if t is large the transform is sharp, and if t is small the transform is broad. Thus the transform has a reciprocal nature with respect to real space. This is also true of the Fourier coefficients of a function; they fall off with n more rapidly if the function is broad, than if it is sharp; see Problems 6.2 and 6.5.

6.2.3 Fourier Series and Transforms in Vector Space

So far we have been dealing with one-dimensional functions; let us now turn to three-dimensional functions involving three variables x, y, z, which are the coordinates along three axes $\mathbf{a}_1$, $\mathbf{a}_2$, $\mathbf{a}_3$, i.e., our function has values at the tip of a general three-dimensional vector $\mathbf{r}$. The function $f(x, y, z)$ can then be written as $f(\mathbf{r})$. Let $f(\mathbf{r})$ be a function that is triply periodic, repeated on a lattice whose edges are defined by $\mathbf{a}_1$, $\mathbf{a}_2$, $\mathbf{a}_3$. We may generalize Eq. (6.6) to three dimensions as

$$f(\mathbf{r}) = \sum_h \sum_k \sum_l C_{hkl} \exp\left[-2\pi i \left(\frac{hx}{a_1} + \frac{ky}{a_2} + \frac{lz}{a_3}\right)\right]. \tag{6.24}$$

In order to obtain the coefficients of this series, we can proceed in the same way as for the one-dimensional series we considered before. Therefore, we should consider the integral

$$\int_{V_c} f(\mathbf{r}) \exp\left[+2\pi i \left(\frac{h'x}{a_1} + \frac{k'y}{a_2} + \frac{l'z}{a_3}\right)\right] dV_\mathbf{r}, \tag{6.25}$$

over the volume V_c of the unit cell. Even if the cell edges are not perpendicular,

$$dV_\mathbf{r}/dx \, dy \, dz = V_c a_1 a_2 a_3$$

Therefore, $dV_\mathbf{r}$ can be replaced by $V_c/(a_1 a_2 a_3) \, dx \, dy \, dz$.

Defining fractional coordinates x', y', and z' by

$$x' = x/a_1, \qquad y' = y/a_2, \qquad z' = z/a_3,$$

we have

6.2 Applications of Fourier Analysis to Diffraction

$$\int_{V_c} f(\mathbf{r})\exp[+2\pi i(h'x' + k'y' + l'z')]\,dV_\mathbf{r}$$

$$= V_c \sum_h \sum_k \sum_l^{+\infty}{}_{-\infty} C_{hkl} \int_{-1/2}^{+1/2} \exp[-2\pi i(h-h')x']\,dx'$$

$$\times \int_{-1/2}^{+1/2} \exp[-2\pi i(k-k')y']\,dy' \int_{-1/2}^{+1/2} \exp[-2\pi i(l-l')z']\,dz'\,.$$

As each integral has the value of the period for $h = h'$, etc., and zero otherwise,

$$\int_{V_c} f(\mathbf{r})\exp[+2\pi i(h'x' + k'y' + l'z')]\,dV_\mathbf{r} = V_c C_{h'k'l'}\,. \qquad (6.26)$$

We can greatly simplify Eq. (6.26) by using the reciprocal vectors $\mathbf{b}_i$ which were so useful in crystallographic calculations. Consider

$$\mathbf{r}^*_{hkl} = h\mathbf{b}_1 + k\mathbf{b}_2 + l\mathbf{b}_3 \quad \text{and} \quad \mathbf{r} = x\mathbf{a}_1 + y\mathbf{a}_2 + z\mathbf{a}_3\,,$$

therefore,

$$\mathbf{r} \cdot \mathbf{r}^*_{hkl} = hx + ky + lz\,.$$

Hence Eq. (6.26) may be rewritten as

$$f(\mathbf{r}) = \sum_h \sum_k \sum_l^{+\infty}{}_{-\infty} C_{hkl}\exp(-2\pi i \mathbf{r}\cdot\mathbf{r}^*_{hkl})\,, \qquad (6.27)$$

with coefficients C_{hkl} determined from Eq. (6.26).

The sum of Eq. (6.27) is a set of plane waves in real space. That is, planes of constant amplitude for each wave are defined by $\mathbf{r}\cdot\mathbf{r}^*_{hkl}$ = constant, and each wave has a maximum amplitude for $\mathbf{r}\cdot\mathbf{r}^*_{hkl}$ an integer.[1] The waves have their fronts perpendicular to $\mathbf{r}^*_{hkl}$, i.e., parallel to (hkl) planes, and each has a wavelength d_{hkl}. We can see that this is so by comparing $2\pi \mathbf{r}\cdot\mathbf{r}^*_{hkl}$ to the corresponding term for this wave traveling in one dimension which we have previously considered, $2\pi x/\lambda$. As $\mathbf{r}^*_{hkl}$ has magnitude $(1/d_{hkl})$, $\mathbf{r}\cdot\mathbf{r}^*_{hkl} = (r/d_{hkl})\cos(\mathbf{r},\mathbf{r}^*)$. Thus, $\mathbf{r}\cdot\mathbf{r}^*$ is merely the projection of $\mathbf{r}$ on the unit normal to the periodic planes, divided by the period or wavelength of these planes, d_{hkl}. Multiplying by 2π gives us the phase angle for that particular position $\mathbf{r}$ in real space for the hkl wave.

In carrying out the triple sum, many harmonics (nh, nk, nl) must be included because the sum runs from $-\infty$ to $+\infty$. Thus $\mathbf{r}^*$ should be thought of as representing such planes as nh, nk, nl, even though we do not normally use such a notation for planes. The spacing of such planes is d_{hkl}/n.

Similar generalizations of Eqs. (6.16) and (6.17) show that *the Fourier transforms for a three-dimensional function* can be written as

---

[1] The equation $\mathbf{r}\cdot\mathbf{r}^*_{hkl}$ = const. is the equation of a plane. To see this, let $\mathbf{r}$ be some vector from the origin to a plane. Then let $\mathbf{n}$ be a unit normal to the plane. We can write that $\mathbf{n}\cdot\mathbf{r} = d_{hkl}$. But $\mathbf{n}$ can be written as $\mathbf{r}^*_{hkl}/|\mathbf{r}^*_{hkl}|$ or $\mathbf{r}^*_{hkl}/1/d_{hkl}$. Hence $\mathbf{r}^*_{hkl}\cdot\mathbf{r} = 1$.

$$f(\mathbf{r}) = \int F(\mathbf{s})e^{-2\pi i \mathbf{s} \cdot \mathbf{r}} dV_s, \qquad (6.28)$$

$$F(\mathbf{s}) = \int f(\mathbf{r})e^{+2\pi i \mathbf{s} \cdot \mathbf{r}} dV_r. \qquad (6.29)$$

Relations similar to Eqs. (6.20) hold for these functions. An important example is for the function $f(-\mathbf{r})$, which is $f(\mathbf{r})$ inverted through the origin:

$$Tf(-\mathbf{r}) = \int_{-\infty}^{+\infty} f(-\mathbf{r})\exp[+2\pi i(\mathbf{r}) \cdot \mathbf{s}] dV_r$$

$$= \int_{-\infty}^{+\infty} f(\mathbf{r}')\exp(-2\pi i \mathbf{r}' \cdot \mathbf{s}) dV_r' = F^*(\mathbf{s}).$$

This result reduces to Eq. (6.20d) for one dimension.

6.2.4 Fourier Transforms of Periodic Functions — The Diffraction Pattern

Let us now turn our attention to useful functions for periodic arrays. One such function is *the Dirac delta function* $\delta(\mathbf{r})$. By definition, $\delta(\mathbf{r}) = 0$ for $\mathbf{r} \neq 0$ and $\delta(\mathbf{r})$ approaches infinity for $\mathbf{r} \to 0$. Furthermore, $\int \delta(\mathbf{r}) dV_r = 1$. Multiplying a function $f(\mathbf{r})$ by $\delta(\mathbf{r} - \mathbf{r}_0)$ samples it at $\mathbf{r}_0$. To see this, consider the function

$$g(\mathbf{r}) = \int f(\mathbf{r})\delta(\mathbf{r} - \mathbf{r}_0) d\mathbf{r}.$$

Since $f(\mathbf{r})$ is slowly varying compared to $\delta(\mathbf{r} - \mathbf{r}_0)$, its value at $\mathbf{r} = \mathbf{r}_0$ (the only place where δ has a value) may be taken out of the integral, i.e.,

$$g(\mathbf{r}) = f(\mathbf{r}_0)\int \delta(\mathbf{r} - \mathbf{r}_0) d\mathbf{r} = f(\mathbf{r}_0). \qquad (6.30)$$

Using this result, we can evaluate the Fourier transform of $\delta(\mathbf{r} - \mathbf{r}_0)$:

$$T\delta(\mathbf{r} - \mathbf{r}_0) = \int \delta(\mathbf{r} - \mathbf{r}_0)e^{+2\pi i \mathbf{s} \cdot \mathbf{r}} dV_r = e^{+2\pi i \mathbf{s} \cdot \mathbf{r}_0}\int \delta(\mathbf{r} - \mathbf{r}_0) dV_r = e^{+2\pi i \mathbf{s} \cdot \mathbf{r}_0}. \qquad (6.31)$$

Thus, *the Fourier transform of a delta function in* $\mathbf{r}$ *space is a plane wave in* $\mathbf{s}$ *space and vice versa*. With this kind of function we can represent a one-dimensional net in $\mathbf{r}$ space as $\sum_{m=0}^{N-1} \delta(\mathbf{r} - m\mathbf{a})$. As an exercise let us take the transform of this net:

$$F(\mathbf{s}) = \int_{-\infty}^{\infty} \sum_{m=0}^{N-1} \delta(\mathbf{r} - m\mathbf{a})e^{+2\pi i \mathbf{s} \cdot \mathbf{r}} dV_r,$$

Making the substitution $\mathbf{r}_1 = \mathbf{r} - m\mathbf{a}$,

6.2 Applications of Fourier Analysis to Diffraction

$$F(\mathbf{s}) = \sum_m e^{+2\pi i \mathbf{s} \cdot m\mathbf{a}} \int \delta(\mathbf{r}_1) e^{+2\pi i \mathbf{s} \cdot \mathbf{r}_1} dV_{\mathbf{r}_1} = \sum_m e^{+2\pi i \mathbf{s} \cdot m\mathbf{a}} \int \delta(\mathbf{r}_1) dV_{\mathbf{r}_1}$$

$$= \sum_{m=0}^{N-1} e^{+2\pi i \mathbf{s} \cdot m\mathbf{a}} = \frac{\sin \pi \mathbf{s} \cdot N\mathbf{a}}{\sin \pi \mathbf{s} \cdot \mathbf{a}} \exp(2\pi i (N-1) \mathbf{s} \cdot \mathbf{a}/2), \tag{6.32}$$

where the last step follows from our discussion in Sect. 2.4. In that section we found that this kind of function is zero (for large N) except for $\mathbf{s} \cdot \mathbf{a} = h$, that is, for $s = h/a \cos(\mathbf{s}, \mathbf{a})$. The transform of a one-dimensional net of points is then a set of planes in $\mathbf{s}$ space perpendicular to $\mathbf{a}$ and spaced $1/a$ apart. Another way to see that this is so is to write each wave $\exp(2\pi i \mathbf{s} \cdot m\mathbf{a})$ as $\exp[2\pi i \mathbf{s} \cdot m\mathbf{n}/1/a]$, where $\mathbf{n}$ is a unit vector along $\mathbf{a}$. This is clearly a plane wave with fronts normal to $\mathbf{a}$ of wavelength $1/a$. The transform is the sum of many such waves, which leads to sharp planes in $\mathbf{s}$ space.

Conversely, the transform of a set of continuous planes can be shown to be a lattice row. In the limit as the number of planes in the set goes to infinity, we can prove that such a sum is also a δ function. Consider the function $h(x)$ which is a one-dimensional set of δ functions with period x_0. From Eq. (6.6), we can write this periodic function as a Fourier sum:

$$h(x) = \sum_{-\infty}^{+\infty} \delta(x - nx_0) = \sum_{-\infty}^{\infty} C_n \exp\left(\frac{-2\pi i n x}{x_0}\right). \tag{6.32a}$$

The coefficients are evaluated from Eq. (6.7) as

$$C_n = \frac{1}{x_0} \int_{-x_0/2}^{x_0/2} h(x) \exp\left(\frac{+2\pi i n x}{x_0}\right) dx.$$

The limits of the integral confine the region of interest of $h(x)$ to the origin, so we can write

$$C_n = \frac{1}{x_0} \int_{-x_0/2}^{x_0/2} \delta(x) \exp\left(\frac{+2\pi i n x}{x_0}\right) dx = \frac{1}{x_0},$$

where the results of Eq. (6.31) have been used. (For functions periodic in three dimensions, C_n is equal to the reciprocal of the volume of the unit cell.) Thus we can write the periodically repeated δ function as an infinite set of waves

$$\sum \delta(x - nx_0) = \frac{1}{x_0} \sum \exp\left(\frac{-2\pi i n x}{x_0}\right). \tag{6.32b}$$

Similarly, in transform space, it is easy to show that

$$\sum \delta\left(s - \frac{n}{x_0}\right) = x_0 \sum \exp(+2\pi i n x_0 s). \tag{6.32c}$$

Note that not only can an infinite sum of waves in one space be written as a δ function, but in transform space it is *also* a δ function. The periodicity in one space is the inverse of the periodicity in the other.

Next we consider the transform of a two-dimensional net. We write the net as

$$\sum_{m=0}^{N_1-1} \sum_{n=0}^{N_2-1} \delta[\mathbf{r} - (m\mathbf{a}_1 + n\mathbf{a}_2)].$$

Then

$$F(\mathbf{s}) = \int \sum_m \sum_n \delta[\mathbf{r} - (m\mathbf{a}_1 + n\mathbf{a}_2)] \exp(+2\pi i \mathbf{s} \cdot \mathbf{r}) dA_\mathbf{r}.$$

$$= \sum_m \exp(+2\pi i \mathbf{s} \cdot m\mathbf{a}_1) \sum_n \exp(+2\pi i \mathbf{s} \cdot n\mathbf{a}_2). \tag{6.33}$$

This function is the intersection of the two sets of planes represented by the two sums. As illustrated earlier in Fig. 2.21, this is a series of lines perpendicular to the original net.

For a δ function repeated in three dimensions (a lattice), we can therefore expect the transform to consist of points. These points are then produced by the intersections of the planes perpendicular to the third axis and the lines of the two-dimensional solution; this has also been illustrated in Fig. 2.22. This network of points is produced by the intersection of planes perpendicular to the $\mathbf{a}_1$, $\mathbf{a}_2$, and $\mathbf{a}_3$ axes and spaced apart by $1/a_1$, $1/a_2$, and $1/a_3$. This network is identical to that called the reciprocal lattice in Chap. 1. Thus we have formally shown that *the transform of a lattice is its reciprocal lattice* and we may refer to transform or s space, and reciprocal-space interchangeably.

It will be instructive to review some of the expressions for diffraction developed in Chap. 2 as examples of Fourier series and transforms. The scattering of x-rays from an atom is given by the scattering factor

$$f(\mathbf{s}) = \int \rho(\mathbf{r}) e^{2\pi i \mathbf{s} \cdot \mathbf{r}} dV_\mathbf{r}, \tag{6.34}$$

[see Eq. (2.14)], where $\rho(\mathbf{r})$ is the electron density of the atom. We recognize Eq. (6.34) as the Fourier transform of the electron density. It is now clear that measurements of $f(\mathbf{s})$ may be Fourier transformed experimentally to obtain $\rho(\mathbf{r})$ which can be directly compared with theory.

Similarly, the amplitude of x-radiation scattered by an assembly of electrons (atoms arranged in some structure, not necessarily periodic) is given by

$$F(\mathbf{s}) = \int \rho(\mathbf{r}) e^{2\pi i \mathbf{s} \cdot \mathbf{r}} dV_\mathbf{r}, \tag{6.35}$$

[see Eq. (2.13)], where now $\rho(\mathbf{r})$ is the electron density of all electrons in the sample. (Here, as in Chap. 2, we are neglecting ρ_0, the electron density of the medium surrounding the sample.) Again in principle, if we know $F(\mathbf{s})$ for all $\mathbf{s}$, we can take its Fourier transform to determine directly $\rho(\mathbf{r})$, which will tell us where all the atoms are, i.e., determine the structure. However, as we already know, $F(\mathbf{s})$ is a complex quantity and we do not measure $F(\mathbf{s})$ directly, but rather $F(\mathbf{s})F^*(\mathbf{s}) = |F(\mathbf{s})|^2$. To see this point another way,

$$F(\mathbf{s}) = T\rho(\mathbf{r})$$

and from Eq. (6.20d),

$$F^*(\mathbf{s}) = T\rho(-\mathbf{r}).$$

Hence, $|F(\mathbf{s})|^2$ is the product of the transforms of the electron density whose distribution we wish to know and this density inverted through the origin. The *measurement of intensity thus adds a center of symmetry even if one is not present in the original structure. This is known as Friedel's law and means that the presence or absence of a center of symmetry is not directly revealed in the diffraction pattern.* Special procedures do exist for obtaining this information by using radiation near an absorption edge of one of the atoms (the anomalous scattering technique discussed in Sect. 3.4) or by a statistical analysis of all the available intensity data (see Sect. 6.6).

Since the phase of $F(\mathbf{s})$ is not known, we cannot directly transform Eq. (6.35) to obtain the desired $\rho(\mathbf{r})$. There are many approaches to solve this phase problem and, in subsequent sections of this chapter, we will discuss several of the most frequently used methods. However, we shall first examine procedures for electron density mapping and refinement of structures using Fourier transformations. For further reference to the applications of Fourier series and transformations to diffraction, the reader is referred to Buerger (1960), Ramachandran and Srinivasan (1970), Lipson and Taylor (1958), and Cowley (1975). We shall see more examples in the subjects treated in Chap. 7 and 8.

We close this section with a note about numerical calculation of Fourier integrals. To determine the coefficients of the series representing a function, or to carry out a Fourier transformation, we may replace the desired integral by a sum, e.g.,

$$A_n \equiv \frac{1}{a} \int_{-a/2}^{a/2} f(x) \cos \frac{2\pi n x}{a} dx \simeq \frac{1}{a} \sum f(x) \cos \frac{2\pi n x}{a} \Delta x \,.$$

Computer programs for evaluating such sums are available (see Crystallogr. Comput. Comm. of the Am. Crystallogr. Assoc.) and a rapid algorithm for carrying out the transformation, called the "fast Fourier transform," is described by Brigham (1974) and available as a subroutine at most computer facilities.

6.3 Electron Density Mapping

6.3.1 Three-Dimensional Density Maps

In Sect. 2.34 it was pointed out that because we measure intensity not amplitude, we lose information about the phase angle of a reflection. In this section we shall assume that the phase problem has been solved, either by trial and error as suggested in Sect. 5.7 or by one of the more general methods to be discussed in subsequent sections. We wish to use the technique of Fourier transformation to give us a "picture" of the electron distribution in such an unknown structure.

In a defect-free crystal the electron density $\rho(x, y, z)$ is triply periodic, repeated in every unit cell. We can thus use Eq. (6.27) to describe it;

$$\begin{aligned} \rho(\mathbf{r}) = \rho(x,y,z) &= \sum_h \sum_k \sum_l C_{hkl} \exp(-2\pi i \mathbf{r} \cdot \mathbf{r}^*_{hkl}) \\ &= \sum_h \sum_k \sum_l C_{hkl} \exp[-2\pi i (hx + ky + lz)] \,. \end{aligned} \quad (6.36)$$

The Fourier coefficients are given by Eq. (6.28) as

$$C_{hkl} = \frac{1}{V_c} \int_{V_c} \rho(\mathbf{r}) \exp(+2\pi i \mathbf{r} \cdot \mathbf{r}^*_{hkl}) dV_r . \qquad (6.37)$$

Referring to Eqs. (2.13) and (2.15), we recognize this expression as the structure factor per unit cell divided by the volume of the cell, F_{hkl}/V_c. Hence, Eq. (6.36) may be written as

$$\rho(\mathbf{r}) = \rho(x,y,z) = \frac{1}{V_c} \sum_h \sum_k \sum_{l}^{\infty}{}_{-\infty} F_{hkl} \exp(-2\pi i \mathbf{r} \cdot \mathbf{r}^*_{hkl}) . \qquad (6.38a)$$

As we saw in Sect. 5.7, we may write $F_{hkl} = A_{hkl} + iB_{hkl}$, so

$$\rho(x,y,z) = \frac{1}{V_c} \sum_h \sum_k \sum_{l}^{\infty}{}_{-\infty} \{ A_{hkl} \cos 2\pi(hx + ky + lz)$$

$$+ B_{hkl} \sin 2\pi(hx + ky + lz) \} , \qquad (6.38b)$$

where the usual condition $F_{hkl} = F^*_{\bar{h}\bar{k}\bar{l}}$, or equivalently $A_{hkl} = A_{\bar{h}\bar{k}\bar{l}}$ and $B_{hkl} = -B_{\bar{h}\bar{k}\bar{l}}$ assures that the density is real. These conditions hold for the condition of no anomalous scatterers. If the radiation used has wavelength close to an absorption edge, then the atomic scattering factors have imaginary components. The reader should determine the expansion for $\rho(x,y,z)$ assuming that $F = F' + iF'' = (F'_r + iF'_i) + i(F''_r + iF''_i)$, where r and i represent real and imaginary parts of the structure factor. Writing $f = f_0 + \Delta f' + i\Delta f''$, the expressions for F'_r and F'_i are

$$F'_r = \sum_n (f_{0n} + \Delta f'_n) \cos 2\pi(hx_n + ky_n + lz_n)$$

and

$$F'_i = \sum_n \Delta f''_n \cos 2\pi(hx_n + ky_n + lz_n) .$$

Similar expressions for F''_r and F''_i are obtained by replacing the cosine functions by sine functions of the same argument.

Returning to the discussion of Eq. (6.38b), we note that in a centro-symmetric crystal, with origin taken at the center of symmetry, $B_{hkl} = 0$ for all hkl and

$$\rho(x,y,z) = \frac{1}{V_c} \sum \sum \sum_{-\infty}^{\infty} F_{hkl} \cos 2\pi(hx + ky + lz) . \qquad (6.38c)$$

Note that the expression for the electron density requires the summation of an infinite number of terms with coefficients F_{hkl}. Actual experiments will of course yield only a finite number of such F values out to some F_{max}. This is roughly equivalent to multiplying the function $F(\mathbf{s})$ by a function that is unity for $|\mathbf{s}| < |\mathbf{s}|_{max}$ and zero for $|\mathbf{s}| > |\mathbf{s}|_{max}$. Terminating the series at F_{max} obviously gives us a distorted picture of the electron density, a distortion known as termination error. We will return to this in Sect. 6.4. Although such termination errors may cause problems, when very careful work is done (see Dawson 1975) fine structure in the electron density maps may reveal electron concentrations in regions between atoms showing

6.3 Electron Density Mapping

direct evidence of bonding in solids. One novel approach to this problem is to use neutron diffraction to establish accurate values of the atomic positions and thermal parameters (since neutron scattering amplitudes are experimentally determined constants, they are not variables in the structure refinement.) Then one may form the electron density map based on the difference between the F's observed using x-rays and those calculated from the parameters that resulted from the neutron study, assuming the f's for x-rays for spherical distributions of electrons. This difference pattern will reveal deviations from the spherical electron distribution and evidence of bonding. Experimental measurements with both x-rays and neutrons should be done at low temperatures to reduce the thermal parameters to a minimum. Short wavelength x-rays obtainable from a synchrotron source may be employed to minimize absorption. Such work is now being done on small molecules, to compare the bonding to theoretical calculations (Coppens and Hall 1982).

A second point to note about Eqs. (6.38) is that they give a value for the electron density at a specific point in the unit cell with coordinates x, y, z. A complete map of this continuous density function is impractical, and instead the density is calculated at all points on a fine mesh within the cell, for example, at intervals of 1/60th of unit cell dimensions. Subsequently, by interpolating between points on the mesh, contours of constant electron density are plotted. Plots of $\rho(x, y, z_1)$ may be made for constant z_1. When drawn on transparent plastic, one may stack sheets of varying z to get a three-dimensional representation. Such plots are often difficult to visualize in three dimensions and an alternate representation may be produced by making two drawings of the structure with slightly different viewing angles using a computer to form a stereopair. Observations of such stereopairs help to give the true three-dimensional perspective of the structure, but details can sometimes be obtained by looking at the two- and one-dimensional representations of the electron density known as sections.

6.3.2 Sections and Projections

For example, *a plane section representing electron density on the plane* $z = z_1$ is given by

$$\rho(x, y, z_1) = \frac{1}{V_c} \sum_h \sum_k \sum_l F_{hkl} \exp[-2\pi i(hx + ky + lz_1)]$$

$$= \frac{1}{V_c} \sum_h \sum_k \left\{ \sum_l F_{hkl} \exp(-2\pi i l z_1) \right\} \exp[-2\pi i(hx + ky)]$$

$$= \frac{1}{V_c} \sum_h \sum_k Q_{hk} \exp[-2\pi i(hx + ky)]. \qquad (6.39)$$

Examples of such sections are given in Fig. 6.2 for NaCl. Similarly, *the electron density section along a line* (defined by $y = y_1, z = z_1$) *is a single Fourier sum*

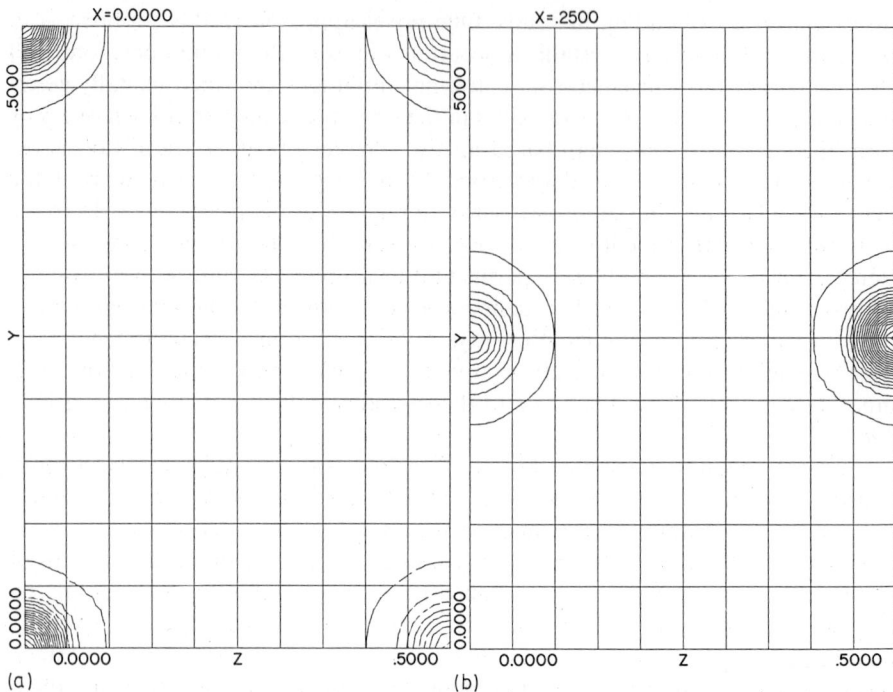

Fig. 6.2. Electron density sections of NaCl for (a) $\rho(0, u, u)$ and (b) $\rho(\tfrac{1}{4}, u, u)$. In this figure, and in Fig. 6.3, the "data" used to compute the Fourier transforms is the set of F_{hkl} calculated for the range of $\sin\theta/\lambda$ available using Mo K_α x-radiation and $0 \le 2\theta \le 160°$, including 138 reflections out to $hkl = 15, 3, 3$. Values for the scattering factors of Na^+ and Cl^- were taken from the International Tables, Volume III, p. 202. The isointensity lines are plotted at intervals of 50, with intensities normalized to 999 at the Cl^- positions. The intensity maximum at the Na^+ position is then 441

$$\rho(x, y_1, z_1) = \frac{1}{V_c} \sum_h P_h \exp(-2\pi i h x) \,. \tag{6.40}$$

where

$$P_h = \sum_k \sum_l F_{hkl} \exp[-2\pi i(ky_1 + lz_1)]$$

contain coefficients F_{hkl} for fixed h, variable k and l such as might be obtained from one layer of a rotation film or a single precession film. An example of electron density along such a line is shown in Fig. 6.3 for NaCl.

Sometimes in preliminary structure analysis, the electron density from all or part of the unit cell is projected onto a single plane. Such projections may reveal the atomic arrangement without necessitating collecting a full set of F_{hkl}. This will be particularly important in large, complex, low-symmetry structures such as those found in biological molecules. Suppose the electron density for the whole cell is projected along the a_3 axis on to the $z = 0$ plane. The *two-dimensional projected density function* is then

6.3 Electron Density Mapping

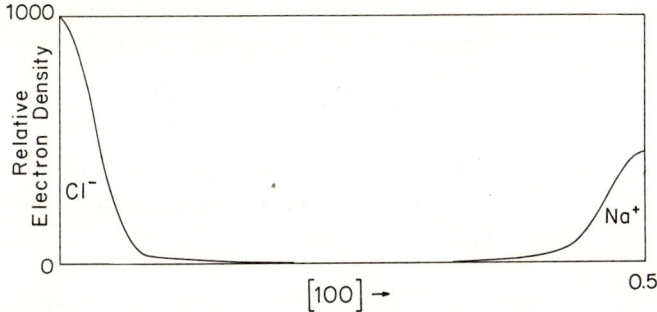

Fig. 6.3. Electron density along the line [100] in NaCl passing through the origins of the Na$^+$ and Cl$^-$ ions

$$\rho(x, y) = \int_{-a_3/2}^{a_3/2} \rho(x, y, z)\, dz$$

$$= \frac{1}{V_c} \sum_h \sum_k \sum_l F_{hkl} \exp[-2\pi i(hx + ky)] \int_{-a_3/2}^{a_3/2} \exp(-2\pi i l z)\, dz$$

$$= \frac{a_3}{V_c} \sum_h \sum_k F_{hk0} \exp[-2\pi i(hx + ky)], \quad (6.41a)$$

since the integral is zero for $l \neq 0$ and a_3 for $l = 0$. All the necessary data for this projection are obtainable from a single section in reciprocal space, such as a single precession film. This is another example of the reciprocal character of **s** space and **r** space. A projection along a direction in one space is related to a section in the other. Since it is the diffraction from such a section in **s** space that is employed in electron microscopy, the image we see is the projection of the electron density along the direction of the electron beam.

Since projections of the unit cell onto a plane must be one of the allowed two dimensionally periodic patterns, there are only 17 allowable plane projection patterns (see Sect. 5.4). The most useful projections are along the directions of axial symmetry elements. Such projections have symmetry related to the space group symmetry of the unit cell and are important clues to the overall symmetry. This subject is discussed in some detail in Chap. 11 of Buerger (1960).

Note that in Eq. (6.41a), $a_3/V_c = a_3/(\mathbf{a}_3 \cdot \mathbf{a}_1 \times \mathbf{a}_2) = a_3/(a_3|\mathbf{a}_1 \times \mathbf{a}_2|\cos\varphi)$, where φ is the angle between $\mathbf{a}_3$ and $\mathbf{a}_1 \times \mathbf{a}_2$. The magnitude a_3 cancels, leaving the reciprocal of the area of the cell perpendicular to the direction of projection. This is a general result which holds for any direction of projection.

Via high-speed digital computation, the calculation of $\rho(x, y, z)$ is possible, even for structures for which thousands of individual F_{hkl}'s must be included. It is important to note that the number of computations increases rapidly with decreasing mesh. Thus for a mesh of 1/60th of each unit cell edge, $60^3 = 216{,}000$ triple sums of type Eqs. (6.38b) would be performed. For a mesh of $\frac{1}{120}$, 1,728,000 computations are necessary. Even though routine, such computations are time consuming and hence costly. Any shortcuts to reduce the number of such computations are important to help stretch the crystallographer's limited computer budget. The most important shortcut comes from analysis of the symmetry of the unit cell. By definition, points related by this symmetry have identical electron densities and only need be computed once. There is then a minimum volume (or area) required to represent all unique points.

As an example of the usefulness of symmetry in reducing the required number of computations, we shall consider the projection along $\mathbf{a}_3$. From Eq. (6.41a), we may write

$$\rho(x,y) = \frac{a_3}{V_c} \sum_{h}^{\infty} \sum_{k}_{-\infty} \{A_{hk0} \cos 2\pi(hx + ky) + B_{hk0} \sin 2\pi(hx + ky)\} . \tag{6.41b}$$

Consider the reciprocal space broken into four regions as indicated in Fig. 6.4. When no appreciable anomalous scattering is present, the diffraction pattern contains a center of symmetry (even if the structure does not). Thus only regions hk and $\bar{h}k$ need be measured. We can see this as follows. From the general properties of a Fourier series, $A_{hk0} = A_{\bar{h}\bar{k}0}$ and $B_{hk0} = -B_{\bar{h}\bar{k}0}$. Combining terms for these two quadrants of Fig. 6.4 gives

$$\rho(x,y) = \frac{a_3}{V_c} \Bigg\{ A_{000} + 2 \sum_{h} \sum_{k>0} A_{hk0}(\cos 2\pi hx \cos 2\pi ky - \sin 2\pi hx \sin 2\pi ky)$$

$$+ B_{hk0}(\sin 2\pi hx \cos 2\pi ky + \cos 2\pi hx \sin 2\pi ky) \Bigg\}.$$

Note that A_{000} which is the total of the electron density in the unit cell [Eq. (6.19)] must then be the total atomic number in the unit cell. A similar expression may be written for the two quadrants $h\bar{k}$ and $\bar{h}k$. Combining all terms, the projected density is

$$\rho(x,y) = \frac{a_3}{V_c} \Bigg\{ A_{000} + 2 \sum_h \sum_k [(A_{hk0} + A_{\bar{h}k0}) \cos 2\pi hx \cos 2\pi ky$$

$$- (A_{hk0} - A_{\bar{h}k0}) \sin 2\pi hx \sin 2\pi ky + (B_{hk0} - B_{\bar{h}k0}) \sin 2\pi hx \cos 2\pi ky$$

$$+ (B_{hk0} + B_{\bar{h}k0}) \cos 2\pi hx \sin 2\pi ky] \Bigg\}. \tag{6.41c}$$

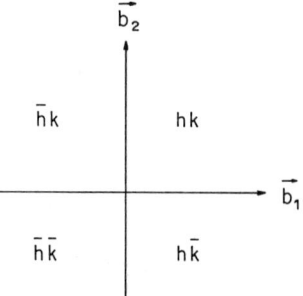

Fig. 6.4. For computing the projected electron densities, reciprocal space may be conveniently broken up into the four quadrants indicated. (After Buerger, M.J., "Crystal Structure Analysis". Copyright 1960, John Wiley & Sons. Reprinted by permission of John Wiley & Sons, Inc.)

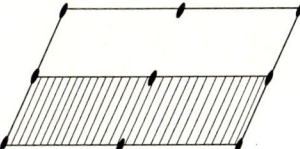

Fig. 6.5. Analysis of the minimum area for which the electron density need be computed in the two-dimensional space group *P*2. (After Buerger, M.J., "Crystal Structure Analysis". Copyright 1960, John Wiley & Sons. Reprinted by permission of John Wiley & Sons, Inc.)

Fig. 6.6. Analysis of the minimum area for which the electron density need be computed in the two-dimensional space group *Cm*. (After Buerger, M.J., "Crystal Structure Analysis". Copyright 1960, John Wiley & Sons. Reprinted by permission of John Wiley & Sons, Inc.)

As an example of the effect of the projection symmetry, consider some structure that projects to the two-dimensional space group p2 shown in Fig. 6.5. Looking in the International Tables (Vol. I. p. 58), we see that for any atom at coordinates x, y there must be one at $\bar{x}$, $\bar{y}$. Thus the structure factors must obey the relation $F_{hk0} = F_{\bar{h}\bar{k}0}$. But we also have the general requirement (for no anomalous scatterers) that $F_{hk0} = F^*_{\bar{h}\bar{k}0}$. That is, $(A_{hk0} + iB_{hk0}) = (A_{\bar{h}\bar{k}0} + iB_{\bar{h}\bar{k}0}) = (A_{hk0} - iB_{hk0})$. Consequently, $B_{hk0} \equiv 0$. Now examining Eq. (6.41c), with $B_{hk0} = 0$, we see that $\rho(x,y) = \rho(\bar{x},\bar{y})$ and $\rho(x,\bar{y}) = \rho(\bar{x},y)$. Thus we need only compute the projected electron density for one-half of the unit cell to know it for all points; for example, only the shaded region in Fig. 6.5 need be computed. Another simple example is for projection onto the two-dimensional space group *Cm* shown in Fig. 6.6 (see p. 60 of the International Tables). Again only one-half of the cell need be calculated, due to the center of symmetry introduced by the measurement of intensity; but in addition, the computation is simplified since $F_{hk0} = F_{\bar{h}k0}$ due to the mirror, i.e., $A_{hk0} = A_{\bar{h}k0}$ and $B_{hk0} = B_{\bar{h}k0}$, eliminating two terms from Eq. (6.41). We only need data from one quarter of reciprocal space in Fig. 6.6. The reader should consider other examples of projections, sections, and full unit cell densities to see how symmetry reduces the number of computations (see Problem 6.7). Further discussion of these points and other types of electron density maps may be found in Buerger (1960, Chaps. 11 and 17), Ramachandran and Srinivasan (1970), and James (1954).

One important application of the Fourier mapping technique is in the process of structure refinement. Let us assume that the structure is approximately known. Using techniques such as those described in Sect. 5.7 or Sects. 6.4–6.7, the approximate positions of the atoms or chemical molecules have been determined, and may be used to compute an estimate of the structure factor F_c. While these calculated values of F_c may be erroneous owing to small errors in the unknown atomic coordinates, the computed phases of F_c can be used as a starting point in the structure refinement. In the simplest case of a centrosymmetric structure, the phase is either ± 1 and for a good first guess of the structure, this sign will be known accurately for most of the strong *hkl* reflections.

Thus we may begin the Fourier analysis and obtain an electron density map using calculated phases and measured $|F_0|$'s. In such a calculation it is wise to omit the F's for weak reflections as their intensity and phase contain the greatest errors. The resultant Fourier synthesis will be incomplete, but it may be used to refine the structure. Atoms will be identified by regions of high electron density. An iterative procedure will lead to a better set of phases and more certainty for weaker reflections which may then be used in subsequent iterations. A corrected model of the crystal structure may now be used as input to the final least-squares refinement discussed in Sect. 6.9.

6.3.3 Optical Diffraction

Before closing this section we shall discuss a mapping technique developed by H. Lipson and C.A. Taylor that is based on the analogy between optical and x-ray diffraction. Consider the scattering of light from a single circular hole in an otherwise opaque sheet of material called the optical mask. In Problem 6.10 the amplitude scattered from such a circular aperature of diameter a is shown to be $J_1(\pi a s)/s$ where s is the radial coordinate in reciprocal space and $J_1(X)$ is the first-order Bessel function of argument x. This is a function which has a peak at $s = 0$ and then oscillates with decreasing amplitude. The first minimum of this function occurs at a scattering angle φ given by

$$\sin \varphi = 0.61 \lambda / a ,$$

where λ is the wavelength of the incident light. Taking typical values of $\lambda \approx 5(10)^{-7}$ m and $a \approx (10)^{-3}$ m, we find $\varphi \approx 2$ min of arc. A long distance is employed so that this position occurs at the edge of the magnified pattern. The scattering from a pattern of holes will then always be within this region. (This should be clear to the reader from the reciprocal nature of the real and transform spaces.)

Consider a beam of light incident on an optical mask with several holes cut into it as shown in Fig. 6.7. Let $a_j(s')$ be the amplitude scattered by a single hole in

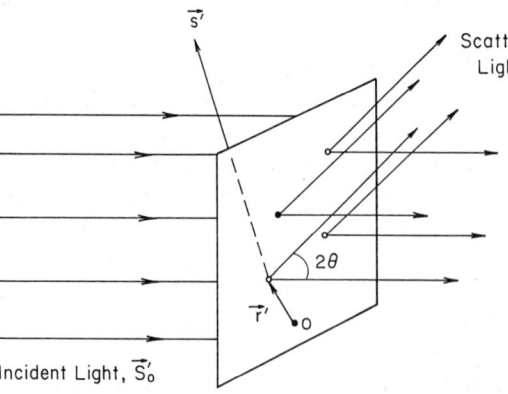

Fig. 6.7. Light scattered from holes in an optical mask will produce a diffraction pattern characteristic of the positions and sizes of the holes. (After Woolfson, M.M., "An Introduction to X-ray Crystallography". Cambridge Univ. Press, London and New York, 1970)

6.3 Electron Density Mapping

the mask located at vector $\mathbf{r}'_j$ from some arbitrary origin. If the holes in the mask are arranged with the symmetry of a projection along the $\mathbf{a}_2$ axis, then the resultant diffraction pattern is equivalent to a section of the Fourier transform of the unit cell contents in the $\mathbf{b}_1 \mathbf{b}_3$ plane. To see this, note that the total amplitude of light scattered from the mask in direction $\mathbf{s}'$ is

$$A(\mathbf{s}') = \sum_j a_j(\mathbf{s}') \exp(2\pi i \mathbf{s}' \cdot \mathbf{r}'_j). \tag{6.42}$$

Comparing Eq. (6.42) to Eq. (2.16), we see that *the optical transform is proportional to the section of the Fourier transform of the unit cell contents* if the scale of the optical mask is adjusted so that

$$\mathbf{r}'_j \cdot \mathbf{s}' = \mathbf{r}_j \cdot \mathbf{s},$$

and the amplitude $a_j(\mathbf{s}')$ is adjusted so that it is equivalent to the atomic scattering factor of the atom represented by the jth hole. This latter requirement can be approximately done by varying the size of the holes in the optical mask.

The optical diffraction pattern described by the plane waves in Eq. (6.42) is generated by placing the object at the focal point of a convergent lens. But then the image is at an infinite distance, and other convergent lenses are used to bring the diffraction pattern into focus in practical distances. A small angle optical diffractometer of the type used by Lipson and Taylor (1958) is sketched in Fig. 6.8. The image may be observed by a microscope focused at the focal plane of lens L_2 or by enlarging a photographic film placed in this focal plane. An example of such a transform is shown in Fig. 6.9. When a possible structure has been guessed, the optical mask corresponding to the projection of this structure along a high symmetry direction is made. The optical transform may then be compared to the x-ray diffraction pattern of the appropriate section in reciprocal space. This is a relatively quick method of establishing that the assumed structure is a good guess, and that techniques for refinement may be attempted.

In the next sections of this chapter we will discuss several of the more commonly used techniques to solve the phase problems, i.e., to determine the structure using only the observed $|F|^2$.

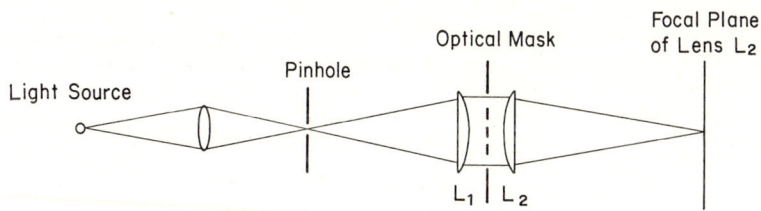

Fig. 6.8. An optical diffractometer used to bring the diffraction pattern from an optical mask into focus at the focal plane of lens L_2. (After Woolfson, M.M., "An Introduction to X-ray Crystallography", Cambridge Univ. Press, London and New York, 1970)

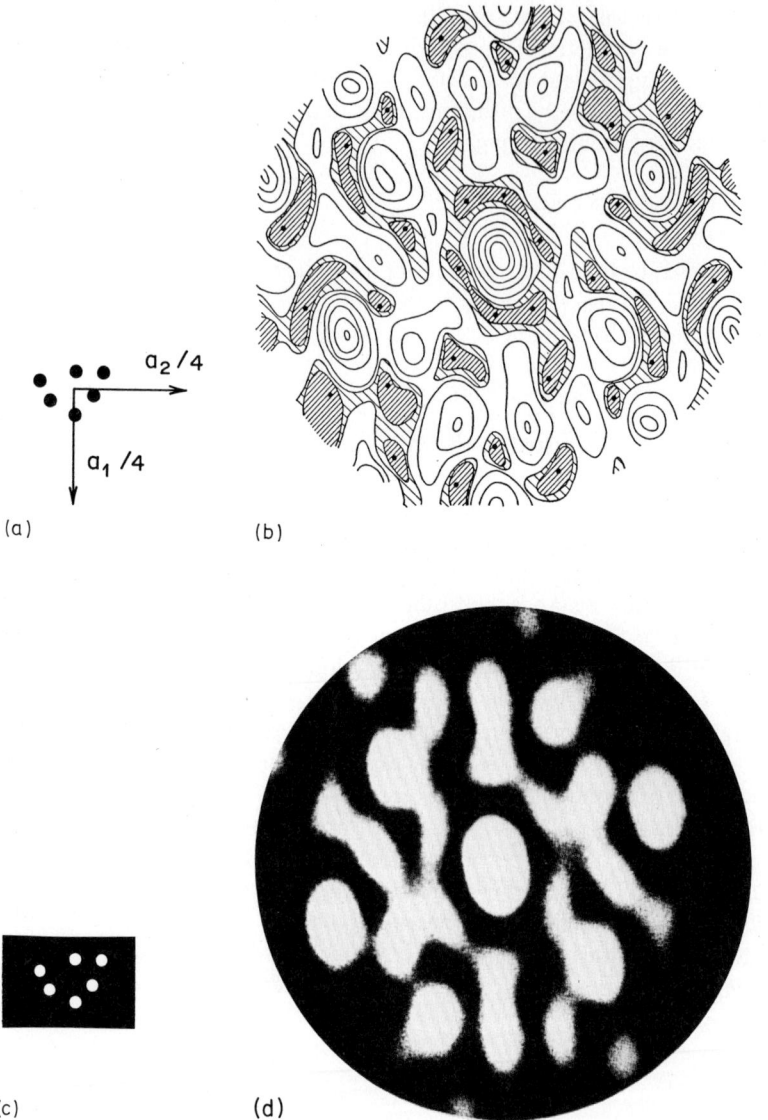

Fig. 6.9a–d. An example of the optical diffraction pattern for a hypothetical model. **a** Six-atom "molecule," **b** The calculated modulus of the transform of **a**. The lowest regions are *shaded* and the *black spots* are points of zero amplitude. **c** Optical mask representing the structure shown in **a**. **d** Optical transform of the mask in **c**. The essential features of the transforms shown in **b** and **d** are equivalent. (Reproduced from Lipson, H., and Taylor, C.A., "Fourier Transforms and X-ray Diffraction". By permission of Bell, London, 1958)

6.4 The Patterson Function

6.4.1 Definition of the Patterson Function

The solution to the crystal structure problem involves assigning phases to the measured $|F_{hkl}|$, since a direct Fourier transformation of F_{hkl} will yield the electron density map. While no completely automatic technique exists for obtaining these phases, there are some approaches that are always possible and may yield useful information. One of these approaches is known as *Patterson synthesis* after A.L. Patterson who initiated this technique in 1934. Patterson sought to obtain information directly from the measured $|F|^2$'s by considering their Fourier Transform.

The amplitude scattered from a sample of electron density $\rho(\mathbf{r})$ is given by Eq. (6.35):

$$F(\mathbf{s}) = \int_{V_r} \rho(\mathbf{r}) e^{+2\pi i \mathbf{s} \cdot \mathbf{r}} dV_\mathbf{r},$$

so the corresponding intensity is

$$F(\mathbf{s})F^*(\mathbf{s}) = \int_{V_r} \rho(\mathbf{r}) \exp(+2\pi i \mathbf{s} \cdot \mathbf{r}) dV_\mathbf{r} \int_{V_{r'}} \rho(\mathbf{r}') \exp(-2\pi i \mathbf{s} \cdot \mathbf{r}') dV_{\mathbf{r}'}.$$

Substituting $\mathbf{r} - \mathbf{r}' = \mathbf{u}$, eliminating $\mathbf{r}$, and rearranging

$$F(\mathbf{s})F^*(\mathbf{s}) = \int_{V_u} \left\{ \int_{V_{r'}} \rho(\mathbf{u}+\mathbf{r}')\rho(\mathbf{r}') dV_{\mathbf{r}'} \right\} e^{+2\pi i \mathbf{s} \cdot \mathbf{u}} dV_\mathbf{u} = TP(\mathbf{u}), \tag{6.43}$$

where

$$P(\mathbf{u}) = \int_{V_{r'}} \rho(\mathbf{u}+\mathbf{r}')\rho(\mathbf{r}') dV_{\mathbf{r}'} \tag{6.44}$$

is known as the *Patterson function*. Thus we see that if we take the inverse transform of the measured intensity $F(\mathbf{s})F^*(\mathbf{s})$, we obtain the Patterson function:

$$T^{-1} F(\mathbf{s})F^*(\mathbf{s}) = \int_{V_s} |F(\mathbf{s})|^2 e^{-2\pi i \mathbf{s} \cdot \mathbf{r}} dV_\mathbf{s} = P(\mathbf{r}). \tag{6.45}$$

From its definition in Eq. (6.44), we can see that the *Patterson function* will have large values only when $\rho(\mathbf{r})$ and $\rho(\mathbf{u}+\mathbf{r})$ are both large, i.e., for $\mathbf{u}$ equal to interatomic vectors. That is, it is a function which shows peaks for all interatomic vectors in the structure. These peaks will be broader than the individual atoms, since $P(\mathbf{u})$ will still have finite values when either $\rho(\mathbf{r})$ or $\rho(\mathbf{u}+\mathbf{r})$ (but not both) is small. Thus a vector joining atoms of diameters d_A and d_B will have a Patterson peak of diameter $(d_A + d_B)$. Furthermore, the areas of the peaks are the products of the electron densities at the two ends of the vector involved.

A one-dimensional example is illustrative. Consider the electron density distributed as in Fig. 6.10a with atoms at x_1, x_2, x_3. The Patterson function for $\rho(x)$ is

$$P(x) = \int \rho(x+u)\rho(u)du. \tag{6.46}$$

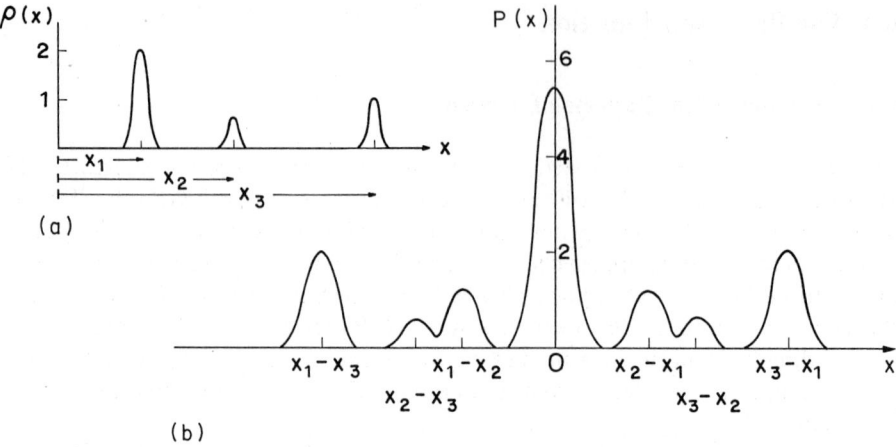

Fig. 6.10a,b. A one-dimensional example of the Patterson function. **a** Electron density. **b** The Patterson function. (After Buerger, M.J., "Vector Space", Copyright 1959, John Wiley & Sons. Reprinted by permission of John Wiley & Sons, Inc.)

To obtain this result for a fixed x, we must take ρ at each point u and multiply it by ρ evaluated at $x + u$, then sum these values for all u. Peaks will occur at values of x equal to $x_2 - x_1$, $x_1 - x_2$, $x_3 - x_1$, $x_1 - x_3$, $x_2 - x_3$, and $x_3 - x_2$, as shown in Fig. 6.10b. The peak at the origin of $P(x)$ is due to all interatomic vectors $x = 0$, i.e., the sum of all terms of the type $\rho^2(u)$ for each atom. For n atoms in the crystal, there are n such peaks at $x = 0$. The total number of peaks is n^2 (every atom with respect to all others and itself), so there are $(n^2 - n)$ non-origin peaks. Since there is an inversion center contained in $P(x)$ [to see this, set $x = -x$ in Eq. (6.46) and let $v = u - x$], there are only $(n^2 - n)/2$ unique peaks. In this simple case the structure can be easily inferred from the Patterson function.

6.4.2 Definition of the Convolution Function

The Patterson function is a particular example of a very general and useful function, the convolution. Let $f(\mathbf{x})$ and $g(\mathbf{x})$ be two integrable functions of $\mathbf{x}$. Then the convolution of $f(\mathbf{x})$ and $g(\mathbf{x})$ is

$$f(\mathbf{x}) \otimes g(\mathbf{x}) \equiv \int f(\mathbf{u}) g(\mathbf{x} - \mathbf{u}) \, dV_\mathbf{u} \, . \tag{6.47}$$

Comparing Eqs. (6.47) and (6.44), we see that they are identical for $f(\mathbf{x}) = \rho(\mathbf{r}')$ and $g(\mathbf{x}) = \rho(-\mathbf{r}')$, i.e.,

$$P(\mathbf{r}) = \rho(\mathbf{r}) \otimes \rho(-\mathbf{r}) \, . \tag{6.48}$$

The Patterson function is the convolution of the electron density with its own inversion in the origin. Let us explore the properties of the convolution. The Fourier transform of the convolution $y(\mathbf{x})$ is

6.4 The Patterson Function

$$Y(\mathbf{s}) = \int y(\mathbf{x}) \exp(+2\pi i[\mathbf{s}\cdot\mathbf{x}])\,dV_x$$

$$= \int\int f(\mathbf{u})g(\mathbf{x}-\mathbf{u})\exp(+2\pi i[\mathbf{s}\cdot\mathbf{x}])\,dV_x\,dV_u\,. \tag{6.49}$$

Letting $\mathbf{x} - \mathbf{u} = \mathbf{w}$,

$$Y(\mathbf{s}) = \int\int f(\mathbf{u})g(\mathbf{w})\exp(+2\pi i[\mathbf{s}\cdot(\mathbf{u}+\mathbf{w})])\,dV_u\,dV_w\,,$$

$$= \int f(\mathbf{u})\exp(+2\pi i[\mathbf{s}\cdot\mathbf{u}])\,dV_u \int g(\mathbf{w})\exp(+2\pi i[\mathbf{s}\cdot\mathbf{w}])\,dV_w$$

$$= F(\mathbf{s})G(\mathbf{s})\,. \tag{6.50}$$

This property can sometimes be useful in evaluating a convolution by obtaining the transforms of each function, multiplying them and then transforming back. The reader can show directly from Eq. (6.50) that a convolution is commutative, i.e., $f \otimes g = g \otimes f$. This result also tells us that the integral of a convolution of two functions is the product of the transforms at $s = 0$, hence the product of the areas of the two functions. The reader can readily prove that if $w(\mathbf{x}) = f(\mathbf{x})g(\mathbf{x})$

$$Tw(\mathbf{x}) = W(\mathbf{s}) = F(\mathbf{s}) \otimes G(\mathbf{s})\,. \tag{6.51}$$

It also follows from Eq. (6.50) that

$$|G(\mathbf{s})|^2 = G(\mathbf{s})G^*(\mathbf{s}) = \int g(\mathbf{x}) \otimes g(-\mathbf{x})\exp(+2\pi i[\mathbf{s}\cdot\mathbf{x}])\,dV_x \tag{6.52a}$$

$$|g(\mathbf{x})|^2 = g(\mathbf{x})g^*(\mathbf{x}) = \int G(\mathbf{s}) \otimes G(-\mathbf{s})\exp(-2\pi i[\mathbf{s}\cdot\mathbf{x}])\,dV_s\,. \tag{6.52b}$$

Equation (6.52) tells us that the square of the modulus of $G(\mathbf{s})$ is the transform of the convolution of the function $g(\mathbf{x})$ and the same function inverted through the origin, as we have pointed out for the Patterson function. We have seen that the diffracted amplitude, which we can think of as $G(\mathbf{s})$, times its complex conjugate is the intensity we measure in a diffraction experiment. The loss of information about the phase of $G(\mathbf{s})$, resulting from the measurement of only $|G(\mathbf{s})|^2$ gives rise to diffraction data that imply a center of symmetry even if none actually exists in the structure.

Finally, some interesting results occur if we combine δ functions and the convolution. Suppose we convolute $g(\mathbf{x})$ and $\sum_n \delta(\mathbf{x} - \mathbf{x}_n)$:

$$y(\mathbf{x}) = \int g(\mathbf{u}) \sum \delta(\mathbf{x} - \mathbf{u} - \mathbf{x}_n)\,du\,. \tag{6.53}$$

Now the integral has values only for zeros of the argument of the δ function, i.e., for $\mathbf{u} = \mathbf{x} - \mathbf{x}_n$. Thus,

$$y(\mathbf{x}) = \sum_n g(\mathbf{x} - \mathbf{x}_n)\,, \tag{6.54}$$

and the function $g(\mathbf{x})$ is repeated at the points $\mathbf{x}_n$ by convoluting it with $\sum \delta(\mathbf{x} - \mathbf{x}_n)$.

As an example of the presence of convolutions in analyses we have already made, consider Eq. (2.18b), the amplitude $A_T(\mathbf{s})$ diffracted from a three-dimensional periodic crystal. The result is the product of the transform of the electron density of the unit cell, $F(\mathbf{s})$, and an interference term which is the transform of the lattice, as shown in Eqs. (6.32). Thus *the amplitude is the transform of the convolution of the electron density in a unit cell with the δ function representing the lattice*:

$$A_T(\mathbf{s}) = T[\rho(\mathbf{r}) \otimes \sum_{l,m,n} \delta(\mathbf{r} - l\mathbf{a}_1 - m\mathbf{a}_2 - n\mathbf{a}_3)] .$$

Similarly, we note that the intensity scattered from such a triply periodic crystal, given by $A_T(\mathbf{s})A_T^*(\mathbf{s})$ in Eq. (2.18c), is the transform of the convolution of two terms: one is the convolution of the electron density in a unit cell with the same density inverted through the origin, $F(\mathbf{s})F^*(\mathbf{s})$, while the other is the convolution of the lattice with the lattice inverted through the origin.

We can now examine the effect of measuring only a finite number of reflections. This limitation is equivalent to multiplying the desired infinite set of F_{hkl}'s by a function that has unit magnitude in the range of hkl measured and zero outside that range, a function that looks like a box in $\mathbf{s}$ space. Since the true electron density is the transform of the infinite set of F_{hkl}'s, the result of transforming only a finite set is the convolution of the true density and the transform of the limiting box. As shown in Fig. 6.1b for one dimension, the transform of a box will have oscillations that will be superimposed on the electron density and Patterson maps. These oscillations are called *termination errors*, and of course, can be reduced by extending the measurements to larger values of $|\mathbf{s}|$. The fact that the x-ray scattering factors fall off with increasing $|\mathbf{s}|$ means the higher order F_{hkl}'s will generally be small and the termination errors due to their neglect will also be small. This is not true for neutron scattering where the scattering amplitudes are independent of $|\mathbf{s}|$, and termination errors are more important in this case. A similar effect occurs in electron microscopy where the aperture limits the number of diffraction spots that can be recombined causing termination errors in the image which is the transform of the diffraction pattern. We see that convolutions appear often in the development of diffraction problems, and this will be discussed further in Sect. 7.4.

6.4.3 Use of the Patterson Function

Let us return to the subject of Patterson functions and look at two further examples, now in two dimensions. In Fig. 6.11a, a simple molecule is shown centered at the origin. Its inverse is indicated by open circles. Following the definition of Eq. (6.48), the convolution is obtained by drawing the structure three times with its center of inversion placed successively at the three atomic centers of the inverted struction, (i.e., displaced by interatomic vectors). An alternative way to arrive at the same result shown in Fig. 6.11a is to follow the definition of convolution given in Eq. (6.46). Each atom of the molecule is successively placed at the origin and the molecule redrawn in this location. The resultant pattern of dotted circles is the same as that shown in the figure. In Fig. 6.11b, the same atomic arrangement is drawn

6.4 The Patterson Function

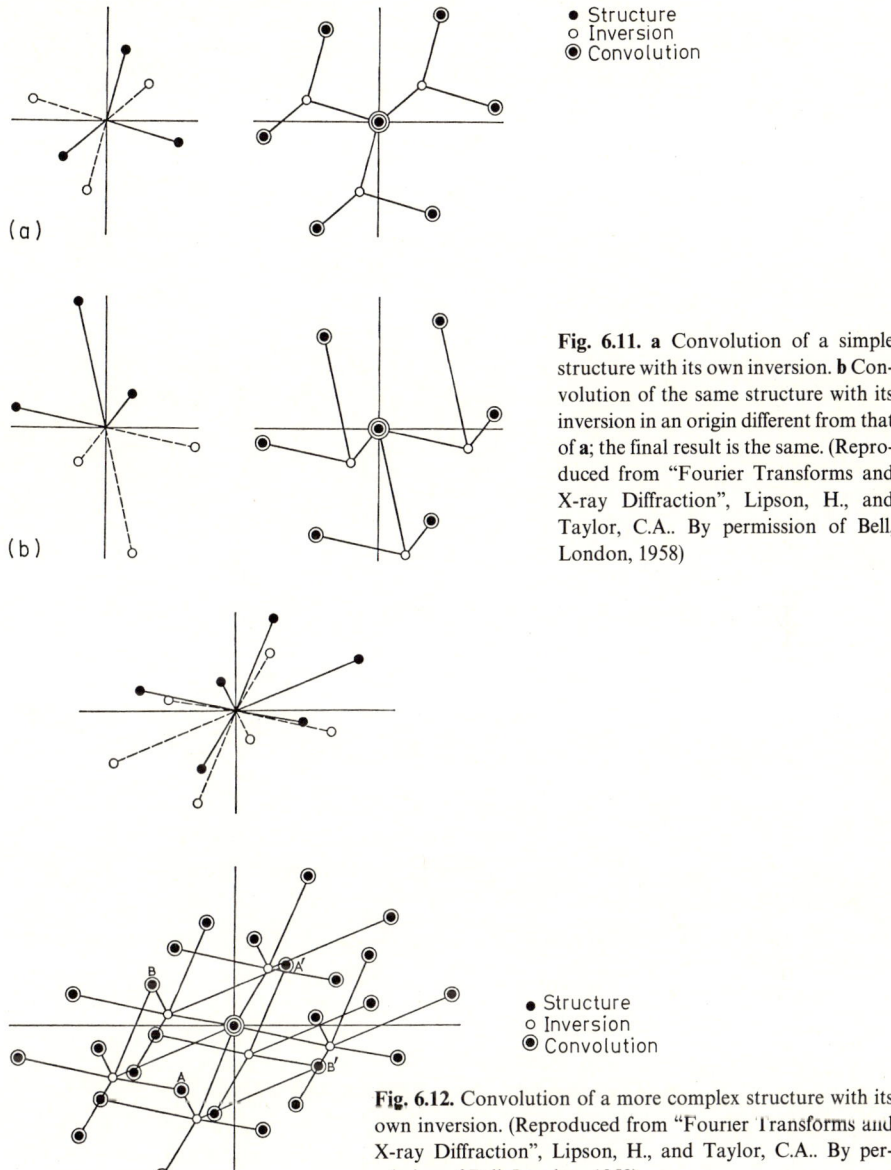

- Structure
- ○ Inversion
- ⊙ Convolution

Fig. 6.11. a Convolution of a simple structure with its own inversion. **b** Convolution of the same structure with its inversion in an origin different from that of **a**; the final result is the same. (Reproduced from "Fourier Transforms and X-ray Diffraction", Lipson, H., and Taylor, C.A.. By permission of Bell, London, 1958)

- Structure
- ○ Inversion
- ⊙ Convolution

Fig. 6.12. Convolution of a more complex structure with its own inversion. (Reproduced from "Fourier Transforms and X-ray Diffraction", Lipson, H., and Taylor, C.A.. By permission of Bell, London, 1958)

rotated and with a new origin; but as shown, the same convolution results. This is to be expected, since we have shown in Sect. 2.4 that displacement of a molecule changes only the phase, not the amplitude of the scattering, and no phase information is included in the Patterson function.

A more complicated structure is shown in Fig. 6.12. Note that several of the peaks arise twice in the construction of the convolution (at A, A', B, and B'), because there are two pairs of atoms with each of these interatomic vectors. Overlap effects such as these contribute to the problem of unraveling the Patterson maps to produce a final structure. Clearly as we approach the three-dimensional crystal structures we must become more analytical in our treatment. We confine our attention here to structures that are periodic and discuss nonperiodic structures in Chap. 7.

For periodic structures we need not integrate Eq. (6.43) over the entire crystal, but only over one unit cell. From Eq. (6.38a), the electron density at vector **r** in a cell is

$$\rho(\mathbf{r}) = \frac{1}{V_c} \sum_h \sum_k \sum_l F_{hkl} \exp\left(-2\pi i \left[\frac{hx}{a_1} + \frac{ky}{a_2} + \frac{lz}{a_3}\right]\right), \qquad (6.55a)$$

while the density at a vector $\mathbf{r}' + \mathbf{r}$ is

$$\rho(\mathbf{r}' + \mathbf{r})$$
$$= \frac{1}{V_c} \sum_h \sum_k \sum_l F_{hkl} \exp\left(-2\pi i \left[\frac{h}{a_1}(x + X) + \frac{k}{a_2}(y + Y) + \frac{l}{a_3}(z + Z)\right]\right), \qquad (6.55b)$$

where

$$\mathbf{r}' = (X/a_1)\mathbf{a}_1 + (Y/a_2)\mathbf{a}_2 + (Z/a_3)\mathbf{a}_3 = U\mathbf{a}_1 + V\mathbf{a}_2 + W\mathbf{a}_3$$

and

$$\mathbf{r} = (x/a_1)\mathbf{a}_1 + (y/a_2)\mathbf{a}_2 + (z/a_3)\mathbf{a}_3 = u\mathbf{a}_1 + v\mathbf{a}_2 + w\mathbf{a}_3 \, .$$

Then, from Eq. (6.44), the Patterson function is

$$P(U, V, W) = \frac{1}{V_c^2} \sum_h \sum_k \sum_l \sum_{h'} \sum_{k'} \sum_{l'} F_{hkl} F_{h'k'l'} \exp(-2\pi i[hU + kV + lW])$$
$$\times \frac{V_c}{a_1 a_2 a_3} \int_0^{a_3} \int_0^{a_2} \int_0^{a_1} \exp(-2\pi i[(h + h')u + (k + k')v$$
$$+ (l + l')w]) \, dx \, dy \, dz \, , \qquad (6.56)$$

where the substitution $dV/dx \, dy \, dz = V_c/a_1 a_2 a_3$ has been used. The integral over x is zero unless $h + h' = 0$ in which case it is equal to a_1. Similarly, for the y and z integration, the conditions $k = -k', l = -l'$ are required. Then

$$P(U, V, W) = \frac{1}{V_c} \sum_h \sum_k \sum_l^{\infty}_{-\infty} F_{hkl} F_{\bar{h}\bar{k}\bar{l}} \exp(-2\pi i[hU + kV + lW]) \, . \qquad (6.57a)$$

But $F_{hkl} F_{\bar{h}\bar{k}\bar{l}} = F_{hkl}^2$, and in the absence of anomalous scatterers, $F_{hkl}^2 = F_{\bar{h}\bar{k}\bar{l}}^2$, so we may write

$$P(U, V, W) = \frac{1}{V_c} \sum_h \sum_k \sum_l^{\infty}_{-\infty} F_{hkl}^2 \cos 2\pi[hU + kV + lW] \, . \qquad (6.57b)$$

There are several general features of Patterson maps that we may note. Consider the region near the origin of interatomic vector space. For zero arguments, the

6.4 The Patterson Function

cosine is one, so

$$P(U, V, W) \xrightarrow[(U,V,W)\to 0]{} \frac{1}{V_c} \sum_{hkl} F_{hkl}^2 = V_c \rho_c^2 ,$$

where ρ_c is the electron density of the unit cell. The expression in terms of F_{hkl}^2 is from Eq. (6.57b) with $U, V, W = 0$, while the expression in terms of ρ_c follows directly from the definition of the Patterson function in Eq. (6.44) where **u** is set equal to zero. This may be used to normalize measured intensities to an absolute scale; then the heights of peaks can be used to identify which atoms are at the ends of the interatomic vector corresponding to peaks. For example, for two atoms of atomic number Z_i and Z_j, the height of their peak H_{ij} is $Z_i Z_j$. As already indicated, the convolution integral spreads a peak out so that it has a diameter equal to the sum of the diameters of the two atoms involved, resulting in a loss of resolution in the Patterson function compared to an electron density map. Since atomic sizes are not all the same, it is better to use the entire intensity under a Patterson peak to estimate $Z_i Z_j$ rather than the height. Furthermore, this loss of resolution makes Patterson projections more poorly defined than the corresponding electron density projections, increasing the difficulty of analysis. Most often in practice one divides the measured F^2 by the $\sum_{\text{cell}} f^2$ which has the effect of reducing the origin peak and sharpening the Patterson peaks. In this case the heights of the Patterson peaks mean little. This procedure also increases termination errors since the decrease of f with $|s|$ has been cancelled, but Patterson projections are often used to locate the heavier atoms in the structure and these are not easily missed even when termination errors are present.

6.4.4 Patterson-Harker Sections

The technique of projection is particularly powerful when one heavy atom dominates the Patterson map and the phases. However, in general, for structures with many atoms the Patterson projections have the disadvantage that Patterson peaks that are distinct in three dimensions often project to the same general location in the projection. *One procedure to eliminate the resulting confusion was developed by Harker. He showed that it is possible to isolate the effects of certain symmetry elements in certain Patterson sections now known as Harker sections.* For example, consider a twofold axis parallel to $\mathbf{a}_2$ in the cell. If there is an atom at position x, y, z, there will be another at $\bar{x}, y, \bar{z}$ and they will have an interatomic vector $2x, 0, 2z$. If instead there was a 2_1 screw axis parallel to $\mathbf{a}_2$, the atom at x, y, z would have its mate at $\bar{x}, y + \frac{1}{2}, \bar{z}$, leading to an interatomic vector $2x, \frac{1}{2}, 2z$. Remember that the coordinates of $P(U, V, W)$ are the coordinates of an interatomic vector. Thus to examine only atoms related by the twofold axis, Harker suggested viewing the section:

$$P(U, 0, W) = \frac{1}{V_c} \sum_h \sum_l \left(\sum_k F_{hkl}^2 \right) \cos 2\pi (hU + lW) .$$

While for those related by a 2_1 screw axis,

$$P(U, \tfrac{1}{2}, W) = \frac{1}{V_c} \sum_h \sum_l \left(\sum_k F_{hkl}^2 (-1)^k \right) \cos 2\pi(hU + lW).$$

If there is a mirror plane perpendicular to the $\mathbf{a}_2$ axis, there will be atoms at x, y, z and $x, \bar{y}, z$, leading to an interatomic vector $0, 2y, 0$. Then the section $P(0, V, 0)$ will reveal these. And so on.

Let us look at an example [after Fairhurst and Cohen (1972)]. When the dentist takes Ag_3Sn from his shelf and mixes it with Hg, the "setting" reaction in that hole in your tooth involves the formation of Ag–Hg and Hg–Sn compounds. We discussed the structure of one of these, Ag_3Sn, in Sect. 5.7. Another of these is from chemical analysis, Ag_2Hg_3. Crystals can be made to float in Hg-rich liquid and the crystals are brittle, and so can be ground to a sphere without damage to the crystal. We are dealing with a material with very heavy atoms. The crystal is almost, but not quite a sphere, and so there are some problems with the absorption correction.

Proceeding in the usual way, the material is found to be cubic but with two-fold axes normal to cubic planes, not fourfold axes. From density measurements, there are 50 atoms per unit cell (30 of Hg, 20 of Ag). Conditions on observed reflections give $h + k + l = 2n$, so the Bravais lattice is body-centered cubic. Possible space groups are $I23$, $I2_13$, and $Im3$ (the first is non-centro-symmetric), described on pp. 307, 308 and 314, respectively, in the International Tables, Volume I.

A Patterson-Harker section at $W = 0$, i.e., $P(U, V, 0)$, is shown in Fig. 6.13. The heights or areas are not very helpful in this case, because when we consider the

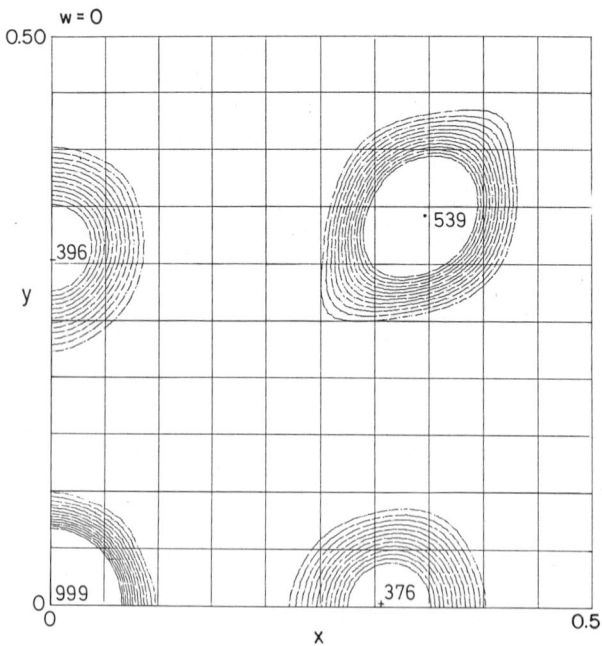

Fig. 6.13. Patterson-Harker section, $w = 0.0$ for Ag_2Hg_3

6.4 The Patterson Function

possible atomic arrangements we see that a great many vectors overlap. This is to be expected for high symmetry systems such as the cubic system.

The shapes of the peaks are slightly distorted, perhaps due to the absorption problem mentioned. Note that there are no peaks exactly at $\frac{1}{2}$, 0, 0 or 0, $\frac{1}{2}$, 0 or $\frac{1}{2}$, $\frac{1}{2}$, 0 or on lines $\frac{1}{2}$, V, 0; U, $\frac{1}{2}$, 0. Let us consider possible atomic distributions for each of the three plausible space groups. For $Im3$:

(30 Hg)	(20 Ag)
(1) rank $24g$ + rank $6b$	(1) $12d + 8c$
(2) $16f + 8c + 6b$	(2) $12d + 6b + 2a$
(3) $12d + 12e + 6b$	(3) $12e + 8c$
(4) $16f + 12d + 2a$	(4) $12e + 6b + 2a$
(5) $16f + 12e + 2a$	
(6) $12d + 12d + 6b$	
(7) $12e + 12e + 6b$	

The Harker section helps us immediately. The rank $6b$ equipoint involves coordinates like 0, $\frac{1}{2}$, $\frac{1}{2}$; $\frac{1}{2}$, 0, $\frac{1}{2}$. The interatomic vector between these is $\frac{1}{2}$, $\frac{1}{2}$, 0, but no such peak appears in the Patterson section. So choices (1), (2), (3), (6), and (7) for Hg and (2) and (4) for Ag are not correct. Choice (5) for Hg involves an atom with coordinates 0, 0, 0 on the rank $2a$ equipoint and one at $\frac{1}{2}$, x, 0 on $12e$. Their interatomic vector would be $\frac{1}{2}$, x, 0. No such peak appears. This elimates choice (5) for Hg. The rank $8c$ equipoint involves coordinates like $\frac{1}{4}$, $\frac{1}{4}$, $\frac{1}{4}$ and $\frac{3}{4}$, $\frac{3}{4}$, $\frac{1}{4}$ leading to the interatomic vector $\frac{1}{2}$, $\frac{1}{2}$, 0 which is not observed. This eliminates the remaining possibilities for Ag and Hg, and we must proceed to the next space group! $I2_13$ does not have the proper equipoints to distribute the 50 atoms. For $I23$, similar considerations eliminate most possibilities. The only possible distribution for Ag (12 + 8) and for Hg (12 + 8 + 8 + 2) require too many atoms along a body diagonal of the cell ($a = 10.0$ Å). We have eliminated all the space groups; but the crystal is still there!!

After some consideration we realize that the electron/atom ratio for this material is close to that for γ-brass structures which are also complex cubics; but generally, these have 52 atoms/cell.

Maybe there are two vacancies in the structure; thus, we start over again. Perhaps there are 22 Ag positions and 30 Hg. $I2_13$ does not have a proper set of equipoints for this, hence it is not the structure. Considerations of packing and interatomic vectors also eliminate the other two possibilities. One more try ... 32 Hg, 20 Ag. There are no combinations of equipoints that will work with these numbers for $Im3$. For $I2_13$:

(20 Ag)	(32 Hg)
12 + 8	(1) $8 + 8 + 8 + 8$
	(2) $24 + 8$
	(3) $12 + 12 + 8$

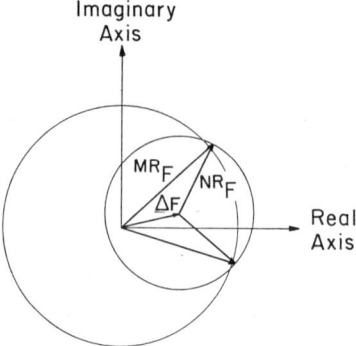

Fig. 6.14. Ambiguity may arise in the determination of phases by the heavy structure. (After Buerger, M.J., "Crystal Structure Analysis". Copyright 1960, John Wiley & Sons. Reprinted by permission of John Wiley & Sons, Inc.) Note that the two *circles* intersect in two locations

their phases, while the magnitude and phase of ΔF are known. All possible values of ^{MR}F and ^{NR}F will be vectors in the complex plane to circles of radii $|^{MR}F|$ and $|^{NR}F|$ as shown in Fig. 6.14. As seen in the figure, there are two solutions possible leaving an ambiguity in the analysis. Two common resolutions of this ambiguity have been used:

(i) Consider yet another isomorphous compound, or chemically add another atom to the structure.
(ii) Choose radiation (or choose the replaceable atom) such that the replaceable atom scatters anomalously. Then plots for the two reflections hkl and $\overline{hkl}$ will eliminate the ambiguity (see Fig. A6.15 and the solution to Problem 6.15).

Heavy atom and replaceable atom techniques have been used frequently in the study of organic compounds where chemical replacements may be achieved with some effort. As an example of such a study, refer to Fig. 6.15 which is a photograph of a model of the haemoglobin molecule. The molecule contains mostly helical chains of amino acids with four "haem" groups (represented by rectangular platelets) containing H, C, and N, and most importantly, Fe^{2+}. The role of this protein is to carry oxygen from the lungs to the body tissue. The oxygen is attached to the ferrous ion. In aqueous solutions, such as body fluids, the iron oxidizes to the trivalent state and gives up the oxygen to the tissue. Note how this structure determination reveals the amino acid chains surrounding the haem groups, creating an oily protective environment to keep the oxygen attached to the haem and away from the water. In structure studies such as these, neutron diffraction may be used to locate the hydrogen atoms that are obscured by heavier atoms in the x-ray analysis.

Hydrogen and deuterium have negative and positive neutron scattering lengths respectively. In an organic structure, substitution of deuterium for hydrogen causes positive peaks in electron density maps, where these were negative. [Neutrons are generally used to detect hydrogen in structures anyway, because its scattering length is similar to that of other atoms, whereas with x-rays it is hardly detectable (Table 3.2).] Thus, with neutrons, hydrogen bonds, or water of hydration can be detected through deuteration. It is also possible to "label" different macro-molecular

6.5 Heavy Atom Techniques

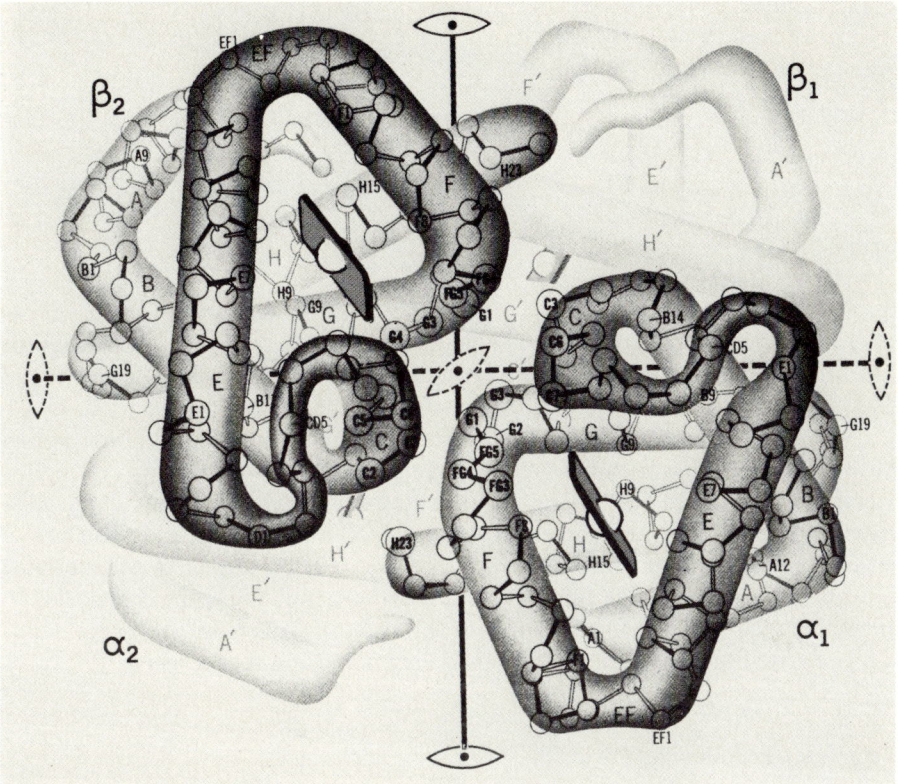

Fig. 6.15. The folding and packing of chains by horse oxyhaemoglobin. The molecule is roughly spherical, 64 × 55 × 50 Å. The four chains each contain a haem group (*flat plate*) in which the Fe^{2+} ion is included. The molecule has nearly 222 symmetry; however, α and β subunits are not identical. The *vertical axis is twofold* while the two *horizontal axes are dashed*, indicating approximate twofold symmetry. (Reprinted from Dickerson, R.E., and Geis, I., "The Structure and Action of Proteins", W.A. Benjamin, Inc., Menlo Park. Copyright 1969 by Dickerson and Geis)

subunits by this replacement and therefore locate them and their orientation. This is simply another example of atom replacement.

Anomalous dispersion — the change in scattering factor close to an absorption edge (Sect. 3.4) provides another means of "tagging" specific atoms, especially because with a synchrotron source, it is possible to choose an x-ray wavelength very close to such an atom's absorption edge so as to maximize the change. (The best choice may be just *below* the edge, so that the energy is not large enough to cause fluorescence from this atom; this might make the background excessive.)

We can then write the structure factor as made up of two parts, that due to "normal" atoms (n) and that part due to anomalous scatterers (a) (see Karle 1980):

$$^{\lambda}F_{hkl} = F^n_{hkl} + {}^{\lambda}F^a_{hkl},$$

where, as usual:

$$F^n_{hkl} = \sum_j f_{j,hkl} \exp(2\pi i \mathbf{r} \cdot \mathbf{r}^*) = |F^n_{hkl}| \exp(i[{}^n\varphi_{hkl}]).$$

but:

$$\begin{aligned}{}^\lambda F^a_{hkl} &= \sum_j ({}^\lambda \Delta f'_{j,hkl} + i\, {}^\lambda \Delta f''_{j,hkl}) \exp(2\pi i \mathbf{r} \cdot \mathbf{r}^*) \\ &= \sum_j {}^\lambda \delta_{j,hkl} \exp(i\Psi),\end{aligned}$$

where:

$$^\lambda \delta_{j,hkl} = \sqrt{(\Delta f'_{j,hkl})^2 + (\Delta f''_{j,hkl})^2}.$$

The anomalous scatterer is like an additional term in the structure factor, with an additional phase, so it is just like a heavy atom; we can represent it on an Argand diagram as we did for such atoms in Sect. 3.4 and Fig. 6.14.

Actually, there is a small anomalous scattering correction for all atoms at any wavelength and it has been possible to make use of even this small correction. For example, this has been done using the sulfur atoms in a biological molecule [protein crambin, Hendrickson and Teeter (1981)], using Ni-filtered CuK_α radiation from a sealed tube. The anomalous scattering from the six sulfur atoms in the molecule contributed only ~ 2–3 pct the average scattering factor. When the change due to anomalous scattering is small, as it is in this case, the difference $\Delta F = F_{hkl} - F_{\overline{hkl}}$ can be derived by considering the Argand diagram in Fig. 6.16. Clearly:

$$\Delta F = 2\delta \sin(\Psi - \varphi)_{hkl}.$$

Therefore, the following Patterson sum can be performed:

$$\begin{aligned}\text{sum} &= \frac{1}{V_c} \sum_h \sum_k \sum_l (\Delta F)^2 \cos 2\pi(hu + kv + lw) \\ &= \frac{1}{V_c} \sum_h \sum_k \sum_l 4\delta^2_{hkl} \sin^2(\Psi - \varphi)_{hkl} \cos \pi(hu + kv + lw)\end{aligned}$$

This will reveal the anomalous scatterers.

How do we proceed to solve the phase problem in cases where we cannot find a usable isomorphous replacement or a heavy (or anomalously scattering) atom? In Sect. 6.4 we discussed techniques for transforming the measured F's to a real space representation, the Patterson map, and searching for the solution in real space. In the next section we will examine techniques for direct determination of the phases by analysis of the intensities, a search for the solution in reciprocal space.

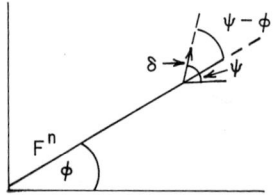

Fig. 6.16. Argand diagram for a weak anomalously scattering atom

6.6 Intensity Statistics and Inequalities

6.6.1 Detection of a Center of Symmetry, Mirrors and Rotation Axes

We have seen in Sect. 5.7 and Problems 5.16 and 5.17 that examinations of systematic extinctions of hkl reflections can be used to identify the presence of glide planes and screw axes; but we have implied that the presence of a center of symmetry, mirror planes, and rotation axes could not be inferred directly from the measured intensities. In this section we will show that for complex structures with many atoms per cell certain *systematic properties of the measured intensities (the "data set") can be used to determine the presence or absence of these symmetry elements* and thus establish the space group. Furthermore, there are other systematic properties of the intensities which may be used to establish the phases by trial and error procedures. These subjects have been described in a very lucid manner by Woolfson (1970) and we will follow his presentation.

Let us first consider the center of symmetry. [Techniques such as etch pits, piezoelectricity, and optical activity may sometimes be used to reveal a lack of centrosymmetry (see Sects. 5.6 and 5.7), but they are not foolproof and often difficult to apply to small crystals.] A general treatment of this problem using the measured intensities was provided by A.J.C. Wilson (1949).

We will treat the atomic coordinates as a large number N of independent variables and examine their statistical properties. Clearly if the intensities are dominated by the locations of one or a few heavy atoms, the assumption of independence will break down and other approaches such as the Patterson techniques will be more appropriate. Denote each of the variables by x_j ($j = 1, N$) with mean values $\bar{x}_j$ and variances σ_j^2. Then from elementary statistics, the sum of these variables is

$$X = \sum_j x_j \tag{6.61a}$$

with mean value

$$\bar{X} = \sum_j \bar{x}_j, \tag{6.61b}$$

and variance

$$\sigma^2 = \sum_j \sigma_j^2 . \tag{6.61c}$$

The central limit theorem of statistics states that the probability distribution for a large number of independent, random variables is the normal distribution, i.e., that the probability that the true sum lies between X and $X + dX$ is given by $P(X)\,dX$ where

$$P(X) = (2\pi\sigma^2)^{-1/2} \exp\{-(X - \bar{X})^2/2\sigma^2\} . \tag{6.62}$$

Consider a centrosymmetric structure. The structure factor may be written as

$$F_h = \sum_{j=1}^{N} f_j \cos 2\pi \mathbf{r}_j \cdot \mathbf{r}_h^* , \tag{6.63a}$$

where we use the short-hand notation $h = hkl$ for convenience. Since for every atom at $\mathbf{r}_j$ there is an identical one at $-\mathbf{r}_j$, the terms in the sum are not independent. However, if we rewrite Eq. (6.63a) as a sum over one asymmetric unit, we have

$$F_h = \sum_{j=1}^{N/2} 2f_j \cos 2\pi \mathbf{r}_j \cdot \mathbf{r}_h^* . \tag{6.63b}$$

Equating $x_j = 2f_j \cos 2\pi \mathbf{r}_j \cdot \mathbf{r}_h^*$, we may use Eq. (6.62). Now

$$\overline{x_j} = \overline{2f_j \cos 2\pi \mathbf{r}_j \cdot \mathbf{r}_h^*} = 2f_j \overline{\cos 2\pi \mathbf{r}_j \cdot \mathbf{r}_h^*} ,$$

since the scattering factor of the atom at $\mathbf{r}_j$ is fixed at f_j. For a large number of atoms in the cell, a random distribution of atomic positions may be assumed. Then,

$$\overline{\cos 2\pi \mathbf{r}_j \cdot \mathbf{r}_h^*} = 0$$

and

$$\overline{x_j} = 0 . \tag{6.64a}$$

The variance of $\overline{x_j}$ may be computed from

$$\sigma_j^2 = \overline{x_j^2} - \overline{x_j}^2 ,$$

but since $\overline{x_j} = 0$,

$$\sigma_j^2 = 4f_j^2 \overline{\cos^2 2\pi \mathbf{r}_j \cdot \mathbf{r}_h^*} = 2f_j^2 . \tag{6.64b}$$

Substituting in Eqs. (6.61), we see that the mean value of the structure factor is

$$\overline{F} = \sum_{j=1}^{N/2} \overline{x_j} = 0 \tag{6.65a}$$

and its variance, denoted Σ, is

$$\Sigma = \sigma^2 = \sum_{j=1}^{N/2} \sigma_j^2 = \sum_{j=1}^{N/2} 2f_j^2 . \tag{6.65b}$$

The distribution function for this centrosymmetric case is then

$$P_{\bar{1}}(F) = (2\pi\Sigma)^{-1/2} \exp(-F^2/2\Sigma) . \tag{6.66}$$

This distribution function is centered around $F = 0$ and is denoted a *centric distribution* shown in Fig. 6.17. We shall define a quantity M which is independent of Σ and is a useful measure of the shape of the distribution. Let

$$M = |\overline{F}|^2 / \overline{|F|^2} . \tag{6.67}$$

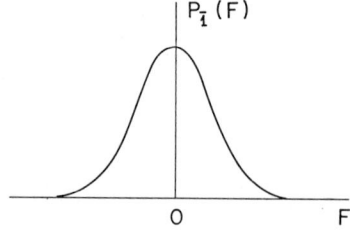

Fig. 6.17. A centric distribution function, $P_{\bar{1}}(|F|)$. (After Woolfson, M.M., "An Introduction to X-ray Crystallography", Cambridge Univ. Press, London and New York, 1970)

6.6 Intensity Statistics and Inequalities

For the centric distribution of Eq. (6.66),

$$\overline{|F|_{\bar{1}}} = \int_0^\infty 2F P_{\bar{1}}(F)\, dF = (2\Sigma/\pi)^{1/2}, \tag{6.68a}$$

and

$$\overline{|F|_{\bar{1}}^2} = \int_{-\infty}^\infty F^2 P_{\bar{1}}(F)\, dF = \Sigma. \tag{6.68b}$$

Thus

$$M_{\bar{1}} = 2/\pi = 0.637. \tag{6.68c}$$

Now let us carry out a similar analysis for noncentrosymmetrics structures. We may write the structure factor as

$$F_h = A_h + iB_h, \tag{6.69a}$$

where

$$A_h = \sum_{j=1}^N f_j \cos 2\pi \mathbf{r}_j \cdot \mathbf{r}_h^* \tag{6.69b}$$

and

$$B_h = \sum_{j=1}^N f_j \sin 2\pi \mathbf{r}_j \cdot \mathbf{r}_h^*. \tag{6.69c}$$

Proceeding as before, we determine the distributions of A and B:

$$P(A) = (\pi\Sigma)^{-1/2} \exp(-A^2/\Sigma), \tag{6.70a}$$

and

$$P(B) = (\pi\Sigma)^{-1/2} \exp(-B^2/\Sigma), \tag{6.70b}$$

where

$$\sigma_A^2 = \sigma_B^2 = \tfrac{1}{2}\Sigma. \tag{6.70c}$$

The quantities A and B may be plotted on an Argand diagram. Then the probability that A lies between A and $A + dA$ *and* also that B lies between B and $B + dB$ is given by

$$P(A)P(B)\, dA\, dB = (\pi\Sigma)^{-1} \exp\{-(A^2 + B^2)/\Sigma\}\, dA\, dB$$
$$= (\pi\Sigma)^{-1} \exp\{-|F|^2/\Sigma\}\, dA\, dB. \tag{6.71}$$

The differential element $dA\, dB$ is an element of area dS on the Argand diagram. Consequently, to solve the probability of finding the structure amplitude between $|F|$ and $|F| + d|F|$, we replace $dA\, dB$ in Eq. (6.71) by the area of an annular ring $2\pi|F|\, d|F|$:

$$P_1(|F|)\, d|F| = (\pi\Sigma)^{-1} \exp(-|F|^2/\Sigma) 2\pi|F|\, d|F|,$$

or

$$P_1(|F|) = (2/\Sigma)|F| \exp(-|F|^2/\Sigma). \tag{6.72}$$

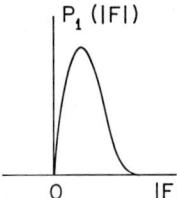

Fig. 6.18. An acentric distribution function, $P_1(|F|)$. (After Woolfson, M.M., "An Introduction to X-ray Crystallography". Cambridge Univ. Press, London and New York, 1970)

This distribution function is *acentric* as shown in Fig. 6.18. To compare with the centric distribution, we wish to eliminate Σ as before. The mean is

$$\overline{|F|}_1 = \int_0^\infty |F| P_1(|F|) d|F| = (\pi\Sigma/4)^{1/2}, \tag{6.73a}$$

and the variance is

$$\overline{|F|^2}_1 = \int_0^\infty |F|^2 P_1(|F|) d|F| = \Sigma \tag{6.73b}$$

Consequently, for the acentric distribution associated with noncentrosymmetric structures,

$$M_1 = \pi/4 = 0.785. \tag{6.73c}$$

The values for $M_{\bar{1}}$ and M_1 are significantly different and one uses these values and the shape of the distribution of $|F|$ to detect noncentrosymmetry; however, several words of caution are in order. Often axial structure factors, $h00$, $0k0$, and $00l$, are atypical and are real even if the structure is noncentrosymmetric. They are best ignored in computing $\overline{|F|}$ and $\overline{|F|^2}$. Also, structure factors at low $(\sin\theta)/\lambda$ are usually statistically unreliable due to extinction and data from $\sin\theta/\lambda < 0.2$ should also be ignored.

When we include the entire data set in the analysis for M, we are assuming that Σ does not vary significantly over the set. However, since Σ depends on $\sin\theta/\lambda$ through the variation in f_j [Eq. (6.65b)], Σ is found to be roughly constant for only small ranges of $\sin\theta/\lambda$. To avoid such problems, crystallographers often use normalized structure factors in these calculations, corrected for the decrease in f_j with $\sin\theta/\lambda$. Such a *unitary structure factor* is *defined* by

$$U_h = F_h \bigg/ \sum_{j=1}^N f_j \tag{6.74a}$$

Since $|F_h| \leq \sum_{j=1}^N f_j$, then

$$|U_h| \leq 1. \tag{6.74b}$$

We can also normalize each individual scattering factor by the same normalizing factor:

$$n_j = f_j \bigg/ \sum_j f_j, \tag{6.74c}$$

defining a set of unitary scattering factors. Then the unitary structure factor becomes

$$U_h = \sum_j n_j \exp(2\pi i \mathbf{r}_j \cdot \mathbf{r}_h^*) . \tag{6.74d}$$

In terms of the unitary structure factor, the centric and acentric distribution functions are

$$P_{\bar{1}}(U) = (2\pi\varepsilon)^{-1/2} \exp(-U^2/2\varepsilon) \tag{6.75a}$$

and

$$P_1(|U|) = (2/\varepsilon)|U|\exp(-|U|^2/\varepsilon) , \tag{6.75b}$$

where

$$\varepsilon = \sum_j n_j^2 . \tag{6.75c}$$

At a synchrotron radiation source it is possible to vary the chosen energy to be within a few ev of the absorption edge of many elements. As was pointed out in Sect. 3.4, changes in the scattering factor of 20–30 pct can occur in this vicinity, and tests for a center of symmetry by comparing F_{hkl}, $F_{\bar{h}k\bar{l}}$ (see Sect. 3.4) should be quite simple in the future with such sources.

Let us turn now to the presence or absence of mirrors and rotation axes. For example, the space group of a monoclinic structure with no systematic absences is P2, Pm, or P2/m. The last of these is centrosymmetric and could be identified by the value of M discussed above. But if the structure is noncentrosymmetric, a distinction must still be made between P2 and Pm. Consider a mirror perpendicular to the $\mathbf{a}_2$ axis versus a twofold rotation axis along $\mathbf{a}_2$. For the mirror, an $\mathbf{a}_2$ projection will show complete overlap of pairs of atoms due to the mirror. Thus in projection there are $N/2$ double-weight peaks and the effective weight of Σ, Σ_m becomes

$$\Sigma_m = \sum_{j=1}^{N/2} (2f_j)^2 = 2\Sigma \tag{6.76a}$$

and

$$\overline{|F_{h0l}|^2} = \Sigma_m = 2\Sigma . \tag{6.76b}$$

Similarly, a twofold axis along $\mathbf{a}_2$ shows

$$\overline{|F_{0k0}|^2} = \Sigma_2 = 2\Sigma , \tag{6.77}$$

since atoms are related in pairs with coordinate (x, y, z) and $(\bar{x}, y, \bar{z})$ so that projected onto the $\mathbf{a}_2$ axis, they overlap at $(0, y, 0)$. An alternate derivation of these results is discussed in Problem 6.16.

6.6.2 Direct Methods — Inequalities

We see from the above that there is information about symmetry elements contained in the average behavior of large numbers of $|F|^2$'s. Can we find similar information about phases by examining the statistical behavior of the $|F|^2$? Such *direct methods of determining the phases* may be derived from one of many inequality expressions. The first application of inequalities to the phase problem was due to Harker and Kasper who treated the problem in 1948. We begin with the Cauchy inequality

$$\left|\sum_{j=1}^{N} a_j b_j\right|^2 \leq \left(\sum_{j=1}^{N} |a_j|^2\right)\left(\sum_{j=1}^{N} |b_j|^2\right), \tag{6.78a}$$

where a_j and b_j are any real or complex quantities. To understand this expression, we may think of the set of numbers $\{a_j\}$ and $\{b_j\}$ defining vectors in an N-dimensional space. The expression in Eq. (6.78a) can then be considered to mean that the dot product of **a** and **b** (the left-hand side is the generalized dot product) is less then or equal to the product of the lengths of the vectors.

We will use the unitary structure factors defined in Eq. (6.74d)

$$U_h = \sum_{j=1}^{N} n_j \exp(2\pi i \mathbf{r}_j \cdot \mathbf{r}_h^*), \tag{6.78b}$$

where the n_j's remain essentially constant over reciprocal space. Let $a_j = \sqrt{n_j}$ and $b_j = \sqrt{n_j} \exp(2\pi i \mathbf{r}_j \cdot \mathbf{r}_h^*)$. Then from Cauchy's inequality,

$$|U_h|^2 \leq \left(\sum_{j=1}^{N} n_j\right)\left(\sum_{j=1}^{N} n_j |\exp(2\pi i \mathbf{r}_j \cdot \mathbf{r}_h^*)|^2\right). \tag{6.79a}$$

This result simplifies since $|\exp(2\pi i \mathbf{r}_j \cdot \mathbf{r}_h^*)| = 1$ and $\sum_j n_j = 1$, i.e.,

$$|U_h|^2 \leq 1. \tag{6.79b}$$

The expression in (6.79b) is not really a new result as it merely restates the fact that $|F_h|^2 \leq \sum_j Z^2$. However, we may now deduce new results by including the presence of symmetry elements.

Consider the case of a center of symmetry. Then Eq. (6.79b) reduces to

$$U_h = \sum_{j=1}^{N} n_j \cos 2\pi \mathbf{r}_j \cdot \mathbf{r}_h^*, \tag{6.80}$$

with U_h real. Now identifying $a_j = \sqrt{n_j}$ and $b_j = \sqrt{n_j} \cos 2\pi \mathbf{r}_j \cdot \mathbf{r}_h^*$ we obtain

$$U_h^2 \leq \left(\sum_j n_j\right)\left(\sum_j n_j \cos^2 2\pi \mathbf{r}_j \cdot \mathbf{r}_h^*\right). \tag{6.81a}$$

Simplifying,

$$\sum_j n_j \cos^2 2\pi \mathbf{r}_j \cdot \mathbf{r}_h^* = \tfrac{1}{2}\sum_j n_j + \tfrac{1}{2}\sum_j n_j \cos 4\pi \mathbf{r}_j \cdot \mathbf{r}_h^* = \tfrac{1}{2} + \tfrac{1}{2}U_{2h}.$$

Thus, inequality (6.81a) becomes

$$U_h^2 \leq \tfrac{1}{2}(1 + U_{2h}). \tag{6.81b}$$

If $|U_h|$ and $|U_{2h}|$ are both sufficiently large, then one can show that U_{2h} must be positive. For example, if $|U_h| = 0.6$ and $|U_{2h}| = 0.5$, from Eq. (6.81b)

$$U_{2h} \geq -0.28,$$

hence U_{2h} must be $+0.5$, not -0.5. The expression in (6.81b) is only a weak

6.6 Intensity Statistics and Inequalities

constraint on U_h and U_{2h} since large values of U occur only rarely in crystals with large N; but the recognition that such constraints do exist produced a revolution in crystal structure analysis that is heading toward a completely automatic solution of the phase problem. Let us consider some other symmetry elements to see what additional information may be obtained.

Consider a twofold axis along $\mathbf{a}_2$. Atoms occur in pairs with coordinates (x, y, z) and $(\bar{x}, y, \bar{z})$. The unitary structure factor becomes

$$U_{hkl} = \sum_j n_j \exp(2\pi i k y_j) \cos 2\pi (h x_j + l z_j) . \tag{6.82a}$$

Again we use the Cauchy inequality with $a_j = \sqrt{n_j} \exp(2\pi i k y_j)$ and $b_j = \sqrt{n_j} \times \cos 2\pi (h x_j + l z_j)$. The result is

$$|U_{hkl}| \leq \tfrac{1}{2}(1 + U_{2h,0,2l}) , \tag{6.82b}$$

from which the sign of a large structure factor $U_{2h,0,2l}$ may be determined if *any* of the structure factors U_{hkl} has sufficiently large magnitude. Similar results can be developed for other symmetry elements, or for combinations of elements to form the various point groups. The reader is referred to Buerger (1960, p. 560) for a list of these *Harker-Kasper inequalities*, and to Problem 6.17 for some further examples.

Still more interesting inequalities result when one considers two reflections simultaneously. For example, let us examine what happens when we form the sum of two unitary structure factors in a centrosymmetric structure. The sum

$$U_h + U_{h'} = \sum_j n_j (\cos 2\pi \mathbf{r}_j \cdot \mathbf{r}_h^* + \cos 2\pi \mathbf{r}_j \cdot \mathbf{r}_{h'}^*) ,$$

$$= \sum_j 2 n_j \cos \pi (\mathbf{r}_h^* + \mathbf{r}_{h'}^*) \cdot \mathbf{r}_j \cos \pi (\mathbf{r}_h^* - \mathbf{r}_{h'}^*) \cdot \mathbf{r}_j$$

leads to the choice $a_j = \sqrt{2 n_j} \cos \pi (\mathbf{r}_h^* + \mathbf{r}_{h'}^*) \cdot \mathbf{r}_j$ and $b_j = \sqrt{2 n_j} \cos \pi (\mathbf{r}_h^* - \mathbf{r}_{h'}^*) \cdot \mathbf{r}_j$. The inequality found is

$$(U_h + U_{h'})^2 \leq (1 + U_{h+h'})(1 + U_{h-h'}) , \tag{6.83a}$$

where the notation $h + h'$ implies the triplet $h + h', k + k', l + l'$, etc. A similar relation may be derived for $U_h - U_{h'}$:

$$(U_h - U_{h'})^2 \leq (1 - U_{h+h'})(1 - U_{h-h'}) . \tag{6.83b}$$

The largest left-hand side provides the most useful relation, so inequality (6.83a) is used when U_h and $U_{h'}$ have the same sign, while (6.83b) is used when they have opposite signs. We may combine these results to form a general relation if we introduce the quantity $s(h)$ meaning the sign of $U(h)$. Then (6.83a) and (6.83b) are equivalent to

$$(|U_h| + |U_{h'}|)^2 \leq \{1 + s(h)s(h')U_{h+h'}\}\{1 + s(h)s(h')U_{h-h'}\} . \tag{6.83c}$$

An interesting further transformation of (6.83c) is possible to produce

$$(|U_h| + |U_{h'}|)^2 \leq \{1 + s(h)s(h')s(h+h')|U_{h+h'}|\}$$
$$\times \{1 + s(h)s(h')s(h-h')|U_{h-h'}|\} . \tag{6.83d}$$

If the U's are sufficiently large, this relationship can show that either or both of the *sign triple products* (stp), $s(h)s(h')s(h+h')$ and $s(h)s(h')s(h-h')$ are positive. Let us illustrate the use of (6.83d) with an example. Let

$$|U_h| = |U_{h'}| = |U_{h+h'}| = |U_{h-h'}| = 0.5 \, .$$

Then the left-hand side of (6.83d) is 1.0, and it can be seen that unless both (stp) are positive, the inequality cannot be satisfied.

6.6.3 Phase Determination Using Direct Methods

We may now schematically describe *the procedure for phase determination in a centrosymmetric structure*. Two or three of the largest U values are assigned a positive sign. They must be + or − in the centrosymmetric structure and the choice is arbitrary. These initial choices of U must be from different parity groups [the signs of h, k, l taken as even, even, even (eee); odd, even, even (oee); oeo, etc., define different parity groups]. The initial choice of reflections is equivalent to establishing an arbitrary origin for the structure. Then the Harker-Kasper inequalities and the stp's are used to identify as many signs as possible. Inevitably one runs out of relationships, and then a symbol (e.g., "p") is assigned to the sign of an important unidentified U and one continues to use the inequalities, determining additional signs in terms of "p." When halted again, a second sign symbol "q" is introduced and the process continues until a sufficiently large number of phases have been identified to justify a trial electron density map. At this point, one carries out the Fourier mapping for all possible combinations of signs for "p," "q," etc., looking for realistic structures in the resultant electron density map. The refinement then proceeds as we have described in Sect. 6.3. This field is still being developed; however, direct methods are now the preferred techniques when dealing with structures containing 20 atoms or less in the asymmetric unit. In such cases, direct methods are used nearly exclusively for centrosymmetric structures and for more than 95% of the noncentrosymmetric structures.[2]

One of the major limitations of the inequality relationships is the necessity to have large U's. For a centrosymmetric unit cell with N identical atoms, the U's will have a centric distribution [Eq. (6.75a)] with zero mean, and variance

$$\varepsilon = \sum_j n_j^2 = \sum_j \left(\frac{f_j}{Nf_j}\right)^2 = N^{-1} \, .$$

Only $\approx 5\%$ of the structure factors would be expected to have magnitudes greater than two standard deviations ($2N^{-1/2}$) and only 0.3% should have magnitudes greater than $3N^{-1/2}$. For $N = 36$, then, only 0.3% of the $|U|$'s will be greater than 0.5. Thus for $N > 40$, it is unlikely to achieve a solution using the inequality relationship above.

Inequality (6.83d) can be used to show that if the U's are sufficiently large, then

$$s(h)s(h')s(h-h') = +1 \, . \tag{6.84}$$

---

[2] The 1985 Nobel Prize for Chemistry was awarded to H. Hauptman and J. Karle for their pioneering work on direct methods of crystal structure analysis (Hauptman and Karle, 1953).

6.6 Intensity Statistics and Inequalities

Actually this relation is probably true even when the $|U|$'s are smaller than necessary to satisfy the inequality relations. This was demonstrated independently by Sayre, Cochran, and Zachariasen. We will follow Sayre's derivation (1952). We consider a structure containing equal nonoverlapping atoms with electron density distribution $\rho(\mathbf{r})$. The "squared structure" of density $\rho^2(\mathbf{r})$ also consists of equal resolved peaks. For the original structure, we have

$$F_h = \sum_j f_h \exp(2\pi i \mathbf{r}_j \cdot \mathbf{r}_h^*), \tag{6.85a}$$

where $(1/V_c)F_h$ are the Fourier coefficients of $\rho(\mathbf{r})$. Similarly, for the squared structure, the corresponding equation is

$$G_h = \sum_j g_h \exp(2\pi i \mathbf{r}_j \cdot \mathbf{r}_h^*), \tag{6.85b}$$

where $(1/V_c)G_n$ are the Fourier coefficients of $\rho^2(\mathbf{r})$. These two expressions are related by

$$F_h = (f_h/g_h)G_h = \theta_h G_h, \tag{6.85c}$$

where θ_h is some constant. It follows from the convolution theorem [Eq. (6.47)] that the Fourier coefficients of $\rho^2(\mathbf{r})$ are given by the self-convolution of the Fourier coefficients of $\rho(\mathbf{r})$, i.e.,

$$\frac{1}{V_c}G_h = \sum_{h'} \frac{1}{V_c} F_{h'} \frac{1}{V_c} F_{h-h'}. \tag{6.86a}$$

Using Eq. (6.85c), we obtain

$$F_h = \frac{\theta_h}{V_c} \sum_{h'} F_{h'} F_{h-h'}. \tag{6.86b}$$

If F_h is large, then any large products in the right-hand side of Eq. (6.86b) are likely to have the same sign as F_h, i.e.,

$$s(h)s(h')s(h-h') \approx 1, \tag{6.87}$$

where $\approx$ is interpreted as probably equal. This expression, often known as *Sayre's relation*, may be made more quantitative, and the probability of its being correct for a given set of circumstances may be evaluated [for example, see Woolfson (1954)].

Sayre's equation [Eq. (6.86b)] applies quite generally and for noncentrosymmetric structure factors can be written as

$$|F_h|\exp(i\phi_h) = \frac{\theta_h}{V_c} \sum_{h'} |F_{h'}||F_{h-h'}|\exp\{i(\phi_{h'} + \phi_{h-h'})\}. \tag{6.88a}$$

If there are a number of significant terms on the right-hand side, then the probable value of $\tan \phi_h$ is given by the ratio of imaginary to real parts of the right-hand side, i.e.,

$$\langle \tan \phi_h \rangle = \frac{\sum_{h'} |F_{h'}||F_{h-h'}|\sin(\phi_{h'} + \phi_{h-h'})}{\sum_{h'} |F_{h'}||F_{h-h'}|\cos(\phi_{h'} + \phi_{h-h'})}. \tag{6.88b}$$

Sayre's relationship or the $\tan \phi_h$ expression in Eq. (6.88b) are very powerful, and have been combined with the inequality relationships discussed earlier to produce a systematic approach to solving the phase problem with a high probability of success. There is still some trial and error involved in this process, and for large N, the digital computer is imperative in trying the many alternate paths to a complete phase determination. Such digital programs are now routinely used extensively by crystallographers to solve structures of up to $N \approx 100$, and as techniques improve, perhaps nearly automatic structure determination will be developed from such direct methods.

Table 6.1. Development of structural crystallography

Step or Result	1948	1986
Choice of problem	Uncertain. Many related complexes examined for one of greatest crystallographic simplicity	If suitable crystals are available examination of related materials is unusual
Unit cell determination	Film methods	Generally by search and index techniques on an automatic diffractometer. Film methods used only when troubles develop (e.g., the structure cannot be solved!)
Intensity data collection	Film methods. Roughly 50–100 intensities per day visually estimated to an accuracy of 15%. Perhaps 500–1000 collected	Automatic diffractometer. Some collect as many as 2500 intensities per day to complete data collection in 1–2 days
Solution of the phase problem	Very uncertain. Three-dimensional Patterson function might require 40 h of computing time. Determination might not progress beyond this point	Reasonably straightforward by direct methods for most small molecule problems, except those involving pseudo symmetry or supercells
Refinement of the structure	Fourier refinement only. Least-squares refinement computationally impossible except on simplest problems	Non-linear least squares. Problems up to 1000 variables or so (about 100–120 independent atoms) are feasible on department computers
Complexity of the structure	Perhaps 20 independent atoms	Up to about 100 independent atoms for least-squares refinement. Fourier refinement permits many more atoms, for example, proteins
Approximate standard deviations in:		
C–C in equal atom structure	0.03	0.005
Fe–C in organometallic crystal	0.02	0.003
Additional information	Very approximate Debye-Waller temperature factors	Thorough description, in terms of assumed model, of thermal motion of atoms in the structure

6.7 Difference Techniques

We close this section with reference to Table 6.1, prepared by Professor J. Ibers of Northwestern University comparing the techniques used in crystal structure analysis in 1948 and 1986. In the intervening years the trend has continued in favor of direct methods. For further discussions of these direct methods the reader is referred to Woolfson (1970, Chap. 8) and to the article by Germain and Woolfson (1968).

6.7 Difference Techniques

6.7.1 Difference Fourier Synthesis

We noted in Sect. 6.4 that termination of the Fourier series due to the finite set of data leads to oscillation in the electron density map called termination errors. These errors may interfere with refinement of the structure. One way to eliminate the effect of series termination near the end of refinement is to use a *difference Fourier synthesis*. For every reflection, form the difference between the observed and calculated structure factor. Then we calculate the function

$$\eta(\mathbf{r}) = \frac{1}{V_c} \sum_h \{(F_o)_h - (F_c)_h\} \exp(2\pi i \mathbf{r} \cdot \mathbf{r}_h^*) \,. \tag{6.89}$$

The quantity $\eta(\mathbf{r}) = \rho_o(\mathbf{r}) - \rho_c(\mathbf{r})$ may be thought of as the difference between the "true" electron density corresponding to the observed F_o's and the "proposed" electron density corresponding to the calculated F_c's. The utility of this synthesis will be revealed by some examples.

Suppose a small error in the x coordinate of an atom has been made. The electron densities projected on the x axis might look as shown in Fig. 6.19a. In the difference pattern shown in Fig. 6.19b, symmetrical negative contours (dotted line) and positive contours reveal the erroneous coordinates. A second important example deals with the thermal vibrations. Suppose the wrong value for the e^{-M} term has been assumed. A small positive or negative bump will reveal such a discrepancy as shown in Fig. 6.20. Similarly, if anisotropic vibrations have been erroneously accounted for, patterns such as shown in Fig. 6.21 will appear in the difference map.

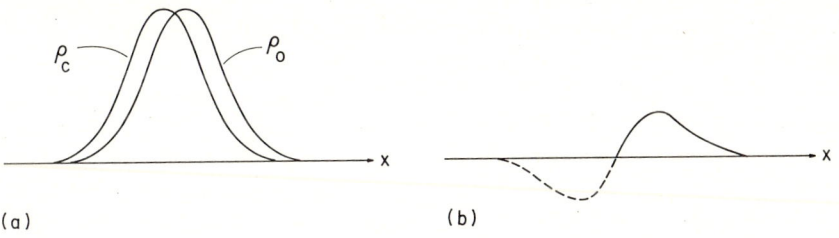

Fig. 6.19.
The effect of a small error in the x coordinate of an atom (**a**), on the difference Fourier pattern (**b**)

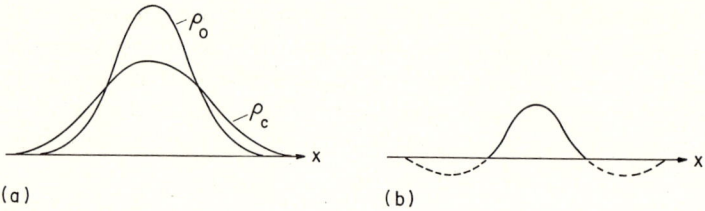

Fig. 6.20. The effect of a small error in the Debye-Waller factor e^{-M} (**a**), on the difference Fourier pattern (**b**). (After Woolfson, M.M., "An Introduction to Crystallography", Cambridge Univ. Press, London and New York 1970)

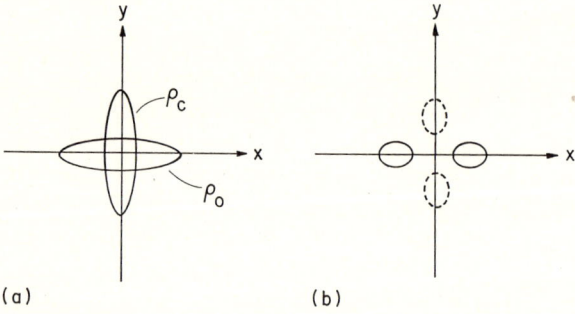

Fig. 6.21. The effect of erroneously neglecting anisotropic vibrations (**a**), on the Fourier difference pattern (**b**). (After Woolfson, M.M., "An Introduction to Crystallography", Cambridge Univ. Press, London and New York 1970)

Care is needed in the use of electron density maps for refinement as there is a tendency for the assumed structure to be forced into these maps. To see this, we rewrite the terms of the electron density (not the difference) as $[|F_o|/|F_c|]|F_c|e^{i\theta_c}$, i.e., the calculated phases are employed with the observed F's. This is the product of two transforms and in the electron density, the effect will then be the convolution of both. The first is a constant, producing a δ function in the map. The second is the *assumed* structure! Thus we get peaks tending to appear at the assumed positions. Refinement then can be slow with difference maps. In general, now, investigators go directly to least-squares solutions for structures with 100 atoms or less (see Sect. 6.9). However, organic structures, such as proteins and enzymes, often contain thousands of atoms and, in such cases, refinement is still done with difference techniques, since computational costs are otherwise too high. Projections on lines and planes are often used to simplify things in such structures.

An application of the difference synthesis to a defect structure was made by Willis in the study of UO_{2+x} (1963a, b, 1964). The crystal structure of the stoichiometric compound is cubic with the structure of the mineral fluorite, space group $Fm3m$ with the U at 000 + fct and 0 at $\frac{111}{444}$ and $\frac{333}{444}$ + fct. Upon oxidation the density of UO_{2+x} increases with x in a way that agrees with the assumption that oxygen is incorporated interstitially. Neutron diffraction studies are necessary to study the oxygen in the presence of the high atomic number uranium. The only detectable changes in the Bragg peaks in neutron diffraction patterns of oxidized single crystals are slight shifts in Bragg angles with x and a change in the diffracted intensities. No new lines are observed ruling out long-range periodicity in the defect structure and suggesting that the space group remains $Fm3m$. The simplest assumption to make is that the extra oxygen occupy one or more interstitial sites at random forming a "statistical" unit cell with fractional occupation of interstitial sites. Data for the compound $UO_{2.12}$ were collected from a single crystal held at 800°C. These data were first analyzed using a general expression for the structure factor:

$$F_{hkl} = \sum_{r=1}^{r=n} m_r b_r \exp[2\pi i(hx_r + ky_r + lz_r)] \exp(-B_r \sin^2\theta/\lambda^2), \tag{6.90}$$

where n is the number of atoms in the statistical cell, m_r the occupation number of the rth atom, b_r the coherent scattering amplitude of the rth atom, x_r, y_r, z_r, the positional coordinates of the rth atom, and B_r the isotropic temperature factor of the rth atom.

Assuming the excess oxygen was only at the "most-likely" holes, $\frac{1}{2}\frac{1}{2}\frac{1}{2}$ + fct, refined to zero occupation at these sites (denoted the O''' sites). Since the extra oxygen must be somewhere, they were searched for by using a difference Fourier map. The map was constructed with coefficients formed by taking the difference between the observed F_o's and those calculated assuming the UO_2 structure. A two-dimensional section of this map for $Z = \frac{3}{8}$ is shown in Fig. 6.22. The extra oxygen atoms show up clearly as density maxima at positions corresponding to the 48(h) (O') and 16(e) (O'') equipoints of the $Fm3m$ space group. Subsequent refinement using these coordinates indicated that on the average a unit cell in $UO_{2.12}$ has 0.08 ± 0.04 O', 0.16 ± 0.06 O'' atoms and no O''', while the number of oxygen at the "normal" fluorite sites is reduced from 2 to 1.87 ± 0.03.

The oxygen interstitials have coordinates that would severely distort neighboring "normal" oxygens if they are present, so it is natural to assume some local

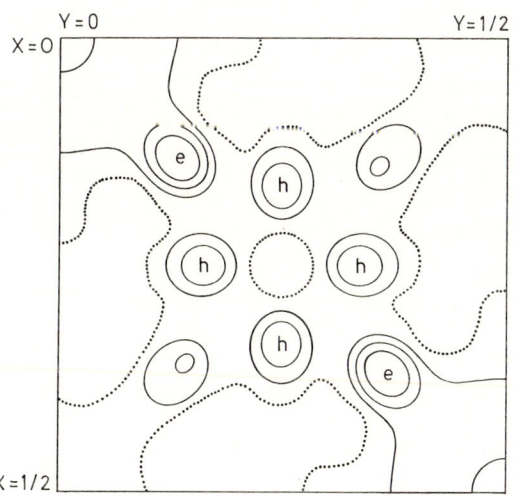

Fig. 6.22. Section of the Fourier difference map of $UO_{2.12}$ at $z = \frac{3}{8}$, with the normal UO_2 structure removed. [Reproduced by permission from Willis, B.T.M., J. Phys. Radium 25, 431 (1964)]

defect complex of "normal" oxygen vacancies with O' and O" interstitials. Since the ratio of the concentrations of normal O : O' : O" was found to be 13 : 8 : 16, a structure containing 2 : 1 : 2 was tried. This implies a model in which an O' enters the lattice displacing two nearby normal O to O" sites. However, if every extra oxygen caused such a defect to occur, the composition of the unit cell for $UO_{2.12}$ would become $UO_{1.76} \, O'_{0.12} \, O''_{0.24}$ resulting in a large error on the occupancy of normal O which had been determined as 1.87 ± 0.03. Willis suggested an alternate 2 : 2 : 2 model in which two O' atoms together produce two vacancies and two O" with the resultant formula $UO_{1.88} \, O'_{0.12} \, O''_{0.12}$.

It is not possible to choose between these models, or for that matter to even know if any local defect structure really exists (even if it seems likely) by looking only at the Bragg reflections. Information about the local atomic arrangement *does* exist in the diffraction pattern, but it appears as diffuse scattering. This subject is treated in Sect. 7.6.

6.7.2 Difference Patterson Synthesis

Let us now turn our attention to another difference technique, *the difference Patterson synthesis*. Suppose that in addition to our crystal with atoms $M \, A \, B \, C \, D$ we have another crystal $N \, A \, B \, C \, D$ identical in every respect except for replacement of M by N. The Patterson for the first crystal can be represented by the matrix of interatomic vectors

$MM \quad MA \quad MB \quad MC \quad MD$

$AM \quad AA \quad AB \quad \text{---} \quad \text{etc.}$

while the Patterson for the second crystal is

$NN \quad NA \quad NB \quad NC \quad ND$

$AN \quad AA \quad AB \quad \text{---} \quad \text{etc.}$

If we form the difference Patterson with these two crystals (using as coefficients of the Fourier series the difference in F^2), we eliminate everything but the first column and row of each matrix; there will be $2n$ peaks in it if n is the number of atoms. The peaks will be of the form:

$MM - NN, \, MA - NA, \, \text{etc.}, \quad \text{or} \quad (M+N)(M-N), \, A(M-N), \, \text{etc.}$

Thus if M and N are heavy atoms, we can eliminate the large origin peak in the Patterson with a difference Patterson. We also reduce the effect of the heavy atoms elsewhere, so that peaks are essentially due to the lighter atoms A, B, C, D weighted by $M - N$. The peaks like (AB), etc., are gone and *the image is simply that of the crystal with peaks of strength proportional to the number of electrons in each atom.* The peaks are broader, though, than in the electron density maps. This procedure is especially useful if there is one heavy atom per cell and we choose it as the origin. Otherwise the effect may be confusing. (We can also eliminate the origin peak by forming the difference of the actual F^2 and those calculated for just the heavy atom. This may help us to see some vectors close to the origin.)

6.7 Difference Techniques

This difference Patterson is especially useful for order-disorder problems in alloys and minerals. For example, consider Cu_3Au. This cubic material exists in a random form in which each lattice point of the fcc cell has on the average $\frac{3}{4}Cu$ and $\frac{1}{4}Au$. An interatomic vector from a corner of the cell to another corner would have the weight $(\frac{3}{4}Z_{Cu} + \frac{1}{4}Z_{Au})^2$ in the Patterson. In the ordered state at low temperatures, all corners are occupied by Au atoms and all three faces by Cu atoms. The same vector would then appear with the weight $(3Z_{Cu}^2 + Z_{Au}^2)/4$. (There are similar vectors, three face-to-face types as well as one corner-to-corner type. You can think of the structure as four interpenetrating simple cubics, three of Cu and one of Au.) At intermediate states of order, the fraction of corner sites occupied by Au is some number $\frac{1}{4} \leq r_{Au} \leq 1$ and the fraction of face sites is similarly written as $\frac{3}{4} \leq r_{Cu} \leq 1$. We may define a degree of order S by the expression

$$S = (r_{Au} - \tfrac{1}{4})/(1 - \tfrac{1}{4}) . \tag{6.91}$$

We see that S varies from 0 for $r_{Au} = \frac{1}{4}$ (random alloy) to 1 for $r_{Au} = 1$ (complete order). The structure factor may then be generally written as

$$F_{hkl} = [r_{Au}f_{Au} + (1 - r_{Au})f_{Cu}]e^{2\pi i(0)} + [(1 - r_{Cu})f_{Au} + r_{Cu}f_{Cu}]$$
$$\times \{e^{\pi i(h+k)} + e^{\pi i(k+l)} + e^{\pi i(l+h)}\} . \tag{6.92}$$

Noting that mass balance requires that the total number of Au atoms is

$$\tfrac{1}{4}r_{Au}N + \tfrac{3}{4}(1 - r_{Cu})N = \tfrac{1}{4}N , \tag{6.93}$$

where N is the total number of atoms in the crystal, expression (6.92) can be simplified to

$$F_{hkl} = \begin{cases} (f_{Au} + 3f_{Cu}), & \text{for } hkl \text{ unmixed} \\ (f_{Au} - f_{Cu})S, & \text{for } hkl \text{ mixed} . \end{cases} \tag{6.94}$$

We see that the regular fcc reflections (*hkl* unmixed) are unaffected by the degree of order and only the superstructure peaks (*hkl* mixed) contain the parameter S. (A further example of order in bcc alloys is discussed in Problem 6.19 and the general subject of order in alloys is pursued in more depth in Sect. 7.6.)

To form a difference Patterson for this structure, we would use the sum

$$\Delta P(uvw) = \sum_{hkl}\sum\sum (F_{\text{ordered}}^2 - F_{\text{disordered}}^2)\cos 2\pi(hu + kv + lw) .$$

But the only difference in the two diffraction patterns is a set of superstructure peaks so we use only those to form the function; the others would cancel anyway!

In this case of order-disorder, there will be negative and positive peaks in the difference Patterson. If the electron density at both ends of an interatomic vector either increases or decreases compared to the random atomic arrangement, there is a positive peak. If the density decreases at one end of the vector but increases at the other, we get a negative peak. As an example of the power of this method we will consider the defect structure of $Fe_{1-x}O$ as determined by Koch and Cohen (1969). The "ideal" structure of $Fe_{1-x}O$ (known in mineral form as wüstite) is that of rocksalt (NaCl). The phase is stable only above about 570°C and although there is a large range of x, the stoichiometric composition FeO does not exist

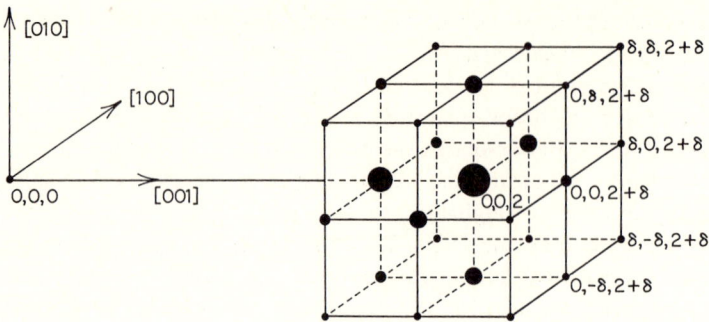

Fig. 6.23. Schematic representation of the superstructure peaks surrounding the fundamental reflections in $Fe_{0.902}O$. [Reproduced by permission from Koch, R., and Cohen, J.B., Acta Crystallogr. *B25*, 275 (1969)]

at normal pressures. The deviation from stoichiometry results from vacancies on cation sites and some of the tetrahedral sites are occupied by cations, as suggested by density measurements. Despite this detailed information there remained one major question — how the vacancies and tetrahedral ions are arranged. Thermodynamic measurements suggested that this arrangement is not random for certain regions of the single phase field, as did measurements of conductivity. Measurements of the chemical diffusion coefficient indicated that as the deviation from stoichiometry increases, the chemical interdiffusion coefficient decreases, rather than increasing as would be expected if random vacancies were being produced.

In quenched samples of $Fe_{1-x}O$ extra diffraction peaks are observed corresponding to a unit cell *about* (but not precisely) 2.5 times the dimensions of the average $Fe_{1-x}O$ cell.[3] Single crystal measurements were made on a slab of $Fe_{0.902}O$ cleaved from a large grain. The superstructure peaks observed around each fundamental NaCl-type reflection are indicated schematically in Fig. 6.23.

Let us discuss the analysis of the fundamental peaks first. For peaks of this type measured with MoK_α, the agreeement with what was expected from a NaCl-type structure was quite good, as illustrated by the temperature factor plot shown in Fig. 6.24. The 111, 222, and 333 peaks appeared to be affected by extinction (see Fig. 6.24) and were not used in subsequent analysis.

The agreement for the fundamental peaks could be improved if some iron ions were introduced on tetrahedral sites. A specific tetrahedral content was assumed and the occupation on octahedral sites adjusted to correspond with the known overall composition. Having thus specified the iron distribution, the scale factor and two isotropic temperature factors (one for oxygen ions and the other for iron ions) were refined with a least-squares program. This process was then repeated for a series of other concentrations of tetrahedral iron ions. A minimum in the R residual [see Eq. (6.110)] was obtained for 3.3% Fe on tetrahedral sites.

---

[3] In some heat treatments, a two-phase mixture may be initiated during the quenching producing an ordered periodic array with satellite peaks corresponding to a repeat distance of 100–500 Å. The discussion in the text is limited to the structure of the quenched *single-phase* $Fe_{1-x}O$.

6.7 Difference Techniques

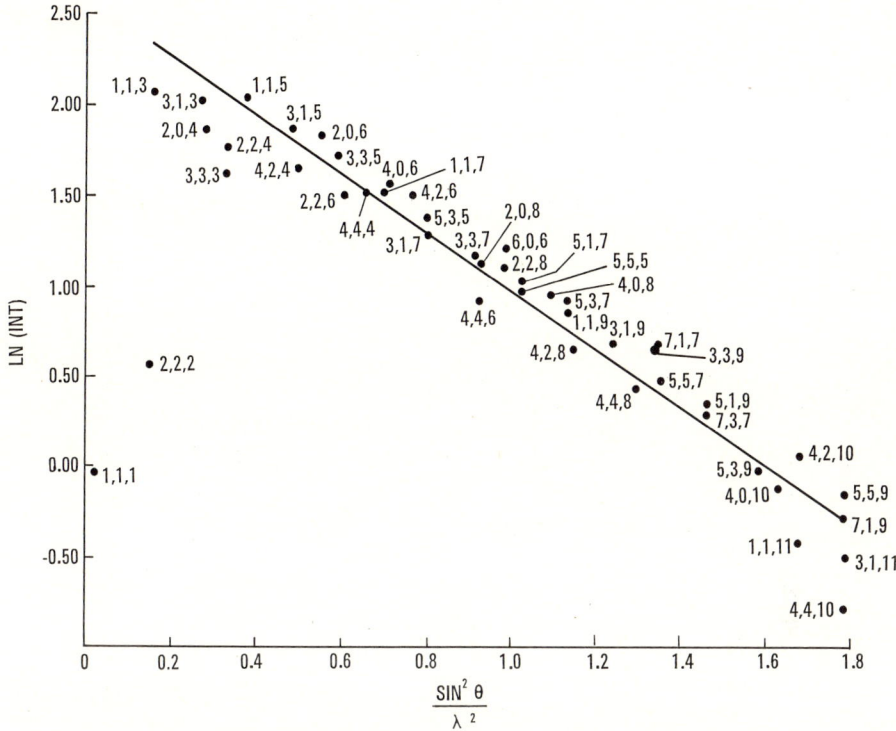

Fig. 6.24. Analysis of the fundamental reflections of $Fe_{0.902}O$ according to a NaCl-type structure. [Reproduced by permission from Koch, F., and Cohen, J.B., Acta Crystallogr. *B25*, 275 (1969)]

The defect structure was determined using the difference Patterson technique. The Patterson function from only the superstructure peaks may be visualized as the Patterson of a difference structure, formed by subtracting the electron density of a disordered crystal from the ordered structure. Since the periodicity does not correspond to an integral multiple of unit cells, analyses were made assuming a super cell of 3× and of 5× the original cell. The results were similar. Two difference Patterson sections are shown in Fig. 6.25 for the 3× cell. An outstanding feature of these Patterson maps is the large positive peaks corresponding to the vectors $\frac{1}{6}, \frac{1}{6}, 0$ and $\frac{1}{3}, 0, 0$. These vectors respectively span the distances between nearest neighbor and next-nearest neighbor octahedral sites. Positive peaks will result if the neighbors of vacancies are also vacancies. Similarly, the negative peak corresponding to the vector $\frac{1}{12}, \frac{1}{12}, \frac{1}{12}$ (the vector between an octahedral site and its tetrahedral neighbor) implies that the vacant octahedral sites tend to be grouped about occupied tetrahedral sites. The symmetry of the array of superstructure peaks (see Fig. 6.23) suggests the appropriate Laue diffraction group is *m3m*. Since no systematic extinctions are present, the ordered structure apparently possesses the point symmetry of *m3m*, $\overline{4}3m$ or 432. Using this information, a specific model for the cation distribution in a 3× cell was deduced. Each complex consisted of 13 vacant octahedral sites (a center site and the 12 nearest neighbor sites around it).

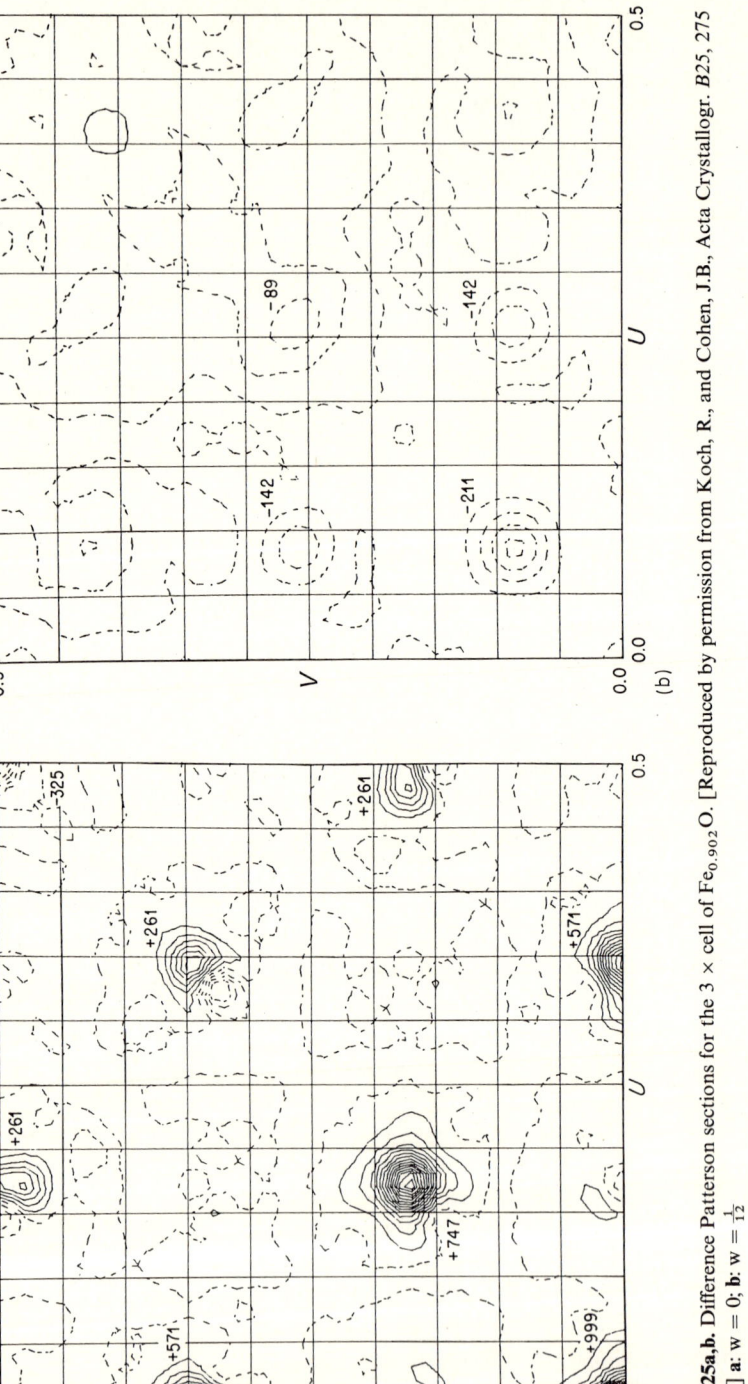

Fig. 6.25a,b. Difference Patterson sections for the 3 × cell of $Fe_{0.902}O$. [Reproduced by permission from Koch, R., and Cohen, J.B., Acta Crystallogr. *B25*, 275 (1969)] **a:** $w = 0$; **b:** $w = \frac{1}{12}$

A total of eight tetrahedral sites are included within each vacancy cluster; some or all of these may be occupied by iron ions. One such complex exists on each of the simple cubic lattice points for the 3× cell. This model was chosen with an eye to the concentration of tetrahedral ions, the known composition, the two positive peaks corresponding to first- and second-neighbor octahedral sites to any octahedral Fe cation and the negative octahedral-tetrahedral peak. Within one 3× unit of the model there are a total of 108 oxygen ions, 95 octahedral iron ions, and from the previously presented data on structure factor calculations for the fundamental peaks, about four tetrahedral ions. This corresponds to an overall formula of $Fe_{99}O_{108}$ or $Fe_{0.918}O$, compared to the known composition of $Fe_{0.902}O$; two more vacant cation sites would yield the correct composition.

As these clusters are large, they can be seen with the electron microscope; S. Iijima (32nd Ann. Proc. Electron Microsc. Soc. Am., 1974) showed that these clusters do indeed exist and that the nonintegral value of the periodicity is a consequence of the arrangement of the defect clusters on a monoclinic lattice. In fact, this is an incommensurate structure, as discussed in Sect. 6.11. A more precise analysis taking this into account, and employing higher dimensional crystallography, led to essentially the same defect array (Yamamoto 1982).

6.8 Helical Structures

Artificial polymers are important materials. These long chain structures of repeating units are often in the form of fibers in a material or folded regularly to form a crystal. One such structure built from ethylene (C_2H_4) is shown in Fig. 6.26. The molecular weights of such polymer chains are in the tens of thousands and often these chains are twisted, i.e., helical due to steric hindrance of large side groups (instead of the

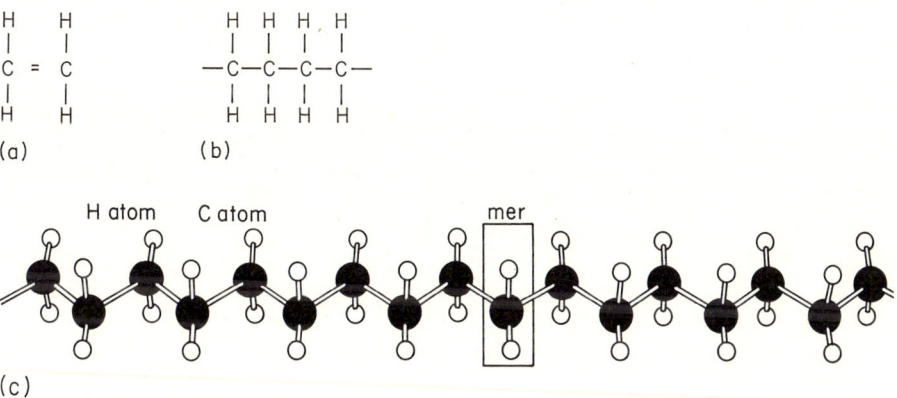

Fig. 6.26. a The molecule ethylene. b A two dimensional sketch of the linear chain molecule of polyethylene. c A schematic drawing of a portion of a chain of polyethylene showing the three dimensional form of the molecule

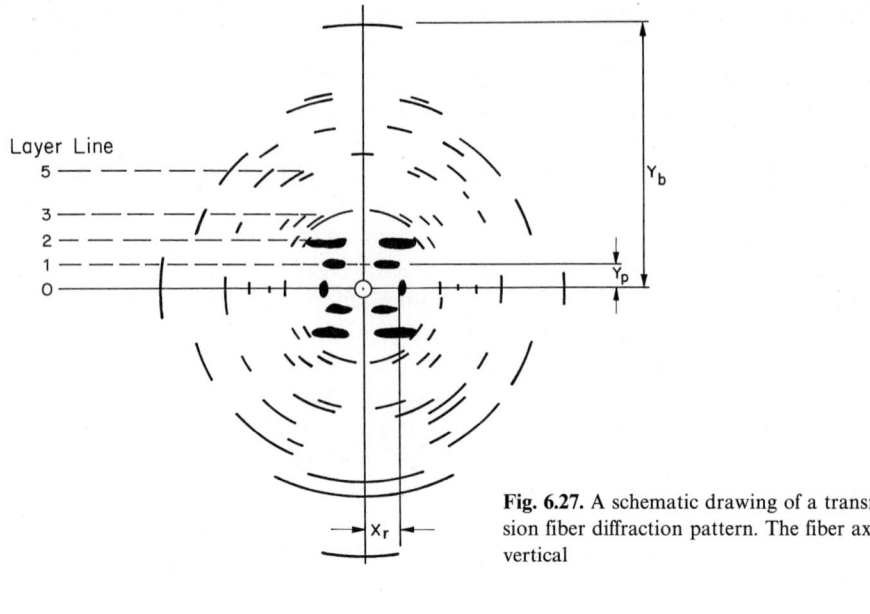

Fig. 6.27. A schematic drawing of a transmission fiber diffraction pattern. The fiber axis is vertical

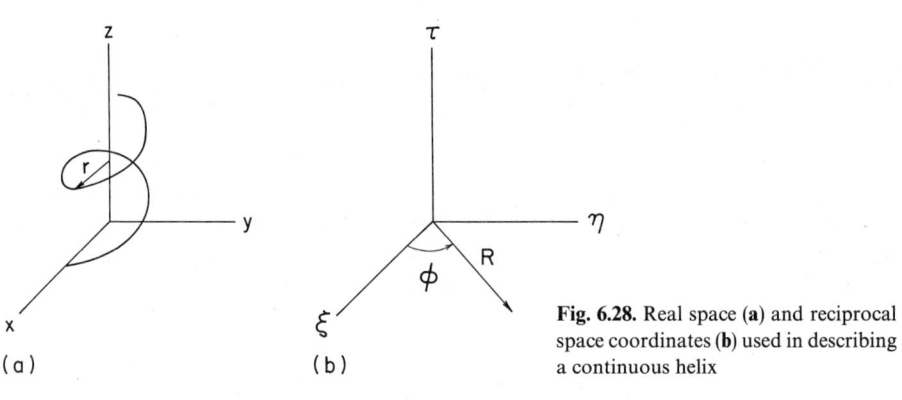

Fig. 6.28. Real space (**a**) and reciprocal space coordinates (**b**) used in describing a continuous helix

H atoms in this figure). Another important polymeric chain is that of the molecule DNA that plays such a key role in body functions. In this case the repeating unit in one helix is bonded to another intertwined helix by compounds known as bases, and each helical structure is stabilized by hydrogen bonding along the chain. Helical sections also are present in hemoglobin, Fig. 6.15.

Because of the significance of these materials, and the vital role diffraction has played in elucidating the structures, we consider them here. A typical diffraction pattern of a fiber taken with its fiber axis vertical is sketched in Fig. 6.27.

A continuous helix is drawn in Fig. 6.28, along with the variables we shall employ. From Fig. 6.28a,

$$x = r\cos 2\pi z/P, \qquad y = r\sin 2\pi z/P,$$

where P is the pitch.

6.8 Helical Structures

To obtain the diffraction pattern, we take the Fourier transform. Let $\rho(\mathbf{r})$ be unity on a continuous helix, 0 elsewhere. Hence,

$$F(s) = \int \exp\left(+2\pi i\left[r\xi\cos\frac{2\pi z}{P} + r\eta\sin\frac{2\pi z}{P} + z\tau\right]\right)dV_\mathbf{r}. \tag{6.95}$$

We shall integrate with respect to z (ignoring here any constant relating $dV_\mathbf{r}$ and dz) and employ cylindrical coordinates in reciprocal space (see Fig. 6.28b) with several variable substitutions:

$$R^2 = \xi^2 + \eta^2, \tag{6.96a}$$

$$\tan\varphi = \eta/\xi, \tag{6.96b}$$

$$\xi/R = \cos\varphi, \tag{6.96c}$$

$$\eta/R = \sin\varphi. \tag{6.96d}$$

Equation (6.95) becomes

$$F(s) = \int \exp\left(+2\pi i\left[Rr\cos\left(\frac{2\pi z}{P} - \varphi\right) + z\tau\right]\right)dz. \tag{6.97}$$

For a long helix with many turns, we can think of the structure as one pitch convoluted with the δ function which repeats it, $\delta(z - nP)$. The transform is then the product of the transforms of the two individual functions. The transform of the δ function is $\delta[\tau - (n/P)]$ and will have values only for $\tau = 1/P$, $2/P$, etc. Therefore,

$$F(s) = \int \exp\left(2\pi i\left[Rr\cos\left(\frac{2\pi z}{P} - \varphi\right) \pm \frac{nz}{P}\right]\right)dz. \tag{6.98}$$

This integral is related to the Bessel function $J_n(w)$:

$$2\pi(i)^n J_n(w) = \int_0^{2\pi} \exp(iw\cos\theta)\exp(in\theta)\,d\theta = 2\pi J_n(w)\exp(in\pi/2). \tag{6.99}$$

We can see this relation more clearly by letting $w = 2\pi Rr$ and $\theta = (2\pi z/P) - \phi$. Solving,

$$2\pi nz/P = n(\theta + \phi) \quad \text{and} \quad (2\pi/P)\,dz = d\theta.$$

Therefore, Eq. (6.98) may be rewritten as

$$F(s) = PJ_n(2\pi Rr)\exp[in(\phi + \tfrac{\pi}{2})]. \tag{6.100}$$

$J_n(w)$ is plotted in Fig. 6.29 for several values of n. Note that J_0 has a value for R and n zero, but is zero for $R = 0$ for $n \neq 0$. Also, the values decrease as R or n increases. Thus *the pattern should consist of layers with broad regions of intensity for each n, spaced n/P apart in reciprocal space* and, on the zero layer, there is intensity near the origin, as seen in Fig. 6.27.

If we consider the helix to be discontinuous perhaps due to heavy side groups (or the bases in DNA) spaced periodically along the helix, we can think of this electron density as the product of the continuous helix and a periodic δ function,

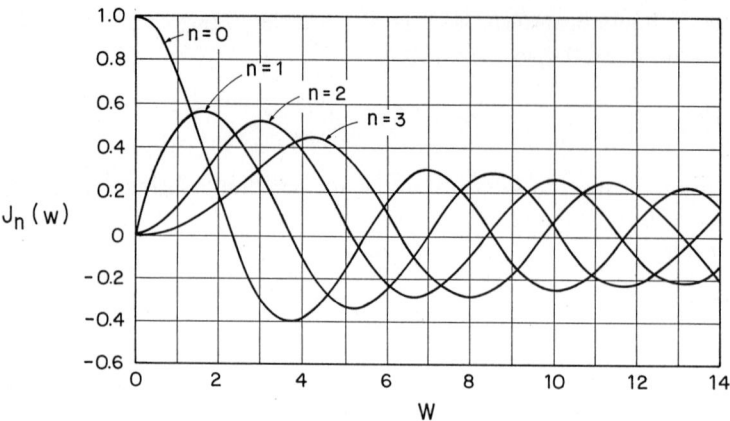

Fig. 6.29. The Bessel functions $J_0(w)$, $J_1(w)$, $J_2(w)$, and $J_3(w)$ for real arguments. Note that $J_n(-w) = (-1)^n J_n(w)$

$\delta(z - mp)$. The transform of this product is the convolution of the two transforms. As a result, we have the convolution of two δ functions, one with values at n/P the other with values at m/p. This can be visualized as each layer m/p in turn serving as an origin for the n/P layers. If P/p is not a ratio of integers, there will be almost continuous scattering. If it is such a ratio, there will be a finite number of layers in the pattern due to overlap.

Consider the pattern sketched in Fig. 6.27. First note that the distance Y_p is related to the pitch, while Y_b is due to discontinuities along the helix. These distances can be related to distances in real space with the same formulas used for layering on a rotation pattern (Sect. 4.3). Thus P/p can be obtained. The repeat of the pattern is clearly evident from the presence of intensity near the vertical line (top of the figure) implying that J_0 is dominant. (This is missing near the origin due to the hole in the film.) Suppose $P/p = a/b$. Then, the "selection rule" for layers in the pattern is

$$\tau = (m/p) + (n/P) = (ma/Pb) + (bn/bP) = (ma + nb)/Pb = l/c, \quad (6.101)$$

where $c = Pb$, and $l = ma + nb$ is an integer, i.e., the pattern repeats for $l = c$. For each $l = ma + nb$, n is obtained by choosing $m = 0, +1, -1$, etc. All the n-order Bessel functions are added for the amplitude for that layer, l. Near the vertical axis, J_1 is important for $l = 1$, J_2 for $l = 2$, etc., so that this can help decide if a layer is $l = 1$ or 2.

The argument of the Bessel function to any peak near the y axis, i.e., distances like x_r, tell us the value of r since the argument is $2\pi Rr$. We proceed as follows. We know the value of this argument for any layer from tables of the dominant J_n. Because the film is flat, the angle 2θ is simply related to coordinates on the film, $\tan 2\theta = (x^2 + y^2)^{1/2}/z$, where z is the fiber to film distance. With this 2θ the value of R may be obtained (see Sect. 4.3) while $r_v^* = 1/c$ and hence r^*. There is a "cross" of spots centered at the hole in the film which arises from the tilt of the material along the helix.

If there are two such helices interwined, interesting results can be produced. If, for example, the two are shifted $\frac{1}{8}$ in the z coordinate, the fourth layer is missing (consider two planes d apart with another inserted at $\frac{1}{8} d$; at the fourth order the scattering from this plane is half a wavelength out of phase with the scattering from the first plane and cancels it). This is how we know DNA is a double stranded helix.

If there are many helices (r) rotated with respect to each other by θ_r and displaced by z_r, each with scattering factor f_r:

$$F(s) = \sum_r \sum_n f_r J_n(2\pi R r) \exp\left[in\left(\phi + \frac{\pi}{2}\right)\right] \exp\left[-in\left(\theta_r + \frac{2\pi l z_r}{c}\right)\right]. \tag{6.102}$$

6.9 Refinement by Least Squares

6.9.1 Single Crystal Data

Suppose that an observable quantity q is a linear function of a set of unknown variables $x_1, x_2, \ldots = \{x_i\}$, so that we may write

$$q = a_1 x_1 + a_2 x_2 + a_3 x_3 + \ldots. \tag{6.103a}$$

If we may make similar experiments in which the constants $a_1, a_2, a_3, \ldots$ are different, but known, we would obtain a set of equations:

$$\begin{aligned} q_1 &= a_{11} x_1 + a_{12} x_2 + a_{13} x_3 + \ldots, \\ q_2 &= a_{21} x_1 + a_{22} x_2 + a_{23} x_3 + \ldots, \end{aligned} \tag{6.103b}$$

etc. If the number of observations (m) is equal to or greater than the number of unknowns (n), and the determinant of the coefficients in Eq. (6.103b) is nonzero, we may solve to obtain the unknowns $\{x_i\}$. However, each experimental observation q_i is associated with some error E_i, so that what is actually measured is $q_i + E_i$. Then a different set of $\{x_i\}$ will be obtained for each set of n equations selected. According to a principle first enunciated by Legendre in 1806, *the most acceptable solution to this problem is the one that minimizes the sum of the squares of the errors*, $\sum_i^m E_i^2$. This least-squares minimization is the basis for a procedure of refinement using either the observed structure amplitudes ($|F|$) or their squares ($FF^* = |F|^2$). We wish to minimize the sum

$$\sum_j E_j^2 = \sum_j (a_{j1} x_1 + a_{j2} x_2 \ldots - q_j)^2. \tag{6.104}$$

This function is a minimum when its partial derivatives with respect to $\{x_i\}$ vanish, i.e., when

$$\partial \sum_j E_j^2 / \partial x_i = 2 \sum_j a_{ji} E_j = 0 \tag{6.105a}$$

for all x_i. Expanding Eq. (6.105a) using the definition of E_j in Eq. (6.104), we have a set of equations

$$\sum_k \sum_j a_{ji} a_{jk} x_k = \sum_j a_{ji} q_j \tag{6.105b}$$

called the normal equations. They comprise a system of n equations in the n unknowns $\{x_i\}$ which may be solved to obtain the desired set of $\{x_i\}$ which best satisfies Legendre's principle.

Owing to various experimental factors, some observations may be considered more reliable than others. For example, counting statistics will be better for stronger peaks if equal time has been spent sampling each hkl reflection, or absorption corrections may be larger for some crystal orientations and affect certain peaks more strongly. We may choose a set of weighting factors $\{w_j\}$ to indicate the relative reliabilities of the observations $\{q_j\}$. When this is done, each quantity in Eq. (6.105b) must be replaced by its product with w_j.

We wish to apply this procedure to find the set of coordinates of atoms in the structure that best fit the *observed structure amplitudes* F_0. The calculated structure amplitude is given by [see Eq. (2.21)]

$$F_c = \sum_r f_r \exp[2\pi i(hx_r + ky_r + lz_r)] . \tag{6.106}$$

This is clearly a nonlinear equation in the unknowns, but it can be linearized by expanding in a Taylor's series and retaining only the first two terms:

$$\Delta F = F_0 - F_c = \sum_r \left(\varepsilon_{x_r} \frac{\partial F_c}{\partial x_r} + \varepsilon_{y_r} \frac{\partial F_c}{\partial y_r} + \varepsilon_{z_r} \frac{\partial F_c}{\partial z_r} \right), \tag{6.107}$$

where the ε's are small changes from the initial guesses of x_r, y_r, z_r. Replacing q_j by ΔF_j for each observed reflection, Eq. (6.104) becomes

$$\sum_j E_j^2 = \sum_j \left[w_j \sum_r \left\{ \varepsilon_{x_r} \frac{\partial F_{cj}}{\partial x_r} + \varepsilon_{y_r} \frac{\partial F_{cj}}{\partial y_r} + \varepsilon_{z_r} \frac{\partial F_{cj}}{\partial z_r} \right\} - \Delta F_j \right]^2 \tag{6.108}$$

Minimizing Eq. (6.108) leads to a set of normal equations in the unknown ε's. While many modern digital computer programs include off-diagonal terms, for the purpose of following the basic nature of the iteration processes, we will neglect them. Then solutions are of the form

$$\varepsilon_{x_r} = \sum_j w_j \left(\frac{\partial F_{cj}}{\partial x_r} \right) \Delta F_j / \sum_j w_j \left(\frac{\partial F_{cj}}{\partial x_r} \right)^2, \tag{6.109}$$

where a weighting factor w_j has been included for each observed F. The calculation is then repeated with a new F_c calculated using corrected values for the coordinates $x_r + \varepsilon_{x_r}$, etc. This iteration proceeds until the ε's are "small" by some statistical standard.

The least-squares refinement procedure is very flexible and has many advantages. Any number of F's in excess of the number of unknowns may be used. Thus it is possible to discard questionable data (for which absorption or extinction errors may be too large) and still obtain a complete solution. Furthermore, the analysis can be expanded to include the scale factor [K in Eq. 4.19] and the temperature factor. In its simplest form we may include a single isotropic temperature factor of the type

$$f_r \to f_{0r} \exp(-B_r \sin^2 \theta/\lambda^2) = f_{0r} \exp(-M_r)$$

(see Sect. 3.4). In general, however, atomic vibrations are described by a symmetric rank two tensor, and the appropriate anisotropic temperature factor should be used, [see, for example, Cruikshank (1956)]:

$$\exp(-M_r) = \exp[-(b_{11}h^2 + b_{22}k^2 + b_{33}l^2 + b_{12}hk + b_{13}hl + b_{23}kl)],$$

with

$$b_{ii} = 2\pi^2 |\mathbf{b}_i|^2 U^{ii}, \qquad b_{ij} = 4\pi^2 |\mathbf{b}_i||\mathbf{b}_j| U^{ij},$$

where the $\mathbf{b}_i$ are reciprocal lattice vectors, and U^{ij} are the components of the tensor describing the anisotropic vibrations. The mean-square vibration in any direction is then given by $U^{ij} l_i l_j$, where $l_i l_j$ are the direction cosines of the direction in real space, and repeated indices imply summation. The initial guesses at the values of b_{ii} and b_{ij} may be taken as all equal to the estimate made as described in Sect. 3.4.[4]

An additional feature of the least-squares fitting procedure is that estimates of the statistical uncertainty (variance) may also be made for all unknowns, so that we may have an estimate of the accuracy with which atomic coordinates have been established. Finally, we should say something about a criterion for stopping the refinement. A *residual or reliability index is usually defined as*

$$R = \sum_j ||F_{0j}| - |F_{cj}|| / \sum_j |F_{0j}| \qquad (6.110\text{a})$$

or, when the values of the structure factors are weighted by $0 \leq w_j \leq 1$, a weighted residual is defined as

$$R_w = \left\{ \sum_j w_j (|F_{0j}| - |F_{cj}|)^2 / \sum_j w_j F_{0j}^2 \right\}^{1/2}. \qquad (6.110\text{b})$$

Residuals in the range 0.05–0.10 generally indicate satisfactory fits, but the expected value of R for a good fit depends on the number and accuracy of observations and the number of parameters determined. A statistical analysis of this problem has been made by Hamilton (1965) and his paper should be consulted for details.

Computer programs for least-squares refinement of crystal structures are available at all laboratories where such studies are made, and a central library listing the various programs and their special features has been established (Crystallographic Computing Committee of the American Crystallographic Association). Powder patterns can also be analyzed this way.

6.9.2 Powder Data

Single crystal methods have been and will continue to be the primary techniques for crystallization analysis. By contrast, the ease of obtaining powder data has always made that technique the method of choice in studies in which samples are to be examined as a function of changing external variables such as temperature,

---

[4] This formulation assumes only rectilinear motions, not librational motions (molecular rotations). See Dawson (1975) for further details.

pressure, magnetic fields, partial pressure of chemically reactive species, etc. Recent advances in the least squares analysis of powder data (known as Rietveld analysis after H.M. Rietveld who originally proposed the technique in 1967) have made possible routine determination of structural parameters with precision rivaling that of good single crystal data. Complex structures are still best determined from single crystal data from which the space group may be most readily derived, but Rietveld analysis of powder data is being increasingly applied to the study of variation of atomic positions as a function of changing external variables (see Cheetham and Taylor 1977; Albinati and Willis 1982).

Instead of using integrated intensities of peaks, the Rietveld method fits the internal structural parameters to the overall profile of the powder pattern assuming the pattern to be the sum of Bragg reflections satisfying some function or some sum of functions (like Gaussian or Lorentzian functions) centered at their respective Bragg angular positions. [For neutron powder data, this shape is simple and a Gaussian fit often suffices, but a more complex peak shape is necessary to account for the influence of the line shape when characteristic x-rays are used as a source, because of the shape of emission lines. However, we may anticipate increased use of the Reitveld method in future years for x-ray powder data obtained using synchrotron radiation as source in which case a single function will suffice.] A

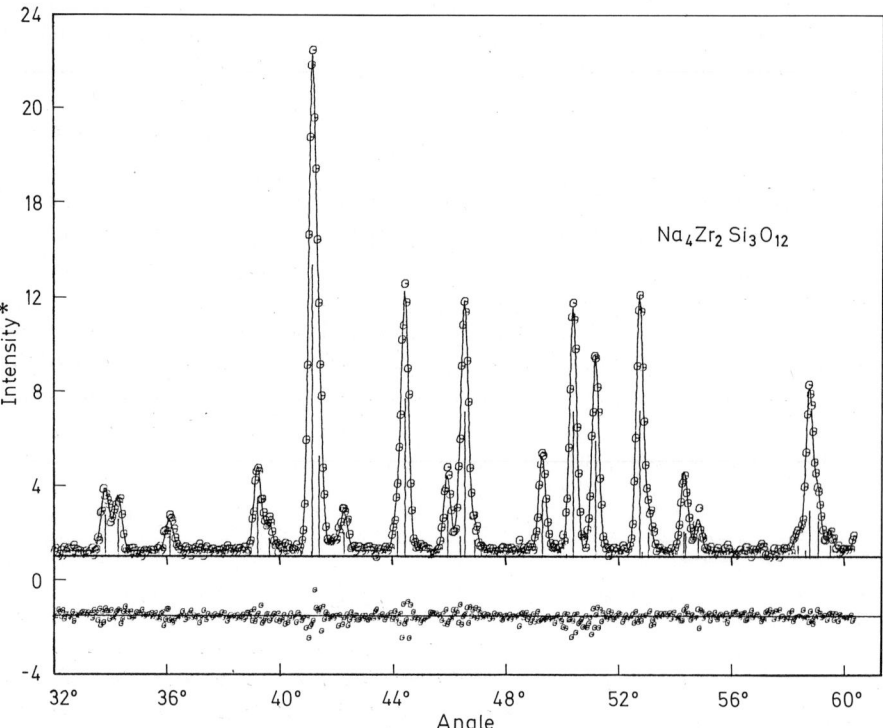

Fig. 6.30. A Rietveld fit to the powder pattern of $Na_4Zr_2Si_3O_{12}$. After B.J. Wuensch, T.J. Schioler and E. Prince, Proc. Conf. on High Temp. Solid Electrolytes, Aug. 1983, Vol II, pp. 55–71, Brookhaven National Laboratory, BNL-51728

Rietveld program uses the actual point counts at closely spaced 2θ intervals and leads to a greater efficiency of information retrieved from the superposed reflections inherent in powder patterns of low symmetry structures.

The program minimizes the quantity:

$$M = \sum_i w_i \left\{ Y_i(\text{obs}) - \frac{1}{c} Y_i(\text{calc}) \right\}^2 \tag{6.111}$$

where Y_i is the background-corrected intensity at a 2θ point i, c is a scale factor, and w_i is a least squares weight, equal to $(\sigma_i^2)^{-1}$, $Y_i(\text{calc})$ is formed by summing the contributions from all Bragg reflections which may overlap at point i and includes the multiplicity of the reflection and the Lorentz factor.

An example of the application of the Rietveld method is shown in Fig. 6.30 which exhibits the measured intensity and best fit line for the neutron powder pattern of $Na_{1+x} Zr_2 Si_x P_{3-x} O_{12}$ with $x = 3$. This is an end member of a series of compounds which includes (for $x = 2$) the superionic conductor called "NaSiCon." This pattern was taken as part of a study of the entire series intended to clarify the mechanism of ionic conductivity in these materials, and would have been prohibitively expensive to carry out had single crystal analysis been the only available technique. Note that the many overlapping peaks (positions indicated by vertical bars) have been uniformly fit as shown by the deviation between theory and experiment displayed along the bottom of the figure.

6.10 The Use of Electron Diffraction

Because the scattering cross-section of atoms for electrons is high, there is considerable multiple scattering in diffraction patterns obtained with electrons. At first glance, this would seem to preclude the use of electron diffraction for structural work. This is not entirely the case. Electron diffraction with low energy electrons is employed to study the structure of surfaces, by measuring the intensity of the rods in reciprocal space from the surface, (see Sect. 3.7) although the calculations are very difficult (Pendry 1974). In fact, the use of glancing angle x-ray diffraction, with its many advantages (discussed in Sect. 3.9) will probably gradually displace this tool.

Nonetheless there are occasions when the structure is required of a small second-phase particle in a matrix. It may be so small that it can be observed only in an electron microscope. There are then two techniques for obtaining structural information:

1) In very thin portions of a transmission specimen it is possible to permit many diffracted beams to pass through the apertures and combine to form an image that has atomic resolution (Cowley 1975) or at least resolution of structural units. If the incident beam is parallel to a principal axis of the crystal, what is observed is a projection of the atom potentials along this direction — the projected electron

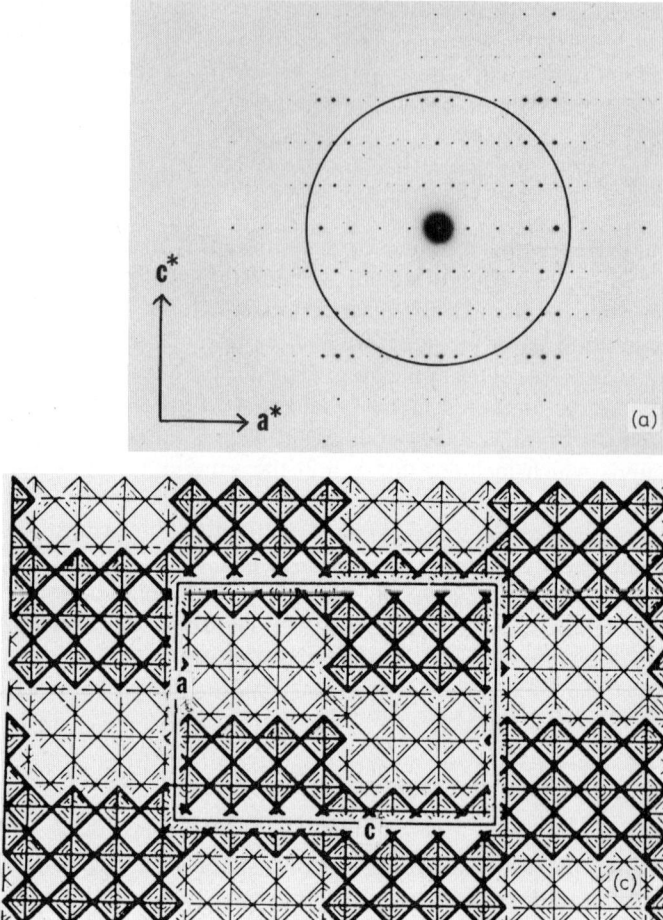

Fig. 6.31a,c

Fig. 6.31a–d. Structure resolution via transmission electron microscopy. **a** Electron diffraction pattern; *circle* indicates spots allowed to pass through the objective aperture to form the image. **b** Image. **c** Model of image. $a = 20.5$ Å, $c = 28.5$ Å. The *shaded squares* are oxygen octahedra containing metal atoms (at two different levels, *dark* and *light*). $T_{12}Nb_{10}O_{29}$, S. Iijima 1971. (Reproduced by permission of J. Cowley, Arizona State University) **d** High resolution structure image of a gold thin film showing a region of (110) surface in profile. The *black 'dots'* where the fringes cross show the precise positions of the atomic columns. (Taken from D.J. Smith and L.D. Marks, Ultramicroscopy *16*, 101–114 (1985))

density. With a thin section and proper focussing conditions, dark regions indicate heavy atom columns and the images reflect the true structure, Fig. 6.31. In some cases actual atom resolution has been obtained.

2) A second possible use of the electron microscope is to obtain information on the crystal symmetry, and in fact the space group itself. Some of this could be done with the normal sharp spot pattern taken along a principal axis. However, in such a case, intensities are extremely sensitive to the exact orientation of the crystal to the incident beam. Instead, "convergent beam electron diffraction" (CBED) is

6.10 The Use of Electron Diffraction

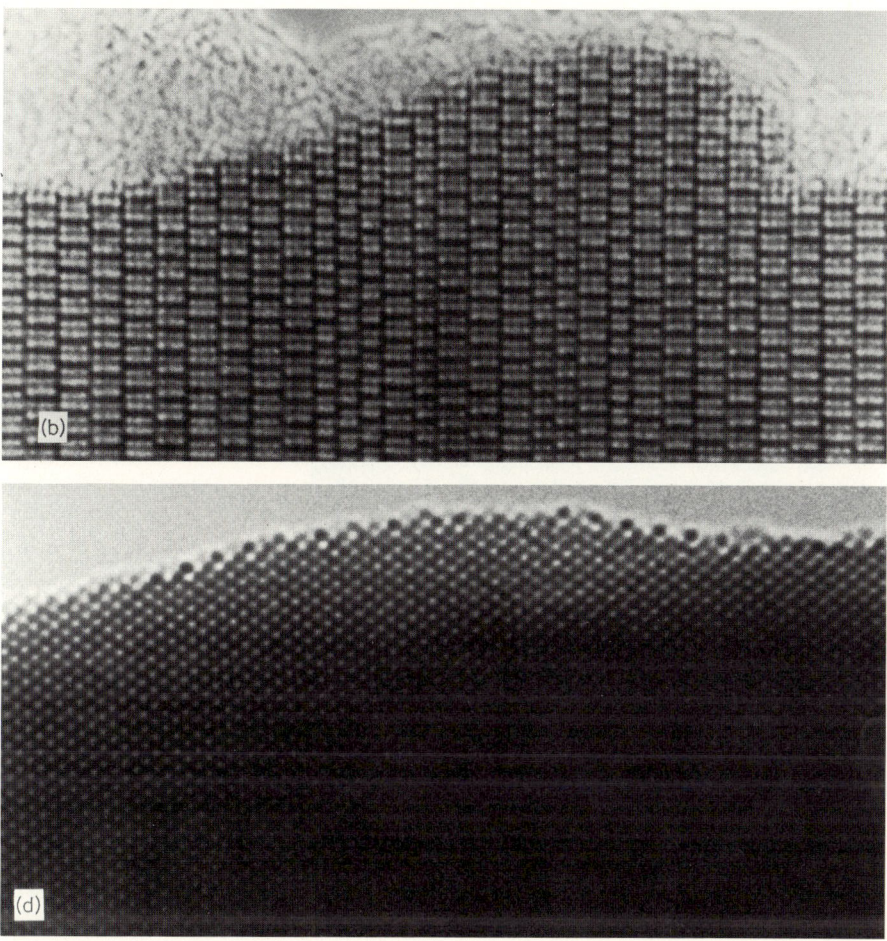

Fig. 6.31b,d

employed. The electron beam is made to converge on the specimen. Instead of rocking the crystal through its reflection (an "omega scan") as might be done with x-rays, a host of incident beam orientations is provided at a small region. The spots produced are broad circles, as in Fig. 6.32. The general pattern of spots around the center provides information about point symmetries due to axes parallel to the direct beam, e.g. rotations and mirrors parallel to this beam. However, it is quite apparent that the large spots in Fig. 6.32 are rich in detail due to multiple scattering, and this detail can be employed to give space group information. To see how we can proceed further consider the case of a mirror plane parallel to the crystal faces, perpendicular to the incident beam as in Figs. 6.33a and b. In our diffraction equations, interchanging **S** and **S**$_0$ does not alter the magnitude of the amplitude. This "reciprocity theorem" implies that we can exchange source and observation points; this is done in (c). The mirror thus produces two-fold symmetry in the spot.

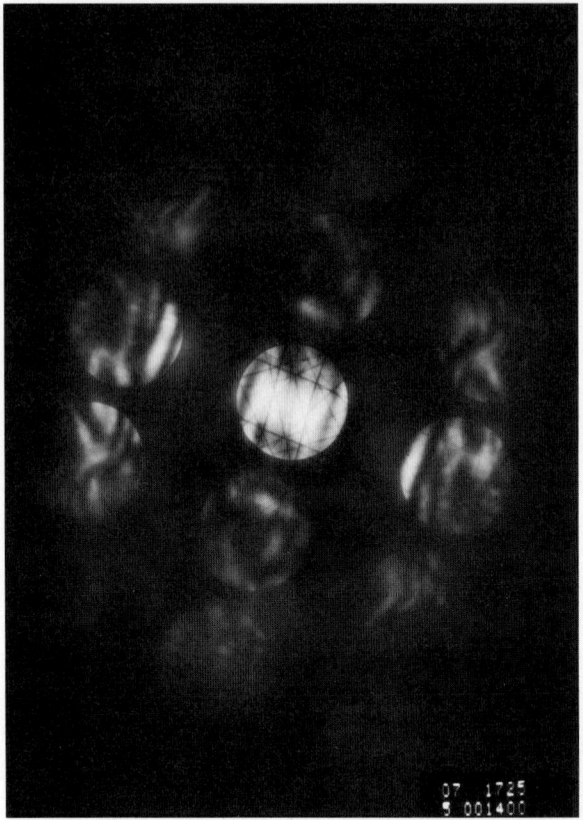

Fig. 6.32. Convergent beam electron diffraction pattern of Al_2O_3 in the [T2TO] direction, taken at 100kV. Reprinted by permission from "Convergent Beam Electron Diffraction of Alloys" compiled by John Mansfield, Adams Hilger Ltd., Boston, 1984. (Taken by M.D. Shannon)

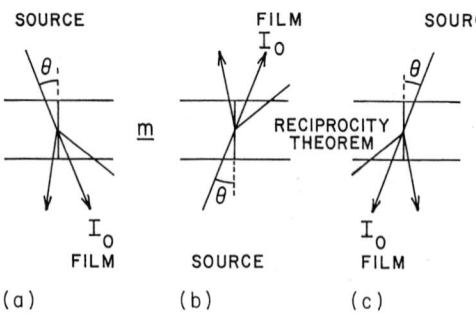

Fig. 6.33a–c. How a mirror plane parallel to the faces of a crystal affects the convergent beam diffraction pattern. The mirror generates **b** and **c**. This is for the direct beam spot, or bright field. After J.W. Steeds in "Quantitative Electron Microscopy" (eds: J.N. Chapman and A.J. Craven) Scottish Universities Summer School in Physics, Glasgow, 1983

A set of tables can be drawn up (Steeds 1983) that indicate the observed symmetries in the direct beam spot (bright field) and the diffraction spots (dark field) and how these relate to the crystal symmetries. When there is strong interaction between the incident beam and the atoms, Friedel's law no longer holds (see Chap. 8) and this helps sort the space groups with even more certainty than with x-rays (if anomalous dispersion is not large in the latter case).

Care is needed, because imperfections can disturb the symmetry. Generally, patterns along two or more axes are required.

6.11 Aperiodic Crystals

6.11.1 Introduction

Chapter 5 of this book was devoted to the development of the classical 230 space groups which describe the symmetry of all three-dimensional periodic geometrical arrays of atoms. It has been clear for many years that most real materials deviate from the assumptions which lead to these idealized descriptions. Many of these deviations are best treated as perturbations of the basic periodic geometrical arrangement of atoms and are discussed elsewhere in this text in some detail. For example, misorientations of small blocks of a crystal relative to one another lead to secondary extinction (Sect. 4.9). Other examples are the effects of point, line, and surface imperfections in crystals (Sect. 7.3), the effect of long-range lattice strain (Sect. 4.7) and the effects of thermal vibrations (Sects. 3.4.4 and 7.2).

When atomic order is not perfect, the partial or complete disorder is usually best described in terms of deviations from the space group characteristic of perfect order or perfect disorder (Sect. 7.6). When a property of the atom other than its position (e.g., magnetic spin) is added to the problem, the number of space groups increases, and magnetic imperfections, when present, are generally treated as deviations from perfect periodicity (Sect. 3.6). Of course, complete relaxation of long-range atomic regularity may occur, and is characteristic of gases and liquids (including super-cooled liquids or glasses), and the structural features of these materials are statistical in nature and not amenable to description in terms of the regularity of space groups and unit cells (Sect. 7.8).

In this section we will discuss a general class of materials called incommensurate structures (mentioned briefly in Sects. 6.7 and 6.8). These materials have generally been described in terms of the classical 3-D space groups with long-range periodic compositional or positional modulations superimposed (Tsakalakos 1984). When the period of this defect structure is equal to an integral number of lattice spacings the modulation is termed commensurate, otherwise it is incommensurate. Examples of such incommensurate structures include long-period superlattices (Sato and Toth 1965), charge density waves (Moneton et al. 1975), and spiral magnetic structures (Corliss et al. 1959). The diffraction patterns of these structures may be represented as that of the basic underlying 3-D space group plus additional sharp reflections which arise as a consequence of the superimposed periodic deviations (see Problems. 2.16 and 6.20). Recently, it has become clear that an alternate and unifying representation of these structures may be achieved by the use of so-called superspace groups, crystallographic space groups in N-dimensional space, where $N > 3$, (de Wolff 1984). In the next subsection we will briefly describe this concept. This esoteric, alternative representation of crystal structure might have remained unknown to most materials scientists were it not for the dramatic discovery by Shechtman et al. (1984) of a material which cannot be described as a deviation from the classical 3-D space groups, but rather *requires* description in terms of a six-dimensional superspace group.

Shechtman and his co-workers examined electron diffraction patterns of rapidly cooled alloys of Al with 10–14% Mn, Fe, or Cr. In these samples they found an aluminum rich matrix and particles of a metallic solid which has a diffraction pattern consisting of sharp peaks like a single crystal but has point group symmetry $m\bar{3}5$ (icosahedral). As we have shown in Sect. 5.3, this is inconsistent with translational periodicity. If the specimen is rotated through the angles of the icosahedral point group (Fig. 6.34a), selected-area diffraction patterns clearly display the six five-fold, ten three-fold, and fifteen two-fold axes characteristic of icosahedral symmetry (Fig. 6.34b). In a series of experiments it was determined that this new solid, dubbed "Shechtmanite," was not an artifact of multiple twinning, but in fact represented the first known example of a material exhibiting long-range orientational order with no translational periodicity. Lacking periodicity, these new solids cannot be classified as crystals, and instead they have been described as "quasicrystalline."

The diffraction pattern of a periodic crystal has an easily recognizable feature, namely spots along a given direction occur at integral intervals. For example going along the h00 direction in reciprocal space we find vectors of length r^*_{100}, $r^*_{200} = 2r^*_{100}$, $r^*_{300} = 3r^*_{100}$, etc. When the diffraction pattern of a quasicrystal is examined, however, this is not true. As an example, Fig. 6.35 shows the relations observed between four different vectors to spots in the first and second ring of spots in the pattern of Fig. 6.34b with five-fold rotational symmetry. Note that

$$\mathbf{l}_4 = \mathbf{l}_1 + \mathbf{l}_2 = -2\cos(\pi/5)\mathbf{l}_3 = -\tau\mathbf{l}_3,$$

where $\tau = 2\cos(\pi/5) = (1 + \sqrt{5})/2$. The diffraction patterns shown in Fig. 6.34b may be indexed in the straightforward manner described in Sect. 4.2 (c.f. Sect. 1.9 and Problem 1.29). A basis system is chosen with three orthogonal 2-fold axes of the icosahedral group forming the basis vectors. This is the coordinate system used in the International Tables of Crystallography for this point group and has the advantages of orthogonality for calculations such as vector summation and distances between points. When the observed electron diffraction patterns are indexed using this basic system, the resultant Miller indices include a mixture of rational and irrational numbers which can always be reduced to the form

$$(h + h'\tau, k + k'\tau, l + l'\tau),$$

where h, k, l, h', k', l' are all integers. Using a simplified notational form for these indices, $(h/h', k/k', l/l')$, Cahn et al. (1986a) have produced the stereographic projection of the icosahedral reciprocal lattice shown in Fig. 6.36.

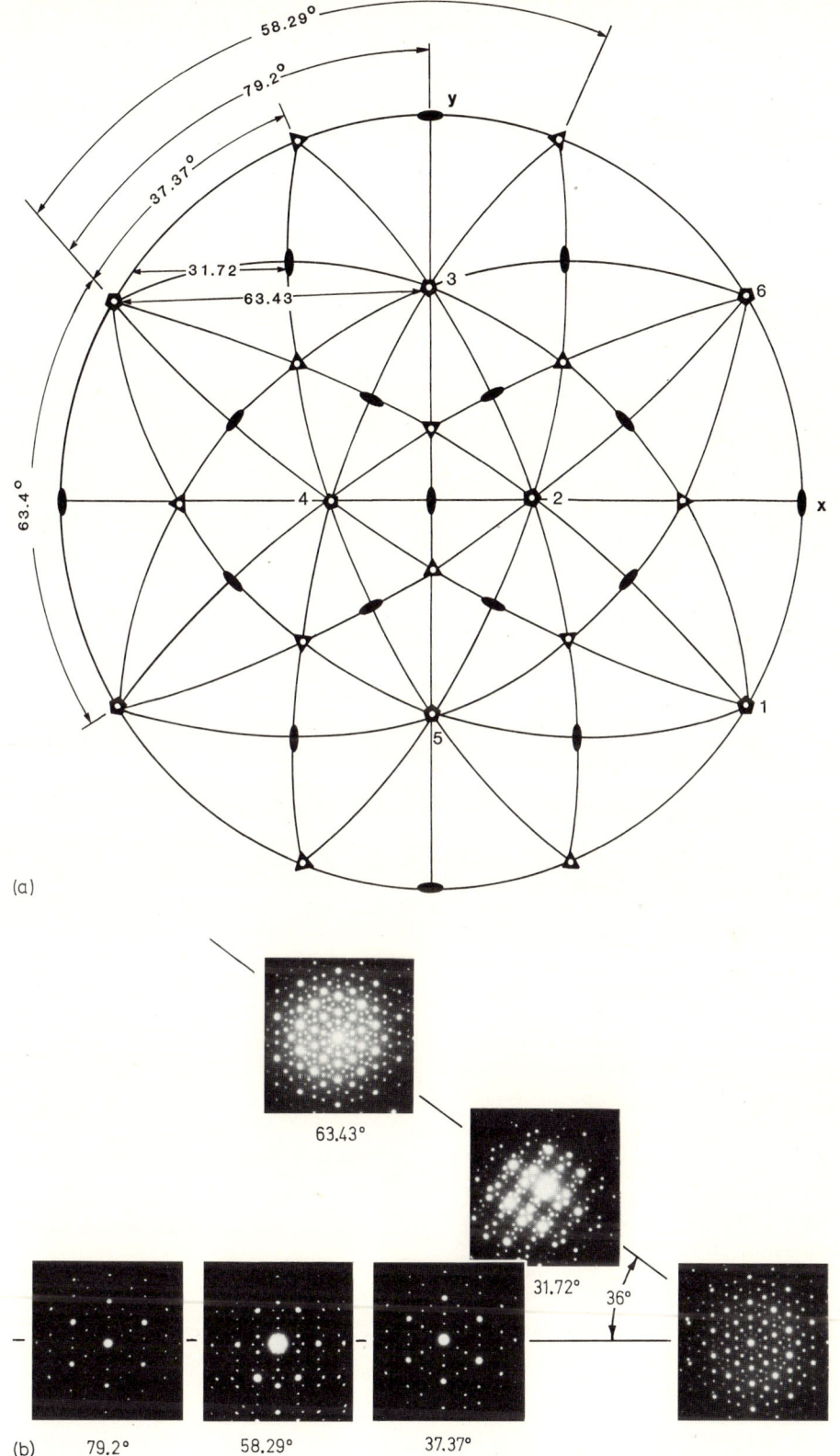

Fig. 6.34. a Icosahedral point group $m\bar{3}\bar{5}$ shown in stereographic projection. **b** Selected-area electron diffraction patterns taken from a single grain of the icosahedral phase. Rotations match those in **a**

350   6. Determination of Crystal Structures

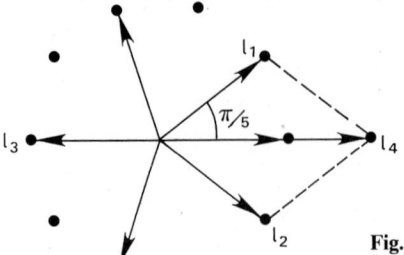

Fig. 6.35. Definition of the irrational number, τ, which arises naturally when a five-fold rotational axis is present

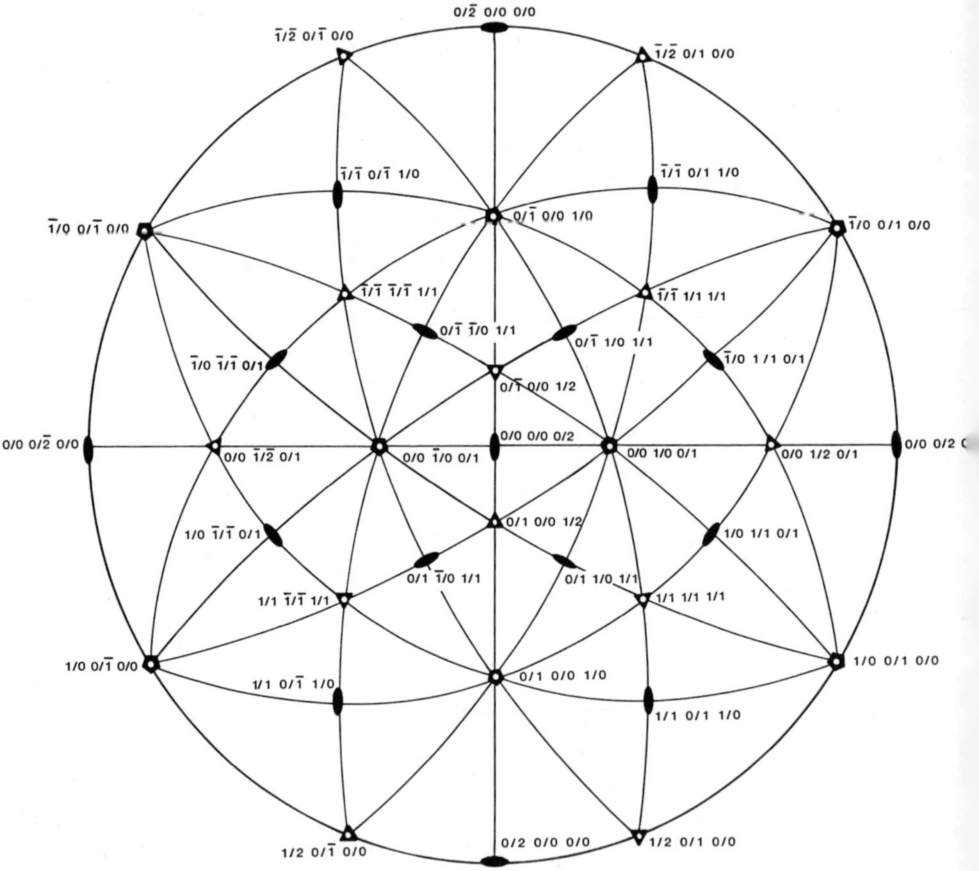

Fig. 6.36. The stereographic projection of Fig. 6.34a indexed using the six index scheme developed by Cahn et al. (1986a)

6.11.2 Diffraction from Quasicrystals

How can we understand the presence of sharp diffraction spots from a structure lacking translational periodicity? In what follows we adopt the presentation of Zia and Dallas (1985), based on the more general treatment of Elser (1986). A quasicrystalline structure may be represented as a sum of δ-functions located at a discrete set of points in D-dimensional space distributed neither randomly nor periodically, but rather located by a complex set of rules. For $D = 2$, for example, these points may be located at vertices of a Penrose tiling pattern (Fig. 5.4). The clearest way to obtain a quasicrystalline structure follows. It is known as the "cut and project" procedure. Consider a regular periodic lattice of points in N-dimensional space *cut* by a D-dimensional ($D < N$) surface located at an arbitrary orientation with respect to the lattice. Take the set of points which lie within an arbitrary distance, w, from the surface and *project* them onto it. The resultant is a generalized Penrose lattice in D dimensions. For simplicity we will consider the square lattice ($N = 2$) with surface $D = 1$, i.e. a line (e.g. the ξ-axis as shown in Fig. 6.37a).

We choose a square lattice with unit spacing in the x, y rectangular coordinate system. The projection line at angle θ to the x-axis is labeled ξ and an orthogonal axis (the projection axis) labeled η. The point at the origin is common to the x, y and ξ, η coordinate systems. If $\cot \theta$ is irrational, there will not be another lattice point on this line and the projected set of points (designated by open circles in Fig. 6.37a) will be arranged aperiodically. To calculate the diffraction pattern from atoms located at such an aperiodic string of points we must first represent them analytically. The square lattice of points may be written as a generalized set of δ-functions:

$$U_0(x, y) = \sum_{j,k} \delta(x - j)\delta(y - k) . \tag{6.112}$$

The projection described above then reduces to (a) writing U_0 in terms of ξ and η,

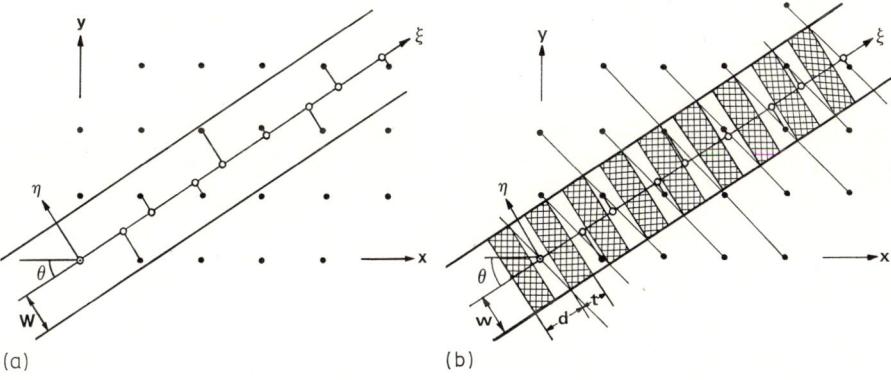

Fig. 6.37. **a** The quasicrystalline string of atoms are located at the *circles* (○), the images of the square lattice points (●) lying within the *solid lines*, projected onto the ξ axis. **b** Figure **a** repeated to demonstrate that all quasicrystalline atoms are located within periodically repeated regions of the one-dimensional space

(b) multiplying by the step function, $R(\eta) \equiv 1$ if $|\eta| < w$ and 0 otherwise, and (c) integrating over η. Thus

$$Q(\xi) = \int d\eta\, R(\eta) U(\xi,\eta), \tag{6.113}$$

where $U(\xi,\eta) \equiv U_0(\xi \cos\theta - \eta \sin\theta, \xi \sin\theta + \eta \cos\theta)$, is an implicit way of writing the sum of quasicrystalline δ-functions represented by open circles in Fig. 6.37a.

To obtain the diffraction pattern, we must calculate $F(s)F^*(s)$ where $F(s)$ is the Fourier transform of $Q(\xi)$. We write $S(p) = \int d\eta\, R(\eta) \exp(2\pi i p \eta)$ and $M(\xi,p) = \int d\eta\, U(\xi,\eta) \exp(2\pi i p \eta)$ as the transforms of R and U respectively. Recall that if f is the Fourier transform of g, then $\int g$ is just $f(0)$. Thus $Q(\xi)$ may be regarded as the transform with respect to η of the function $R(\eta) U(\xi,\eta)$ evaluated at $p = 0$. Since we have the product of two functions $R(\eta)$ and $U(\xi,\eta)$, the transform of their product is a convolution of their transforms. Then

$$Q(\xi) = \int dp\, S(-p) M(\xi,p). \tag{6.114}$$

Finally, substituting from Equation (6.114)

$$F(s) = \int d\xi\, Q(\xi) \exp(2\pi i s \xi)$$

$$= \int dp\, S(-p) V(s,p), \tag{6.115}$$

where $V(s,p)$ is the complete two-dimensional Fourier transform of U.

Now we may write explicit functions for $S(p)$ and $V(s,p)$. We note that U is just U_0 rotated, i.e. a sum over δ-functions at rotated points. Therefore, V is also a sum over rotated δ-functions, i.e.

$$V(s,p) = \sum \delta[s - (n\cos\theta + m\sin\theta)]\delta[p - (-n\sin\theta + m\cos\theta)] \tag{6.116}$$

The Fourier transform of the step function $R(\eta)$ is [see discussion leading to Eq. (6.23b)] $S(p) = 2w(\sin pw/pw)$, and so we have

$$F(s) = \sum_{m,n} S(-p)\delta[s - (n\cos\theta + m\sin\theta)]\delta[p - (-n\sin\theta + m\cos\theta)]. \tag{6.117}$$

This equation displays the spectrum as a sum of densely distributed δ-functions, appropriately weighted by the function $S(-p)$. For each point on the line in reciprocal space with coordinate $s = n\cos\theta + m\sin\theta$, there is a diffraction peak with amplitude $S(-p)$, where $p = -n\sin\theta + m\cos\theta$. Thus two indices, m and n, must be used to specify each point in the 1–D reciprocal space reflecting the "hidden-symmetry" of the 2–D lattice. Equation (6.117) also shows that the location of the δ-functions are those from projection of the complete square but rotated lattice in s, p space, but with lesser strengths for the more distant lattice points. Thus, given a finite sensitivity in the measuring instrument, only those peaks with

6.11 Aperiodic Crystals

a large enough weight will be detected. The result is a discrete spectrum of Bragg-like peaks.

An interesting physical interpretation of these remarks has been suggested by Prince (1986). Using the construction lines shown in Fig. 6.37b, the space for $|\eta| < w$ of Fig. 6.37a is divided onto two periodically repeating regimes, shaded and unshaded. We note that all projected points fall somewhere in the unshaded regimes. Thus, we may describe the quasicrystal lattice as an array of points displaced by a complex formula from periodically repeating lattice positions. This same general description characterizes the phenomena of thermal- and size-effect-displacements of atoms in periodic crystals. In these two cases also, the diffraction pattern consists of sharp δ-functions weighted by a function [of the type $\exp(-\beta \sin^2 \theta/\lambda^2)$] which contains information about the displacements (see Sects. 3.4.4 and 7.2).

While the mathematics become more complex, the cut and project technique described above may be applied to any space of dimension N, with projection onto any surface of dimension $D < N$. In particular, when $N = 6$ and $D = 3$, projection along the five-fold axis of the icosahedral point group (this point group *is* compatible with periodic translation in 6–D space) gives rise to 3–D diffraction patterns containing all the symmetry observed in the Al–Mn studies of Schechtman et al. (see Cahn et al. 1986b). Using these ideas the structure of the Al–Mn icosahedral phase has been clarified by Cahn et al. (1986b). Taking advantage of the relative differences in scattering amplitudes of Al and Mn when neutron and x-ray powder patterns were compared, these workers were able to establish the dimensions of the cut and projections in 6–D space (analogous to the dimension w in Fig. 6.37a) and hence obtain the clue to a structure model. The early, unrefined structure reveals Mn icosahedra rotated relative to one another by 180° about a common 3-fold axis and joined to one another along this axis with the three atoms on each of the two 3-fold planes forming an octahedral arrangement. This structure is not space filling, and the detailed arrangements of atoms in the interstices between octahedra, as well as the specific location of all Al atoms may require "single-quasicrystal" data.

The excitement generated by the discovery of icosahedral phases led to more than one hundred related publications in the first year following the publication by Shechtman et al. This field is developing rapidly with new experimental and theoretical developments appearing at every time. Other examples of icosahedral symmetry have been discovered, and two other quasicrystal point groups have been discovered: decagonal (Bendersky 1985) and duedecagonal (Nissen et al. 1985). In principle an infinite variety of such point groups are possible, and we must look to theory to understand why they are seen so infrequently in nature. As larger quasicrystals of these materials are obtained, property measurements will be made, perhaps leading to exciting new applications for these new solids. The reader is referred to the current literature for up-to-date information about this rapidly developing field.

References

Albinati, A., and Willis, B.T.M., Acta Crystallogr. *B28*, 371 (1982)
Bendersky, L., Phys. Rev. Lett. *55*, 1461 (1985)
Brigham, E.O., "The Fast Fourier Transform", Prentice Hall, Englewood Cliffs, New Jersey (1974)
Buerger, M.J., "Vector Space", Chaps. 10–12, Wiley, New York (1959)
Buerger, M.J., "Crystal Structure Analysis", Wiley, New York (1960)
Cahn, J.W., Shechtman, D., and Gratias, D., J. Mater. Res. *1*, 13 (1986a)
Cahn, J.W., and Gratias, D., J. de Phys., Colloque C3, 415 (1986b)
Champeney, D.C., "Fourier Transforms and Their Physical Applications", Academic Press, New York (1973)
Cheetham, A.K., and Taylor, J.C., J. Sol. St. Chem. *21*, 253 (1977)
Coppens, P., and Hall, M.B. (eds), "Electron Densities and the Chemical Bond", Plenum, NY (1982)
Corliss, L.M., Hastings, J.M., and Weiss, R.J., Phys. Rev. Lett. *46*, 1135 (1959)
Cowley, J.M., "Diffraction Physics", North-Holland/American Elsevier, New York (1975)
Cruikshank, D.W.J., Acta Crystallogr. *9*, 747 (1956)
Crystallographic Computing Committee of the American Crystallographic Association, 335 E. 45th St., New York, 10017
Dawson, B., in "Advances in Structure Research by Diffraction Methods" (W. Hoppe and R. Mason, eds.), Volume 6. Pergamon, New York (1975)
deWolff, P. in "Modulated Structure Materials" (ed. T. Tsakalakos) Martinus Nijhoff, Boston (1984)
Elser, V., Phys. Rev. *B32*, 4892 (1985)
Fairhurst, C. and Cohen, J.B., Acta Crystallogr. *B28*, 371 (1972)
Germain, G., and Woolfson, M.M., Acta Crystallogr. *B24*, 91 (1968)
Hamilton, W.C., Trans. Amer. Cryst. Assoc. *1*, 20 (1965)
Harburn, G., Taylor, C.A., and Welberry, T.R., "Atlas of Optical Transforms." Cornell Univ. Press, Ithaca, New York (1975)
Harker, D., and Kasper, J.S., Acta Crystallogr. *1*, 70 (1948)
Hauptman, H. and Karle, J., Am. Cryst. Assoc. Monograph, No. 3, Polycrystal Book Service, 1953.
Hendrickson, W.A., and Teeter, M.M., Nature, *240*, 107 (1981)
James, R.W., "The Optical Principles of the Diffraction of X-Rays," Chap. 7. London (1954)
Karle, J., in "International Journal of Quantum Chemistry: Quantum Biology Symposium" 7, 357 (1980)
Koch, F., and Cohen, J.B., Acta Crystallogr. *B25*, 275 (1969)
Lipson, H., and Taylor. C.A., "Fourier Transforms and X-Ray Diffraction." Bell, London (1958)
Moncton, D.E., Axe, J.D., and D. Salvo, F.J., Phys. Rev. Lett. *34*, 734 (1975)
Nissen, H.U., Ishimasa, T. and Schologle, R., Helvetia Phys. Acta *58*, 819 (1985)
Patterson, A.L., Phys. Rev. *46*, 372 (1934)
Pendry, J.B., "Low-Energy Electron Diffraction", Academic Press, New York (1974)
Prince, E., Acta Cryst., May, 1987.
Ramachandran, G.N., and Srinivasan, R., "Fourier Methods in Crystallography." Wiley (Interscience), New York (1970)
Sato, H., and Toth, R.S., in "Alloying Behavior and Effects in Concentrated Solid Solutions" (ed. T.B. Massalski), Gordon and Breach, New York (1965)
Sayre, D., Acta Crystallogr. *5*, 60 (1952)
Shechtman, D., Blech, I., Gratias, D., and Cahn, J.W., Phys. Rev. Lett. *53*, 1951 (1984)
Steeds, J.W., in "Quantitative Electron Microscopy" (eds. J.N. Chapman and A.J. Cohen) Universities Summer School in Physica, Glasgow (1983)
Stuart, R.D., "An Introduction to Fourier Analysis." Wiley, New York (1961)
Taylor, C.A., and Lipson, H., "Optical Transforms." Cornell Univ. Press, Ithaca, New York (1964)
Tsakalakos, T. (ed.), "Modulated Structure Materials," Martinus Nijhoff, Boston (1984)
Willis, B.T.M., J. Phys. Radium *25*, 431 (1964)
Willis, B.T.M., Nature *197*, 755 (1963a)
Willis, B.T.M., Proc. R. Soc. London *274*, 122 (1963b)

Willis, B.T.M., and Pryor, A.W., "Thermal Vibrations in Crystallography." Cambridge Univ. Press, Oxford (1975)
Wilson, A.J.C., Acta Crystallogr. 2, 318 (1949)
Woolfson, M.M., Acta Crystallogr. 7, 61 (1954)
Woolfson, M.M., "An Introduction to X-Ray Crystallography." Cambridge Univ. Press, Oxford (1970)
Yamamoto, A., Acta Crystallogr. B3, 9, 1451 (1982)
Zia, R.K.P. and Dallas, W.J., J. Phys. A18, L341 (1985)

Problems

6.1. Prove that Eq. (6.7) is the correct expression for the complex Fourier coefficients, C_m.

6.2. Find the first four sine and cosine coefficients of the Fourier series representing the periodic function $f(x)$, defined on the interval $|x| < a$ by

$$f(x) = \begin{cases} 1, & |x| < a/2 \\ -1, & a/2 < |x| < a \end{cases}.$$

Plot the sum of these terms in the range $|x| < a$.

6.3. Consider the periodic function $f(x)$ defined on the interval $|x| < a/2$ by

$$f(x) = \begin{cases} 1, & |x| < X/2 \\ 0, & X < |x| < a/2 \end{cases}.$$

Evaluate the coefficients of the Fourier series representing the function.

6.4. Prove the results stated in Eqs. (6.20a)–(6.20e). *Hint*: Let $cx \to x'$.

6.5. Take the Fourier transform of the one-dimensional Gaussian function $f(x) = A\exp(-B^2 x^2)$. Discuss the result.

6.6. Prove Eq. (6.32c), i.e., show that $C_n = x_0$ when a set of δ functions in s space is expanded as

$$\sum \delta(s - n/x_0) = \sum C_n \exp(+2\pi i n x_0 s).$$

6.7. Calculate the diffraction pattern from two very narrow slits spaced a distance a apart. *Hint*: Consider these slits as δ functions of unit weight.

6.8. Calculate the diffraction pattern of two slits of width b spaced a distance a apart. (*Hint*: Use the result of Problem 6.7 and think of the scattering density as a convolution of δ functions and a single slit of finite width b.)

6.9. Evaluate the Fourier transform of the two-dimension function $f(x, y)$ which is a constant ρ over the domain l, w, i.e.,

$$f(x, y) = \begin{cases} \rho, & |x| \leq 1/2, \quad |y| \leq w/2 \\ 0, & |x| > 1/2, \quad |y| > w/2 \end{cases}.$$

This Fourier transform is the diffraction pattern of a rectangular hole.

6.10. Find the diffraction pattern from a circular aperture

$$f(x, y) = \begin{cases} 1 & \text{for } (x^2 + y^2)^{1/2} < a/2 \\ 0 & \text{elsewhere.} \end{cases}$$

6.11. Find the irreducible areas (minimum area in reciprocal space) that must be sampled to evaluate the Fourier sums for plane groups $p2mmm$, $p4$, $p4mm$ and $P3$.

6.12. Develop the Patterson patterns for the plane groups $P2mm$, $P2mg$, and $P2gg$. Note the presence of double weight peaks which can be used to separate these groups.

6.13. A structure is found to contain two identical atoms per unit cell. The unit cell is orthorhombic, space group $Pmmm$.
(a) What is the only possible location of these two atoms?
(b) Indicate with the aid of formulas what Patterson section would enable you to obtain the actual coordinates.

6.14. Consider a monoclinic crystal. The diffraction symbol is $P \ldots 2_1$.
(a) What space groups are possible?
(b) For each possible space group show what the plane group is if the structure is *projected* successively along a_1, a_2, and a_3.
(c) Draw the Patterson pattern for each projection, and show how the two structures can be distinguished.
(d) Are there any Harker sections that could quickly distinguish the two space groups?

6.15. In Sect. 6.5 the technique of isomorphous replacement was presented for the case of noncentrosymmetric structures. Two ambiguous solutions were shown to result. This ambiguity may be resolved by:
(a) adding another atom in another place in the unit cell and repeating the measurements, looking for a common set of phase angles;
(b) using an atom that will scatter anomalously. Plots for hkl and $\overline{hkl}$ will yield unique answers.
Make drawings for cases (a) and (b) to show how the ambiguity in phase is resolved.

6.16. (a) Show that when a mirror is present perpendicular to a_2, the average value of square amplitudes

$$\overline{|F_{h0l}|^2} = 2\Sigma,$$

where Σ is defined in Eq. (6.65b). Use the fact that with a mirror present there are pairs of atoms with coordinates (x, y, z) and $(x, \bar{y}, z)$, and use the definition of F_{h0l}.
(b) Repeat the procedures of part (a) for a twofold axis along a_2 for which pairs of atoms occur at (x, y, z), and $(\bar{x}, y, \bar{z})$, and show that for this case

$$\overline{|F_{0k0}|^2} = 2\Sigma.$$

Problems

6.17. Show that when a twofold axis along $\mathbf{a}_2$ and a center of symmetry are combined, the following two relations both apply:
(a) $U_{hkl}^2 \leq \frac{1}{4}(1 + U_{0,2k,0})(1 + U_{2h,0,2l})$,
(b) $U_{hkl}^2 \leq \frac{1}{4}(1 + U_{0,2k,0} + U_{2h,0,2l} + U_{2h,2k,2l})$.

6.18. Consider polytetrafluoroethylene (Teflon):
$$\begin{array}{c} \text{F} \quad \text{F} \\ | \quad | \\ \text{C} - \text{C} \\ | \quad | \\ \text{F} \quad \text{F} \end{array}$$
is the monomer, the C—C distance in the backbone of the polymer is ≈ 1.54 Å; the C—F distance is 1.49 Å; the C—C angle is 116.0°; the molecules form a 13_6 helix, meaning 13 carbons for six turns of the helix.
(a) Work out the pitch and the diameter of the helix in angstrom. (Do not forget the fluorines!)
(b) The spacing of discrete side groups can be taken as the C—C spacing along the helix. Plot the Miller index of the layer spacing (from 0 to ± 15) versus the order n of Bessel function that satisfies the selection rule.
(c) Assuming that there is only a single helix and that the only important Bessel function on each layer is the lowest order, sketch the pattern. Assume Cu K_α radiation, and a specimen to film distance of 10 cm. Show enough layers to form a repeat. Give the intensity and the width of spots on all layers.

6.19. It sometimes happens that similar atoms can exchange positions from unit cell to unit cell, i.e., there is an "average" atom at each position. This can make solution of a structure more difficult, but in simple structures the "degree of order" can be easily evaluated and is a useful measure of the atomic arrangement. An alloy of 50 at. % A and 50 at. % B forms a simple CsCl structure at low temperatures, with A at the corners of the cubic cell and B at the centers. At higher temperatures, atoms interchange and some A atoms appear on centered positions (β sites) and some B's are in A positions (α sites). Let

$r_\alpha =$ the fraction of α sites occupied by A atoms (i.e., rightly occupied).
$w_\alpha =$ the fraction of α sites wrongly occupied (i.e., by B atoms).
$r_\beta =$ fraction of β sites occupied rightly by B atoms.
$w_\beta =$ fraction of β sites wrongly occupied by A atoms.

In this alloy, because of the equal compositions of A and B, $w_\alpha = w_\beta$, $w_\beta = 1 - r_\alpha$, $r_\alpha = r_\beta$, and $w_\alpha = 1 - r_\beta$. Define a long-range order parameter $S = a + br_\alpha$ where $S = 1$, when $r_\alpha = 1$, and $S = 0$, when $r_\alpha = x_A$ (the atomic fraction).
(a) Show that $S = 2r_\alpha - 1$, and $S = r_\alpha - w_\beta$.
(b) Derive an expression for the structure factor, using the terms r_α, r_β, w_α, w_β. Simplify this, if possible, with one of the definitions of S in (a).
(c) Discuss the diffraction pattern and how it will change with increasing temperature up to $S = 0$. Keep in mind that f_A and f_B are really $f_A \exp(-M_A)$ and $f_B \exp(-M_B)$. How could you actually obtain a value for S?

6.20. At one stage of precipitation in some alloys we can imagine the solute distributed sinusoidally with wave number q. Consider a one-dimensional lattice of points p apart and average density of scattering material ρ:

(a) Calculate the appearance of the diffraction pattern and sketch your result.
(b) How will the result be affected by the value of q? Sketch the result.
(c) Suppose the wave is finite, i.e., of length L. How will this length affect the result? That is, if the wave is not the length of the crystal but smaller, what effect will be produced?
(d) If there is only half a wavelength of the wave, how will the pattern be affected?
(e) If any extra diffraction effects are too close to the main peak to be detected, using Parseval's theorem, how much will the intensity of the fundamental peaks be affected?
(f) If the wave is not a pure cosine wave but say a square wave, how will the pattern be affected?
(g) If there are three waves in each of the three [[100]] directions in a cubic crystal, what will the pattern look like? (Besides being important in materials, this problem has many important analogies. Frequency and time form Fourier sets like real and reciprocal space. If s is frequency and r is time, this problem is one on amplitude modulation.)

6.21. The table below shows the 24 most intense peaks from a Patterson map, obtained by a computer search of a 3-D Patterson. The Patterson map is actually an "origin removed Patterson". That is, the F^2 were divided by $\sum f^2$ and unity was subtracted to minimize the origin (but not eliminate it). The limits of the Patterson are 0 to $\frac{1}{2}$ in x, 0 to $\frac{1}{2}$ in y, and 0 to $\frac{1}{2}$ in z. The compound is an iron porphyrin which is illustrated in Fig. P6.21. The space group is $C2/c$ with 4 molecules per unit cell. Each molecule contains one Fe and four nitrogens and the rest are carbons and hydrogens.
(a) What is the equipoint of the iron atoms?
(b) Derive the positions of the Fe atoms and the nitrogen atoms. (*Hint*: Fe–N distances is ~ 2.1 Å.)
(c) Identify the first ten peaks in the map — that is, what are the atom pairs involved in each peak.

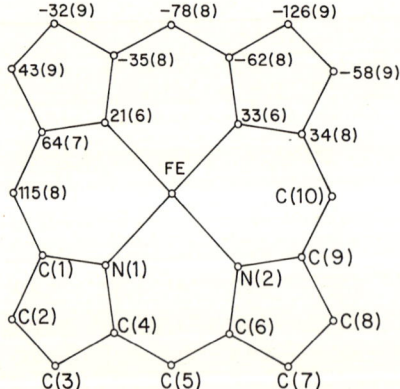

Fig. P6.21

Problems

Caution: The drawing is a *projection* down one axis — don't expect to use a ruler on the drawing to get accurate distances! The atoms are numbered. Other numbers are fractional displacements normal to the drawing in units of 0.001 Å. The uncertainty is in parentheses.

Table P6.21. List of peak positions and their heights (RHO). ED = Electron density (in e/Å^3 if input data are in electrons) = (2/vol)(RMAX/999)RHO

No.	Along crystal axes						Along orthogonal axes			
	z	y	x	RHO	ED		x(Å)	y(Å)	z(Å)	RMAX[a]
1	0.5040	0.2400	0.0000	999	(0.588)	Edge peak	0.000	4.662	9.291	10.395
2	0.5040	0.2550	0.5040	926	(0.545)	Edge peak	9.508	4.953	9.046	14.027
3	0.1631	0.0000	0.0097	601	(0.354)		0.182	0.000	3.002	3.008
4	0.4554	0.2618	0.1011	577	(0.340)		1.907	5.086	8.345	9.957
5	0.1083	0.4950	0.3862	568	(0.334)		7.285	9.615	1.809	12.198
6	0.2424	0.4950	0.4094	556	(0.327)		7.724	9.615	4.269	13.051
7	0.1072	0.0000	0.0452	500	(0.294)		0.853	0.000	1.954	2.132
8	0.3898	0.4950	0.3553	491	(0.289)		6.702	9.615	7.012	13.658
9	0.0000	0.0300	0.0000	480	(0.283)	Edge peak	0.000	0.583	0.000	0.583
10	0.1188	0.0000	0.1085	458	(0.270)		2.047	0.000	2.137	2.959
11	0.0000	0.4650	0.5040	450	(0.265)	Edge peak	9.508	9.032	−0.245	13.117
12	0.0424	0.4950	0.4002	432	(0.254)		7.550	9.615	0.587	12.239
13	0.3946	0.2387	0.4562	413	(0.243)		8.607	4.636	7.053	12.054
14	0.5040	0.2367	0.3358	410	(0.241)		6.335	4.598	9.127	12.025
15	0.2817	0.4384	0.3971	391	(0.230)		7.490	8.517	5.100	12.395
16	0.2798	0.4950	0.1082	376	(0.221)		2.041	9.615	5.106	11.076
17	0.1674	0.4950	0.4370	373	(0.219)		8.244	9.615	2.874	12.988
18	0.2302	0.0000	0.0573	366	(0.216)		1.080	0.000	4.215	4.351
19	0.1620	0.4950	0.5040	360	(0.212)	Edge peak	9.508	9.615	2.741	13.797
20	0.3279	0.2486	0.0638	346	(0.204)		1.204	4.829	6.013	7.806
21	0.3016	0.2355	0.3736	342	(0.201)		7.048	4.574	5.377	9.975
22	0.4466	0.2645	0.2200	326	(0.192)		4.150	5.138	8.126	10.471
23	0.0556	0.0000	0.1232	313	(0.184)		2.323	0.000	0.965	2.516
24	0.0366	0.0000	0.0604	312	(0.183)		1.140	0.000	0.645	1.310

[a] True interatomic distance from origin in Å

7. What Else Can We Learn from a Diffraction Experiment Besides the Average Structure?

7.1 Introduction

So far we have been centering our considerations on the characterization of crystal structures. This is an important aspect of diffraction studies and it has enabled us to learn most of the fundamental concepts. We have in fact come across a great many uses for diffraction in addition to structure determination as the following list indicates:

1. Chemical analysis (in selected regions with the electron microscope, as well as in larger areas with standard x-ray fluorescence techniques)
   (a) From the diffraction pattern from a substance. (Sect. 4.7)
   (b) From the lattice parameter (Sect. 4.7)
   (c) From absorption (Sect. 3.7)
2. Phase boundaries
   (a) From lattice parameters (Sect. 4.7)
   (b) From integrated intensities (Sect. 4.9)
   (c) From absorption (Sect. 3.7)
3. Volume fraction of a phase (Sect. 4.9)
4. Residual stresses (Sect. 4.7)
5. Theoretical densities (Sect. 4.7)
6. Coefficients of thermal expansion (Sect. 4.7)
7. Indexing microscopic features of deformation, precipitation (Sect. 1.4)
8. Texture (Sect. 4.7)

In the last few decades it has been found that perturbations in a periodic structure can also be detected and described in some detail from diffraction patterns, and in this chapter we shall look at this area. We shall be concerned mainly with a qualitative understanding, but in a few cases a detailed treatment will be given to indicate the methods that are actually employed to obtain quantitative information on the perturbations of the structure.

As we proceed, we shall discuss some of the applications of the electron microscope to this area, as this has become a powerful complimentary tool for the study of imperfections. At the outset, it is necessary to realize that although imperfections may play a primary role in the behavior of a material, they may not involve a large fraction of the atoms. Severely distorted crystals may contain 10^{16} dislocations/m^2 and even if the core of each dislocation contains ten atoms per plane, less than one

7.2 Thermal-Diffuse Scattering (TDS)

atom in ten is significantly displaced from the average lattice. The lower long-range strain fields from these defects may, however, affect all atoms, but only a small fraction of the atoms are strongly involved. This is sometimes bothersome in x-ray diffraction, because when we average over all the atoms under the beam, most of them are only affected in a minor way. On the other hand, the small range of effects from such an imperfection is the very reason we can see individual imperfections in the electron microscope. However, in the case of a concentrated close-packed solid solution, the local displacements of atoms from the average lattice due to the different size of the solute and solvent atoms will involve almost all the atoms. Diffraction is then the only tool for examining the effects in detail, and it will be difficult to see these displacements in a microscope. Finally, there are some materials that are not even crystalline and here as well, diffraction is the main tool for studying the atomic arrangements.

7.2.1 The Equations for TDS

There is one "imperfection" that involves all the atoms; they are all vibrating. This vibration can be thought of, at any instant, as due to a great many displacement waves (called phonons) traveling through the solid. The displacement of each atom is the sum total of the effects of all these waves.

We saw in Sect. 3.4 that thermal vibrations reduced the intensity of the diffraction peaks, but because there is conservation of energy, there must be other scattering somewhere. When we looked at this before, we considered each atom individually. This time let us look at the intensity from all the atoms in a solid. We shall again consider only one atom per lattice point and that all atoms are identical:

$$I_{eu}(\mathbf{s}) = \sum_n \sum_m f_n f_m \exp(2\pi i \mathbf{s}) \cdot (\mathbf{r}_n - \mathbf{r}_m) \, .^1$$

Let $\mathbf{r}_n = \mathbf{r}_n^0 + \Delta\mathbf{r}_n$, where $\mathbf{r}_n^0$ is a vector to a lattice point in the average structure, and $\Delta\mathbf{r}_n$ is the instantaneous thermal displacement. Then,

$$I_{eu}(\mathbf{s}) = \sum_n \sum_m f_n f_m \exp(2\pi i \mathbf{s}) \cdot (\mathbf{r}_n^0 - \mathbf{r}_m^0) \exp(2\pi i \mathbf{s}) \cdot (\Delta\mathbf{r}_n - \Delta\mathbf{r}_m) \, .$$

The displacements $\Delta\mathbf{r}_n$ vary with time and are in different stages of this variation at different places in the specimen. It takes a finite time to make a measurement of the intensity and our incident beam exposes a considerable number of atoms.

---

[1] We write the intensity in electron units, i.e., scattering per electron. The actual intensity is this times $(I_0 e^4 \eta^2 \; wh/m^2 c^4 R^2)$ $([1 + \cos^2 2\theta]/2)$ $(A_0/2\mu)$ for unpolarized (filtered) radiation and a flat specimen with an incident beam of intensity I_0, area A_0, and receiving slits of width w and height h.

Accordingly, we shall average the intensity over space and time, and indicate this with carats, $\langle \ \rangle$,

$$\langle I_{eu}(\mathbf{s})\rangle = \sum_n \sum_m f_n f_m \exp(2\pi i \mathbf{s}) \cdot (\mathbf{r}_n^0 - \mathbf{r}_m^0) \langle \exp(2\pi i \mathbf{s}) \cdot (\Delta \mathbf{r}_n - \Delta \mathbf{r}_m) \rangle . \quad (7.1\text{a})$$

We shall assume that the displacements from the average structure are small and expand the averaged term:

$$\langle \exp(2\pi i \mathbf{s}) \cdot (\Delta \mathbf{r}_n - \Delta \mathbf{r}_m) \rangle = 1 + 2\pi i \langle \mathbf{s} \cdot (\Delta \mathbf{r}_n - \Delta \mathbf{r}_m) \rangle$$
$$- 2\pi^2 \langle \{\mathbf{s} \cdot (\Delta \mathbf{r}_n - \Delta \mathbf{r}_m)\}^2 \rangle + \ldots . \quad (7.1\text{b})$$

The term in the expansion linear in $\mathbf{s} \cdot (\Delta \mathbf{r}_n - \Delta \mathbf{r}_m)$ vanishes because there *is* an average lattice. We shall retain only the other two terms. This cannot be precisely correct or there would be no thermal expansion of a crystal, but it will suffice for our considerations of this phenomenon. This is referred to as the harmonic approximation.

For the last term on the right of Eq. (7.1b), the cross product terms $\Delta \mathbf{r}_m \Delta \mathbf{r}_n$ average to zero *only* if the vibrations of the atoms are independent. Because the atoms are "bound" together and because the displacements are of the order of a few per cent of the interatomic distances, this is not really to be expected. It was our assumption in Sect. 3.4, but we shall not use it here.

Let us consider the atomic displacements as the result of many plane waves, transverse and longitudinal, with wave vectors $\mathbf{k}$ and frequencies $\nu_\mathbf{k}$:

$$\Delta \mathbf{r}_n = \sum_\mathbf{k} \mathbf{A}_\mathbf{k} \cos[2\pi \nu_\mathbf{k} t - (\mathbf{k} \cdot \mathbf{r}_n^0) - \delta_\mathbf{k}] . \quad (7.2)$$

The symbol $\mathbf{A}_\mathbf{k}$ represents the maximum amplitude and is a vector, as the wave can represent a longitudinal displacement (parallel) to $\mathbf{k}$ or one that is transverse, that is, perpendicular to $\mathbf{k}$. These waves are illustrated in Fig. 7.1. It is to be emphasized that $\Delta \mathbf{r}_n$ varies with time. The term involving $\mathbf{k} \cdot \mathbf{r}_n^0$ represents the displacement of the atom at $\mathbf{r}_n$ along the direction of propagation of the wave, represented by the vector $\mathbf{k}$ of magnitude $(2\pi/\lambda_\mathbf{k})$. The other part of the phase angle $\delta_\mathbf{k}$ changes often during our observation; the waves may start and damp out rapidly during this time. Each such plane wave is the mathematical representation of a phonon of wavelength $\lambda_\mathbf{k}$. We can add such waves without considering their interaction, if we assume that the atoms are moving according to a linear differential equation, and that there is

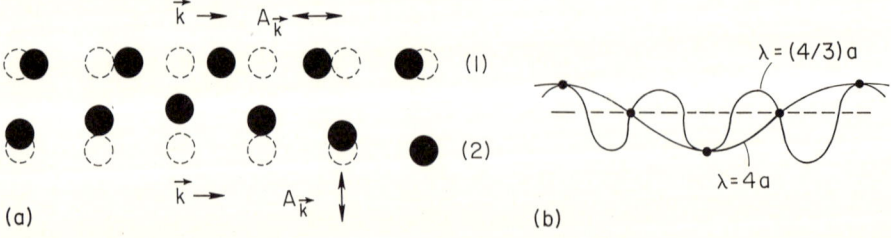

Fig. 7.1. a (1) A longitudinal wave. Positions in the average structure are shown *dotted*. (2) A transverse wave. **b** The same atomic displacement represented by two equivalent waves

7.2 Thermal-Diffuse Scattering (TDS)

a restoring force proportional to displacement, as we saw for electrons in Sect. 3.4 when the scattering factor was discussed.

Using Eq. (7.2) for the last term on the right in Eq. (7.1b), squaring and then taking the time average, cross products in the sums for different waves $\mathbf{k}$, $\mathbf{k}'$ vanish as the phases of the waves are independent, and only the squares of terms with the same $\mathbf{k}$ for the nth and mth atom remain:

$$1 - 2\pi^2 \left\langle \sum_{\mathbf{k}} \{\mathbf{s} \cdot \mathbf{A}_{\mathbf{k}}[\cos(2\pi v_{\mathbf{k}} t - \mathbf{k} \cdot \mathbf{r}_n^0 - \delta_{\mathbf{k}}) - \cos(2\pi v_{\mathbf{k}} t - \mathbf{k} \cdot \mathbf{r}_m^0 - \delta_{\mathbf{k}})]\}^2 \right\rangle.$$

The two cosine-squared terms average with time to one-half, and

$$-\langle 2\cos(2\pi v_{\mathbf{k}} t - \mathbf{k} \cdot \mathbf{r}_n^0 - \delta_{\mathbf{k}}) \cos(2\pi v_{\mathbf{k}} t - \mathbf{k} \cdot \mathbf{r}_m^0 - \delta_{\mathbf{k}}) \rangle$$
$$= -2\langle \cos A \cos B \rangle = -\langle \{\cos(A+B) + \cos(A-B)\} \rangle.$$

The term $\cos(A+B)$ averages to zero over the time of measurement as atoms vibrate at 10^{13} Hz(cps), so that we are left only with the term in $(A-B)$. Thus,

$$1 - 2\pi^2 \langle \{\mathbf{s} \cdot (\Delta \mathbf{r}_n - \Delta \mathbf{r}_m)\}^2 \rangle$$

$$= 1 - 2\pi^2 \left\langle \sum_{\mathbf{k}} (\mathbf{s} \cdot \mathbf{A}_{\mathbf{k}})^2 \{1 - \cos[\mathbf{k} \cdot (\mathbf{r}_m^0 - \mathbf{r}_n^0)]\} \right\rangle$$

$$\cong \exp -\left\{2\pi^2 \left\langle \sum_{\mathbf{k}} (\mathbf{s} \cdot \mathbf{A}_{\mathbf{k}})^2 (1 - \cos[\mathbf{k} \cdot (\mathbf{r}_m^0 - \mathbf{r}_n^0)]) \right\rangle \right\}$$

$$= \exp(-2M) \exp \left\langle 2\pi^2 \sum_{\mathbf{k}} (\mathbf{s} \cdot \mathbf{A}_{\mathbf{k}})^2 \cos[\mathbf{k} \cdot (\mathbf{r}_m^0 - \mathbf{r}_n^0)] \right\rangle.$$

Putting this back in Eq. (7.1a):

$$\langle I_{eu}(\mathbf{s}) \rangle = \sum_n \sum_m f_m f_n \exp(-2M) \exp[2\pi i \mathbf{s} \cdot (\mathbf{r}_n^0 - \mathbf{r}_m^0)]$$

$$\times \exp \left\langle 2\pi^2 \sum_{\mathbf{k}} (\mathbf{s} \cdot \mathbf{A}_{\mathbf{k}})^2 \cos[\mathbf{k} \cdot (\mathbf{r}_m^0 - \mathbf{r}_n^0)] \right\rangle.$$

Expanding the last exponent and retaining only the first two terms,[2]

$$\langle I_{eu}(\mathbf{s}) \rangle = \sum_n \sum_m f_m f_n \exp(-2M) \exp[2\pi i \mathbf{s} \cdot (\mathbf{r}_n^0 - \mathbf{r}_m^0)]$$
$$+ \sum_n \sum_m f_m f_n \exp(-2M) \exp[2\pi i \mathbf{s} \cdot (\mathbf{r}_n^0 - \mathbf{r}_m^0)]$$
$$\times \left\langle 2\pi^2 \sum_{\mathbf{k}} (\mathbf{s} \cdot \mathbf{A}_{\mathbf{k}})^2 \cos[\mathbf{k} \cdot (\mathbf{r}_m^0 - \mathbf{r}_n^0)] \right\rangle. \quad (7.3)$$

The first term in Eq. (7.3) is just the Bragg scattering which peaks at $\mathbf{s} = \mathbf{r}_{hkl}^*$ depressed by $\exp(-2M)$. For the second term, each cosine can be written as

---

[2] $2M = 2\pi^2 \langle \sum_{\mathbf{k}} (\mathbf{s} \cdot \mathbf{A}_{\mathbf{k}})^2 \rangle$ is equivalent to $(16\pi^2 \sin^2\theta/\lambda^2)\langle \Delta x^2 \rangle$ as in Sect. 3.4. This can be seen if it is recalled that $|\mathbf{s}| = 2\sin\theta/\lambda$ and $\langle \Delta x^2 \rangle$ is the mean-square displacement along $\mathbf{s}$ due to a wave, which is half the mean-square amplitude of the wave. (This can easily be proven by averaging the value of $\cos^2\alpha$ (where α is the angle between $\mathbf{s}$, $\Delta \mathbf{r}_n$) for all possible values of α.)

$$\tfrac{1}{2}\{\exp[i\mathbf{k}\cdot(\mathbf{r}_m^0 - \mathbf{r}_n^0)] + \exp[-i\mathbf{k}\cdot(\mathbf{r}_m^0 - \mathbf{r}_n^0)]\} \ .$$

Thus, the second term in Eq. (7.3) has terms $\exp 2\pi i(\mathbf{s} \pm \mathbf{k}/2\pi)\cdot(\mathbf{r}_m^0 - \mathbf{r}_n^0)$ and will have peaks for $\mathbf{s} + \mathbf{k}/2\pi = \mathbf{s} + \mathbf{n}/\lambda_\mathbf{k} = \mathbf{r}_{hkl}^*$, $\mathbf{s} - \mathbf{n}/\lambda_\mathbf{k} = \mathbf{r}_{hkl}^*$ (where $\mathbf{n}$ is a unit vector in the direction of $\mathbf{k}$). There will be a set of satellites for each phonon of wavelength $\lambda_\mathbf{k}$. As $\mathbf{r}_{hkl}^*$ is a vector of the reciprocal lattice, this implies that there are peaks *not* when $\mathbf{s} = \mathbf{r}_{hkl}^*$, but rather when $\mathbf{s}$ is *displaced* from $\mathbf{r}_{hkl}^*$ by $\mathbf{n}/\lambda_\mathbf{k}$.

There is a *finite number of waves* in the sum over $\mathbf{k}$. *The shortest wavelength is twice the interatomic distance* because, as shown in Fig. 7.1b for a transverse wave, any shorter wavelength is equally well represented by a longer one. Thus the maximum $\mathbf{k}$ vector is halfway between diffraction spots in reciprocal space. Also, all such waves must have nodal points at the surface of the solid; because of this nodal requirement the number of waves is limited or quantized. Treatments of waves in solids indicate that in any direction there is one (nearly) longitudinal and two (nearly) transverse waves for each k (Warren, 1969). If we set the number as p along the $\mathbf{a}_1$ direction, and there are N_1 cells in this direction, then $N_1 a_1/p$ represents the possible waves, with p ranging from zero to $\pm N_1/2$. In all three directions then, there are $N_1 N_2 N_3 = N$ waves. Taking into account the longitudinal and transverse modes, there are $3N$ waves. There are a great many discrete waves of different wavelengths, but a minimum wavelength twice the interatomic distance, and a maximum wavelength.

If we differentiate Eq. (7.2) to get the velocity $[v = d(\Delta \mathbf{r}_n)/dt]$ of an atom during vibration, we can see that the average kinetic energy $\tfrac{1}{2}mv^2$ will be proportional to the sum of the frequencies (squared), and also that the amplitude of each satellite will be inversely proportional to the satellite's frequency. For the kinetic energy $\langle E_{\text{kin}} \rangle$, we can write

$$\sum \langle E_{\text{kin}} \rangle = \frac{1}{2}\left\langle \left(\sum_n m\left[\frac{d(\Delta \mathbf{r}_n)}{dt}\right]^2 \right).\right\rangle \tag{7.4a}$$

With the aid of Eq. (7.2),

$$\sum \langle E_{\text{kin}} \rangle = \tfrac{1}{4}Nm \sum_\mathbf{k} 4\pi^2 v_\mathbf{k}^2 (\mathbf{A}_\mathbf{k})^2 \ . \tag{7.4b}$$

The total (kinetic plus potential) energy which is $k_B T$ at high temperatures, for each "mode", that is for each phonon of wave-vector $\mathbf{k}$, is simply twice the value from Eq. (7.4b). We can see that $\langle (\mathbf{A}_\mathbf{k})^2 \rangle$ will be inversely proportional to $v_\mathbf{k}^2$

$$\tfrac{1}{2}Nm\, 4\pi^2 v_\mathbf{k} \mathbf{A}_\mathbf{k}^2 = k_B T \tag{7.4c}$$

or:

$$\mathbf{A}_\mathbf{k}^2 = 2k_B T/Nm\omega_\mathbf{k}^2 \ . \tag{7.4d}$$

(More accurate expressions for the energy of a vibrational mode at any temperature are available, but this will suffice to illustrate the main points.) The frequency is largest for the satellites farthest from the Bragg peak (of smallest wavelength) and hence, these are the weakest in intensity. This fact can be readily understood if we realize that there is a coupling of the motion of atoms. Such coupling is obviously

7.2 Thermal-Diffuse Scattering (TDS)

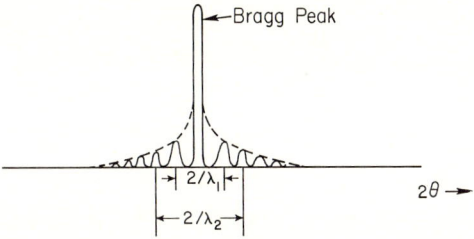

Fig. 7.2. The envelope of satellites under a Bragg peak due to thermal vibrations

strongest between near-neighbors; it is as if a small block of crystal was vibrating as a unit in some way. We expect the amplitude of diffraction at any instant from such a unit to be the Fourier transform of the product of the shape of the block and the average structure. The transform will then be a convolution of the transform of the structure and the transform of the block. We have seen in Sect. 6.5 that the transform of the block will have the form of a damped sine wave ($\sin kx/kx$). This function has its largest value for small k, i.e., for the intensity closest to Bragg peaks. (More detailed treatments of this dependence of amplitude on frequency can be found in almost any text on solid state physics.)

An "envelope" of satellites then appears under each Bragg peak, as shown in Fig. 7.2, increasing in intensity with increasing diffraction angle due to the s dependence [see Eq. (7.3)] and with temperature, since the amplitude, A, increases with T.

7.2.2 Measuring TDS

At each point in reciprocal space for this temperature-diffuse scattering (TDS), we know the k value, from the intensity we can get frequency, and from the frequency and wavelength we can get the velocity of an elastic wave at frequencies up to 10^{13} Hz — a much higher frequency than is possible with any mechanical device.

Rewriting Eq. (7.3):

$$\langle I_{eu}(s) \rangle = \tfrac{1}{2} \sum_n \sum_m f_m f_n \exp(-2M) 2\pi \sum_k (\mathbf{s} \cdot \mathbf{A_k})^2 \left[\exp 2\pi i (\mathbf{s} + \mathbf{k}/2\pi) \cdot (\mathbf{r}_m^0 - \mathbf{r}_n^0) \right.$$
$$\left. + \exp 2\pi i (\mathbf{s} - \mathbf{k}/2\pi) \cdot (\mathbf{r}_m^0 - \mathbf{r}_n^0) \right].$$

Now, let $\mathbf{A_k} = A_{kj} \mathbf{e}_{kj}$, where j represents the type of wave (transverse or longitudinal) and $\mathbf{e}$ is a unit vector in the direction of vibration. Therefore, $\sum_k \to \sum_k \sum_j$. Next, we replace the sum over k by an integral over reciprocal space, $b^3 dh_1 dh_2 dh_3$, because of the high density of phonons, N/b^3. Thus:

$$\langle I_{eu}(s) \rangle = \tfrac{1}{2} f^2 \exp(-2M) 2\pi^2 \sum_j (\mathbf{s} \cdot \mathbf{A}_{kj} \mathbf{e}_{kj})^2 \int\int\int \left(\sum_m \sum_n [\exp 2\pi i (\mathbf{s} + \mathbf{k}/2\pi) \right.$$
$$\left. \cdot (\mathbf{r}_m^0 - \mathbf{r}_n^0) + \exp 2\pi i (\mathbf{s} - \mathbf{k}/2\pi) \cdot (\mathbf{r}_m^0 - \mathbf{r}_n^0)] \frac{N}{b^3} b^3 dh_1 dh_2 dh_3 \right).$$

The integral limits are: $h_i - \tfrac{1}{2}$ to $h_i + \tfrac{1}{2}$, to include all the waves. As we saw in Chap. 2, the integral of a sum of such plane waves is N, and so:

$$\langle I_{\text{eu}}(\mathbf{s})\rangle = \tfrac{1}{2}f^2 \exp(-2M) 2\pi^2 \sum_j (\mathbf{s}\cdot\mathbf{A}_{kj}\mathbf{e}_{kj})^2 (N+N)N\ .$$

With Eq. (7.4c):

$$\langle I_{\text{eu}}(\mathbf{s})\rangle = f^2 \exp(-2M) \frac{Nk_B T}{M\omega_{kj}^2} 4\pi^2 \sum_j (\mathbf{s}\cdot\mathbf{e}_{kj})^2$$

$$= f^2 \exp(-2M) \frac{Nk_B T}{M\omega_{kj}^2} \sum_j s^2 \cos^2\alpha_j\ , \tag{7.5}$$

where α is the angle between $\mathbf{s}$ and the vibration direction. We can explore the entire vibrational spectrum to compare with theory. In particular, it is possible to calculate the lattice specific heat, Debye temperature, and interatomic force constants from such data. More complete discussion can be found in James (1950), Warren (1969), Bacon (1962), and Willis and Pryor (1975).

How do we know whether we are examining a longitudinal or transverse mode? Recall that the intensity associated with thermal vibrations is proportional to $(\mathbf{s}\cdot\mathbf{A}_k)$, Eq. (7.3). By properly choosing the position of the scattering vector in reciprocal space, we can choose whether to sample a longitudinal or transverse mode. For example, for the direction shown in Fig. 7.3c, the dot product is zero for the transverse waves with vibration directions perpendicular to $(\mathbf{S}-\mathbf{S}_0)/\lambda$ and with vibration direction in the plane of the drawing, or perpendicular to it. But the dot product is large for longitudinal modes in this direction. Thus, in Eq. (7.5), there is only one term in the sum over j.

If the integrated intensity of a Bragg peak is to be compared to a calculation to better than 5–10%, a correction is required in the experimental data for this envelope of diffuse scattering under each peak (Walker and Chipman 1970); if the peaks are at very large values of $|\mathbf{s}|$, then the error in neglecting this can be large. This correction involves some knowledge of the Debye temperature or elastic constants, and some approximate estimate of these will be needed (see Helmholdt and Vos 1977).

Actually, we might expect some energy exchange between the waves of vibrations and the photons of an x-ray beam. However, the energy of an x-ray photon is several kiloelectron volts (keV), while phonon energies ($k_B T$) are the order of 0.025 eV at room temperature. The change in energy (and hence wavelength) is too slight to detect with the energy resolution of present detectors. If a perfect crystal (with its narrow reflection range) is employed to energy-analyze the diffracted beam, then the resolution is adequate, especially if a high-angle reflection of the analyzer is employed, because the width of a Bragg peak decreases with order. The intensity of the elastic scattering is too low to do this with normal sources, but it is possible with synchrotron sources.

Due to their large mass, neutrons in a "thermalized" neutron beam are of the same magnitude of energy as the phonons. The energy exchange is then easily detected as a change in the energy of the scattered neutrons either by time-of-flight analysis as discussed in Sect. 4.7, or by placing a third crystal after the monochromator and specimen crystal to analyze the wavelength of the scattered neutrons. A typical experimental arrangement in reciprocal space is shown in Fig. 7.3a. We can

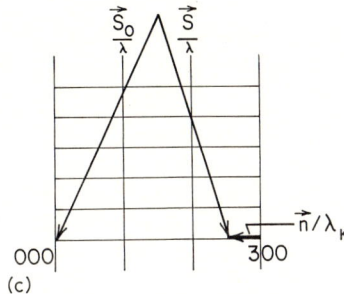

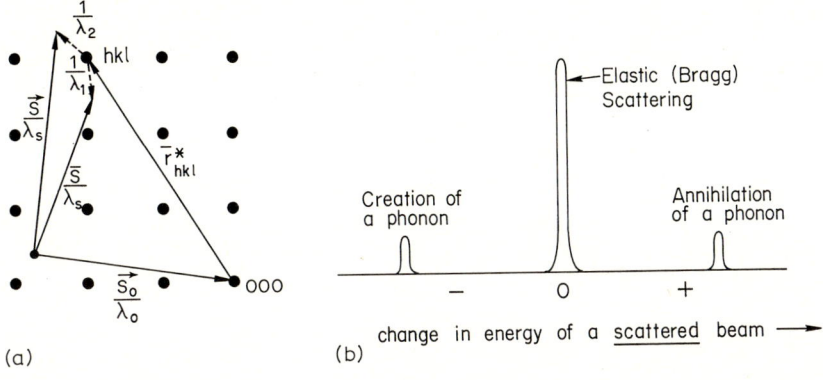

Fig. 7.3. **a** An inelastic scattering experiment represented in reciprocal space. For the wave represented by $1/\lambda_1$ energy is gained by the crystal (phonons are created) and λ_1 is the wavelength of the energy given up to the crystal vibrations, whereas for $1/\lambda_2$, energy is lost from the crystal; λ_2 the wavelength of the "annihilated" phonon. (After Bacon, G.E., "Neutron Diffraction", 2nd edition, Oxford University Press, London and New York, 1962.) **b** Energy analysis of a scattered neutron beam. **c** Orientation for examining [100] longitudinal vibration modes only

write the conservation of energy and momentum as follows. From the equation for a wave's momentum $(h/2\pi)\mathbf{k}$, and dividing it by $h/2\pi$, conservation of momentum requires that:

$$(\mathbf{S}/\lambda_s) - (\mathbf{S}_0/\lambda_0) = \mathbf{r}^*_{hkl} + (\mathbf{n}/\lambda_\mathbf{k}),$$

where n is a unit vector in the direction of propagation of $\mathbf{k}$, λ_s is the wavelength of scattered radiation, λ_0 is the incident wavelength, and $\lambda_\mathbf{k}$ is the wavelength of this phonon. Now $\lambda_\mathbf{k}$ is fixed by the position in reciprocal space. From energy conservation, the frequency is obtained:

$$(h^2/2m\lambda_0^2) - (h^2/2m\lambda_s^2) = 2\pi h v_k .$$

The scattered wavelength may be shorter than λ_0 if energy is taken from the crystal vibrations (phonon of λ_2 in Fig. 7.3a annihilated), or longer if phonons are created (phonon of λ_1 in Fig. 7.3a).

An experiment consists of choosing a position in reciprocal space representing a possible phonon wavelength, and examining the intensity versus energy or wavelength of the scattered neutrons at each such position. Alternatively, one may fix the energy by a setting of the analyzer crystal, and vary the position in reciprocal space, there will be an increase in intensity for those energies which differ from the initial beam's energy by the energy of a possible phonon, as in Fig. 7.3b. This peaking versus energy is due to the fact that there are discrete vibrational modes; only certain phonons can exist, as was already discussed.

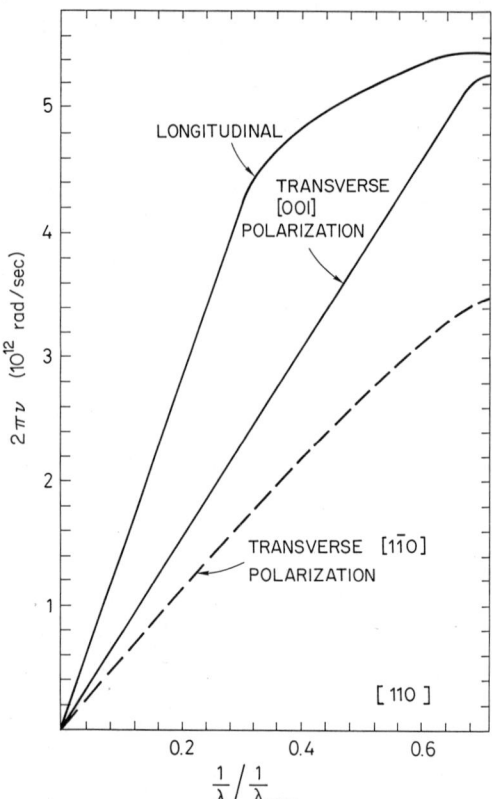

Fig. 7.4. Frequency vs. wave number for Cu_3Au at 683°K. (*Dotted line* is estimated.) There is a linear portion for long wavelengths, (whose slope is the elastic constant) but quite pronounced curvature at short wavelengths. This "dispersion curve" is shown for two transverse and one longitudinal wave. These are the only modes in simple directions for a pure material or random solid solution with one (average) atom per primitive unit cell

We know that for a sound wave (that is, a vibration of long wavelength), the velocity $v = (C/\rho_m)^{1/2}$ where C is some elastic constant and ρ_m is the mass density. Also $v = \nu\lambda = \nu/(1/\lambda)$. Thus constant velocity implies that if we plot ν (or $2\pi\nu = \omega$) vs. $1/\lambda$ (or $2\pi/\lambda = k$) this will be a straight line; the velocity is a constant, so that if $1/\lambda$ increases, ν increases.

Typical results, however, are as shown in Fig. 7.4. There is a linear portion for long wavelengths (whose slope is the elastic constant) but quite pronounced curvature at short wavelengths. This "dispersion curve" is shown for two transverse and one longitudinal wave. These are the only modes in simple directions for a pure material or random solid solution with one (average) atom per primitive unit cell. (Recall that for an fcc cell there is an equivalent primitive cell, Problem 2.8.) For compounds or elements with p atoms per cell, there are these three "acoustical" branches as in Fig. 7.4 (so called because their initial slopes are elastic constants) for which the frequency generally increases toward λ_{max}, and for which nearest atoms move in the same direction at long wavelength. But in addition there are $3p - 3$ "optical" branches. These have a different shape, being higher in frequency than the acoustical modes everywhere and not dropping to the origin. They represent vibrations in which neighboring atoms move in opposite directions. Because they are much higher in frequency than the acoustic modes, they contribute little to the intensity near Bragg peaks, although they can be detected by the neutron inelastic methods. (For such a material, the atomic scattering factor is replaced by the structure factor in all the preceeding equations.)

7.2.3 The Uses of TDS

The elastic constants are an important piece of information that can be obtained from these studies. (While this may seem to be difficult to do, other methods may be as difficult at high temperatures, or if the crystal is very soft or small.) For further details on this subject, see James (1950), Warren (1969), Bacon (1962), Dawson (1975), Maradudin et al. (1963), and Willis and Pryor (1975).

There are a number of phase transitions that have recently been found to be related to features of the phonon spectrum and this has increased the interest in these studies in the last few years to see just how general is this relationship. In particular, *certain vibrational modes may somehow be "frozen" in or condense, leaving static displacements* that are apropriate to the new phase. We shall illustrate this with one particular example. Alloys of Ti and Zr with Nb undergo a first-order transition from a bcc solid solution to a rhombohedral structure known as the "omega phase." A projection of the bcc structure along a $\langle 111 \rangle$ direction is shown in Fig. 7.5a. During the transformation, the atoms A and B move closer together (B moves into the plane of the drawing, A moves out of it). These motions are shown in Fig. 7.5b. Such movements can be initiated with a $\frac{2}{3} \langle 111 \rangle$ longitudinal phonon and, in the dispersion curves of niobium, there is a minimum near this position for the longitudinal acoustic mode, i.e., a very low frequency or long lifetime rather than a smooth increase in v with $1/\lambda$. It is now envisaged that through thermal activation and interactions between waves (anharmonic interactions), regions of this omega phase similar to those sketched in Fig. 7.5c are stabilized and grow. In other systems a phonon gradually "condenses." The frequency "softens", tending toward zero with decreasing temperature, and hence the amplitude increases [Fig. (7.5)], resulting in a static wave whose displacements produce the new phase over a large region of the crystal. It is not yet known if many transitions are related to these "soft modes."

In a close-packed structure, like a metallic-alloy solid solution, the solute *atoms that replace the solvent may be of a different size*. This *causes local static displacements, quite analogous to the dynamic movements* we have been discussing. The displacement $\Delta \mathbf{r}_n$ is independent of time, but will depend on the local environment, i.e., on the number and type of atoms around it. These *static displacements will give rise to effects analogous to vibrations* — reductions in Bragg peak intensities and diffuse scattering under the Bragg peaks. From these, information can be obtained on the atomic volumes of the species in an alloy to compare to the pure elements (see Warren 1969; Averbach 1955; Guinier 1959; Sect. 7.6). Because these are static effects, there can be an average displacement unlike the case of thermal vibrations in the harmonic approximation we have discussed. This leads to additional scattering in the diffraction pattern, which will be discussed in Sect. 7.6.

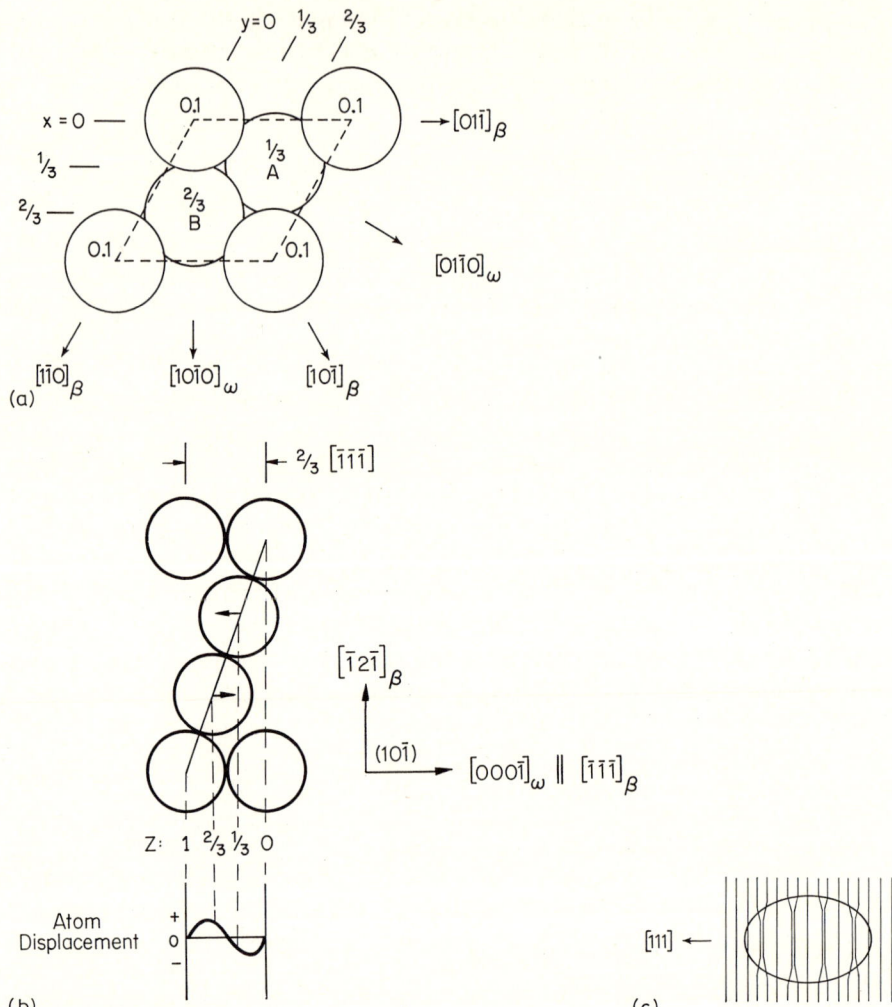

Fig. 7.5. a The bcc structure looking along a $[\bar{1}\bar{1}1]$ direction. The z coordinates normal to the page are indicated by the numbers within the atoms. **b** $(10\bar{1})$ section through **a**. A $\frac{2}{3}[111]$ longitudinal displacement wave causes the displacements indicated by arrows. [After Sass, S.L., J. Less Common Metals *28*, 157 (1972). Reprinted with permission of Elsevier-Sequoia S.A.] **c** A particle of the ω phase. [Reprinted with permission from deFontaine, D., "Phase Transitions" (H.K. Henisch and R. Roy, eds.), p. 169, Copyright 1973, Pergamon, New York]

Finally we can gain some insight into the difference between x-ray and neutron experiments by considering the convolutions described in Sect. 6.4. Here we assume that the electron density depends on time (t) as well as position (**r**) in a solid. Hence the Patterson function may be written as a two-dimensional convolution:

$$P(\mathbf{R}, T) = \int\int \rho(\mathbf{r}, t)\rho(\mathbf{r} + \mathbf{R}, t + T)\, d\mathbf{r}\, dt \ . \tag{7.6a}$$

7.2 Thermal-Diffuse Scattering (TDS)

The square of the structure factor is again the Fourier transform of this Patterson function, but it now depends on the frequency exchange (v) between the beam and phonons as well as $(\mathbf{S} - \mathbf{S}_0)/\lambda = \mathbf{s}$:

$$F^2(\mathbf{s}, v) = \int\int P(\mathbf{R}, T)\exp[(2\pi i(\mathbf{s}\cdot\mathbf{R} + vt)]\,d\mathbf{R}\,dT. \qquad (7.6b)$$

In an x-ray experiment, if we cannot separate the beams that have lost energy, we measure

$$\int F^2(\mathbf{s}, v)\,dv = \int\int\int P(\mathbf{R}, T)\exp(2\pi i\mathbf{s}\cdot\mathbf{R})\exp(2\pi ivt)\,d\mathbf{R}\,dT\,dv\,.$$

Now, from our discussion on delta functions in Sect. 6.2,

$$\int \exp(2\pi ivt)\,dv = \delta(t)$$

and hence

$$\int F^2(\mathbf{s}, v)\,dv = \int P(\mathbf{R}, 0)\exp(2\pi i\mathbf{s}\cdot\mathbf{R})\,d\mathbf{R}\,. \qquad (7.6c)$$

With x-rays we are measuring a section of the Patterson at time $t = 0$. (Here again we see the relationship mentioned in Sect. 6.3 that a projection in measuring space is a section in real space.)

With neutron scattering, or with a synchrotron source and energy analysis of the diffracted beam, we can separate the elastic scattering:

$$F^2(\mathbf{s}, 0) = \int\left[\int P(\mathbf{R}, T)\,dT\right]\exp(2\pi i\mathbf{s}\cdot\mathbf{R})\,d\mathbf{R}\,. \qquad (7.6d)$$

Our section at zero frequency change in measuring space is a time average of the Patterson function. For further discussion of this matter see Cowley (1975).

We close by pointing out that we have included only two terms in Eq. (7.1b). Especially at high temperatures, higher terms may be important. There can be "higher" order TDS, from the higher order terms in this equation. Figure 7.6 shows how two phonons ("second order TDS") can contribute significantly to the intensity. See Reid (1985) for further details on this. Also, there can be anharmonic effects that arise due to phonon creation and annihilation, such as two phonons com-

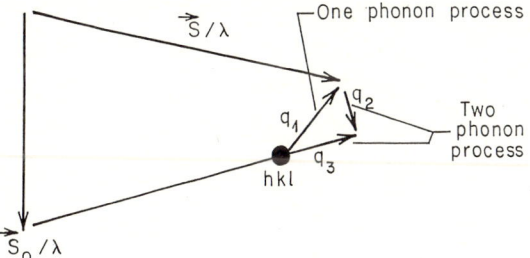

Fig. 7.6. The TDS intensity at the tip of $\mathbf{s}/\lambda$ is due to one phonon processes (q_1) and higher order phonon processes, such as the two-phonon processes shown (q_2, q_3). (This is referred to as "second-order TDS")

bining to form another (see Pirie et al. 1941). Anharmonicity can contribute 3–20 pct as much as harmonic processes, the higher values being along high symmetry directions.

7.3 Distortion and Mosaic Size

7.3.1 Introduction

A great many interesting effects fall into this category. Except in rare cases, most crystals contain defects; line defects such as dislocations, faults in the stacking sequence (planar stacking faults), and point defects such as vacancies and interstitials. Dislocations grouped in some array will produce a mosaic structure, i.e., regions of a crystal tilted slightly with respect to another region. Each region is known as a "subgrain" or "cell." A similar effect is produced by a planar fault. Dislocations produce distortions even at large distances from the defect itself. So also do substitutional atoms of a different size than the host atoms, or clusters of solute, or interstitials. Around a vacant site or "vacancy," the structure may collapse slightly.

We shall develop *expressions for the diffracted intensity that include averaging over the sample irradiated; but with the electron microscope, we would be interested only in the effects near one defect* and the averages are not necessary (on the other hand, it is harder to be quantitative with the microscope because of multiple scattering.) Let us first consider the effects seen in the microscope and some general features of the intensity. Then we shall undertake a more detailed treatment for the averaged intensity. Recall from Sect. 3.3 that the electron diffraction pattern forms near the objective lens as the sum of diffracting waves shown schematically in Fig. 7.7. With an aperture at the focal or diffraction plane, we can look either at the image produced by the direct beam (bright field) excluding diffracted spots, or the image produced by one of the diffraction spots (dark field) excluding the direct beam. Or we can include both spots and look at the interference between them. Changes in diffracted intensity will affect the images in all cases.

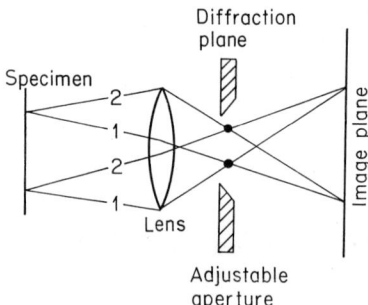

Fig. 7.7. Bright field or dark field in the electron microscope is chosen by placing the adjustable aperture over the transmitted beam (not shown) or some of the diffraction spots, respectively. 1 and 2 are two different diffraction peaks. (The aperture can be moved, or changed in size)

7.3.2 Imaging in the Transmission Electron Microscope

We have seen in Sect. 6.3 that if every unit cell is the same, and the structure is periodic, the electron density can be written as

$$\rho(xyz) = \frac{1}{V_c} \sum_h \sum_k \sum_l F_{hkl} \exp(-2\pi i[hx + ky + lz]) \, .$$

Suppose we are looking through the microscope along the [001] or a_3 axis in a crystal of thickness n unit cells. What we see on the image plane is the projection or sum of this electron density along [001].

Such a projection was discussed in Sect. 6.3 where it was shown that the xy plane of such a projection is related to the structure factors:

$$\rho(xy) = \frac{a_3}{V_c} \sum_h \sum_k F_{hk0} \exp(-2\pi i[hx + ky]) \, . \tag{7.7}$$

This is what we see through the microscope as shown in Fig. 7.7. *If diffraction peaks are cut out by an aperture, the image cannot be perfect* no matter how good the lenses, as the required sums in Eq. (7.7) are infinite. Of course, the scattering factors decrease with increasing $\sin\theta/\lambda$ so that the first few diffraction peaks are the most important in the image. If this were not the case a lens would not serve any purpose!

According to Eq. (7.7), the projected electron density depends on the amplitudes of the various scattered beams (the $F_{hk0's}$). In this equation it has been assumed that the unit cell is the same everywhere; i.e., F_{hk0} depends on h and k but not on position x, y, z. When there are distortions that vary from point to point in the structure, this is *not* the case; instead that scattered amplitude can be written more generally as

$$A_{eu}(\mathbf{s}) = \sum_n F_n \exp(2\pi i \mathbf{s} \cdot \mathbf{r}_n) \, . \tag{7.8a}$$

Here, the structure factor depends on the position of the nth unit cell, $\mathbf{r}_n$, in the structure.

Recall from our study of the interference function in Sect. 2.3 that if we are not exactly on a diffraction spot (i.e., $\mathbf{s} = \mathbf{r}^*_{hkl} + \Delta\mathbf{s}$), the sum A_{eu} on the complex plane is a circle whose radius depends inversely on $\Delta\mathbf{s}$ (for small $\Delta\mathbf{s}$), and which opens and closes as the number of terms in the sum increases (Fig. 2.10). Thus near the edge of a thin piece of material in the electron microscope, there will be fringes due to the thickness variation (Fig. 7.8). Near an imperfection, $\mathbf{r}_n$ is not the vector of the perfect lattice; due to some distortion, we can write as before $\mathbf{r}_n = \mathbf{r}_n^0 + \Delta\mathbf{r}_n$. Now, $\mathbf{s} = \mathbf{r}^*_{hkl} + \Delta\mathbf{s}$. Equation (7.8a) can then be written as

$$A_{eu}(\mathbf{s}) = \sum_n F_n \exp(2\pi i[\mathbf{r}^*_{hkl} \cdot \Delta\mathbf{r}_n]) \exp(2\pi i[\Delta\mathbf{s} \cdot \mathbf{r}_n^0])$$

$$\times \exp(2\pi i[\mathbf{r}^*_{hkl} \cdot \mathbf{r}_n^0]) \exp(2\pi i[\Delta\mathbf{s} \cdot \Delta\mathbf{r}_n]) \, . \tag{7.8b}$$

The last exponent is essentially unity because generally $\Delta\mathbf{r}_n$ and $\Delta\mathbf{s}$ are small. The next to last is just unity because $\mathbf{r}_n^0 \cdot \mathbf{r}^*_{hkl}$ is an integer (Chap. 2).

The first exponential term represents an additional phase factor over that which would be present without the distortion. We shall prove in Chap. 8 that the material

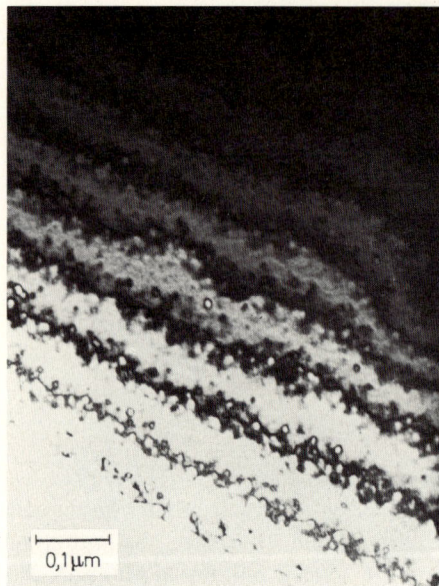

Fig. 7.8. Ni-10 at % Ti. 100,000 ×, 100kV. Fringes near the edge of a foil. (Photograph by S. Sass)

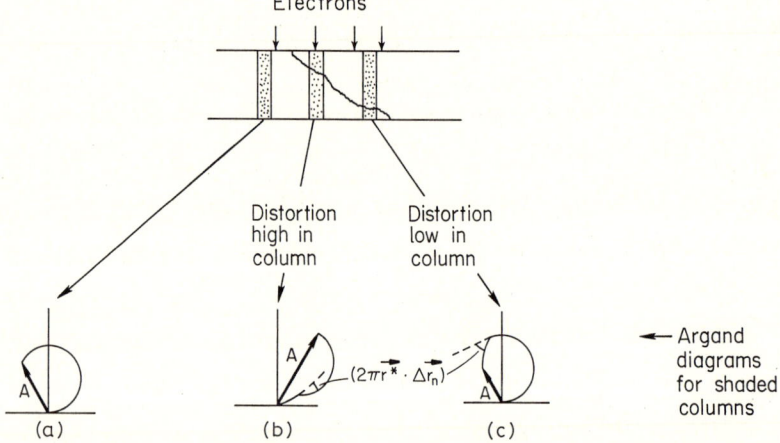

Fig. 7.9a–c. The effect of a distortion on the scattered amplitude A. The phase amplitude diagrams whose sum is indicated by the *arrows* show how the total amplitude scattered by the foil is affected by a fault or distortion. As this diffracted amplitude oscillates with the position of the distortion, so does that of the transmitted beam. The image will then contain fringes which reverse contrast in bright and dark field

can be considered as individual columns of unit cells, so that Eq. (7.8), instead of being a three-dimensional sum for a three-dimensional solid, can be summed for a given Δs in individual columns around the imperfection. We *assume* this result here. Due to the additional phase factor, the imperfection may increase or decrease the resulting diffracted amplitude depending on where the imperfection is in the column. If it extends over some large region, there may be fringes as indicated in Fig. 7.9. In

7.3 Distortion and Mosaic Size

bright field, the pattern will be dark where the diffracted amplitude (blocked by the aperture) is large, and light where it is small.

This first exponential will have no effect if $\mathbf{r}_{hkl}^* \cdot \Delta \mathbf{r}_n = 0$. Thus, *we can tell the direction of distortion, such as the Burger's vector of a dislocation, by examining dark field images made with different diffracted beams* and looking for the one diffraction spot for which the contrast in the image vanishes. For a spherical distortion ($\Delta \mathbf{r}_n$ the same in all directions), there will always be a "plane of no contrast" perpendicular to the strongest diffraction spot in the sum forming the image as shown in Fig. 7.10a, b. (The condition $\mathbf{r}_{hkl}^* \cdot \Delta \mathbf{r}_n = 0$, with $|\Delta \mathbf{r}_n|$ a constant is the equation of a plane as shown in Sect. 2.3.)

Consider the second exponential. If a region of the foil is tilted to reduce Δs, the diffraction will increase and a black band or "extinction contour" will appear in

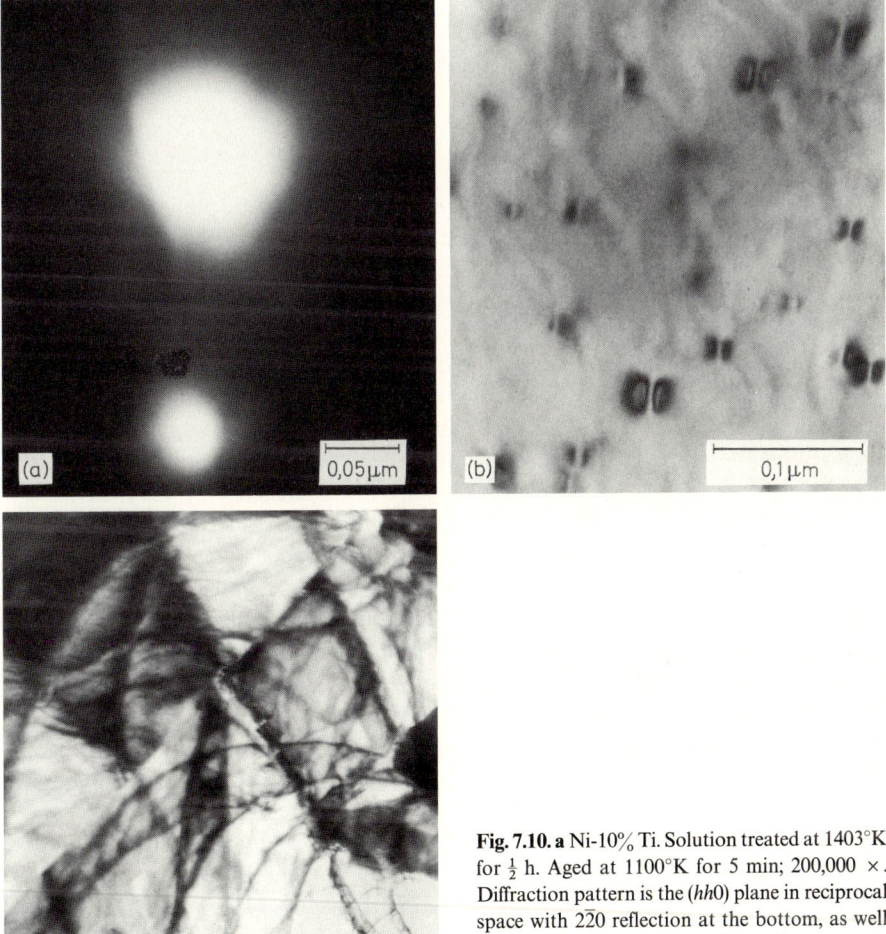

Fig. 7.10. **a** Ni-10% Ti. Solution treated at 1403°K for $\frac{1}{2}$ h. Aged at 1100°K for 5 min; 200,000 ×. Diffraction pattern is the (*hh*0) plane in reciprocal space with $2\bar{2}0$ reflection at the bottom, as well as the central beam. Note plane of no contrast in **b** perpendicular to $2\bar{2}0$. **c** Bend contours in Ni-10% Ti. 75,000 ×, 100 kV. (Photos by S. Sass)

376   7. What Else Can We Learn from a Diffraction Experiment Besides the Average Structure?

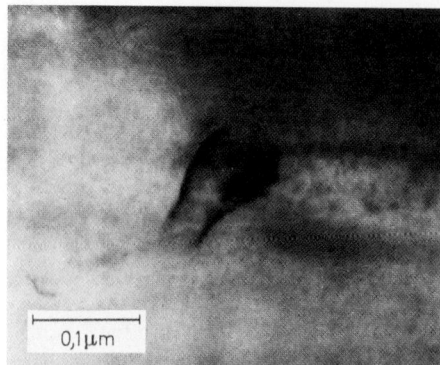

Fig. 7.11. Ni-10% Ti solution treated at 1150°C for $\frac{1}{2}$ hr. 150,000 ×, 100 kV. Asymmetric image of a dislocation. (Photograph by S. Sass)

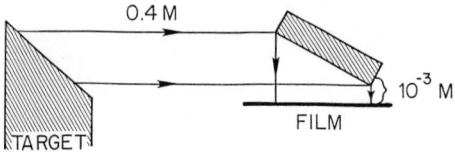

Fig. 7.12. The Berg-Barrett method for imaging imperfections with x-rays

bright field, Fig. 7.10c. On one side of an edge dislocation, the planes are tilted one way, on the other side the other way. The term $2\pi \mathbf{r}_{hkl}^* \cdot \Delta \mathbf{r}_n$ is replaced with $2\pi \Delta \mathbf{s} \cdot \mathbf{r}_n^0$ in Fig. 7.9b; on one side of the dislocation the change in phase is as shown leading to an increase in amplitude, whereas on the other side it is a rotation in the opposite sense, closing the circle so that the resultant amplitude is smaller than for defect-free material. This light-dark contrast is illustrated in Fig. 7.11.

The same kind of image effects can be produced with x-rays. A film is placed near a sample (1 mm or less) and the sample is bathed with the beam, as in Fig. 7.12. Unfortunately, even high resolution films can be enlarged only a few hundred times, so that the density of distortion sites must be quite low or highly clustered if individual ones or groups are to be examined in this way. This technique is employed to measure the dislocation densities in the almost perfect materials used in the semiconductor industry, but is not useful for the densities of imperfections in normal materials. It is difficult here to be quantitative about our discussion of images in the electron microscope, of these "Berg-Barrett" x-ray pictures, or of the *local* scattered amplitude because in highly perfect regions our "kinematic" (single scattering) theory does not apply. This problem will be examined in Chap. 8.

Here we now turn our attention to what we can learn from the shape of a diffraction peak from a bulk material irradiated over a large area containing many imperfections. That is, what we can learn from the shape of a peak recorded in the normal way with a film or diffractometer. This problem really has two parts: (i) what we can learn about the state of the material from the peak shape, and (ii) how we can correct for the effects produced by the finite receiving slits, finite wavelength spread, and other measuring effects. We will consider (i) first. We examine what we can expect to occur in general to the diffraction pattern due to mosaic or distortions.

7.3.3 Effects on Diffraction Peaks

The presence of mosaic implies that the crystal consists of many small crystals diffracting independently. We have seen in Chap. 2 that the *breadth of a diffraction peak depends inversely on the number of unit cells, N.* On a precession film or electron diffraction pattern, *all* spots change in the same manner with N. *But these are direct maps of reciprocal space.* In 2θ space, for example, the effect will be different. Suppose, that normal to the $00l$ diffracting planes (we can always choose a new unit cell so that these are the indices) the mosaic is N cells thick. With h_3 a continuous variable in reciprocal space, the broadening is Δh_3. Then, since in this case

$$h_3|\mathbf{b}_3| = 2\sin\theta/\lambda,$$

differentiation yields

$$(\Delta h_3 b_3 / \cos\theta)\lambda = \Delta 2\theta.$$

As $\Delta h_3 b_3$ is $1/(Na_3)$,

$$\Delta 2\theta = \lambda/Na_3 \cos\theta.$$

The broadening varies inversely with $\cos\theta$ in 2θ space and is larger at large scattering angles.

Consider now, distortions. First, consider the case where the distortion is the same everywhere, as if the unit cell is larger (or smaller) or changes it shape. This will obviously change the diffraction angles, but not the shape of the peaks.

On the other hand, suppose that the *displacement varies from point to point* in the solid. Then there is a displacement of the form $\mathbf{r}_n + \Delta \mathbf{r}_n$ as mentioned above. (While this is similar in form to the thermal displacements we discussed earlier, it is to be emphasized that those we are considering *now* are static.) In this case the *effect on the intensity increases with* $\mathbf{s}$ [see Eq. (7.8b)] and therefore, can be seen to vary with distance from the origin of reciprocal space. This implies that the broadening will vary with 2θ more than the $\cos\theta$ dependence due to size.

Differentiating Bragg's law:

$$0 = 2\Delta d \sin\theta + 2d\cos\theta \Delta\theta$$

$$\Delta 2\theta = -\frac{2\Delta d}{d}\tan\theta.$$

This equation actually describes the peak shift due to a constant strain. We will apply it in a different way, to the case of a distortion which varies from place to place in a crystalline material, by considering the variance in d, $\left\langle \left(\frac{\Delta d}{d}\right)^2 \right\rangle^{1/2}$ and how this produces a broadening. In this case, the width 2θ is

$$\Delta 2\theta = 2\left\langle \left(\frac{\Delta d}{d}\right)^2 \right\rangle^{1/2} \tan\theta.$$

As we expected the variation with 2θ is larger than $1/\cos\theta$. If both effects are present:

$$\Delta 2\theta = \frac{\lambda}{Na_3 \cos\theta} + 2\left\langle \left(\frac{\Delta d}{d}\right)^2 \right\rangle^{1/2} \tan\theta.$$

Therefore, $\Delta 2\theta \cos\theta$ is linear vs. $\sin\theta$:

$$\Delta 2\theta \cos\theta = \frac{\lambda}{Na_3} + 2\left\langle\left(\frac{\Delta d}{d}\right)^2\right\rangle \sin\theta.$$

From such a plot, the size can be obtained from the intercept and the variance in strain from the slope. From such studies of the peak width we cannot learn anything about the size or strain *distributions*. Even more importantly, there is instrumental broadening and shape due to the emission line from an x-ray tube. It is impossible to obtain simple functions to describe the combined effect (although this is possible at a well collimated white radiation source like a storage ring) and for *both* these reasons breadth measurements cannot be very quantitative. In Sect. 7.4, we shall see how to correct for this in a completely general way. This will allow us to obtain the entire diffraction peak free of such effects. As we will need to record the entire shape of a diffraction peak to do this, not just its breadth — why not use all the information available and look more closely at the effect of mosaic and distortion on the entire diffraction peak.

7.3.4 A More Quantitative Evaluation of the Effect of Mosaic Size and Distortion

We shall follow the treatment by Warren and co-workers (see Warren, 1969). The material will be considered to be *cubic with a cell edge* a_0. The intensity can be written from Eq. (2.18):

$$I_{eu} = \sum_{n_1}\sum_{n_2}\sum_{n_3}\sum_{m_1}\sum_{m_2}\sum_{m_3} F^2 \exp(2\pi i[\mathbf{s}\cdot(\mathbf{r}_n - \mathbf{r}_m)])$$

$$= \sum_n\sum_m F^2 \exp(2\pi i\{h_1(n_1-m_1) + h_2(n_2-m_2) + h_3(n_3-m_3)$$

$$+ \mathbf{s}\cdot(\Delta\mathbf{r}_n - \Delta\mathbf{r}_m)\}).$$

The Δ's are displacements of unit cells from the positions in the average lattice.

We will henceforth assume that each and every cell has the same structure factor, F. On a diffractometer, which is the tool we would employ for precisely recording intensities, we generally record power as a function of 2θ; *we need an expression not of the total integrated power as in Sect. 4.9, but power as a function of 2θ* ($P'_{2\theta}$).

From Chap. 4, for the case of a powder sample [see the development of Eq. (4.44)],

$$(P)_{eu} = \frac{mnR^2\lambda^3}{4V_c}\int\int\int \frac{I_{eu}}{\sin\theta} dh_1\, dh_2\, dh_3$$

and

$$(P')_{eu} = P_{eu}/2\pi R\sin 2\theta,$$

where P'_{eu} is the power per unit length of diffracting cone, m is the multiplicity, and n is the number of regions oriented to diffract. We shall let $M = mn$ represent the number of domains or regions which diffract. As we have mentioned, we can always *define a unit cell, such that the hkl diffraction peak is a 00l' peak* from this new cell.

7.3 Distortion and Mosaic Size

Table 7.1. The various Miller indices and other terms employed in Sect. 7.3 and 7.5

	hkl	Actual indices in the true cubic unit cell with edge a_0 and continuous variables in reciprocal space h_1, h_2, h_3.
	h_0	$(h^2 + k^2 + l^2)^{1/2}$
	$00l'$	Indices in a unit cell with edges a_1', a_2', a_3', chosen so that hkl can be reindexed as $00l'$. The variables in reciprocal space are h_1', h_2', h_3'.
	L	na_3', a column length normal to the diffracting plane.
	HKL_0	Indices in the hexagonal axial system for fcc shown in Fig. 7.19a. The variables in reciprocal space are H_1, H_2, H_3.
	N_n	Number of cells in a column of a coherently reflecting region with an nth neighbor cell.
$\bar{N}_{h_1'}, \bar{N}_{h_2'}, \bar{N}_{h_3'}$		Average dimensions of a coherently diffracting region along the crystal axes a_1', a_2', a_3'.

The distance from the origin in reciprocal space to the diffracting point is the same regardless of the coordinate system chosen, so that

$$l'/a_3' = (h^2 + k^2 + l^2)^{1/2}/a_0 \equiv h_0/a_0 \,. \tag{7.9}$$

Table 7.1 contains a summary of all the indices to be used here and in subsequent sections. We continue with this new cell in order to simplify the mathematics; the variable is now only h_3', which is along the normal to the diffracting planes. Since $h_3'|\mathbf{b}_3'| = 2\sin\theta/\lambda$, then $dh_3' = (\cos\theta\,d2\theta)/|\mathbf{b}_3'|\lambda$. Hence, replacing the integration with respect to h_3 with this expression and dropping the integration over 2θ,

$$(P_{2\theta}')_{eu} = \frac{MR\lambda^2}{16\pi V_c |\mathbf{b}_3'|\sin^2\theta} \int\int I_{eu}\,dh_1'\,dh_2'$$

$$= \frac{MR\lambda^2 F^2}{16\pi V_c |\mathbf{b}_3'|\sin^2\theta}\int_{+1/2}^{+1\ 1/2}\int_{+1/2}^{+1\ 1/2}\sum_{n_1}\sum_{n_2}\sum_{n_3}\sum_{m_1}\sum_{m_2}\sum_{m_3}$$

$$\times \exp(2\pi i[h_1'(n_1 - m_1) + h_2'(n_2 - m_2) + h_3'(n_3 - m_3)$$

$$+ \mathbf{s}\cdot(\Delta\mathbf{r}_n - \Delta\mathbf{r}_m)])\,dh_1'\,dh_2'\,.$$

The factors $\sin^2\theta$ and F^2 are removed from the integral because they vary slowly compared to the exponential terms. This preintegral factor will be written as $K(\theta)$. The integration is a projection along h_1' and h_2'. Consider only one peak, say the 001. Then the limits of the integrals with respect to h_1' and h_2' are from $\frac{1}{2}$ to $\frac{3}{2}$. This kind of integral was evaluated in Sect. 6.2 (the transform of a rectangle). Also,

$$\Delta\mathbf{r}_n = x_n\mathbf{a}_1' + y_n\mathbf{a}_2' + z_n\mathbf{a}_3'\,,\qquad \mathbf{s}\cong l'\mathbf{b}_3'$$

and

$$\mathbf{s}\cdot(\Delta\mathbf{r}_n - \Delta\mathbf{r}_m) = l'(z_n - z_m)\,.$$

Therefore,

$$(P_{2\theta}')_{eu} = K(\theta)\sum_{n_1}\sum_{n_2}\sum_{n_3}\sum_{m_1}\sum_{m_2}\sum_{m_3}\frac{\sin\pi(n_1 - m_1)}{\pi(n_1 - m_1)}$$

$$\times \frac{\sin\pi(n_2 - m_2)}{\pi(n_2 - m_2)}\exp(2\pi i l'[z_n - z_m])\exp(2\pi i h_3'[n_3 - m_3])\,.$$

The sine terms are *each* zero unless $m_1 = n_1$, $m_2 = n_2$ and then they are unity, so that the sum over $m_1 m_2, n_1 n_2$ becomes $\bar{N}_{h_1} \bar{N}_{h_2}$, where the $\bar{N}$'s are the average number of columns in a plane perpendicular to $[00l']$ in a mosaic region. The remaining terms $\sum_{m_3} \sum_{n_3}$ are the sums between pairs of cells $n_3 - m_3$ apart in a column. Let $n = n_3 - m_3$, and N_n be the average number of cells with an nth neighbor along the direction $[00l']$, the average being over all columns in the mosaic regions under the beam. Take the average value of the term involving z over all these pairs of cells n cells apart in all the columns in all the mosaic regions:

$$(P'_{2\theta})_{\text{eu}} = K(\theta) \bar{N}_{h'_1} \bar{N}_{h'_2} \sum_{n=-\infty}^{+\infty} N_n \langle \exp(2\pi i l' z_n) \rangle \exp(2\pi i n h'_3).$$

Multiplying and dividing by $\bar{N}_{h'_3}$, the average number of cells in a column, expressing the exponents in trigonometric form and keeping in mind that cross terms for (n) and $(-n)$ cancel:

$$(P'_{2\theta})_{\text{eu}} = K(\theta) \bar{N}_{h'_1} \bar{N}_{h'_2} \bar{N}_{h'_3} \left\{ \sum_{n=-\infty}^{+\infty} \frac{N_n}{\bar{N}_{h'_3}} \langle \cos 2\pi l' z_n \rangle \right.$$
$$\left. \times \cos 2\pi n h'_3 - \frac{N_n}{\bar{N}_{h'_3}} \langle \sin 2\pi l' z_n \rangle \sin 2\pi n h'_3 \right\}. \quad (7.10)$$

Equation (7.10) is in a form suitable for Fourier analysis, as it is a Fourier series like those discussed in Chap. 6, with

$$A_n = (N_n / \bar{N}_{h'_3}) \langle \cos 2\pi l' z_n \rangle, \qquad B_n = -(N_n / \bar{N}_{h'_3}) \langle \sin 2\pi l' z_n \rangle.$$

However, the coefficients A_n, B_n are those due to the sample only, without instrumental broadening; a correction for this will be discussed later.

7.3.5 The Meaning of the Fourier Coefficients

Let us look at the various terms in this equation. *Those involving displacement depend on l' as we predicted in our qualitative considerations, while N_n, the particle size term does not depend on l'*. The *peak will be asymmetric if there is a net displacement*, i.e., if the sine term involving z_n does not average to zero; the peak will be asymmetric toward low angles for a positive net displacement, and in the opposite direction for a negative net z_n. *If there are many sine terms, i.e., if the strain exists over a large distance, the peak will actually shift.*

Because the part of the Fourier coefficients due to particle size does not depend on l' while that due to distortions does, *these parts can be separated*. Several orders of a peak, e.g., $hkl = 200, 400, 600$, are analyzed and a plot is made of A_n versus l' for each n and extrapolated to $l' = 0$. The ordinate is $N_n / \bar{N}_{h_3}$. (Unless the material is elastically isotropic, the displacements will vary with crystallographic direction; thus we cannot use peaks representing different directions.) However, with a powder or a polycrystalline material, the third order often overlaps with another peak. For example, with an fcc material the 600 occurs at the same angle as the 442. We can

7.3 Distortion and Mosaic Size

eliminate the need for a third-order term if the strains are small so that the term $\cos 2\pi l' z_n$ in A_n can be expanded ($\cos x = 1 - x^2/2 + x^4/24 + \ldots$). Including only the first two terms and taking the logarithm of the expansion,

$$\ln(\langle \cos 2\pi l' z_n \rangle) \cong -2\pi^2 (l')^2 \langle z_n^2 \rangle.$$

Now, $z_n a_3'$ is the change in length of a column of length na_3, so that the strain ε_n is

$$\varepsilon_n = z_n a_3'/na_3' = z_n/n \quad \text{or} \quad n\varepsilon_n = z_n.$$

Inserting this value for z_n in Eq. (7.10) using Eq. (7.9), and the above approximation for the cosine term:

$$\ln A_n = \ln(N_n/\bar{N}_{h_3'}) - 2\pi^2 [h_0^2 (a_3')^2/a_0^2] n^2 \langle \varepsilon_n^2 \rangle, \tag{7.11a}$$

or

$$\ln A_L = \ln(N_n/\bar{N}_{h_3'}) - (2\pi^2 h_0^2 L^2/a_0^2) \langle \varepsilon_L^2 \rangle, \tag{7.11b}$$

where $L = na_3'$ is the true distance between cells in a column normal to the diffracting planes. *If there are no strains, only small particles, $\ln A_L$ versus h_0^2 for each L will be a horizontal line. If there is no particle broadening but only strains, all the lines for various L's intersect at $L = 0$, $\ln A_L = 0$.*

According to Eq. (7.11b) *only two orders of a peak are needed to separate the effects of particle size and strain* and obtain $N_n/\bar{N}_{h_3'}$ and $\langle \varepsilon_L^2 \rangle$ vs. L. This is because $\ln A_L$ versus h_0^2 is a straight line according to this equation (with a slope yielding $\langle \varepsilon_L^2 \rangle$): it *is* possible to do the analysis with a powder or polycrystalline specimen. But Eq. (7.11b) resulted when we neglected higher terms than the second in the expansion of the cosine. How valid is this? a_3'/a_0 is typically 5, $\langle \varepsilon_L^2 \rangle^{1/2}$ is at most 0.005. Substituting these values in the expansion of $\cos 2\pi l' z_n = \cos 2\pi (h_0 a_3'/a_0) n\varepsilon_n$, nh_0 can be 6 with the third term in the expansion still only 6% of the second. Thus, for the first few harmonics n, $\ln A_L$ versus h_0^2 is in fact, linear.[3]

We have seen that if there are net strains over large distances, the peak will shift. If the origin for the analysis is chosen at the position of a well-annealed standard (with sharp peaks), then for small n

$$B_n \equiv -\frac{N_n}{\bar{N}_{h_3'}} \left(\sin 2\pi \frac{h_0 a_3'}{a_0} n \langle \varepsilon_n \rangle \right) \rightarrow \left(-\frac{N_n}{\bar{N}_{h_3'}} \frac{2\pi h_0 L}{a_0} \langle \varepsilon_n \rangle \right).$$

On the other hand, if the center of the actual peak is used, in the analysis the measured strain is the value above or below the long range (or mean) strain:

$$\langle \varepsilon_L^2 \rangle_{\text{meas}} = \langle (\varepsilon_L - \varepsilon_{\text{long range}})^2 \rangle = \langle \varepsilon_L^2 \rangle - 2 \langle \varepsilon_L \varepsilon_{\text{long range}} \rangle + \varepsilon_{\text{long range}}^2,$$

$$\langle \varepsilon_L^2 \rangle_{\text{meas}} = \langle \varepsilon_L^2 \rangle - \varepsilon_{\text{long range}}^2, \tag{7.12}$$

because

$$-2 \langle \varepsilon_L \varepsilon_{\text{long range}} \rangle = -2\varepsilon_{\text{long range}} \langle \varepsilon_L \rangle = -2\varepsilon_{\text{long range}}^2.$$

---

[3] A_L itself, as well as $\ln A_L$ is approximately linear with h_0^2: $A_L \cong N_n/\bar{N}_{h_3'}[1 - (2\pi^2 h_0^2 L^2/a_3'^2) \langle \varepsilon_L^2 \rangle]$, Delhez and Mittemeijer. (1976).

The average $\langle \varepsilon_L \rangle$ is the long-range strain. *In an analysis with the center of the actual peak, we are therefore measuring the deviation from the mean or long-range strain.* Some typical data are presented in Fig. 7.13b.

It is easy to see that $N_n/\bar{N}_{h'_3} \cong (1 - |n|/\bar{N}_{h'_3})$ for small n, or if the column lengths are the same everywhere. Take a column of, say, five cells and evaluate N_i, the number of cells with an ith neighbor in the $+n$ direction. (The terms in the sum involving $-n$ include counts in the other direction.) For this column, $N_0 = 5$, $N_1 = 4$, $N_2 = 3$, $N_3 = 2$, $N_4 = 1$, $N_5 = 0$. The term $N_1/\bar{N}_{h'_3}$, is $\frac{4}{5}$, which is the same as $(1 - \frac{1}{5})$.

This implies that $N_n/\bar{N}_{h'_3}$ decreases rapidly with increasing n or L if $\bar{N}_{h'_3}$ small; the peak will be broader, the smaller the regions, as we anticipated. If the values of $N_n/\bar{N}_{h'_3}$ obtained from the intersections of the curves with the ordinate in Fig. 7.13b [see Eq. (7.11)] are plotted versus L, the initial slope is

$$(d(N_n/\bar{N}_{h'_3})/dn)_{n \to 0} = -1/\bar{N}_{h'_3} \quad \text{or} \quad (d(N_n/\bar{N}_{h'_3})/dL)_{L \to 0} = -1/(a'_3 \bar{N}_{h'_3}) = 1/D_{\text{eff}} .$$
(7.13)

Thus *if a plot of $N_n/\bar{N}_{h'_3}$ versus L normalized so that its intercept at $L = 0$ is unity, the slope gives the average value (D_{eff}) of the length of the columns normal to the diffraction planes* — a measure of the effective mosaic size in that direction. Such a plot is shown in the insert of Fig. 7.13b. (The data in Fig. 7.13b were obtained after correction for broadening due to instrumental effects; we shall explore this correction below.)

Often there is a small bend or hook in the data near $n = 0$. This occurs because with a broad peak it is difficult to estimate background, and hence, as A_0 is the area of the peak (see Sect. 6.2), its value will be too small. The values from beyond $n = 3 - 4$ should be extrapolated to $n = 0$ in doing this normalization. As $\ln(N_n/\bar{N}_{h'_3})$ vs. L is generally linear for $n > 3$, such a plot is a simple procedure for this extrapolation.

An analysis of the coefficients of only one peak can yield approximate values for both strain and mosaic size if two or more orders are not available (Nandi et al. 1984). Finally, the second derivative of the size coefficients is the particle size distribution (Warren 1969).

7.4 Slit Corrections

7.4.1 Theoretical Considerations

We turn now to consider how we can correct a peak's shape for the various effects that cause the peak from a sample with even few defects to have a finite breadth. There are a variety of causes of this broadening: the finite sizes of the slits at the x-ray tube and the receiving slits, the range of wavelengths in the beam, the fact that in a diffractometer a flat specimen is only tangent to the focusing circle at one point, the size of the source.

7.4 Slit Corrections

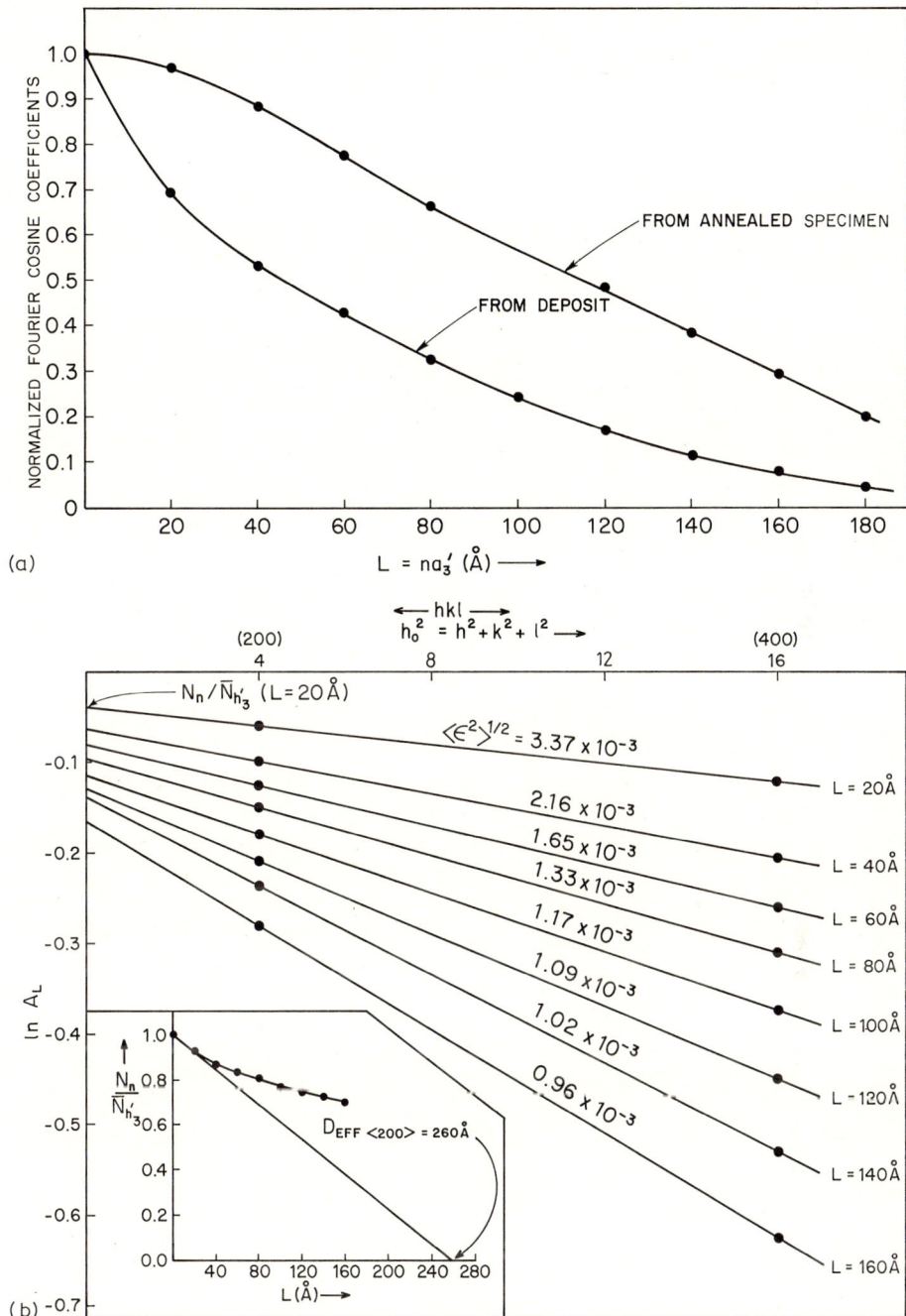

Fig. 7.13. **a** Fourier cosine coefficients of 400 peaks, from Ag electrodeposited from (KCN + AgCN) solution, and from a well-annealed standard. **b** Separation of particle size and strain and (*insert*) the determination of particle size, using the corrected Fourier cosine coefficients, A_L from **a**. [From Hinton, R.W., Schwartz, L.H., and Cohen, J.B., J. Electrochem. Soc. *110*, 103 (1963); reprinted with permission from the Journal of the Electrochemical Society]

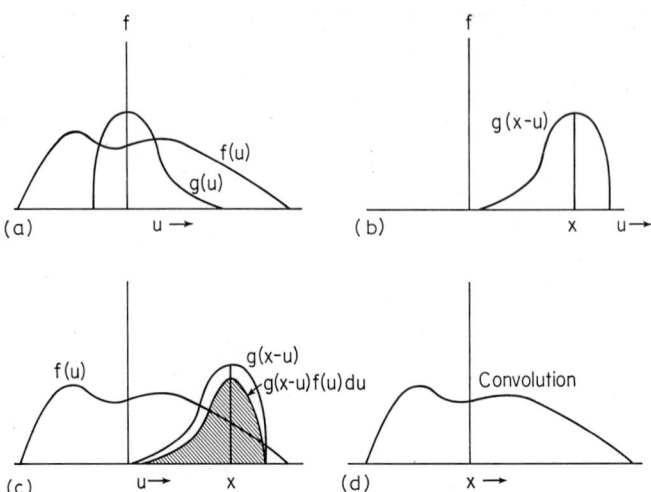

Fig. 7.14. The convolution of two functions g and f

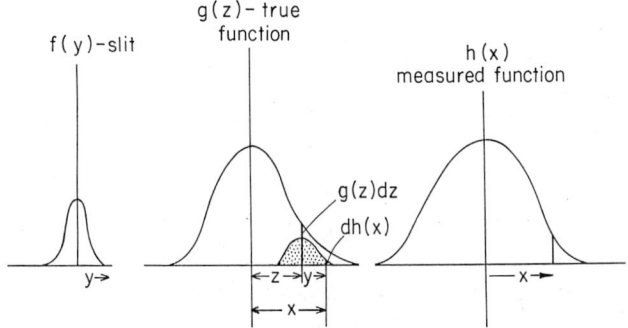

Fig. 7.15. A slit $f(y)$ redistributes an element of a true function, $g(z)\,dz$, so that it has the shape $f(y)$ and contributes $dh(x)$ to the measured function, $h(x)$. (From Warren, B.E., "X-ray Diffraction". Addison-Wesley, Reading, Massachusetts, 1969)

The function that we shall find useful in this discussion is the convolution discussed in Sect. 6.4. Equation (6.47) is repeated here (see Fig. 7.14).

$$f(\mathbf{x}) \otimes g(\mathbf{x}) = \int_{-\infty}^{+\infty} f(\mathbf{u})g(\mathbf{x} - \mathbf{u})\,dV_{\mathbf{u}}. \tag{6.47}$$

Physically, this function can be thought of as follows: $g(\mathbf{u})$ is transferred to position $\mathbf{x}$, and inverted through (or folded about) this position, Fig. 7.14b. Then, at each position the product of $f(\mathbf{u})$ and $g(\mathbf{u})$ is obtained, Fig. 7.14c. The sum of all such products for all $\mathbf{u}$ is the convolution evaluated for $\mathbf{x}$. This is illustrated in Fig. 7.14d.

Our own problem will also serve to illustrate the meaning. Suppose a slit in a measuring system can be described by a function $f(y)$, and the true function to be measured is $g(z)$, Fig. 7.15. The slit takes any element $g(z)\,dz$ as it "sees" it, and

7.4 Slit Corrections

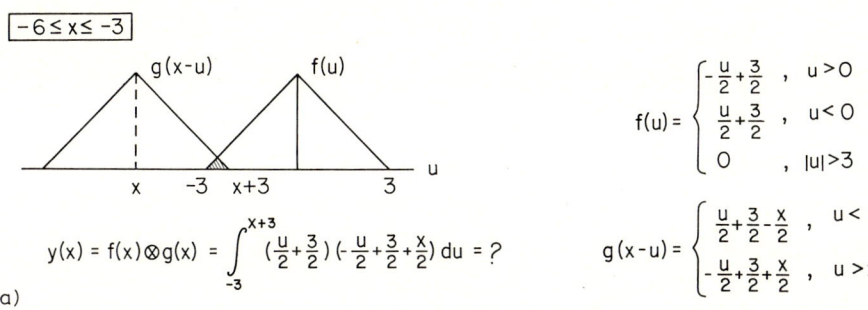

(a) $\boxed{-6 \le x \le -3}$

$$y(x) = f(x) \otimes g(x) = \int_{-3}^{x+3} \left(\tfrac{u}{2}+\tfrac{3}{2}\right)\left(-\tfrac{u}{2}+\tfrac{3}{2}+\tfrac{x}{2}\right) du = ?$$

$$f(u) = \begin{cases} -\tfrac{u}{2}+\tfrac{3}{2}, & u > 0 \\ \tfrac{u}{2}+\tfrac{3}{2}, & u < 0 \\ 0, & |u| > 3 \end{cases}$$

$$g(x-u) = \begin{cases} \tfrac{u}{2}+\tfrac{3}{2}-\tfrac{x}{2}, & u < x \\ -\tfrac{u}{2}+\tfrac{3}{2}+\tfrac{x}{2}, & u > x \end{cases}$$

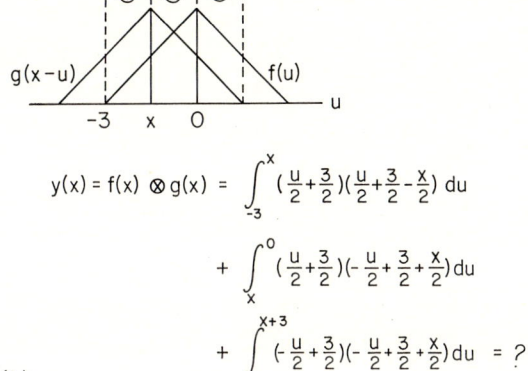

(b) $\boxed{-3 \le x \le 0}$

$$y(x) = f(x) \otimes g(x) = \int_{-3}^{x} \left(\tfrac{u}{2}+\tfrac{3}{2}\right)\left(\tfrac{u}{2}+\tfrac{3}{2}-\tfrac{x}{2}\right) du$$

$$+ \int_{x}^{0} \left(\tfrac{u}{2}+\tfrac{3}{2}\right)\left(-\tfrac{u}{2}+\tfrac{3}{2}+\tfrac{x}{2}\right) du$$

$$+ \int_{0}^{x+3} \left(-\tfrac{u}{2}+\tfrac{3}{2}\right)\left(-\tfrac{u}{2}+\tfrac{3}{2}+\tfrac{x}{2}\right) du = ?$$

Fig. 7.16. The convolution of two triangles

spreads it out into a function of the form $f(y)$. Some point on this spread function is a contribution $d[h(x)]$ to the measured function.

$$d[h(x)]/g(z)\,dz = f(y)/A_y,$$

where A_y is the area of $f(y)$. Then,

$$d[h(x)] = (1/A_y)g(z)f(y)\,dz.$$

As can be seen in the figure, $y = x - z$. Hence,

$$h(x) = \frac{1}{A_y} \int_{-\infty}^{+\infty} g(z)f(x-z)\,dz. \tag{7.14}$$

Our resulting function $h(x)$ is proportional to a convolution of the slit function and the true function.[4] This is actually a very good way to visualize convolutions in setting up any analytical treatment. For example, suppose we wish to convolute two triangles $f(x_1)$ and $f(x_2)$ as in Fig. 7.16. The various stages of the convolution,

---

[4] If the x-ray source has finite dimensions, then we must convolute source and sample function and the *result* must then be convoluted with the receiving slit, and so on for all phenomena that cause broadening. The function $f(y)$ really represents all these instrumental effects.

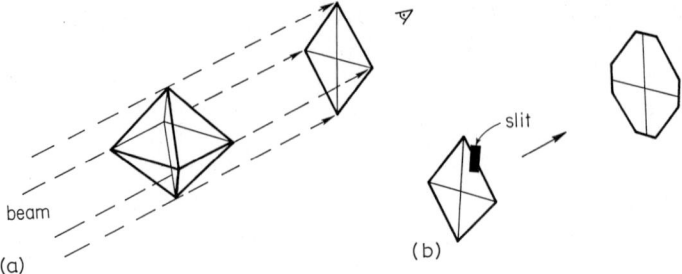

Fig. 7.17. Effect of a slit on the projection of an octahedron. The octahedron is projected by the beam in **a** to form a diamond shape. In **b** a rectangular slit sampling this diamond produces an octagonal figure

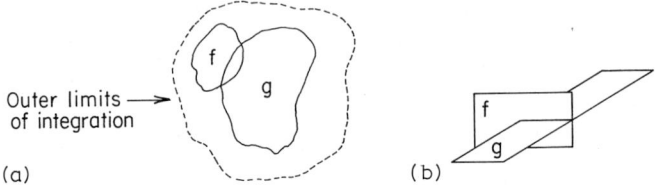

Fig. 7.18. a Two-dimensional convolution in one plane. **b** A convolution of two-dimensional functions in three-dimensional space

with one triangle as a slit moving past the other, are shown in the figure, along with the required integrals for part of the evaluations.

[Two-dimensional convolutions can be important. Suppose for example you have an octahedral crystal. An incident beam "views" this as a diamond-shaped two-dimensional figure shown in Fig. 7.17. If the beam is square, the convolution results in an octagon of intensity on a film. We can write such a two-dimensional convolution as follows:

$$\int f(\mathbf{u}_1 \mathbf{u}_2) g(\mathbf{x}_1 - \mathbf{u}_1, \mathbf{x}_2 - \mathbf{u}_2) d\mathbf{u}_1 d\mathbf{u}_2 . \tag{7.15}$$

As shown in Fig. 7.18a, this is an integration over the area of overlap of the two functions. (The extent of the convolution is shown by dotted lines.) If the two functions to be convoluted are two dimensional, the result can be three-dimensional if the two are not functions of the same two dimensions; see Fig. 7.18b.]

We now write the functions in the convolution Eq. (7.14) as Fourier series:

$$\begin{aligned} h(x) &= \frac{1}{A_y} \int_{-\infty}^{+\infty} g(z) f(x - z) \, dz \\ &= \frac{1}{A_y} \int_{-\infty}^{+\infty} \sum G_n \exp\left(-\frac{2\pi i n z}{a}\right) \sum F'_n \exp\left(-\frac{2\pi i n'(x - z)}{a}\right) dz \\ &\cong \frac{1}{A_y} \int_{-a/2}^{+a/2} \sum_n G_n \sum_{n'} F'_n \exp\left(\frac{2\pi i [n' - n] z}{a}\right) \exp\left(-\frac{2\pi i n' x}{a}\right) dz . \end{aligned} \tag{7.16}$$

7.4 Slit Corrections

The limits on a convolution are really $-\infty$ to $+\infty$; by truncating the limits to represent practical evaluations, we are effectively multiplying our convolution by a function which is unity in the interval $-a/2$ to $+a/2$ and zero elsewhere. Then *the Fourier transform of $h(x)$ is its true transform convoluted with the transform of the unit function*. As the latter exhibits oscillations, these will appear in the transform of $h(x)$ when $h(x)$ is evaluated with finite limits. Because a transform and the determination of Fourier coefficients are closely related processes, the Fourier coefficients of $h(x)$ will have such oscillations. Care is needed not to truncate too close to a peak to minimize these.

The above integral involving $n - n'$ is zero unless $n = n'$, in which case it is a. Therefore,

$$h(x) = \frac{a}{A_y} \sum F_n G_n \exp\left(-\frac{2\pi i n x}{a}\right) = \sum H_n \exp\left(-\frac{2\pi i n x}{a}\right). \tag{7.17}$$

[This is simply a restatement of the result in Sect. 6.4, that the transform of a convolution (H_n) is the product of the transforms of the two functions $F_n G_n$).] Thus, *to obtain the coefficients* of the true function $g(x)$,

$$G_n = (H_n/aF_n)A_y \tag{7.18}$$

7.4.2 Evaluating the Slit Correction

We need only have the Fourier coefficients for our measured function and those for a peak from a sample that has few defects to obtain the coefficients of the true diffraction peak without the effects of slits, source size, etc. To facilitate computation, this equation is generally separated into sine and cosine terms G_n^i and G_n^r, respectively):

$$G_n^r + iG_n^i = [(H_n^r + iH_n^i)/(F_n^r + iF_n^i)](A_y/a). \tag{7.19}$$

Multiplying top and bottom of the right-hand side by $F_n^r - iF_n^i$,

$$G_n^r = \left(\frac{H_n^r F_n^r + H_n^i F_n^i}{(F_n^r)^2 + (F_n^i)^2}\right)\frac{A_y}{a}, \tag{7.20a}$$

$$G_n^i = \left(\frac{H_n^i F_n^r - H_n^r F_n^i}{(F_n^r)^2 + (F_n^i)^2}\right)\frac{A_y}{a}. \tag{7.20b}$$

This entire procedure was first developed by Stokes (1948). His original paper is still well worth reading!

If the measured functions are symmetrical, the sine terms of the series vanish and we have

$$G_n^r = (H_n^r/aF_n^r)A_y. \tag{7.20c}$$

In other words, if the slit and measured functions are symmetrical, so is the true function.

With these coefficients G, we can synthesize the true peak. If you wish to have the true function in absolute terms, instead of just its shape, it is vital to keep track of all the constants in front of the sums in evaluating the coefficients of each function! An easy way to do this is to choose a function for which the Fourier coefficients can be readily evaluated analytically and with the same period as the function to be evaluated numerically. The numerical solution can then be normalized to the analytical solution.

We have seen in Sect. 6.2 that it is possible to obtain the Fourier coefficients of any real function by replacing the integral for these coefficients by a sum for practical evaluation on a computer:[5]

$$G_n = \frac{1}{a} \sum g(x) \exp\left(\frac{2\pi i n x}{a}\right) \Delta x .$$

For evaluation, let x/a be $t/120$ (the function is divided into 120 parts, $\Delta x = a/120$). Then

$$G_n = \frac{a}{120} \frac{1}{a} \sum_{t=-60}^{t=+60} g(t) \exp\left(\frac{2\pi i n t}{120}\right)$$

or

$$G_n = \frac{1}{120} \sum_{t=-60}^{+60} g(t) \exp\left(\frac{2\pi i n t}{120}\right),$$

where the sum is over integers t in the interval a. Now the true peak $g(t) \equiv g(x)$ can be written as

$$g(t) = \sum_{n'} G_{n'} \exp\left(-\frac{2\pi i n' t}{120}\right).$$

Therefore,

$$G_n = \frac{1}{120} \sum_{t=-60}^{+60} \sum_{n'} G_{n'} \exp\left(\frac{2\pi i [n-n']t}{120}\right)$$

$$= \frac{1}{120} \sum_{n'} G_{n'} \left[\sum_{t} \exp\left(\frac{2\pi i [n-n']t}{120}\right)\right].$$

The exponential term is zero unless $(n - n')/120$ is an integer. Thus, G_n is the sum of $G_{n'}$ values for $(n - n')$ equal to an integral multiple of 120. That is, $n - n' = 0$, 120, 240, etc. Hence $n' = n$, $n' = n - 120$, $n' = n + 120$, etc.

$$G_n = G_{n'} + (G_{n'+120}) + (G_{n'+240}) + \cdots + (G_{n'-120}) + (G_{n'-240}) + \cdots .$$

If the coefficients fall to nearly zero values as n increases in the range of n from -60 to $+60$, there will be only one coefficient in this sum for each n; this then is the required test that our numerical analysis is correct. Furthermore, *only as many independent coefficients as there are data points can be obtained*, in this case, 120. Additional coefficients are related to these by the periodicity. The function is only defined by

---

[5] Alternately, a least-squares fit to the series can be employed; uncertain regions of the peak can then be ignored (Kidron and DeAngelis 1971).

7.4 Slit Corrections

these coefficients at the discrete data points and all 120 coefficients are, of course, required to properly reproduce the function.

It often happens that it is not convenient to use the same period for the "slit" function and the measured function because the latter is so much broader; if we *did* use the same period with the sharper function, we would have many intervals (t) with zero height.

A Fourier series representing a diffraction peak, say an $00l'$ peak for an orthorhombic cell, is a series representing a function in reciprocal space, so that the period is b'_3. We can always find such a cell. Therefore,

$$\frac{2\pi nx}{a} \to \frac{2\pi nx}{b'_3} \to 2\pi na'_3 x \,.$$

This result shows that we do not have to employ the same interval for slit and broadened functions; it is only necessary to compare coefficients at equal values of $L = na'_3$. The fictitious cell edge, a'_3, is calculated for $00l'$ peak from the high and low-angle positions of where the tails join the background, θ_H and θ_L, rather than from the distance between orders, i.e.,

$$[(l' + \tfrac{1}{2}) - (l' - \tfrac{1}{2})]b'_3 = (2\sin\theta_H/\lambda) - (2\sin\theta_L/\lambda) \,, \tag{7.21}$$

where $b'_3 = 1/a'_3$. In this way the large regions between two orders of a peak where there are no data are not included in the analysis.

The Fourier coefficients when summed will produce peaks with period b'_3, even though all of these but the first really do not exist. However, it is only the actual peak that concerns us in this analysis and we may ignore the others.

Thus we see how to actually obtain the Fourier coefficients of our true peak without the effects of slits and other broadening factors. Our "slit" function is simply the same peak recorded from a sample that is well annealed. If we wish to do so, we can then synthesize a diffraction peak from a sample with distortions, without these "slit" effects, with the Stokes-corrected Fourier coefficients, following the procedures described in Sect. 6.2 for such a synthesis. Then the effects of various treatments of the specimen could be compared by, say, comparing the breadth of the peaks. This is really unnecessary, however, for we have just seen in Sect. 7.3 that we can learn a great deal about the specimen from the corrected Fourier coefficients themselves! Note also that it is only necessary to obtain the coefficients and normalize them in this procedure; it is *not* necessary to know the actual magnitudes of the coefficients, so that it is *not* necessary to keep track of constants. Each peak (the one in which we are interested and the "slit" function) is corrected for the angular-dependent functions in $K(\theta)$, F^2, and trigonometric terms, and then the Fourier coefficients are obtained for each.

It is a good idea to *record at least four to five times the breadth of a peak on either side of it*, and *to compare the background of a standard and the pattern to be analyzed*; the background in both should be the same and this comparison will help in any extrapolation of overlapping peaks on the broadened pattern. These precautions will minimize oscillations in the coefficients.

The peaks to be analyzed should be at least 20% broader than the standard peaks, or else the analytical procedures will result in considerable scatter. (This means, e.g., that a particle size larger than about 2000 Å cannot be measured this way. Because of the limited instrumental broadening available at a synchrotron source, sizes up to 4000 Å may be possible).

7.4.3 Errors in Fourier Coefficients

Associated with the actual determination of each Fourier coefficient is an error due to counting statistics. Suppose that a peak is measured by counting for a fixed time T, at a series of points, $R = 2r + 1$, across the profile. The variance for each count, N, is $\sigma^2(N_k) = N_k/T$. Furthermore, let g be the average background and L_0 the background corrected integrated intensity (which is A_0 the lead Fourier coefficient). With an equal number of intervals on both sides of the peak, the variance in the cosine coefficient $\sigma^2(A_n)$ is (Wilson 1967, 1968, 1969; Schlosberg and Cohen 1983):

$$\sigma^2(A_n) = \frac{1}{TL_0^2}\left\{A_n^2(Rg - L_0) + [A_{2n}L_0]/2 + [L_0 + Rg]/2\right.$$

$$\left. + g\left[1/2 \sum_{k=-r}^{k=+r} \cos 4\pi nk/R - 2A_n \sum_k \cos(2\pi nk/R)\right]\right\}.$$

The covariance between A_n and A_m is:

$$\text{cov}(A_n, A_m) = \frac{1}{TL_0^2}\left\{A_n A_m(Rg - L_0) + [A_{n+m}L_0]/2 + [A_{n-m}L_0]/2\right.$$

$$+ g\left[1/2 \sum_k \cos(2\pi[n+m]k/R) + 1/2 \sum_k \cos(2\pi[n-m]k/R)\right]$$

$$\left. - A_m \sum_k \cos(2\pi nk/R) - A_n \sum_k \cos(2\pi mk/R)\right\}.$$

Furthermore:

$$\sigma^2(B_n) = \frac{1}{TL_0^2}\left\{B_n^2(Rg - L_0) - [A_{2n}L_0]/2 + [L_0 + Rg]/2\right.$$

$$\left. - \left[1/2g \sum_k \cos(4\pi nk/R) + 2B_n(G_R - G_L)/R \sum_k k\sin(2\pi nk/R)\right]\right\}.$$

Here G_R, G_L are the background intensities at the ends of the peak, $+r$ and $-r$ respectively.

Thus, not only can the Fourier coefficients be determined from actual data, but their errors may be assigned. These equations follow (with some manipulation) from the equations:

$$A_n = \sum_k [I_k - g]\cos(2\pi nk/R),$$

and from the standard equation for variance:

$$\sigma^2(A_n) = \sum_k \left(\frac{\partial A_n}{\partial k}\right)^2 \sigma^2(I_k).$$

These errors have been propagated to the effects on particle size and microstrain (Schlosberg and Cohen 1983). Therefore, software can be written to determine D and $\langle \varepsilon_L^2 \rangle^{1/2}$ to an operator-specified precision. A first quick pass over a peak provides sufficient information to determine the proper counting times to achieve this precision. Such a procedure is superior to calculating the error *after* a measurement, in which case it may be that too short (or too long) a counting time had been employed. There are also some interesting algorithms for determining the background position if the peak is well above this level (Huang and Parrish 1984).

7.4.4 Some Uses of Particle Size and Strain

With a minimum of assumptions, we are able, with this analysis, to obtain the rms strain $\langle \varepsilon_L^2 \rangle^{1/2}$ *as a function of column length in a given crystallographic direction and the mosaic size* (D_{eff}) *in this direction*. The errors in both quantities are typically ±10 pct. From this information we can proceed to learn about the degree of anisotropy and the imperfections. In a deformed material, $(1/D_{\text{eff}})^2$ can be taken as a *dislocation density*; that is, D_{eff} is the spacing of such imperfections. From the formulas for strain around a dislocation, the data on $\langle \varepsilon_L^2 \rangle^{1/2}$ can also be used to obtain a density. These two should agree if the dislocations are randomly arranged, but if $(1/D_{\text{eff}})^2$ is smaller than the value obtained from the strains, the dislocations may be clustered. The densities calculated in this way are in agreement with those observed in the electron microscope (see Cohen and Hilliard 1966). (If there are cells or clusters of dislocations, D_{eff} is a measure of the cell size not the dislocation spacing, but this can generally be ascertained by comparing the densities obtained with $\langle \varepsilon^2 \rangle$ and D as mentioned.) *This technique is especially useful for high concentrations of dislocations and this is exactly where the electron microscope is least useful* because of overlap of images. By combining the two tools, a whole range of concentrations can be studied, with a region of overlap sufficient to allow comparison of the two.

Line broadening can also be employed to examine inhomogeneities in composition and diffusion. When the composition varies from point to point, so do the lattice parameters and this broadens the peak (Rudman 1960; Carpenter et al. 1971).

7.5 Stacking Disorder

7.5.1 The Geometry of Stacking Faults

There are many cases in materials where faults occur in the stacking of planes of atoms. As an example, consider a close packed fcc material, e.g., copper or silver. As we saw in Sect. 1.6, the stacking sequence of the (111) planes in *ABCABC*. That is, the atoms of each successive layer fit in the holes of the one below, with the fourth layer directly over the atoms in the first. It is also possible to have close packing with the third layer over the first, *BCBC* ... — the hexagonal stacking sequence.

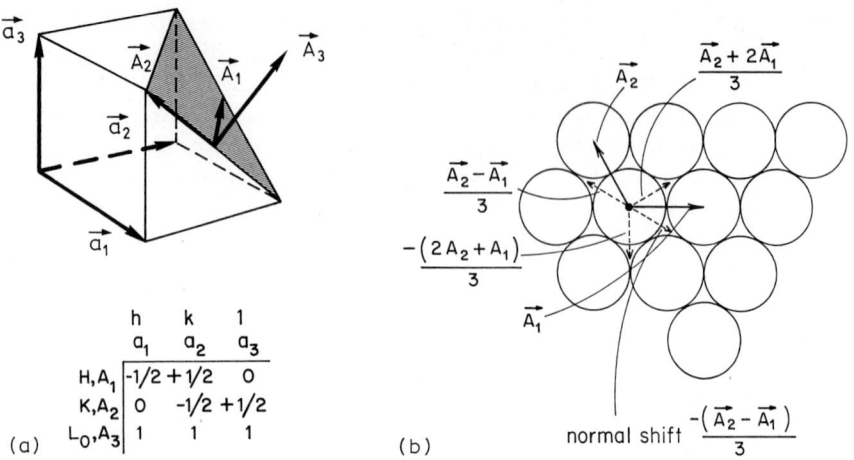

	h	k	l
	a_1	a_2	a_3
H,A_1	-1/2	+1/2	0
K,A_2	0	-1/2	+1/2
L_0,A_3	1	1	1

(a) (b) normal shift $-(\overrightarrow{A_2}-\overrightarrow{A_1})/3$

Fig. 7.19. a Hexagonal axes in a cubic system. **b** Vectors to stacking positions

Table 7.2

	Shift (Δr_n)	$s \cdot \Delta r_n = r^* \cdot \Delta r_n$	$360°(r^* \cdot \Delta r_n) = 360° \times (H\mathbf{B}_1 + K\mathbf{B}_2 + L_0\mathbf{B}_3) \cdot (\Delta r_n)$
Faulted positions	$(\mathbf{A}_2 + 2\mathbf{A}_1)/3$	$(K + 2H)/3$	$-240° = +120°$
	$(\mathbf{A}_2 - \mathbf{A}_1)/3$	$(K - H)/3$	$+120°$
	$(2\mathbf{A}_2 + \mathbf{A}_1)/3$	$-(2K + H)/3$	$+120°$
(Normal shift for the next layer)	$-(\mathbf{A}_2 - \mathbf{A}_1)/3$	$-(K - H)/3$	$-120°$

During the growth of an fcc crystal, or as a result of deformation, one layer may be in a wrong position:

$A\ B\ C\ A\ B\ C\ B\ C\ A\ B\ C\ \ldots$.

This stacking includes a one-layer fault, or in effect a small hexagonal region, $\ldots BCBC \ldots$. It will be easier in the discussion that follows (and for all kinds of layer faults) if we switch now to an axial system with one axis normal to the plane of the fault. Assume the fault is on the (111). Then the hexagonal axes in Fig. 7.19a will be convenient; A_3 is normal to the (111). In Fig. 7.19b, A_1 and A_2 are shown along with the possible positions for a faulted or correctly placed layer. The phase angle associated with each position is given in Table 7.2, for a *hkl* = 200 peak (or $HKL_0 = \bar{1}02$ with the hexagonal axes). [The matrix for transforming indices established by the methods described in Chap. 1, Eq. (1.9f), is given under Fig. 7.19a.] *All three faulted positions cause the same change in phase angle, but this is different than the change for the normal position of a layer*. In the microscope, with the fault running at an angle to the surface of the foil, there would be a *series of fringes* in the image to indicate the presence of the fault, Fig. 7.20. This is caused by the opening and closing of the amplitude circle on the complex plane due to the different phase shift when a fault occurs, as in Fig. 7.8.

7.5 Stacking Disorder

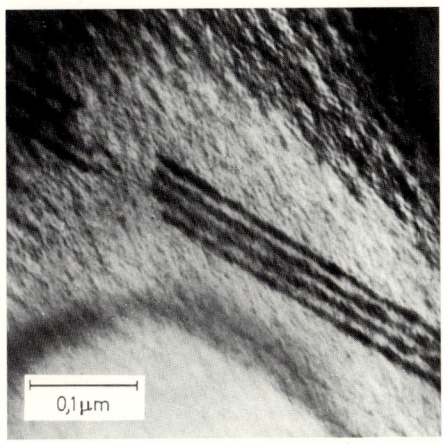

Fig. 7.20. Ni-10% Ti. 150,000 ×, 100 kV. Fringes at a stacking fault. (Photograph by S. Sass)

The resultant *effect on the diffraction pattern can be qualitatively predicted. If we have a fault on several consecutive* (111) *planes stacked ABCAB, then we will produce a twinned region CBACBA* as shown below (i.e., a region related by 180° rotation around the normal to the last plane without a fault):

one fault:	A	B	C	A	B	C	<u>B</u>	C	A	B	C...
two faults:	A	B	C	A	B	C	<u>B</u>	<u>A</u>	B	C...	
three faults:	A	B	C	A	B	C	<u>B</u>	<u>A</u>	<u>C</u>	A	B	C...
(etc.)											

(Note that the faulted layer cannot have the same letter as the previous layer, i.e., it cannot be directly above the previous layer, nor can it have the letter for the layer in the correct sequence.)

If we fault every other layer, we will produce a hexagonal close packed region:

one fault:	A	B	C	A	B	C	<u>B</u>	C	A	B	C...
hexagonal region:	A	B	C	A	B	C	<u>B</u>	C	<u>B</u>	C	A	B	C...

The $H0L$ plane of reciprocal space for an fcc system (with the hexagonal axial system), its twin (rotated 180° around B_3), and a hexagonal structure are shown in Fig. 7.21. As more and more faults occur in a fcc material, so that there are more and more regions that resemble twins and hexagonal regions, the $1\bar{1}\bar{1}$ *and* $\bar{1}11$ *peaks will shift away from the origin to the higher* 2θ *values of a hexagonal material, but the* 200 *and* $\bar{2}00$ *will move toward the origin*; on a powder pattern the 111 and 200 peaks will be closer together than in an annealed specimen. The peaks will first develop tails in these directions when the fault density is low and then gradually change position. But not all the components of a peak in a powder pattern will be affected. The 111 is not affected in this case because the shift $\Delta \mathbf{r}_n$ is normal to $\mathbf{r}^*$, and thus there is no phase change for this $\mathbf{r}^*$; $\Delta \mathbf{r}_n \cdot \mathbf{r}^* = 0$. All peaks but the 111 streak in a [111] direction, because the faults separate the material into small coherently diffracting regions in this direction.

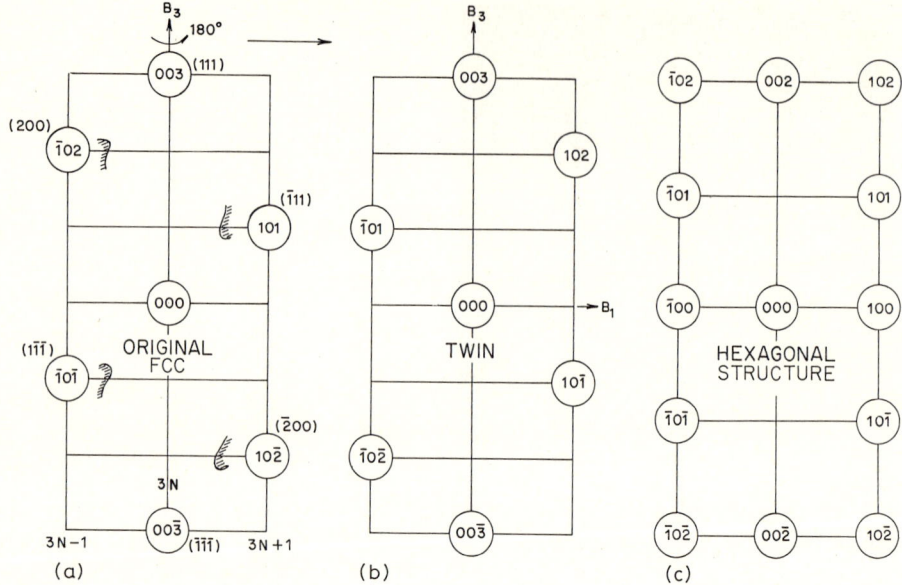

Fig. 7.21. In **a**, indices in the *circles* are hexagonal; those in *parentheses* are the Miller indices of the peak with the (original) cubic axial system. On the right, note that the B_3 axis for the hexagonal cell is $\tfrac{3}{2}$ that for the cubic cell because the hexagonal close-packed structure consists of two (111) layers but the cubic cell is made up of three of these. [After Patterson, M.S., J. Appl. Phys. *23*, 805 (1952).] Schematic diffraction peaks are shown in **a**, with "tails" in the direction expected due to faulting

We can carry out a quantitative treatment by summing over layers in an identical way to the treatment for distortion and particle size. We will assume that:

1. The faults are only on the (111) plane, not any other plane in this form.
2. The fault is as wide as the mosaic region.
3. There is no segregation to the fault if the material is an alloy, so that the scattering per layer is the same in the fault as in other layers.
4. There is no change in the interplanar spacing at the fault, which might occur due to segregation or because there are a different number of neighbors (neighbors in the second layer below a fault are reduced from three to one and are at a different distance).
5. The effect of any strains has been removed by the multiple-order Fourier technique discussed in the previous section.
6. The number of faulted layers is small.

All of these assumptions have been examined in the literature (see Cohen and Hilliard 1966). Assumptions (2–4) can be removed and the effects included in the equations. Assumption (6), as we shall see, is found to be correct from experimental measurements using the equations to be developed. Such an assumption greatly simplifies the mathematics to follow, and as it can be checked, it is a worthwhile starting point. It has been found that to a good approximation, the results for faults on more than one {111} plane in the form are additive.

7.5.2 Fourier Analysis of Faulting

We can write, following the previous treatment of distortion, that[6]

$$I_{eu}(s) = (\bar{f})^2 \bar{N}_{H_3} \frac{\sin^2 \pi s \cdot (\bar{N}_{H_1}) A_1}{\sin^2 \pi s \cdot A_1} \frac{\sin^2 \pi s \cdot (\bar{N}_{H_2}) A_2}{\sin^2 \pi s \cdot A_2}$$

$$\times \sum_n \frac{N_n}{\bar{N}_{H_3}} \langle \exp(-2\pi i [s \cdot \Delta r_n]) \rangle \exp(-2\pi i [n H_3/3]) . \tag{7.22}$$

The interference terms arise from the sums of exponential terms, as in Eq. (2.18), i.e., the terms $\sum_{m_1} \sum_{m_2} \sum_{n_1} \sum_{n_2} \ldots$. The terms $\bar{N}_{H_1}$, $\bar{N}_{H_2}$ are the number of columns in the plane of the fault; $\bar{N}_{H_3}$ is the average number of (111) layers in the coherent regions. N_n is the number of layers with a layer n layers away. The factor 1/3 in the last exponent arises because, in using our hexagonal axial scheme, each atom is related to the first by a vector of the form

$$m_1 A_1 + m_2 A_2 + (m_3 A_3/3) .$$

Each layer contains only one atom per unit cell in the layer with scattering factor $\bar{f}$ if the material is an alloy. That is, the axes A_1, A_2, $A_3/3$ define a layer with only one atom on it so that the structure factor is $\bar{f}$. For simplicity we choose a new axis, $A_3' = A_3/3$.

As the simplest procedure, we shall now *assume* that the *faults are randomly distributed*. Therefore, the presence of one fault does not effect the stacking of the next layers. Then, recalling that Δr_n is the difference in displacement at two layers separated by $(m_3 - n_3) A_3$,

$$\langle \exp(-2\pi i [s \cdot \Delta r_n]) \rangle = \langle \exp(-2\pi i [s \cdot \Delta r_{0 \to 1}]) \exp(-2\pi i [s \cdot \Delta r_{1 \to 2}]) \ldots \rangle$$

$$= \langle \exp(-2\pi i [s \cdot \Delta r_{0 \to 1}]) \rangle \langle \exp(-2\pi i [s \cdot \Delta r_{1 \to 2}]) \rangle \ldots$$

$$= \langle \exp(-2\pi i [s \cdot \Delta r_{0 \to 1}]) \rangle^n .$$

Let α be the probability that a (111) plane is faulted and, considering Table 7.2,

$$\langle \exp(-2\pi i [s \cdot \Delta r_{0 \to 1}]) \rangle^n = \{(1 - \alpha) \exp[-2\pi i (H - K)/3]$$

$$+ \alpha \exp[+2\pi i (H - K)/3]\}^n . \tag{7.23a}$$

There will be no effect due to the fault if $(H - K) = 3N$, where N is an integer. There will be an effect for $(H - K) = 3N \pm 1$, which is the only other possible value for the fcc system. Let

$$\langle \exp(-2\pi i [s \cdot \Delta r_{0 \to 1}]) \rangle = Z \exp(2\pi i y)$$

$$= (1 - \alpha) \exp[-2\pi i (H - K)/3]$$

$$+ \alpha \exp[+2\pi i (H - K)/3] . \tag{7.23b}$$

---

[6] The sign of the exponential terms is negative [rather than positive as in Eq. (7.10)] and its predecessors, because we are considering shifts that correspond to stacking layers in the negative A_3 direction, that is, in Fig. 7.19b, the "correct" position is *below* the one drawn.

Then,

$$Z \exp(2\pi i y + 2\pi i (H-K)/3) = 1 - \alpha + \alpha \cos 4\pi[(H-K)/3]$$
$$+ i\alpha \sin 4\pi[(H-K)/3], \quad (7.24)$$

and multiplying Eq. (7.24) by its complex conjugate:

$$Z \cdot Z^* \equiv Z^2 = (1-\alpha)^2 + 2(1-\alpha)(\alpha)\cos 4\pi(H-K)/3 + \alpha^2. \quad (7.25a)$$

As $\cos 2x = 1 - 2\sin^2 x$,

$$Z^2 = 1 - 4\alpha(1-\alpha)\sin^2[2\pi(H-K)/3]. \quad (7.25b)$$

For small x, $(1-x)^{1/2} \cong 1 - x/2$, so that

$$Z \cong 1 - 2\alpha \sin^2 2\pi(H-K)/3. \quad (7.26)$$

Also taking the imaginary parts in Eq. (7.24),

$$2\pi y + 2\pi(H-K)/3 = \arcsin\{\alpha \sin[4\pi(H-K)/3]/Z\},$$

and for small α, the angle is small and $Z \cong 1$ so that

$$2\pi y \cong -2\pi[(H-K)/3] + \alpha \sin[4\pi(H-K)/3]. \quad (7.27)$$

Recalling that $N_n/\bar{N}_{H_3'} \cong (1 - |n|/\bar{N}_{H_3})$ and substituting (7.23a) and (7.23b) in (7.22), we have,

$$I_{eu}(s) \cong \bar{f}^2 \frac{\sin^2 \pi s \cdot \bar{N}_{H_1} A_1}{\sin^2 \pi s \cdot A_1} \frac{\sin^2 \pi s \cdot \bar{N}_{H_2} A_2}{\sin^2 \pi s \cdot A_2} \bar{N}_{H_3}$$

$$\times \sum_{-\infty}^{+\infty} \left(1 - \frac{|n|}{\bar{N}_{H_3}}\right) Z^{|n|} \exp(2\pi i n [H_3' + y]). \quad (7.28)$$

At this point we can see from the exponential term that there will indeed be a peak shift (the term y) just as we expected. The absolute values for n comes from the fact that in Eq. (7.23), Z is not affected by the sign of n. Furthermore, as the coefficients of the series are the same for $+n$ and $-n$, the exponential form can be replaced by a cosine series.

7.5.3 Peak Shifts Due to Faults

To make our equations useful, we shall basically employ the same approach followed in considering particle size and strain. We must transform our equation to a form suitable for integration in reciprocal space for a $00l'$ reflection so that we can integrate and obtain $P'_{2\theta}$ the power versus angle, which is what we can measure; the result should be a one-dimensional sum. That is, we must take any peak indexed in the hexagonal system and reindex it as a $00l'$ peak for an orthorhombic system, and then obtain $P'_{2\theta}$. So far our equations are for those peaks that are affected by faults, but some components of peaks in a powder pattern are not affected, like the 111 fcc peak (the $11\bar{1}$ component is affected). If we are going to examine a powder pattern or a polycrystalline solid, we must take into account these unaffected components.

7.5 Stacking Disorder

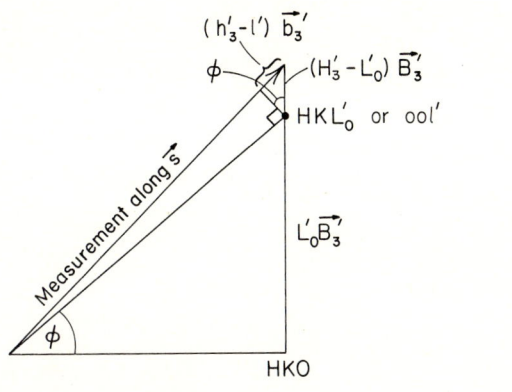

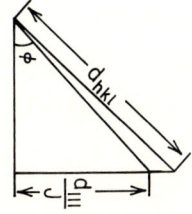

Fig. 7.22. Reciprocal space (**a**) and real space (**b**), showing the diffraction vector $(\mathbf{S} - \mathbf{S}_0)/\lambda = \mathbf{s}$ near an HKL'_0 peak. (Reprinted with permission of Warren, B.E., in "Progress in Metal Physics", Vol. 8. Copyright 1959, Pergamon, New York)

From Fig. 7.22,

$$|\mathbf{b}'_3| h'_3 = 2\sin\theta/\lambda, \tag{7.29}$$

$$l'|\mathbf{b}'_3| = 1/d_{hkl}, \tag{7.30}$$

$$[(H'_3) - L'_0]|\mathbf{B}'_3|\sin\phi = [h'_3 - l']|\mathbf{b}'_3|, \tag{7.31}$$

$$\sin\phi \cong |\mathbf{B}'_3| L'_0/(1/d_{hkl}). \tag{7.32}$$

Also,

$$|\mathbf{B}'_3| = \frac{1}{d_{111}}. \tag{7.33}$$

Furthermore, define:

$$q = |\mathbf{b}'_3|/|\mathbf{B}'_3|\sin\phi. \tag{7.34}$$

From Eq. (7.32), with $l = 1$:

$$q = d_{111}^2/L'_0 d_{hkl}^2. \tag{7.35}$$

For a powder pattern, as shown in Sect. 7.3 for the total power per unit length in a diffraction cone (with M the number of diffracting regions), and employing Eq. (7.28):

$$(P)_{eu} = \frac{MR\lambda^3 (\overline{f})^2 \overline{N}_{H_3}}{8\pi V_c \sin\theta \sin 2\theta} \int\int\int \frac{\sin^2 \pi \overline{N}_{H_1} H_1}{\sin^2 \pi H_1} \frac{\sin^2 \pi \overline{N}_{H_2} H_2}{\sin^2 \pi H_2}$$

$$\times \sum_n \left(1 - \frac{|n|}{\overline{N}_{H_3}}\right) Z^{|n|} \cos 2\pi n [H'_3 + y] \, dH_1 \, dH_2 \, dH_3.$$

However, what we want is not the total power, but instead power per unit angle

2θ. With Eqs. (7.29) and (7.31),

$$dH'_3 = \cos\theta(d\,2\theta)/\lambda|\mathbf{B}'_3|\sin\phi. \tag{7.36}$$

Therefore, integrating over H_1 and H_2 only for one peak to obtain the intensity versus 2θ, $P'_{2\theta}$:

$$(P'_{2\theta})_{eu} = \frac{M\bar{N}_{H_1}\bar{N}_{H_2}\bar{N}_{H_3}R\lambda^2(\bar{f})^2}{16\pi V_c|\mathbf{B}'_3|\sin\phi\sin^2\theta} \times \sum_n \left(1 - \frac{|n|}{\bar{N}_{H_3}}\right) Z^{|n|}\cos 2\pi n[H'_3 + y]. \tag{7.37}$$

We now rewrite this equation in the form describing one peak around the value of $H'_3 = L'_0$, indicating all terms, like $\bar{f}$ and the trigonometric terms which are functions of θ and constant terms, by $K'(\theta)$,

$$(P'_{2\theta})_{eu} = \frac{K'(\theta)}{|\mathbf{B}'_3|\sin\phi} \sum_n \left(1 - \frac{|n|}{\bar{N}_{H_3}}\right) Z^{|n|}\cos 2\pi n[H'_3 - L_0 + (y + L_0)]; \tag{7.38}$$

Eq. (7.38) is in terms of the variable H'_3 in hexagonal reciprocal space. It is now necessary to transform to h'_3 in our substitute orthorhombic cell.

From the definition of q and with $B'_3 = 1/d_{111}$ and $l'h'_3 = 1/d_{hkl}$, $q = (d_{111}/ld_{hkl})\sin\Phi$. In real space (Fig. 7.22b) $\sin\Phi = d_{111}/L'_0 d_{hkl}$. Therefore: $q = L'_0/l$, which is a ratio of integers. Let this ratio be m/n, so that $n = m/q$. With Eqs. (7.31) and (7.34), Eq. (7.38) becomes

$$(P'_{2\theta})_{eu} = \frac{K'(\theta)}{|\mathbf{B}'_3|\sin\phi} \sum \left(1 - \frac{|m|}{|q|\bar{N}_{H_3}}\right) Z^{|m/q|}\cos 2\pi nq\left[H'_3 - l' + \frac{(y + L'_0)}{q}\right].$$

Now,

$$\int A_n \cos 2\pi nqx\, dn = \frac{1}{q}\int A_{m/q}\cos 2\pi mx\, dm. \tag{7.39}$$

Therefore, with Eq. (7.34)

$$(P'_{2\theta})_{eu} = \frac{K'(\theta)}{|\mathbf{b}'_3|} \sum \left(1 - \frac{|m|}{|q|\bar{N}_{H_3}}\right) Z^{|m/q|}\cos 2\pi m\left[(H'_3 - l') + \left(\frac{y + L'_0}{q}\right)\right]. \tag{7.40}$$

We are now ready to put the equation in a form suitable for powders and to examine the terms in more detail. Let

$$\delta = \frac{1}{X}\sum_X \frac{y + L'_0}{q}, \tag{7.41}$$

where X is the number of powder components in any one peak in a powder pattern; i.e., the multiplicity. Normally, the peak occurs for $(h'_3) - l' + \delta_0 = N$. With faults, it occurs for $(h'_3) - l' + \delta = N$ (an integer). Thus,

$$\Delta h'_3 = \text{(position with no faults)} - \text{(position with faults)} = (\delta_0 - \delta).$$

As δ increases, the peak in the cosine series occurs at smaller values of h'_3.

Substituting Eq. (7.27) for y in (7.41),

$$\delta = \frac{1}{X}\sum_X \left\{\frac{L'_0}{q} - \frac{(H-K)}{3q} + \frac{\alpha}{2\pi q}\sin\left[4\pi\frac{(H-K)}{3}\right]\right\}. \tag{7.42}$$

7.5 Stacking Disorder

The first two terms represent δ_0, and so,

$$\Delta h'_3 = -\frac{1}{X}\sum_X \frac{\alpha}{2\pi q} \sin\left[4\pi\frac{(H-K)}{3}\right]. \tag{7.43}$$

With Eq. (7.35),

$$\Delta h'_3 = -\frac{1}{X}\sum_X \frac{\alpha}{2\pi}\frac{d^2_{hkl}L'_0}{d^2_{111}} \sin\left[4\pi\frac{(H-K)}{3}\right]. \tag{7.44}$$

Now $\Delta h'_3 |\mathbf{b}_3| = \cos\theta(\Delta 2\theta)/\lambda$ in radians, so, in degrees:

$$\Delta 2\theta° = (360/2\pi)\Delta h'_3 \lambda/|\mathbf{a}'_3|\cos\theta,$$

and with:

$$|\mathbf{a}'_3| = d_{hkl}, \qquad \lambda = 2d_{hkl}\sin\theta,$$

we obtain

$$(\Delta 2\theta°)_{hkl} = (360/\pi)\tan\theta \Delta h'_3.$$

Hence,

$$(\Delta 2\theta°)_{hkl} = -\frac{180}{\pi^2}\frac{d^2_{hkl}}{d^2_{111}}\alpha\tan\theta\left[\frac{1}{X}\sum_X L'_0 \sin\left(4\pi\frac{(H-K)}{3}\right)\right]. \tag{7.45}$$

Now,

$$1/d^2_{hkl} = h^2_0/a^2_0 \quad \text{and} \quad 1/d^2_{111} = 3/a^2_0.$$

Our final result is

$$\Delta 2\theta° = -(180/\pi^2)(3/h^2_0)\alpha\tan\theta G_{hkl}. \tag{7.46}$$

Some terms for G_{hkl} are given in Table 7.3 for the 111 and 200 peaks of the powder pattern. (Terms for $\overline{hkl}$ give the same result as hkl so that only half the multiplicity need be considered.)

For the 111:

$$G_{111} = \frac{1}{X}\sum_X L'_0 \sin 4\pi\left(\frac{H-K}{3}\right) = -\frac{\sqrt{3}}{8},$$

Table 7.3

(hkl)	H [a]	K [a]	$3L'_0$	$4\pi(H-K)/3$ [b]	$\sin 4\pi(H-K)/3$
111	0	0	3	0	0
11$\bar{1}$	0	$\bar{1}$	1	+240°	$-\sqrt{3}/2$
$\bar{1}$11	1	0	1	+240°	$-\sqrt{3}/2$
1$\bar{1}$1	$\bar{1}$	1	1	$-480° = +240°$	$-\sqrt{3}/2$
200	$\bar{1}$	0	2	$-240°$	$\sqrt{3}/2$
020	1	$\bar{1}$	2	$+480° = -240°$	$\sqrt{3}/2$
002	0	1	2	$-240°$	$\sqrt{3}/2$

[a] Taken from Fig. 7.19a.
[b] Expressed in degrees.

and

$$\Delta 2\theta^\circ_{111} = +(45\sqrt{3}/2\pi^2)\alpha \tan\theta. \tag{7.47}$$

For the 200:

$$G_{200} = \sqrt{3},$$

and

$$\Delta 2\theta^\circ_{200} = -(90\sqrt{3}/2\pi^2)\alpha \tan\theta. \tag{7.48}$$

These equations are simple enough when we finally get them! The 200 powder peak shifts to low angles and the 111 to high angles, just as we anticipated, and *from the shifts, we can measure α. We want to minimize errors due to the positioning of the standard and deformed samples* in the diffractometer. (Recall from Sect. 4.7 that if the specimen is slightly displaced from the center of the goniometer there will be a peak shift, which is largest for low-angle peaks like the 111 and 200.) *We also want to eliminate any back lash in the goniometer gears.* We do the latter by recording the 111 and 200 peaks continuously and minimize the effect of sample position *by taking the change in the separation of these peaks for two specimens*:

$$\begin{aligned}\Delta 2\theta^\circ_{200} - \Delta 2\theta^\circ_{111} &= (2\theta^{\text{deformed}}_{200} - 2\theta^{\text{deformed}}_{111}) - (2\theta^{\text{annealed}}_{200} - 2\theta^{\text{annealed}}_{111}) \\ &= (2\theta^{\text{deformed}}_{200} - 2\theta^{\text{annealed}}_{200}) - (2\theta^{\text{deformed}}_{111} - 2\theta^{\text{annealed}}_{111}) \\ &= -(45\sqrt{3}/2\pi^2)\alpha(2\tan\theta_{200} - \tan\theta_{111}). \end{aligned} \tag{7.49}$$

The difference in the change in 111 and 200 peak positions for each specimen is involved and any effect of sample displacement is small, as these peaks are close together.

7.5.4 Peak Broadening Due to Faults

Next, consider the Fourier coefficients, which represent the broadening of a peak:

$$Z^{|m/q|} = 1 + |m/q|\log Z + [|m/q|\log Z]^2/2 + \cdots.$$

For small m and $Z \cong 1$, only the first two terms are important, and it is from the values for small m that, as we have seen, we get our "particle" size — in this case the spacing between faults. Hence,

$$Z^{|m/q|} \cong 1 + (|m/q|)\log\{1 - 2\alpha\sin^2[2\pi(H-K)/3]\}.$$

$\text{Log}(1-x) = -x$ for small x so that for small $|m/q|$,

$$Z^{|m/q|} \cong 1 - |m/q|\{2\alpha\sin^2[2\pi(H-K)/3]\}.$$

Thus, from Eq. (7.40),

$$A_{|m/q|} = \frac{1}{X}\sum_X \left\{\left(1 - \frac{|m|}{|q|\bar{N}_{H_3}}\right)\left[1 - \frac{|m|}{|q|}2\alpha\sin^2 2\pi\frac{(H-K)}{3}\right]\right\},$$

and, using Eqs. (7.34) and (7.35) for q as well as

7.5 Stacking Disorder

$$B'_3 = \frac{1}{d_{111}} \quad \text{and} \quad |L| = md_{hkl} = \frac{|m|}{|\mathbf{b}'_3|},$$

$$A_{|m/q|} \cong 1 - \frac{|m|}{|\mathbf{b}'_3|} \frac{1}{X} \sum_X \left\{ \frac{|\sin\varphi|}{\bar{N}_{H'_3} d_{111}} \right\} - |m| \frac{1}{X} \sum_X \frac{|L'_0| d^2_{hkl}}{d^2_{111}} 2\alpha \sin^2 2\pi \frac{(H-K)}{3},$$

and with Eq. (7.26)

$$A_{|m/q|} = 1 - L \left\{ \frac{1}{X} \sum_X \frac{|\sin\varphi|}{\bar{N}_{H'_3} d_{111}} + \frac{1}{X} \sum_X \frac{3|L'_0|}{a_0 h_0} (1-Z) \right\}.$$

Call the first sum $1/D$. It is, after all, a particle size term, due to effects other than faults; i.e., consider

$$D = \bar{N}_{H'_3} d_{111}/|\sin\varphi| = \bar{N}_{H'_3}/|\mathbf{B}'_3| \sin\Phi .$$

The numerator is the average thickness of a mosaic region in the [111] direction, and $1/\sin\varphi$, as can be seen in Fig. 7.22a, gives the projected size in the measuring direction **s**.

Evaluating the second sum using Table 7.3, we have for the 111 peak:

$$A_L = 1 - L\{(1/D) + (1.5\alpha/a_0)(\sqrt{3}/4)\},$$

and for the 200:

$$A_L = 1 - L\{(1/D) + (1.5\alpha/a_0)\} .$$

Thus from the initial slope of the Fourier cosine coefficients[7] versus L we obtain

$$(1/D_{\text{eff}})_{111} = (1/D) + (1.5\alpha/a_0)(\sqrt{3}/4) , \tag{7.50a}$$

and

$$(1/D_{\text{eff}})_{200} = (1/D) + (1.5\alpha/a_0) . \tag{7.50b}$$

If D is large,

$$(D_{\text{eff}})_{111}/(D_{\text{eff}})_{200} = 4/\sqrt{3} = 2.3 .$$

The 200 peak will be broader than the 111 because the effective particle size for the 200 is smaller. With α measured from peak shifts, D can be obtain from D_{eff}. *The unusual shifts and the difference in broadening of the 111 and 200 are characteristic of faults.* If there were residual tensile strains, say normal to the diffracting planes, this would cause *both the 111 and 200 to have larger than normal d values* and hence to appear at lower angles; because $\Delta\theta = -(\Delta d/d)\tan\theta$, the shift would be greater for the 200 than the 111, and these would move closer together. However, so would the 222 and 400, whereas, as the reader can verify for himself by evaluating the terms for these peaks, *faults on the (111) will cause these two high-angle peaks to separate*, in contrast to the 111–200 peaks.

Typical values of α in filings of copper and gold and their alloys are of the order of 0.01 to 0.05. Our assumption of small α was valid! The corresponding changes

---

[7] If there is strain present, these are the cosine coefficients *after* extrapolating out strain in a plot of $\ln A_L$ vs. h_0^2 as in Fig. 7.13b.

in separation of the 111 and 200 peaks are about 0.04° 2θ to 0.2° 2θ with Cu K_α. These values of α imply that there is one fault every 100 to every 20 (111) planes ($1/\alpha$ is the average number of planes between faults). After severe deformation all four $\{111\}$ planes are probably involved and the effects are additive; it is more likely that each of the $\{111\}$ planes is faulted every 400 to every 80 planes ($4/\alpha$).

We have examined the mathematics in detail for this case because the same techniques can be followed for faults of different kinds on different planes — and other than random distributions can be examined. Basically, we have seen how to work from a convenient reciprocal space (hexagonal in this case) to a single measuring variable (the $00h'_3$ line) and how to average our expressions for powder patterns. We have also seen that faults can cause rather specific effects on the pattern. There is a basic philosophy in this kind of derivation — to attempt to use whatever approximations are required to get the equations in a form usable with actual data, and then to see if the approximations were justified. Faulting occurs not only in metals, but in minerals and oxides. For example, many oxide structures consist of oxygen octahedra containing metal ions, the structure being built up by corner sharing by these octahedra. Deviations from stoichiometry are accommodated at planes where two octahedra share edges.

There are many elegant treatments of layer faults in the literature. For example, in Warren's work (1969), the reader will find a treatment that includes twinned regions or collections of faults. Recently (Cowley 1975, 1976) a general theory has been developed involving the Patterson function particularly suitable for studies with single crystals and hence, for effects seen in the electron microscope. See also Howard (1977).

7.6 Local Ordering and Clustering

7.6.1 Effects of Concentration Fluctuations on the Diffraction Pattern

In covalently or ionically bonded structures it is generally unlikely that a given atom type will mix with another on a given equipoint, but this can occur in alloys and minerals. This problem of degree of order, the occupancy of an equipoint or sublattice was discussed in Sect. 6.4 and Problem 6.19. The intensity of certain sharp peaks, the extra or superlattice reflections were affected. In Sect. 3.6 we found that if different types of atoms are randomly distributed, there are no extra peaks, but instead a weak continuous background — the Laue monotonic scattering. (There we considered isotopes and neutron scattering, but this is the same situation, for x-rays or neutrons, if there is a random distribution of solute.) If the atoms are arranged in a less than random fashion, we shall see in this section that this scattering becomes more localized in the form of broad maxima that sharpen into extra or super-lattice reflections as order increases, or become the peaks of a precipitate.

7.6 Local Ordering and Clustering

Considering N atoms, one per (primitive) unit cell in the cubic system:

$$I_{eu}(\mathbf{s}) = \sum_m \sum_n f_m f_n \exp(2\pi i[\mathbf{s} \cdot (\mathbf{r}_m - \mathbf{r}_n)]) . \tag{7.51a}$$

This can be rewritten with $\mathbf{r}_{mn} = \mathbf{r}_m - \mathbf{r}_n$ as

$$I_{eu}(\mathbf{s}) = \sum_m \sum_n (\bar{f})^2 \exp(2\pi i[\mathbf{s} \cdot \mathbf{r}_{mn}]) + \sum_m \sum_n [f_m f_n - (\bar{f})^2] \exp(2\pi i[\mathbf{s} \cdot \mathbf{r}_{mn}]) . \tag{7.51b}$$

The average intensity for all the atoms in the specimen (assuming no displacements of atoms from the lattice points) is:

$$I_{eu}(\mathbf{s}) = \underbrace{\sum_m \sum_n (\bar{f})^2 \exp(2\pi i[\mathbf{s} \cdot \mathbf{r}_{mn}])}_{I_{eu_1}} + \underbrace{\sum_m \sum_n [\langle f_m f_n \rangle - (\bar{f})^2] \exp(2\pi i[\mathbf{s} \cdot \mathbf{r}_{mn}])}_{I_{eu_2}} . \tag{7.51c}$$

The first term, I_{eu_1}, represents the Bragg peaks with "average" atoms at each lattice site. We now evaluate I_{eu_2}.

Consider a binary alloy. Let P_{mn}^{AA} be the conditional probability that there is an A atom at the end of the vector $\mathbf{r}_{mn}$ *if* there is an A atom at the origin of this interatomic vector. Similarly, let P_{mn}^{AB} be the conditional probability that there is a B at n *if* there is an A at m, and so on for P_{mn}^{BB}, P_{mn}^{BA}.

Because we have chosen to work with a binary system, the atomic fraction c_A and c_B (of A and B atoms), are related.

$$c_A + c_B = 1 . \tag{7.52a}$$

Also since the vector from an A atom must end on an A or B atom (and similarly for a B atom):

$$P_{mn}^{AA} + P_{mn}^{AB} = 1 , \tag{7.52b}$$

$$P_{mn}^{BA} + P_{mn}^{BB} = 1 . \tag{7.52c}$$

The number of AB pairs does not depend on whether we start with the A or B atom in the pair; hence,

$$Nc_A P_{mn}^{AB} = Nc_B P_{mn}^{BA} . \tag{7.52d}$$

[The term Nc_A is the total number of starting A atoms in all the pairs and multiplying this by P_{mn}^{AB} yields the total number of AB (not BA) pairs.]

With these probabilities we can evaluate $\langle f_m f_n \rangle$, the average value of $f_m f_n$ for all atoms with the same $\mathbf{r}_{mn}$ under the x-ray beam,

$$\langle f_m f_n \rangle = c_A f_A P_{mn}^{AA} f_A + c_B f_B P_{mn}^{BB} f_B + (c_A P_{mn}^{AB} + c_B P_{mn}^{BA}) f_A f_B .$$

Eliminating all the pair probabilities but P_{mn}^{AB} with Eqs. (7.52b)–(7.52d),

$$\langle f_m f_n \rangle = c_A f_A^2 + c_B f_B^2 - c_A P_{mn}^{AB} (f_A - f_B)^2 .$$

Now

$$(\bar{f})^2 = (c_A f_A + c_B f_B)^2$$

and adding and subtracting $c_A c_B (f_A - f_B)^2$,

$$(\bar{f})^2 = c_A f_A^2 + c_B f_B^2 - c_A c_B (f_A - f_B)^2 \ .$$

Therefore:

$$\langle I_{eu_2}(\mathbf{s}) \rangle = \sum_m \sum_n c_A (c_B - P_{mn}^{AB})(f_A - f_B)^2 \cos 2\pi \mathbf{s} \cdot \mathbf{r}_{mn} \ .$$

The cosine term is all that is necessary; for every vector $\mathbf{r}_{mn}$ in the sum, there occurs the opposite vector $\mathbf{r}_{nm}$ or $-\mathbf{r}_{mn}$ and according to Eq. (7.52d) the P_{mn}^{AB} term is the same for *both* vectors. We define the "Warren short-range order parameter" α_{mn}:

$$\alpha_{mn} \equiv 1 - (P_{mn}^{AB}/c_B) \equiv 1 - (P_{mn}^{BA}/c_A) \ . \tag{7.53a}$$

Therefore,

$$\langle I_{eu_2}(\mathbf{s}) \rangle = \sum_m \sum_n c_A c_B (f_A - f_B)^2 \alpha_{mn} \cos 2\pi \mathbf{s} \cdot \mathbf{r}_{mn} \ . \tag{7.53b}$$

This diffuse intensity is simply the Fourier transforms of the difference Patterson function described in Sect. 6.7, i.e., the scattering due to the difference between the actual local configuration and the random structure.

If there is no long-range order, α_{mn} *approaches zero beyond a small number of near neighbors from each atom.* That is, P_{mn}^{AB} approaches its random value c_B so that α_{mn} approaches zero. The double sum in Eq. (7.53) can then be replaced by N (the number of atoms in the volume diffracting)[8] times a single sum over the neighbors of an average atom;

$$\langle I_{eu_2}(\mathbf{s}) \rangle = N c_A c_B (f_A - f_B)^2 \sum_m \alpha_m \cos 2\pi \mathbf{s} \cdot \mathbf{r}_m \ . \tag{7.54}$$

The sum over m really implies a triple sum in terms of the coordinates lmn of each vector around the average atom. In a cubic system these vectors can be written as $[l + m + n]\mathbf{a}_0$. For example, from a corner to a face site in an fcc cell one vector is $\mathbf{a}_0[\frac{1}{2}, \frac{1}{2}, 0]$.

$$\langle I_{eu_2}(\mathbf{s}) \rangle = N c_A c_B (f_A - f_B)^2 \sum_{l=-\infty}^{\infty} \sum_{m=-\infty}^{\infty} \sum_{n=-\infty}^{\infty} \alpha_{lmn} \cos 2\pi (h_1 l + h_2 m + h_3 n) \ . \tag{7.55}$$

Whether the alloy is random or not, $\alpha_0 = 1$ (P_0^{AB}, the probability of a B at an A atom's position, is zero). But the other α's are zero if the alloy is random, because the $P_{mn}^{AB} = c_B$ in Eq. (7.53a) for all vectors. Then only the first term in the series exists, which is the Laue monotonic scattering. If there is local order (local $ABAB$ regions) or clustering (local regions rich in A or B) for an fcc structure, which means $\alpha_{1/2, 1/2, 0}$, $\alpha_{1, 0, 0} \neq 0$, this introduces cosine terms modulating the Laue monotonic, raising it in some places and lowering it in others. The total scattering due to local order (the first term in the Fourier series proportional to α_0) remains constant regardless of the arrangement. The sharpness of the modulation will depend on the number of α's different than zero, the extent of local order around atoms.

Each α can be positive or negative. If there is local order, then $\alpha_{1/2, 1/2, 0}$ is negative, because $P_{1/2, 1/2, 0}^{AB} > c_B$. As an example, consider perfectly ordered Cu_3Au, with

---

[8] For a solid flat specimen in a diffractometer, from Sect. 4.9, the volume irradiated for a *beam of unit cross section* is $1/2(\mu/\rho)\rho$. If N_0 is the number of atoms per unit volume, $N = N_0/2(\mu/\rho)\rho$.

7.6 Local Ordering and Clustering

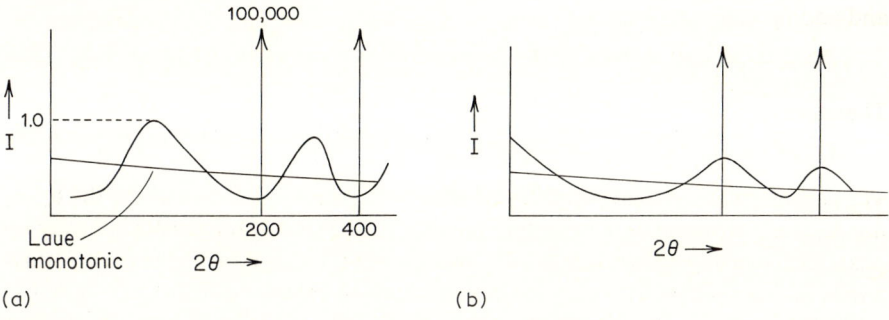

Fig. 7.23.a Scattering from an alloy single crystal due to local order. **b** Scattering due to clustering of like atoms. Bragg peaks indicated by *vertical arrows*

gold at the corners and copper at the faces of a cubic unit cell; the first neighbors to a corner gold are all copper atoms and $\alpha_{1/2,1/2,0}$, $\alpha_{0,1/2,1/2}$, etc. $= 1 - \frac{1}{0.75} = -\frac{1}{3}$. The terms α_{100}, α_{010}, α_{001}, etc., corresponding to second neighbor atoms at corners of the unit cell, are unity, etc. The first cosine term with coefficient $\alpha_{1/2,1/2,0}$ is negative at $\mathbf{s} = 0$, and peaks occur for h_1, h_2, h_3 such that $2\pi(h_1 l + h_2 m + h_3 n) = \pi$, i.e., at the positions 100, 110, 300 .... If there is clustering, $P^{AB}_{1/2,1/2,0} < X_B$ and the intensity peaks near the origin, and for $2\pi(h_1 l + h_2 m + h_3 n) = 0, 2\pi, \ldots$, i.e., right at the Bragg peaks 000, 200, 400, ... These two situations are sketched in Fig. 7.23, along with some typical intensities for a single crystal with $c_A c_B (f_A - f_B)^2 \cong 500$ (e.g., such as Cu$_3$Au). *With a powder sample, the* diffuse scattering is of the order of 1% of the peaks. In this case, Eq. (7.55) would have to be averaged for all orientations of **r** and **s** (see Sect. 7.8 for the equation).

The diffuse scattering has a half breadth of several degrees 2θ as compared to about $0.02°$ 2θ for the Bragg peaks, and broad receiving slits can be used to make it easier to detect this scattering, without appreciably affecting its shape.

If the data are gathered in a volume of reciprocal space, *Fourier cosine inversions* (Sect. 6.2) *will yield the α's.* Before the data can be inverted, however, there are several important steps. Although the scattering can often be detected with a slow scan, chart recording, broad slits, and a pulse-height analyzer, if any accuracy at all is to be obtained for the α's, it is necessary to use a monochromator or solid-state detector and to point count. *The background of air scattering and electronic noise must be measured and subtracted. Compton scattering must be corrected for by calculation, and this requires that the intensities be measured absolutely in electron units.* How this is done is described in Appendix C.

For a cubic system, such as fcc, Eq. (7.55) can be simplified further; by expanding the trigonometric term

$$\langle I_{eu_2}(\mathbf{s}) \rangle = N c_A c_B (f_A - f_B)^2 \sum_{l}^{+\infty} \sum_{m} \sum_{n} \alpha_{lmn} [\cos 2\pi h_1 l \cos 2\pi h_2 m \cos 2\pi h_3 n$$

$$- \cos 2\pi h_1 l \sin 2\pi h_2 m \sin 2\pi h_3 n - \sin 2\pi h_1 l \sin 2\pi h_2 m \cos 2\pi h_3 n$$

$$- \sin 2\pi h_1 l \cos 2\pi h_2 m \sin 2\pi h_3 n] .$$

Now for a cubic material $\alpha_{lmn} = \alpha_{l\bar{m}\bar{n}} = \alpha_{lm\bar{n}} = \alpha_{\bar{l}mn} = \alpha_{\bar{l}\bar{m}n}$ etc. This eliminates the last three terms, that is these terms for lmn, $\bar{l}mn$, $\bar{l}\bar{m}n$, etc. cancel at any $h_1 h_2 h_3$. Thus we can write:

$$\langle I_{eu_2}(s) \rangle = Nc_A c_B (f_A - f_B)^2 \sum_l \sum_m \sum_n \alpha_{lmn} \cos 2\pi h_1 l \cos 2\pi h_2 m \cos 2\pi h_3 n \, . \quad (7.56)$$

We immediately see that this diffuse intensity has a period of two along the $\mathbf{b}_1$, $\mathbf{b}_2$ and $\mathbf{b}_3$ axes, because l, m, n in fcc has values $p/2$. The short-range order intensity repeats itself in a unit cell with edges, $2b_i$. But it is not even necessary to make measurements in that volume. It is easy to show that only a volume for h_i from 0–1 (or 1–2 etc.) is sufficient for fcc. We have to prove $I_{eu_2}(1 + h_1, h_2, h_3) = I_{eu_2}(1 - h_1, h_2, h_3)$. This can be done by inserting the term $(h_1 + 1)$ for h_1 in Eq. (7.56), and expanding this cosine term, and repeating this for $(1 - h_1)$. Then $\cos 2\pi(1 + h_1)l$ can be expanded to $\cos 2\pi l h_1 \cos 2\pi l - \sin 2\pi l h_1 \sin 2\pi l$ and $\cos 2\pi(1 - h_1)l$ can be written as $\cos 2\pi l h_1 \cos 2\pi l + \sin 2\pi l h_1 \sin 2\pi l$. The quantity l is an integer for fcc and thus the sine terms vanish, and the two terms are equal. This is a formal proof that (100) is a mirror plane in the reciprocal space of these structures.

By quite similar arguments, (110) type planes can be shown to be mirror planes in reciprocal space and the symmetry is shown in Fig. 7.24a. Furthermore, for an fcc structure, $(2l + 2m + 2n)$ is an even number. The phase factor in Eq. (7.51), is $\exp 2\pi i[lh_1 + mh_2 + nh_3]$. Therefore, from this parity condition, for any point $(1 \pm h_1, 1 \pm h_2, 1 \pm h_3)$, I_{eu_2} is symmetrically related to the intensity at h_1, h_2, h_3. Thus the minimum volume is reduced still further, as in Fig. 7.24b. (*Note the combination of symmetry elements for this case.*) The intensity needs to be measured only in this small volume, and symmetry can be employed to produce the intensity in a full unit cell.

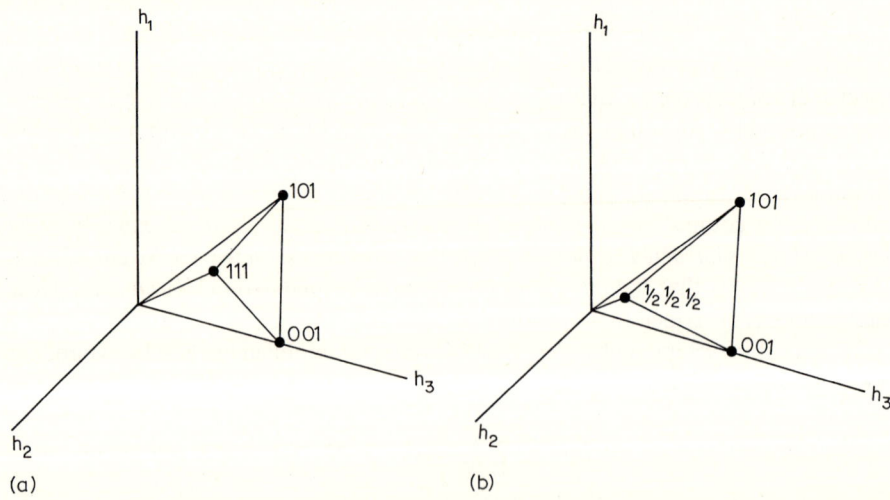

Fig. 7.24 a The volume reduced by symmetry for I_{SRO}: $(h_1, h_2, h_3) = (h_1, h_3, h_2)$; $(h_1, h_2, h_3) = (h_3, h_2, h_1)$; and (h_1, h_2, h_3). **b** The volume in **a** reduced by the symmetry for I_{SRO}: $(h_1, h_2, h_3) = (1 - h_1, 1 - h_3, 1 - h_2)$. (From Gragg, J.E., Jr., Ph.D. Thesis, Northwestern University, 1970)

7.6.2 Including the Effects of Atomic Displacements

If the atoms have different sizes, the static displacements from lattice sites will cause additional scattering. We have mentioned this kind of scattering earlier in Sect. 7.2. The scattering from this effect and from local order can be separated, and information can then be obtained about the difference in interatomic separation as a function of distance between atoms, as well as the α's. While qualitative work can be done with powders, it is difficult to do these separations unless single crystals are employed. We, therefore, shall now examine a treatment for single crystals developed by Georgopoulos and Cohen (1977, 1984) which includes these "size effects". We start with Eq. (7.51a).

Allowing the atoms to be displaced from their ideal lattice positions by small amounts, the vectors $\mathbf{r}_m$ and $\mathbf{r}_n$ can be expressed as:

$$\mathbf{r}_m = \mathbf{R}_m + \boldsymbol{\delta}_m, \qquad \mathbf{r}_n = \mathbf{R}_n + \boldsymbol{\delta}_n, \tag{7.58a}$$

where $\mathbf{R}_m$, $\mathbf{R}_n$ are ideal lattice vectors and $\boldsymbol{\delta}_m$, $\boldsymbol{\delta}_n$ are small displacement vectors from the ideal lattice sites. Then, Eq. (7.51a) can be written as:

$$I_T = \sum_m \sum_n f_m f_n \exp[i\mathbf{k} \cdot \mathbf{R}_{mn}] \exp[i\mathbf{k} \cdot \boldsymbol{\delta}_{mn}], \tag{7.58a}$$

where:

$$\mathbf{R}_{mn} = \mathbf{R}_m - \mathbf{R}_n \quad \text{and} \quad \boldsymbol{\delta}_{mn} = \boldsymbol{\delta}_m - \boldsymbol{\delta}_n. \tag{7.58b}$$

Using the conditional pair probability P_{mn}^{AB} the spatially averaged intensity can be expressed as:

$$\begin{aligned}
\langle I_T \rangle = \sum_m \sum_n \big(& c_A f_A^2 P_{mn}^{AA} \langle \exp[i\mathbf{k} \cdot \boldsymbol{\delta}_{mn}^{AA}] \rangle \\
& + c_A f_A f_B P_{mn}^{AB} \langle \exp[i\mathbf{k} \cdot \boldsymbol{\delta}_{mn}^{AB}] \rangle \\
& + c_B f_A f_B P_{mn}^{BA} \langle \exp[i\mathbf{k} \cdot \boldsymbol{\delta}_{mn}^{BA}] \rangle \\
& + c_B f_B^2 P_{mn}^{BB} \langle \exp[i\mathbf{k} \cdot \boldsymbol{\delta}_{mn}^{BB}] \rangle \big) \exp[i\mathbf{k} \cdot \mathbf{R}_{mn}].
\end{aligned} \tag{7.59}$$

where the carats imply an average over either AA, or AB or BB pairs.

If the displacements of atoms from their ideal lattice sites are assumed small compared with the interatomic distance, the exponentials in Eq. (7.59) can be expanded as follows, denoting by i, j the species on the m or n site (i and j stand for A or B):

$$\langle \exp[i\mathbf{k} \cdot \boldsymbol{\delta}_{mn}^{ij}] \rangle = \Big\langle 1 + i\mathbf{k} \cdot \boldsymbol{\delta}_{mn}^{ij} - \frac{1}{2}(\mathbf{k} \cdot \boldsymbol{\delta}_{mn}^{ij})^2$$
$$- \frac{1}{3!}(\mathbf{k} \cdot \boldsymbol{\delta}_{mn}^{ij})^3 + \frac{1}{4!}(\mathbf{k} \cdot \boldsymbol{\delta}_{mn}^{ij})^4 + \cdots \Big\rangle. \tag{7.60a}$$

Grouping all odd-powered terms together and factoring out $i\mathbf{k} \cdot \boldsymbol{\delta}_{mn}^i$:

$$\langle \exp[i\mathbf{k} \cdot \boldsymbol{\delta}_{mn}^{ij}] \rangle = \Big\langle \Big[1 - \frac{1}{2}(\mathbf{k} \cdot \boldsymbol{\delta}_{mn}^{ij})^2 + \frac{1}{4!}(\mathbf{k} \cdot \boldsymbol{\delta}_{mn}^{ij})^4 + \cdots \Big]$$
$$+ i\mathbf{k} \cdot \boldsymbol{\delta}_{mn}^{ij} \Big[1 - \frac{1}{3!}(\mathbf{k} \cdot \boldsymbol{\delta}_{mn}^{ij})^2 + \cdots \Big] \Big\rangle. \tag{7.60b}$$

Now both terms in square brackets are approximately Taylor expansions of $\exp[-\frac{1}{2}(\mathbf{k}\cdot\boldsymbol{\delta}_{mn}^{ij})^2]$. Hence:

$$\begin{aligned}\langle\exp[i\mathbf{k}\cdot\boldsymbol{\delta}_{mn}^{ij}]\rangle &= \exp[-\tfrac{1}{2}\langle(\mathbf{k}\cdot\boldsymbol{\delta}_{mn}^{ij})^2\rangle]\cdot(1+i\langle\mathbf{k}\cdot\boldsymbol{\delta}_{mn}^{ij}\rangle)\\ &= \exp[-\tfrac{1}{2}\langle(\mathbf{k}\cdot\boldsymbol{\delta}_{m}^{i})^2\rangle]\exp[-\tfrac{1}{2}\langle(\mathbf{k}\cdot\boldsymbol{\delta}_{n}^{j})^2\rangle]\\ &\quad\times\exp[\langle(\mathbf{k}\cdot\boldsymbol{\delta}_{m}^{i})(\mathbf{k}\cdot\boldsymbol{\delta}_{n}^{j})\rangle](1+i\langle\mathbf{k}\cdot\boldsymbol{\delta}_{mn}^{ij}\rangle).\end{aligned} \quad (7.61)$$

The first two exponentials on the right-hand side of Eq. (7.61) involve lattice averages of squared displacements of atoms and depend only on the atomic species involved. Hence, they are nothing but Debye-Waller like factors, M_i. Equation (7.61) can then be rewritten as:

$$\langle\exp[i\mathbf{k}\cdot\boldsymbol{\delta}_{mn}^{ij}]\rangle = (1+i\langle\mathbf{k}\cdot\boldsymbol{\delta}_{mn}^{ij}\rangle+\langle(\mathbf{k}\cdot\boldsymbol{\delta}_{m}^{i})(\mathbf{k}\cdot\boldsymbol{\delta}_{n}^{j})\rangle)\exp(-M_i)\exp(-M_j), \quad (7.62)$$

expanding $\exp\langle(\mathbf{k}\cdot\boldsymbol{\delta}_{m}^{i})(\mathbf{k}\cdot\boldsymbol{\delta}_{n}^{j})\rangle$, multiplying and keeping terms up to the second order in displacements.

Substituting (7.62) in (7.59) and grouping together terms with the same powers of δ, one obtains (with f corrected for the mean square displacements, that is corrected for the term M_i):

$$\begin{aligned}\langle I_T\rangle = \sum_m\sum_n[&c_Af_A^2 P_{mn}^{AA} + c_Af_Af_B P_{mn}^{AB} + c_Bf_Bf_A P_{mn}^{BA} + c_Bf_B^2 P_{mn}^{BB}\\ &+ ic_Af_A^2\langle\mathbf{k}\cdot\boldsymbol{\delta}_{mn}^{AA}\rangle P_{mn}^{AA} + ic_Af_Af_B\langle\mathbf{k}\cdot\boldsymbol{\delta}_{mn}^{AB}\rangle P_{mn}^{AB}\\ &+ ic_Bf_Af_B\langle\mathbf{k}\cdot\boldsymbol{\delta}_{mn}^{BA}\rangle P_{mn}^{BA} + ic_Bf_B^2\langle\mathbf{k}\cdot\boldsymbol{\delta}_{mn}^{BB}\rangle P_{mn}^{BB}\\ &+ c_Af_A^2\langle(\mathbf{k}\cdot\boldsymbol{\delta}_{m}^{A})(\mathbf{k}\cdot\boldsymbol{\delta}_{n}^{A})\rangle P_{mn}^{AA} + c_Af_Af_B\langle(\mathbf{k}\cdot\boldsymbol{\delta}_{m}^{A})(\mathbf{k}\cdot\boldsymbol{\delta}_{n}^{B})\rangle P_{mn}^{AB}\\ &+ c_Bf_Af_B\langle(\mathbf{k}\cdot\boldsymbol{\delta}_{m}^{B})(\mathbf{k}\cdot\boldsymbol{\delta}_{n}^{A})\rangle P_{mn}^{BA}\\ &+ c_Bf_B^2\langle(\mathbf{k}\cdot\boldsymbol{\delta}_{m}^{B})(\mathbf{k}\cdot\boldsymbol{\delta}_{n}^{B})\rangle P_{mn}^{BB}]\exp[i\mathbf{k}\cdot\mathbf{R}_{mn}].\end{aligned} \quad (7.63)$$

We will use Eq. (7.52), the definition of α_{mn} [Eq. (7.53a)] and the following relationships that follow from these expressions:

$$c_A P_{mn}^{AB} = c_B P_{mn}^{BA} = c_A c_B(1-\alpha_{mn}), \quad (7.64a)$$

$$P_{mn}^{AA} = c_A + c_B\alpha_{mn}, \quad (7.64b)$$

$$P_{mn}^{BB} = c_B + c_A\alpha_{mn}. \quad (7.64c)$$

Also, as the material has an average lattice:

$$c_A P_{mn}^{AA}\langle\boldsymbol{\delta}_{mn}^{AA}\rangle + c_A P_{mn}^{AB}\langle\boldsymbol{\delta}_{mn}^{AB}\rangle + c_B P_{mn}^{BA}\langle\boldsymbol{\delta}_{mn}^{BA}\rangle + c_B P_{mn}^{BB}\langle\boldsymbol{\delta}_{mn}^{BB}\rangle = 0. \quad (7.65a)$$

This can be written with Eq. (7.64) as:

$$-[\langle\mathbf{k}\cdot\boldsymbol{\delta}_{mn}^{AB}\rangle+\langle\mathbf{k}\cdot\boldsymbol{\delta}_{mn}^{BA}\rangle] = \frac{c_A/c_B+\alpha_{mn}}{1-\alpha_{mn}}\langle\mathbf{k}\cdot\boldsymbol{\delta}_{mn}^{AA}\rangle + \frac{c_B/c_A+\alpha_{mn}}{1-\alpha_{mn}}\langle\mathbf{k}\cdot\boldsymbol{\delta}_{mn}^{BB}\rangle \quad (7.65b)$$

with Eqs. (7.64) and (7.65):

7.6 Local Ordering and Clustering

$$\langle I_T \rangle = (c_A f_A + c_B f_B)^2 \sum_m \sum_n \exp[i\mathbf{k} \cdot \mathbf{R}_{mn}]$$

$$+ c_A c_B (f_A - f_B)^2 \sum_m \sum_n \alpha_{mn} \exp[i\mathbf{k} \cdot \mathbf{R}_{mn}]$$

$$+ ic_A c_B (f_A - f_B)^2 \sum_m \sum_n [\eta(c_A/c_B + \alpha_{mn})\langle \mathbf{k} \cdot \boldsymbol{\delta}_{mn}^{AA} \rangle$$

$$- \zeta(c_B/c_A + \alpha_{mn})\langle \mathbf{k} \cdot \boldsymbol{\delta}_{mn}^{BB} \rangle] \exp[i\mathbf{k} \cdot \boldsymbol{\delta}_{mn}]$$

$$+ c_A c_B (f_A - f_B)^2 \sum_m \sum_n (\eta^2(c_A/c_B + \alpha_{mn})\langle (\mathbf{k} \cdot \boldsymbol{\delta}_m^A)(\mathbf{k} \cdot \boldsymbol{\delta}_n^A)\rangle$$

$$+ 2\eta\zeta(1 - \alpha_{mn})\langle (\mathbf{k} \cdot \boldsymbol{\delta}_m^A)(\mathbf{k} \cdot \boldsymbol{\delta}_n^B)\rangle$$

$$+ \zeta^2(c_B/c_A + \alpha_{mn})[\langle (\mathbf{k} \cdot \boldsymbol{\delta}_m^B)(\mathbf{k} \cdot \boldsymbol{\delta}_n^B)\rangle]\exp[i\mathbf{k} \cdot \mathbf{R}_{mn}], \quad (7.66a)$$

where:

$$\eta = f_A/(f_A - f_B), \quad \zeta = f_B/(f_A - f_B). \quad (7.66b)$$

(Note that there are no zero order terms in the sums involving displacements because there are no displacements of an atom with respect to itself.)

7.6.3 Simplifying the Basic Equation

We proceed in the following way:

1) The first term in Eq. (7.66) represents the Bragg reflections and will not concern us any further. In addition we note that, in the case of short range order, coefficients like α_{mn}, $\langle \mathbf{k} \cdot \boldsymbol{\delta}_{mn} \rangle$ and $\langle (\mathbf{k} \cdot \boldsymbol{\delta}_m)(\mathbf{k} \cdot \boldsymbol{\delta}_n)\rangle$ vanish when the sites m and n are distant. Hence, the sums over m and n can be replaced by N times a single sum over all interatomic vectors, N being the total number of atoms. This expresses the fact that lattice averages are independent of the choice of origin of coordinates.

2) Consider the displacements of an atom at the end of interatomic vector $\mathbf{R}_{lmn}$, as shown in Fig. 7.25. When considered from the opposite direction $\mathbf{R}_{\bar{l}\bar{m}\bar{n}}$, $\boldsymbol{\delta}$ changes sign. Therefore, we can combine the two terms involving the average displacement (for each type of pair).

$$Nic_A c_B (f_A - f_B)^2 \eta (c_A/c_B + \alpha_{lmn})[\langle \mathbf{k} \cdot \boldsymbol{\delta}_{lmn}^{ij} \rangle e^{i\mathbf{k} \cdot \mathbf{R}_{lmn}} + \langle \mathbf{k} \cdot \boldsymbol{\delta}_{\bar{l}\bar{m}\bar{n}}^{ij} \rangle e^{-i\mathbf{k} \cdot \mathbf{R}_{lmn}}]$$

$$= -2Nc_A c_B (f_A - f_B)^2 \eta (c_A/c_B + \alpha_{lmn})\langle \mathbf{k} \cdot \boldsymbol{\delta}_{lmn}^{ij} \rangle \sin \mathbf{k} \cdot \mathbf{R}_{lmn}).$$

(In the following equations we will use the sine form, but keep the lmn, $\overline{lmn}$ terms separate, so that the sums over lmn run from minus to plus infinity.)

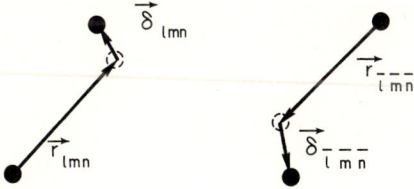

Fig. 7.25. Displacements associated with an interatomic vector $\mathbf{r}$ and its inverse $\bar{\mathbf{r}}$. The *dotted* positions represent locations without displacements

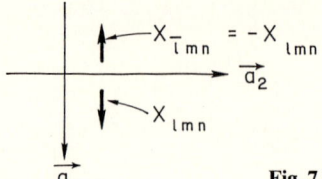

Fig. 7.26. Displacement vectors X_{lmn} and $X_{\bar{l}mn}$

3) Finally, we substitute:

$$\mathbf{R}_{mn} = l\mathbf{a}_1 + m\mathbf{a}_2 + n\mathbf{a}_3,$$

$$\delta_{lmn}^{AA} = X_{lmn}^{AA}\mathbf{a}_1 + Y_{lmn}^{AA}\mathbf{a}_2 + Z_{lmn}^{AA}\mathbf{a}_3, \quad (7.67)$$

$$k = 2\pi(h_1\mathbf{b}_1 + h_2\mathbf{b}_2 + h_3\mathbf{b}_3),$$

where the $\mathbf{a}_1$ and the $\mathbf{b}_1$ are the real and reciprocal basis vectors respectively. With these substitutions the sine term becomes $\sin 2\pi(lh_1 + mh_2 + nh_3)$.

4) Also we will make use of cubic symmetry as we did for local order. To do this we need to explore how the displacements are affected by symmetry. We consider fcc ($m3m$) symmetry in Eq. (7.66). The substitution of Eqs. (7.67) into Eq. (7.66) leads to displacement terms involving $\langle X \rangle$, $\langle Y \rangle$, $\langle Z \rangle$, $\langle X_0 X_{lmn} \rangle$, $\langle X_0 Y_{lmn} \rangle$ etc, times trigonometric terms. To simplify these trigonometric terms as we did for local order intensity we have to consider how this symmetry affects these displacement moments.

For the term $\langle X_{lmn}^{ij} \rangle$, from symmetry across $\{100\}$ type planes (see Fig. 7.26):

$$\langle X_{lmn}^{ij} \rangle = -\langle X_{\bar{l}mn}^{ij} \rangle = \langle X_{l\bar{m}n}^{ij} \rangle = \langle X_{lm\bar{n}}^{ij} \rangle, \quad (7.68a)$$

while from symmetry across $\{110\}$ type planes,

$$\langle X_{lmn}^{ij} \rangle = \langle X_{lnm}^{ij} \rangle = \langle Y_{mln}^{ij} \rangle = \langle Z_{nml}^{ij} \rangle. \quad (7.68b)$$

For the term $\langle X_0^i X_{lmn}^j \rangle$, from symmetry across $\{100\}$:

$$\langle X_0^i X_{lmn}^j \rangle = \langle X_0^i X_{\bar{l}mn}^j \rangle = \langle X_0^i X_{l\bar{m}n}^j \rangle = \langle X_0^i X_{lm\bar{n}}^j \rangle \quad (7.68c)$$

while from symmetry across $\{110\}$ planes:

$$\langle X_0^i X_{lmn}^j \rangle = \langle X_0^i X_{lnm}^j \rangle = \langle Y_0^i Y_{mln}^j \rangle = \langle Z_0^i Z_{nml}^j \rangle. \quad (7.68d)$$

and similarly for the cross terms.

These four steps allow us to re-write Eq. (7.66):

$$I_D = Nc_A c_B (f_A - f_B)^2 [I_{SRO} + \eta h_1 Q_x^{AA} + \zeta h_1 Q_x^{BB} + \eta h_2 Q_y^{AA} + \zeta h_2 Q_y^{BB}$$
$$+ \eta h_3 Q_z^{AA} + \zeta h_3 Q_z^{BB} + \eta^2 h_1^2 R_x^{AA} + 2\eta\zeta h_1^2 R_x^{AB} + \eta^2 h_1^2 R_x^{BB}$$
$$+ \eta^2 h_2^2 R_y^{AA} + 2\eta\zeta h_2^2 R_y^{AB} + \zeta^2 h_2^2 R_y^{BB} + \eta^2 h_3^2 R_z^{AA} + 2\eta\zeta h_3^2 R_z^{AB}$$
$$+ \zeta^2 h_3^2 R_z^{BB} + \eta^2 h_1 h_2 S_{xy}^{AA} + 2\eta\zeta h_1 h_2 S_{xy}^{AB} + \zeta^2 h_1 h_2 S_{xy}^{BB}$$
$$+ \eta^2 h_2 h_3 S_{yz}^{AA} + 2\eta\zeta h_2 h_3 S_{yz}^{AB} + \zeta^2 h_2 h_3 S_{yz}^{BB} + \eta^2 h_3 h_1 S_{zx}^{AA}$$
$$+ 2\eta\zeta h_3 h_1 S_{zx}^{AB} + \zeta^2 h_3 h_1 S_{zx}^{BB}], \quad (7.69)$$

7.6 Local Ordering and Clustering

where I_D is the diffuse intensity in electron units and:

$$I_{SRO} = \sum_l \sum_m \sum_n \alpha_{lmn} \cos 2\pi l h_1 \cos 2\pi m h_2 \cos 2\pi n h_3 ,$$

$$Q_x^{AA} = -2\pi \sum_l \sum_m \sum_n (c_A/c_B + \alpha_{lmn}) \langle X_{lmn}^{AA} \rangle \sin 2\pi l h_1 \cos 2\pi m h_2 \cos 2\pi n h_3 ,$$

$$R_x^{AA} = 4\pi^2 \sum_l \sum_m \sum_n (c_A/c_B + \alpha_{lmn}) \langle X_0^A X_{lmn}^A \rangle \cos 2\pi l h_1 \cos 2\pi m h_2 \cos 2\pi n h_3 ,$$

$$S_{xy}^{AA} = 4\pi^2 \sum_l \sum_m \sum_n (c_A/c_B + \alpha_{lmn}) \langle X_0^A Y_{lmn}^A \rangle \sin 2l h_1 \sin 2\pi m h_2 \cos 2\pi n h_3$$

(7.70)

etc. for the rest of the terms in (7.69). Every component has a periodicity of two. The various terms are shown plotted along an $h00$ line in Fig. 7.27. Each contribution varies differently with distance from the origin, and this difference will shortly permit us to separate all the Fourier series in Eq. (7.69).

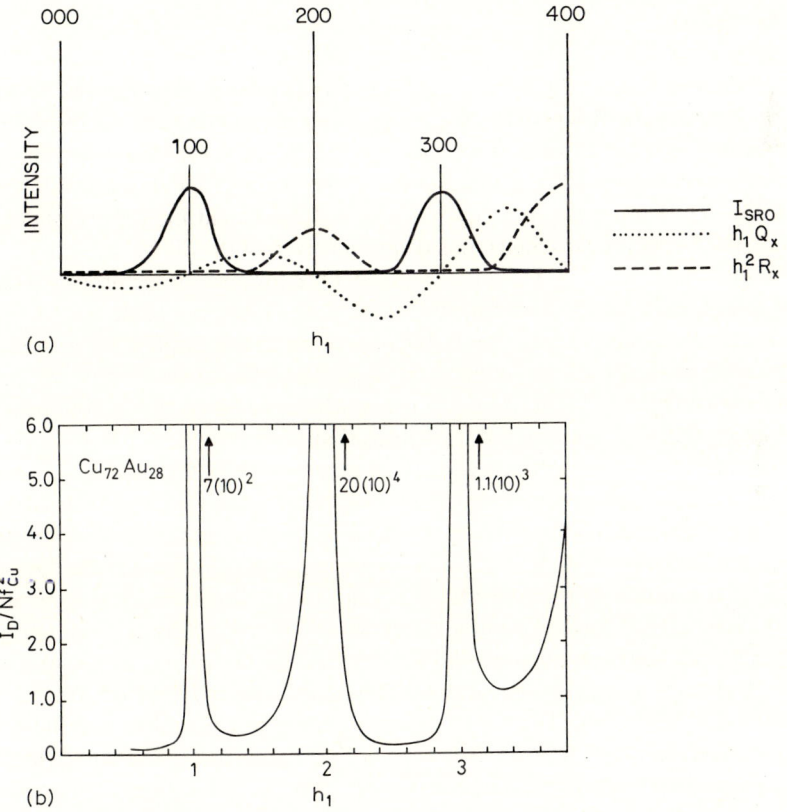

Fig. 7.27. a Schematic of the diffuse intensity components. (For Q_x, R_x only one term in the Fourier series is illustrated.) **b** Intensity for Cu_3Au in electron units (divided by Nf^2_{Cu}) along the $[h00]$ direction in reciprocal space. Data corrected for TDS and Compton scattering. Alloy was heat treated to give partial long range order. (At full order all Au atoms are at corners, Cu at faces of a cubic unit cell.) [After Schwartz, L.H., and Cohen, J.B., J. Appl. Phys. **36**, 598 (1965)]

What have we accomplished so far? Quite a lot! We have managed to write the diffuse intensity as a sum of various components, each of which contains a particular piece of information about the local arrangements of the atoms in the solid solution. The first term is a function of the short range order parameters, which tell us about the chemical environment about each atomic species, i.e., on the average how many atoms of each kind and at what distance from an A or B atom. The remaining twenty four terms in the intensity expansion contain very detailed information about the local displacements. The first order terms give us the preferred interatomic distance of AA or BB pairs, separated by a given lattice vector. The second order terms give us correlations of atomic displacements: $\langle X_0^A Y_{lmn}^B \rangle$, for instance, gives the probable displacement of a B atom at the tip of the vector $[lmn]$ in the Y direction when an A atom at the origin has been displaced by X in the $i = 1$ direction. In the next section we will see how we can extract this information from measurements of the total diffuse scattering.

7.6.4 Data Analysis

We remind the reader that the various terms in Eq. (7.69) are each Fourier series with period 2, whose coefficients are short range order parameters and the displacement averages. So, it seems that, if we managed to break down the total diffuse intensity into its components contained in Eq. (7.69), a three-dimensional Fourier transform of each component in reciprocal space should yield the corresponding parameters. But how can we achieve this decomposition of the diffuse intensity?

Following a suggestion by Tibballs (1975) we look for a set of reciprocal space points, at which each of the components, (I_{SRO}, Q, R, S) has the same value; the intensity is not the same at these points because terms are multiplied by η, h_i etc. After measuring the total diffuse intensity at these reciprocal space points, we can form a system of linear equations like Eq. (7.69), in which the twenty five components are the constants to be determined. It is obvious that we need twenty five measurements — or more — for the solution of the linear system to be feasible.

Now we will systematically construct the reciprocal space point set referred to in the previous paragraph, identify it by its member that lies in the first minimum volume (i.e. the one closest to the origin of reciprocal space) and call it the "associated set" of this point. It is clearer to illustrate this procedure with a concrete example: Take 0.4, 0.5, 0.6 as the point and assume fcc symmetry. By examining Eq. (7.69), we can spot some obvious candidates for the associated set: all reciprocal space points which lie integral reciprocal lattice vectors away from 0.4, 0.5, 0.6, such as 2.4, 0.5, 0.6; 1.4, 1.5, 1.6, etc. All such points accessible experimentally are members of the associated set of 0.4, 0.5, 0.6. But we can find many more, not so obvious. Besides translational symmetry, reciprocal space possesses additional symmetry elements, such as mirror planes, rotational axes and inversion centers. These additional symmetries and their effects on the various intensity components in Eq. (7.69) are listed in Tables 7.4–7.6 for simple cubic, bcc and fcc symmetry respectively. We now illustrate the role of these symmetry elements on the various terms in Eq. (7.69). To do this we make use of Eqs. (7.68).

7.6 Local Ordering and Clustering

Table 7.4. Symmetry relations among the intensity components for simple cubic symmetry

Mirror planes: $\left\{\dfrac{n}{2}, 0, 0\right\}$, $n = 0, \pm 1, \pm 2, \ldots$

$I_{SRO}(\tfrac{1}{2} + h_1 h_2 h_3) = I_{SRO}(\tfrac{1}{2} - h_1 h_2 h_3)$
$Q_x(\tfrac{1}{2} + h_1 h_2 h_3) = -Q_x(\tfrac{1}{2} - h_1 h_2 h_3)$
$R_x(\tfrac{1}{2} + h_1 h_2 h_3) = R_x(\tfrac{1}{2} - h_1 h_2 h_3)$
$S_{xy}(\tfrac{1}{2} + h_1 h_2 h_3) = -S_{xy}(\tfrac{1}{2} - h_1 h_2 h_3)$

Relations among components

$I_{SRO}(h_1 h_2 h_3) = I_{SRO}(h_2 h_3 h_1) = I_{SRO}(h_3 h_1 h_2)$
$I_{SRO}(h_1 h_3 h_2) = I_{SRO}(h_3 h_2 h_1) = I_{SRO}(h_2 h_1 h_3)$
$Q_x(h_1 h_2 h_3) = Q_x(h_1 h_3 h_2) = Q_y(h_2 h_1 h_3) = Q_z(h_3 h_2 h_1)$
$R_x(h_1 h_2 h_3) = R_x(h_1 h_3 h_2) = R_y(h_2 h_1 h_3) = R_z(h_3 h_2 h_1)$
$S_{xy}(h_1 h_2 h_3) = S_{xy}(h_2 h_1 h_3) = S_{yz}(h_3 h_2 h_1) = S_{zx}(h_1 h_3 h_2)$

Table 7.5. Symmetry relations among the intensity components for body centered cubic symmetry

Mirror planes: $\{n, 0, 0\}$, $n = 0, \pm 1, \pm 2, \ldots$

$I_{SRO}(n + h_1 h_2 h_3) = I_{SRO}(n - h_1 h_2 h_3)$
$Q_x(n + h_1 h_2 h_3) = -Q_x(n - h_1 h_2 h_3)$
$R_x(n + h_1 h_2 h_3) = R_x(n - h_1 h_2 h_3)$
$S_{xy}(n + h_1 h_2 h_3) = -S_{xy}(n - h_1 h_2 h_3)$

Rotation axes: 2-fold $\langle n + \tfrac{1}{2}, n + \tfrac{1}{2}, 0 \rangle$, $n = 0, \pm 1, \pm 2, \ldots$

$I_{SRO}(\tfrac{1}{2} + h_1, \tfrac{1}{2} + h_2, h_3) = I_{SRO}(\tfrac{1}{2} - h_1, \tfrac{1}{2} - h_2, h_3)$
$Q_x(\tfrac{1}{2} + h_1, \tfrac{1}{2} + h_2, h_3) = -Q_x(\tfrac{1}{2} - h_1, \tfrac{1}{2} - h_2, h_3)$
$R_x(\tfrac{1}{2} + h_1, \tfrac{1}{2} + h_2, h_3) = R_x(\tfrac{1}{2} - h_1, \tfrac{1}{2} - h_2, h_3)$
$S_{xy}(\tfrac{1}{2} + h_1, \tfrac{1}{2} + h_2, h_3) = -S_{xy}(\tfrac{1}{2} - h_1, \tfrac{1}{2} - h_2, h_3)$

Relations among components:

Same as for simple cubic symmetry

Table 7.6. Symmetry relations among the intensity components for face centered cubic symmetry

Mirror planes: Same as for *BCC* symmetry

Inversion centers: $(n + \tfrac{1}{2}, n + \tfrac{1}{2}, n + \tfrac{1}{2})$, $n = 0, \pm 1, \pm 2, \ldots$

$I_{SRO}(\tfrac{1}{2} + h_1, \tfrac{1}{2} + h_2, \tfrac{1}{2} + h_3) = I_{SRO}(\tfrac{1}{2} - h_1, \tfrac{1}{2} - h_2, \tfrac{1}{2} - h_3)$
$Q_x(\tfrac{1}{2} + h_1, \tfrac{1}{2} + h_2, \tfrac{1}{2} + h_3) = -Q_x(\tfrac{1}{2} - h_1, \tfrac{1}{2} - h_2, \tfrac{1}{2} - h_3)$
$R_x(\tfrac{1}{2} + h_1, \tfrac{1}{2} + h_2, \tfrac{1}{2} + h_3) = R_x(\tfrac{1}{2} - h_1, \tfrac{1}{2} - h_2, \tfrac{1}{2} - h_3)$
$S_{xy}(\tfrac{1}{2} + h_1, \tfrac{1}{2} + h_2, \tfrac{1}{2} + h_3) = -S_{xy}(\tfrac{1}{2} - h_1, \tfrac{1}{2} - h_2, \tfrac{1}{2} - h_3)$

Relations among components:

Same as for simple cubic symmetry

For the component $Q_x^{ij}(h_1, h_2, h_3)$, the symmetry of this term is determined by the symmetry of the displacement field moment $\langle X_{lmn}^{ij} \rangle$. Thus, from the symmetry across {100} type planes,

$$Q_x^{ij}(h_1, h_2, h_3) = -Q_x^{ij}(-h_1, h_2, h_3) = Q_x^{ij}(h_1, -h_2, h_3) = Q_x^{ij}(h_1, h_2, -h_3). \tag{7.71a}$$

Similarly, from symmetry across {110} type planes,

$$Q_x^{ij}(h_1, h_2, h_3) = Q_x^{ij}(h_1, h_3, h_2) = Q_y^{ij}(h_2, h_1, h_3) = Q_z^{ij}(h_3, h_2, h_1). \tag{7.71b}$$

For the component $R_x^{ij}(h_1, h_2, h_3)$, the symmetry of this term is determined by the symmetry of the displacement field moment $\langle X_0^i X_{lmn}^j \rangle$. Thus, from symmetry across {100} type planes,

$$R_x^{ij}(h_1, h_2, h_3) = R_x^{ij}(-h_1, h_2, h_3) = R_x^{ij}(h_1, -h_2, h_3) = R_x^{ij}(h_1, h_2, -h_3). \tag{7.72a}$$

Similarly, from symmetry across {110} type planes,

$$R_x^{ij}(h_1, h_2, h_3) = R_x^{ij}(h_1, h_3, h_2) = R_y^{ij}(h_2, h_1, h_3) = R_z^{ij}(h_3, h_2, h_1). \tag{7.72b}$$

For the component $S_{xy}^{ij}(h_1, h_2, h_3)$, the symmetry of this term is determined by the symmetry of the displacement field moment $\langle X_{lmn} Y_{lmn} \rangle$. Thus,

$$S_{xy}^{ij}(h_1, h_2, h_3) = -S_{xy}^{ij}(-h_1, h_2, h_3) = -S_{xy}^{ij}(h_1, -h_2, h_3)$$
$$= S_{xy}^{ij}(h_1, h_2, -h_3). \tag{7.73a}$$

Similarly, from symmetry across {110} type planes, one obtains,

$$S_{xy}^{ij}(h_1, h_2, h_3) = S_{xy}^{ij}(h_2, h_1, h_3) = S_{yz}^{ij}(h_3, h_2, h_1) = S_{zx}^{ij}(h_1, h_3, h_2). \tag{7.73b}$$

[The reader should now prove to himself (from the symmetries of the terms) that the total diffuse intensity in Eq. (7.69) indeed has $m3m$ point symmetry.]

In addition to the previously discussed symmetries of the diffuse intensity components, there exists an additional "symmetry" element for these terms within the reciprocal unit cell which is in fact a consequence of the existing symmetries. Thus, since we have stated that these components are periodic in reciprocal space with the period of the reciprocal lattice unit cell, the term $Q_x^{ij}(h_1, h_2, h_3)$, for example, must satisfy the condition $Q_x^{ij}(h_1 + 1, h_2, h_3) = Q_x^{ij}(h_1 - 1, h_2, h_3)$. But, as a result of the previously discussed symmetry elements of this term, $Q_x^{ij}(h_1 - 1, h_2, h_3) = -Q_x^{ij}(-h_1 + 1, h_2, h_3)$. Hence, $Q_x^{ij}(h_1, h_2, h_3)$ is antisymmetric across the plane $h_1 = 1$.

Similar arguments for the other reciprocal lattice axes and diffuse intensity components lead to the result that

$$Q_x^{ij}(h_1 + 1, h_2, h_3) = -Q_x^{ij}(-h_1 + 1, h_2, h_3) \tag{7.74a}$$

$$Q_x^{ij}(h_1, h_2 + 1, h_3) = Q_x^{ij}(h_1, -h_2 + 1, h_3) \tag{7.74b}$$

$$Q_x^{ij}(h_1, h_2, h_3 + 1) = Q_x^{ij}(h_1, h_2, -h_3 + 1) \tag{7.74c}$$

$$R_x^{ij}(h_1 + 1, h_2, h_3) = R_x^{ij}(-h_1 + 1, h_2, h_3) \tag{7.75a}$$

$$R_x^{ij}(h_1, h_2 + 1, h_3) = R_x^{ij}(h_1, -h_2 + 1, h_3) \tag{7.75b}$$

$$R_x^{ij}(h_1, h_2, h_3 + 1) = R_x^{ij}(h_1, h_2, -h_3 + 1) \tag{7.75c}$$

7.6 Local Ordering and Clustering

$$S_{xy}^{ij}(h_1 + 1, h_2, h_3) = -S_{xy}^{ij}(-h_1 + 1, h_2, h_3) \tag{7.76a}$$

$$S_{xy}^{ij}(h_1, h_2 + 1, h_3) = -S_{xy}^{ij}(h_1, -h_2 + 1, h_3) \tag{7.76b}$$

$$S_{xy}^{ij}(h_1, h_2, h_3 + 1) = S_{xy}^{ij}(h_1, h_2, -h_3 + 1) \tag{7.76c}$$

$$I_{SRO}(h_1 + 1, h_2, h_3) = I_{SRO}(-h_1 + 1, h_2, h_3) \tag{7.77a}$$

$$I_{SRO}(h_1, h_2 + 1, h_3) = I_{SRO}(h_1, -h_2 + 1, h_3) \tag{7.77b}$$

$$I_{SRO}(h_1, h_2, h_3 + 1) = I_{SRO}(h_1, h_2, -h_3 + 1). \tag{7.77c}$$

Recalling the parity condition for the indices of the fcc lattice vectors, $(2l + 2m + 2n)$ even, point $(1 \pm h_1, 1 \pm h_2, 1 \pm h_3)$ is symmetrically related to the point (h_1, h_2, h_3). Whether the relationship is symmetric or antisymmetric obviously depends on the particular diffuse intensity component. Moreover, any point (h'_1, h'_2, h'_3) which is symmetrically related to the point (h_1, h_2, h_3) yields an additional point $(1 \pm h'_1, 1 \pm h'_2, 1 \pm h'_3)$, although not necessarily by the same type of symmetry (symmetric or antisymmetric). Note that we have restricted ourselves to points of the form $(1 \pm h_1, 1 \pm h_2, 1 \pm h_3)$, for example, because the point (h_1, h_2, h_3) and all points symmetrically related to it have been shown to be restricted to the first positive octant ($0 \leq h_i \leq 1$) of the reciprocal lattice unit cell.

As a result of all these symmetry considerations, it should be apparent that any intensity component can be constructed in a unit cell in reciprocal space from the measurements in the volume for I_{SRO}. Using Tables 7.4–7.7, we can find more points belonging to the associated set of 0.4, 0.5, 0.6. Table 7.7 lists part of this set. Tables 7.4–7.7 also include what we have just seen — that the various symmetry elements present in each case can impose certain constraints on the intensity components.

Table 7.7. The Associated set of 0.4, 0.5, 0.6 and the constraints due to face centered cubic symmetry

Constraints	Associated set			Associated set		
	h_1	h_2	h_3	h_1	h_2	h_3
$Q_x^{AA} = -Q_z^{AA}$	1.4	0.5	0.4	2.6	1.6	1.5
$Q_x^{BB} = -Q_z^{BB}$	1.5	0.6	0.4	2.5	2.4	0.6
$R_x^{AA} = R_z^{AA}$	1.6	0.6	0.5	2.6	2.5	0.4
$R_x^{AB} = R_y^{AB}$	1.5	1.4	0.4	2.5	2.4	1.4
$R_x^{BB} = R_y^{BB}$	1.5	1.4	0.4	3.4	1.5	0.4
$S_{zx}^{AA} = S_{yz}^{AA}$	1.6	1.5	0.6	2.6	2.4	1.5
$S_{zx}^{AB} = S_{yz}^{AB}$	2.4	0.6	0.5	3.6	1.5	0.6
$S_{zx}^{BB} = S_{yz}^{BB}$	2.6	0.5	0.4	3.5	1.6	1.4
	2.4	1.4	0.5	3.4	2.5	0.4
	2.5	1.6	0.6	3.5	2.6	0.4
	2.6	1.5	0.4	3.4	2.5	1.6
	2.4	1.5	1.4	4.4	1.4	0.5
	2.5	1.6	1.4			

Let us continue with an example of the use of the these constraints to form the associated set. From the definition of Q_x in Eq. (7.70), the following can be easily verified:

$$Q_x(h_1 h_2 h_3) = Q_x(h_1 h_3 h_2) = Q_z(h_3 h_2 h_1)$$

which, for our base point 0.4, 0.5, 0.6 implies:

$$Q_z(0.4, 0.5, 0.6) = Q_x(0.6, 0.5, 0.4) = Q_x(0.6, 0.4, 0.5).$$

From Table 7.6, using the inversion center symmetry:

$$Q_x(0.6, 0.4, 0.5) = Q_x(0.5 + 0.1, 0.5 - 0.1, 0.5 + 0.0)$$
$$= -Q_x(0.5 - 0.1, 0.5 + 0.1, 0.5 - 0.0) = Q_x(0.4, 0.6, 0.5)$$

which gives:

$$Q_z(0.4, 0.5, 0.6) = -Q_x(0.4, 0.6, 0.5) = -Q_x(0.4, 0.5, 0.6).$$

Similar constraints can be found for other intensity components. For our base point 0.4, 0.5, 0.6, they are all listed in the first column of Table 7.7. Some of these constraints can result from a combination of the symmetry elements shown in Table 7.5. Note that each constraint reduces the number of unknowns in Eq. (7.69) by one.

Now the strategy for the diffuse scattering measurement has been laid out. For each base point in reciprocal space at which we wish to separate the diffuse intensity into its components, we construct the associated set, carry out the measurements and form a linear system of equations like (7.69). This system is then solved by a linear least squares procedure. Care should be exercised, because the matrix of this system is often ill–conditioned, i.e. the coefficients of the intensity components are almost multiples of one another, since η and ζ do not vary much as a function of $\sin\theta/\lambda$. Procedures for handling this problem are described by Wu et al. (1983).

[In the case of diffuse neutron scattering Eq. (7.69) is of course still valid, but the various components are no longer linearly independent. Since the neutron scattering length is usually independent of the scattering angle, the quantities η and ζ defined in Eq. (7.69b) are constant and the matrix of the linear system of equations is singular. Equation (7.69) must be modified to remove this singularity, by combining all the AA, AB and BB terms of each component into one. The number of unknowns is thus reduced to ten. The short range order intensity component is still the same, but the new Q, R and S terms contain less information, since they are now inseparable combinations of AA, AB and BB pair contributions. A different separation procedure is possible in this case (Borie and Sparks 1971; Gragg and Cohen 1971).]

7.6 Local Ordering and Clustering

7.6.5 The Minimum Volume for a Measurement

In order to be able to Fourier invert each intensity component to recover the short range order coefficients and the various displacement moments, the proper separation procedure must be carried out for base points in a uniform net, covering one minimum volume of reciprocal space.

This volume is the minimum volume for I_{SRO}, but not for the other terms. Consider Q_x^{ij} and R_x^{ij}. As can be seen from the results of the previous discussion, the diffuse intensity components $Q_x^{ij}(h_1, h_2, h_3)$ and $R_x^{ij}(h_1, h_2, h_3)$ have identical symmetry elements, save for the symmetric/antisymmetric nature of the operations — a factor that has no influence on the minimum repeat volume. Thus, we can expect these two components to have identical minimum repeat volumes.

The evolution of the minimum repeat volume for these two diffuse intensity components can be traced as follows:

(i) From the periodic nature of these components and their respective symmetries across planes such as $h_1 = 0$, the value of these terms can be restricted to the first positive unit cell in reciprocal space.
(ii) From the respective symmetries of these terms across such planes as $h_1 = 1$, the volume in which they must be known can be further restricted to the first octant of the first positive unit cell in reciprocal space. This volume is shown in Fig. 7.28a.
(iii) From the symmetries of these two components across the plane $h_2 = h_3$, their value can be further restricted to the volume shown in Fig. 7.28b. Note that these two terms do not have any symmetric relationship across the planes $h_1 = h_2$ and $h_1 = h_3$.
(iv) For the fcc lattice, these terms can be restricted to the volume shown in Fig. 7.28c by the symmetry element $(h_1, h_2, h_3) = (1 - h_1, 1 - h_3, 1 - h_2)$. This is a combined symmetry element resulting from the generating elements $(h_1, h_2, h_3) = (h_1, h_3, h_2)$ and $(h_1, h_2, h_3) = (1 - h_1, 1 - h_2, 1 - h_3)$. All other symmetry elements for these terms for the fcc lattice have no effect in reducing this volume further. Thus, this volume is the minimum repeat volume for the diffuse intensity components $Q_x(h_1, h_2, h_3)$ and $R_x(h_1, h_2, h_3)$ in the fcc lattice reciprocal space. This volume can be filled from the data on Q_x (or R_x) in the minimum volume for I_{SRO} and the relationships between Q_y, Q_z and Q_x. Once this volume is filled it can be employed to fill an entire unit cell in reciprocal space and transformed to find the coefficients of each series.

7.6.6 The Significance of the Results

How well do these methods of analysis work? How reliable are the calculated values for the short range order parameters and the displacement moments? Of course, the error in intensity measurements is well-known, since it follows Poisson statistics. The least-squares procedure will propagate these errors through to give error estimates for the intensity components and Wu et al. (1983) have shown how to further propagate these errors, through the Fourier transform, to produce error

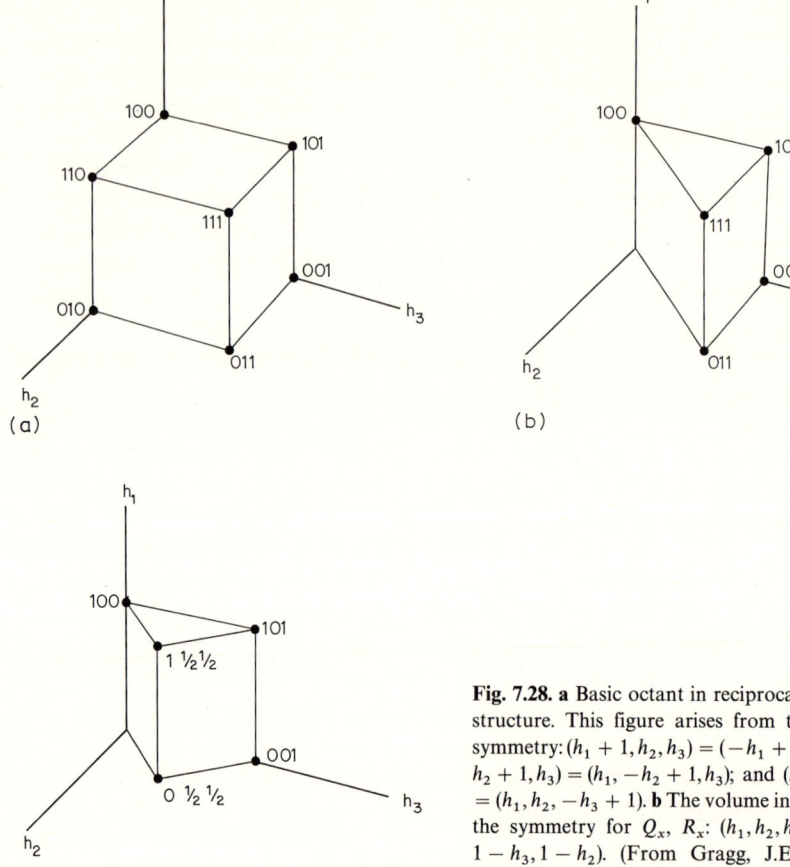

Fig. 7.28. **a** Basic octant in reciprocal space; m3m structure. This figure arises from the following symmetry: $(h_1 + 1, h_2, h_3) = (-h_1 + 1, h_2, h_3); (h_1, h_2 + 1, h_3) = (h_1, -h_2 + 1, h_3);$ and $(h_1, h_2, h_3 + 1) = (h_1, h_2, -h_3 + 1)$. **b** The volume in **a** reduced by the symmetry for Q_x, R_x: $(h_1, h_2, h_3) = (1 - h_1, 1 - h_3, 1 - h_2)$. (From Gragg, J.E., Jr., Ph.D. Thesis, Northwestern University, 1970)

estimates for the short range order parameters and the displacement moments. Typically α's can be measured to a statistical precision of ~ 5 pct and the displacements to about 7–15 pct. But this is only part of the problem. Many other sources of errors exist besides counting statistics. First and foremost is the fact that Eq. (7.69) is an approximation, valid only if the local atomic displacements are small and third and higher order displacements negligible; the high order displacement terms are, in fact, included in the second order terms, R and S, in an approximate manner (Hayakawa et al. 1975; Wu et al. 1983), but the small displacement assumption may not be a legitimate one in certain alloy systems.

Other errors of experimental nature also enter. The net diffuse intensity is only part of the measured value. Other contributions, such as Compton and parasitic scattering, sample fluorescence etc., must be measured or estimated and subtracted. Uncertainties in the composition of the specimen can also introduce errors in the analysis. The diffuse intensity must be placed on an absolute scale (see Appendix C), that is, the intensity of the incident beam (which is orders of magnitude higher

7.6 Local Ordering and Clustering

than the measured intensities) must be accurately measured. It is more difficult to estimate the contribution of all the possible errors to the uncertainty in the final parameters, and each must be accurately assessed.

7.6.7 Experimental Details

The experimental apparatus for diffuse scattering measurements is centered around a single crystal four-circle diffractometer, preceeded by an incident beam monochromator. Fig. 7.29 shows a typical laboratory setup. Figure 7.30 shows a diffuse scattering station set up at the Cornell University High Energy Synchrotron Source, CHESS. Note that in the latter case, the diffractometer has been mounted such that diffraction is measured in the vertical plane, due to the polarization of the synchrotron beam parallel to the floor.

There are no fundamental differences between a laboratory and a synchrotron setup, except for the design of the incident-beam-monochromator. The same basic components are present, perhaps with small adaptations. Due to the duration of the measurement and the variability of the incident beam flux (much more severe at the synchrotron), a beam flux monitor is always used. The sample is often placed under an evacuated beryllium dome to eliminate air scattering.

We will now examine various procedures that must be followed in the actual measurement of diffuse scattering. They are enumerated below to make the presentation more systematic; it is not implied that they are followed in the indicated sequence.

1. The crystallographic orientation of the sample with respect to the diffractometer axes must be measured accurately. Since this is a fairly time-consuming procedure

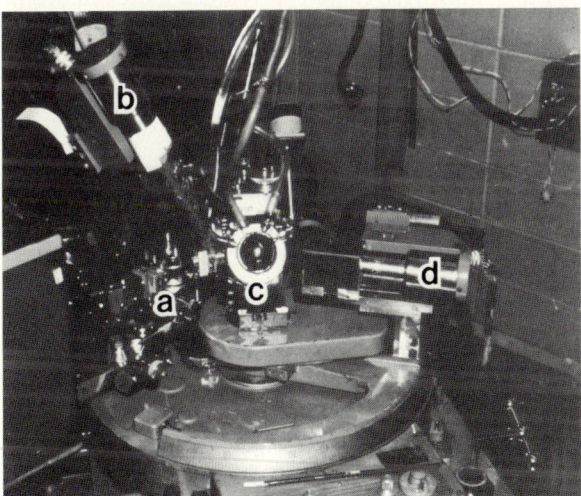

Fig. 7.29. Laboratory diffuse scattering appartus. a, monochromator; b, monitor detector; c, sample chamber; d, diffracted intensity detector

Fig. 7.30. Diffuse scattering station at CHESS. Components indicated the same as in Fig. 7.29

to perform manually, it is highly desirable to have the computer search for Bragg reflections, determine their exact locations and compute the orientation matrix automatically.
2. The set of reciprocal space points, at which measurement of the diffuse intensity is needed, must be generated and converted to diffractometer settings. In order to save time spent in moving the instrument from one setting to the next, the settings should be ordered in a sequence that minimizes diffractometer motion.
3. The measurements should be performed by counting each data point until a fixed monitor count is reached, rather than for a fixed time. This is much more important at a synchrotron than with an x-ray generator.

7.6.8 Checking the Results

Fortunately, there are some independent, global checks that one can apply after the data analysis is completed. They are all based on physical arguments, which the numerical analysis does not have to obey automatically.

First, since the short range order intensity is just a re-distribution of the Laue monotonic scattering (the uniform diffuse scattering from a random solid solution), it cannot be negative anywhere in reciprocal space. The same holds true for the second order size effect scattering [the combination of the R and S terms of Eq. (7.69)], since it is merely the re-distribution of the intensity lost from the Bragg reflections through the Debye-Waller factor. Any successful experiment should produce positive values for these components, and the few occasional negative values should be within the statistical error.

From the definition of the short range order parameters, recall that the first coefficient, α_{000}, must have the value of unity, since the probability of having a B

atom on top of an *A* atom is obviously zero. The value of α_{000} is a very sensitive check on the quality of a diffuse scattering measurement and the validity of the data analysis. Typically now, it is possible to obtain α_{000} values within 5–10 pct of unity. Other alphas have errors typically of 0.02. Also, one can establish ranges of values that the α's can take, by setting the pair probabilities involved equal to zero or unity. Any value far outside these physical bounds [more than twice the standard deviation (Wu et al. 1983)] again indicates that the results are not reliable. The displacements must be realistic as well.

Once the short range order parameters have been obtained, they can be employed in computer simulations, involving up to 108,000 "atoms" on a three-dimensional lattice (Gehlen and Cohen 1965; Gragg et al. 1971). The "atoms" (represented in the computer by ones and zeroes) are rearranged to satisfy the measured short range order parameters. The resulting local atomic configurations can then be examined in detail. It has been shown that the pair probabilities strongly confine arrangements in triplets, quadruplets etc., which makes this technique realistic. Also in some systems it has been possible to confirm the computer generated structures by transmission electron microscopy and field-ion microscopy. We show in Fig. 7.31 an example of such a study. For a recent review of other results from such studies see Cohen and Georgopoulos (1984) and Cohen (1986).

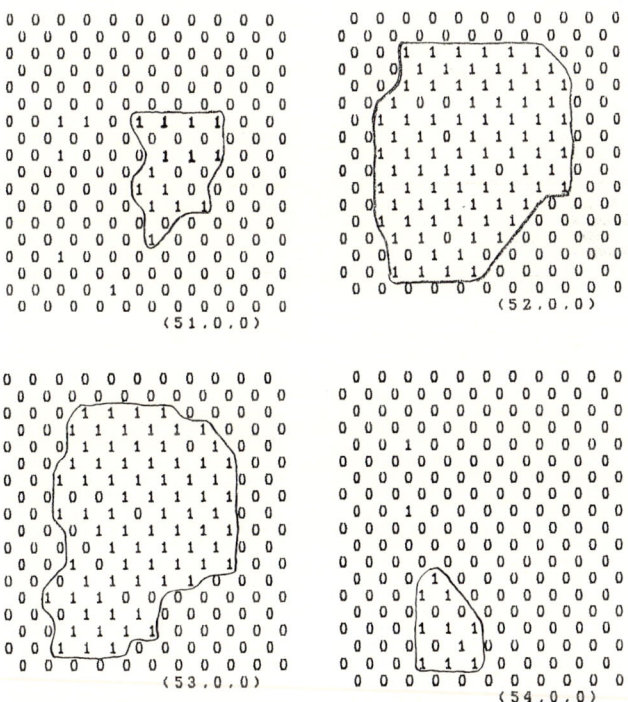

Fig. 7.31. Successive {100} planes in a simulation of the diffuse scattering from G.P. zones in an Al-Cu alloy: 1 = Cu, 0 = Al. Note the thick Cu rich platelike zone. For further details see Matsubara, E. and Cohen, J.B., Acta Metallurgica *33*, 1945, 1957 (1985)

Peisl (1975) and Larson (1985) have used the size effect scattering to examine the displacements around solute atoms in dilute solutions, and around point defects. Thermodynamic information can also be obtained by calculating pair interaction potentials from the observed α's (Clapp and Moss 1966, 1968; Wu et al. 1982, 1984).

It is worth mentioning that a diffuse scattering measurement is not necessarily a long one. The diffuse intensity must be sampled with reasonable statistical accuracy (say 1–2%) at a large number of reciprocal lattice points, typically 3000–5000. Given the fact that diffuse scattering is fairly weak, an experiment in the laboratory can easily last one to two weeks, sometimes longer. But with synchrotron radiation, a high quality measurement takes only 24–36 h. In view of this and the complexity of the data analysis procedures, it is hardly worth emphasizing the necessity of using a computer for automation of the experiment, as well as the generation of the associated sets and the final data analysis. Over the past few years, extensive software development has occurred for this purpose and it is now possible to collect data and analyze it almost completely automatically, with minimal operator intervention.

7.6.9 More Complex Systems

While it is the power of the synchrotron that permits us to make such studies in a reasonable time, the tunability of synchrotron radiation also permits studies of ternary (or quarternary) systems. The basic equations for more complex systems have been derived (Hayakawa et al. 1975).

Cenedese et al. (1984) have provided a number of important suggestions as to how to proceed with such systems. For example, for the ternary ABC system, with two energies, one a few eV just below the absorption edge of element A (where this element's scattering factor changes by as much as 20 pct as pointed out in Sect. 3.4), and one a few hundred eV below this edge:

$$\begin{aligned} I(\lambda_1, h_i) - I(\lambda_2, h_i) &= c_A c_B |f_A(h_i) - \bar{f}_B(h_i)|^2_{\lambda_1, \lambda_2} \alpha^{AB}(h_i) \\ &+ c_A c_C |f_A(h_i) - \bar{f}_c(h_i)|^2_{\lambda_1, \lambda_2} \alpha^{AC}(h_i) \\ &+ (c_A)^2 |f_A(h_i)|^2_{\lambda_1, \lambda_2} [h_i Q_i^{AA}(h_i) + (h_i)^2 R_i^{AA}(h_i) \\ &+ h_i h_j S_{ij}^{AA}(h_i)] + c_A c_B [\bar{f}_B^*(h_i) f_A(h_i) \\ &+ f_A^*(h_i) \bar{f}_B(h_i)]_{\lambda_1, \lambda_2} \times [h_i Q_i^{AB}(h_i) \\ &+ (h_i)^2 R_i^{AB}(h_i) + h_i h_j S_{ij}^{AB}(h_i)] + c_A c_C (f_c^*) f_A(h_i) \\ &+ f_A^*(h_i) f_c(h_i))_{\lambda_1, \lambda_2} \times [h_i Q_i^{AC}(h_i) + (h_i)^2 R_i^{AC}(h_i) \\ &+ h_i h_j S_{ij}^{AC}(h_i)], \end{aligned} \quad (7.78)$$

where $(f_A)_{\lambda_1, \lambda_2}$ denotes the difference $(f_A)_{\lambda_1} - (f_A)_{\lambda_2}$, and:

$$f_B = [(f_B)_{\lambda_1} + (f_B)_{\lambda_2}]/2 \simeq (f_B)_{\lambda_1} \simeq (f_B)_{\lambda_2}. \quad (7.79)$$

In Eq. (7.78) there are 29 periodic functions, which may be separable by the least squares approach if f_A, f_B, f_C have different angular dependences. These modulations could then be introduced as input in the expression of $I(\lambda_1 h_i) + I(\lambda_2, h_i)$ (or either one), and the remaining unknown modulations separated. A similar approach is possible with neutrons, if isotopes are available, and if the scattering features are not too sharp. Indeed, successful analysis of a ternary Fe-Cr-Ni alloy has already been carried out with such substitutions using neutrons (Cenedese et al. 1984), although not at this writing with x-rays.

In Chap. 3, it was mentioned that additional diffuse scattering occurs associated with the decrease in f', $\Delta f'$. Thus, "resonant Raman" scattering can be quite important in experiments that make use of anomalous scattering, because it changes its intensity as the edge is approached, and can be an appreciable fraction of a measurement if this is not a very strong intensity. Therefore it has to be calculated and subtracted. Because of its importance, we explore its nature a bit further. If the incident energy is less than, say, the K edge of the element of interest, the 1s electron is raised to an excited state, but not quite high enough in energy to be free of the atom (i.e., not to the Fermi level), an L electron drops to the 1s state and the excited 1s also decays, emitting energy. All three phenomena are part of a single resonant process, not separate and the net effect is two fold: (1) the emitted energy is $\sim h\nu_{incident} - h\nu_{L\,edge}$, (2) the energy has a structure with a separation in components like the $K_\alpha K_\beta$ peaks, because the L and M levels are involved, as for fluorescence.

As the incident energy is increased toward the K edge this resonant energy moves toward the K edge also, maintaining its separation from $h\nu_{incident}$, and it sharpens and increases in magnitude. (Once the K edge is crossed, and fluorescence occurs, the fluorescent energies are fixed and no longer follow the incident energy.) With an energy sensitive detector it is possible to separate the portion of the resonant Raman scattering associated with the L shell, but as the edge is approached, the M portion is too close to the edge to separate and must be calculated, if the elastic portion only is of interest, as is done for other inelastic processes like Compton scattering. If this intensity is large, it may be better to choose a wavelength further from the edge, even if the dispersion correction is less. Key references for this effect are Eisenberger et al. (1976) and Tulkki (1983).

7.7 Small-Angle Scattering (SAS)

7.7.1 The Shape Transform

In the previous section we noted that when there is clustering, the diffuse scattering has its maxima at the Bragg peaks — including the origin. This latter region has certain interesting features that have led to a whole field of study referred to as "small-angle scattering". In this region, effects of displacement (including TDS) are small, as is the Compton scattering. (There can be some trouble with multiple scattering, illustrated in Fig. 7.32, but even this can be avoided by orienting a thin single crystal so that there is no diffraction or, with neutrons, a wavelength can be chosen that is too large to satisfy Bragg's law.) One of the pioneers in this field,

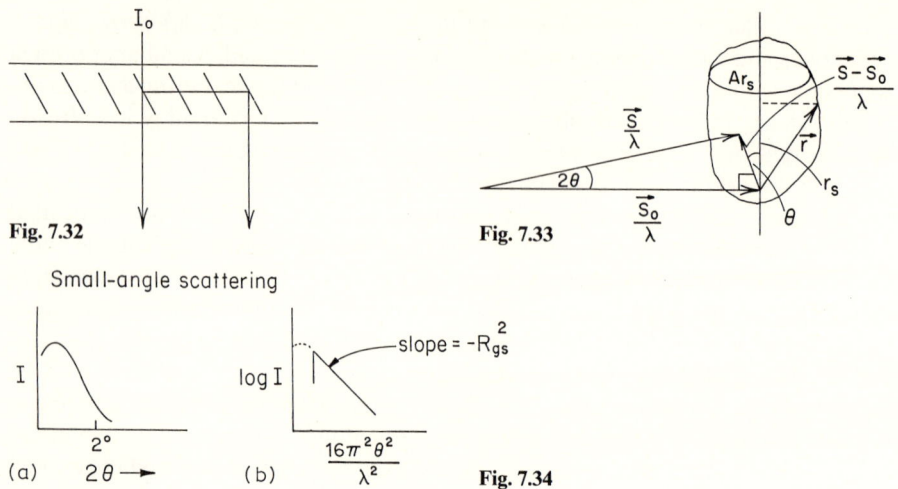

Fig. 7.32. Multiple diffraction can give rise to small-angle scattering

Fig. 7.33. Scattering near the origin of reciprocal space. (After Guinier, A., "X-ray Diffraction in Crystals, Imperfect Crystals, and Amorphous Bodies". Copyright by Freeman, San Francisco, CA, 1963)

Fig. 7.34. a Small-angle scattering pattern. **b** The determination of the radius of gyration (R_g) from **a**

A. Guinier, has demonstrated that some very interesting information can be obtained in this region (see Guinier and Fournet 1959). We could evaluate the scattering in terms of α's, but suppose the clustering corresponds to small regions of constant composition different that of the matrix. This is the situation where we have small particles, such as precipitates in a solid solution, or polymers in a liquid solution, or porosity in an otherwise homogeneous solid. In our development of kinematic diffraction theory we have seen (Sect. 2.4) that the amplitude is the Fourier transform of the electron density. [Of course, if the matter is surrounded by another medium (even air), it is the difference in density that is involved; but when it is just air, we neglect this component.[9]] For one particle, the scattering is then the transform of the difference in electron density between the matrix and particle:

$$A_{eu}(\mathbf{s}) = \Delta\rho \int \exp(2\pi i[\mathbf{s}\cdot\mathbf{r}])\,dv_r\,.$$

The transform can be evaluated for specific shapes as was done in Sect. 6.2 and this will yield the analytical expression for the scattering pattern characteristic of the shape, rods of scattering for plates, etc. Let us examine a different approach. Let $\mathbf{s}\cdot\mathbf{r} = sr_s$, where r_s is the magnitude of $\mathbf{r}$ along $\mathbf{s} = (\mathbf{S} - \mathbf{S}_0)/\lambda$. At small angles, $s \cong 2\theta/\lambda$ and r_s is essentially the component of $\mathbf{r}$ along a line perpendicular to $\mathbf{S}_0$, as seen in Fig. 7.33. If A_{r_s} is the cross-sectional area of the particle along this line.

---

[9] That is, we measure and subtract the air scattering.

7.7 Small-Angle Scattering (SAS)

$$A_{eu}(s) = \Delta\rho \int \exp(2\pi i[sr_s])A_{r_s}\,dr_s.$$

Expanding the exponential, because s is small in the small-angle region:

$$A_{eu}(s) = \Delta\rho\left[\int A_{r_s}\,dr_s + \int 2\pi i s r_s A_{r_s}\,dr_s - \int 2\pi^2 s^2 r_s^2 A_{r_s}\,dr_s\right].$$

If we choose as our origin for the particle its center of gravity, the second term is zero. The first is the volume per particle V times $\Delta\rho$. The third integral is the second moment of the particle, related to the *radius of gyration* (R_{gs}):

$$R_{gs}^2 = \frac{1}{V}\int r_s^2 A_{r_s}\,dr_s. \tag{7.80}$$

Thus,

$$A_{eu}(s) = \Delta\rho[V - V2\pi^2 s^2 R_{gs}^2] \cong \Delta\rho\exp(-2\pi^2[2\theta/\lambda]^2 R_{gs}^2)V,$$

and for the intensity for n independent particles:

$$I_{eu}(s) = |A_{eu}(s)|^2 \cong n(\Delta\rho)^2 V^2 \exp(-4\pi^2[2\theta/\lambda]^2 R_{gs}^2) \tag{7.81}$$

By plotting $\ln I$ vs. $16\pi^2\theta^2/\lambda^2$ (Fig. 7.34), R_{gs} can be obtained, and by evaluating Eq. (7.80) for the shape of the particle, judged, say from electron microscopy or by directional streaking on a film, measurements of R_{gs} can be employed to obtain dimensions of the particle.

If the particles have a range of sizes and take on all possible orientations in space, then we can think of particle dimensions not only along s playing a role, but the dimensions along the two orthogonal axes perpendicular to s. If we call these axes x and y and recall the distance to a point (r) is related to its coordinates, i.e., $r^2 = x^2 + y^2 + s^2$, then $\langle R_g^2 \rangle = \langle R_{gx}^2 \rangle + \langle R_{gy}^2 \rangle + \langle R_{gs}^2 \rangle$. Furthermore, if all three averages are equal, that is, if all orientations are equally possible, then $\langle R_g^2 \rangle/3$ replaces R_{gs}^2 in Eq. (7.81), where

$$\langle R_g^2 \rangle = \int r^2\,dv/V. \tag{7.82}$$

Some typical equipment for work in the small-angle region is shown in Fig. 7.35.

Note that in Fig. 7.34a there is a "bump" in the intensity. If the particles are not randomly arranged (perhaps because the density of particles is high), there will be some *interparticle interference*. A range of sizes may also be present and these will be indicated by a range of slopes in a ln-plot like that in Fig. 7.34b. In some cases, it is possible to determine the size distribution; see Sect. 7.7.5.

It is difficult to make measurements near the direct beam in reflection geometry. Therefore, most small angle studies involve transmission. With the geometry in Fig. 7.35, the sample is fixed and measurements are made vs. 2θ (with film, a normal moving detector, or a position sensitive detector). In this case, the measured power (P) can be written as proportional to $I_0 A_0 t \exp[-(\mu/\rho)\rho t]$, where t is the sample thickness and the remaining terms have their usual meaning. By setting $dP/dt = 0$,

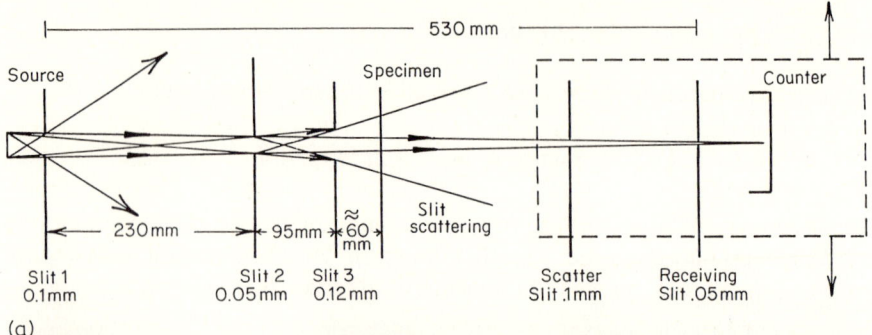

Fig. 7.35. a The slit system shown defines a beam about 0.05° 2θ wide and scattering from the slits is small beyond $2\theta = 0.08° \, 2\theta$, so that a d spacing of 500 Å could easily be observed with Cu K_α. (Divergence of the beam and scattering from the slits is also shown.) **b** Typical appearance of the equipment

it is easy to show that the optimal thickness is $t = 1/\mu$. This gives the maximum intensity; however as the variation of diffracted power with t is slowly varying, thickness 2–3 times as great or 1/2 to 1/3 are still useful. With x-rays $\mu \propto \lambda^3$ so that $t \propto \lambda^{-3}$. With longer wavelengths selected, to "push" the scattering pattern to higher 2θ away from the direct beam, the thickness can be prohibitively small. With neutrons, μ is independent of λ, and low, so that quite thick specimens can be examined. In fact volume fractions as low as 10^{-6} can be detected with "SANS" (small-angle neutron scattering). If the particles causing the scattering are large, there can be apparent small angle scattering which is really broadening due to refraction through the particles. This contribution can be estimated (Compton and Allison 1935; Weiss 1951) as well as the parasitic intensity, such as multiple scat-

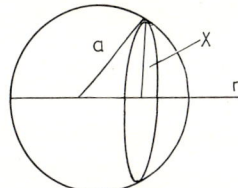

Fig. 7.36. A spherical particle of radius a, and the definition of x for calculating the radius of gyration

tering (Schelten and Schmatz 1980), that due to oxide films (Roth 1977) and that due to dislocations (Atkinson and Hirsch 1958) which contribute because of the change in local atom density due to the strain field. [Recently, Berk and Hardmann-Rhyne (1985) have demonstrated that in favorable cases for SANS the multiple scattering effects can be analyzed in detail, providing information about particle size and size distribution in samples with high densities of particles and sizes overlapping the optical scattering range.]

It is also interesting to note that the neutron or x-ray beam does not have to be very monochromatic. By differentiating Bragg's law $\Delta\lambda = 2d\cos\theta\,\Delta\theta$ or $\Delta\theta = \tan\theta\,(\Delta\lambda/\lambda)$. The amount of broadening $\Delta\lambda$, is small at low θ even for moderate $\Delta\lambda/\lambda$. Thus single crystal monochromators are not necessary; filtered x-rays or a neutron beam monochromated by a chopper with $\Delta\lambda/\lambda = 0.25$ (Sect. 4.7) are adequate.

Let us evaluate the radius of gyration for a sphere; with the aid of Fig. 7.36

$$R_{gs}^2 = \frac{1}{(4/3)\pi a^3} \cdot 2 \cdot \int_0^a \pi x^2(a^2 - x^2)\,dx = a^2/5\,.$$

For a random array, with Eq. (7.82), and using spherical coordinates

$$R_g^2 = \int_0^{2\pi}\int_0^{2\pi} r^2[r^2\sin\phi\,dr\,d\phi\,d\theta]/\tfrac{4}{3}\pi a^3 = 3a^2/5\,.$$

Thus, for spheres, a Guinier plot gives the average size. For other shapes, such as a cylinder or ellipsoid, *which have two dimensions*, measurements in an electron microscope of the aspect ratio are necessary, unless these are well aligned so that the different dimensions can be determined by measurements in different directions in reciprocal space.

In determining a radius of gyration when particles of varying size are present, heavy weighting is placed on those with largest size. To see this consider that there are k different sizes present, each with fraction w_k and volume V_k. In developing Eq. (7.81) the term VR_g^2 appears for the amplitude.

Hence:

$$\langle R_g^2 \rangle = \frac{\sum_k w_k \Delta\rho^2 V_k^2 (R_g^k)^2}{\sum_k w_k \Delta\rho^2 V_k^2}\,. \tag{7.83}$$

Now $\Delta\rho^2 V^2$ is the square of the number of electrons per particle, which is proportional to R^6. Therefore:

$$\langle R_g^2 \rangle = \frac{\langle R^8 \rangle}{\langle R^6 \rangle}\,,$$

a ratio which is heavily weighted toward large R.

7.7.2 The Guinier Region

Perhaps we can get a still better feeling for the Guinier equation by examining the more general situation, where the density of scattering ability is not a constant through our small particles. We start with our general equation for the diffracted intensity for such a particle:

$$I_{eu}(s) = \sum_n \sum_m f_n f_m \exp[2\pi i s \cdot r_{nm}]. \tag{7.84}$$

Now assume that all possible orientations are present, that is we will average this equation by allowing r_{nm} to take on all possible orientations around the diffraction vector, s.[10] As in Fig. 7.37 we allow r_{mn} to lie anywhere on the surface of a sphere of radius, r_{mn}:

$$I_{eu}(s) = \frac{\sum_m \sum_n \int_0^\pi f_n f_m \exp[2\pi i s \cdot r_{mn} \cos \alpha] 2\pi r_{mn} \sin \alpha\, r_{mn}\, d\alpha}{4\pi r_{mn}^2}.$$

With $k = 4\pi \sin \theta / \lambda$,

$$I_{eu}(s) = \sum_m \sum_n f_m f_n \sin k r_{mn} / k r_{mn}. \tag{7.85}$$

Expanding the sine term:

$$I_{eu}(k) = \sum_m \sum_n f_m f_n - \sum_m \sum_n f_m f_n 4\pi^2 s^2 r_{mn}^2 / 6 \ldots$$

Now $r_{mn}^2 = (r_m - r_n)^2 = r_m^2 + r_n^2 - 2 r_m r_n$.

Therefore:

$$\sum_m \sum_n f_m f_n r_{mn}^2 = \sum_m \sum_n f_m f_n r_m^2 + \sum_m \sum_n f_m f_n r_n^2 - 2 \sum_m \sum_n f_m f_n r_m r_n.$$

Defining the center of gravity of the particle as $\sum f_m r_m = 0$, the last term vanishes. Also:

$$\sum_m \sum_n f_m f_n r_m^2 = \sum_n f_n \sum_m f_m r_m^2 = \rho \sum_m f_m r_m^2,$$

where ρ is the number of electrons in the particle [f approaches Z (the atomic number) as $S \to 0$]. With $R_g^2 = \sum f_m r_m^2 / \sum f_m$

$$I_{eu}(s) = N(\rho^2 - (4\pi^2/3)\rho^2 R_g^2 s^2 = N\rho^2 \exp[-(4\pi^2/3)s^2 R_g^2].$$

This is the same Guinier expression we derived before!

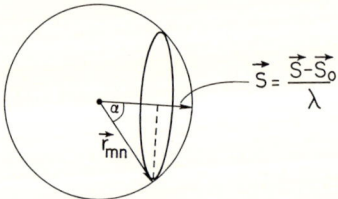

Fig. 7.37. The vector r_{mn} can take on all orientations around the diffraction vector s, that is occupy all positions with equal probability on the surface of a sphere of radius r_{mn}

---

[10] Note that this is the same kind of average to be taken to obtain an equation for I_{eu_2} the diffuse scattering from a powder, or for a gas of molecules.

7.7.3 Small-Angle Scattering from Polymers

One interesting area where this radius is of interest is in the measurement of molecular size, in biology or in polymer science. Of course, in a polymer there can be crystalline and amorphous regions and their difference in density will give rise to small angle scattering. Here we wish to consider an entirely amorphous substance, and we start by considering the case of a long chain molecule in a solvent. Typically, the radius of gyration is 100–500 Å, and small angle scattering is employed to measure this dimension using light[11], x-rays, or neutrons. The backbone of a long coiled chain is illustrated in Fig. 7.38. There are $(n + 1)$ mass points, and n bonds. The length of a C—C— bond is l_i, and the center of mass is c so that R_i^c is the distance from c and $r_{ij} = R_i - R_j$.
Therefore:

$$R_g^2 = \frac{1}{n+1} \sum_{i=0}^{n} (R_i^c)^2 . \tag{7.86a}$$

Both **R** and **r** are vectors, so:

$$\mathbf{R}_i^c = \mathbf{R}_0^c + \mathbf{r}_{0i} . \tag{7.86b}$$

Therefore, with $\mathbf{r}_{00} = 0$:

$$R_g^2 = (R_0^c)^2 + \frac{2}{n+1} R_0^c \sum_{i=1}^{n} r_{0j} + \frac{1}{n+1} \sum_{i=1}^{n} r_{0i}^2 . \tag{7.68c}$$

From Eq. (7.86b):

$$\sum_{i=1}^{n} R_i^c = 0 ,$$

and therefore

$$R_0^c = -\frac{1}{n+1} \sum_{i=1}^{n} r_{0i} .$$

So:

$$(R_0^c)^2 = \frac{1}{(n+1)^2} \sum_{i=1}^{n} \sum_{j=1}^{n} r_{0i} r_{0j} \tag{7.86d}$$

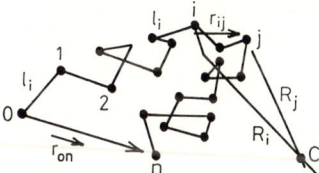

Fig. 7.38. Definition of terms to describe the backbone configuration of a long chain organic molecule

---

[11] With light, the "scattering ability" is related to index of refraction; because of the long λ, the angles for this "small-angle" scattering are $\sim 90°$.

and:

$$R_g^2 = \frac{1}{(n+1)^2}\sum\sum r_{0i}r_{0j} - \frac{2}{(n+1)^2}\sum\sum r_{0i}r_{0j} + \frac{1}{n+1}\sum r_{0i}^2 \quad (7.68e)$$

$$= \frac{1}{n+1}\sum_{i=1}^{n} r_{0i}^2 - \frac{1}{(n+1)^2}\sum_{i=1}^{n}\sum_{j=1}^{n} r_{0i}r_{0j}. \quad (7.68f)$$

For two vectors **a**, **b**:

$$\mathbf{a}\cdot\mathbf{b} = \frac{a^2 + b^2 - |\mathbf{a}-\mathbf{b}|^2}{2}.$$

In our case

$$\mathbf{r}_{0i}\cdot\mathbf{r}_{0j} = \frac{r_{0i}^2 + r_{0j}^2 - r_{ij}^2}{2}.$$

Therefore:

$$R_g^2 = \frac{1}{n+1}\sum r_{0i}^2 - \frac{n}{(n+1)^2}\sum_{i=1}^{n} r_{0i}^2 + \frac{1}{2(n+1)^2}\sum_{i=1}^{n}\sum_{j=1}^{n} r_{ij}^2. \quad (7.68g)$$

With the identity:

$$\frac{1}{n+1} - \frac{n}{(n+1)^2} = \frac{n+1}{(n+1)^2} - \frac{n}{(n+1)^2} = \frac{1}{(n+1)^2},$$

$$R_g^2 = \frac{1}{(n+1)^2}\sum r_{0i}^2 + \frac{1}{2(n+1)^2}\sum_i\sum_j r_{ij}^2.$$

As $r_{0i}^2 = r_{0j}^2$:

$$R_g^2 = \frac{1}{2(n+1)^2}\sum_{i=1}^{n} r_{0i}^2 + \frac{1}{2(n+1)^2}\sum r_{0j}^2 + \frac{1}{2(n+1)^2}\sum_i\sum_j r_{ij}^2$$

$$= \frac{1}{2(n+1)^2}\sum_{i=0}^{n}\sum_{j=0}^{n} r_{ij}^2. \quad (7.86h)$$

With a defined as the C—C bond distance, and also $\bar{\mathbf{r}}_{ij} = \sum_{n=1}^{j}\mathbf{l}_n$, then:

$$\mathbf{r}_{ij}\cdot\mathbf{r}_{ij} = \sum_{m=i}^{j}\sum_{n=i}^{j}\mathbf{l}_m\cdot\mathbf{l}_n = |i-j|a^2 + \sum_{\substack{m=i\\m\neq n}}^{j}\sum_{n=i}^{j}\mathbf{l}_m\cdot\mathbf{l}_n$$

$$= |i-j|a^2$$

on the average, for a freely joined polymer (so that a bond can take on all orientations). Then:

$$R_g^2 = \frac{a^2}{2(n+1)^2}\sum_{i=0}^{n}\sum_{j=0}^{n}|i-j|$$

and for large n:

$$R_g^2 = \frac{a^2}{2n^2}\sum_{i=0}^{n}\sum_{j=0}^{n}|i-j|. \quad (7.86i)$$

7.7 Small-Angle Scattering (SAS)

Now with $m = |i - j|$ and writing out the matrix represented by the double sum:

$$\sum_{i=0}^{n} \sum_{j=0}^{n} |i - j| = 2 \sum_{m=0}^{n} (n + 1 - m)m . \tag{7.87a}$$

Since:

$$\sum_{m=0}^{n} m = \frac{n(n + 1)}{2}, \tag{7.87b}$$

and

$$\sum m^2 = \frac{n(n + 1)(2n + 1)}{6}, \tag{7.87c}$$

we obtain:

$$R_g^2 = \frac{1}{6}\left(\frac{n + 2}{n + 1}\right)na^2 = \frac{1}{6}na^2 \quad \text{for large } n . \tag{7.88}$$

Measuring R_g thus gives us a value for the length of the molecule (n). [For further details see Flory (1953, 1969).]

By varying the density of scattering ability of the solution in which the polymer is dissolved, it is possible to decide on such things as whether the molecule is homogeneously dense, or whether there is a dense core and outer layer of different density (by matching the solution scattering to that of what is believed to be outer layer). In the study of solid polymers, tagging a molecule (or specified chemically unique parts of a molecule) may make them "visible" in a small angle diffraction experiment. Some of the hydrogen in the molecules or its segments can be replaced by deuterium for SANS, or a heavy atom (iodine, for example) may be incorporated for SAXS. A simple check is possible in such cases to confirm that single molecule scattering is occurring and that intermolecular scattering or molecular clustering is not occurring. According to Eq. (7.81), the intensity extrapolated to $s = 0$ is $N\Delta\rho^2 V^2$, where N is the (known) number of tagged molecules in the irradiated volume, and V is the volume per tagged particle. As N is known, the value $(\Delta\rho V)^2$ is obtained, a measure of the total electron count per particle, related to the molecular weight. (In the case of a polymer with varying chain length, this can be shown to be an average weight, lying between the number and weight average molecular weights.) Therefore, this extrapolated value, if measured on an absolute scale, should give the electron count (or nucleon count for SANS) expected for a single molecule.

Additional information may be extracted from the SAS pattern by noting that, as pointed out in Sect. 7.7.1, the amplitude is essentially a shape transform. The intensity is the product of two such transforms, and hence intensity is the transform of the convolution of the shape and itself in real space. Let p be zero outside the particle and unity inside, then the intensity is proportional to the transform of:

$$V(x) = \int p(u)p(x + u) \, dV_u . \tag{7.89}$$

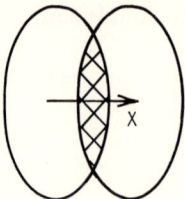

Fig. 7.39. A particle and its "ghost", displaced from it by x. Their convolution is the common (shaded) area

This is the common volume between a particle and its "ghost", displaced by x, as in Fig. 7.39. From our study of the convolution, (Sect. 6.4) the value of the convolution at $x = 0$, which is just V, is the integral of the intensity in reciprocal space. This integral is referred to as "invariant", as it is independent of the size or shape of the particles causing the scattering.[12] (This is really the same statement that we found in the previous section, that $\alpha_{000} = 1$.)

$$\int I(s)\, dV_s = N\Delta\rho^2 V. \tag{7.90a}$$

Therefore:

$$\frac{I(0)}{\int I(s)\, dV_s} = \frac{N\Delta\rho^2 V^2}{N\Delta\rho^2 V} = V. \tag{7.90b}$$

This is the volume per particle. This yields another measure of the particle dimension, in addition to R_g, and is especially useful in fixing dimensions for biological molecules which are uniform in size (but less so with polymers which vary in length).

It is no wonder that the "Guinier region" of small angle scattering is of wide interest. Accordingly, it is appropriate to examine just how precise is the Guinier equation. Consider a solid sphere. The shape transform can be written in spherical coordinates with $p = 1$ within the sphere, zero outside as:

$$\int \exp(-2\pi i \mathbf{s} \cdot \mathbf{r})\, dV_r = \int_0^{2\pi} \int_0^{\pi} \int_0^a \exp(-2\pi i s r \cos\alpha) r^2 \sin\alpha\, d\alpha\, d\phi\, dr. \tag{7.91}$$

Write the exponential in terms of cosine and sine, let $u = 2\pi s r \cos\alpha$ so that $du = -2\pi s r \sin\alpha\, d\alpha$. Then as α varies from 0 to π, u varies from $+2\pi s r$ to $-2\pi s r$. Thus the integral in Eq. (7.91) can be written:

$$\int_0^{2\pi} \int_{2\pi s r}^{-2\pi s r} \int_0^a [\cos u - i \sin u][-\tfrac{1}{2}\pi s] r\, d\phi\, dr\, du.$$

The integral with respect to ϕ is 2π. Integrating with respect to u and then r, the result is:

$$\tfrac{4}{3}\pi a^3 \left\{ \frac{3[\sin 2\pi s a - 2\pi s a \cos 2\pi s a]}{(2\pi s a)^3} \right\}. \tag{7.92}$$

---

[12] Determining the integral in the small angle region is not as easy as it might appear. Data must be extrapolated under the direct beam to $s = 0$, and out to infinity. The latter can be done analytically because, as we shall see, at large s intensity varies with s^{-4}

7.7 Small-Angle Scattering (SAS)

This expression is proportional to the amplitude; the intensity is related to the square of this term. This square has zeroes for $(2K + 1)/4$ (where $K \geq 1$ is an integer) and maxima (beyond $s = 0$ for which it is unity) at $K/2$ for $K \geq 2$, with intensities $0.1/K^4$. The Guinier approximation follows this function quite closely for $2\pi s R_g \leq 1.2$. This criterion is often employed to be certain that the data is in the range where the determination of R_g is valid.

7.7.4 The Porod Region

There is another region of small angle scattering that has many applications, and we can see why by employing Eq. (7.92). We square this equation and recall that $I_{eu}(s) = \Delta \rho^2$ times this square:

$$I_{eu}(s) = [\Delta \rho 4\pi a^3]^2 (3)^2 \left(\frac{\sin 2\pi as - 2\pi as \cos 2\pi as}{(2\pi sa)^3} \right)^2. \tag{7.93}$$

Forming the square, using $2 \sin \alpha \cos \alpha = \sin 2\alpha$ (for the middle term) and $1 - \cos^2 \alpha = \sin^2 \alpha$ (for the first):

$$I_{eu}(s) = \frac{16\Delta \rho^2}{64\pi^4 s^6} [1 + (4\pi^2 a^2 s^2 - 1) \cos^2 2\pi as - 2\pi as \sin 4\pi as].$$

With $\cos^2 \alpha = (1 + \cos 2\alpha)/2$, and multipling by $(2/\pi)/(2/\pi)$.

$$I_{eu}(s) = \frac{\Delta \rho^2}{8\pi^3} \left[\frac{1}{\pi s^6} + \frac{4\pi a^2}{s^4} + \left(\frac{4\pi a^2}{s^4} - \frac{1}{\pi s^6} \right) \cos 4\pi as - \frac{4a}{s^5} \sin 4\pi as \right].$$

Usually we have a size distribution, (each size with some frequency w_a) and then we can neglect the trigonometric terms, which cancel due to contributions from the various sizes of particles. Also $1/s^6 \ll 1/s^4$. Therefore,

$$I_{eu}(s) = \frac{\Delta \rho^2 \sum_a w_a 4\pi a_a^2}{8\pi^3 \, s^4} = \frac{\Delta \rho^2}{8\pi^3} \frac{S}{s^4}. \tag{7.94}$$

Here, S is the total surface area in the irradiated volume. We can find the region where our approximations are valid by plotting $\ln I$ vs. $\ln s$ and choosing the region where the slope is -4, or we can form the quantity $s^4 I_{eu}(s)$ vs. s, Fig. 7.40, to obtain $\Delta \rho^2 S$, from the horizontal region, where $kR_g \geq 3$. The oscillations at low s arise due to the trigonometric terms, and will be present *only* if any size distribution is narrow in width. In such a case, the shaded areas should be comparable in magnitude. While we have derived this result for spherical particles, it may be shown to be much more

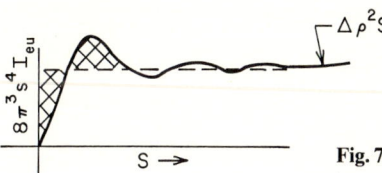

Fig. 7.40. A Porod plot. The *shaded areas* should be equal if there is a narrow size distribution

general. However, if the particles are very small in one direction, or angular, the shaded areas are not equal, and a horizontal line is never quite reached; a perfect cube (without rounded corners) is an example of an angular shape (see Tchoubar-Vallant and Méring 1965, 1966; Schiller and Méring 1967).

It would seem that absolute intensities would be required to employ Eq. (7.94) to obtain the surface area "S". However, recall that the "invariant", the integrated small-angle intensity, is equal to $N\Delta\rho^2 V$. Dividing by this, all constants are eliminated:

$$\left(\frac{\Delta\rho^2}{8\pi^3}\frac{S}{s^4}\right)/N\Delta\rho^2 V = \frac{S/V_{total}}{8\pi^3 s^4}, \tag{7.95}$$

where $V_{total} = NV$ is the total volume for all particles under the beam. In this way we measure the surface to volume ratio, without absolute intensities.[13]

Surface areas are particularly interesting in the field of heterogeneous catalysis. Fine powders of silica gel and alumina with 10–100 m^2/g surface areas are typical in this field. Often such a powder, with pores of 100 Å, has particles of a metal 10–100 Å on the pore walls. In such a case, there are three surface areas — metal-powder, powder-air, metal-air. All three separate areas can be measured (Goodisman et al. 1981; Espinat et al. 1984).

As we are measuring a surface to volume ratio, the inverse of this is related to an average dimension called the Porod radius, R_p. This is related to a moment of the distribution, $\langle R^3 \rangle / \langle R^2 \rangle$, and is therefore numerically different than R_G.

7.7.5 Size Distributions

It is possible to obtain not only moments of any size distribution, such as R_G and R_p, but information on the size distribution itself. Consider the ghost function, $V(\mathbf{r})$ and take the average $\langle V(\mathbf{r}) \rangle$, for all possible directions of $\mathbf{r}$. We define the "characteristic function", $\gamma(\mathbf{r})$ as $\langle V(\mathbf{r}) \rangle / V$, and $P(r)\,dV = 4\pi\Delta\rho^2 V\gamma(r)\,dV$. Now

$$I_{eu}(s) = \Delta\rho^2 \int V(\mathbf{r})\exp[-2\pi i \mathbf{s}\cdot\mathbf{r}]\,dV_{\mathbf{r}}.$$

With all possible orientations, the exponential term is averaged, as before:

$$I_{eu}(s) = \Delta\rho^2 V \int \gamma(r)\frac{\sin kr}{kr} 4\pi r^2\,dr = \Delta\rho^2 \int P(r)\frac{\sin kr}{kr}\,dV.$$

Therefore, the sine transform of $(kI_{eu}/\Delta\rho^2)$ gives $P(r)r$ or $\gamma(r)$. With $G(l)$ the distribution of chord lengths through a particle (or pore) and Fig. 7.41:

$$\gamma(r) = (1/\bar{l})\int_0^D (l-r)G(l)\,dl.$$

Here D is the maximum length and $\bar{l} = \int_0^D lG(l)\,dl$. Therefore:

---

[13] Sometimes people use $Q = q = k = 2\pi s$, and then the equation is $2\pi(S/V)/Q^4$. Also, in the literature on this topic, instead of 2θ, the scattering angle is sometimes taken as θ; then $k = 4\pi\sin(\theta/2)/\lambda$.

7.7 Small-Angle Scattering (SAS)

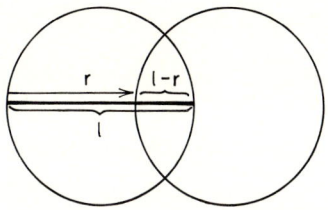

Fig. 7.41. The definition of chord length, l

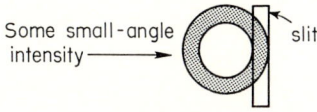

Fig. 7.42. How a rectangular slit sees a typical small angle scattering pattern

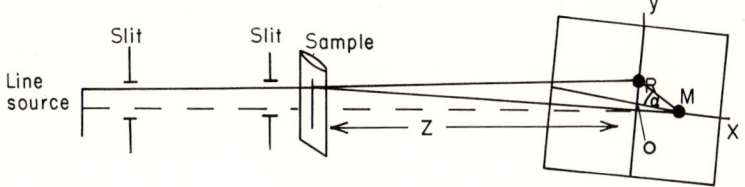

Fig. 7.43. The geometry of the effect of beam (and slit) size on the small angle scattering

$$\frac{d^2\gamma(r)}{dr^2} = \frac{1}{\bar{l}}G(l) \ . \qquad (7.96)$$

Thus, the second derivative of the characteristic function gives the chord length distribution. See also Brill and Schmidt (1968), Federova and Schmidt (1978), Yang (1984).

7.7.6 Slit Corrections

Before we leave this topic, it is necessary to point out that because such studies are made at small scattering angles ($< 2°2\theta$), the dimensions of the slit can strongly affect the results, as shown in Fig. 7.42. This is especially true if rectangular slits are employed to increase the measured intensity. With the aid of Fig. 7.43 which details this effect, we define $k_x = 2\pi OM/2\lambda$, $k_y = 2\pi OR/2\lambda$. Measurements are made along x but include scattering at positions y. Let I_0 be the vertical intensity distribution (along y) at the sample:

$$I_{\text{meas}}(k_x) \propto \int I_0(y) I(\sqrt{k_x^2 + k_y^2}) \, dy \ . \qquad (7.97)$$

We have ignored the effect of beam width. With the usual collimation, this is generally small, and we have seen how to deconvolute this effect in Sect. 7.4.

Now suppose $I(k)$ is a Gaussian like the Guinier expression, i.e., of the form

$$Ale^{-BR_g^2 k^2} = Ale^{-BR_g^2(k_x^2 + k_y^2)} \ .$$

Then:

$$I_{meas}(k_x) \propto \left\{ \int I_0(y) A \exp[-BR_g^2 k_y^2] \, dk_y \right\} \exp[-BR_g^2 k_x^2].$$

The term in parentheses is a constant and thus the result is of the form of the Guinier expression. The Guinier relationship is unaffected by vertical collimation, which is one of the reasons it is so widely used in small-angle scattering.

What about the Porod region, where $I \propto c/k^4$?
In this region:

$$I_{meas}(k_x) \propto \int_{-\infty}^{+\infty} \frac{1}{(k_x^2 + k_y^2)^2} \, dk_y.$$

From Fig. 7.43, $k_y = k_x \tan \alpha$. Then:

$$I_{meas}(k_x) \propto \frac{2}{k_x^3} \int_0^{\pi/2} \frac{d\alpha}{(1 + \tan^2 \alpha)\cos^2 \alpha} = \frac{2}{k_x^3} \int_0^{\pi/2} \cos^2 \alpha \, d\alpha = \frac{\pi}{2k_x^3}$$

or:

$$k_x^3 I_{meas}(k_x) = \frac{\pi}{2} k_x^4 I(k_x). \tag{7.98}$$

We also may wish to use the invariant $\int k_x^2 I(k_x) \, dk_x$ with slit collimation. By similar methods it can be shown that this becomes:

$$\tfrac{1}{2} \int k_x I_{meas}(k_x) \, dk_x. \tag{7.99}$$

There are many procedures described in the literature for correcting the entire scattering curve for slit effects [Glatter, O. and Kratky, O., "Small Angle X-ray Scattering", Academic Press, New York (1982)].

We close this section by referring the reader to a paper of P.B. Moore [J. Appl. Cryst. *13*, 168 (1980)], where equations are given for estimating errors in the various quantities we have discussed due to errors in the data.

7.8 Liquids and Amorphous Solids

7.8.1 Single Component Substances

From liquids and amorphous solids, the diffraction pattern appears as in Fig. 7.44. There are some details, but the pattern is devoid of the sharp peaks characteristic of a crystalline solid. Actually, the exact distinction between crystalline and amorphous is quite difficult to specify. After extreme deformation or rapid electrolytic or vapor deposition or rapid quenching, normally crystalline materials often give patterns similar to those for amorphous material, in some cases due to a very small

7.8 Liquids and Amorphous Solids

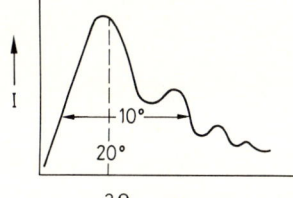

Fig. 7.44. Scattering from a liquid or amorphous solid

mosaic size and large distortions — but sometimes also due to a quite different aperiodic network.

What can we determine from such a pattern? First, consider a material of only one atom type. We recall that the diffracted intensity is a Fourier transform of the convolution of the atomic density and the same density inverted through the origin, i.e.

$$I_{eu}(s) = KT[\rho(\mathbf{r}) \otimes \rho(-\mathbf{r})] \ .$$

Let $\mathbf{u} + \mathbf{r} + \mathbf{w}$, then:

$$I_{eu}(s) = K \int \int \rho(\mathbf{w})\rho(\mathbf{w} + \mathbf{r}) \, dV_w \exp(2\pi i \mathbf{s} \cdot \mathbf{r}) \, dV_r \ .$$

The density product peaks when both density values exist, i.e. is, when there are atoms at $\mathbf{r}$ and $\mathbf{r} + \mathbf{w}$. This is the "Patterson function" discussed in Sect. 6.4. We assumed there that the densities were periodic, but if they are not, this is about as far as we can go. Because liquids are isotropic, we can work with *radial Patterson functions*, i.e., $\langle \rho(r + w) \otimes \rho(w) \rangle$ averaged over all atoms as origin.[14] Referring to Fig. 7.37:

$$I_{eu}(s) = K \int_0^r \int_0^\pi \langle \rho(r+w) \otimes \rho(w) \rangle \exp[2\pi i r \cos\alpha(2\sin\theta/\lambda)] 2\pi r \sin\alpha \, dr \, d\alpha \ .$$

With $k = 4\pi \sin\theta/\lambda$,

$$\langle I_{eu}(s) \rangle = K \int_0^r \langle \rho(r+w) \otimes \rho(w) \rangle 4\pi r^2 \frac{\sin kr}{kr} dr \qquad (7.100a)$$

or

$$\frac{4\pi \sin\theta}{\lambda} \langle I_{eu}(s) \rangle = K \int_0^r \langle \rho(r+w) \otimes \rho(w) \rangle 4\pi r \sin kr \ . \qquad (7.100b)$$

We can further simplify this expression by writing the Patterson function or convolution as $Nf^2 + Nf^2\rho(r)$ where $\rho(r) \, dr$ is the number of atoms in a shell of thickness dr at a distance r from another atom. (The first term is just the term associated with the origin of the Patterson function.) Then we can write:

---

[14] This is also the averaging procedure to follow with Eq. (7.55) to obtain the diffuse intensity in a powder pattern due to local order.

$$\left[\frac{[k\langle I_{eu}(s)\rangle]}{Nf^2} - K\right] = K\int_0^r \rho(r)4\pi r \sin kr\, dr. \qquad (7.101)$$

A term involving the average atomic density ρ_a is now added and subtracted:

$$\left[\frac{(k\langle I_{eu}(s)\rangle)}{Nf^2} - K\right] = K\int_0^\infty [\rho(r) - \rho_a]4\pi r \sin kr\, dr$$

$$+ \int_0^r \rho_a 4\pi r \sin kr\, dr. \qquad (7.102)$$

In Eq. (7.102), the upper limit of the first term on the right-hand side can be taken as infinity instead of the sample dimension, because $[\rho(r) - \rho_a]$ vanishes rapidly with increasing r. The second term is simply the transform of the sample; as it has a macroscopic size, this transform makes a contribution only very close to the origin in reciprocal space and we can neglect it. [For further details about this, see Warren (1969).]

By taking the Fourier sine transform of the left-hand side of Eq. (7.102), $4\pi r[\rho(r) - \rho_a]$ or, the radial distribution function ("RDF"), $\rho(r)$ can be obtained. But first it is necessary to determine K, and eliminate parasitic and Compton scattering. The term K includes the direct beam, which may be evaluated by the methods in Appendix C. However, there is an alternative procedure. *At large angles, the intensity is the sum of independent scatterers, KNf^2 plus Compton scattering.* Therefore, *calculated values can be matched to experimental data minus background in this region* (where there are no oscillations in the data) to evaluate K. Then it is possible to take the transform. But a serious problem remains; whereas the *intensity* decreases rapidly with increasing θ, the function on the left side of Eq. (7.102) does not because the intensity is multiplied by k. The sine transform which has limits zero to infinity is in effect terminated by the limit of data. While this limit should always be at as great a value of $\sin \theta/\lambda$ as possible (and hence experiments should be done with short wavelengths), there is still a truncation effect. There are two current approaches to this problem. One is to multiply the left-hand side by a *damping function*, $\exp(-4\pi a \sin \theta/\lambda)^2$, so that this term does become small at the largest measured point; this broadens $[\rho(r) - \rho_a]$ slightly. Such an approach is discussed in detail by Warren (1969) and Warren and Mozzi (1975). The other approach is an example of the basic properties of transforms.

The "missing" data at high angles will contribute primarily at low values of r as a result of the transform relationship of these two spaces. Now, two atoms cannot come closer than their diameter, so that in the first term on the right-hand side of Eq. (7.102) *the transform versus r at small r should be a straight line of slope* $-4\pi\rho_a$, since $\rho(r) = 0$ in this region. In the actual transform, oscillations will appear here (and elsewhere) due to the termination, and *the difference between them and the line with slope* $-4\pi\rho_a$ *can be retransformed to an intensity*. The amount of this synthesized intensity *beyond the maximum measuring point* can then be added to the actual data and the sum can be reinverted. The oscillations occur throughout the pattern, not only in the small r range, but as we are retransforming only the information at small r, several iterations may be required. This procedure also tends to correct for an error in K; such an error causes the left-hand side of Eq. (7.102) to differ from the

7.8 Liquids and Amorphous Solids

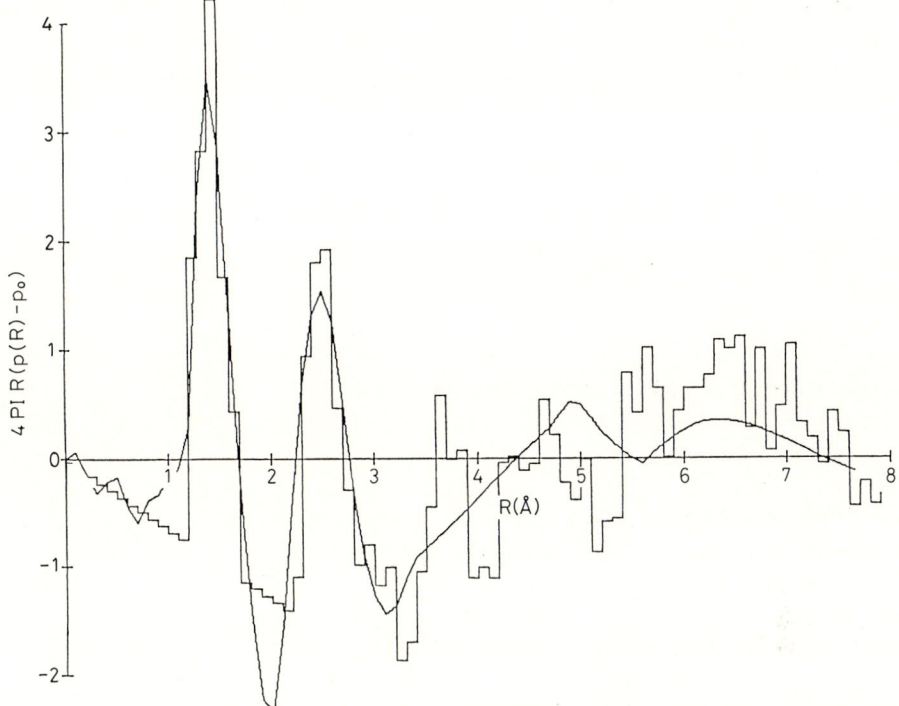

Fig. 7.45. The transform of the left side of Eq. (7.102) along with a histogram calculated from a histogram of the crystal. The specimen was isotactic polystyrene. For further details see Wecker, S.M., Davidson, T., and Cohen, J.B., J. Mater. Sci. 7, 1249 (1972). The units of ρ are atoms per cubic centimeter. To obtain the value in a peak the integral $\{\int_{r_1}^{r_2} 4\pi r^2 [\rho(r) - \rho_0] \, dr\} + \rho_0$ is evaluated

true value in a slowly varying manner with θ, and hence affects the transform at small r. There may be other spurious oscillations throughout the pattern which can be gradually eliminated in the same manner. This procedure was originally developed by Kaplow and co-workers (Azaroff et al. 1974).

A typical result after these corrections is shown in Fig. 7.45. *The integral over one of the peaks yields the number of atoms in that range of r* (the coordination number) minus the average value.

A powder pattern of a crystalline material can be treated in this same way, yielding some very interesting information on binding forces and elastic constants; see Kaplow et al. (1964).

What can we learn if the amorphous or liquid material is made up of *many atom types*?

7.8.2 Multicomponent Substances

In this case, we will follow notes of Wagner (1985). We assume that there are n elements and N total atoms seen by the incident beam (N_i of each species, i in the irradiated volume). The average atomic density in the alloy is $\bar{\rho}_a = N/V$, the average atom density of the ith species is $\bar{\rho}_i = N_i/V$ and average concentration $\bar{c}_i = N_i/V$. We use the bar to indicate the average value of a quantity over the entire sample, whereas these quantities can have local values.

$$\sum N_i = N, \tag{7.103a}$$

$$\sum \bar{\rho}_i = \bar{\rho}, \tag{7.103b}$$

$$\sum \bar{c}_i = 1. \tag{7.103c}$$

Also:

$$c_i = \bar{\rho}_i/\bar{\rho}, \tag{7.103d}$$

and the density of i and j atoms around an i atom:

$$\rho_i(r) = \sum_k^{N_i} \delta(r - r_{ik}). \tag{7.103e}$$

With this equation:

$$\int \rho_i(r)\, dr = N_i. \tag{7.104}$$

The amplitude of the scattered radiation from i is,

$$A_i(\mathbf{s}) = f_i \int_V \rho_i(r) \exp[-2\pi i \mathbf{s} \cdot \mathbf{r}]\, dr \tag{7.105a}$$

or:

$$A_i(\mathbf{s}) = f_i P_i(\mathbf{s}). \tag{7.105b}$$

[Of course for x-rays the f_i are also functions of s.]
The definition of $P_i(\mathbf{s})$ is contained in Eq. (7.105a) which can also be written in terms of our now familiar shape function (Sect. 7.7):

$$P_i(\mathbf{s}) = \int_{-\infty}^{\alpha} \rho_i(r) p(r) \exp[2\pi i \mathbf{s} \cdot \mathbf{r}]\, dr. \tag{7.105c}$$

With $p(r)$ equal to unity within the sample, zero elsewhere. Therefore the scattered intensity is:

$$I_{eu}(\mathbf{s}) = A(\mathbf{s})A^*(\mathbf{s}) = \sum_i^n \sum_j^n f_i f_j^* P_i(\mathbf{s}) P_j^*(\mathbf{s}). \tag{7.106}$$

Next we incorporate the partial atomic distribution functions, of the form $\rho_{ij}(\mathbf{r})$, which is the density of j atoms at distance $\mathbf{r}$ from an i atom.

Now, $N_i \rho_{ij}(\mathbf{r}) = N_j \rho_{ji}(\mathbf{r})$ or $\bar{c}_i N \rho_{ij}(\mathbf{r}) = \bar{c}_j N \rho_{ji}(\mathbf{r})$. Thus:

$$\bar{c}_i \rho_{ij}(\mathbf{r}) = \bar{c}_j \rho_{ji}(\mathbf{r}). \tag{7.107}$$

The reader will recognize this relationship as akin to that for the conditional pair probabilities we employed in discussing local order. In fact, there will be many echos of the treatment of this topic (introduced in Sect. 7.6.4) because considering a

7.8 Liquids and Amorphous Solids

multicomponent liquid is a generalization of the local order in alloys; here we have large variations in local position (there are no lattice sites), as well as chemical fluctuations.

In a binary (AB) alloy there are therefore three independent distribution functions, ρ_{AA}, ρ_{AB}, ρ_{BB}, (but ρ_{BA} is not independent of ρ_{AB} as shown in Eq. (7.107). In general for an n component system, using Eq. (7.107), there are $n(n+1)/2$ independent partial densities.

The quantity $P_i(\mathbf{s})$ is the transform of the product of $\rho_i(\mathbf{r})$ and $P(\mathbf{r})$, our shape function. Therefore, $P_i(\mathbf{s})P_j^*(\mathbf{s})$ is the transform of the convolution $[\rho_i(\mathbf{r})P(\mathbf{r}) \otimes \rho_j(-\mathbf{r})P(-\mathbf{r})]$. We can think of $P_i(\mathbf{s})P_j^*(\mathbf{s})$ as the product of the transforms of the "ghost" function $P(\mathbf{r})$ [Eq. (7.89)] times a function which describes the convolution $\rho_i(\mathbf{r}) \otimes \rho_j(-\mathbf{r})$, that is the Patterson function, which we can write as $N_i[\delta_{ij}\delta(\mathbf{r}) + \rho_{ij}(\mathbf{r})]$. Therefore:

$$P_i(\mathbf{s})P_j^*(\mathbf{s}) = \int n_i V(\mathbf{r})[\delta_{ij}\delta(\mathbf{r}) + \rho_{ij}(\mathbf{r})]\exp[-2\pi i \mathbf{s} \cdot \mathbf{r}]\,d\mathbf{r}\,. \tag{7.108}$$

Now $\delta(\mathbf{r}) = 0$ unless $r = 0$, and then $V(\mathbf{r}) = 1$, and with $V(\mathbf{s})$ the transform of $V(\mathbf{r})$:

$$P_i(\mathbf{s})P_j^*(\mathbf{s}) = N\left\{\bar{c}_i\delta_{ij} + \bar{c}_i\bar{c}_j\rho_a V(\mathbf{s}) + \int c_i c_j \left[\frac{\rho_{ij}(\mathbf{r})}{\bar{c}_j} - \rho_a\right]\exp[-2\pi i \mathbf{s} \cdot \mathbf{r}]\,d\mathbf{r}\right\}. \tag{7.109}$$

For samples of the usual size, the second term is zero except very close to $\mathbf{s} = 0$, as discussed in Sect. 7.8.1, so we will ignore this term. We define the integral as $\bar{c}_i\bar{c}_j[I_{ij}(\mathbf{s}) - 1]$.

Therefore:

$$I_{eu}(\mathbf{s}) = N\left\{\sum_i \bar{c}_i f_i^2 - \sum_{i\neq j}\sum \bar{c}_i\bar{c}_j f_i f_j^* + \sum_i\sum_j \bar{c}_i\bar{c}_j f_i f_j^* I_{ij}(\mathbf{s})\right\} \tag{7.110a}$$

$$= N\left\{\langle f^2\rangle - \langle f\rangle^2 + \sum_i\sum_j \bar{c}_i\bar{c}_j f_i f_j^* I_{ij}(\mathbf{s})\right\}. \tag{7.110b}$$

Furthermore, we can obtain the spherical average of the exponent in Eq. (7.105c) as we did for monatomic liquids, and change to coordinate $k = 4\pi\sin\theta/\lambda$. Also, we are after the fluctuation term $\{I_{ij}(k) - 1\}$. Accordingly, we write:

$$kI'(k) - \frac{k[I(k)/N - [\langle f^2\rangle - \langle f\rangle^2]]}{\langle f\rangle^2} - 1 = \sum_i\sum_j \bar{c}_i\bar{c}_j[f_i f_j^*/\langle f\rangle^2][I_{ij}(k) - 1] \tag{7.111a}$$

or:

$$kI'(k) = \sum_i\sum_j W_{ij}[I_{ij}(k) - 1]\,. \tag{7.111b}$$

Therefore, to obtain information on $\rho_{ij}(r)$ vs. r, a sine transform of $kI'(k)$ is taken. The resultant quantity is related to the sums for all ij of the convolution of the transform of W_{ij} and $4\pi r\left(\dfrac{\rho_{ij}(r)}{\bar{c}_j} - \rho_a\right)$.

In some substances, it is reasonable to think that the structure is built up of certain strongly bonded units. For example, in the case of glass, SiO_4^{-4} tetrahedra are likely to be involved because of the strong covalent bonding in this unit and the fact that it appears intact in so many structures. Certain peaks are specifically associated with Si—Si, Si—O, and O—O bonds whose interatomic distances are distinct and known from determinations of crystal structures in which this unit is found. Having made these assignments, it is possible to evaluate the transform W_{ij} and deconvolute the effect on each peak (see Mozzi and Warren (1969) who were also able to extract information on the range of distortions of the tetrahedra from such data). However, if neutrons are used, the scattering factor is often not a function of $\sin\theta/\lambda$ and the term $W_{ij}(r)$ is a constant. Another example is a polymer whose radial distribution function is shown in Fig. 7.45. As the data were taken with x-rays, the primary scattering element is carbon. Hence, the first peak in this figure is due to carbon neighbors to an atom along a chain, and integrates to two atoms (after adding ρ_a, of course). In a more general case such as a liquid or amorphous alloy, it will be necessary to try various assignments for the peaks and synthesize the intensity or else make simplifying assumptions. This is discussed further in chapters by Kaplow (Azaroff et al. 1974; Cohen and Hilliard 1966).

7.8.3 The Keating Approach

If it is desired to obtain separate information on the partial radial distribution functions $\rho_{ij}(r)$, there are two possible procedures, which we shall examine for a binary system. Rewriting Eq. (7.111a):

$$I'(k) = W_{AA}[I_{AA}(k) - 1] + W_{BB}[I_{BB}(k) - 1] + (W_{AB} + W_{BA})[I_{AB}(k) - 1]. \tag{7.112}$$

In the case of local order in crystalline alloys, Sect. 7.6.4, we employed various symmetries to separate terms in a similar equation. Of course, this is not possible in this case of a liquid or amorphous solid. But measurements can be made at different x-ray energies at each k, close to and far from an absorption edge of the elements involved, to change the scattering factors as was suggested by Keating (1963). At least three energies are needed for the three terms and more if a least squares solution is preferred. The W_{ij}'s can be evaluated for these energies and the set of equations could be solved for the terms in brackets (at each k). Large changes in the neutron scattering factors are possible (in the W_{ij}) by studying samples with different isotopes using neutron scattering and such solutions have been obtained in this case (Enderby 1967; Edwards et al. 1975). However with x-rays, measurements have not yet been made with this approach very close to the edge where the changes in f'''s are large. (If this is done with small changes, the simultaneous equations are ill-conditioned.) In such a case, very accurate values of the scattering factors are needed [for example, those by Cromer and Liberman (1981)], and measured values may even be better (see Sect. 3.72 for one method). Account will have to be taken of the resonant Raman scattering (Sec 7.6). Even with neutrons, systems must be carefully chosen to make the changes large. Once the separate

7.8 Liquids and Amorphous Solids

terms $I_{ij}(k)$ are obtained, a sine transform will reveal the $\rho_{ij}(r)$. With ternary systems there are six independent probabilities, and it will be interesting to see if solutions with real data are possible.

7.8.4 The Anomalous Dispersion Approach

There is a second approach suggested by Shevchik (1977) which we can understand best by expanding Eq. (7.110a):

$$\frac{I_{eu}(s)}{N} = \bar{c}_A f_A^2 + \bar{c}_B f_B^2 - \bar{c}_A^2 f_A^2 - \bar{c}_B^2 f_B^2 - \bar{c}_A \bar{c}_B (f_A f_B^* + f_B f_A^*)$$
$$+ \bar{c}_A^2 f_A^2 I_{AA}(s) + \bar{c}_B^2 f_B^2 I_{BB}(s) + \bar{c}_A \bar{c}_B (f_A f_B^* + f_B f_A^*) I_{AB}(s). \quad (7.113)$$

We next consider the difference in two patterns, taken at a synchrotron source with two wavelengths, one very close to the absorption edge of A (within a few eV), one 100–200 eV away. This second measurement is made at an energy removed from the edge, but not too far away from the first energy, to keep the f_B essentially unchanged. We also choose both energies just below the edge, eliminating $\Delta f_A''$ and also to avoid large change in background at the two positions, due to fluorescence if the edge is crossed. If we take the difference in these two measurements then:

$$\Delta(I_{eu}(k)/N = 2\bar{c}_A \bar{c}_B [\Delta f_A'(E_2) - \Delta f_A'(E_1)] \left\{ f_A^0 \left(1 + \frac{\bar{c}_A}{\bar{c}_B} I_{AA}\right) + f_B^0 (I_{AB} - 1) \right\}.$$
$$(7.114)$$

This is simply the weighted sum of the transforms of the two partial pair distributions (ρ_{AA}, ρ_{AB}) around an A atom. Therefore the Fourier sine transform of $\Delta(I_{eu}(s)/N)$ gives a radial distribution function (RDF) related to this weighting. Fuoss et al. (1981) were able to use this procedure to show that in amorphous GeSe there was threefold coordination, similar to that in the crystalline phase, even though the bonding is ionic in the crystal and covalent in the amorphous state.

One way to obtain information from this difference would be to make the sine transform to real space, then transform back each peak in this weighted RDF separately. This transform is Eq. (7.114) for one shell only and various functions for $e_{ij}(r)$ could be tried to see which give the best fit in each shell. For example, a Gaussian function could be tried for each and, by least squares techniques, the position and width for each distribution could be obtained, and as well the vibrational amplitude of each species. This has not been attempted at this writing, nor have such techniques been applied to ternaries or quaternaries, but the extension to these cases is clear and the results could be quite interesting.

7.8.5 The Bhatia-Thornton Equations

There is another interesting approach to the study of this class of materials, developed by Bhatia and Thornton (1970). We define $\rho_N(r) = \sum_{i=1}^{n} \rho_i(r)$ and a local deviation in concentration of the ith element, $c_i(r) = \rho(r) - \bar{c}_i \rho_N(r)$. These equations have the following properties:

$$c_i(k) = \rho_i(k) - \rho_{iN}(k) \tag{7.115a}$$

$$\int \rho_N(r)\, dr = \sum_{i=1}^{n} \int \rho_i(r)\, dr = \sum_{i=1}^{n} N_i = N \tag{7.115b}$$

$$\int c_i(r)\, dr = \int [\rho_i(r) - \bar{c}_i \rho_N(r)]\, dr = N_i - \bar{c}_i N = 0 \tag{7.115c}$$

$$\lim_{r \to \infty} c_i(r) = \bar{\rho}_i - \bar{c}_i \sum \bar{\rho}_i = \bar{\rho}_i - \bar{c}_i \bar{\rho}_N = 0 \tag{7.115d}$$

$$\sum_{i=1}^{n} c_i(r) = \sum_{i=1}^{n} \rho_i(r) - \rho_N(r) \sum \bar{c}_i = \rho_N(r) - \rho_N(r) = 0. \tag{7.115e}$$

Due to this last equation, there are only $(n-1)$ independent $c_i(r)$ functions. Also, this equation implies that the sum of the transforms of $c_i(r)$, $\sum c_i(k)$, is zero.

Continuing for a binary system:

$$\rho_N(r) = \rho_1(r) + \rho_2(r). \tag{7.116}$$

Therefore:

$$c_1(r) = \rho_1(r) - \bar{c}_1 \rho_N(r) = \rho_1(r) - c_1 \rho_1(r) - c_2 \rho_2(r)$$
$$= \bar{c}_2 \rho_1(r) - \bar{c}_1 \rho_2(r) = -c_2(r). \tag{7.117}$$

The terms $\rho_1(r)$ and $\rho_2(r)$ can be written in terms of $\rho_N(r)$ and a term $c(r)$, by first defining $c_1(r) \equiv c(r)$ and $c_2(r) \equiv -c(r)$:

$$\rho_1(r) = \bar{c}_1 P_N(r) + c(r) \tag{7.118a}$$

$$P_1(k) = \bar{c}_1 P_N(k) + c(k) \tag{7.118b}$$

and similarily for $\rho_2(r)$, $P_2(k)$. Therefore, the scattering amplitude is:

$$A(k) = \sum_{i=1}^{n} f_i P_i(k) = f_1 P_1(k) + f_2 P_2(k)$$
$$= [\bar{c}_1 f_1 + \bar{c}_2 f_2] P_N(k) + [f_1 - f_2] c(k) \tag{7.119a}$$

or:

$$A(k) = \langle f \rangle P_N(k) + \Delta f(k). \tag{7.119b}$$

With $I_{eu}(k) = A(k) A^*(k)$:

$$I_{eu}(k) = \langle f \rangle \langle f^* \rangle P_N(k) P_N^*(k) + \Delta f \Delta f^* c(k) c^*(k)$$
$$+ \langle f \rangle \Delta f^* P_N(k) c^*(k) + \langle f \rangle \Delta f P_N^*(k) c(k). \tag{7.120}$$

Defining:

$$S_{NN}(k) = \frac{1}{N} P_N(k) P_N^*(k) \tag{7.121a}$$

7.8 Liquids and Amorphous Solids

$$S_{NC}(k) = \frac{1}{N} P_N(k) c^*(k) \tag{7.121b}$$

$$S_{CC}(k) = \frac{1}{N} c(k) c^*(k) \tag{7.121c}$$

then,

$$I_{eu}(k) = \langle f \rangle^2 S_{NN}(k) + (\Delta f)^2 S_{CC}(k) + 2\Delta f \langle f \rangle S_{NC}(k) . \tag{7.122}$$

By proper choice of elements (with x-rays) or isotopes (with neutrons) one of these three terms may be made dominant by adjusting the f's. See, for example, Wagner (1980).

Of course the two treatments we have followed are related. For example:

$$S_{NN}(k) = \frac{1}{N} P_N(k) P_N^*(k) = \frac{1}{N} \sum_i \sum_j P_i(k) P_j^*(k) \tag{7.123a}$$

$$= \sum_i \sum_j \{c_i \delta_{ij} + \bar{c}_i \bar{c}_j \rho_a V(k) + \bar{c}_i \bar{c}_j [I_{ij}(k) - 1]\} . \tag{7.123b}$$

As before the second term makes no contribution beyond very small angles, and:

$$S_{NN}(k) = \sum_i \sum_j \bar{c}_i \bar{c}_j I_{ij}(k) \tag{7.124a}$$

which for a binary system is:

$$S_{NN}(k) = \bar{c}_1^2 I_{11}(k) + \bar{c}_2^2 I_{22}(k) + 2\bar{c}_1 \bar{c}_2 I_{12}(k) . \tag{7.124b}$$

Similarly:

$$S_{CC}(k) = \bar{c}_1 \bar{c}_2 \{1 + \bar{c}_1 \bar{c}_2 [I_{11}(k) + I_{22}(k) - 2I_{12}(k)]\} \tag{7.124c}$$

and:

$$S_{NC}(k) = \bar{c}_1 \bar{c}_2 \{[\bar{c}_1 I_{11}(k) + \bar{c}_2 I_{12}(k)] - [\bar{c}_2 I_{22}(k) + \bar{c}_1 I_{21}(k)]\} . \tag{7.124d}$$

Thus, in the general case, we can obtain the S functions from measurements of the I_{ij} and do need to rely on weighting one contribution or another.

To understand the meaning of these Bhatia-Thornton (s) terms we need their limits. Now $I_{ij}(k)$ approaches unity for large k, because $\sin kr / kr$ approaches zero at this limit. Then:

$$S_{NN}(k) \to \bar{c}_1^2 + \bar{c}_2^2 + 2\bar{c}_1 \bar{c}_2 = (\bar{c}_1 + \bar{c}_2)^2 = 1 \tag{7.125a}$$

$$S_{CC}(k) \to \bar{c}_1 \bar{c}_2 \tag{7.125b}$$

$$S_{NC}(k) \to 0 . \tag{7.125c}$$

Therefore, S_{NN} oscillates about unity, and it describes the overall structure or topology, or short-range order of atoms *independent* of their type. The term S_{CC} oscillates about $\bar{c}_1 \bar{c}_2$ for a nonrandom distribution of atoms 1 and 2, and describes the chemical short-range order. Finally, S_{NC} oscillates about zero, and *is* zero if the surroundings of the A and B atoms are identical, such as might be expected for a binary substitutional alloy with atoms of identical size. It is the equivalent to the size effect scattering we discussed for crystalline alloys in Sect. 7.6.

448   7. What Else Can We Learn from a Diffraction Experiment Besides the Average Structure?

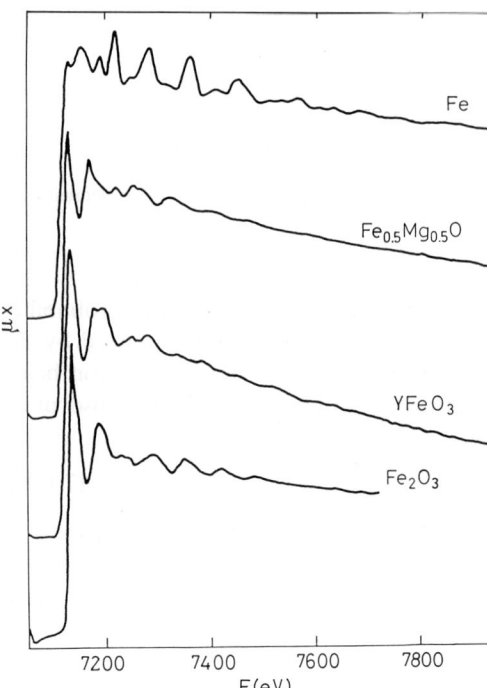

Fig. 7.47. The x-ray absorption coefficient (multiplied by the thickness) as a function of energy in the vicinity of the K absorption edge of Fe for a number of iron-containing compounds. Reprinted with permission from Knapp, G.S. and Georgopoulos, P., Crystal Growth Properties and Applications 7, 75 (1982)

suggest that EXAFS is caused by, and is sensitive to, the atomic environment of the absorbing atoms.

7.9.2 The Physics of the EXAFS Oscillations

Let us look a little more closely at the physics of the photoelectric absorption process. It is of course, a quantum effect and therefore it is only necessary to consider an initial state of the photon-atom system and a final state. (What happens in between is indeterminate.) The process is characterized by a transition probability from the initial to the final state. In the case of photoelectric absorption, the initial state is described by the wavefunction of the photon plus the atomic wavefunctions of all the electrons of the atom. In the final state, the photon has disappeared and the wavefunction of one of the core electrons is delocalized, since this electron now has aquired the energy of the photon and cannot occupy an atomic orbital any more. Remember that electrons are fermions and as such obey the Pauli exclusion principle. If we change a quantum number (and energy has the principal quantum number), we must find an empty state in which to put the electron. The transition probability for photoelectric absorption is measured by the absorption coefficient μ, which is given by the Fermi "golden rule":

$$\mu \sim M^2 \rho(E_F), \tag{7.129}$$

7.9 Extended X-ray Absorption Fine Structure (EXAFS)

Here M is the matrix element for the transition; it is a measure of the coupling between a photon and an electron and expresses the ease with which they can exchange energy. $\rho(E_F)$ is the density of empty electronic states at the final energy of the photoelectron. If there are no empty states with energy E_F, the absorption coefficient is zero and the transition is not allowed. This explains very nicely the presence of edges in the absorption spectrum. If the energy of the final state is lower than the highest occupied atomic orbital, the electron cannot make the transition and the photon is not absorbed. As soon as the energy of the final state exceeds that of the highest occupied state (the vacuum level for isolated atoms, the Fermi energy for atoms in condensed matter), the density of states suddenly becomes non-zero and the absorption coefficient shows a sharp increase.

With this brief introduction to the physics of absorption, let us now see if we can explain the presence of EXAFS. The photoelectron has an energy of hundreds of eV in the EXAFS region and is not likely to be perturbed by small bond energy modulations of the order of one eV. On the other hand, it has been shown that the fine structure present in the *immediate* vicinity of the absorption edge (within ~ 50 eV) is, in fact, due to the details of the density of states, which in this energy range deviates markedly from that of a free electron. We will not concern ourselves with the near-edge structure (called XANES, x-ray absorption near-edge spectroscopy) any further, because it is not a diffraction phenomenon. We will only mention that there is indirect structural information in the near-edge fine structure. This is because the atomic arrangement around the absorbing atom is reflected in the local density of states and hence in the absorption fine structure. As yet, there is no general theory describing this relationship and structural inferences have been drawn by a "finger printing" method by which an unknown spectrum is compared in some way with spectra from compounds of known structure. For example, if the near absorption edge region is examined for an element in some compound whose valence is unknown, from comparisons with the edge position in compounds containing this element with known valence, the valence in the unknown can be obtained (Tang et al. 1985). At best, we can hope to distinguish only gross features of the atomic structure.

A theory which explains all the essential features of EXAFS was proposed by Stern (1974). He assumes that the density of empty states is featureless, appropriate to that of a free electron. The matrix element is modulated by scattering of a photoelectron by the environment of the absorbing atom. In effect, there is interference between the outgoing portion of the photoelectron wavefunction and the part scattered back by the atoms in the vicinity of the absorbing atom. This interference can be constructive or destructive, thereby enhancing or reducing the matrix element and with it the absorption coefficient, compared to that of an isolated atom.

So EXAFS is shown to be due to electron diffraction. Each atom in the sample is a mini-diffraction instrument, both an electron source and electron detector. The scattering angle is fixed at 180 degrees, but the photoelectron energy can be varied by changing the photon energy, much like energy-dispersive x-ray diffraction instruments or time-of-flight neutron diffractometers. It is important to note that the electrons can also be considered as waves characterized by a wavelength or, better, a wavenumber $k = 2\pi/\lambda$ which is related to the electron kinetic energy by

$$E - E_b = h^2 k^2 / 2m, \tag{7.130}$$

where the kinetic energy is equal to the photon energy minus the binding energy, E_b, of the electron in the atom, k is the wavenumber and m is the electron rest mass. From this relation, we can interchangeably consider the absorption coefficient a function either of photon energy or photoelectron wavenumber.

7.9.3 Quantitative Analysis of EXAFS

Let us attempt now to describe EXAFS in a quantitative fashion, using familiar diffraction concepts. To simplify matters, rather than considering the total absorption coefficient, we define a new quantity $\chi(k)$ as

$$\chi(k) = \frac{\mu(k) - \mu_0(k)}{\mu_0(k)}, \tag{7.131}$$

where $\mu_0(k)$ is the absorption coefficient of a free atom. The term $\chi(k)$ is a natural quantity to use, because it isolates and normalizes the condensed matter effects. Let us see what factors control $\chi(k)$.

1) Since EXAFS is a diffraction phenomenon, one should expect to find a trigonometric term in $\chi(k)$ with a phase factor of $2kR$, where R is the distance between the absorbing atom and its neighbor. (The total path length of the photoelectron is $2R$.) Due to this trigonometric term $\chi(k)$ should show an oscillatory behavior with more closely spaced oscillations the larger the distance R. The presence of such a term makes EXAFS a sensitive tool for measuring near neighbor distances! This term should be of the form $(\sin kR/kR)$, as we are seeing the effects of an entire spherical shell of atoms around the absorber.
2) The amplitude of the oscillations should obviously depend linearly on the number of neighboring atoms at a given distance. The more neighbors, the larger the probability of backscattering.
3) The atomic number of the backscattering species determines their electron scattering strength. A quantity analogous to the atomic scattering factor for x-rays should be included in the EXAFS formula. This quantity is the absolute backscattering cross section (like f^2) of the atoms and it is a function of the photoelectron energy.
4) The outgoing wave is spherical and essentially planar by the time it returns. Therefore, an additional $1/R$ term is required to take into account the outgoing spherical wave.
5) Finally, since EXAFS is a scattering phenomenon, any uncertainty in the interatomic distance should manifest itself in a reduction of the amplitude of oscillations, since the interference that gives rise to these oscillations is smeared. This effect will be stronger at larger values of the photoelectron wavenumber than at smaller values, as a given spread in interatomic distances will produce larger changes in the phase of the backscattered waves, thereby making the interference effects less pronounced. This is analogous to the Debye-Waller factor in diffraction theory, where similar

7.9 Extended X-ray Absorption Fine Structure (EXAFS)

arguments apply. The only difference is that the Debye-Waller factor is a measure of the mean squared deviation of an atom from its ideal lattice site, whereas for EXAFS the corresponding term is a measure of the mean squared deviation of the interatomic distance from its average value. Just as in x-ray diffraction, this Debye-Waller factor reflects a range of interatomic distances due either to thermal vibrations or to static atomic displacements, because the absorbing atoms do not all have identical surroundings.

Putting all these effects together and noting that the overall amplitude will be a composite of partial amplitudes due to atomic shells at different distances around the absorbing atom, one can write $\chi(k)$ as follows:

$$\chi(k) = \sum_j \frac{N_j}{kR_j^2} f_j(k,\pi) e^{-2R_j/\lambda} \sin[2kR_j + \varphi_j(k)] e^{-2k^2\langle\sigma_j^2\rangle}, \qquad (7.132)$$

where the N_j is the number of atoms in the jth shell, $f_j(k,\pi)$ the backscattering cross section of the atoms in that shell, R_j the shell radius and $\langle\sigma_j^2\rangle$ the mean squared relative displacements of the shell atoms from their ideal distance from the absorbing atom. The notion of a shell of atoms is a useful one in this context, since as mentioned above, only radial distance information is contained in EXAFS. Keep in mind also that the local atomic arrangements are fairly regular even in amorphous materials, hence interatomic distances tend to be grouped around discrete values.

Two other terms are also involved, which do not appear to have an obvious counterpart in x-ray diffraction, but in fact they do. The first one of them is the exponential term involving a quantity λ, which is the mean free path for inelastic scattering of the photoelectron. This term expresses the fact that, as the photoelectron travels through matter, it can experience an inelastic collision with an atom in its path. After an event like this the photoelectron energy changes and it can no longer interfere with itself, so it is lost to the EXAFS process. The exponential term merely states that the probability for inelastic scattering increases with distance that the photoelectron must travel. There is a similar quantity for x-ray photons, called the coherence length of the photon. It is not commonly mentioned in x-ray diffraction theory because its value is larger that the scattering length (the distance over which a photon has a high probability of being scattered). For electrons, however, typical inelastic mean free paths are only 4–5 Å. This is a limitation or an advantage of EXAFS, depending upon one's point of view: We can obtain information about the immediate vicinity of the absorbing species. Interference effects due to distant neighbors are strongly attenuated.

The second term mentioned above is a phase shift appearing in the argument of the sine. This is a phase shift that the photoelectron experiences upon leaving and reentering the absorbing atom and also as it scatters off the neighbors. Its value can be obtained from theoretical calculations, just as $f_j^2(k,\pi)$. Its diffraction counterpart is the imaginary dispersion correction to the scattering factor, which is also the out-of-phase component of the scattered amplitude.

The power of the EXAFS technique as a tool for structure determination lies in its simplicity, but most importantly on this fact, which has been experimentally verified. The backscattering amplitudes and phase shifts are properties of the absorbing and backscattering atoms and to a good approximation do not depend on the environment of these atoms (Lytle et al. 1974). This means that amplitudes and phase shifts can be calculated theoretically, tabulated and used in the EXAFS equation without prior knowledge of the particular distances, number of neighbors etc. involved (which is, after all, the kind of information we want!). This is an important fact. [If it were not true, the only way that one could obtain structural information would be through extremely involved self-consistent electronic calculations, in which a structure would have to be assumed, the photoelectric absorption problem solved from first principles, and the process repeated while varying the structural parameters until a satisfactory fit to the experimental spectrum was achieved.]

7.9.4 Experimental Procedures for EXAFS

Experimentally, EXAFS measurements are very simple, at least conceptually. All that is needed is a monochromatic x-ray beam (with no harmonics!) of variable photon energy, an x-ray detector and the specimen in the form of a thin slab (later we will give criteria concerning the proper thickness). The intensity of the beam is measured with and without the specimen in the beam path and from Eq. (7.128a) the product μx can be determined. Equation (7.131) shows that this is all that is needed; the thickness cancels in the ratio. In practice the experiment is done slightly differently. A semi-transparent x-ray detector is placed in the beam ahead of the specimen. This detector gives a reading proportional to the intensity of the incident beam, whereas at the same time the detector following the specimen measures the intensity of the transmitted beam. This arrangement has the advantage that beam intensity instabilities are normalized out, since both detectors are active at the same time. The quantity μx is not directly available, unless the efficiencies of both detectors are factored in, but again, from Eq. (7.131), we see that this is not necessary. As will be shown in Sect. 7.9.6 the ideal sample absorbs ~ 90 pct of the beam. Therefore, the first (semi-transparent detector should be chosen (or the gas adjusted) to absorb 10 pct. Then both detectors will see about the same count rate, minimizing any differences in dead time losses. At a synchrotron source, it is possible to achieve 1–2 eV resolution with perfect crystals, and to record spectra from many types of samples in a matter of minutes. With a curved focusing crystal to select a wavelength, resolution of 4–6 eV with standard x-ray sources is possible, and with a rotating anode generator patterns often take only 1–3 h. This resolution depends on slit and target dimensions (the slit at the focus *after* the monochromator) and the diffraction angle, which can be varied by varying the diffracting plane of the monochromator (Georgopoulos and Knapp 1981). Some calculated resolutions are shown in Fig. 7.48. It can be seen that heavier elements require high angle reflections. With a resolution of 20 eV, it is possible to deal with edges (K or L) up to ~ 30 keV. In air, the lightest element possible would correspond to an edge of 3 keV, e.g. K. In an electron microscope, by energy loss spectroscopy (the energy loss of the electrons after passing through a sample) it is possible to see EXAFS down to the element oxygen. [At some synchrotrons there is an adequate photon flux up to 50–60 keV; however

7.9 Extended X-ray Absorption Fine Structure (EXAFS)

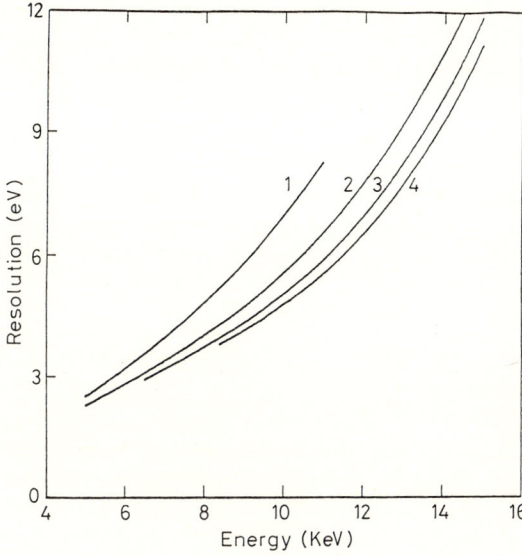

Fig. 7.48. Calculated energy resolution for a bent Si monochromator mounted at the sample position in a normal diffractometer. (The EXAFS sample and detectors are placed after the reciving slit.) *Curve 1)* 111 reflection, 2) 220, 3) 400, 4) 511. From Tang, C., Ph.D. Thesis, Northwestern University, 1983)

the lifetime of the electron hole is very short, and the uncertainty principle leads to large energy broadening, so that the useful limit is still ~ 30 keV.]

There are many interesting possibilities with this tool. By looking at the electron yield (Auger and photoelectrons) it is possible to limit the region sampled to surface layers (due to the low penetration depth of electrons). Thus surface EXAFS or SEXAFS is being used to explore such things as adsorbed molecule configurations. By polarizing the x-ray beam, photoelectrons from bonds normal and parallel to the surface can be sampled, revealing details about the chemical bonding of these adsorbed molecules.

Finally it is worth mentioning that interpretation of EXAFS oscillations allows the experimental determination of $\Delta f'$, $\Delta f''$, as discussed in Sect. 3.7.

7.9.5 Data Analysis

Our first task after obtaining μx (or something proportional to it) as a function of energy is to calculate $\chi(k)$ from Eq. (7.131). Immediately we see that a new unknown appears, μ_0. This is the absorption coefficient of the free atom and in practice it cannot be obtained experimentally (imagine trying to produce a dense monoatomic vapor of something as refractory as tungsten!). Instead, we use the experimental spectrum to remove the EXAFS oscillations without affecting the absorption jump and the overall shape of the absorption curve. $\chi(k)$ can then be obtained. For example, data points χ_i may be obtained such that:

$$\chi_i = \tfrac{1}{4}\chi_{i-1} + \tfrac{1}{2}\chi_i + \tfrac{1}{4}\chi_{i+1} \, . \tag{7.133}$$

Repeating this procedure about 50 times will smooth out the oscillations. Then we can proceed as follows:

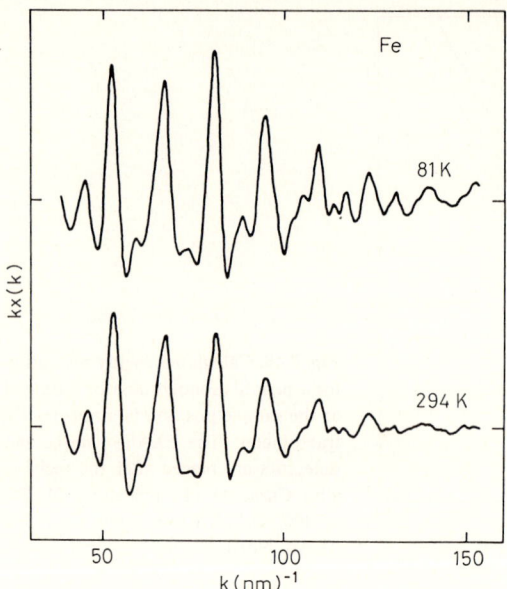

Fig. 7.49. The EXAFS spectrum $k\chi(k)$ vs. k for alpha iron. Note the more severe damping of the oscillations at the higher temperature. Reprinted with permission from Knapp, G.S. and Georgopoulos, P., Crystal Growth Properties and Applications, 7, 75 (1982)

1) The data below the edge, which represents all the electron shells but the shell responsible for the absorption jump (which we assume here to be a K shell), can be fit to an analytical function. One of the most popular is the Victoreen formula, which is of the form:

$$\mu(E) = CE^{-3} + DE^{-4} \,. \tag{7.134}$$

This equation is then used above the edge to subtract from the smoothed one leaving only the absorption due to the atom and shell of interest.

2) $\chi(k)$ is obtained; see Fig. 7.49.

3) Sometimes χ is multiplied by k^n to emphasize some of the weak fluctuations. Either $\chi(k)$ or $k^n\chi(k)$ is then Fourier transformed, as shown in Fig. 7.50. This resembles a radial distribution function (Sect. 7.8) but is called a "radial structure function". In fact it is very similar to the weighted RDF obtained by the difference in x-ray scattering using two wavelengths close to an absorption edge (Sect. 7.8). However the k values are 40 Å^{-1}, twice those possible via scattering and should in principle provide good information at small R. (On the other hand, the near edge region is avoided, which leads to errors and oscillations at large R.) EXAFS patterns are certainly easier to obtain than the results from scattering, but it remains to be seen which technique provides better information.

4) A window function (unity inside, zero outside) is placed around one peak, say for the first neighbor shell, and the data is back transformed to k space.

5) Nonlinear least–squares fitting of this peak to Eq. (7.132) is used to obtain N_j, R_j, and σ_j. Usually this analysis includes adjusting E_b (which causes changes in k). Care is needed as there is some correlation between E_b and R, and strong correlation between σ_j and N_j. If two or more shells overlap, correlations will be very strong

7.9 Extended X-ray Absorption Fine Structure (EXAFS)

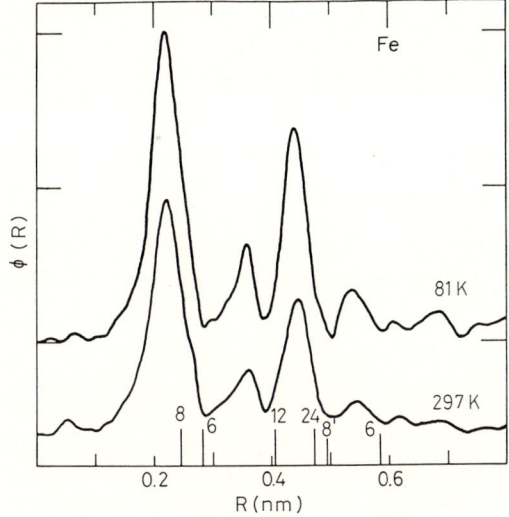

Fig. 7.50. The modulus of the Fourier transforms of Fig. 7.49. The *vertical bars* indicate the distances and co-ordination numbers of the first few neighbor shells. Reprinted with permission from Knapp, G.S. and Georgopoulos, P., Crystal Growth Properties and Applications, 7, 75 (1982)

and the analysis is difficult. In this procedure, tabulated values of f_j and ϕ_j are employed (Teo and Lee 1979).

This process can be repeated with other than the first shell, but often this is of little meaning due to extensive peak overlap.

7.9.6 Statistical Error Limits in EXAFS Data

Transmission Case

An EXAFS experiment involves an accurate measurement of the linear absorption coefficient. Since in the laboratory, but sometimes also at a synchrotron, we are limited by the total amount of x-ray flux, we must try to optimize the signal to noise ratio to make the most out of the available flux. We are also interested in spending the least amount of time possible to achieve a given quality in our spectra.

Following Eq. (7.128a), the linear absorption coefficient is given by:

$$\mu x = \ln N_0 - \ln N, \tag{7.135}$$

where x is the thickness of the specimen and N_0 and N are the total number of photons in the incident and transmitted beam respectively (the flux times the counting time). Differentiating:

$$\delta(\mu x) = \delta N/N. \tag{7.136a}$$

Since the time of arrival of photons at the detector follows Poisson statistics, we have:

$$\delta N/N = N^{-1/2}, \tag{7.136b}$$

hence:

$$\delta(\mu x) = N^{-1/2}. \tag{7.137}$$

Substituting (7.135) into (7.137):

$$x\delta\mu = N_0^{-1/2} \exp(\mu x/2). \tag{7.138a}$$

Before proceeding, we note a couple of points. First of all, we are interested in the relative error in μ, something like $\delta\mu/\mu$, since this is exactly the EXAFS function $\chi(k)$. Also although we are measuring the total absorption coefficient of our specimen, it is only the contribution of the K (or L) edge of the element of interest that we are after. Hence, we are seeking an expression for $\delta\mu_K/\mu_K$. Noting that $\delta\mu_K = \delta\mu$, since only μ_K varies in the energy range we are scanning, we can rewrite Eq. (7.138a) as follows:

$$\delta\mu_K/\mu_K = N_0^{-1/2}[\exp(\mu x/2)/\mu x](\mu/\mu_K). \tag{7.138b}$$

Equation (7.138b) teaches us a number of things. First, our statistical errors in the determination of $\chi(k)$ get worse as μ/μ_K gets large for a given N_0. This means that dilute samples are harder to study than concentrated samples, in which the K (or L) edge is responsible for almost all the absorption. Also, for a given total absorption coefficient μ, we see that the term $\exp(\mu x/2)/\mu x$ is not constant but depends on the sample thickness. Let us then calculate what thickness minimizes this expression. (That there is minimum is obvious, since this function becomes infinite at both $x = 0$ and x tending to infinity.) Differentiating with respect to x:

$$\frac{d}{dx}\left(\frac{\exp(\mu x/2)}{\mu x}\right) = \frac{(\mu^2 x/2)\exp(\mu x/2) - \mu\exp(\mu x/2)}{(\mu x)^2} = 0 \tag{7.139}$$

which holds if $x = 2/\mu$. So, for optimum results the samples must be made two absorption lengths thick. Note that the total absorption coefficient is involved here, not just the contribution of the edge we are studying.

Fluorescence Case

The fluorescence intensity is given by:

$$N = N_0 \frac{\varepsilon\Omega}{4\pi}\left(\frac{\mu_K}{\mu + \mu(E_F)\sin\beta}\right)\left\{1 - \exp\left[-\left(\frac{\mu}{\sin\beta} + \mu(E_F)\right)x\right]\right\}, \tag{7.140a}$$

where ε is the fluorescence yield, Ω is the solid angle that the sample subtends on the detector, $\mu(E_F)$ is the absorption coefficient of the sample for fluorescent photons, β is the angle that the incident beam makes with the sample surface and x is the thickness. By using thick samples, the term in parentheses is negligible. Also, by letting the incident beam strike the sample at small angles, $\sin\beta$ is small and Eq. (7.140a) considerably simplified:

$$N = N_0 \frac{\varepsilon\Omega}{4\pi} \frac{\mu_K}{\mu}. \tag{7.140b}$$

Before proceeding with the error analysis we note that fluorescence detection is not appropriate for concentrated samples. In this case, μ is approximately equal to

7.9 Extended X-ray Absorption Fine Structure (EXAFS)

μ_K and N does not change appreciable with energy. This does not mean that there is no fluorescence; it just means that the EXAFS oscillations are severely attenuated. So, for practical purposes, fluorescence detection is useful only for dilute samples.

Differentiating Eq. (7.140b):

$$\delta N = N_0 \frac{\varepsilon\Omega}{4\pi} \frac{\delta\mu_K}{\mu} \qquad (7.141a)$$

since we are only interested in cases where $\mu \gg \mu_K$. Again, by using Eq. (7.136) we can show that:

$$\frac{\delta\mu_K}{\mu_K} = N_0^{-1/2} \left(\frac{\varepsilon\Omega}{4\pi}\right)^{-1/2} \left(\frac{\mu}{\mu_K}\right)^{1/2}. \qquad (7.141b)$$

Comparing Eqs. (7.138b) and (7.141b), we see that in the fluorescence case the loss of statistical accuracy is slower than in transmission as the sample gets more dilute, going down only by the square root of μ/μ_K rather than linearly. Hence, it is obvious that concentrated samples should be studied in transmission, whereas for dilute samples fluorescence is more advantageous. The breakeven point is a function of the fluorescence yield and the maximum solid angle we can use with the fluorescence detector. Some examples of the limits of detection are given in Table 7.8. One caution — in the case of fluorescence, the detector sees other scattering as well — diffuse scattering from a single crystal, or parts of Debye-Scherrer cones with powders. This can mask or confuse the oscillations. Thus, in this case, the actual error limits are probably somewhat larger.

It has been the purpose of this chapter to reveal the vast amount of information that can be obtained from materials with broad diffraction peaks or no peaks at all,

Table 7.8. Minimum concentrations for various experimental methods[a]

Method	Count rates (s^{-1})	μ_K/μ for a 20-h run	Examples
Transmission			
Flat crystal	$\sim 10^3$	0.7	Elements
Curved crystal with 1 kW generator	$\sim 10^5$	~ 0.07	$\sim 2\%$ Fe in H_2O
Curved crystal with rotating anode	$\geq 10^6$	~ 0.02	$\sim 0.4\%$ Fe in H_2O
Synchroton (bending magnet)	$10^8 - 10^{10}$/time on machine (%)		
Fluorescence			
Curved crystal with 1 kW generator	$\sim 10^5$	0.5×10^{-2}	0.003 Cu in Al
Curved crystal with rotating anode	$\geq 10^6$	0.5×10^{-3}	0.001 Cu in Al

[a] Adapted from G.S. Knapp and P. Georgopoulos, Crystal Growth, Properties, and Applications, 7, 75 (1982)

and from unusual background scattering, and something of the techniques involved. If we knew only about determining crystal structures, we might throw such materials away, or throw up our hands in frustration if such materials were important to our research. But diffraction is, in fact, a powerful tool, whose limits are still uncharted and the reader is encouraged to try it to see what information it might provide. He will generally be pleasantly surprised!

References

Atkinson, H.H., and Hirsch, P.B., Phil. Mag. *3*, 213 (1958)
Averbach, B.L., in "Theory of Alloy Phases", ASM, Metals Park, Ohio, 1955
Azaroff, L.V., Kaplow, R., Kato, N., Weiss, R.J., Wilson, A.J.C., and Young, R.A., "X-ray Diffraction", McGraw-Hill, New York, 1974
Bacon, G.E., "Neutron Diffraction", Clarendon Oxford, 1962
Berk, N.F., and Hardmann-Rhyne, K.A., J. Appl. Cryst. *18*, 467 (1985); 473 (1985)
Bhatia, A.B. and Thornton, D.E., Phys. Rev. *B2*, 3004 (1970)
Borie, R., and Sparks, C.J., Acta Cryst. *A27*, 198 (1971)
Brille, O.L., and Schmidt., P.W., J. Appl. Cryst. *39*, 2274 (1968)
Carpenter, J.A., Tenney, D.R., and Houska, C.R., J. Appl. Phys. *42*, 4305 (1971)
Cenedese, P., Fley, F., and LeFebvre, S., Acta Cryst. *A40*, 228 (1984)
Chen, H.D., Comstock, R.J., and Cohen, J.B., Ann. Rev. Mater. Sci. *9*, 51 (1979)
Clapp, P.C., and Moss, S.C., Phys. Rev. *142*, 418 (1966); *171*, 754 (1968)
Cohen, J.B., in "Phase Transformations", (H.I. Aaronson, ed.), ASM, Novelty Park, Ohio (1970)
Cohen, J.B., and Hilliard, J.E., (eds.), "Local Atomic Arrangements Studied by X-ray Diffraction", Gordon and Breach, and AIME, New York, 1966
Cohen, J.B., in "Adv. in Solid State Phys. *39*, 131 (1986)
Cohen, J.B., and Georgopoulos, P., in "Phase Transformations in Solids", (ed. T. Tsakalakos), MRS. Symposia Proceedings, Vol. 21, North Holland, New York (1984)
Compton, A.H. and Allison, S.K., "X-rays in Theory and Experiment", Van Nostrand, New York (1935)
Cowley, J.M., "Diffraction Physics", North Holland, Amsterdam, 1975 (American Elsevier, New York); Acta Cryst. *A32*, 83, 88 (1976)
Cromer, D.T. and Liberman, D.A.J., Acta. Cryst. *A37*, 267 (1981)
Dawson, B., "Advances in Structure Research", Vol. 6, Pergamon, New York, 1975
Delhez, R., and Mittemeijer, E.J., J. Appl. Crystallogr. *9*, 232 (1976)
Edwards, F.G., Enderby, J.E., Howe, P.A., and Page. D.I., J. Phys. C, Solid State Physics *8*, 3483 (1975)
Eisenberger, P., Platzman, P.M. and Winick, H., Phys. Rev. Letters *36*, 623 (1976)
Enderby, J.E., North, D.M., and Egelstaff, P.A., Phil. Mag. *14*, 961 (1966); Adv. Phys. *16*, 171 (1967)
Espinat, D., Morawek, B., Larue, J.F. and Renouprez, A.J., J. Appl. Cryst. *17*, 269 (1984)
Federova, I.S., and Schmidt, P.W., J. Appl. Cryst. *11*, 405 (1978)
Flory, P.J., in "Principles of Polymer Chemistry", Cornell University Press, Ithaca, NY (1953)
Flory, P.J., "Statistical Mechanics of Chain Molecules", Wiley, New York, 1969
Fuoss, P.H., Eisenberger, P., Warburton, W.K., and Bienenstock, A., Phys. Rev. Letters *23*, 1537 (1981)
Gehlen, P., and Cohen, J.B., Phys. Rev. *139A*, 844 (1965)
Georgopoulos, P., and Cohen, J.B., in "Modulated Structure Materials", (ed. T. Tsakalakos), NATO ASI Series E., No. 83, Martinus Nijhoff, Boston (1984)
Georgopoulos, P., and Cohen, J.B., J. de Physique *38C7*, 191 (1977)
Glatter, O and Kratky O., "Small Angle X-Ray Scattering", Academic Press, New York (1982)
Georgopoulos, P., and Knapp, G.S., J. Appl. Cryst. *14*, 3 (1981)
Goodisman, T., Brumberger, H., and Cupelo, R., J. Appl. Cryst. *14*, 305 (1981)
Gragg, J.E., Jr., and Cohen, J.B., Acta Metall. *19*, 507 (1971)

References

Gragg, J.E., Jr., Bardhan, P., and Cohen, J.B., in "Critical Phenomena in Alloys, Magnets and Superconductors", (R.E. Mills, E. Ascher, and R.I. Jaffee, eds), Chap. 6, Part 3, McGraw-Hill, New York, 1971
Gragg, J.E., Jr., Hayakawa, M., and Cohen, J.B., J. Appl. Cryst. *6*, 59 (1973)
Guinier, A., Adv. Solid State Phys. *9* (1959)
Guinier, A., "X-ray Diffraction", Freeman, San Francisco, 1960
Guinier, A., and Fournet, G., "Small-Angle Scattering of X-rays", Wiley, New York, 1959
Hayakawa, M., and Cohen, J.B., Acta Cryst. *A31*, 635 (1975)
Hayakawa, M., Bardhan, P., and Cohen, J.B., J. Appl. Cryst. *88*, 87 (1975)
Helmholdt, R.B., and Vos, A., Acta. Cryst. *A33*, 38 (1977)
Hilliard, J.E., in "Phase Transformations", (H.I. Aaronson, ed), ASM, Novelty Park, Ohio (1970)
Howard, C.J., Acta Cryst *A33*, 29 (1977)
Huang, T.C. and Parrish, W., "Advances in X-ray Analysis 27, 45 (1984)
James, R.W., "The Optical Principles of the Diffraction of X-rays", Bell, London, 1950
Kaplow, R., Averbach, B.L., and Strong, S.L., J. Phys. Chem. Solids *25*, 1195 (1964)
Keating, D.T., J. Appl. Phys. *24*, 923 (1963)
Kidron, A., and DeAngelis, R.J., "Symposium on Computer Aided Engineering" (G.M. Gladwell, ed.), p. 285. Univ. of Waterloo Press (1971)
Krivoglaz, M.A., "Theory of X-ray and Thermal Neutron Scattering by Real Crystals", Plenum, New York, 1969
Lake, J.A., Acta Cryst. *23*, 191 (1967)
Larson, B.C., J. Appl. Cryst. *8*, 150 (1975)
Lytle, F.W., Sayers, D.E., and Stern, E.A., Phys. Rev. *10*, 4826, 4836 (1974)
Maradudin, H.A., Montroll, E.W., and Weiss, G.H., "Theory of Lattice Dynamics in the Harmonic Approximation", Solid State Phys., Supp. 3 (F. Seitz and D. Turnbull, eds.), Academic Press, New York, 1963
Matsubara, E., and Georgopoulos, P., J. Appl. Cryst. *18*, 377 (1985)
Moore, P.B., J. Appl. Cryst. *13*, 168 (1980)
Mozzi, R.L. and Warren, B.E., J. Appl. Cryst. *2*, 164 (1969)
Nandi, R.K., Kuo, H.K., Schlosberg, W., Wissler, G., Cohen, J.B., and Crist, B., J. Appl. Cryst. *17*, 22 (1984)
Ohshima, K., Harada, J., Morinaga, M., Georgopoulos, P., and Cohen, J.B., J. Appl. Cryst. *19*, 188 (1986)
Peisl, H., J. Appl. Cryst. *8*, 143 (1975), this entire issue contains many interesting papers on local order
Pirie, J.D., Reid, J.S., and Smith, T.J., Phys. *C4*, 289 (1971)
Reid, J.S., Acta Cryst. *A41*, 513 (1985)
Roth, M., J. Appl. Cryst. *10*, 172 (1977)
Ruppersberg, H. and Egger, H., J. Chem. Phys. *63*, 4095 (1975)
Schelten, J. and Schmatz, W., J. Appl. Cryst. *13*, 385 (1980)
Schiller, C. and Mering, J., C.R. Acad. Science Paris *264*, 247 (1967)
Schlosberg, W.H., and Cohen, J.B., J. Appl. Cryst. *16*, 304 (1983)
Shevchick, N.J., Phil. Mag. *35*, 805, 1289 (1977)
Stern, E.A., Phys. Rev. *10*, 3027 (1974)
Stokes, A.R., Proc. Phys. Soc. London *61*, 382 (1948)
Tang, C., Georgopoulos, P., Fine, M.E., Cohen, J.B., Nygren, M., Knapp, G.S., and Aldred, A., Phys. Rev. *B31*, 1000 (1985)
Tchoubar-Vallat, D., and Mering, J., C.R. Acad. Science Paris, *261*, 3361 (1965); *263*, 1030 (1966)
Teo, B.K. and Lee, P.A., J. Am. Chem. Soc. *101*, 2815 (1979)
Tibballs, J.E., J. Appl. Cryst. *8*, 111 (1975)
Tulkki, J., Phys. Rev. *A27*, 3375 (1983)
Wagner, C.N.J., J. NonCryst. Solids *42*, 3 (1980)
Walker, C.B., and Chipman, D.R., Acta Cryst. *A26*, 447 (1970)
Warren, B.E., "X-ray Diffraction", Addison-Wesley, Reading, Massachusetts, 1969
Warren, B.E., and Mozzi, R.L., J. Appl. Cryst. *8*, 674 (1975)
Weiss, R.J., Phys. Rev. *83*, 379 (1951)
Willis, B.T.M., and Pryor, A.W., "Thermal Vibrations in Crystallography", Cambridge Univ. Press, Oxford, (1975)

Wilson, A.J.C., Acta Cryst. *A23*, 888 (1967); *A24*, 478 (1968); *A25*, 584 (1969)
Wu, T.B., Matsubara, E., and Cohen, J.B., J. Appl. Cryst. *16*, 407 (1983)
Yang, M., Ph.D. Thesis, Northwestern University, Evanston, IL, 1984

Problems

7.1. The data in Tables 7.9–7.12 at the end of the Problems are for two orders of a peak, for a worked specimen and a reference. The material has a bcc structure.
(a) Determine the Fourier coefficients of all four peaks. (A computer program for this can be found in Cohen and Hilliard, 1966, Chap. 8.)
(b) Obtain Stokes-corrected coefficients.
(c) Replot the peaks after the Stokes correction.
(d) Determine particle or "mosaic" size, D_{eff}, and rms microstrains versus L (show all plots).
(e) Do the corrected Fourier coefficients satisfy the Stokes criterion of falling to zero in the interval?
(f) Estimate the dislocation density.

7.2. Suppose that there are no faults, just cells and that the cell boundaries are on (111) planes. Show that the ratio of D_{eff} measured from a bcc powder material in the [110] direction (from 110, 220 peaks) to that in the [100] (from 200, 400 peaks) is (1.42)/(1).

7.3. From the equations in Sect. 7.5 derive simple forms for the Fourier cosine coefficients for peak shape and for peak shifts for the 311, 222, and 400 peaks from a fcc material containing random stacking faults on the (111) planes, with probability α. Include also an equation for the angular separation of the 400 and 222 peaks, in degrees 2θ.

7.4. Consider a layered structure such as graphite which normally has a layer spacing a_3. Assume that there is a probability α that some of the layer spacings open up to $a_3 + \varepsilon$. (There is considerable interest in such materials as superconductors where the opening can occur due to insertion or "intercalation" of another species.) Assume that the faults occur at random and that both α and ε are small.
(a) Derive a simple equation for the Fourier coefficients.
(b) Derive an equation for the sin θ/λ position of a peak. Compare to the application of Bragg's law, where $d = a'_3/l$ and a'_3 is an average spacing, $(1 - \alpha)a_3 + \alpha(\varepsilon + a_3)$.
(c) Derive an expression for the integral breadth $[\int I d2\theta]/I_{max}$ in radians for a 00l reflection in terms of θ, λ, l, ε, a_3, α. Simplify for small α, ε. Disregard the particle size, D_{eff}.
(d) Let $\lambda = 1.54$ Å, $a_3 = 3.50$ Å, $\varepsilon = 0.7$ Å, $\alpha = 0.15$. Compute the sin θ/λ positions for 001, 002, 003, 004. What are the a'_3 values from $\lambda = (2a'_3/l) \sin \theta$ (using the calculated sin θ/λ)?
(e) From the data in (d), calculate the integral breadths in radians for the four reflections. How closely do the results follow the usual assumption that distor-

tion produces a breadth proportional to $\tan\theta$? Again disregard the particle size $[(\Delta d/d) \propto \tan\theta]$.

7.5. When Cu_3Au is fully ordered, there are gold atoms at the corners and copper atoms at the faces of a cubic unit cell. Thus every Au atom has all Cu first neighbors. Disordered boundaries can be introduced on $\{100\}$ planes by a $\frac{1}{2}[[110]]$ shift *in* the $\{100\}$ planes. The first neighbors to a Au atom are still all Cu atoms across such a boundary, even though along the $[[100]]$ direction normal to the boundary the ordering pattern has been changed from Au, Au, Au ... at corners to Cu, Cu, ... Cu. Because first neighbors are the same, such a boundary has a low energy. These occur during ordering of Cu_3Au, because the ordering starts at many places at once, each place "unaware" of whether the other has Cu or Au at corners. Draw several unit cells with one such boundary. Derive the effect of such boundaries on the peak shapes and positions from a powder specimen.

7.6. Calculate values in cps for the incoherent scattering at $2\theta = 20°$ and $2\theta = 150°$, for Cu K_α ($I_0 A_0 = 10^9$ cps), for (a) carbon (b) cobalt, and (c) polystyrene (C_8H_8). Assume reflection geometry, an infinitely thick specimen, with receiving slits 0.1 mm × 0.1 mm at 145 mm from the specimen. Let $\lambda = 1.54$ Å. (See Appendix C.)

7.7. A powder sample contains N atoms of fcc "ridiculum." Imagine that 10% of the atoms are selected at random and removed, without disturbing the positions of the others.
(a) By what percentage is the integrated intensity of the crystalline reflections reduced?
(b) What percentage of the Bragg peaks is the diffuse intensity?

7.8. For a fcc alloy $I^0_{SRO}(h_i)$ can be written in the form

$$I^0_{SRO} = \sum_l \sum_m \sum_n \alpha_{lmn} \cos 2\pi l h_1 \cos 2\pi m h_2 \cos 2\pi n h_3 .$$

(a) Show that the result in (a) indicates that the periodicity of I^0_{SRO} is the reciprocal unit cell for a fcc structure.
(b) Similarly, prove that

$$Q_x = \sum_l \sum_m \sum_n \gamma^x_{lmn} \sin 2\pi l h_1 \cos 2\pi m h_2 \cos 2\pi n h_3 .$$

7.9. For a bcc structure, considering the restrictions on lmn for the interatomic vectors, show that

$$Q_x(h_1 h_2 h_3) = -Q_x(1 - h_1, 1 \pm h_2, h_3) .$$

7.10. By means of sketches, show that to maintain cubic symmetry, displacements need to be along the interatomic vector r_{lmn}, only if $l = m, n = 0$; $l, m = n = 0$, or $l = m = n$.

7.11. Consider the solid solution $(Fe_xMg_{1-x})O$. This has the same structure as NaCl; the oxygen sublattice is full, and Fe and Mg are on the cation sublattice. Derive an expression for the intensity due to local order in this system, in electron units.

7.12. Gases of molecules give rise to interesting diffraction patterns, somewhat like those from liquids.
(a) Show that if a molecule can take on all positions in space, then

$$F^2 = \sum_m \sum_n f_m f_n \frac{\sin kr_{mn}}{kr_{mn}},$$

where $k = 4\pi \sin \theta/\lambda$. The sum is over atoms in the molecule. [*Hint*: Consider the structure factor of a molecule and the averaging in Eq. (7.100).]
(b) Patterns of gases are most easily obtained by electron diffraction; the electron beam is incident on a small jet of the gas. Suppose such a pattern is taken for Br_2 at 300°K and 1 atm pressure. The first peak in the pattern is at $\sin \theta/\lambda = 0.25$ Å^{-1}. Derive an expression for the intensity in electron units, and then
(c) What is the Br–Br distance in the molecule from the given position of the first peak?
(d) Plot the intensity per molecule versus $\sin \theta/\lambda$.
(e) How will the pattern change with pressure?

7.13. A dispersion curve was shown in Fig. 7.4. The units for the ordinate are terahertz (10^{12} rad/s) and for the abscissa, reduced reciprocal lattice vector from a Bragg point, i.e., the distance divided by the distance to the next same point.

From this figure calculate the initial slope and convert this to the elastic constant C_{44} given the acoustic velocity V in a [110] direction in a cubic material is

$$V = \sqrt{C_{44}/\rho},$$

where ρ is the density. Compare the result to mechanical measurements of C_{44} [see Siegel, S., *Phys. Rev.* **57**, 537 (1940)].

7.14. Inelastic neutron scattering experiments are often carried out with a "time-of-flight" spectrometer. A pulse of neutrons of fixed wavelength is incident on a specimen at a fixed angle. The scattered neutrons (having a range of energies due to exchange of energy with the phonons) is examined at fixed angle, with an analyzer that measures their velocity spectrum. Draw this situation in reciprocal space and indicate several reciprocal lattice vectors. Assume a fcc structure ($a_0 = 7.7$ Å), $2\theta = 20°$, and an incident beam of 4.0 Å at 20° to a [100] direction. Assume that both **S** and **S**$_0$ lie in a (100) plane in reciprocal space. Discuss what portion of the Brillouin zone is sampled, and possible advantages and disadvantages of the method. [The Brillouin zone is defined in the next problem, part (b).]

7.15. (a) Draw to scale a ($\bar{1}10$) plane in reciprocal space, for a fcc structure, with $a_0 = 4.0$ Å. Include several 002, 111, 222, and 220 type spots around the 000 point.
(b) On the drawing made in (a) place all the traces of the Brillouin zones, so that the drawing is filled with such zones. (These zones in this structure are the planes perpendicular to [111] and [001] directions and bisecting the distances to the actual diffraction spots.) These are the boundaries due to the minimum phonon wavelengths. Note from Sect. 7.2 why the planes are halfway to the spots.
(c) Suppose that you wish to measure a dispersion curve like that in Fig. 7.4 of a single crystal of this material with a "triple axis" spectrometer, consisting of a monochromator, specimen, and finally, an analyzer crystal, each on their own

goniometer. The analyzer can be set to measure any given wavelength scattered at some position in the reciprocal space of the specimen. Draw a sketch of this situation in reciprocal space, with incident and scattered waves from all three crystals interconnected.

(d) This kind of work is now generally done with the "constant q" method ($q = 2\pi/\lambda_k$). The vector **q** is from a diffracting spot to a point in one Brillouin zone representing a phonon of specific wavelength λ and a specific direction. Let $\lambda_{incident} = 2.33$ Å and let $\mathbf{S}_0$ to the specimen be fixed *in direction*. The scattered wavelength and its direction are allowed to vary, but **q** is fixed. By moving the analyzer around the specimen and changing its Bragg angle, **S** and $\lambda_{scattered}$ are varied. If there is a gain or loss of neutron energy (which occurs when the neutron interacts with a transverse or longitudinal phonon of that (**q**) there will be peaks of energy as shown in Fig. 7.3b. Sketch this method in reciprocal space, showing how **S** changes with λ. Do this for energy gain and loss.

(e) Choose at least two equivalent positions on your drawing of part (b) to measure the longitudinal mode and one transverse mode in a [110] direction. The transverse modes are polarized in the [001] and [110] directions.

(f) What determines if these modes can be analyzed by energy gain or loss? Sketch the gain and loss situations for each of your points in (e).

(g) Show analytically that for both the incident and diffracted beams to a specimen any uncertainty in wavelength of the incident and diffracted neutrons due to divergencies is perpendicular to $\mathbf{r}^*_{hkl}$. (*Hint*: consider the parts of Bragg's law due to $\mathbf{S}_0$, **S** in vector form and differentiate with respect to $1/\lambda$.)

(h) The result of the effect in (g) is that there is an ellipse of resolution, with its long axis essentially perpendicular to $\mathbf{r}^*_{hkl}$. In an experiment you want this function to pass through the dispersion curve with its short axis to minimize broadening of the energy peak. Examine your choices of equivalent points in (e) for an acoustic mode like that in Fig. 7.4, in both the steep and "flat" portions. Are any of your equivalent points better than others? Consider as well as resolution, that the intensity is largest the further from the origin (see Sect. 7.2).

Consider the ellipse of resolution as rising only from the beams incident and diffracted from the specimen. Neglect any effect due to the monochromator or analyzer.

7.16. The physical properties of branched polymer molecules are very different than linear ones. Part of the reason is that the radius of gyration of a branched molecule is different than a linear one. Calculate the radius of gyration of a star shaped macromolecule containing four arms and having a total of n units. Assume the units are freely jointed so that

$$\langle r_{ij}^2 \rangle = a_K^2,$$

where K is the number of units between the ith and jth unit.

Useful Hint: Number the units in each arm from 1 to $n/4$, starting from the center and moving toward the ends.

What is the ratio of the radius of gyration of this molecule to a linear one containing this same number of units?

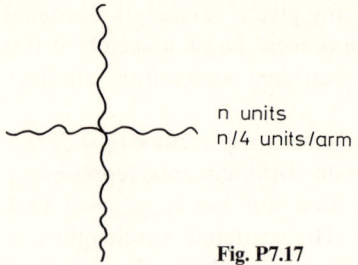

Fig. P7.17

7.17. In a research project on high temperature fatigue of copper, it has been found that pores develop at grain boundaries as the deformation proceeds. These pores eventually link up and cause intergranular failure. But it is difficult to find an adequate number of these pores using electron microscopy to obtain good statistical information on their size, (to study their growth) especially when they are small. Accordingly, neutron small-angle scattering experiments were initiated. The specimen was 99.999 pct. Cu annealed (final grain size = 200 μ) and then fatigued in a vacuum at 405°C at a maximum strain amplitude of 0.044 pct., to 25 pct of its expected life. The incident neutron beam had an intensity of 5.3×10^5 n/s/cm². This was obtained by comparing the intensity scattered from vanadium foils with that calculated from the vanadium incoherent scattering cross-section.[16] The data have been corrected for background. This was done by measuring the intensity without the specimen (I_1) and with cadmium (I_2), a high neutron absorber. If T is the transmission of the copper specimen,

$$I_{corrected} = \frac{I_{measured} - T(I_1 - I_2) - I_2}{T}. \qquad (P7.17.1)$$

The wavelength was 10.78 Å (with a 10% spread), obtained with a "velocity selector".

At relatively high scattering angle ($q = 4\pi \sin\theta/\lambda > 0.06$ Å^{-1}), the intensity is constant and equal to the incoherent scattering cross-section for Cu(0.68 barns).

The relevant scattering equations after the above corrections are:

$$I(q) = I_0 t s e \frac{d\sigma}{d\Omega} \Delta\Omega + I_\infty \qquad (P7.17.2a)$$

$$I_\infty = I_0 t s e N \sigma_1 \frac{\Delta\Omega}{4\pi}, \qquad (P7.17.2b)$$

where

I_0 = incident beam intensity = 5.3×10^5 n/s/cm²;
t = time of measurement at each point (actually a multidetector was employed to obtain all points at once) = 3790 s;
s = sample cross-section = 0.451 cm²;

---

[16] Vanadium gives no coherent scattering.

Problems

e = sample thickness = 0.901 mm;
$d\sigma/d\Omega$ = differential scattering cross-section per unit volume (cm^2/cm^3);
$\Delta\Omega$ = solid angle seen by the detector (the detector cells were 1×1 cm^2 at 282 cm from the specimen);
N = atomic density of the sample;
σ_1 = total incoherent scattering cross-section per atom = 0.59 barns;
σ_a = true absorption cross-section of Cu at 10.78 Å = 21.96 barns.

Other pertinent data: Density of Cu = 8.47×10^{22} atoms/cm^3; $T = e^{-N(\sigma_1+\sigma_a)e}$ = 0.848 (measured) vs. 0.840 (calculated).

The data given below are in intervals of $q = 2 \times 10^{-3}$ Å^{-1}. These have been corrected for the counting efficiency of the detector and Eq. (P7.17.1) and similar data for an unfatigued but otherwise identical specimen has been subtracted to further eliminate any effects not due to fatigue. (The data were averaged over a circle of constant q on the 2-D detector for each q.)

Intervals of q of 2×10^{-3} Å^{-1}		I_{net} (counts)	Intervals of q of 2×10^{-3} Å^{-1}		I_{net} (counts)
1	0.0000		18	0.03425	9.4
5	0.00825	6100.9	19	0.03625	6.2
6	0.01025	1608.9	20	0.03825	4.9
7	0.01225	717.9	21	0.04025	4.7
8	0.01425	331.9	22	0.04225	5.5
9	0.01625	183.8	23	0.04425	2.7
10	0.01825	114.5	24	0.04625	3.2
11	0.02025	67.1	25	0.04825	1.5
12	0.02225	50.3	26	0.05025	1.8
13	0.02425	34.1	27	0.05225	0.8
14	0.02625	26.2	28	0.05425	1.5
15	0.02825	17.4	29	0.05625	$-.1$
16	0.03025	12.0	30	0.05825	$+.2$
17	0.03225	11.1			

Other relevant formulae:

$$\frac{d\sigma}{d\Omega} \text{ in Porod region} = \frac{2\pi \Delta\rho^2 S}{q^4 V}. \qquad (P7.17.3)$$

Here $\Delta\rho$ is related to the scattering length density (Cu in this case) and has units cm/cm^3; scattering lengths can be found in the International Tables and put in these units (see Appendix A for pertinent pages). In this case it is 6.44×10^{10} cm^{-2}. Here S is the total surface in the specimen and V is the volume irradiated.

The invariant in these units is:

$$\frac{K}{2\pi^2} \int q^2 I(q)\, dq \qquad (P7.17.4)$$

(K represents all the constants in Eq. (P7.17.2a). Point collimation can be assumed.

(a) Is Porod's law valid in the measured region?
(b) Assuming the net counts are from the pores evaluate the "Porod radius" (Surface per unit volume of particle irradiated is a dimension = $\langle R^3 \rangle / \langle R^2 \rangle$). (Do any required extrapolations of data as *best* you can.)
(c) What is the surface area of the pores per cm^3 of sample.
(d) From the results of (b) and (c) assuming the pores are uniformly sized spheres, and with the information that there is 120 cm^2 of grain boundary area per cm^3 of specimen, (determined metallographically), calculate:
(i) number of voids per unit of grain boundary area.
(ii) average distance between voids along the grain boundaries.

7.18. You are given a sample of Al-0.5 at pct Cu, to examine EXAFS above the Cu absorption edge. The only detector you have is a NaI scintillation counter. Its active area is 4 cm in diameter and (due to constraints of the apparatus) cannot be positioned any closer than 20 mm from the sample.

Would you attempt to measure EXAFS in the transmission mode, or via fluorescence? Why? (The fluorescence yield, ε_k, is 0.4.)

Table 7.9. Intensity data: 110 peak — Cr radiation, V Filter, worked sample

$\sin\theta/\lambda$ [a]	$I_{(net)}$ [b]	$\sin\theta/\lambda$ [a]	$I_{(net)}$ [b]	$\sin\theta/\lambda$ [a]	$I_{(net)}$ [b]
0.21907	0	0.24423	27	0.25439	2.6
0.22002	0.010	0.24486	59	0.25517	2.2
0.22167	0.036	0.24549	142	0.25595	1.8
0.22331	0.066	0.24581	212	0.25672	1.5
0.22492	0.10	0.24612	281	0.25826	1.2
0.22658	0.16	0.24628	301	0.25979	1.0
0.22820	0.24	0.24634	305	0.26284	0.72
0.22983	0.34	0.24637 [c]	306	0.26588	0.55
0.23144	0.46	0.24640	306	0.26889	0.43
0.23306	0.61	0.24644	304	0.27189	0.33
0.23467	0.83	0.24659	295	0.27486	0.26
0.23627	1.1	0.24675	272	0.27781	0.20
0.23676	1.4	0.24707	207	0.28074	0.14
0.23787	1.7	0.24738	141	0.28365	0.11
0.23867	2.0	0.24801	62	0.28654	0.089
0.23947	2.4	0.24863	31	0.28940	0.071
0.24026	3.1	0.24926	19	0.29224	0.053
0.24106	4.2	0.24989	13	0.29507	0.040
0.24185	5.8	0.25051	10	0.29787	0.027
0.24265	8.1	0.25129	6.9	0.29925	0.020
0.24312	10	0.25207	5.3	0.30064	0.014
0.24344	13	0.25222	4.3	0.30202	0.007
0.24391	20	0.25362	3.3	0.30340	0.003
				0.30397	0

[a] $\lambda = 2.2896$ Å
[b] Intensities above background were corrected by dividing by the Lorentz polarization factor.
[c] Center of range.

Table 7.10. Intensity data: 110 peak, reference sample

$\sin\theta/\lambda$	$I_{(net)}$	$\sin\theta/\lambda$	$I_{(net)}$	$\sin\theta/\lambda$	$I_{(net)}$
0.22035	0	0.24644	568	0.25979	0.63
0.22167	0.011	0.24659	790	0.26132	0.55
0.22331	0.024	0.24669	833	0.26284	0.49
0.22494	0.037	0.24672	840	0.26436	0.45
0.22658	0.053	0.24675[a]	845	0.26588	0.41
0.22820	0.071	0.24678	839	0.26739	0.38
0.22983	0.090	0.24691	725	0.26889	0.34
0.23144	0.11	0.24707	470	0.27039	0.31
0.23306	0.15	0.24738	168	0.27188	0.29
0.23467	0.19	0.24769	32	0.27337	0.27
0.23627	0.24	0.24801	15	0.27486	0.25
0.23787	0.30	0.24863	7.7	0.27634	0.23
0.02947	0.38	0.24926	4.4	0.27781	0.22
0.24106	0.53	0.24989	3.6	0.27928	0.19
0.24185	0.64	0.25051	2.6	0.28220	0.15
0.24266	0.78	0.25129	2.0	0.28365	0.13
0.24344	1.0	0.25207	1.7	0.28510	0.12
0.24376	1.2	0.25222	1.5	0.28797	0.093
0.24423	2.5	0.25362	1.3	0.28940	0.078
0.24454	4.9	0.25439	1.1	0.29082	0.062
0.24486	9.6	0.25517	1.0	0.29366	0.030
0.24549	27	0.25672	0.86	0.29507	0.013
0.24581	53	0.25826	0.73	0.29645	0
0.24612	197				

[a] Center of range.

Table 7.11. Intensity data: 220 peak — Co radiation, Fe filter, worked sample

$\sin\theta/\lambda$ [a]	$I_{(net)}$	$\sin\theta/\lambda$ [a]	$I_{(net)}$	$\sin\theta/\lambda$ [a]	$I_{(net)}$
0.46561	0	0.49173	19.1	0.50135	1.3
0.46749	0.026	0.49219	27.4	0.50243	1.1
0.46882	0.033	0.49242	31.7	0.50349	0.97
0.47014	0.039	0.49265	34.9	0.50455	0.85
0.47146	0.052	0.49288	37.5	0.50559	0.76
0.47276	0.068	0.49300	37.6	0.50663	0.69
0.47406	0.090	0.49311[b]	37.9	0.50765	0.62
0.47535	0.11	0.49323	37.6	0.50867	0.57
0.47663	0.14	0.49334	37.2	0.50968	0.51
0.47790	0.17	0.49357	35.5	0.51067	0.45
0.47916	0.21	0.49380	33.0	0.51166	0.39
0.48041	0.28	0.49403	30.2	0.51264	0.33
0.48165	0.36	0.49448	22.7	0.51361	0.28
0.48289	0.46	0.49494	16.1	0.51456	0.24
0.48411	0.63	0.49539	11.1	0.51551	0.20
0.48532	0.89	0.49584	8.2	0.51645	0.16
0.48653	1.3	0.49640	4.8	0.51738	0.13
0.48773	1.8	0.49696	3.5	0.51921	0.084
0.48833	2.2	0.49752	2.6	0.52099	0.051
0.48891	2.7	0.49808	2.2	0.52274	0.032
0.48950	3.5	0.49862	1.9	0.52360	0.023
0.49009	4.9	0.49918	1.7	0.52445	0.015
0.49068	7.7	0.50027	1.5	0.52581	0
0.49126	12.6				

[a] $\lambda = 1.7889$ Å.
[b] Center of range.

Table 7.12. Intensity data: 220 peak, reference sample

$\sin\theta/\lambda$	$I_{(net)}$	$\sin\theta/\lambda$	$I_{(net)}$	$\sin\theta/\lambda$	$I_{(net)}$
0.47513	0	0.49311	65	0.49696	0.65
0.47663	0.02	0.49334	134	0.49808	0.48
0.47790	0.04	0.49343[a]	143	0.49918	0.41
0.47916	0.05	0.49346	142	0.50027	0.35
0.48041	0.07	0.49357	121	0.50135	0.30
0.48165	0.09	0.49380	56	0.50243	0.26
0.48289	0.11	0.49403	34	0.50349	0.22
0.48411	0.15	0.49426	54	0.50455	0.19
0.48533	0.20	0.49437	69	0.50559	0.16
0.48653	0.27	0.49448	73	0.50663	0.14
0.48773	0.33	0.49471	48	0.50765	0.11
0.48891	0.45	0.49494	20	0.50867	0.09
0.49009	0.69	0.49516	8.0	0.50968	0.08
0.49126	1.2	0.49539	4.3	0.51067	0.06
0.49184	1.9	0.49562	2.2	0.51166	0.05
0.48242	3.6	0.49584	1.4	0.51264	0.03
0.49265	7.7	0.49617	0.84	0.51423	0
0.49288	19				

[a] Center of range.

8. The Dynamical Theory of Diffraction[1]

8.1 Introduction

If an incident beam is scattered from atoms at some depth in a crystal, it is not at all clear why we have been assuming that the scattered beam can leave the crystal without rescattering. Consider a small crystal in the shape of a cube, 10^{-3} m on an edge, entirely bathed by an x-ray beam. The peak intensity from such a crystal can be written as follows, for unpolarized radiation and receiving slits of dimensions $w \times h$ (see Sect. 4.9).

$$I(\text{cps}) \cong (I_0 A_0)(e^4/m^2 c^4)(wh/R^2)[(1 + \cos^2 2\theta)/2] F^2 N_1^2 N_2^2 N_3^2 \ .$$

Taking reasonable values for all the constants ($w = 0.001$, $h = 0.01$ m) with the cubic unit cell edge being 10 Å so that $N_i = 10^6$ and $F^2 \cong 1000$, with $I_0 A_0 = 10^9$ cps

$$I \cong (10^9)(10^{-29})[(0.01 \times 0.001)/0.02](1)10^3(10^6)^6 \cong 10^{16} \text{ cps} \ .$$

The calculated intensity is greater than the incident beam! Obviously the kinematic theory we employed for this calculation has its shortcomings. As von Laue once remarked, it is a happy circumstance that most crystals are not perfect enough to deserve any better than the kinematic theory! As long as the coherent regions are small, a few hundred Ångstroms or so, the *scattered intensity* from each such region is *low* and the *intensity* from each such region is *added*, not the amplitude, as was done to obtain the above equation. Then this problem does not occur. But intellectual curiosity alone demands that we look more closely at highly perfect crystals, and if that is not enough, there are practical reasons for us to be concerned:

(1) In doing structural work, we often obtain a first estimate of the temperature factor by plotting $(\ln I)/\sum f^2 (\text{LP})$ versus $\sin^2 \theta/\lambda^2$ [see Eq. (3.25)]; the data can be from fine powders or single crystals. It is not unusual for the strong low-angle peaks to be considerably below the line determined from the others, even on this logarithmic plot. (By deforming the crystal slightly, by abrasion or rapid quenching, this effect can be eliminated.) In the kinematic theory, we have assumed that a once-scattered beam is not rescattered. However, if a crystal or powder particle is

---

[1] This chapter is based on a review by Batterman, B.W., and Cole, H., Rev. Mod. Phys. 36, 681–717 (1964). Comments by the authors on this version are gratefully acknowledged. Figures captions marked with an asterisk (∗) are adopted from this review and are reproduced with the permission of "Rev. Mod. Phys." and the authors.

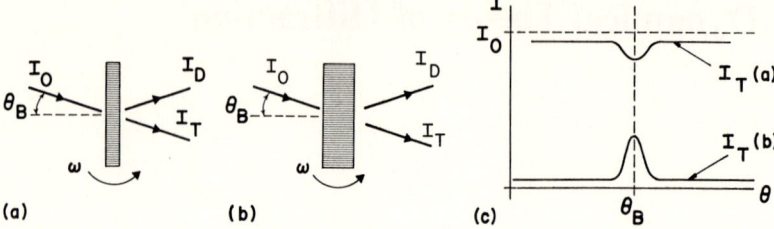

Fig. 8.1a,b. Anomalous transmission. **a** Thin crystal Laue diffraction. **b** Thick crystal Laue diffraction. **c** Transmitted intensity. [Batterman, B.W., and Cole, H., Rev. Mod. Phys. **36**, 681–717 (1964)]*

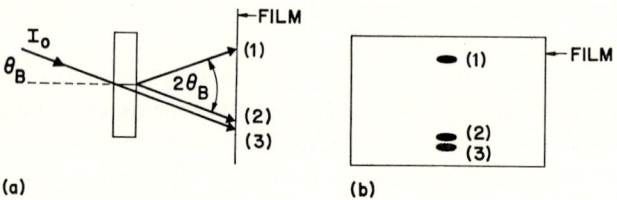

Fig. 8.2a,b. Geometry of the beams when anomalous transmission is occurring. **a** Ray diagram required to explain the spots on the film. **b** The energy in the diffracted beam*

highly perfect and the *once-diffracted beam can find a plane above it that is also oriented to diffract*, part of this beam will then be returned in the initial direction of the incident radiation. As shown by Eq. (3.25c) *the scattered radiation is 90° out of phase with the incident beam.* Hence *after two scattering events, it is* 180° *out of phase with the incident radiation and partially cancels it.* The cancellation is only partial because the scattered beam is not as intense as the incident beam. The result is that the incident beam's power is reduced by more than the usual absorption as it penetrates the crystal, and this in turn reduces the total diffracted intensity.[2]

(2) There are many highly perfect crystals these days, particularly as a result of the demand for these in the semiconductor industry. Some very unusual diffraction effects can be obtained with these. One of these effects was first discovered by Borrmann (1941) and is illustrated in Fig. 8.1. A single crystal is shown cut to have parallel faces, with the diffracting planes perpendicular to these faces. In (a) the crystal is thin ($\mu t \ll 1$), and in (b) it is thick ($\mu t \gg 10$). Diffraction is in transmission or Laue geometry. In (a) as the crystal is rotated through the Bragg angle, there is a dip in the transmitted beam I_T below the value given by $I_0 \exp(-\mu t)$. This is to be expected from kinematic theory, as energy is removed from the transmitted beam for diffraction. In (b), however, the transmitted intensity shows a small *peak*. A further experiment shown in Fig. 8.2 helps us to understand this unusual effect. The

---

[2] The reader should study Sects. 3.4 and 3.7. Here the close connection between scattered intensity, index of refraction and absorption are detailed. In particular it is shown that (a) scattering from an electron is 180° out of phase with the incoming radiation, (b) scattering from a collection of electrons leads to an additional 90° shift, (c) the index of refraction is imaginary due to scattering alone, and (d) the imaginary portion causes absorption. We shall be seeing many of these results again in this chapter.

8.1 Introduction

thick crystal is set at its Bragg angle and a film is placed on the far side to receive the emerging beams. The developed film shows *three* spots, (1) being the diffracted beam, (2) being about as dark as (1) and separated by the correct 2θ, and (3) being a very weak spot, but in-line with the incident beam. The separation between (2) and (3) is proportional to the sample's thickness. The peak in Fig. 8.1 is due to ray (2). From the way Fig. 8.2 is drawn [to account for this effect of thickness on the separation of rays (2) and (3)], *the radiation forming spots* (1) *and* (2) *emerges from the crystal at a point opposite to where the incident beam impinges*. We shall find that *the radiation apparently travels along atomic planes* with little absorption and that this *anomalous transmission from* a thick crystal is really a phenomenon due to diffraction. In fact, we shall see that ray (2) in Fig. 8.2 is best described as a forward diffracted beam, rather than a transmitted beam.

The surprising thing is that any appreciable intensity at all gets through a crystal whose absorption factor is $\exp(-10)$! The theory we shall explore will indicate that there is a standing wave pattern in the crystal for the *total wave* field and that it is better to consider this pattern than our former separation of the beams into incident and diffracted rays. This standing wave pattern is shown in Fig. 8.3. The solid line represents such a pattern with *nodal points at atomic planes*. Obviously, such a beam will not be absorbed by photoelectric emissions as it passes through a crystal. There is a second solution with *antinodes at the atoms* (shown dotted). This wave field suffers enhanced absorption and scattering. While it is elminated in a transmission experiment, it is important in reflection geometries. In fact, there are four fields, if both states of x-ray polarization are considered. For some of you, this result may not be new; this is the same thing that occurs to electrons at a Brillouin zone boundary. The difference in energy of the fields is the energy gap.

This unusual effect in highly perfect regions and the differences in scattered intensity in perfect and imperfect regions is the basis of the various "topographical" techniques now employed to detect imperfections with x-rays or electrons. (We shall discuss these more fully later.) In fact, because the electron scattering factor is $\approx 10^4$

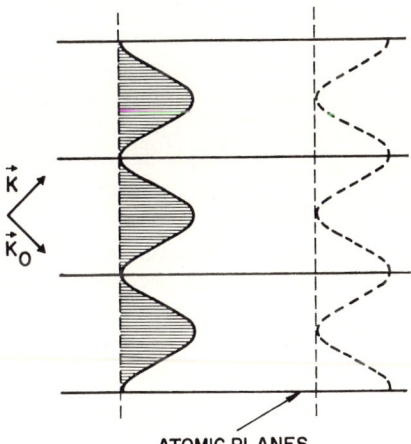

ATOMIC PLANES

Fig. 8.3. Standing wave pattern produced by two coherent, traveling plane waves with wave vectors $\mathbf{K}_0$ and $\mathbf{K}$. Nodes of *dotted curve* coincide with antinodes of *solid curve*. When such a pattern exists relative to the atomic planes, the normal photoelectric absorption is radically altered*

times that for x-rays, such effects are even more important in electron microscopy and diffraction. It is impossible in *this* field to obtain quantitative results without dynamical theory.

The first attempt to explain the reduction in diffracted intensities from highly perfect crystals was made by Darwin (1914) very shortly after the discovery of x-rays. He employed the multiple scattering approach we discussed above in a qualitative way. We shall follow a different method, developed by Ewald (1916, 1917, 1958) in which we shall solve Maxwell's equations for the waves that can exist in a medium with a periodic complex dielectric coefficient. Darwin's approach is described by Warren (1969, Chap. 14). It is often used in electron diffraction (see Hirsch et al. 1965). Actually, there was not much interest in this at the time of its discovery as when Moseley (Sect. 3.1) checked Darwin's ideas by integrating intensities from a rock salt crystal he found that the kinematic theory was adequate. [Soon afterwards when others began measuring intensities and troubles began to arise, particularly with calcite and diamonds, Darwin was invited to lecture on his ideas to a small group — but he had forgotten all his equations!]

Before we continue, it is worth pointing out that "multiple scattering" can be a problem even in amorphous materials; see Warren (1969, pp. 145–149). Even at $100°\ 2\theta$ with Cu K_α, about 7% of the intensity from polystyrene is due to such events. It can be important in small angle scattering, as shown in Fig. 8.4.

In the kinematic theory, diffraction occurs when, as shown in Fig. 8.5, the Ewald sphere touches a spot in reciprocal space; each such spot represents a set of parallel planes. The geometry of this figure leads to Bragg's law as we have already seen. The crucial difference, when we consider dynamical theory, will be in the concept of the Ewald sphere. Rather than a single such sphere, even for a monochromatic beam, there will be many. The locus of the centers of these form a "dispersion surface." Wave vectors from this surface to reciprocal lattice points will be seen to be the permitted solutions.

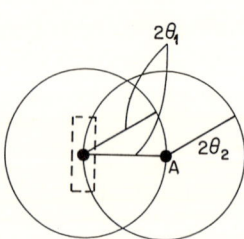

Fig. 8.4. A Reflection from a powder at angle $2\theta_1$ acts as a source at A for a second reflection $2\theta_2$. This leads to some portion of the diffracted cone of semiapex angle $2\theta_2$ appearing in the small-angle region (*dotted outline*). The direct beam emerges at the *filled dot*, and the large *circles* represents the Debye-Scherrer cones of diffraction

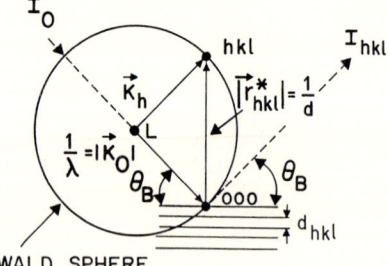

Fig. 8.5. Ewald sphere and Bragg's law. The incoming wave with wave vector $\mathbf{K}_0$ is diffracted by the atomic planes with interplanar spacing "d" *

8.2 The Dielectric Constant in a Crystal

We shall consider the case of x-radiation, by examining solutions to Maxwell's equations inside a crystal. For electrons, we would examine Schrödinger's wave equation for the waves that are solutions to this formulation in the presence of a periodic potential. The resultant effects are quite similar, so that with the treatment for x-rays, we shall learn most of the essential features. For the electron case the reader is referred to Hirsch et al. (1965) and Cowley (1975).

8.2 The Dielectric Constant in a Crystal[3]

8.2.1 Away from any Resonance

In a crystal, we are dealing with a periodic array of electrons. When the electric field **E** due to the incoming radiation impinges on this array, the electric displacement vector **D** or electric flux intensity that is induced can be written, for small displacements, as

$$\mathbf{D} = \varepsilon_0 \mathbf{E} + \mathbf{P} = \kappa \varepsilon_0 \mathbf{E} \,. \tag{8.1}$$

In this equation, **P** is the polarization per unit volume in the crystal (that is, the sum of charges times their displacements, per unit volume). **D**, **P** have units of charge per unit area, **E** is force per unit charge (n-m/coulomb or volts/meter). Thus, ε_0 has units of charge per unit area divided by force per unit charge. The term κ is the *dielectric constant* (a rank 2 tensor) and ε_0 is the permittivity of vacuum. The $\kappa \varepsilon_0$ term may be thought of as the proportionality constant between the charges inside the specimen and the total field through the specimen from these charges. Equation (8.1) simply states that the total field at a point in a crystal is the sum of the applied field and the field due to displacement of charges in the crystal. This total field **D** need not be parallel to **E**, since κ is a tensor.

From Eq. (8.1), it can be seen that

$$\kappa = 1 + (\mathbf{P}/\varepsilon_0 \mathbf{E}) \,. \tag{8.2}$$

As was demonstrated in Sect. 3.4, a sinusoidal field **E** acting on an electron held by a simple restoring force to an ion core leads to a displacement:

$$\mathbf{X} = (e/m)\mathbf{E}/(\omega_0^2 - \omega^2) \,. \tag{8.3}$$

If ω is very much greater than ω_0, which is the case for x-rays and most of the electrons in an atom, then these electrons move 180° out of phase with **E**, but they move all together and

$$\mathbf{P} = \rho(\mathbf{r}) e \mathbf{X} \,, \tag{8.4}$$

where $\rho(\mathbf{r})$ is the electron density. Then, substituting Eq. (8.3) into (8.4) and the result

---

[3] For simplicity, in all subsequent figures the subscripts hkl will be denoted h and we will use the subscript 0 for 000.

into (8.2) with frequency

$$\omega = 2\pi\nu = 2\pi c/\lambda, \tag{8.5}$$

$$\kappa(\mathbf{r}) = 1 - [(e^2/mc^2)\lambda^2/4\pi^2\varepsilon_0]\rho(\mathbf{r}). \tag{8.6}$$

The quantity $(e^2/4\pi\varepsilon_0 mc^2)$ is known as the classical electron radius r_e and has a value of 2.818×10^{-15} m. Thus

$$\kappa(\mathbf{r}) = 1 - r_e(\lambda^2/\pi)\rho(\mathbf{r}). \tag{8.7a}$$

Employing a Fourier series for $\rho(\mathbf{r})$, Eq. (8.6) becomes

$$\kappa(\mathbf{r}) = 1 - r_e\frac{\lambda^2}{\pi V}\sum_{hkl} F_{hkl}\exp(-2\pi i\mathbf{r}\cdot\mathbf{r}^*_{hkl})$$

$$= 1 - \Gamma\sum_{hkl} F_{hkl}\exp(-2\pi i\mathbf{r}\cdot\mathbf{r}^*_{hkl}). \tag{8.7b}$$

The value of Γ is of the order of 10^{-7}–10^{-8} for x-rays and is dimensionless.

8.2.2 Near an Absorption Edge

If the x-ray has a frequency near a resonant value (ω_0) for some electron and if there is absorption, the situation is more involved, as was indicated in Sect. 3.4. *In x-ray studies resonance and absorption are generally included by the corrections $\Delta f'$ and $\Delta f''$*:

$$F_{hkl} = \sum_n (f + \Delta f' + i\Delta f'')_n \exp(2\pi i\mathbf{r}_n\cdot\mathbf{r}^*_{hkl}). \tag{8.8}$$

For convenience, we shall group the real $(f + \Delta f')$ and imaginary parts $(\Delta f'')$ and we shall write

$$F_{hkl} = F'_{hkl} + iF''_{hkl}. \tag{8.9}$$

It should be realized that F'_{hkl} and F''_{hkl} can both be complex due to the spatial atomic arrangement (the exponential terms). However, consider the term $hkl = 000$. Then F'_0 and F''_0 are both real and

$$\kappa_0 = 1 - \Gamma[F'_0 + iF''_0], \tag{8.10a}$$

$$\kappa_0 = 1 - \Gamma\sum_n (f_0 + \Delta f'_0)_n - i\Gamma\sum_n (\Delta f''_0)_n. \tag{8.10b}$$

The second term is the order of 10^{-5} due to Γ, the term in $i\Gamma$, 10^{-6}. As these are the leading terms in the Fourier series representation of the dielectric constant, κ_0 *represents the average value of the dielectric constant and it is only slightly different from unity.* This result is not surprising; *the squre root of κ is the index of refraction* n[4] and in Sect. 3.4, it was shown that n differs from unity by only a few parts per million for x-rays.

---

[4] See, for example, Seitz, F., "The Modern Theory of Solids," Chap. 17. McGraw-Hill, New York, 1940.

In Sect. 3.7, it was shown that the absorption term for the amplitude through a material of thickness Δz is related to the imaginary part of the index of refraction n_i:

$$\exp(-\omega n_i \Delta z/c) = \exp[-2\pi(\Delta z/c)(c/\lambda)n_i]. \tag{8.11a}$$

For the absorption of *intensity*, the term is

$$\exp(-4\pi \Delta z n_i/\lambda) = \exp(-\mu \Delta z).$$

Thus

$$\mu = 4\pi n_i/\lambda. \tag{8.11b}$$

Since $n = \sqrt{\kappa}$ and $(1 - X)^{1/2} = 1 - X/2$ if X is small, then with κ_i, the imaginary part of the dielectric constant, from Eq. (8.10b)

$$\mu = 2\pi\kappa_i/\lambda = (2\pi/\lambda)\Gamma \sum_n (\Delta f_0'')_n,$$

or

$$\mu = (2\pi/\lambda)\Gamma F_0''. \tag{8.11c}$$

The zeroth order term ($hkl = 000$) is employed since it is the beam at $0°\ 2\theta$ that is involved in absorption. Thus, from a measurement of μ, we can obtain F_0''.

8.3 Waves that Satisfy Bragg's Law

8.3.1 Maxwell's Equations Inside a Crystal

We shall employ only two of Maxwell's equations:

$$\nabla \times \mathbf{E} = -\partial \mathbf{B}/\partial t = -\mu \partial \mathbf{H}/\partial t, \tag{8.12a}$$

$$\nabla \times \mathbf{H} = (\partial \mathbf{D}/\partial t) + (\text{current/constant}). \tag{8.12b}$$

Equation (8.12a) states that the "circulation" of the field $\mathbf{E}$, that is, its tangential component around a closed path, is proportional to the rate of change of $\mathbf{B}$ with time [from Stokes' theorem $\oint \mathbf{E} \cdot \mathbf{dx} = \int (\nabla \times \mathbf{E}) \cdot \mathbf{dA}$, where A is a surface whose normal is $\mathbf{A}$]. For example, if we have a wire with a circular electric field around it, this field exerts a force on the electrons in the wire, causing them to accelerate in a circular path, inducing a magnetic field that varies with time *along* the wire. This equation also represents the well-known induction in a coil due to a moving magnetic field. The second equation states that the circulation around a path (say, around a wire) of the magnetic field is related to the current in the wire and to any changing electric fields.

Atoms with their associated electrons vibrate at frequencies $\approx 10^{13}$ cps, whereas x-rays have a frequency of 10^{18} cps. In view of this large difference, it is reasonable to assume that the electrical conductivity due to the motion of atomic charges is zero at x-ray frequencies so that the current in Eq. (8.12b) is zero and that magnetically the crystal has the same behavior as empty space, i.e., that $\mu = \mu_0$. Because of atomic vibrations, the dielectric constant is really time dependent, but we shall neglect its time derivatives (i.e, we neglect time derivatives of the structure factor). Then

$$\nabla \times \mathbf{E} = -\partial \mathbf{B}/\partial t = -\mu_0 \partial \mathbf{H}/\partial t \, . \tag{8.13a}$$

This equation implies that **B** and **H** are collinear. Also,

$$\nabla \times \mathbf{H} = \partial \mathbf{D}/\partial t = \varepsilon_0 \partial(\kappa \mathbf{E})/\partial t = \kappa \varepsilon_0 \partial \mathbf{E}/\partial t \, . \tag{8.13b}$$

Next, we assume that the fields, **E**, **D**, and **H** can be expressed as sums of plane waves. We also write down Bragg's law for a scattered wave with **K** as the diffracted wave vector $\mathbf{S}/\lambda$ and $\mathbf{K}_0$ as the incident wave vector $\mathbf{S}_0/\lambda$:

$$\mathbf{K}_{hkl} = \mathbf{S}_{hkl}/\lambda = \mathbf{K}_0 + \mathbf{r}^*_{hkl} \, . \tag{8.14}$$

With this conservation of momentum equation, we also assume that the crystal is perfect, i.e, that $\mathbf{r}^*_{hkl}$ is the same everywhere in the crystal. To include absorption, all wave vectors **K** will be assumed to be complex:

$$\mathbf{K} = \mathbf{K}' - i\mathbf{K}'' \, . \tag{8.15a}$$

That this accounts for absorption can be seen by writing the expression for a plane wave:

$$\mathbf{A} = \mathbf{A}_0 \exp(-2\pi i \mathbf{K} \cdot \mathbf{r}) = \mathbf{A}_0 [\exp(-2\pi i \mathbf{K}' \cdot \mathbf{r}) \exp(-2\pi \mathbf{K}'' \cdot \mathbf{r})] \, . \tag{8.15b}$$

The last term has a real exponent and if $\mathbf{K}''$ is positive, represents a decrease in amplitude with increasing **r**.

Now, **K**, **K**', and **K**'' need not be collinear; *the planes of constant absorption* (perpendicular to **K**'') *need not be the same as planes of constant phase* (perpendicular to **K**') as shown in Fig. 8.6. From Eq. (8.11c), $|\mathbf{K}''|/|\mathbf{K}'| \approx 10^{-6}$, so that the direction of **K** is essentially that of the real part. Although we will keep **K** as a complex quantity, for its direction we shall consider only its real part.

Let **A** stand for **E**, **D**, or **H**, so that we may express a field as a plane wave:

$$\mathbf{A} = \mathbf{A}_{hkl} \exp(+2\pi i v t) \exp(-2\pi i \mathbf{K}_{hkl} \cdot \mathbf{r}) \, .$$

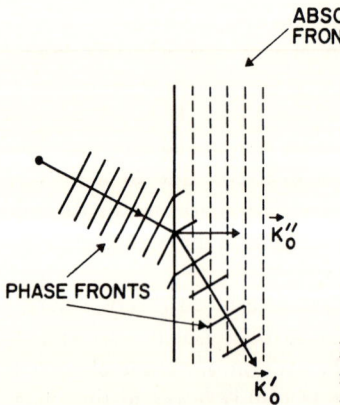

Fig. 8.6. The imaginary part of the wave vector, $\mathbf{K}''_0$, which represents absorption, is normal to the crystal surface. The real part of the wave vector, $\mathbf{K}'_0 = \mathbf{S}_0/\lambda$, describes the propagation*

Employing Eq. (8.14),

$$\mathbf{A} = \mathbf{A}_{hkl}\exp(+2\pi i v t)\exp(-2\pi i\mathbf{K}_0\cdot\mathbf{r})\exp[-2\pi i(\mathbf{r}^*_{hkl}\cdot\mathbf{r})] \ .$$

But $\mathbf{r}^*_{hkl}\cdot\mathbf{r}$ is an integer in a crystal and hence, the last factor is unity. Therefore, *all* solutions for *all* reciprocal lattice vectors are equally possible and we can write more generally:

$$\mathbf{A} = \exp(+2\pi i v t)\sum_{hkl}\mathbf{A}_{hkl}\exp(-2\pi i\mathbf{K}_{hkl}\cdot\mathbf{r}) \ . \tag{8.16}$$

Consider now the terms needed in Eq. (8.12a):

$$\nabla\times\mathbf{E} = -2\pi i\exp(+2\pi i v t)\sum_{hkl}\mathbf{K}_{hkl}\times\mathbf{E}_{hkl}\exp(-2\pi i\mathbf{K}_{hkl}\cdot\mathbf{r}) \tag{8.17a}$$

and

$$\partial\mathbf{H}/\partial t = +2\pi i v\exp(+2\pi i v t)\sum_{hkl}\mathbf{H}_{hkl}\exp(-2\pi i\mathbf{K}_{hkl}\cdot\mathbf{r}) \tag{8.17b}$$

According to Eq. (8.13a), Eqs. (8.17a) and (8.17b) are equal when the constant μ_0 is added in front of Eq. (8.17b). As these are both Fourier series, this means that their individual coefficients must be equal:

$$\mathbf{K}_{hkl}\times\mathbf{E}_{hkl} = v\mu_0\mathbf{H}_{hkl} \ . \tag{8.18a}$$

Similarly, from Eq. (8.13b)

$$\mathbf{K}_{hkl}\times\mathbf{H}_{hkl} = -v\mathbf{D}_{hkl} \ . \tag{8.18b}$$

Neglecting the very small imaginary parts of $\mathbf{K}$, Eq. (8.18a) states that $\mathbf{H}_{hkl}$ is perpendicular to $\mathbf{K}_{hkl}$. Eq. (8.18b) then implies that $\mathbf{K}$, $\mathbf{H}$, and $\mathbf{D}$ form an orthogonal set. However, $\mathbf{E}_{hkl}$ is in the plane of $\mathbf{D}_{hkl}$ and $\mathbf{K}_{hkl}$, but not necessarily along $\mathbf{D}_{hkl}$ as we saw earlier in considering polarization in a crystal. This is shown in Fig. 8.7; $\mathbf{E}_{hkl}$ may have a longitudinal component. Employing Eqs. (8.1) and (8.7b):

$$\sum_{hkl}\mathbf{D}_{hkl}\exp(-2\pi i\mathbf{K}_{hkl}\cdot\mathbf{r})$$

$$= \varepsilon_0\left[1 - \Gamma\sum_{h'k'l'}F_{h'k'l'}\exp(-2\pi i\mathbf{r}^*_{h'k'l'}\cdot\mathbf{r})\right]\times\sum_{hkl}\mathbf{E}_{hkl}\exp(-2\pi i\mathbf{K}_{hkl}\cdot\mathbf{r}) \ . \tag{8.19}$$

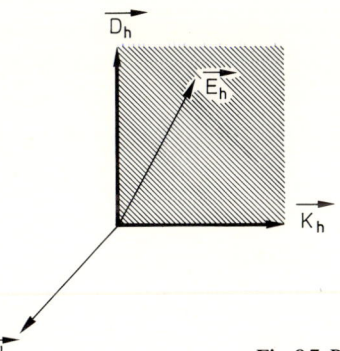

Fig. 8.7. Relationship between fields **D**, **H**, **E**, and propagation vector **K**

The primes merely distinguish the two sums. For the right-hand side of Eq. (8.19), we make use of the following. From Eq. (8.14)

$$\mathbf{K}_{h'k'l'} + \mathbf{r}_{hkl}^* = \mathbf{K}_{h'+h,k'+k,l'+l} \, .$$

Also, let

$$h' + h, \quad k' + k, \quad l' + l \equiv M \, ;$$

then

$$\sum_{h'k'l'} \sum_{hkl} F_{h'k'l'} \mathbf{E}_{hkl} \to \sum_{hkl} \mathbf{E}_{hkl} \sum_{M-hkl} F_{M-hkl} \to \sum_{p} \sum_{(hkl-p)} \mathbf{E}_p F_{hkl-p} \, .$$

In the last change we have let $hkl \to p$, $M \to hkl$. We then write:

$$\sum_{hkl} \mathbf{D}_{hkl} \exp(-2\pi i \mathbf{K}_{hkl} \cdot \mathbf{r}) = \varepsilon_0 \sum_{hkl} \mathbf{E}_{hkl} \exp(-2\pi i \mathbf{K}_{hkl} \cdot \mathbf{r})$$

$$- \varepsilon_0 \Gamma \sum_{hkl} \left(\sum_{p} F_{hkl-p} \mathbf{E}_p \right) \exp(-2\pi i \mathbf{K}_{hkl} \cdot \mathbf{r}) \, . \quad (8.20)$$

In the last term, the sum should be over $hkl - p$, but a sum over $(hkl - p)$ that runs to plus and minus infinity is the same as a sum over hkl.

Again, since the Fourier components must be equal,

$$\mathbf{D}_{hkl} = \varepsilon_0 \mathbf{E}_{hkl} - \varepsilon_0 \Gamma \sum_{p} F_{hkl-p} \mathbf{E}_p$$

$$= \varepsilon_0 (1 - \Gamma F_0) \mathbf{E}_{hkl} - \varepsilon_0 \Gamma \sum_{p \neq hkl} F_{hkl-p} \mathbf{E}_p \, . \quad (8.21)$$

Because Γ is so small, the field $\mathbf{D}_{hkl}$ is predominantly the first term $\varepsilon_0 \kappa_0 \mathbf{E}_{hkl}$, but is modified slightly by small contributions from other Fourier components of the electric field.

We now examine Maxwell's equations again. In particular, we take the cross product of $\mathbf{K}_{hkl}$ with each side of Eq. (8.18a) and substitute Eq. (8.18b) to combine Eqs. (8.18):

$$\mathbf{K}_{hkl} \times (\mathbf{K}_{hkl} \times \mathbf{E}_{hkl}) = \nu \mu_0 (\mathbf{K}_{hkl} \times \mathbf{H}_{hkl}) = -\nu^2 \mu_0 \mathbf{D}_{hkl} \, . \quad (8.22)$$

We substitute for $\mathbf{D}_{hkl}$ from Eq. (8.21):

$$\mathbf{K}_{hkl} \times (\mathbf{K}_{hkl} \times \mathbf{E}_{hkl}) = -\nu^2 \mu_0 \varepsilon_0 \left[\mathbf{E}_{hkl} - \Gamma \sum_{p} F_{hkl-p} \mathbf{E}_p \right] \, . \quad (8.23)$$

Now, $\mu_0 \varepsilon_0 \equiv 1/c^2$ and $\nu^2/c^2 = (1/\lambda)^2 = k^2$, the wavelength in a vacuum. Hence, Eq. (8.23) may be written as

$$\mathbf{K}_{hkl} \times (\mathbf{K}_{hkl} \times \mathbf{E}_{hkl}) + k^2 \mathbf{E}_{hkl} - k^2 \Gamma \sum_{p} F_{hkl-p} \mathbf{E}_p = 0 \, . \quad (8.24)$$

We know that the triple cross product may be written as

$$\mathbf{K}_{hkl} \times (\mathbf{K}_{hkl} \times \mathbf{E}_{hkl}) = -(\mathbf{K}_{hkl} \cdot \mathbf{K}_{hkl}) \mathbf{E}_{hkl} + (\mathbf{K}_{hkl} \cdot \mathbf{E}_{hkl}) \mathbf{K}_{hkl} \, ;$$

thus

$$-(\mathbf{K}_{hkl} \cdot \mathbf{K}_{hkl}) \mathbf{E}_{hkl} + (\mathbf{K}_{hkl} \cdot \mathbf{E}_{hkl}) \mathbf{K}_{hkl} + k^2 \mathbf{E}_{hkl} - k^2 \Gamma \sum_{p} F_{hkl-p} \mathbf{E}_p = 0 \, .$$

8.3 Waves that Satisfy Bragg's Law

Separating the term in the sum for $p = hkl$ and combining it with the first and third terms, we have our final form:

$$[k^2(1 - \Gamma F_0) - (\mathbf{K}_{hkl} \cdot \mathbf{K}_{hkl})]\mathbf{E}_{hkl} - k^2\Gamma$$
$$\times \sum_{p \neq hkl} F_{hkl-p}\mathbf{E}_p + (\mathbf{K}_{hkl} \cdot \mathbf{E}_{hkl})\mathbf{K}_{hkl} = 0 \,. \tag{8.25a}$$

This is the fundamental set of equations describing the field inside the crystal which must hold for each component of the vectors and for real and imaginary parts. Consider Eq. (8.25a) in the form

$$\mathbf{E}_{hkl} = \left[k^2\Gamma \sum_{p \neq hkl} F_{hkl-p}\mathbf{E}_p - (\mathbf{K}_{hkl} \cdot \mathbf{E}_{hkl})\mathbf{K}_{hkl} \right] \Big/ \underbrace{[k^2(1 - \Gamma F_0)}_{A} - \underbrace{\mathbf{K}_{hkl} \cdot \mathbf{K}_{hkl}]}_{B} \,. \tag{8.25b}$$

Since Γ is so small, the square root of the term A can be written as $k(1 - \frac{1}{2}\Gamma F_0)$. In this term k is the reciprocal of the wavelength in vacuum and $(1 - \frac{1}{2}\Gamma F_0)$ is the square root of the average dielectric constant — the average index of refraction (n). We should really draw Ewald's sphere with radius n/λ not $1/\lambda$ inside the crystal, but of course the difference is really quite small. Note that no field $\mathbf{E}_{hkl}$ will have any appreciable value unless $\mathbf{K}_{hkl}$ differs only slightly from n/λ. *Only those wave vectors very close to the Ewald sphere will have a large amplitude, as in kinematic theory.*

8.3.2 Applications of the Equations for the Fields Inside a Crystal

Suppose that only one point, $hkl = 000$ is close to the Ewald sphere, i.e., there are no Bragg peaks. Then $\mathbf{E}_p = 0$ for $p \neq 000$ and there are no contributing terms in the sum over p. Equation (8.25a) becomes

$$[k^2(1 - \Gamma F_0) - K_0^2]\mathbf{E}_0 = -(\mathbf{K}_0 \cdot \mathbf{E}_0)\mathbf{K}_0$$

Thus the component of $\mathbf{E}_0$ along $\mathbf{K}_0$ is very small, but there *is* some component.

We can see this more generally by taking the dot product of $\mathbf{K}_{hkl}$ and Eq. (8.25a). Then we have

$$k^2\mathbf{K}_{hkl} \cdot \mathbf{E}_{hkl}(1 - \Gamma F_0) = k^2\Gamma \sum_{p \neq hkl} F_{hkl-p}(\mathbf{K}_{hkl} \cdot \mathbf{E}_p) \,. \tag{8.26}$$

As $\Gamma F \approx 10^{-5}$, the component of $\mathbf{E}_{hkl}$ along $\mathbf{K}_{hkl}$ is indeed quite small!

We now consider the case of two points close to the Ewald sphere, 000 and hkl. We define the plane of incidence by the two vectors $\mathbf{K}_0$ and $\mathbf{K}_{hkl}$, and consider $\mathbf{E}_0$ and $\mathbf{E}_{hkl}$ resolved into their scalar components normal to this plane (which we call the "σ" polarization state) and those parallel to this plane (the "π" state).

For the σ polarization state and two points 000 and hkl near the Ewald sphere, Eq. (8.25a) becomes (note the scalar form):

$$[k^2(1 - \Gamma F_0) - (\mathbf{K}_0 \cdot \mathbf{K}_0)]E_0^\sigma - k^2\Gamma F_{\overline{hkl}}E_{hkl}^\sigma = 0 \,, \tag{8.26a}$$

$$-k^2\Gamma F_{hkl}E_0^\sigma + [k^2(1 - \Gamma F_0) - (\mathbf{K}_{hkl} \cdot \mathbf{K}_{hkl})]E_{hkl}^\sigma = 0 \,. \tag{8.26b}$$

The last term in Eq. (8.25a) is zero for this state of polarization. For the π state, the components of **E** lie in the plane of incidence. As we have shown that **E** is essentially perpendicular to **K**, if the angle between $\mathbf{K}_0$ and $\mathbf{K}_{hkl}$ is 2θ, this is also the angle between $\mathbf{E}_0$ and $\mathbf{E}_{hkl}$. In this case Eq. (8.25a) is a difference between the projection of the hkl and 000 fields:

$$[k^2(1 - \Gamma F_0) - \mathbf{K}_0 \cdot \mathbf{K}_0]E_0^\pi - k^2(\cos 2\theta)\Gamma F_{\overline{hkl}}E_{hkl}^\pi = 0, \tag{8.26c}$$

$$-k^2(\cos 2\theta)\Gamma F_{hkl}E_0^\pi + [k^2(1 - \Gamma F_0) - \mathbf{K}_{hkl} \cdot \mathbf{K}_{hkl}]E_{hkl}^\pi = 0. \tag{8.26d}$$

This result is only approximate as we have ignored the component of the field **E** along the wave vector.

We can combine both equations by writing T for the factor due to the polarization (1 or $\cos 2\theta$). The simultaneous solution for the fields $\mathbf{E}_0$ and $\mathbf{E}_{hkl}$ will be nontrivial only if the following determinant is zero:

$$0 = \begin{vmatrix} k^2(1 - \Gamma F_0) - \mathbf{K}_0 \cdot \mathbf{K}_0 & -k^2 T\Gamma F_{\overline{hkl}} \\ -k^2 T\Gamma F_{hkl} & k^2(1 - \Gamma F_0) - \mathbf{K}_{hkl} \cdot \mathbf{K}_{hkl} \end{vmatrix}. \tag{8.27}$$

We shall have much to say about Eq. (8.27), but first let us examine the case of *three* points near the Ewald sphere, 000, $h_1 k_1 l_1$, and $h_2 k_2 l_2$. For simplicity we shall consider only those components perpendicular to the plane of incidence and, furthermore, that all three reciprocal lattice vectors are in one plane. From Eq. (8.25a), we can then write three equations. Let

$$W_{hkl} = k^2(1 - \Gamma F_0) - \mathbf{K}_{hkl} \cdot \mathbf{K}_{hkl}.$$

Then

$$W_{000}\mathbf{E}_0 - k^2\Gamma F_{\bar{h}_1\bar{k}_1\bar{l}_1}\mathbf{E}_{h_1 k_1 l_1} - k^2\Gamma F_{\bar{h}_2\bar{k}_2\bar{l}_2}\mathbf{E}_{h_2 k_2 l_2} = 0, \tag{8.28a}$$

$$-k^2\Gamma F_{h_1 k_1 l_1}\mathbf{E}_0 + W_{h_1 k_1 l_1}\mathbf{E}_{h_1 k_1 l_1} - k^2\Gamma F_{h_1 k_1 l_1 - h_2 k_2 l_2}\mathbf{E}_{h_2 k_2 l_2} = 0, \tag{8.28b}$$

$$-k^2\Gamma F_{h_2 k_2 l_2}\mathbf{E}_0 - k^2\Gamma F_{h_2 k_2 l_2 - h_1 k_1 l_1}\mathbf{E}_{h_1 k_1 l_1} + W_{h_2 k_2 l_2}\mathbf{E}_{h_2 k_2 l_2} = 0. \tag{8.28c}$$

Assume that $F_{h_1 k_1 l_1} = 0$ because of the structure. It is immediately clear from Eq. (8.28c) that there is a solution for $\mathbf{E}_{h_1 k_1 l_1}$. There is intensity for this point in reciprocal space, *even though the structure* factor is zero, but it depends on $F_{h_2 k_2 l_2}$ and $F_{h_2 k_2 l_2 - h_1 k_1 l_1}$ not on $F_{h_1 k_1 l_1}$. We can see what this means with the following equations and their expression in Fig. 8.8.

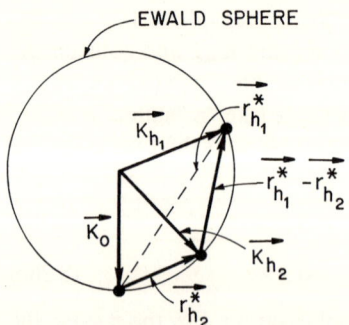

Fig. 8.8. The diffracted beam $\mathbf{K}_{h_2}$, acts as a source for diffraction from $h_1 k_1 l_1$; even though $F_{h_1 k_1 l_1} = 0$, intensity appears because the diffraction vector is $(\mathbf{r}^*_{h_1 k_1 l_1} - \mathbf{r}^*_{h_2 k_2 l_2})$, not $\mathbf{r}^*_{h_1 k_1 l_1}$.

$$\mathbf{K}_{h_2} \equiv \mathbf{K}_{h_2 k_2 l_2} = \mathbf{K}_0 + \mathbf{r}^*_{h_2 k_2 l_2}, \tag{8.29a}$$

$$\mathbf{K}_{h_1} \equiv \mathbf{K}_{h_1 k_1 l_1} = \mathbf{K}_0 + \mathbf{r}^*_{h_1 k_1 l_1} = \mathbf{K}_{h_2 k_2 l_2} - \mathbf{r}^*_{h_2 k_2 l_2} + \mathbf{r}^*_{h_1 k_1 l_1}$$
$$= \mathbf{K}_{h_2 k_2 l_2} + (\mathbf{r}^*_{h_1 k_1 l_1} - \mathbf{r}^*_{h_2 k_2 l_2}). \tag{8.29b}$$

It is as if the diffracted beam $\mathbf{K}_{h_2 k_2 l_2}$ is the source for a second vector in reciprocal space, $(\mathbf{r}^*_{h_1 k_1 l_1} - \mathbf{r}^*_{h_2 k_2 l_2})$. This is a form of "multiple scattering."

In electron microscopy it is possible to tilt a specimen to have only two reciprocal lattice points near the Ewald sphere, or perhaps three. This is satisfactory for dislocation imaging because the image of a dislocation is so wide due to its strain field. However, to form a detailed image as we have seen in Chaps. 6 and 7, as many beams as possible should be active and inside the aperture (see Cowley 1975). Because of the strong scattered intensity in electron diffraction, it is practically impossible to obtain useful information without this dynamical theory (see Hirsch et al. 1965; Cowley 1975).

8.3.3 The Role of the Actual Index of Refraction

We now return to Eq. (8.27) for two points near the Ewald sphere. The diagonal terms, as we have already seen, are the difference (of the squares) of the magnitude of k in a vacuum corrected for the average index of refraction, and the actual magnitude K. If there is no difference, there is no solution, but we have also seen that the difference must be small [Eq. (8.25b)] for any appreciable scattered field. Thus, the index of refraction for $\mathbf{K}_0$ and $\mathbf{K}_{hkl}$ must be different from the average value, and it is this *difference* that is the important parameter. Hence, let us define two new quantities:

$$2k\xi_0 \equiv \mathbf{K}_0 \cdot \mathbf{K}_0 - k^2(1 - \Gamma F_0), \tag{8.30a}$$

$$2k\xi_{hkl} \equiv \mathbf{K}_{hkl} \cdot \mathbf{K}_{hkl} - k^2(1 - \Gamma F_0). \tag{8.30b}$$

The right-hand side of Eq. (8.30a) may be written as

$$[(\mathbf{K}_0 \cdot \mathbf{K}_0)^{1/2} + k(1 - \Gamma F_0)^{1/2}][(\mathbf{K}_0 \cdot \mathbf{K}_0)^{1/2} - k(1 - \Gamma F_0)^{1/2}]$$
$$\cong 2k[(\mathbf{K}_0 \cdot \mathbf{K}_0)^{1/2} - k(1 - \tfrac{1}{2}\Gamma F_0)],$$

since $|\mathbf{K}_0| \cong k$. Therefore,

$$\xi_0 = (\mathbf{K}_0 \cdot \mathbf{K}_0)^{1/2} - k(1 - \tfrac{1}{2}\Gamma F_0), \tag{8.30c}$$

$$\xi_{hkl} = (\mathbf{K}_{hkl} \cdot \mathbf{K}_{hkl})^{1/2} - k(1 - \tfrac{1}{2}\Gamma F_0). \tag{8.30d}$$

Let us look at the real and imaginary parts of ξ_0:

$$(\mathbf{K}_0 \cdot \mathbf{K}_0)^{1/2} = [(K'_0)^2 - (K''_0)^2 - 2iK'_0 K''_0 \cos\beta]^{1/2}, \tag{8.31}$$

where β is the angle between $\mathbf{K}'_0$ and $\mathbf{K}''_0$. Now, $K''_0/K'_0 \ll 1$ as we have already seen. Therefore, Eq. (8.31) can be written [with $(1 - X)^{1/2} \cong 1 - X/2$] as

$$K_0 = K'_0[1 - (K''^2_0/K'^2_0) - 2i(K''_0/K'_0)\cos\beta]^{1/2}$$
$$\cong K'_0 - iK''_0 \cos\beta.$$

are shown. Consider the point A_2. From this point, lines are drawn to 000 and hkl. The real values of ξ can be found by dropping perpendiculars from the point A_2 to the loci, as shown. From the definitions in Eqs. (8.30c) and (8.30d), the values of $\xi_{0\alpha}$ and $\xi_{hkl\alpha}$ are both positive, whereas the corresponding values for the β branch are always negative. Tie points A_1 and A_3 are on opposite sides of a line through Q; their ξ' values are equal in magnitude but opposite in sign, i.e., $\xi'_{hkl\alpha} = -\xi'_{hkl\beta}$.

For points A_4 and A_5, $\xi'_0 = \xi'_{hkl}$. Later we shall show that this distance is the width of total reflection from a perfect crystal in Bragg geometry. Accepting that this is the case, we can immediately determine this distance (w) from Fig. 8.11; from the projection of $\xi_{hkl\alpha}$ on LQ, and setting $\xi_0 = \xi_{hkl}$ in Eq. (8.33):

$$w = 2(\tfrac{1}{2})kT\Gamma|F_{hkl}|/\cos\theta . \tag{8.34}$$

Note that for the π state of polarization ($T < 1$), the width decreases. *The value of the width is typically of the order of a few seconds of* arc. To see this true width, we must somehow eliminate wavelength dispersion. Recall from Sect. 3.1 that even for characteristic radiation K_{α_1}, $\Delta\lambda/\lambda \cong 1/2000$. From Bragg's law $(\Delta\lambda/\lambda)\tan\theta = \Delta\theta$, and even for a 111 K_{α_1} reflection from silicon, $\Delta\theta$ is ≈ 26 sec. The way around this problem is shown in Fig. 8.12a. In this $(1, -1)$ or *parallel* arrangement of two perfect crystals cut to the same plane, K_{α_1} from the source diffracts at a lower angle than K_{α_2} from crystal (1). As a result *both components diffract at the same position* from crystal (-1). In this way *dispersion can minimized*. (Nonetheless, the reflection curve obtained by rotating either crystal through the Bragg angle θ, known as the *rocking curve* is the convolution of the reflection curves from both crystals and the measured width is larger than the calculated value; see Problem 8.3. We have seen how to deconvolute such a situation in Sect. 7.4 and we can use as a "standard" the shape

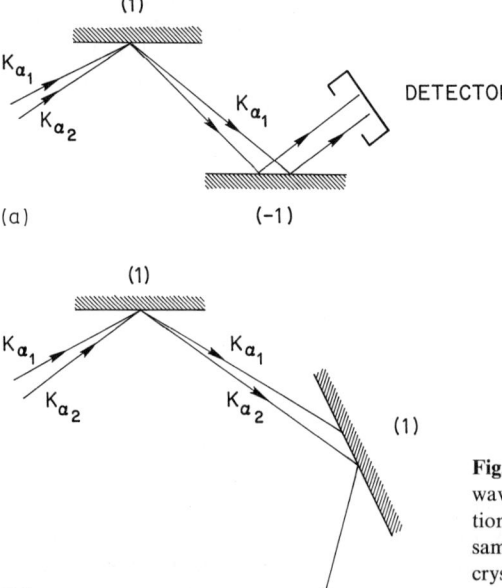

Fig. 8.12. Two crystals in **a** are set for eliminating wavelength dispersion in the $(1, -1)$ configuration. Note that two wavelengths diffract at the same angle of the second crystal. **b** The two crystals set for dispersion $(1, 1)$, so that the two wavelengths do not diffract at the same angle

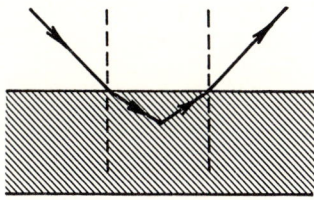

Fig. 8.13. The index of refraction of x-rays is less than one and therefore a ray bends away from the surface normal as it enters the crystal. Bragg's law is satisfied inside the crystal, so that the external angle is larger

of the curve from only one crystal,[5] but we shall see other methods later.) In Fig. 8.12b, in the (1, 1) or anti-parallel position, wavelength components are dispersed at the second crystal and this is one way of studying the wavelength dispersion. An excellent discussion of these rocking curves from "double crystal spectrometers" can be found in James (1950).

Incidentally, because n is less than one, the x-rays traveling in a crystal can be drawn as in Fig. 8.13. For Bragg reflection inside the crystal, the angle θ outside must be somewhat larger than inside. This has been discussed in Sect. 3.9 and Problems 3.22, 3.23 where it was also shown that any correction disappears in symmetric transmission geometry. We shall shortly be able to calculate the correction for the Bragg case in a simpler manner than that in Problem 3.23.

We have so far seen that the tie points determine the direction of the **K**'s and the imaginary part of the ξ's determine absorption. In addition, these ξ's characterize the field amplitudes. From Eq. (8.30a), (8.30b), and (8.26a–d):

$$\mathbf{E}_{hkl}/\mathbf{E}_0 = -2\xi_0/kT\Gamma F_{\overline{hkl}} = -kT\Gamma F_{hkl}/2\xi_{hkl} \,. \tag{8.35}$$

At points A_4 or A_5 in Fig. 8.11, $|\mathbf{E}_{hkl}| = |\mathbf{E}_0|$. This is so because with Eq. (8.33), $\xi_0 = \frac{1}{2}kT\Gamma|F_{hkl}|$. More about this later. First we need to consider how to decide *which* tie points are operative. The conditions of angle of incidence and orientation of the crystal surface with respect to the diffracting plane are involved. To see this, we must consider the boundary conditions on the fields and wave vectors at the crystal surfaces.

8.4 Boundary Conditions

8.4.1 Fields at the Entrance Source

Consider Fig. 8.14. When *fields cross a surface*, the usual condition for continuity is that their *components parallel to the surface are continuous*. However, we have seen that the longitudinal components of **E** and **H** in a crystal are negligible. Furthermore, because the index of refraction is so close to unity for *x-rays*, there is no

---

[5] In such a case, a receiving slit is needed to explore the beam from the first stationary crystal. Slits are *not* needed, however, for the rocking curve.

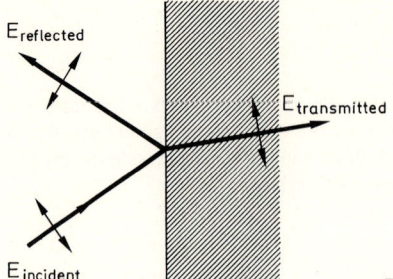

Fig. 8.14. Relation between fields at a boundary

appreciable reflection. This results because the change in angle on entering the crystal is slight. Thus there is no effect of a surface on the field vectors.

8.4.2 The Wave Vectors at a Surface

Let us then examine the wave vectors. We shall match a set of plane waves outside a boundary to a set inside when there is no change in amplitude or frequency upon crossing the surface. Let $\mathbf{k}_i$ be the wave vector outside and $\mathbf{k}_p$ the wave vector inside. For one wave with unit amplitude,

$$\exp(-2\pi i \mathbf{k}_i \cdot \mathbf{\tau}) = \exp(-2\pi i \mathbf{k}_p \cdot \mathbf{\tau}). \tag{8.36}$$

The origin is assumed to be in the boundary and $\mathbf{\tau}$ is a vector in the boundary. To satisfy this equation at every point in the surface, i.e. for every $\mathbf{\tau}$

$$\mathbf{k}_i \cdot \mathbf{\tau} = \mathbf{k}_p \cdot \mathbf{\tau}, \tag{8.36a}$$

which restricts $\mathbf{k}_i$ to

$$\mathbf{k}_i = \mathbf{k}_p + a\mathbf{n}, \tag{8.36b}$$

where $\mathbf{n}$ is a unit vector normal to the surface and a is a constant; *the wave vectors can only differ by a component perpendicular to the boundary.* For a set of waves with amplitudes, E:

$$\sum_i \mathbf{E}_i \exp(-2\pi i \mathbf{k}_i \cdot \mathbf{\tau}) = \sum_p \mathbf{E}_p \exp(-2\pi i \mathbf{k}_p \cdot \mathbf{\tau}). \tag{8.37}$$

The only sets of waves that are of interest are the ones that satisfy Bragg's law, i.e., those that are close to the Ewald sphere. Let the "inside incident" waves be given by $\mathbf{K}_{0j}$, where j is an index whose value will depend on how many reciprocal lattice points are near the sphere of reflection and how many tie points are active (on different branches for example). Assume that only the 000 and hkl points are near the Ewald sphere. According to Bragg's law, the inside diffracted waves are

$$\mathbf{K}_{hklj} = \mathbf{K}_{0j} + \mathbf{r}^*_{hkl}. \tag{8.38}$$

For the outside waves, we can have $\mathbf{k}_{0q}$ and $\mathbf{k}_{0q} + \mathbf{r}^{*\prime}_{hkl}$. Then we can write Eq. (8.37) as follows:

8.4 Boundary Conditions

$$\sum_q \exp(-2\pi i \mathbf{k}_{0q} \cdot \mathbf{\tau})[\mathbf{E}^i_{0q} + \mathbf{E}^i_{hklq} \exp(-2\pi i \mathbf{r}^{*\prime}_{hkl} \cdot \mathbf{\tau})]$$

$$= \sum_j \exp(-2\pi i \mathbf{K}_{0j} \cdot \mathbf{\tau})[\mathbf{E}_{0j} + \mathbf{E}_{hklj} \exp(-2\pi i \mathbf{r}^*_{hkl} \cdot \mathbf{\tau})]. \tag{8.39a}$$

Suppose there is no diffraction. Then:

$$\sum_q \mathbf{E}^i_{0q} \exp(-2\pi i \mathbf{k}_{0q} \cdot \mathbf{\tau}) = \sum_j \mathbf{E}_{0j} \exp(-2\pi i \mathbf{K}_{0j} \cdot \mathbf{\tau}), \tag{8.39b}$$

Now, suppose there is one outside wave, $\mathbf{k}^i_0$. Keep in mind that we have already found that the fields are unaffected by the boundary so that $\mathbf{E}^i_0 = \sum_j \mathbf{E}_{0j}$; *the outside incident amplitude* $\mathbf{E}^i_0$ *must add up to the inside incident amplitudes of* $\mathbf{E}_{0\alpha}$ *and* $\mathbf{E}_{0\beta}$ *for the two branches* (or to four amplitudes if both polarization states are considered). Similarly, the *inside diffracted amplitudes must add up to the outside diffracted amplitude* $\mathbf{E}^i_{hkl}$. Equation (8.39) then requires that

$$\mathbf{k}^i_0 = \mathbf{K}_{0j} + a_{0j}\mathbf{n}, \tag{8.40a}$$

and

$$\mathbf{r}^{*\prime}_{hkl} = \mathbf{r}^*_{hkl} + b_{hkl}\mathbf{n}. \tag{8.40b}$$

Equation (8.40a) states that *the inside wave vectors can differ from* $\mathbf{k}^i_0$ *only by a vector normal to the surface*. Assume that the outside incident wave is not being absorbed (hence, its wave vector is real). Then the imaginary parts of $\mathbf{K}_{0\alpha}$ and $\mathbf{K}_{0\beta}$ are as shown in Fig. 8.6 — perpendicular to the surface.

8.4.3 Selecting Tie Points

In Fig. 8.15 we show how the tie points are selected from these conditions. The surface is indicated by $s-s$. *The normal to the surface is drawn at a distance such that* $LP/(1/\lambda)$ *is the angular displacement from the kinematic Bragg angle*. This is the angle $\Delta\theta$ not $\Delta 2\theta$, as shown in Fig. 8.15. The vector from P to 000 represents the outside wave vector $\mathbf{k}^i_0$. Where the normal intersects the dispersion surface, points A and B determine the tie points; then the wave vector differs from $\mathbf{k}^i_0$ by a component

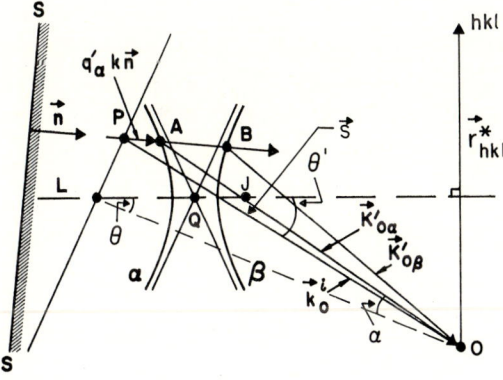

Fig. 8.15. Selection of the tie points by use of entrance point P and surface normal $\mathbf{n}$. PO represents the outside incident wave vector. The deviation from the correct Bragg angle, $\Delta\theta$, is given by LP/k. The real parts of each of the inside incident wave vectors, $\mathbf{K}_{0\alpha}$ and $\mathbf{K}_{0\beta}$, thus have the same surface tangential component as has the outside incident wave vector, $\mathbf{k}_0$. In triangle $LJ0$: $180° - \theta' + \theta + \alpha = 180°$. Therefore, $\theta' = \theta + \alpha$ and $\theta' - \theta = \Delta\theta = \alpha^*$

normal to the surface which we have found as the required situation. Note that **n** is directed inwards from the surface. We can write at point A in Fig. 8.15 that

$$\mathbf{K}_{0\alpha} = \mathbf{k}_0^i - \mathbf{PA} = \mathbf{k}_0^i - q_\alpha k \mathbf{n}. \tag{8.41a}$$

A similar expression can be written for point B. As the K's are complex, q must be complex. With Bragg's law we can write

$$\mathbf{K}'_{0\alpha} = \mathbf{k}_0^i - q'_\alpha k \mathbf{n}, \qquad \mathbf{K}''_{0\alpha} = q''_\alpha k \mathbf{n}; \tag{8.41b}$$

$$\mathbf{K}'_{hkl\alpha} = \mathbf{k}_0^i + \mathbf{r}^*_{hkl} - q'_\alpha k \mathbf{n}, \qquad \mathbf{K}''_{hkl\alpha} = q''_\alpha k \mathbf{n}. \tag{8.41c}$$

Because the reciprocal lattice vector is real, the imaginary components of both wave vectors (and hence their absorption coefficients) are identical.

Figure 8.15 is drawn for the Laue or transmission case, i.e., when both $\mathbf{K}_{hkl\alpha}$ and $\mathbf{K}_{hkl\beta}$ are directed *into* the crystal. For Bragg or reflection geometry, Fig. 8.16, the surface is horizontal and the vector **n** can cut only one branch or pass *between* the two branches. In the latter case only the imaginary part of $\mathbf{K}_0$ is involved, and only an attenuated field can exist in the crystal. *In this case, total reflection occurs over the range corresponding to the separation of the sheets of the dispersion surface*: From Fig. 8.17, the width in seconds can be calculated with Eq. (8.34):

$$\Delta\theta(\sec) = \frac{P_1 P_2}{k} = \left[\frac{w}{\sin\theta}/(1/\lambda)\right] 3600 \times \frac{360}{2\pi} = 3600 \times \frac{360}{2\pi} \frac{k \, \Gamma |T| |F_{hkl}|}{k \, \cos\theta \sin\theta}$$

$$= \frac{2\Gamma |T| |F_{hkl}|}{\sin 2\theta} \times 3600 \times \frac{360}{2\pi}. \tag{8.42}$$

This is not exactly the width in an actual measurement, but is sufficiently close for our purposes here. Further details are given in Problem 8.3.

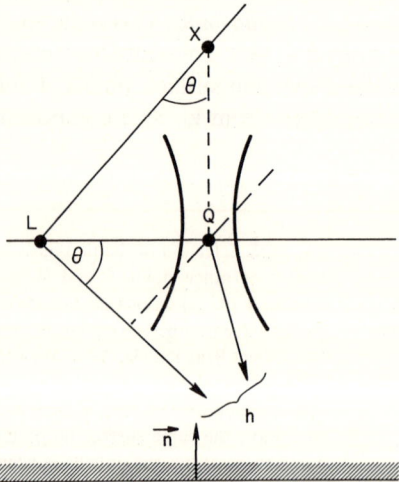

Fig. 8.16. The shift of a Bragg reflection from the kinematic position is LX

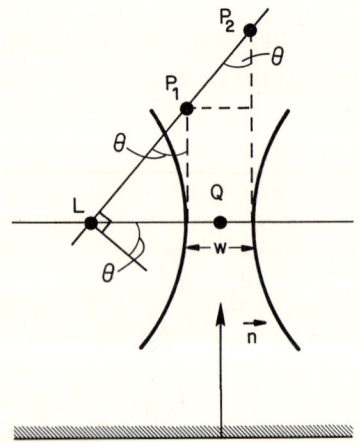

Fig. 8.17. The width of a Bragg reflection is $P_1 P_2$

8.4 Boundary Conditions

Table 8.1. Width $\Delta\theta$ from Eq. (8.42) and peak shifts due to index of refraction from Eq. (8.43) with Cu K$_\alpha$

Material	hkl	2θ (degrees)	$\sin\theta/\lambda$	f_{hkl}	$\|F_{hkl}\|$	$\sin 2\theta$	$\cos 2\theta$	$\Delta\theta^\sigma_{(T=1)}$ (sec)	$\Delta\theta^\pi_{(T=\cos 2\theta)}$ (sec)	Shift (sec)
Ge: $\Gamma = 1.178 \times 10^{-7}$	111	27.28	0.1531	27.36	154.77	0.4588	0.8885	16.39	14.56	14
	333	90.17	0.4593	17.34	98.09	1.000	−0.0030	4.77	0.01	6
	220	45.34	0.2500	23.76	190.08	0.7111	0.7029	12.98	9.12	9
Si: $\Gamma = 1.3305 \times 10^{-7}$	111	28.47	0.1595	10.58	59.85	0.4767	0.8791	6.89	6.05	6
	333	95.05	0.4784	6.45	36.49	0.9961	−0.0880	2.01	0.18	3
	220	47.34	0.2604	8.72	69.76	0.7354	0.6776	5.21	3.53	4

As θ increases (up to 45°) F decreases and $\sin 2\theta$ increases and both changes lead to a marked decrease in width with increasing angle, as shown in Table 8.1. Unlike kinematic theory, the width does not depend on the size of the crystal but on the structure factor.

8.4.4 The Center of a Bragg Reflection

The center of a Bragg reflection is clearly displaced from the value predicted from kinematic theory as shown in Fig. 8.16. The displacement $(LX)/k$ can be readily seen from this figure, recalling that $LQ = \frac{1}{2}k\Gamma F_0'$:

$$LX/k = [\tfrac{1}{2}k\Gamma F_0'/k \sin\theta] \times 3600 \times \frac{360}{2\pi}. \tag{8.43}$$

The shift of a low-angle peak is on the order of 10 arcsec. Results are given in the last column of Table 8.1 for two materials and several peaks.

8.4.5 The Exit Conditions

We see that for Laue geometry there are two inside incident waves for each polarization for a single incident wave, one from the α and one from the β branch. So far we have discussed boundary conditions inside the crystal. But we generally make our observations outside the crystal! There are similar conditions for the exit surface as illustrated in Fig. 8.18. An inward directed normal is shown, drawn to one of the active tie points. The extension of this normal to the asymptotes of the dispersion surface establishes the exit wave vectors $\mathbf{k}_0^e$ and $\mathbf{k}_{hkl}^e$. Thus for a single polarized incident ray, two forward diffracted beams and two hkl diffracted beams emerge from the far side of the crystal, one of each from the α branch and one of each from the β branch.

These exit incident and exit diffracted waves make slightly different angles with respect to each other. Thus this phenomenon is analogous to the splitting of light due to the difference in indices of refraction in some crystals. We return to this later.

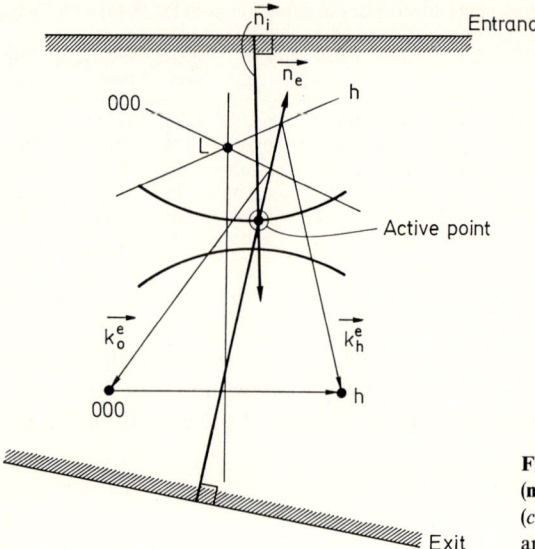

Fig. 8.18. Exit conditions. The exit normal ($\mathbf{n}_e$) drawn back through the active tie point (*circled*) to the positions 000 and h fixes k_h^e and k_{000}^e, the exit values of the wave vectors

8.5 Field Amplitudes

If we know ξ_0 for a tie point, both its real and imaginary parts, then from Eqs. (8.14), (8.30c), and (8.30d) we can obtain the wave vectors $\mathbf{K}_0$ and $\mathbf{K}_{hkl}$ and Eq. (8.35) gives us the ratios of field amplitudes. The imaginary components allow us to take into account absorption and hence, to evaluate the fields everywhere in the crystal. It would be useful then to relate the ξ's to the experimental conditions; in particular, to how far off the Bragg angle is the incident ray, and at what angle does it strike the surface. The real parts are most conveniently handled geometrically since we can draw them. Referring to Fig. 8.19, vectors from 000 or hkl to a point on the α branch of the dispersion surface A and to P are parallel to within a few seconds of arc. Treating them as parallel:

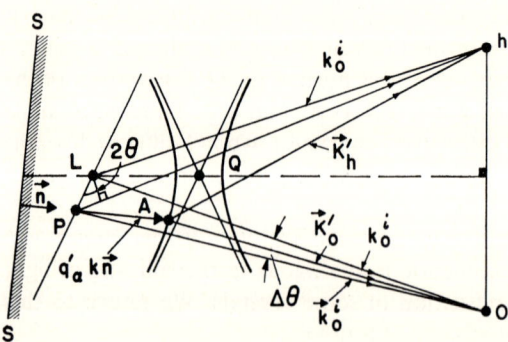

Fig. 8.19. Geometrical constructions used to derive expression for dispersion surface parameters ξ in terms of entrance surface boundary conditions*

8.5 Field Amplitudes

$$K'_0 = k_0^i - q'k\mathbf{n}\cdot\mathbf{S}_0 , \tag{8.44a}$$

$$K'_{hkl} = Ph - q'k\mathbf{n}\cdot\mathbf{S} . \tag{8.44b}$$

Let $\mathbf{n}\cdot\mathbf{S}_0$ be γ_0 and $\mathbf{n}\cdot\mathbf{S} = \gamma_{hkl}$. Also, from Fig. 8.19,

$$Ph = k_0^i + LP\sin 2\theta = k_0^i + (kLP/k)\sin 2\theta$$
$$= k - k|\Delta\theta|\sin 2\theta . \tag{8.45}$$

(In this figure $\Delta\theta = LP/k$ and is negative.) We now combine Eqs. (8.44), (8.45), and (8.32a):

$$\xi'_0 = K'_0 - k(1 - \tfrac{1}{2}\Gamma F'_0) = \tfrac{1}{2}k\Gamma F'_0 - q'k\gamma_0 .$$

Similarly,

$$\xi'_{hkl} = K'_{hkl} - k + \tfrac{1}{2}k\Gamma F'_0$$
$$= k - k|\Delta\theta|\sin 2\theta - k + \tfrac{1}{2}k\Gamma F'_0 - q'k\gamma_{hkl}$$
$$= \tfrac{1}{2}k\Gamma F'_0 - q'k\gamma_{hkl} - k|\Delta\theta|\sin 2\theta .$$

For the imaginary parts we simply combine Eqs. (8.32b), (8.41b), and (8.41c) with $\cos\beta = \gamma$ because the imaginary part of K is along the normal to the surface:

$$\xi''_0 = -K''_0\gamma_0 + \tfrac{1}{2}k\Gamma F''_0 = -q''kn\gamma_0 + \tfrac{1}{2}k\Gamma F''_0 ,$$

$$\xi''_{hkl} = -q''kn\gamma_{hkl} + \tfrac{1}{2}k\Gamma F''_0 .$$

Thus we can write in combined form that

$$\xi_0 = \tfrac{1}{2}k\Gamma F_0 - qk\gamma_0 , \tag{8.46a}$$

$$\xi_{hkl} = \tfrac{1}{2}k\Gamma F_0 - qk\gamma_{hkl} - k\Delta\theta\sin 2\theta . \tag{8.46b}$$

We now combine these last equations with Eq. (8.33). First we eliminate q by solving for it in Eq. (8.46a):

$$q = (\tfrac{1}{2}\Gamma F_0/\gamma_0) - (\xi_0/k\gamma_0) .$$

Then with Eqs. (8.33) and (8.46b):

$$\xi_0\{\tfrac{1}{2}k\Gamma F_0 - k\gamma_{hkl}[(\tfrac{1}{2}\Gamma F_0/\gamma_0) - (\xi_0/k\gamma_0)] - k|\Delta\theta|\sin 2\theta\} = \tfrac{1}{4}k^2 T^2\Gamma^2 F_{hkl}F_{\overline{hkl}} . \tag{8.47}$$

Multiplying both sides by γ_0/γ_{hkl}, and defining b as

$$b \equiv \gamma_0/\gamma_{hkl} , \tag{8.48a}$$

and

$$\delta \equiv \frac{b|\Delta\theta|\sin 2\theta + \tfrac{1}{2}\Gamma F_0(1-b)}{\Gamma|T||b|^{1/2}|F_{hkl}F_{\overline{khl}}|^{1/2}} , \tag{8.48b}$$

then

$$\xi_0 = \tfrac{1}{2}k|T||b|^{1/2}\Gamma(F_{hkl}F_{\overline{hkl}})^{1/2}\left[\delta \pm \left(\delta^2 + \frac{b}{|b|}\right)^{1/2}\right] \tag{8.49a}$$

and similarly

$$\xi_{hkl} = \tfrac{1}{2} k |T| \frac{\Gamma}{|b|^{1/2}} (F_{hkl} F_{\overline{hkl}})^{1/2} \left[\delta \pm \left(\delta^2 + \frac{b}{|b|} \right)^{1/2} \right]^{-1}. \tag{8.49b}$$

These equations appear a bit complicated but will simplify considerably for *symmetric* Laue or Bragg geometry as we shall see. For the former, $b = +1$, for the latter $b = -1$. (In fact b negative implies a Bragg reflection in general.) From Eqs. (8.49) we can obtain the real and imaginary parts of the ξ's from the incident conditions and hence, from Eqs. (8.32) and (8.14), the real and imaginary parts of the wave vectors can be obtained. Then from Eq. (8.35) the field ratios can be calculated.

The term δ can be considered as nothing more than a big constant times $\Delta\theta$. When $\Delta\theta$ is a few seconds, δ is a small number. Note in Eq. (8.48b) that for Bragg geometry, increasing θ means decreasing δ, as b is negative.

8.5.1 The Laue Case

We shall consider the Laue case first; then $b/|b| = 1$. It has been found that the following substitution is useful:

$$\delta = \tfrac{1}{2}(e^v - e^{-v}) = \sinh(v). \tag{8.50a}$$

Then the last terms in Eqs. (8.49) may be written as

$$\delta \pm (\delta^2 + 1)^{1/2} = \delta \pm \cosh(v) = \pm e^{\pm v}. \tag{8.50b}$$

Thus from Eq. (8.49a):

$$\xi_0 = \pm \tfrac{1}{2} k |b|^{1/2} |T| \Gamma |F_{hkl} F_{\overline{hkl}}|^{1/2} e^{\pm v}. \tag{8.51}$$

And from Eq. (8.35):

$$\frac{E_{hkl}}{E_0} = \mp \left[\frac{|T| |b|^{1/2}}{T} \right] \frac{|F_{hkl} F_{\overline{khl}}|^{1/2}}{F_{\overline{hkl}}} e^{\pm v}. \tag{8.52}$$

Since ξ_0 is positive for the α branch and there is a minus sign in front of Eq. (8.35), the $[-e^{+v}]$ case is for this branch, the $[+e^{-v}]$ for the β branch. Equation (8.52) indicates that E_{hkl} and E_0 are 180° out of phase.

For the symmetric Laue case, $b = +1$ at the angle of reflection, when P in any of the figures is at L; hence $\Delta\theta = 0$, $\delta = 0$, and $v = 0$. Then as we have seen,

$$\xi_0 = \xi_{hkl} = \pm \tfrac{1}{2} k |T| \Gamma |F_{hkl} F_{\overline{khl}}|^{1/2}$$

and

$$E_{hkl}/E_0 = \mp \{|T|/T\} |F_{hkl} F_{\overline{hkl}}|^{1/2} / F_{\overline{hkl}}. \tag{8.53}$$

Since ξ_0' is positive for the α branch, F times its complex conjugate is formed and the positive square root is employed in Eq. (8.35). From this equation we can see that if the crystal is centrosymmetric so that $F_{hkl} = F_{\overline{hkl}}$, then $E_{hkl} = E_0$.

8.5 Field Amplitudes

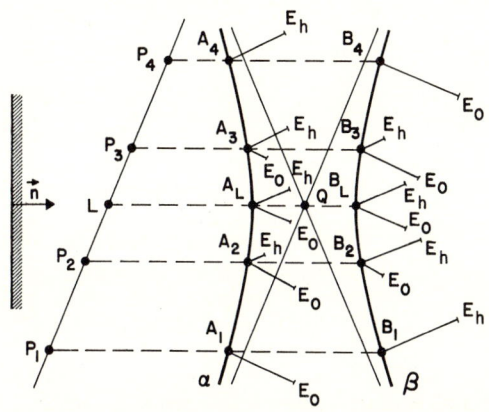

Fig. 8.20. Schematic representation of ratios of field strengths E_h and E_0 for different tie points, using the symmetric Laue entrance conditions*

Let us examine in the symmetric Laue case how the field ratio varies along the dispersion surface. Let $b = +1$ in Eq. (8.49a). Then,

$$\xi'_0 = \pm \tfrac{1}{2} k |\Delta\theta| \sin 2\theta \pm \tfrac{1}{2} [k^2 \Delta\theta^2 \sin^2 2\theta + k^2 T^2 \Gamma^2 F_{hkl} F_{\bar{h}\bar{k}\bar{l}}]^{1/2} \qquad (8.54)$$

For $\Delta\theta$ large, $\Delta\theta^2 \sin^2 2\theta \gg T^2 \Gamma^2 F_{hkl} F_{\bar{h}\bar{k}\bar{l}}$ and

$$\xi_0 \to \pm \tfrac{1}{2} k |\Delta\theta| \sin 2\theta \pm \tfrac{1}{2} k |\Delta\theta| \sin 2\theta$$

or

$$\xi'_0 \to 0, \quad \pm k |\Delta\theta| \sin 2\theta \qquad \text{(plus for the } \alpha \text{ branch)}$$

$$\xi''_0 \to 0.$$

Thus for the points A_1 and B_4 in Fig. 8.20, $\xi_0 \to 0$ and $E_{hkl}/E_0 \to 0$. There is no diffracted wave for these tie points and only a single wave E_0 propagates through the crystal, as illustrated in the figure. The wave vector for this wave is, from Eq. (8.32a):

$$K'_0 = \xi'_0 + k(1 - \tfrac{1}{2} \Gamma F'_0) = k(1 - \tfrac{1}{2} \Gamma F'_0).$$

This is just the vacuum value of the wave vector times the index of refraction, or the wave vector in the crystal. With Eq. (8.32b),

$$K''_0 \cos \beta = \tfrac{1}{2} k \Gamma F''_0 - \xi''_0 \to \tfrac{1}{2} k \Gamma F''_0.$$

The absorption of intensity is then

$$[\exp(-2\pi \mathbf{K}''_0 \cdot \mathbf{r})]^2 = \exp(-2\pi k \Gamma F''_0 \Delta z),$$

where Δz is the distance along $\mathbf{K}''_0$. Therefore,

$$\mu_t = \mu_0 = 2\pi k \Gamma F''_0.$$

We have already seen this result in Eq. (8.11c).

We now return to our discussion of Eq. (8.35). As shown in Fig. 8.20, *as we approach the center of the dispersion curve from points A_1 or B_4, $E_{hkl} \approx E_0$. Note that for $\Delta\theta$ large and negative, only tie points on the α branch are active, whereas*

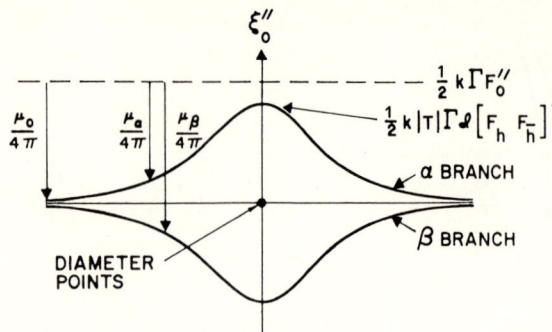

Fig. 8.21. Schematic representation of absorption associated with each tie point. The absorption depends on the difference between the *solid curves* and the *upper horizontal dashed line*. The symbol $\mathscr{I}$ means imaginary part. From Eq. (8.54) this term is the imaginary component ξ_0'' when $\Delta\theta = 0^*$

for $\Delta\theta$ large and positive, only the β branch is active. That is, for points B_1 and A_4, $\xi_{hkl} \to 0$ which implies an infinite $\mathbf{E}_{hkl}$ in Eq. (8.35), for any $\mathbf{E}_0$. Thus these points cannot be active and for them, $\mathbf{E}_0 = 0$.

Referring to Eq. (8.54) when $\Delta\theta = 0$, ξ_0 is proportional to $(\pm)$ square root of a complex constant $|F_{hkl}F_{\bar{h}\bar{k}\bar{l}}|$. As we move away from the diameter tie points, a rapidly increasing real part is added to this term before taking the square root. Thus ξ_0'' (the imaginary component), becomes less and less important, becoming essentially zero well off the Bragg peak as we have already seen (Fig. 8.21). The dotted horizontal line is the term $\frac{1}{2}k\Gamma F_0''$; as indicated in Eq. (8.32b), it is the difference between this line and the solid curves that is related to the absorption coefficient, through K_0''. There is always some absorption, so $\frac{1}{2}K\Gamma F_0''$ is always larger than ξ_0'', but the terms in Eq. (8.32b) are almost equal for the α branch at the diameter points, so that there is very low absorption for this branch at these locations. This accounts for Borrmann's observation of low absorption in a highly perfect single crystal. From the boundary conditions,

$$\mathbf{E}_0^i = \mathbf{E}_{0\alpha} + \mathbf{E}_{0\beta}, \tag{8.55a}$$

$$0 = \mathbf{E}_{hkl\alpha} + \mathbf{E}_{hkl\beta} = (E_{hkl\alpha}/E_{0\alpha})\mathbf{E}_{0\alpha} + (E_{hkl\beta}/E_{0\beta})\mathbf{E}_{0\beta}. \tag{8.55b}$$

Thus, considering the form of Eqs. (8.55a) and (8.52), we can write the equation for the dispersion surface as

$$E_{0\alpha} = E_0^i e^{-v}/2\cosh v, \qquad E_{0\beta} = E_0^i e^{+v}/2\cosh v. \tag{8.56}$$

Employing these relationships for the magnitudes:

$$\mathbf{E}_{0\alpha} = \mathbf{E}_0^i(e^{-v}/2\cosh v)\exp(+2\pi i v t)\exp(-2\pi i \mathbf{K}_{0\alpha}' \cdot \mathbf{r})\exp(-2\pi \mathbf{K}_{0\alpha}'' \cdot \mathbf{r}), \tag{8.57a}$$

$$\mathbf{E}_{0\beta} = \mathbf{E}_0^i(e^{+v}/2\cosh v)\exp(+2\pi i v t)\exp(-2\pi i \mathbf{K}_{0\beta}' \cdot \mathbf{r})\exp(-2\pi \mathbf{K}_{0\beta}'' \cdot \mathbf{r}). \tag{8.57b}$$

With Eqs. (8.52) and (8.55b),

$$\mathbf{E}_{hkl\alpha} = (-\mathbf{E}_0^i/2\cosh v)(|T|/T)|b|^{1/2}(|F_{hkl}F_{\bar{h}\bar{k}\bar{l}}|^{1/2}/F_{\bar{h}\bar{k}\bar{l}})$$
$$\times \exp(+2\pi i v t)\exp(-2\pi i \mathbf{K}_{hkl\alpha}' \cdot \mathbf{r})\exp(-2\pi \mathbf{K}_{hkl\alpha}'' \cdot \mathbf{r}), \tag{8.57c}$$

$$\mathbf{E}_{hkl\beta} = (+\mathbf{E}_0^i/2\cosh v)(|T|/T)|b|^{1/2}(|F_{hkl}F_{\bar{h}\bar{k}\bar{l}}|^{1/2}/F_{\bar{h}\bar{k}\bar{l}})$$
$$\times \exp(+2\pi i v t)\exp(-2\pi i \mathbf{K}_{hkl\beta}' \cdot \mathbf{r})\exp(-2\pi \mathbf{K}_{hkl\beta}'' \cdot \mathbf{r}). \tag{8.57d}$$

8.6 Poynting's Vector and Energy Flow in a Crystal

The wave vectors are related to the incident conditions through Eqs. (8.41), (8.46), (8.51), and (8.11c).

Figures 8.15, 8.18, 8.20, and 8.21 should be studied closely as they contain most of the important features of what we have learned so far. Summarizing:

(a) *Far from the Bragg angle*, a tie point close to the asymptote related to the 000 point is selected on one branch and *only* one wave is present (for each polarization state). *This wave travels substantially in the direction of the incident beam, and suffers normal photoelectric absorption.*

(b) For a field whose tie point is a diameter point on the α branch or *near the diameter, two waves are excited*, one in the "transmitted" direction and one in the "diffracted" direction. These suffer *less than normal absorption* and have essentially *equal amplitude and wavelength. This is also true of the β branch, except that this branch suffers more than normal absorption* at the diameter point.

We now turn to an examination of the energy flow associated with these beams.

8.6 Poynting's Vector and Energy Flow in a Crystal

8.6.1 What is the Poynting's Vector

The instantaneous Poynting vector of an electromagnetic field is given by $\mathbf{L} = \mathbf{E} \times \mathbf{H}$, and it represents the electric and magnetic energy flowing across a unit area perpendicular to a beam's direction $\mathbf{S}$ in unit time.

To see that Poynting's vector does really represent energy flow, we let u represent the energy density in the field and let $\mathbf{L}$ represent the energy flux (across an area perpendicular to the flow). The force on a charge moving with velocity $\mathbf{v}$ due to the field is $\mathbf{F} = q(\mathbf{E} + \mathbf{v} \times \mathbf{B})$ and the rate of work done is $\mathbf{F} \cdot \mathbf{v} = q\mathbf{E} \cdot \mathbf{v}$. If there are N charges, the total rate of work is $Nq\mathbf{E} \cdot \mathbf{v} = \mathbf{E} \cdot \mathbf{j}$ where $\mathbf{j}$ is the current. Therefore, with V the volume, A the area around the material, and $\mathbf{n}$ the unit normal to the surface, conservation of energy requires

$$-\frac{\partial}{\partial t}\int_V u\, dV = \int_A \mathbf{L} \cdot \mathbf{n}\, dA + \int_V \mathbf{E} \cdot \mathbf{j}\, dV.$$

Employing Gauss's theorem to rewrite the first integral on the right hand side,

$$-\frac{\partial}{\partial t}\int_V u\, dV = \int \mathbf{\nabla} \cdot \mathbf{L}\, dV + \int \mathbf{E} \cdot \mathbf{j}\, dV$$

or

$$-\partial u/\partial t = \mathbf{\nabla} \cdot \mathbf{L} + \mathbf{E} \cdot \mathbf{j}, \qquad (8.58a)$$

or

$$\mathbf{E} \cdot \mathbf{j} = -(\partial u/\partial t) - \mathbf{\nabla} \cdot \mathbf{L}.$$

From Maxwell's equations,

$$\mathbf{j} = \varepsilon_0 c^2 (\nabla \times \mathbf{B}) - \varepsilon_0 \partial \mathbf{E}/\partial t \, .$$

Thus,

$$\mathbf{E} \cdot \mathbf{j} = \varepsilon_0 c^2 \mathbf{E} \cdot (\nabla \times \mathbf{B}) - \varepsilon_0 \mathbf{E} \cdot \partial \mathbf{E}/\partial t \, .$$

Now,

$$\mathbf{E} \cdot \nabla \times \mathbf{B} = \nabla \cdot (\mathbf{B} \times \mathbf{E}) + \mathbf{B} \cdot (\nabla \times \mathbf{E}) \, .$$

Thus,

$$\mathbf{E} \cdot \mathbf{j} = \varepsilon_0 c^2 \nabla \cdot (\mathbf{B} \times \mathbf{E}) + \varepsilon_0 c^2 \mathbf{B} \cdot (\nabla \times \mathbf{E}) - (\partial/\partial t)(\tfrac{1}{2}\varepsilon_0 \mathbf{E} \cdot \mathbf{E}) \, .$$

Also from Maxwell's equations [(Eq. (8.12a)],

$$\mathbf{B} \cdot (\nabla \times \mathbf{E}) = \mathbf{B} \cdot (-\partial \mathbf{B}/\partial t) = -\partial/\partial t (\mathbf{B} \cdot \mathbf{B}/2) \, ;$$

hence,

$$\mathbf{E} \cdot \mathbf{j} = \nabla \cdot (\varepsilon_0 c^2 \mathbf{B} \times \mathbf{E}) - (\partial/\partial t)[(\varepsilon_0 c^2/2)\mathbf{B} \cdot \mathbf{B} + (\varepsilon_0/2)\mathbf{E} \cdot \mathbf{E}] \, . \tag{8.58b}$$

Comparing this to Eq. (8.58a), since the energy density u is $\varepsilon_0/2[c^2 \mathbf{B} \cdot \mathbf{B} + \mathbf{E} \cdot \mathbf{E}]$ we have,[6]

$$\mathbf{L} = +\varepsilon_0 c^2 \mathbf{E} \times \mathbf{B} = +\varepsilon_0 \mu_0 c^2 \mathbf{E} \times \mathbf{H} = +\mathbf{E} \times \mathbf{H} \, .$$

Now it can be shown that the time average of $\mathbf{L}$ is given by[7]

$$\langle \mathbf{L} \rangle = \tfrac{1}{2} \mathrm{Re}[\mathbf{E} \times \mathbf{H}^*] \, , \tag{8.59}$$

(where Re means the real part). To prove this, let $\mathbf{E} = \mathbf{E}_0 e^{i\omega t}$, and similarly for $\mathbf{H}$. Then let both $\mathbf{E}_0$ and $\mathbf{H}_0$ be complex. The real part of $\mathbf{E}_0 e^{i\omega t}$ is what is involved in any energy flow:

$$\mathrm{Re}(\mathbf{E}_0 e^{i\omega t}) = \mathrm{Re}[(\mathbf{E}_r + i\mathbf{E}_i)(\cos \omega t + i \sin \omega t)] = \mathbf{E}_r \cos \omega t - \mathbf{E}_i \sin \omega t \, .$$

Now,

$$\mathrm{Re}\mathbf{E} \times \mathrm{Re}\mathbf{H} = (\mathbf{E}_r \times \mathbf{H}_r)\cos^2 \omega t + (\mathbf{E}_i \times \mathbf{H}_i)\sin^2 \omega t$$
$$- [(\mathbf{E}_r \times \mathbf{H}_i) + (\mathbf{E}_i \times \mathbf{H}_r)]\sin \omega t \cos \omega t \, .$$

The first two terms have a time average of $\tfrac{1}{2}$ and the last term vanishes when averaged over time. Thus with only the real parts,

$$\langle \mathbf{L} \rangle \equiv \langle \mathrm{Re}\mathbf{E} \times \mathrm{Re}\mathbf{H} \rangle = \tfrac{1}{2} \langle \mathbf{E}_r \times \mathbf{H}_r + \mathbf{E}_i \times \mathbf{H}_i \rangle \, . \tag{8.60a}$$

If we now form $\mathbf{E} \times \mathbf{H}^*$,

$$\mathbf{E} \times \mathbf{H}^* = (\mathbf{E}_0 e^{i\omega t})(\mathbf{H}_0^* e^{-i\omega t}) = \mathbf{E}_0 \times \mathbf{H}_0^* = (\mathbf{E}_r + i\mathbf{E}_i) \times (\mathbf{H}_r - i\mathbf{H}_i)$$
$$= (\mathbf{E}_r \times \mathbf{H}_r + \mathbf{E}_i \times \mathbf{H}_i) + i(\mathbf{E}_i \times \mathbf{H}_r - \mathbf{E}_r \times \mathbf{H}_i) \, . \tag{8.60b}$$

---

[6] In this section, we use $\mathbf{B} = \mu_0 \mathbf{H}$, where μ_0 is the permeability of empty space. This use of the symbol μ_0 should not be confused with the use of μ_0 for the linear absorption coefficient elsewhere in Chap. 8.
[7] The asterisk implies the complex conjugate for fields. (But keep in mind that $\mathbf{r}^*$ is a vector in reciprocal space, not a complex conjugate!)

8.6 Poynting's Vector and Energy Flow in a Crystal

Except for the factor $\frac{1}{2}$, the real parts are the same as Eq. (8.60a). Thus Eq. (8.59) is proved.

8.6.2 The Use of Poynting's Vector in Dynamical Diffraction

For both **E** and **H**, we can employ sums of terms like Eq. (8.57) for all the points selected by the incident boundary conditions. Let w describe the particular dispersion surface, i.e., α, β, σ, π states, etc. Then, expanding **E** and **H** as in Eq. (8.16), and substituting for $\mathbf{K}_{hklw}$ and $\mathbf{K}_{hklw'}$ as in Eq. (8.38),

$$\mathbf{E} \times \mathbf{H}^* = \sum_w \sum_{w'} \exp[-2\pi i(\mathbf{K}'_{0w} - \mathbf{K}'_{0w'}) \cdot \mathbf{r}] \exp[-2\pi(\mathbf{K}''_{0w} + \mathbf{K}''_{0w'}) \cdot \mathbf{r}]$$
$$\times \sum_{hkl} \sum_{h'k'l'} (\mathbf{E}_{hklw} \times \mathbf{H}^*_{h'k'l'w'}) \exp[-2\pi i(\mathbf{r}^*_{hkl} - \mathbf{r}^*_{h'k'l'}) \cdot \mathbf{r}] \,. \quad (8.61)$$

We now average this over a unit cell, to obtain a more revealing form. Assume that the second exponent, the attenuation, does not vary appreciably across such a unit cell. Also the term $\mathbf{K}'_{0w} - \mathbf{K}'_{0w'}$, approximately the diameter of the dispersion surface, is a very small quantity and if we are integrating over a unit cell we can ignore its variation. The integration of the last exponential term over a unit cell is zero unless $hkl = h'k'l'$. Thus over a unit cell,

$$\langle\!\langle \mathbf{L} \rangle\!\rangle = \tfrac{1}{2}\mathrm{Re}\left\{\sum_w \sum_{w'} \exp[-2\pi i(\mathbf{K}'_{0w} - \mathbf{K}'_{0w'}) \cdot \mathbf{r}] \exp[-2\pi(\mathbf{K}''_{0w} + \mathbf{K}''_{0w'}) \cdot \mathbf{r}] \right.$$
$$\left. \times \sum_{hkl} (\mathbf{E}_{hklw} \times \mathbf{H}^*_{hklw'})\right\}. \quad (8.62)$$

Equation (8.18a) may be written as

$$\mathbf{H}^*_{hklw'} = \mathbf{K}_{hklw'} \times \mathbf{E}^*_{hklw'}/\nu\mu_0 \,.$$

Therefore,

$$\mathbf{E}_{hklw} \times \mathbf{H}^*_{hklw'} = \mathbf{E}_{hklw} \times \mathbf{K}_{hklw'} \times \mathbf{E}^*_{hklw'}/\nu\mu_0 \,.$$

Neglecting $\mathbf{K}''$ and the correction for the index of refraction, $\mathbf{K}_{hklw} \approx |k|\mathbf{S}_{hkl}$, where $\mathbf{S}_{hkl}$ is a unit vector, then

$$\mathbf{E}_{hklw} \times \mathbf{H}^*_{hklw'} = (\mathbf{E}_{hklw} \times \mathbf{S}_{hkl} \times \mathbf{E}^*_{hklw'})k/\nu\mu_0 \,.$$

Now $k/\nu = 1/c = (\varepsilon_0\mu_0)^{1/2}$, so

$$\mathbf{E}_{hklw} \times \mathbf{H}^*_{hklw'} = (\varepsilon_0/\mu_0)^{1/2}(\mathbf{E}_{hklw})(\mathbf{E}^*_{hklw'})\mathbf{S}_{hkl} \,, \quad (8.63)$$

where the fact that $\mathbf{E}_w$ and $\mathbf{E}^*_w$ are perpendicular has been employed. (This is so, because $\mathbf{E}_w$ and $\mathbf{E}^*_w$ are 90° out of phase as can be seen by sketching **E**, **H**, **K** for both states.)

With Eqs. (8.62) and (8.63) for the case of two active reciprocal lattice points (hkl and 000) and two tie points (one polarization state):

$$(\mu_0/\varepsilon_0)^{1/2} \langle\!\langle \mathbf{L} \rangle\!\rangle = \tfrac{1}{2}\exp(-4\pi \mathbf{K}''_{0\alpha}\cdot\mathbf{r})(E_{0\alpha}^2 S_0 + E_{hkl\alpha}^2 S_{hkl})$$
$$+ \tfrac{1}{2}\exp(-4\pi \mathbf{K}''_{0\beta}\cdot\mathbf{r})(E_{0\beta}^2 S_0 + E_{hkl\beta}^2 S_{hkl})$$
$$+ \exp(-2\pi(\mathbf{K}''_{0\alpha} + \mathbf{K}''_{0\beta})\cdot\mathbf{r})\{E_{0\alpha}E_{0\beta}S_0 + E_{hkl\alpha}E_{hkl\beta}S_{hkl}\}$$
$$\times \cos 2\pi[(\mathbf{K}'_{0\alpha} - \mathbf{K}'_{0\beta})\cdot\mathbf{r}]. \tag{8.64}$$

We can write this as

$$(\mu_0/\varepsilon_0)^{1/2} \langle\!\langle \mathbf{L} \rangle\!\rangle = \mathbf{S}_\alpha + \mathbf{S}_\beta + \mathbf{S}_{\alpha\beta}. \tag{8.65}$$

There are three energy flows, one for each of the two branches and one representing coupling. Each of these flows has its own absorption term.

We shall be mainly concerned with the directions of these flows, so, for simplicity, assume there is no absorption. Thus $K''_{0\alpha} = K''_{0\beta} = 0$. In this case the energy flows $\mathbf{S}_\alpha$ and $\mathbf{S}_\beta$ are independent of depth below the surface, while $\mathbf{S}_{\alpha\beta}$ has a sinusoidal dependence on this depth. As can be seen in Fig. 8.15, the boundary conditions demand that $(\mathbf{K}'_{0\alpha} - \mathbf{K}'_{0\beta})$, and hence, $\mathbf{S}_{\alpha\beta}$ is constant in planes parallel to the crystal surface. This term varies sinusoidally with depth, with wavelength Λ:

$$\Lambda = 1/|\mathbf{K}'_{0\alpha} - \mathbf{K}'_{0\beta}|. \tag{8.66}$$

For the symmetric Laue case at the center of the reflection, this is the inverse of the distance between branches, w in Eq. (8.34). From this equation, this Λ is the order of 10^{-5} m for x-rays and 10^{-8} m for electrons. The energy flow shifts around the average direction $\mathbf{S}_\alpha + \mathbf{S}_\beta$ periodically with depth. This pheonomenon is called "Pendellösung" a term coined by Ewald because of the similarity between this energy flow and the energy transfer between weakly coupled pendulums. Λ is known as the "extinction distance" in electron microscopy and since the scattering factors for electrons are much larger than for x-rays, this distance is small for electrons.

This coupling can be made clearer by substituting Eq. (8.57) into Eq. (8.64), letting $a = 1/(2\cosh v)$ and $F_{hkl} \cong F_{\bar{h}\bar{k}\bar{l}}$. Then, still ignoring absorption:

$$\langle\!\langle \mathbf{L} \rangle\!\rangle = \tfrac{1}{2}a^2|E_0^i|^2[e^{-2v}\mathbf{S}_0 + b\mathbf{S}_{hkl}] + \tfrac{1}{2}a^2|E_0^i|^2[e^{+2v}\mathbf{S}_0 + b\mathbf{S}_{hkl}]$$
$$+ a^2|E_0^i|^2(\mathbf{S}_0 - b\mathbf{S}_{hkl})\cos 2\pi Z/\Lambda, \tag{8.67}$$

where Z is the depth below the surface. At $\cos 2\pi Z/\Lambda = 1$, the component of $\mathbf{S}_{\alpha\beta}$ in the direction of $\mathbf{S}_{hkl}$ cancels the corresponding contributions from $\mathbf{S}_\alpha$ and $\mathbf{S}_\beta$ and *all the energy flows in the direction $\mathbf{S}_0$. When $\cos 2\pi Z/\Lambda = -1$, the beam is strong in the direction of $\mathbf{S}_{hkl}$*, while in the direction $\mathbf{S}_0$ it is proportional to $\sinh^2 v/\cosh^2 v$, which is zero at the center of a reflection. *The energy swaps back and forth as a function of depth* extinguishing completely in the $\mathbf{S}_{hkl}$ direction, and very nearly so in the $\mathbf{S}_0$ direction with period Λ. Note that *this swapping requires the excitation of the points on both α and β branches*. Because $b \equiv \gamma_0/\gamma_{hkl}$, Eq. (8.67) tells us that the direction of the $\mathbf{S}_{\alpha\beta}$ energy flow is $\gamma_{hkl}\mathbf{S}_0 - \gamma_0\mathbf{S}_{hkl}$. Considering Fig. 8.22, it is clear that the component of this flow along $\mathbf{n}$ is zero, as the component of $\gamma_{hkl}\mathbf{S}_0$ along $\mathbf{n}$ is $(\gamma_{hkl}\mathbf{S}_0\gamma_0)$ and the component of $\gamma_0\mathbf{S}_{hkl}$ along $\mathbf{n}$ is $(\gamma_0\mathbf{S}_{hkl}\gamma_{hkl})$. Thus, *$\mathbf{S}_{\alpha\beta}$ is always parallel to the surface and does not contribute to flow through* the crystal.

Consider Eq. (8.67) for the center of a diffraction peak in symmetric transmission so that $v = 0, b = 1$. The total energy in the diffracted beam is $a^2(\mathbf{E}_0^i)^2\{\mathbf{S}_0 + \mathbf{S}_{hkl} +$

8.6 Poynting's Vector and Energy Flow in a Crystal

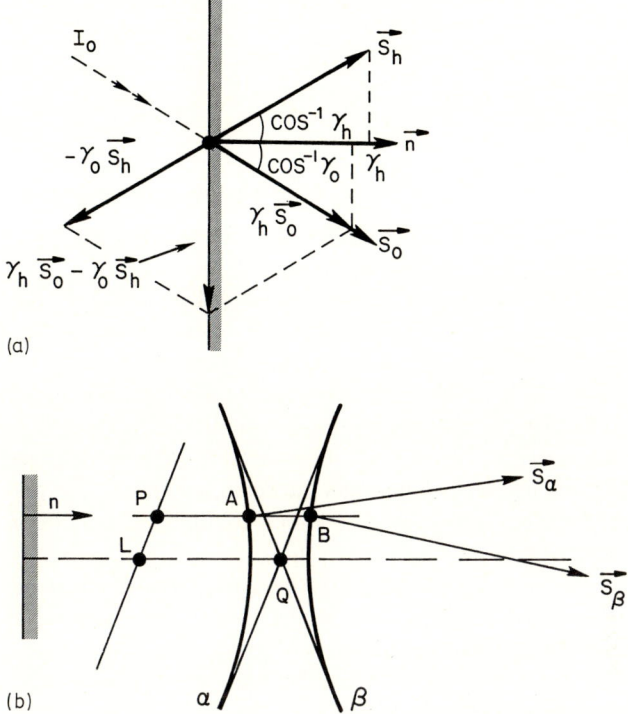

Fig. 8.22. a Vector diagram used in discussing Pendellösung effect. **b** Averaged energy flow associated with the points A and B. Poynting's vectors $\mathbf{S}_\alpha$ and $\mathbf{S}_\beta$ for the flow at each tie point are perpendicular to the real part of the dispersion curve at that tie point. At the exit surface $\mathbf{S}_\alpha$ (and $\mathbf{S}_\beta$), splits into incident and diffracted beams. Thus with both states of polarization there will be eight exit beams (four for the symmetric Laue case)*

$(\mathbf{S}_0 - \mathbf{S}_{hkl})\cos 2\pi Z/\Lambda\}$. Therefore, the magnitude of the diffracted beam is proportional to $(1 + \cos 2\pi Z/\Lambda)$ if $\mathbf{S}_0 - \mathbf{S}_{hkl}$ and $\mathbf{S}_0 + \mathbf{S}_{hkl}$ are of the same magnitude. As $(1 + \cos 2\pi Z/\Lambda) = \sin^2 \pi Z/\Lambda$, for small arguments the diffracted intensity is proportional to $(\pi Z/\Lambda)^2$. From Eqs. (8.36) and (8.66), $1/\Lambda^2$ is proportional to F^2. Thus, for crystals or mosaic regions (such that $Z \cong 10^{-6}$–10^{-7} m for x-rays), the kinematic theory applies!

The net direction of energy flow can be found by averaging Eq. (8.64) or (8.65) over an extinction distance. Then, neglecting absorption.

$$\langle\!\langle\!\langle \mathbf{L} \rangle\!\rangle\!\rangle = \mathbf{S}_\alpha + \mathbf{S}_\beta = (E_{0\alpha}^2 \mathbf{S}_0 + E_{hkl\alpha}^2 \mathbf{S}_{hkl}) + (E_{0\beta}^2 \mathbf{S}_0 + E_{hkl\beta}^2 \mathbf{S}_{hkl}) . \tag{8.68}$$

This represents a complete decoupling of the four plane waves for the one polarization state. That is, $E_{0\alpha}$ and $E_{hkl\beta}$ interact over a unit cell and this interaction has been taken into account in obtaining Eq. (8.64). The *α and β wave fields interact over an extinction distance* and this has been taken into account in deriving Eq. (8.68). The averaged Poynting vector is then just the sum of the Poynting vectors for each of the four plane waves.

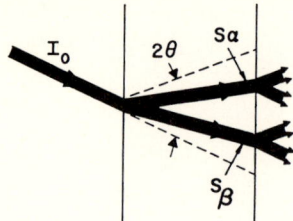

Fig. 8.23. Schematic representation of the separation of the energy flows in the crystal for a narrow beam off the Bragg angle*

Kato (1958) has shown that *the direction of energy flow $\langle\!\langle\!\langle \mathbf{L} \rangle\!\rangle\!\rangle$ corresponding to a tie point on a particular sheet of the dispersion surface is normal to the surface at that point.* This is illustrated in Fig. 8.22b. Note that the energy flows *along* the atomic planes *for the tie points at the diameter*. It is because of this that calculation of images in transmission electron microscopy by simply summing scattering over columns of cells is successful (see Hirsch et al. 1965). Because of the scattering angle, one might at first think that in passing through a foil the beam would "see" several such columns, but the energy does not spread to other columns. This column approximation was assumed in Sect. 7.3.

For any tie point not on the diameter there are two flows, one directed above the planes and one below. In an actual experiment with x-rays, the incident beam has a cross section which is often smaller than the thickness of the crystal. For a crystal slightly off the Bragg angle, *a decoupling will take place* between the two wave fields because of their physical separation, as shown in Fig. 8.22b and 8.23. The *interaction occurs only near the entrance surface*; outside of that region the energy flows independently along paths determined by the Poynting vectors associated with the tie points. The direction of, say, $\mathbf{S}_\alpha$ depends on $E_{0\alpha}$ and $E_{hkl\alpha}$ which in turn depends through Eq. (8.35) on the deviation from Bragg conditions. [From Eq. (8.64) it can be seen that the direction of $\mathbf{S}_\alpha$ depends only upon the ratios of intensities in primary and diffracted beams.]

8.6.3 The Case of Spherical Waves

So far we have assumed that the waves are planar. Kato (1960) has considered this assumption. The waves are really wave packets and as such made up of a small range, ΔK. Thus, a small region of the dispersion surface is excited. To apply plane wave theory as we have done, the angular width of a reflection, $\Delta\theta$, must far exceed the range of angles associated with the wave packet or incident range of angles, $\Delta\Omega$. In electron diffraction $\Delta\theta \approx 10^{-2}$ rad and the condenser lens can reduce $\Delta\Omega$ to 10^{-4} rad so that plane wave theory is adequate for first considerations. This is not the case for x-rays as we can readily see. Suppose we have a target a distance L from a specimen and a slit of width B is placed near the specimen to define the angular resolution. Then $\Delta\Omega \approx B/L$. But if the slit is narrow enough, diffraction phenomena set in and the width $\Delta\Omega \approx \lambda/B$. These two values are identical when $\Delta\Omega \approx (\lambda/L)^{1/2}$. For reasonable values, $\lambda = 1$ Å, $L = 0.5$ m, $\Delta\Omega \approx 10^{-5}$ rad. However, $\Delta\theta \approx 10^{-6}$ rad for x-rays! For plane wave theory to apply, distances of 100 m or more are

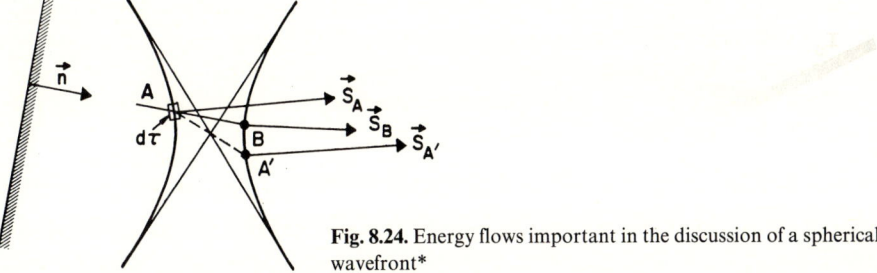

Fig. 8.24. Energy flows important in the discussion of a spherical wavefront*

needed (the length of a football field) and this is hardly possible without an unacceptable loss of intensity! The result of this for x-rays in Laue geometry[8] at least is that different tie points are not independently excited as would be the case for plane waves. In a typical experiment *the entire dispersion surface is illuminated*! There is a range of incident angles sufficient so that S_α (or S_β) covers the entire dispersion surface and hence, the entire range of 2θ. This is called the "fan" of beams. *Interference* is possible *between any pair of points where fields superpose*. Kato showed that the incident beam could be represented by a superposition of plane waves each of which produces a set of waves in the crystal and which can be treated as we have done. These plane waves superimpose to create wave packets in the crystal. The wave packet with vectors terminating in the region $d\tau$ in Fig. 8.24 separates physically and each component propagates independently. For this reason the Pendellösung phenomenon between two points, say A and B in Fig. 8.24, does not occur appreciably in a thick crystal; however, the normals to points A and A' are parallel and interference from waves associated with these points will cause oscillations. The fields from these two points *do not* separate in traveling through the crystal. This results in the well-known thickness changes seen near the wedge-shaped edges of a foil in transmission electron microscopy, Fig. 7.3 (this was explained qualitatively with kinematic theory in Chap. 7).

8.7 Absorption in More Detail

8.7.1 Along the Normal to the Surface

We have just seen that the direction of energy flow depends only on the ratio of field amplitudes which, according to Eq. (8.35), depends only on the tie point. We have related absorption (the K'' values) to incident conditions through Eqs. (8.41), (8.46), (8.51), (8.11c). We would expect, though, that since the flow depends on the tie point, its absorption should not depend on how this point was selected. We now

---

[8] Bragg geometry is discussed in Sect. 8.10.

show that the imaginary parts of ξ can be written in terms of the real parts and also that we can express the ratios of the fields [Eq. (8.35)] in similar terms. This will prove that the absorption is indeed fixed solely by the tie point.

The expression for $\xi_0 \xi_{hkl}$ is the fundamental equation for the dispersion surface, Eq. (8.35). Writing this in terms of real and imaginary parts:

$$(\xi_0' + i\xi_0'')(\xi_{hkl}' + i\xi_{hkl}'') = C^2[\text{Re}(hkl) + i\,\text{Im}(hkl)], \tag{8.69}$$

where $C = (1/2)Tk\Gamma$ and Re and Im are the real and imaginary parts of $(F_{hkl}F_{\overline{hkl}})$. Then

$$\xi_0'\xi_{hkl}' - \xi_0''\xi_{hkl}'' = C^2\,\text{Re}(hkl), \tag{8.70a}$$

$$\xi_0'\xi_{hkl}'' + \xi_0''\xi_{hkl}' = C^2\,\text{Im}(hkl), \tag{8.70b}$$

It can be shown from Eqs. (8.32) that $\xi_0''\xi_{hkl}''$ is small compared to $\xi_0'\xi_{hkl}'$ and hence, the former will be neglected. From Eqs. (8.41b) and (8.41c), $\mathbf{K}_0'' = \mathbf{K}_{hkl}''$. Inserting the definitions of ξ_0'', ξ_{hkl}'' from Eqs. (8.32a) and (8.32b) into Eqs. (8.70) results in

$$\xi_0'\xi_{khl}' = C^2\,\text{Re}(hkl) \tag{8.71a}$$

$$-K_0''(\xi_0'\gamma_{hkl} + \xi_{hkl}'\gamma_0) = C^2\,\text{Im}(hkl) - (\xi_0' + \xi_{hkl}')\tfrac{1}{2}k\Gamma F_0''. \tag{8.71b}$$

Thus, we have expressed K_0'' in terms of ξ_0' and structure factors, but our expression still contains the γ's. In Eq. (8.35) we multiply each right-hand term by its complex conjugate and multiply the two together to obtain

$$|E_{hkl}/E_0|^4 = \xi_0 \xi_0^* F_{hkl} F_{hkl}^* / \xi_{hkl} \xi_{hkl}^* F_{\overline{hkl}} F_{\overline{hkl}}^*. \tag{8.72}$$

The *ratio of the structure factors generally will only differ from unity by a small amount for a noncentrosymmetric crystal*, so we shall assume this value. Also, $\xi_0'' < \xi_0'$. Then,

$$|E_{hkl}/E_0|^2 \cong \xi_0'/\xi_{hkl}'. \tag{8.73}$$

We now combine Eqs. (8.71) and (8.73) and let the attenuation coefficient along the normal to the surface be $\mu_n = 4\pi K_0''$ and $\mu_0 = 2\pi k\Gamma F_0''$. From Eq. (8.71b),[9]

$$\mu_n = 4\pi K_0'' = \frac{1}{(\xi_0'\gamma_{hkl} + \xi_{hkl}'\gamma_0)}[(\xi_0' + \xi_{hkl}')\mu_0 - 4\pi C^2\,\text{Im}(hkl)].$$

Factoring ξ_{hkl}',

$$\mu_n = \frac{1}{\xi_{hkl}'[(\xi_0'/\xi_{hkl}')\gamma_{hkl} + \gamma_0]}\left[\xi_{hkl}'\left(\frac{\xi_0'}{\xi_{hkl}'} + 1\right)\mu_0 - 4\pi C^2\,\text{Im}(hkl)\right]$$

$$= \frac{1}{[(\xi_0'/\xi_{hkl}')\gamma_{hkl} + \gamma_0]}\left[\left(\frac{\xi_0'}{\xi_{hkl}'} + 1\right)\mu_0 - \frac{4\pi C^2\,\text{Im}(hkl)}{\xi_{hkl}'}\right].$$

Substituting Eq. (8.73) for ξ_0'/ξ_{hkl}' and factoring $(E_{hkl}^2 + E_0^2)/E_0^2$ and μ_0:

---

[9] Recall that planes of constant absorption are parallel to the surface; that is $\mathbf{K}_0''$ is perpendicular to the surface.

8.7 Absorption in More Detail

$$\mu_n = \mu_0 \frac{E_{hkl}^2 + E_0^2}{\gamma_{hkl} E_{hkl}^2 + \gamma_0 E_0^2} \left[1 - \frac{2C^2 \operatorname{Im}(hkl) E_0^2}{k \Gamma F_0''(E_{hkl}^2 + E_0^2) \zeta_{hkl}'} \right]$$

$$= A \left[1 - \frac{\frac{1}{2} T^2 k \Gamma \operatorname{Im}(hkl)}{F_0''} \frac{E_0^2}{(E_{hkl}^2 + E_0^2) \zeta_{hkl}'} \right]$$

$$= A \left[1 - \frac{TC \operatorname{Im}(hkl)}{F_0''} B \right].$$

With Eq. (8.71a) and substituting for C,

$$\mu_n = A \left[1 \mp \frac{T \operatorname{Im}(hkl)}{F_0'' \operatorname{Re}(hkl)^{1/2}} \frac{|E_{hkl}||E_0|}{(E_{hkl}^2 + E_0^2)} \right]. \qquad (8.74)$$

8.7.2 Along Poynting's Vector

The absorption coefficient in the direction of **L**, the energy flow, is less than the value normal to the surface, as absorption depends only on distance along this normal. Therefore in the direction of $\langle\!\langle\!\langle \mathbf{L} \rangle\!\rangle\!\rangle$:

$$\mu_L = \mu_n \cos(\mathbf{n}, \mathbf{L}). \qquad (8.75)$$

From Eq. (8.68), for one tie point $\langle\!\langle\!\langle \mathbf{L} \rangle\!\rangle\!\rangle = E_0^2 \mathbf{S}_0 + E_{hkl}^2 \mathbf{S}_{hkl}$ and

$$|\mathbf{L}| \cos(\mathbf{n}, \mathbf{L}) = \langle\!\langle\!\langle \mathbf{L} \rangle\!\rangle\!\rangle \cdot \mathbf{n} = E_0^2 \gamma_0 + E_{hkl}^2 \gamma_{hkl}. \qquad (8.76)$$

From Fig. 8.25 with **u** a unit vector in the diffracting planes,

$$\mathbf{L} \cdot \mathbf{u} = S \cos \Delta = (E_0^2 + E_{hkl}^2) \cos \theta. \qquad (8.77)$$

Combining Eqs. (8.76) and (8.77):

$$\cos(\mathbf{n}, \mathbf{L}) = [(\gamma_0 E_0^2 + \gamma_{hkl} E_{hkl}^2)/(E_0^2 + E_{hkl}^2)](\cos \Delta / \cos \theta). \qquad (8.78)$$

Substituting this into Eq. (8.75) and combining with Eq. (8.74):

$$\mu_L = \mu_0 \frac{\cos \Delta}{\cos \theta} \left[1 \mp \frac{T \operatorname{Im}(hkl) |E_0||E_{hkl}|}{F_0''(\operatorname{Re}(hkl))^{1/2} (E_0^2 + E_{hkl}^2)} \right]. \qquad (8.79)$$

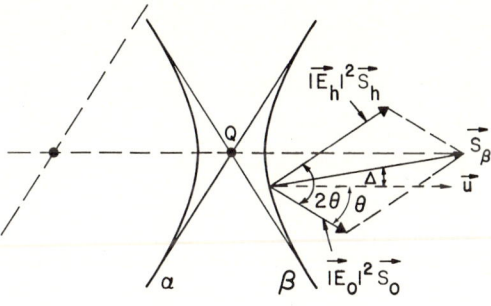

Fig. 8.25. Geometry used in derivation of energy flow as a function of angle, Δ, between flow direction and atomic planes. The Poynting's vector **S** is the vector sum of the individual flows*

Let

$$p = \tan \Delta / \tan \theta . \tag{8.80}$$

The value of this p is zero for the diameter points (as $\Delta = 0$). It becomes plus or minus unity in the wings of the dispersion surface as there Δ approaches θ (see Fig. 8.25). From these facts

$$p = \frac{|E_{hkl}|^2 - |E_0|^2}{|E_{hkl}|^2 + |E_0|^2} . \tag{8.81}$$

Substituting Eq. (8.81) into Eq. (8.79):

$$\mu_L = \mu_0 \frac{\cos \Delta}{\cos \theta} \left[1 \mp \frac{T \text{Im}(hkl)}{2 F_0''(\text{Re}(hkl))^{1/2}} (1 - p^2)^{1/2} \right] . \tag{8.82}$$

The *absorption coefficient* is indeed independent of incident conditions and depends only on Δ, the angle between ⟪⟪L⟫⟫ the direction of energy flow, and the crystal planes. This in turn is *solely a property of the dispersion surface which depends only on the internal structure through the structure factor.*

We can make Eq. (8.82) somewhat simpler as follows. First we define a new quantity, ε:

$$\varepsilon = \text{Im}(hkl)/2F_0''(\text{Re}(hkl))^{1/2} . \tag{8.83}$$

Secondly,[10]

$$F_{hkl} = F'_{hkl} + iF''_{hkl} = (F'_{hkl})_r + i(F'_{hkl})_i + i[(F''_{hkl})_r + i(F''_{hkl})_i] ,$$

$$F_{\bar{h}\bar{k}\bar{l}} = F'_{\bar{h}\bar{k}\bar{l}} + iF''_{\bar{h}\bar{k}\bar{l}} = (F'_{hkl})_r - i(F'_{hkl})_i + i[(F''_{hkl})_r - i(F''_{hkl})_i] .$$

Therefore,

$$F_{hkl} F_{\bar{h}\bar{k}\bar{l}} = \text{Re}(hkl) + i\text{Im}(hkl)$$

$$= (F'_{hkl})_r^2 - (F''_{hkl})_i^2 + 2i[(F'_{hkl})_r(F''_{hkl})_r + (F'_{hkl})_i(F''_{hkl})_i] . \tag{8.84a}$$

If the crystal structure is centrosymmetric and the origin is taken at a center of symmetry, then $(F'_{hkl})_i = (F''_{hkl})_i = 0$ and

$$\text{Re}(hkl) = (F'_{hkl})_r^2 , \quad \text{Im}(hkl) = 2(F'_{hkl})_r (F''_{hkl})_r . \tag{8.84b}$$

In the noncentrosymmetric case, also assuming that F'' is small,

$$\text{Re}(hkl) = (F'_{hkl})^2 , \tag{8.84c}$$

$$\text{Im}(hkl) = 2[(F'_{hkl})_r(F''_{hkl})_r + (F'_{hkl})_i(F''_{hkl})_i] . \tag{8.84d}$$

For the centrosymmetric case, Eq. (8.83) becomes

$$\varepsilon = F''_{hkl}/F_0'' . \tag{8.85}$$

---

[10] Recall from Sect. 8.2 that F' represents the contribution from the real part of the *atomic* scattering factor, and F'' represents the contribution from the imaginary part. Both F' and F'' may, in general, be complex.

8.7.3 The Origin of the Borrmann Effect

In the primary beam's direction, $\mu_0 = \mu_L \cos\theta/\cos\Delta$ and

$$\mu_0(\text{effective}) = \mu_0[1 \mp |T|\varepsilon(1-p^2)^{1/2}]. \tag{8.86}$$

From this expression, we can see the Borrmann effect once again. For those reflections where all atoms scatter in phase, ε becomes $\sum \Delta f''(2\theta)/\sum \Delta f''(0)$. Consulting tables of scattering factors will show that this quantity departs *only slightly from unity*. Hence, μ_0(effective) for the σ polarization state ($T = 1$) becomes zero or $2\mu_0$ for the energy flow along the crystal planes (for which $p = 0$). The wave field for the α branch is associated with the upper sign, and it traverses the crystal with very low absorption. The β field is heavily absorbed. *As the energy flows at greater angles to the planes, absorption of the α (β) branch increases (decreases) arriving at the normal value when $p = \pm 1$ for which the energy flows in either primary or diffracted beam directions, because two tie points are not excited.* A schematic of this variation in absorption is given in Fig. 8.26. With a thick crystal, only those rays traveling parallel to net planes will survive, even though there are, initially, energy flows over the entire range of 2θ. To examine this more closely, let us makes some calculations for a typical case with the aid of Eq. (8.86). We shall assume $\Delta\theta = 0$ and consider the case of a Ge crystal and Cu K$_\alpha$ radiation. For the 220, $F = +8f$, and for the 333, $F = 4f(1 + i)$. For the 220 reflection, $\varepsilon = F''_{hkl}/F''_0$; for the 333, this quantity requires the more general expression. The lattice parameter $a = 5.66$ Å, so that $d_{220} = 2.01$ Å, and $d_{333} = 1.09$ Å. The values of $\sin\theta/\lambda$ that we need for the scattering factors are, for the 220, 0.25, and for the 333, 0.46. For a first approximation, we shall consult the International Tables Vol. III, Table 3.32 for the scattering factors:[11]

$$\varepsilon_{220} \cong 8(1.05)/8(1.1) = 0.955.$$

For the 333, $\Delta f'' = 0.95$, $\Delta f' = -1.3$. Now $(F'_{hkl})_r = (F'_{hkl})_i = 4(f + \Delta f')$, $(F''_{hkl})_r = (F''_{hkl})_i = 4\Delta f''$ and $\varepsilon \cong 0.86$. Also $\mu_0 = 3.50 \times 10^4\,\text{m}^{-1}$.

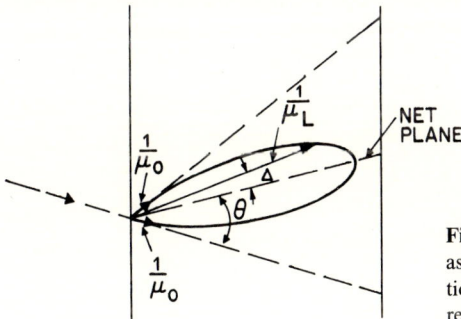

Fig. 8.26. Schematic representation of absorption associated with tie points in the α branch as a function of the angle Δ that the energy flow makes with respect to the atomic planes*

---

[11] More accurate values of $\Delta f'$, $\Delta f''$ are available in volume IV, Table 2.3, but this table does not include an estimate of the angular variation.

Consider first the 220 reflection ($T^\pi = \cos 2\theta = 0.707$), for $\Delta z = 8 \times 10^{-5}$ m.

	$\alpha_\sigma{}^{12}$	α_π	β_σ	β_π
μ_0(effective):	0.158	1.14	6.84	5.86
I/I_0^{12}:	0.88	0.40	0.004	0.009 .

At four times this thickness:

I/I_0^{13}:	0.60		0.026 .	

And at eight times the thickness:

I/I_0^{13}:	0.36		0.0007 .	

For the 333 reflection, $2\theta = 90°$, $\cos 2\theta = 0$:

	α_σ	α_π	β_σ	β_π
μ_0(effective):	1.4760	3.5000	5.6520	3.5000
$I/I_0(\Delta z = 2 \times 10^{-4}\,\text{m})$:	0.39	0.0009	2×10^{-6}	0.0009 .

In these calculations, we have ignored the effect of temperature on the scattering factors. As a better approximation, ε can be taken to be $\approx \exp(-M)$ because when the atoms scatter in phase, $\varepsilon = \sum \Delta f''(2\theta)/\sum \Delta f''(0)$ [see also Wagenfield (1962)]. But it is clear from these calculations that even with *quite a thick crystal* of Ge, with a 220 reflection, *the α_σ branch will pass through, while the other states are all essentially eliminated*. It is thus possible to devise a simple monochromator (Cole et al. 1961) that produces a polarized beam, as shown in Fig. 8.27a. Furthermore, with such a monochromator it is easy to check the perfection of a sample, as shown in Fig. 8.27b. A word of caution: perfect single crystals reflect over narrow angular ranges. Therefore, such a monochromator will select only a portion of the incident beam; the resultant intensity will be about two orders of magnitude lower than from a more mosaic crystal and is typically 10^6 cps with a normal x-ray tube. However, with the new high-intensity generators or with the high-intensity x-rays from synchrotrons, a much larger intensity ($10^8 - 10^{10}$ cps) is possible.

We return now to consider the absorption in the x-ray fan still further. We restrict our consideration to the symmetric Laue case, $b = 1$, and a centrosymmetric crystal. We once again relate the absorption to the incident conditions. From Eq. (8.52), with $|F'_{hkl}F'_{\overline{hkl}}|^{1/2}/F_{hkl} = 1$:

$$E_{hkl}/E_0 = e^{\pm v} . \tag{8.87}$$

From Eqs. (8.48b) and (8.84a) with $(F''_{hkl})^2 \ll (F'_{hkl})^2$:

$$\delta = \sinh v = \frac{\Delta\theta \sin 2\theta}{\Gamma |T| F'_{hkl}} - i \frac{\Delta\theta \sin 2\theta}{\Gamma |T| F'^2_{hkl}} F''_{hkl} . \tag{8.88}$$

Now, $v = v' + iv''$. Thus,

---

[12] Each of these states splits into two beams on leaving the crystal. The calculation here refers to the absorption *inside* the crystal.

[13] I/I_0 is given for *each* component. That is, I_0 is not the total direct beam.

8.7 Absorption in More Detail

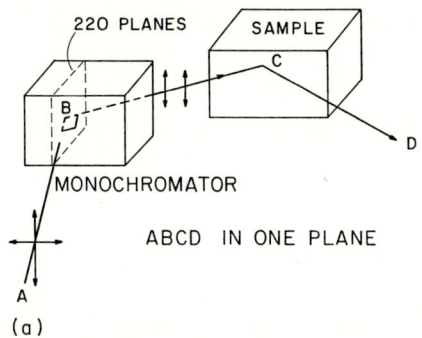

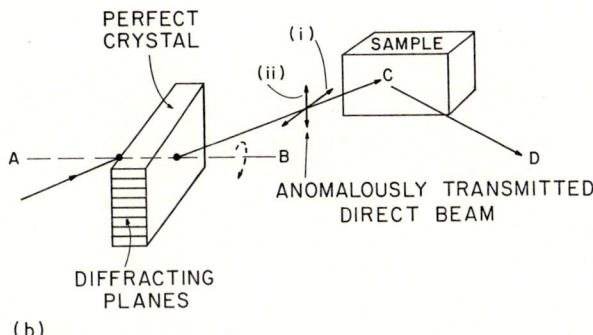

Fig. 8.27. a *Two-sided arrows* indicate direction of polarization for direct beam as it passes through a perfect crystal monochromator of Ge set for a 220 reflection. **b** How to measure perfection of a sample with an anomalously transmitted direct beam. Rotation about AB will change polarization from (i) (fields in ACD plane) to (ii) (fields perpendicular to ACD). Integrated intensities from the sample will have values which depend on the polarization by the specimen and this polarization varies with perfection

$$\delta = \sinh v = \sinh v' \cos v'' + i \cosh v' \sin v'' . \tag{8.89}$$

Comparing the imaginary terms in Eqs. (8.88) and (8.89), we can see that $\cosh v' \sin v''$ must be small compared to $\sinh v' \cos v''$. But v' is not necessarily small, so v'' must be small. Thus,

$$\delta' \cong \sinh v' = \Delta\theta \sin 2\theta / (\Gamma |T| F'_{hkl}) . \tag{8.90}$$

Comparing Eq. (8.87) with Eq. (8.79):

$$\frac{|E_0||E_{hkl}|}{E_0^2 + E_{hkl}^2} = \frac{1}{e^{-v'} + e^{+v'}} = \frac{1}{2 \cosh v'} = \frac{1}{2[1 + (\delta')^2]^{1/2}} . \tag{8.91}$$

From Eqs. (8.79), (8.83), and (8.91),

$$\mu_0(\text{effective}) = \mu_0 \left\{ 1 \mp \frac{|T|\varepsilon}{[1 + (\delta')^2]^{1/2}} \right\} . \tag{8.92}$$

This equation relates the absorption to incident conditions through Eq. (8.90).

8.8 Physical Interpretation of Anomalous Absorption

8.8.1 For Energy Flow Parallel to the Diffracting Planes

Assuming that $\delta' = \Delta = p = 0$, i.e., we are at a diameter point, and also that $F_{hkl} = F_{\bar{h}\bar{k}\bar{l}}$, the two plane waves from a diameter tie point will be equal in magnitude. For further simplicity, we will consider both fields to be real. The total electric field is then the sum of two plane waves:

$$\mathbf{E}_T = \mathbf{E}_0 \exp(+2\pi i v t)\exp(-2\pi i \mathbf{K}_0 \cdot \mathbf{r}) \pm \mathbf{E}_{hkl} \exp(+2\pi i v t)\exp(-2\pi i \mathbf{K}_{hkl} \cdot \mathbf{r}). \tag{8.93a}$$

Let $\mathbf{K}_{hkl} = \mathbf{K}_0 + \mathbf{r}^*_{hkl}$. Then,

$$E_T^2 = E_0^2 + E_{hkl}^2 \pm 2\mathbf{E}_0 \cdot \mathbf{E}_{hkl} \cos 2\pi \mathbf{r}^*_{hkl} \cdot \mathbf{r}. \tag{8.93b}$$

As $E_0 = E_{hkl}$ under the conditions we are considering, and $\mathbf{E}_0 \cdot \mathbf{E}_{hkl} = E_0^2 T$, Eq. (8.93b) may be written as

$$E_T^2 = 2E_0^2(1 \pm T \cos 2\pi \mathbf{r}^*_{hkl} \cdot \mathbf{r}). \tag{8.94}$$

From this equation we see that we have standing waves $\mathbf{r}^* \cdot \mathbf{r} = $ const parallel to the diffracting planes and with a wavelength, $1/r^*_{hkl} = d_{hkl}$. The sign is introduced to take into account the sign of ξ_0. For the α branch and σ state ($T = 1$), the negative sign applies because ξ_0 is positive and hence, E_{hkl} in Eq. (8.93) is negative [see Eq. (8.35)]. In this case for $T = 1$ the intensity is zero at the crystal planes themselves; the nodal points are located there. As a result, this wave cannot suffer appreciable photoelectric absorption. On the other hand, for the β_σ branch the *antinodal* points are at the planes ($\mathbf{r}^*_{hkl} \cdot \mathbf{r} = $ an integer) and high absorption will occur. These situations are shown schematically in Fig. 8.28. Note that for the π states, the field cannot

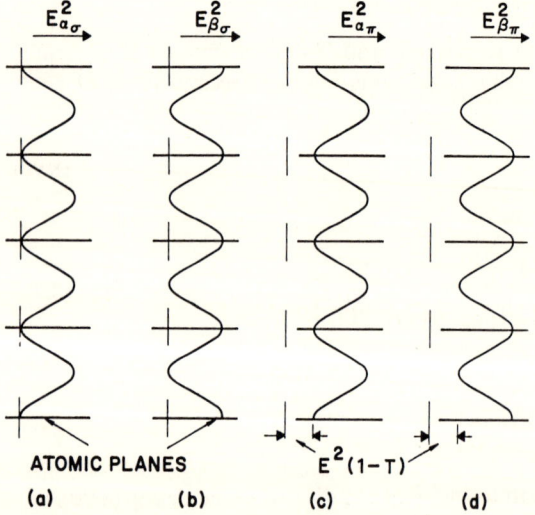

Fig. 8.28. Possible configurations of the standing wave patterns assuming that the nodes of the α branch, σ polarization, coincide with the actomic planes*

8.8 Physical Interpretation of Anomalous Absorption

go to zero for the α branch ($T \neq 1$) and thus there is always higher absorption for this state, as we have already seen from our calculations.

However, in the previous paragraph, we have not *really* proved that the standing waves have nodal or antinodal points *at* the planes, only that these points are on planes *parallel* to the crystal planes, and that these points are separated by the spacing of the crystal planes. Let us start again. We can write Eq. (8.93a) as

$$\mathbf{E} = \exp(+2\pi i v t)\exp(-2\pi i \mathbf{K}_0 \cdot \mathbf{r})\mathbf{E}_0[1 \pm (E_{hkl}/E_0)\exp(-2\pi i \mathbf{r}^*_{hkl} \cdot \mathbf{r})] \ . \qquad (8.95)$$

The field intensity is then proportional to R:

$$R = |1 \pm (E_{hkl}/E_0)\exp(-2\pi i \mathbf{r}^*_{hkl} \cdot \mathbf{r})|^2 \ . \qquad (8.96)$$

The value of $\mathbf{r}$ appears twice in this equation, once in the exponent and once in the field ratio, as this depends on the structure factor, which in turn depends on the choice of origin. But our result should be independent of the choice of origin. Assume a nodal plane exists, and then let $\mathbf{r}$ originate in this plane. From Eq. (8.96), $R = 0$ when $r = 0$ only if $E_{hkl}/E_0 = -1$. For the Laue case and $b = +1$, Eq. (8.52) applies. And if there is a center of symmetry which is chosen as origin, the term $|F_{hkl}F_{\bar{h}\bar{k}\bar{l}}|^{1/2}/F_{\bar{h}\bar{k}\bar{l}} = 1$. Then from Eq. (8.52), $E_{hkl}/E_0 = \mp(|T|/T)e^{\pm v}$. The field ratio is -1 when $T = 1$, $v = 0$, which occurs for the exact Bragg conditions, and for the α branch. Thus the nodal points *are* on the planes.

The *strongest anomalous transmission will occur when all of the atoms in the crystal lie on these nodal planes*. It follows that *only those reflections for which all atoms scatter in phase* (and therefore *all* lie in the plane) can *satisfy this condition*. These are sometimes referred to as *full reflections*. That is why we found in Sect. 8.7 that the transmission is much better for the Ge 220 reflection than for the 333. (Examine these two planes in the diamond structure.)

8.8.2 When the Energy Flow is not Parallel to the Diffracting Planes

We now generalize our treatment for the case where the energy flow is *not* parallel to the diffracting planes. In Fig. 8.29, let $\mathbf{r}_0$ be the position of a nucleus. Let $\mu(\mathbf{r}_a)\,dV$ be the contribution to the absorption by the volume element dV at the end of $\mathbf{r}_a$. Then, assuming the actual absorption at any point depends on the field intensity,

$$\mu(\mathbf{r}_0) = A \int E^2(\mathbf{r})\mu(\mathbf{r}_a)\,dV. \qquad (8.97)$$

Employing Eq. (8.93b):

$$E_T^2 = E_0^2 + E_{hkl}^2 \pm 2\mathbf{E}_0 \cdot \mathbf{E}_{hkl}\exp(-2\pi i \mathbf{r}^*_{hkl} \cdot \mathbf{r}) \ , \qquad (8.98a)$$

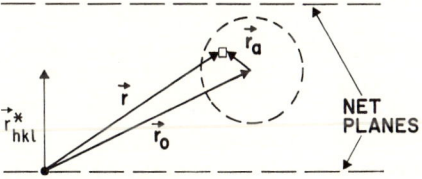

Fig. 8.29. Geometry used to calculate the absorption of an atom in a standing wave pattern*

and noting that

$$\int \mu(\mathbf{r}_a) dV = \mu_0, \tag{8.98b}$$

$$\mu(\mathbf{r}_0) = A\{(E_0^2 + E_{hkl}^2)\mu_0 \pm 2\mathbf{E}_0 \cdot \mathbf{E}_{hkl} \exp(2\pi i \mathbf{r}_{hkl}^* \cdot \mathbf{r}_0)$$
$$\times \int_{\substack{\text{at} \\ \text{vol}}} \exp(-2\pi i \mathbf{r}^* \cdot \mathbf{r}_a) \mu(\mathbf{r}_a) dV\}. \tag{8.99}$$

The averge of $\mu(\mathbf{r}_0)$ over all positions in the field will give the average value μ_0. Hence, the constant $A = (E_0^2 + E_{hkl}^2)^{-1}$. Abbreviating the last integral as μ_{0hkl}, taking the atom in the plane $\mathbf{r}^* \cdot \mathbf{r}_0 = 0$, and replacing $\mathbf{E}_0 \cdot \mathbf{E}_{hkl}$ by $\pm T E_0 E_{hkl}$:

$$\mu = \mu_0 \left[1 \pm T \frac{2 E_0 E_{hkl}}{(E_0^2 + E_{hkl}^2)} \frac{\mu_{0hkl}}{\mu_0} \right]. \tag{8.100}$$

We can arrive at an alternative expression for μ by combining Eqs. (8.79) and (8.83):

$$\mu = \mu_0 \left[1 \mp |T| \frac{2 E_0 E_{hkl}}{(E_0^2 + E_{hkl}^2)} \varepsilon \right]. \tag{8.101}$$

From Eqs. (8.100) and (8.101), we see that ε is related to μ_{0hkl}, which is the weighted distribution of absorbing power within the atom. We see also that ε should depend on temperature since thermal vibrational affects the atomic positions. The larger the amplitude of thermal vibration the more deeply the atom penetrates into regions of higher field intensity on the α branch and the greater will be the absorption. We naively expect ε to have the form $\varepsilon = \varepsilon_0^{-M}$ since $\varepsilon \cong F_{hkl}''/F_0''$ according to Eq. (8.85); this has been confirmed by experiment (Okkerse 1962); but quantum mechanical calculations of this problem have not yet been made.

The anomalous absorption effect has been verified in electron diffraction and in neutron diffraction as well as with x-rays. Duncomb (1962) observed enhanced x-ray emission in extinction contours, from thin films in the transmission electron microscope, indicating the presence of the antinodal wave field. Knowles (1956) chose a crystal for which there was a reaction between a neutron and nucleus resulting in γ-ray emission from the host atoms. During diffraction the neutron intensity for the α branch is reduced at the nucleus and a corresponding reduction in γ-ray output was observed in silicon. Patel and Batterman (1963) have used the destruction of anomalous transmission due to strain to detect as little as 1 ppm of oxygen.

An ingenious use of the interference between transmitted and diffracted beams and the relative absorption of beams has been made by Bonse and Hart (1965b, 1966a, c) in devising an *x-ray interferometer*. This device is illustrated in Fig. 8.30. There is a beamsplitter S, two "mirrors" M, and an analyzer A. Actually these are generally all cut from one crystal, for ease in alignment and rigidity. These crystals are made thick enough so that the beam from only one branch (and polarization state) emerges. The maxima in intensity of the waves that get through the crystal are between atomic planes (as we have seen) and these are only from the σ state. If the mirrors M are equidistant from S and A, the two coherent waves from M overlap

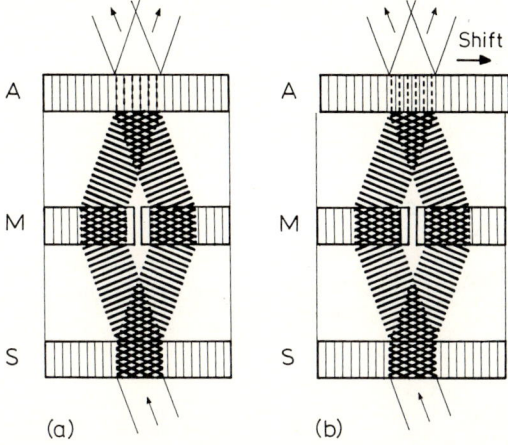

Fig. 8.30a,b. A Laue x-ray interferometer consisting of a beamsplitter S, "mirrors" M, and an analyzer A. The minimum intensities of waves shown *dotted* in **a** are on extensions of atom planes in S. If A, M, and S are all cut into one block in **a** these planes are not displaced at A. In **b** some deformation causes such a displacement enhancing absorption and decreasing the transmitted intensity. [From Bonse, U., and Hart, M., A. Phys. *140*, 455 (1966)]

on the entrance surface of A and set up the standing wave field between continuations of planes in S. A phase shift due to a specimen in one beam path (refracting that beam) or a change in d spacing of A, M, or S, perhaps due to deformation, can cause the intensity maxima to be on atom planes, enhancing absorption (Fig. 8.30b). *Translations of 1 Å, rotations of 10^{-3} sec, or dilations $\Delta d/d$ of 10^{-8} can be measured.* Neutron interferometers are also possible (Overhauser and Collela 1980). Other applications will be discussed when we consider "topography" in Sect. 8.11.

8.9 Exit Beams Again—The Symmetric Laue Case

We have already seen how to choose the *exit wave vectors* and that in the Laue case there are in general eight such beams.

The choice of boundary conditions for the field amplitudes depends on whether the outgoing beams overlap or not. If the *cross section of the incident beam is smaller than the path length* in the crystal, there is no overlap at the exit surface. (We cannot really use infinite plane waves, Sect. 8.6). In such a case, *we must treat the waves separately* and *intensities rather than amplitudes add* for the exit beams.

In Fig. 8.31 (for the symmetric Laue case and a parallel-sided slab of crystal), the incident ray $\mathbf{k}_0^i$ is assumed to be slightly off the Bragg angle, but within the range of reflection $\Delta\theta$. Two flows, $\mathbf{S}_\alpha$ and $\mathbf{S}_\beta$, are produced for each polarization state. If these are not absorbed completely, they split at the exit surface into the diffracted beam $\mathbf{k}_{hkl}^e$ and forward diffracted beam $\mathbf{k}_0^e$. Then for one polarization state,

$$(E_{hkl}^e)^2 = E_{hkl\alpha}^2 + E_{hkl\beta}^2 , \tag{8.102a}$$

and similarly, for the forward diffracted beam. If the fields do overlap, however,

$$(E_{hkl}^e)^2 = |\mathbf{E}_{hkl\alpha} + \mathbf{E}_{hkl\beta}|^2 . \tag{8.102b}$$

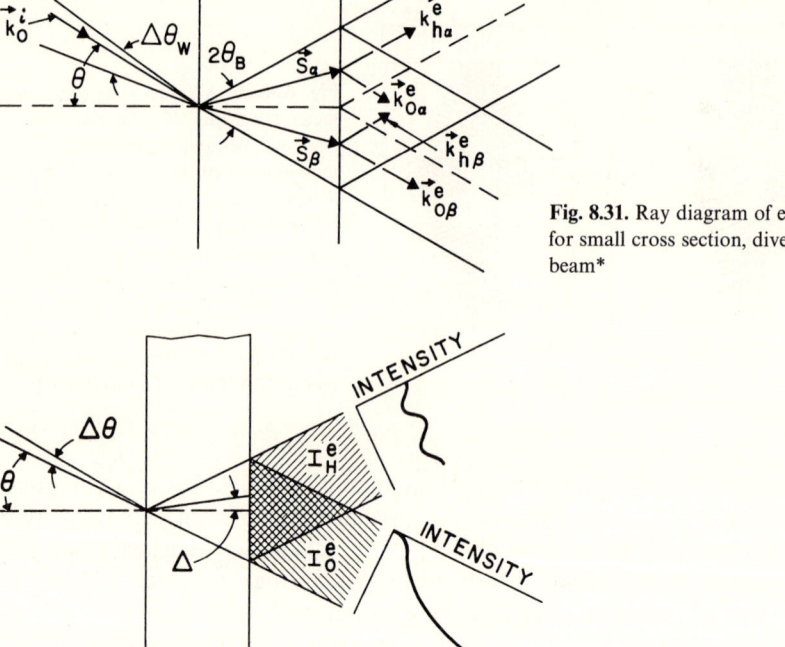

Fig. 8.31. Ray diagram of energy flows for small cross section, divergent input beam*

Fig. 8.32. Energy distribution, as calculated by Kato (1960) across exit beams for small cross section, divergent input beam ($\mu_0 t \approx 2$)*

Now *the ratios of the field amplitudes are fixed by the tie points selected, and their actual values are set by the entrance conditions*. Therefore, *the only way the fields inside the exit surface can differ from those inside the entrance surface is through absorption*. If *the crystal is very thin* with respect to an extinction distance, then from Eq. (8.102b) and the entrance conditions Eq. (8.55b), *there will be no diffracted beam*. When the *crystal is half as thick as this distance, all of the energy for the ray at the exact Bragg angle will be flowing in the diffracted beam's direction*. This flow gets smaller as we deviate from the exact Bragg angle. Thus, there will be an intense beam with narrow angular divergence, giving a certain integrated intensity. *At the full extinction distance, there is again no diffracted beam*. This oscillation continues until the β field is absorbed, or until the beams separate physically. Thus *as the crystal thickens* we go from Eq. (8.102b) to Eq. (8.102a), for the field amplitudes, from *the sum of amplitudes, to the sum of intensities*.

We now wish to explore (a) the intensity distribution across the 2θ fan, (b) the shape of a "rocking" curve, and (c) the integrated intensity.

Assuming a small cross section for the incident beam, we quote Kato's result (Kato 1960). All the important parameters are illustrated in Fig. 8.32. It is assumed that the *incident* beam's intensity is uniform across its cross section. Recall that

8.9 Exit Beams Again—The Symmetric Laue Case

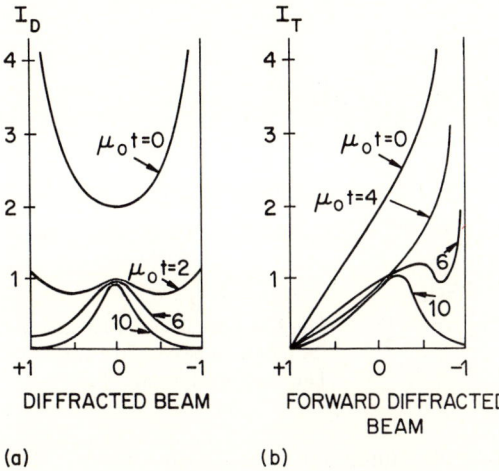

Fig. 8.33a,b. After Kato (1960) at the exit surface. **a** Energy distribution across diffracted beam (across the beam in Fig. 8.32). **b** Forward diffracted beam*

$p = \tan \Delta/\tan \theta = \pm 1$ at either end of the x-ray fan. In Fig. 8.33, the intensity distributions are shown. Note the "hot edges" of the diffracted beam. As can be seen in Fig. 8.20, there is a "pileup" of Poynting vectors in the S_0 and S_{hkl} directions because of the curvature of the dispersion surface, and for thin crystals there is little attenuation. That is, the energy flows are all nearly parallel far away from the center of the dispersion surface. Often in a transmission Laue pattern from a thin crystal doubled spots appear as a result of this. As the crystal thickens and absorption becomes more important (see Fig. 8.26), the central region remains strong but the wings of the beam are seriously absorbed. The forward diffracted beam is asymmetric even for thick crystals, $\mu_0 t = 10$.

Using the boundary conditions for the case where the exit beams are separated, Eqs. (8.102a) and (8.57) with r at the exit surface, r_e:

$$I^e_{0w}/I_0 = (E_{0w}/E^i_0)^2 = \tfrac{1}{4}(e^{\pm 2v'}/|\cosh v|^2)\exp(-4\pi \mathbf{K}''_{0w} \cdot \mathbf{r}_e) \tag{8.103a}$$

(where the subscript w relates to the α or β branches).
Furthermore, with $|F_{hkl} F_{\overline{hkl}}|^{1/2}/F_{hkl} \cong 1$,

$$I^e_{hklw}/I_0 = (E_{hklw}/E^i_0)^2$$
$$= \tfrac{1}{4}(|b|/|\cosh v|^2)\exp(-4\pi \mathbf{K}''_{hklw} \cdot \mathbf{r}_e) . \tag{8.103b}$$

I_0 *is the intensity of a given polarization state.* If the beam is unpolarized, multiplying the right-hand side by $\tfrac{1}{2}$ gives the result with reference to the total beam. We can express these results in terms of the incident conditions with Eqs. (8.90) and (8.92) and the fact that $\delta \pm (\delta^2 + 1)^{1/2} = \pm e^{\pm v}$:

$$I^e_{0w}/I_0 = \frac{1}{4}\left\{1 \mp \frac{\delta'}{[1+(\delta')^2]^{1/2}}\right\}^2 \exp\left\{-\frac{\mu_0 t}{\gamma_0}\left(1 \mp \frac{|T|\varepsilon}{[1+(\delta')^2]^{1/2}}\right)\right\}, \tag{8.104a}$$

$$I^e_{hklw}/I_0 = \frac{1}{4}\left[\frac{1}{1+(\delta')^2}\right]\exp\left\{\frac{-\mu_0 t}{\gamma_{hkl}}\left(1 \mp \frac{|T|\varepsilon}{[1+(\delta')^2]^{1/2}}\right)\right\}. \tag{8.104b}$$

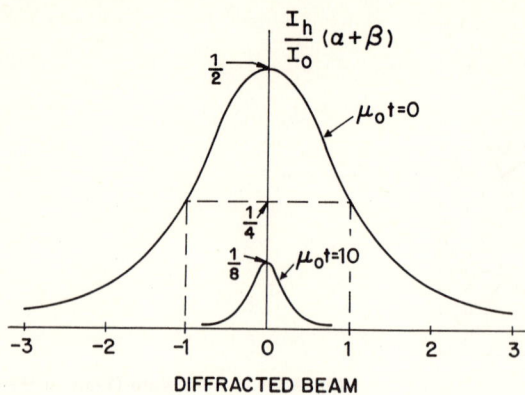

Fig. 8.34. Peak shape on a δ' scale, for the diffracted beam in the symmetric Laue case as a function of $\mu_0 t$*

In these equations, t is the thickness and the minus sign refers to the α branch. The term δ' is related to θ by Eq. (8.90), so that converting to a θ scale is straightforward. For a very thin crystal ($\mu_0 t \ll 1$), the peak shape of the diffracted beam is simply the sum of the intensities for the α and β branches. For one polarization state and a thin crystal, summing the intensities for the α and β branches:

$$I_{hkl}/I_0 = \tfrac{1}{2}\{1/[1 + (\delta')^2]\} . \tag{8.105}$$

This is shown as the upper curve in Fig. 8.34. For $\Delta\theta = \delta' = 0$, $I_{hkl}/I_0 = \tfrac{1}{2}$, and the crystal is simply a beamsplitter diverting one half of the incident beam into the diffracted beam. It should be realized that this result is only valid when the beams separate physically so that we can add intensities.

From Eq. (8.105) or Fig. 8.34, it is easy to see that the half-width (the positions for which $I/I_0 = \tfrac{1}{4}$) is $\Delta\delta' = 2$ or from Eq. (8.90),

$$\Delta\theta(\text{rad}) = 2|T||\Gamma|F'_{hkl}|/\sin 2\theta . \tag{8.106}$$

This is the width of a Bragg reflection as we saw before in Eq. (8.42).

For $\mu_0 t > 10$, radiation associated with the β branch of both polarization states is completely absorbed, as well as the π state of the α branch. As a result, I/I_0 at the maximum will at most be $\tfrac{1}{8}$ that for an *unpolarized* total incident beam:

$$\frac{I_{hkl}}{I_0} = \frac{1}{8}\frac{1}{(1 + (\delta')^2)}\exp\left\{-\frac{\mu_0 t}{\gamma_{hkl}}\left(1 - \frac{|T|\varepsilon}{[1+(\delta')^2]^{1/2}}\right)\right\} . \tag{8.107}$$

The maximum value of $\tfrac{1}{8}$ occurs for $\delta' = 0$ and $\varepsilon = 1$. The absorption term serves to sharpen the peak. This result is shown as the lower curve in Fig. 8.34. Generally, Eq. (8.107) has appreciable values only for $(\delta')^2 \ll 1$. Then $[1 + (\delta')^2]^{1/2} \approx 1 + [(\delta')^2/2]$. Therefore,

$$1/[1 + (\delta')^2]^{1/2} \approx 1/\{1 + [(\delta')^2/2]\} \approx 1 - [(\delta')^2/2] \approx \exp[-(\delta')^2/2] .$$

Hence, Eq. (8.104b) may be written *for the α branch* as

$$I_{hkl\alpha}/I_0 = \tfrac{1}{8}\exp[-(\mu_0 t/\gamma_{hkl})(1 - |T|\varepsilon)]\exp[-(\mu_0 t/\gamma_{hkl})|T|\varepsilon(\delta')^2/2] . \tag{8.108}$$

In this case, the half-width can be determined by recognizing that on the δ' scale

8.9 Exit Beams Again—The Symmetric Laue Case

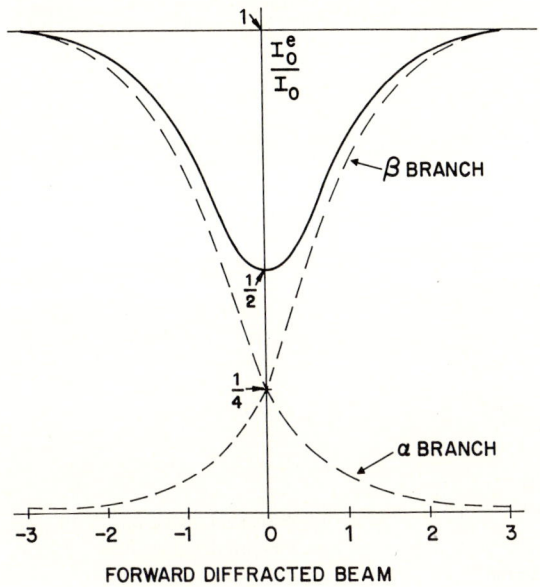

Fig. 8.35. Peak shape, on a δ' scale, for the forward diffracted beam, symmetric Laue case, for zero absorption. The contribution from the points on the individual branches of the dispersion surfaces are shown *dotted* *

the peak drops to half its value when $\frac{1}{2}(\mu_0 t/\gamma_{hkl})|T|\varepsilon(\delta')^2 = \ln 2$, or

$$\Delta\theta(\text{rad}) = \frac{2TF'_{hkl}}{\sin 2\theta}\left[\frac{2\ln 2}{|T|\varepsilon\mu_0 t/\gamma_{hkl}}\right]^{1/2}. \tag{8.109}$$

This width is *smaller* than that for a specimen with a low absorption.

We turn now to the forward diffracted beam. For μt approaching zero, Eq. (8.104a) for one polarization state is

$$\frac{I_0^e}{I_0} = \frac{1}{4}\left\{1 - \frac{\delta'}{[1+(\delta')^2]^{1/2}}\right\}^2 + \frac{1}{4}\left\{1 + \frac{\delta'}{[1+(\delta')^2]^{1/2}}\right\}^2. \tag{8.110}$$

This is the sum of α and β intensities. The behavior of each term is shown as dotted lines in Fig. 8.35; the solid curve is the sum, which is merely the (upside down) diffracted beam in Fig. 8.34 for low $\mu_0 t$. As $\mu_0 t$ increases, the β contribution vanishes more quickly than the α term and the wings die out faster than the middle, as a result of the variation of absorption with angle. The result is shown as the middle curve in Fig. 8.36. When $\mu_0 t > 10$, only the σ polarization state of the α branch contributes and for an unpolarized incident beam,

$$\frac{I_0^e}{I_0} = \frac{1}{8}\left(1 - \frac{\delta'}{[1+(\delta')^2]^{1/2}}\right)\exp\left[-\frac{\mu_0 t}{\gamma_0}\left(1 - \frac{\varepsilon}{[1+(\delta')^2]^{1/2}}\right)\right]. \tag{8.111}$$

This is *almost* the same as Eq. (8.107); there is a slight asymmetry however, due to the preexponential term such that the forward diffracted beam is shifted to lower angles; see Fig. 8.36. As $\mu_0 t$ increases still further, the values of δ' for which Eq. (8.111) has any appreciable value become much less than unity and the forward diffracted and diffracted beams become similar in shape again.

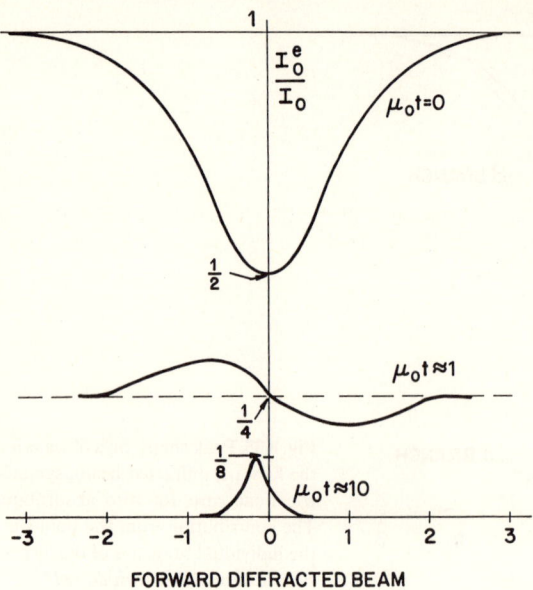

Fig. 8.36. Peak shape, on a δ' scale, for the forward diffracted beam, symmetric Laue case as a function of $\mu_0 t$*

Integrated intensities can be measured by rocking the crystal through an angular range. Thus we can define the power in a reflection as

$$P_{hkl} = \int \frac{AI_{hkl}}{A_0 I_0} d\theta .$$ (8.112)

The term A_0 is the cross-sectional area of the incident beam and A is the area of the diffracted beam. For the symmetric Laue case, the areas are the same and we may write

$$P_{hkl} = \frac{d\theta}{d\delta'} \int \frac{I(\delta')}{I_0} d\delta' ;$$

from Eq. (8.90),

$$\frac{d\theta}{d\delta'} = \frac{|T||\Gamma||F'_{hkl}|}{\sin 2\theta} .$$

Thus,

$$P_{hkl} = \frac{|T||\Gamma||F'_{hkl}|}{\sin 2\theta} \int \frac{I(\delta')}{I_0} d\delta' .$$ (8.113)

For $\mu_0 t \ll 1$, employing Eq. (8.105), for each polarization state,

$$P_{hkl} = \frac{|T||\Gamma||F'_{hkl}|}{\sin 2\theta} (\pi/4) .$$ (8.114)

Note particularly that the integrated intensity varies as $|T||F|$, whereas in kinematic theory it varies as $T^2 F^2$.

8.10 Intensity in Bragg Reflections

For a thick crystal, numerical integration of Eq. (8.104b) is required for precision, but we shall use Eq. (8.108), which upon integration yields

$$P_{hkl} = \frac{|T||\Gamma||F'_{hkl}|}{4\sin 2\theta}\left[\frac{2\pi}{|T|\varepsilon\mu_0 t/\gamma_{hkl}}\right]^{1/2}\exp\left[-\frac{\mu_0 t}{\gamma_{hkl}}(1-|T|\varepsilon)\right]. \tag{8.115}$$

For a more precise evaluation, see Batterman (1962). Actually, for $\mu_0 t = 10$ this equation is valid to about 3%; as $\mu_0 t$ increases, the agreement with a more accurate numerical calculation is even better. For thick crystals, this equation may be used to obtain values of ε; see Problem 8.5.

8.10 Intensity in Bragg Reflections

8.10.1 Neglecting Absorption

The orientation of the dispersion surface for a Bragg reflection is shown in Fig. 8.37. Recall that we have already found in Sect. 8.3 that there is a range of angles for which the inward directed normal does not intersect the dispersion surface; for this range there are no possible fields inside the crystal because there are no possible wave vectors. Total reflection must occur. The situation in Fig. 8.37 is one is which the normal *is* cutting the dispersion surfaces — but note that *both* intersections (T_1 and T_2) are on *one* branch. Incident and exit wave vectors are also shown.

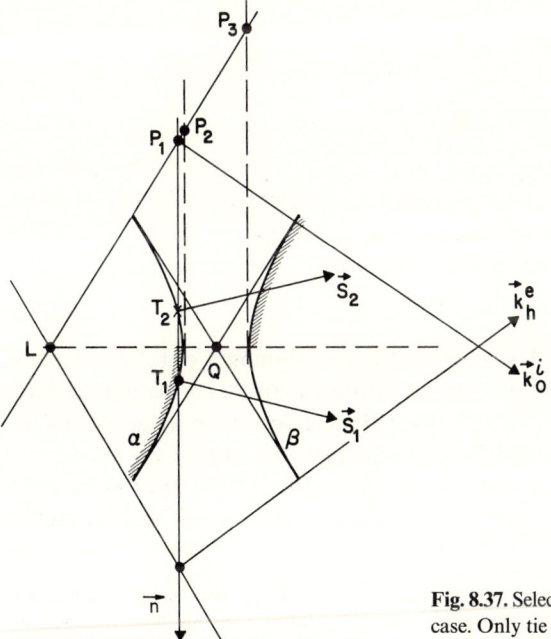

Fig. 8.37. Selection of tie points in the symmetric Bragg case. Only tie points in the *hatched region* correspond to permitted solutions*

The two Poynting vectors are interesting; one, S_2, is outward along the surface but a little above it. But this vector represents the flow averaged over time and an extinction distance. It represents one of the flows *throughout* the crystal. Hence this direction of flow implies, due to the loss by absorption that must occur inside the crystal, that an infinitely thick crystal would require an infinitely strong incident field to give any finite field at the surface! Thus, the tie point T_2 cannot be active. In Fig. 8.37, only the shaded regions are active.[14] As there is only one excited tie point (for each polarization state), the boundary conditions in the Bragg case are

$$\mathbf{E}_0^i = \mathbf{E}_{0j}, \tag{8.116a}$$

$$\mathbf{E}_{hkl}^e = \mathbf{E}_{hklj}. \tag{8.116b}$$

The subscript j represents the tie point. The diffracted beam is more clearly like a reflection, not a flow reaching the surface and splitting up, as was the case in Laue geometry. Outside the region of total reflection (the range P_2 to P_3 in Fig. 8.37), an incident beam strikes the crystal, an internal flow and a reflected beam are generated. The ratio E_{hkl}/E_0 is the same in both beams. The internal flow is completely absorbed if the crystal is thick enough. Multiplying the expressions in Eq. (8.35) together and employing Eq. (8.49) with $b = -1$:

$$\left(\frac{E_{hkl}^e}{E_0^i}\right)^2 = \frac{\xi_0}{\xi_{hkl}} \frac{F_{hkl}}{F_{\overline{hkl}}} = |b|(\delta \pm (\delta^2 - 1)^{1/2})^2 \frac{|F_{hkl}|}{|F_{\overline{hkl}}|}. \tag{8.117}$$

As $|F_{hkl}| = |F_{\overline{hkl}}|$:

$$\frac{|E_{hkl}^e|^2}{|E_0^i|^2} = \frac{|E_{hklj}|^2}{|E_{0j}|^2} = |b||\delta \pm (\delta^2 - 1)^{1/2}|^2. \tag{8.118}$$

With $b = -1$, and assuming there is no absorption, from Eq. (8.48b):

$$\delta = \delta' = (-\Delta\theta \sin 2\theta + \Gamma F_0')/(|T|\Gamma F_{hkl}'). \tag{8.119}$$

If $\Delta\theta$ is large and negative, δ' is large and positive; such a point is on the low-angle side of the α branch. In Eq. (8.118), the α branch is represented by the negative sign and thus $(|E_{hkl}|^2/|E_0^i|^2) \to 0$; only the transmitted beam exists as we have just seen. As δ' reduces to $+1$, then from Eq. (8.118):

$$|E_{hkl}^e/E_0|^2 = 1. \tag{8.120a}$$

This corresponds to point P_2 in Fig. 8.37, and is the beginning of total reflection. At this point, the Poynting vector is along the atomic planes and the physical surface. The fact that energy flow is indeed along the surface in Bragg geometry has been verified by allowing the incident beam to fall near the edge of the crystal and detecting the radiation leaving the side surface (Batterman 1962; Borrmann 1951; Borrmann et al. 1955). [If the incident beam is defined by slits, this implies that its

---

[14] This is similar to the situation in Laue geometry, Sect. 8.5, but in that case an infinite internal field occurred for one branch well away from the diffraction peak. Also, if the crystal has finite thickness, the flow associated with T_2 can be active. It is as if there is a reflection from the bottom surface of the crystal.

8.10 Intensity in Bragg Reflections

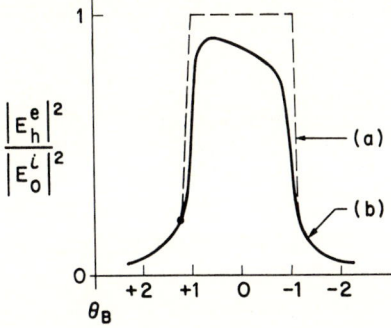

Fig. 8.38. Shape of the Bragg reflection from a perfect crystal, on a δ' scale. The *upper dotted curve* is for zero absorption (Darwin curve) and the *lower curve* is with absorption (Darwin-Prins curve)*

cross section reduces to zero when **S** is parallel to the surface. This is clearly impossible and is a result of assuming that there is no absorption, as we shall shortly see.]

At $\delta = -1$, from Eq. (8.119).

$$\Delta\theta \sin 2\theta = \Gamma F'_0 + |T|\Gamma F'_{hkl}, \qquad (8.120b)$$

This occurs at the diameter point on the β branch. The curve represented by this discussion is shown dotted in Fig. 8.38. Equation (8.118) continues to give unity for the intensity for all values of δ between ± 1, even though there are no tie points on the dispersion surface and thus the top of the peak is flat. Note that at the center of the peak, $\delta = 0$ and from Eq. (8.119)

$$\Delta\theta(\text{rad}) = \Gamma F'_0/\sin 2\theta .$$

This is like the equation we found before for the shift in the peak from the position in kinematic theory, Eq. (8.43). This curve without absorption is often referred to as the "Darwin curve."

8.10.2 Including Absorption

We now turn to reflection including absorption. We can write δ in its complex form from Eq. (8.48b) as

$$\delta = \frac{-\Delta\theta \sin 2\theta + \Gamma F'_0}{\Gamma |T||F'_{hkl}|}\left[1 - i\frac{|F''_{hkl}|}{|F'_{hkl'}|}\right].$$

Therefore,

$$\delta' = \frac{-\Delta\theta \sin 2\theta + \Gamma F'_0}{\Gamma |T||F'_{hkl}|} + \frac{\Gamma F''_0|F''_{hkl}|}{|F'_{hkl}|\Gamma|T||F'_{hkl}|}. \qquad (8.121a)$$

As $F''/F' \ll 1$, the last term can be ignored. Furthermore,

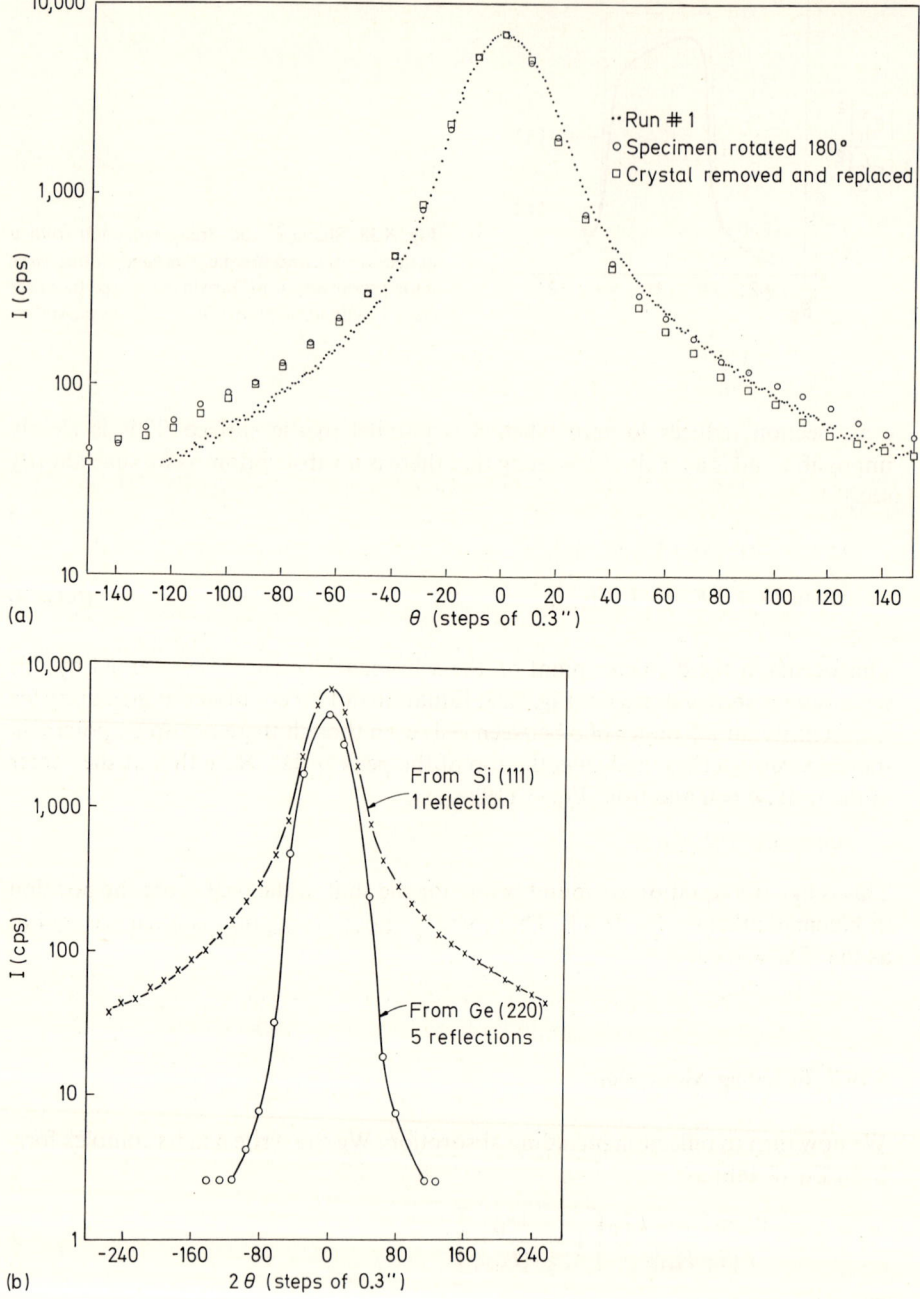

Fig. 8.39. a A real rocking curve from two Si crystals cut to (111) planes and set in (1, −1) configuration. Run 1: ...; specimen rotated 180°: ○; crystal removed and replaced: □. **b** Ge 220 reflection after five reflections. Data taken by H. Chen, P. Georgopoulos, and T. Tsakalakos at Northwestern University

$$\delta'' = \frac{\Delta\theta \sin 2\theta |F''_{hkl}|}{\Gamma |T| |F'_{hkl}| |F'_{hkl}|} - \frac{\Gamma F'_0 |F''_{hkl}|}{|F'_{hkl}| \Gamma |T| |F'_{hkl}|} + \frac{\Gamma F''_0}{\Gamma T |F'_{hkl}|},$$

thus,

$$\delta'' = -\frac{|F''_{hkl}|}{|F'_{hkl}|}\left[\delta' - \frac{F''_0 |F'_{hkl}|}{|T||F'_{hkl}||F''_{hkl}|}\right] \cong -\frac{|F''_{hkl}|}{|F'_{hkl}|}\left(\delta' - \frac{1}{|T|\varepsilon}\right). \tag{8.121b}$$

The "Darwin–Prins" curve can then be generated directly from Eqs. (8.121) and (8.117), and is shown as the solid curve in Fig. 8.38. This curve is less than the Darwin curve over most of its range. The physical reason for the asymmetry will be discussed below.

The integrated intensity must be determined by numerical methods; see Miller (1935). The value of ε can obtained from theory (Hountas et al. 1973) or from the integrated intensity in Laue geometry as mentioned in Sect. 8.9 or Problem 8.5.

We note here, that for weaker reflections, i.e., as F'_{hkl} gets smaller, the α and β branches come closer together since the range depends on F [Eq. (8.34)]. Also for $\delta' = 0$, which is *near* the center of the reflection curve, $\delta'' = F''_0/|T|F'_{hkl}$. *The greater δ'', the greater the deviation from total reflection and the closer is the behavior to kinematic theory.*

The shape of this rocking curve is quite sensitive to the dislocation density in a crystal and its arrangement, and can be used to measure these quantities (Hordon and Averbach 1961).

In contrast to the case without absorption, there is never total reflection since $|E_{hkl}|^2/|E_0|^2 < 1$, and so there must always be some energy flow into the crystal to conserve energy. The energy *never* really flows parallel to the surface. This can be seen formally by combining Eqs. (8.80) and (8.81), giving

$$\tan \Delta = \left[\frac{E^2_{hkl}/E^2_0 - 1}{E^2_{hkl}/E^2_0 + 1}\right]\tan\theta. \tag{8.122}$$

Since $E^2_{hkl}/E^2_0 < 1$, Δ can never be zero. We can approach the diameter points in Fig. 8.37 but never actually reach them.

A real rocking curve in the $(1, -1)$ configuration from silicon single crystals cut to a (111) plane is shown in Fig. 8.39a. *Note the narrow width, but the very long tails which are shown in Problem 8.6 to be of the form $B/(\Delta\theta)^n$*. Bonse and Hart (1965a) have shown that by *using multiple reflections the tails can be drastically reduced*; the result is a peak only a few seconds wide but with little "tail," as shown in Fig. 8.39b for five reflections. With two crystals, cut for multiple reflections in Fig. 8.40, it

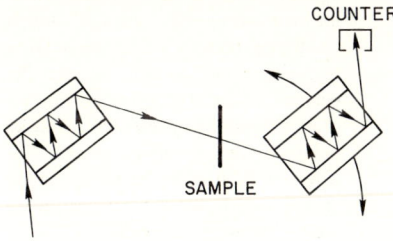

Fig. 8.40. Slotted perfect crystal for small-angle scattering. By rotating the crystal after the specimen, scattering from the specimen vs. 2θ can be explored very close to the direct beam

is possible to have a small-angle scattering system without slits, with which *one can get to within* 10–20 s *of the direct beam* (Bonse and Hart 1965a)! This allows measurements of particle size by small-angle scattering (Sect. 7.7), into the range of the optical microscope. While there is some loss in intensity due to repeated scattering, this is not very large; the order of 20% of the original beam is still obtained. But again, it should be kept in mind that perfect crystals accept only a small portion of the incident beam because of their narrow range of reflection and the intensity is therefore low compared to other small-angle systems. This approach should be employed only when it is needed (see Bonse and Hart 1966b; Brady 1972). One important feature of this technique is that the beam is so narrow that the "slit broadening," discussed in Sect. 7.7 is minimal.

Such "multibounce" crystals can be quite useful at a synchrotron source, as a monochromator. As the width of a reflection is less for higher orders, harmonics can be rejected by a slight distortion of one face. The results is that the intensity is coming from the side of a reflection in first order, and beyond the width of a second order from $\lambda/2$. (Recall that since the synchrotron is a "white" source, harmonics can be as intense as the desired wavelength.)

There is very little difference in the Bragg case for intensity versus θ with and without absorption. Also, we have seen that in the Bragg case there is almost total reflection, and energy flow is nearly parallel to the surface during this reflection. It is therefore of some interest to consider the penetration of the fields in such a case. We consider the symmetric Bragg case, with no absorption and a centrosymmetric crystal, $F_{hkl} = F_{\overline{hkl}}$. To consider attenuation, we examine ξ_0'' and then K_0'' as we have done before. We choose Eq. (8.49a) as it relates ξ to the incident conditions. For the case we are considering,

$$\xi_0 = (k/2)|T|\Gamma F_{hkl}[\delta \pm (\delta^2 - 1)^{1/2}] \,. \tag{8.123}$$

The term ξ_0 is real only outside the range of total reflection, $|\delta| > 1$. Inside this range,

$$\xi_0' = (k/2)|T|\Gamma F_{hkl}\delta \,, \tag{8.124a}$$

$$\xi_0'' = (-k/2)|T|\Gamma F_{hkl}(1 - \delta^2)^{1/2} \,. \tag{8.124b}$$

(Only the minus sign is necessary as we shall see in a moment.) Both of these quantities are plotted in Fig. 8.41. Note particularly that ξ_0' varies linearly with $\Delta\theta$ within the range of reflection; the imaginary term is zero outside the range of reflection but is parabolic within this range.

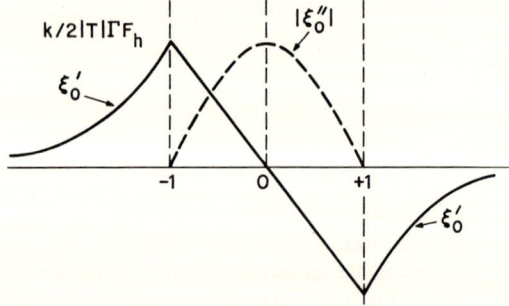

Fig. 8.41. Plot of real part ξ_0' and imaginary part ξ_0'', of ξ_0, as a function of δ' for the no-absorption case and symmetric Bragg conditions. The curved parts of ξ_0' correspond to points on the dispersion surface. The linear portion represents points between the diameter points of the dispersion sheets. The values of ξ_0'' give the extinction depth, or skin depth, for the penetration of radiation during total reflection*

8.10 Intensity in Bragg Reflections

From Eq. (8.32b), remembering that for no absorption, $F_0'' = 0$:

$$\xi_0'' = -K_0'' \sin\theta. \tag{8.125}$$

Hence, an "extinction factor" can be written for intensity as

$$\exp(-4\pi \mathbf{K}_0'' \cdot \mathbf{r}) = \exp[-2\pi k\{|T||\Gamma||F_{hkl}|(1-\delta^2)^{1/2}/\sin\theta\}Z]. \tag{8.126}$$

Here, Z is depth in the crystal. *This attenuation factor is very much larger than* the normal absorption factor for $\delta < 1$ (i.e., during a reflection) as it contains the *real part of the structure factor. Only a thin surface contributes even without normal absorption and that is why the Darwin reflection curve* (no absorption) *and the Darwin–Prins curve* (with absorption) *differ so little.*

8.10.3 Shift in Position of the Standing Wave

In discussing the fields inside a crystal for the Laue case, we found that the intensity at a position $\mathbf{r}$ depended on Eq. (8.96):

$$R = |1 \pm (E_{hkl}/E_0)\exp(-2\pi i \mathbf{r}_{hkl}^* \cdot \mathbf{r})|^2.$$

Recall that for a Bragg reflection, the low 2θ side of a peak is associated with the α branch, whereas the high 2θ side is from the β branch. Recall also, that for the α branch, the E ratio is negative and there are nodal points at $\mathbf{r}$ corresponding to the atomic planes, i.e., $R \to 0$. As δ and θ *change, from the low-angle side, R first decreases* (as $|E_{hkl}/E_0|$ approaches unity) and then increases, as we pass to the β branch. *The position of the standing waves shifts so that there are antinodal points on planes at the end of the reflection* (the high-angle side). This variation in R is shown in Fig. 8.42. The drop in absorption as the Bragg peak is approached occurs as soon as there is even a weak diffracted beam causing the E_0 and E_{hkl} fields to interact. The dashed line represents the case for atoms somewhat below the surface. The R value is lower there because the energy flow is parallel to the surface and less intensity reaches the lower depths. Note that the shape of the curve explains the shape of the intensity curve in Fig. 8.38.

When we consider normal absorption, then we know the absorption coefficient varies from $\mu_0(1-|T|\varepsilon)$ to $\mu_0(1+|T|\varepsilon)$ across the reflection, [see Eq. (8.86)]. With this fact, a direct proof of the appearance of nodal points in the Bragg case has been obtained by Batterman (1964). He employed an approach similar to the one we

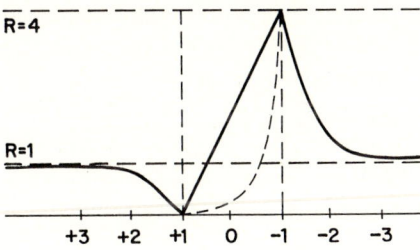

Fig. 8.42. Field intensity at an atom for a full reflection with no absorption in the Bragg case as a function of δ'. The nodes which start at the atoms, α branch ($\delta = +1$), move linearly with glancing angle, until antinodes are at the atoms, β branch ($\delta = -1$). The *solid curve* is for surface atoms, the *dashed curve* is for atoms deeper in the crystal*

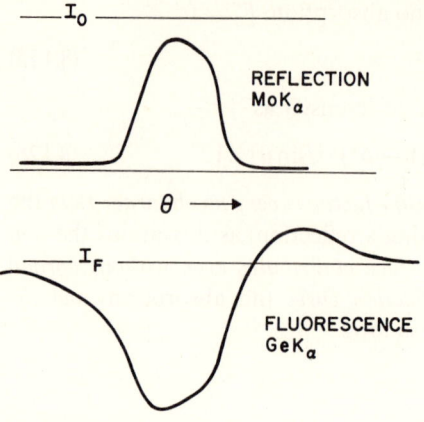

Fig. 8.43. Fluorescence measured during a rotation through a Bragg peak, used as a measure of the field strengths shown in Fig. 8.42. The background in the *bottom figure* is the fluorescence far from a reflection*

described for neutron diffraction. He measured the K shell fluorescence from a perfect Ge crystal produced by Mo K_α radiation and investigated the variation in this fluorescence through a reflection. The results are shown in Fig. 8.43 and correlate extremely well with Fig. 8.42. Far on the low-angle side of the peak, the intensity of the field is away from the atoms. The primary beam penetrates more deeply before it is absorbed, but the fluorescence is absorbed from large depths on leaving the crystal and its intensity is *lower* than the value far from a Bragg reflection. The surface generation on the high-angle side leads to higher intensity for the fluorescence. Batterman (private communication) has suggested that this asymmetric fluorescence could be employed to probe for interstitials, and recent application of the technique to the study of ion-implanted interstitials in silicon has been made by Anderson et al. (1976). The wave fields extend some distance *above* the surface, and so it has been possible to determine the position and spacing of adsorbed films on a perfect crystal surface (Cowan et al 1980).

8.10.4 Remarks Concerning Integrated Intensities

It was pointed out in Sect. 8.6 that dynamical theory should really be more precisely formulated in terms of spherical waves. This approach is reviewed by Authier (1970). We have chosen the simpler plane wave approximation for a first introduction to this topic. However, it turns out that the equations for widths of peak and integrated intensities are the same, so that the formulae given in this chapter can be employed with confidence.

One word of caution is necessary concerning an actual measurement of integrated intensities. In view of the low absorption through thick crystals, it is important to correct for the absorption due to all scattering processes such as that due to the Compton effect. For example, with Ag K_α and a 220 reflection in transmission through a thick slab of silicon, errors as large as 50% are encountered if

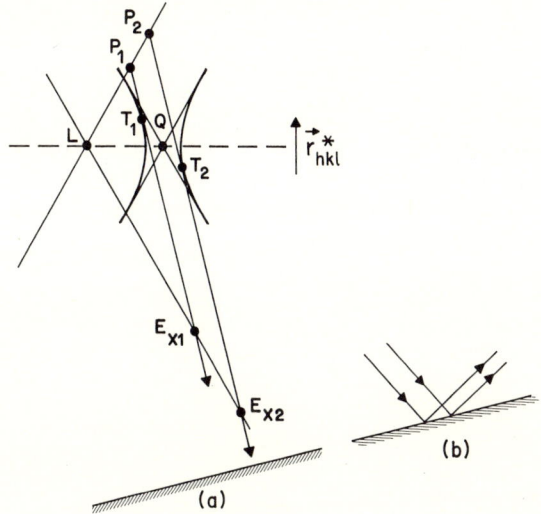

Fig. 8.44. a Asymmetric Bragg reflection. Note that in this case the angular width in the incident beam $P_1 P_2$ is less than the angular width in the diffracted beam $E_{x1} E_{x2}$. At the same time, b the cross section of the diffracted beam is less than that of the incident beam*

Compton scattering is not substracted. As the diffraction angle increases, this effect becomes less important because the Compton scattering comes from all electron shells; the contribution from the outer shells decreases rapidly with angle, in contrast to the photoelectric contribution which is mainly due to the K shell. Details on this problem and a correction for it can be found in Giardina and Merlini (1973).

Consider crystals where the reflecting planes are tilted with respect to the surface, as shown in Fig. 8.44a. Here the physical surface is such that the incident beam is more nearly perpendicular to this boundary. However, the energy flow is still parallel to the surface. The angular range $P_1 P_2$ of total reflection is now narrower than in the symmetric case, whereas the range or divergence in the diffracted beam is larger ($E_{x1} - E_{x2}$). As shown in Fig. 8.44b, the diffracted beam is physically narrower, even though its divergence is larger. If the beam is more nearly parallel to the surface, the opposite will occur. Thus such asymmetrically cut crystals have possibilities for "condensing" or "collimating" a beam.

Interesting papers on the effect of elastic strain and plastic displacement can be found in Penning and Polder (1961), Okkerse and Penning (1965), and Azaroff et al. (1974). The work of Hirsch et al. (1965) is an excellent book on electron diffraction and the applications of dynamical theory to it. Cowley (1975) is a valuable treatise on many-beam theory for this same field. An excellent review of dynamical theory in greater depth can be found in Cowley (1975), James (1963), and Authier (1970). We now turn to a brief discussion of imaging dislocations and other strain fields. This "topographical" method is an application of the differences in kinematic and dynamic theory. There are perfect and imperfect regions, so both theories are important. There are many discussions of this topic for electron microscopy [Hirsch et al. (1965) and Cowley (1975), for example] so we shall briefly discuss the x-ray method — both the techniques and the images.

8.11 X-ray Topography

8.11.1 The Various Techniques

Fig. 8.45 shows several of the techniques for topography, or "x-ray microscopy". Note that both transmission and back reflection can be employed. Pertinent references are also given. We first examine the techniques, then we shall discuss

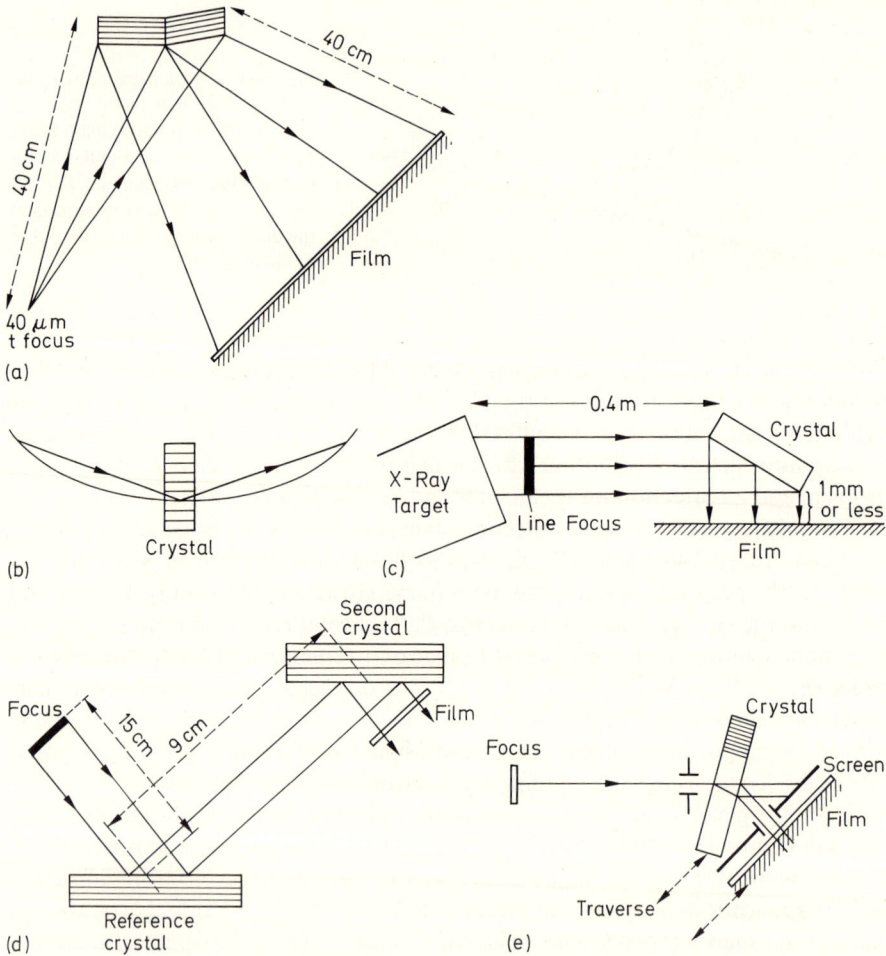

Fig. 8.45. a The Schulz technique [J. Metals *200*, 1082 (1954)]. **b** Guinier-Tennevin method [Acta Crystallogr. *2*, 133 (1940); see also Weissmann, S., J. Appl. Phys. *27*, 389 (1956)]. **c** Berg-Barrett technique [Berg, W., Naturwissenschaften *19*, 391 (1931); Barrett, C. S., Trans. AIME *101*, 15 (1945); see also, Newkirk, J.B., Trans. AIME *215*, 483 (1959); Turner, A.P.L., Vreeland, T., Jr., and Pope, D.P., Acta Crystallogr. A *24*, 452 (1968)]. **d** Modified double crystal Berg-Barrett method [Bonse, U., Z. Phys. *153*, 287 (1958)]. **e** Scanning method ($\mu_0 t < 1$: Lang, A.B., J. Appl. Phys. *29*, 597 (1958); *30*, 1748 (1959). $\mu_0 t \gg 1$: Borrmann, G. et al., Z. Naturforsch. *13a*, 423 (1958)]. (Figures reprinted with permission from Bonse, U., Hart, M., and Newkirk, J.B., "Encyclopaedic Dictionary of Physics", Supplementary Volume 1, p. 391. Copyright, Pergamon Press Ltd., New York, 1961)

contrast. In the Schulz technique (a) white radiation is employed and differences in perfection (not tilt) cause most of the contrast unless the film is far from the specimen. Guinier and Tennevin (b) used the principle of a focusing circle to cause a sharp diffraction line with line focus and a "good" crystal. (Monochromatization occurs as a result of the reflection.) The presence of subgrains breaks the line into spots. Weissman developed techniques to trace such a reflection with film back to the specimen. Then the location of tilts and distortions could be correlated with optical or electron microscopy.

In the Berg-Barrett method (c), the crystal is set to reflect the characteristic radiation, with a film close to the crystal. Note how the line focus is placed; the thin dimension of the target is perpendicular to the illustrations.[15] In this way vertical divergence, which causes image distortion, is reduced but a large area of the sample is exposed. Also there is a near normal incidence on the film; oblique incidence broadens the image unless the thin emulsions of high-resolution nuclear plates are employed. Variations of the reflecting power due to imperfections cause the image; distortions cause the scattering to be kinematic and hence, greater than from perfect regions. Individual dislocations can be resolved. In the illustration, the planes shown in the specimen produce a zero layer reflection. There is some shortening of the image because of the angle of the planes, but because of the low angle of incidence a wide area of the specimen can be covered and the film can be close to the surface. Non-zero layer spots can be used but cause more distortion. The depth of penetration can be adjusted by varying the angle of incidence. Contrast is limited by stray radiation. While the technique does not depend on collimation, some is generally employed to reduce the beam size to the specimen and to avoid general scattering. A filter on the film may be helpful to reduce fluorescence *if* the diffracting wavelength is longer than that of the fluorescence and the filter has an absorption edge between the two; otherwise, lowering the x-ray tube voltage below the voltage necessary to cause such fluorescence may help. The film is often held in a cassette that can be evacuated to "pull" it to the front of the cassette; this helps keep the film flat and allows a closer approach to the specimen. This is necessary to avoid spreading of the image due to the $K_{\alpha_1} - K_{\alpha_2}$ doublet.

A reflection can be quickly located with a fluorescent screen and "tuned" for the maximum intensity with successive exposures on Polaroid film a few centimeters from the specimen. Then the final film can be taken employing a fine emulsion (which as a result takes much longer to expose). Eastman Kodak Recordak Micro-File Film (type 5455) has been found to give quite good pictures in a few minutes and can be magnified perhaps $100 \times$. High-resolution plates can take 10–20 h to expose but can be magnified $500-400 \times$.

As the film is close to the specimen, if there are simultaneous reflections, these will overlap and degrade the image. These can be avoided in the following way (Turner et al. 1968). A stereographic projection of the crystal is made and around each pole or plane normal, a circle of radius $90° - \theta$ is drawn. The incident and

---

[15] The point focus windows on an x-ray tube accomplish this.

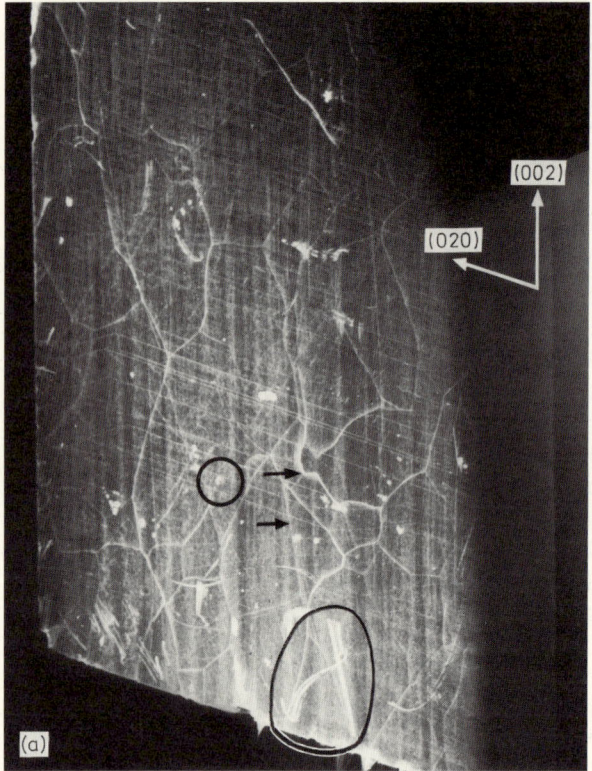

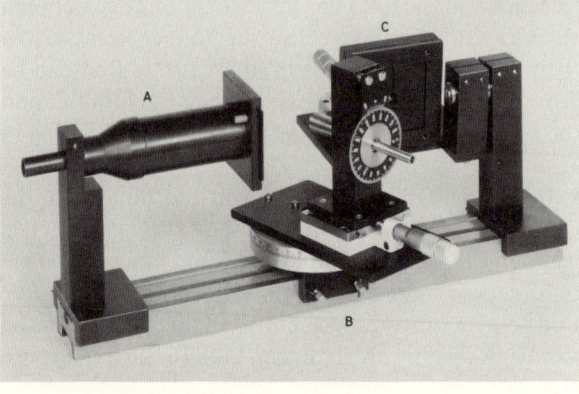

Fig. 8.46. a Typical Berg-Barrett photograph of cleaved LiF, 202 reflection CoK$_\alpha$ radiation. Kodak Type R film enlarged 20 ×. The surface is a {100} plane. (Photographs by T.B. Wu and I.C. Noyan, taken at Northwestern University.) **b** Typical equipment: *A*, divergent beam tube with slits; *B*, goniometer with ω and ϕ tilts; *C*, film holder (which can be evacuated to hold the film flat. (Courtesy of Blake Industries, Scotch Plains, NJ)

diffracted beams for one pole will lie on this circle at opposite ends of a line through the small circle's (true) center. Reflections are chosen such that the diffracted beam will occur near the equator, and 5–10° from other reflection circles. Of course, the crystal may be turned around its normal to produce such orientations. The incident beam should make as large an angle with the normal to the surface as is possible. It is sometimes worthwhile to try such plots for different wavelengths. There are other reasons to consider different x-ray tubes; depth in the crystal can be varied.

By choosing $2\theta \approx 90°$, the π state scattering is minimized. Also as shown in Fig. 8.44, proper choice of the diffracting plane can minimize the angular divergence of a reflection and narrow features in the image.

With some care, and for high-angle reflections whose angular widths are quite narrow, double images of dislocations can be observed due to the tilts on opposite sides of dislocations. A typical Berg–Barrett photograph is shown in Fig. 8.46 along with standard equipment. Use of softer radiation is sometimes convenient as thinner emulsions can be employed, and because of the decreased penetration, more heavily dislocated crystals can be examined.

Finally, the Berg–Barrett technique's sensitivity can be enhanced by first monochromating the beam, as shown in Fig. 8.45d; the monochromator should be of the same material as the specimen, and cut to the same planes, and set in $(1, -1)$ geometry. The reflection is then independent of wavelength dispersion. If the specimen is set on a steep portion of its reflection, tilts in the specimen of <0.1 arc sec or strains $\Delta d/d < 10^{-8}$ can be seen, corresponding to distances 100 μm from a dislocation. By using asymmetrically cut crystals in the diffracted beam, the image may be magnified before recording it (Boettinger et al. 1979).

Intense radiation from a synchrotron makes it possible to perform realtime topography using TV cameras coupled to a radiation sensitive phosphor. Also, the resolution and intensity are sufficient so that each spot in a Laue pattern is an excellent topograph. A fine example of this possibility is shown in Fig. 8.47. With this range of diffraction vectors in a single film, a great deal can be learned about the source of contrast from the conditions described in Sect. 7.3.

The Lang ($\mu_0 t \approx 1$) and Borrmann ($\mu_0 t > 10$) methods are shown in Fig. 8.45e. The beam is collimated just enough so that only K_{α_1} will diffract (not K_{α_2}); one slit of 10–100 μm is employed at some distance from the target, as shown. If w is the width of the focus, W the slit width, and A the target to slit distance, $(W + w)/A$ must be less than $\theta_{\alpha_1} - \theta_{\alpha_2}$, for then the beam is collimated well enough so that both component wavelengths cannot diffract at the same angular setting. The specimen and film are traversed past the slit to give an image of a large specimen. For highest sensitivity however, no scan is performed and this is called a "section topograph". Recall that waves propagate in all directions inside the Borrmann fan, and that there is interference between these waves, either from the same branch if the crystal is thin, or from parallel beams from different branches. Parallel fringes appear on a film from a perfect crystal in the form of a flat plate, and their visibility decreases as imperfection increases. These effects are illustrated in Fig. 8.48.

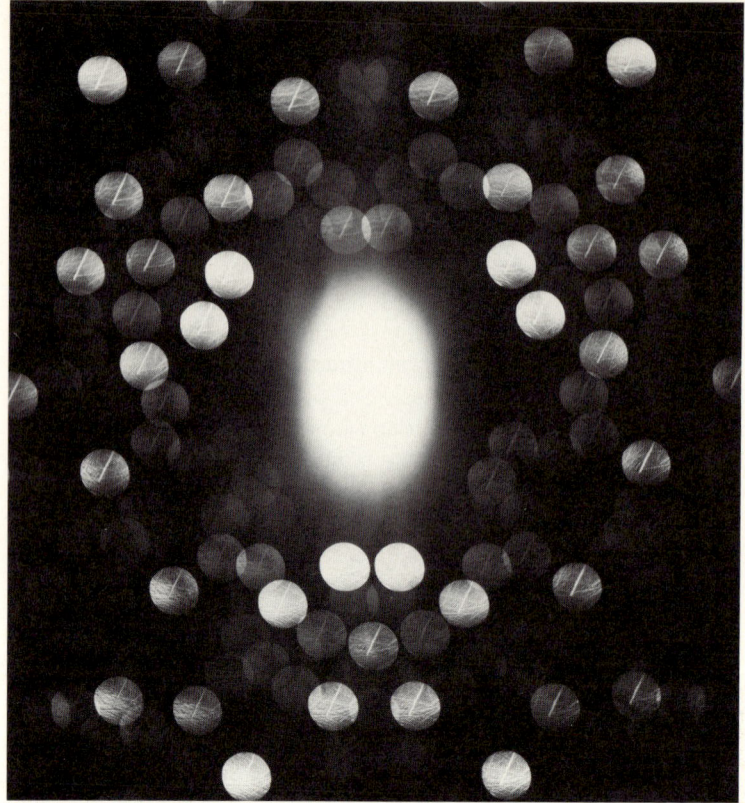

Fig. 8.47. Laue pattern of a garnet single crystal wafer used as a substrate for magnetic garnet epitaxial films, for magnetic bubble memory. Taken in transmission at the Stanford University synchrotron source, 3.7 GeV, 18 mA, 15 second exposure through a 6mm aperature. The 0.5 mm wafer was 10 cm from the Kodak Type R film. Each Laue spot is a topograph. (W. Parrish, IBM Corp.)

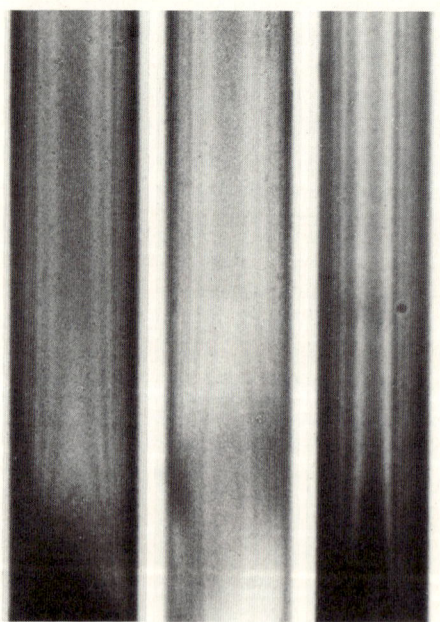

Fig. 8.48. Section topographs of silicon. Elastic deformation has eliminated Pendellösung fringes at the bottom of the photographs. The topographs at the *left* and *center* were taken with the 333 diffraction vector; the former was closer to the strain center than the latter. The topograph at the *right* was taken with the 040 diffraction vector. (Courtesy of S. Stock, Georgia Institute of Technology)

8.11 X-ray Topography

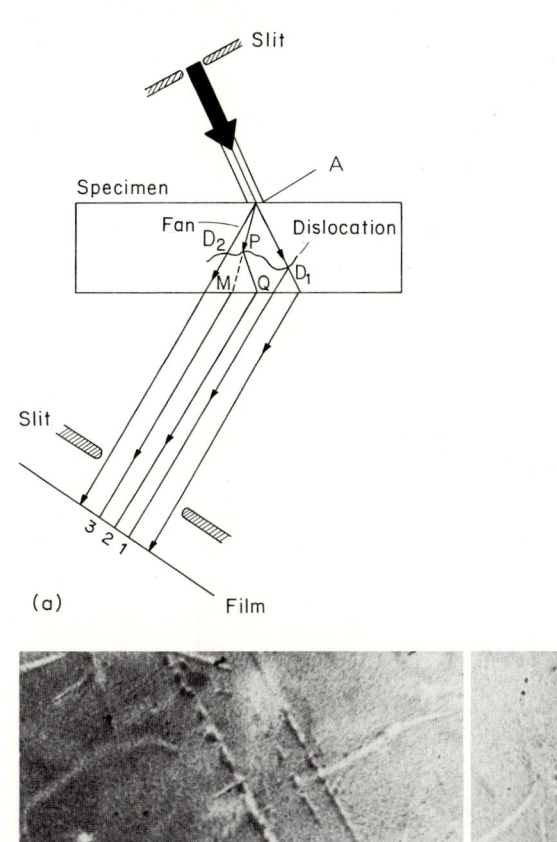

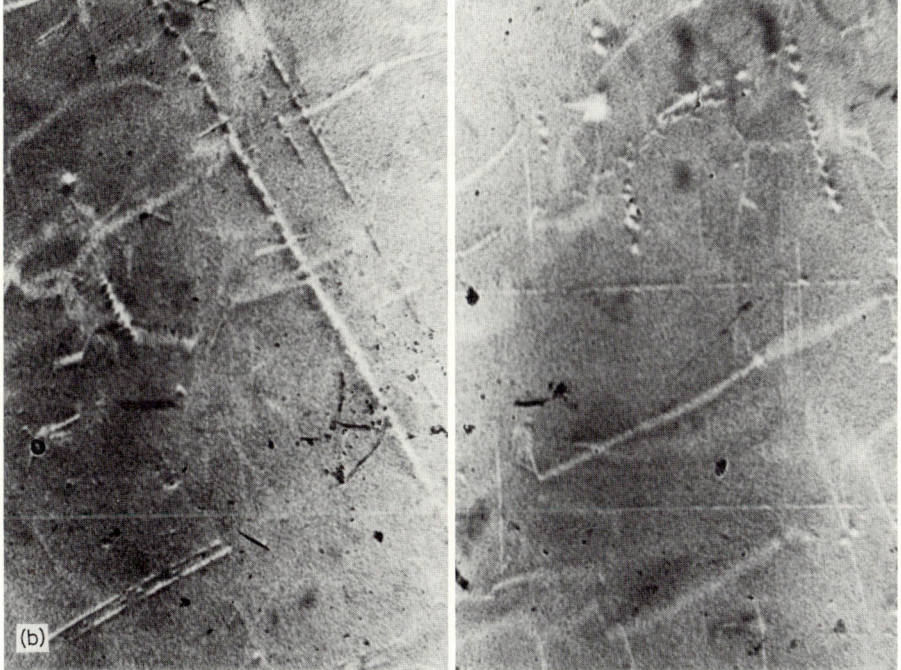

Fig. 8.49. a,b Typical transmission topographs. Ge crystal: 156×. (Film photographed in an optical microscope by A. Purdes, M. James, and B. Ditchek taken at Northwestern University.) **c** The equipment: *1*, beam guide tube; *2*, slit housing; *3*, ω and 2θ gears; *4*, film holder; *5*, movable table; *6*, slit between samples and film defining diffracted beam; *7*, goniometer for specimen; *8* and *9*, ω and 2θ drives; *10*, track for detector (**Fig. 8.49c** see next page)

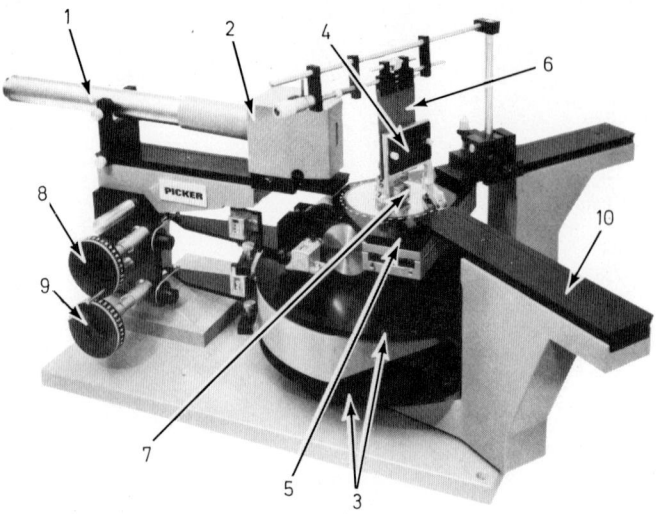

Fig. 8.49c

8.11.2 Contrast Mechanisms

We now turn briefly to contrast mechanisms in traverse patterns. We shall assume that the incident beam has a much wider angular divergence than the reflection. Then most of the beam does not make the right angle for reflection, unless regions of strain $[\Delta\theta = (\Delta d/d)\tan\theta]$ or regions of tilt $(\Delta\theta)$ occur that bring the region into reflection. Such regions diffract kinematically, proportional to the volume of the region, and because of the angular range in the incident beam, the resultant contrast is essentially an integrated intensity.

This "direct image" is the principle mechanism in Berg–Barrett films or for thin crystals in transmission. But in transmission techniques with thick crystals (the Borrmann technique), the contrast is much richer. Consider a dislocation intersecting the Borrmann fan at D_1 and D_2 in Fig. 8.49a. The diffraction of the direct beam by the strained region near D_1 gives rise to the *direct image*, 1. Next consider the path, AP, of a particular wave field which passes very close to the dislocation at P. This path will be perturbed: Near the core of the dislocation, the strain is so great that interbranch scattering may occur. Wave fields propagating along AP will excite wave fields propagating in a parallel direction along AM, and new fields having a tie-point on the other branch of the dispersion surface (and propagating along PQ). This results in a decrease in intensity along PM and this lighter region is called the *dynamic image*, 3. The increase in intensity along PQ leads to the intermediary image, 2. Here, the intensity in this image is often fringed, due to the interference of the original and new wave fields. Farther from the dislocation core, the wave fields are only curved or bent by the local strain, and the image deviates from its original position, leading to a further decrease in intensity of the dynamical image.

8.11 X-ray Topography

Table 8.2. Comparisons of various x-ray topographic techniques[a]

Technique	Schulz, Guinier and Tennevin (Fig. 8.45a, b)	Berg-Barrett (Fig. 8.45c)	Double crystal (Fig. 8.45d)	Scanning transmission (Fig. 8.45e)	
				$\mu_0 t \sim 3$	$\mu_0 t < 1$
Apparatus	Simple	Simple	Complicated	Complicated	Complicated
Exposure time	10–25 h	Short (~ 1 h)	Short (~ 1 h)	10–30 h	2–10 h
Defect for which technique is most suited with kind of contrast[b]	Grain misorientation, subgrains (1)	Subgrains (1) Dislocations (2)	Subgrains (1) Dislocations (1) Stacking faults	Dislocations (2), (3)	Subgrains (1) Dislocations (2) Stacking faults
Best geometric resolution	50 μm	1 μm	1 μm	1 μm	1 μm
Sensitivity to deformations	Low	Low	High	High	Low
Sensitive to the sense of deformations	Tilts: Yes Inhomogeneous deform: No	Subgrains: Yes Dislocation: No	Yes	Yes	No
Thickness t of specimen contributing to topograph[c]	Schulz: ≦5 μm Guinier and Tennevin 50 → 1000 μm	≦5 μm	≦5 μm (back reflection) ≦300 μm (transmission)	0.1 → 5 mm	0 → 2 mm
Dislocation image width[d]	—	1 → 5 μm	Up to 150 μm	Up to 150 μm	1 → 10 μm
Upper limit of dislocation density (lines/m²)	—	5×10^8	10^9	5×10^7	5×10^8

[a] Reproduced with permission from Bonse, U.K., Hart, M., and Newkirk, J.B., "Encyclopaedic Dictionary of Physics," Supplementary Volume 1, p. 391. Copyright Pergamon Press Ltd., New York, 1961.
[b] (1) and (2) homogeneous dilation and tilt contrast, and (3) dynamical contrast.
[c] This is determined in the Bragg case by the extinction depth, and in the Laue case, by the value of μ_0 (the absorption coefficient) for the material and the value of t imposed by the technique.
[d] Based on the assumption that this is determined by normal image overlap.

The relative importance of each image depends on the position within the Borrmann triangle. Thus in very thick crystals only the dynamic shadow is generally present in so-called "Borrmann topographs". Note in Fig. 8.49a that by adjusting the exit slit, the position of a defect within a crystal can be ascertained. It is also worth pointing out that if the incident beam is very wide, so also are the exit beams and it is difficult to place a film in only the diffracted or direct beam close to the specimen. This is another reason for collimation, as well as the monochromatization mentioned earlier.

A typical transmission topograph taken with a diffracted beam is shown in Fig. 8.49b. The contrast would be reversed in the direct beam; the equipment is shown in Fig. 8.49c.

It is possible to determine Burger's vectors (**b**) as these disappear in reflections, such that $\mathbf{r}^* \cdot \mathbf{b}$ and $\mathbf{r} \cdot \mathbf{n} = 0$, where **n** is normal to the slip plane. For such reflections, the interplanar arrangement is undisturbed by the dislocation as can be readily seen by sketching an edge dislocation and these conditions.

In Table 8.2 the various techniques are compared. It is possible to see coherent precipitates, dislocations, or effects of strains due to vapor deposits on a crystal and also magnetic domains (due to magnetostriction). Unfortunately, due to the widths of images and the limits of magnification of x-ray film, the upper limit of imperfection density it is possible to image is $\approx 10^9/m^2$. The eye can resolve 10^{-4} m. At $1000\times$ magnification, two dislocation lines $2 \times (10)^{-4}$ m apart correspond to a density of $\approx 10^{13}/m^2$, 3–4 orders of magnitude higher than the values in the table. If ways can be found to reduce image widths and yet give significant contrast, these techniques would then be of more general use. Unfortunately now, few materials have low enough imperfection densities to be useful. With the new high-intensity x-ray generators (200–2000 mA), or the high intensity of radiation from synchrotrons (Tuomi et al. 1974; Hart, 1975) now being used in these studies, narrower images mgiht be produced, for example, by working near the edge of the x-ray fan. Nevertheless, extensive applications exist for the highly perfect materials, silicon, germanium, GaAs, etc., that are widely used in the electronics industry, or in research employing a few other readily obtainable highly perfect materials, such as LiF, Cu, and Zn.

For further information see Newkirk and Wernick (1962), Amelinckx et al. (1970), Authier (1970), Tanner (1976), and Tanner and Bowen (1980).

References

Amelinckx, S., Gevers, R., Remaut, G., and Van Landuyt, J. (eds.), "Modern Diffraction and Imaging Techniques in Materials Science." North-Holland, Amsterdam, 1970

Anderson, S.K., Golovechenko, J.A., and Mair, G., Phys. Rev. Letters *37*, 1141 (1976)

Authier, A., "Advances in Structure Research by Diffraction Methods" (R. Brill and R. Mason, eds.), Vol. 3. Pergamon, New York, 1970

Azaroff, L.V., Kaplow, R., Kato, N., Weiss, R.J., Wilson, A.J.C., and Young, R.A., "X-Ray Diffraction." McGraw-Hill, New York, 1974

Batterman, B.W., Phys. Rev. *126*, 1461 (1962)

Batterman, B.W., Phys. Rev. *133*, A759 (1964)

Boettinger, W.J., Burdette, H.E. and Kuriyama, M., Rev. Scientific Instr. *50*, 26 (1979)
Bonse, U., and Hart, M., Appl. Phys. Lett. 7, 238 (1965a)
Bonse, U., and Hart, M., Z. Phys. *188*, 154 (1965b)
Bonse, U., and Hart, M., Z. Phys. *189*, 269 (1966a)
Bonse, U., and Hart, M., Z. Phys. *189*, 151 (1966b)
Bonse, U., and Hart, M., Z. Phys. *190*, 455 (1966c)
Borrmann, G., Z. Phys. *42*, 157 (1941)
Borrmann, G., Naturwissenschaften *38*, 330 (1951)
Borrmann, G., Hildebrant, G., and Wagner, H., Z. Phys. *142*, 406 (1955)
Brady, G.W., J. Chem. Phys. *57*, 91 (1972)
Cole, H., Chambers, F. W., and Wood, C., J. Appl. Phys. *32*, 1942 (1961)
Cowan, P.L., Golovchenko, J.A., and Robbins, M.F., Phys. Rev. Letters *44*, 1680 (1980)
Cowley, J.M., "Diffraction Physics." North Holland, Amsterdam, 1975 (American Elsevier, New York)
Darwin, C.G., Philos. Mag. *27*, 315, 675 (1914)
Duncumb, P., Philos. Mag. 7, 2101 (1962)
Ewald, P.P., Ann. Phys. *49*, 1, 117 (1916); *54*, 519 (1917); Acta Crystallogr. *11*, 888 (1958)
Giardina, M.D., and Merlini, A., Z. Naturforsch. *28a*, 1360 (1973)
Hart, M., J. Appl. Crystallogr. *8*, 436 (1975)
Hirsch, P.B., Howie, A., Nicholson, R.B., Pashley, D.W., and Whelan, M.J., "Electron Microscopy of Thin Crystals", Butterworth, London, 1965
Hordon, M.J., and Averbach, B.L., Acta Metall. *9*, 237 (1961)
Hountas, A., Filippakis, S.E., Papathanassopoulos, C., and Tsakalakos, Th., J. Phys. C. *6*, 1693 (1973)
James, R.W., "The Optical Principle of the Diffraction of X-Rays," p. 304, Bell, London, 1950
James, R.W., Solid State Phys. *15*, 55 (1963)
Kato, N., Acta Crystallogr. *11*, 885 (1958)
Kato, N., Acta Crystallogr. *13*, 349 (1960)
Knowles, J.W., Acta Crystallogr. *9*, 61 (1956)
Miller, F., Phys. Rev. *47*, 209 (1935)
Newkirk, J.B., and Wernick, J.H. (eds.), "Direct Observation of Imperfection in Crystals." Interscience, New York, 1962
Okkerse, B., Philips Res. Reports *126*, 464 (1962)
Okkerse, B., and Penning, P., Philips Res. Reports *18*, 82 (1965)
Overhauser, A.W., and Collela, R., Amer. Scientist *68*, 70 (1980)
Patel, J.R., and Batterman, B.W., J. Appl. Phys. *34*, 2716 (1963)
Penning, P., and Polder, D., Philips Res. Reports *16*, 419 (1961)
Tanner, B.K., "X-ray Diffraction Topography", Pergamon Press, New York, (1976)
Tanner, B.K. and Bowen D.K., eds. "Characterization of Crystal Growth Defects by X-ray Methods", Plenum Press, New York, (1980)
Tuomi, T., Naukkarinen, K., and Rabe, P., Phys. Status Solidi, *A 25*, 93 (1974)
Turner, A.P., Vreeland, R., Jr., and Pope, D.P., Acta Crystallogr. *A 24*, 452 (1968).
Wagenfield, H., J. Appl. Phys. *33*, 2907 (1962)
Warren, B.E., Acta Crystallogr. *12*, 837 (1959); J. Appl. Phys. 7, 1111 (1959)
Warren, B.E., "X-Ray Diffraction." Addison-Wesley, Reading, Massachusetts, 1969

Problems

8.1. Verify the results in Table 8.1 by direct calculation.

8.2. Repeat the calculations in Table 8.1 for Mo K_α and compare the results.

8.3. It is estimated that the convolution of two reflections from perfect crystals in $(1, -1)$ geometry leads to a reflection width ≈ 1.32 times that for a single reflection.

Furthermore, although the width of the Darwin curve is $2\delta'$, because of the tails the actual width at half intensity is $2.12\delta'$. Using the data in Table 8.1 and Fig. 8.39a, compare experimental and theoretical rocking curves. (Notice that the widths can be employed to measure scattering factors.)

8.4. Away from Bragg peaks the absorption coefficient $\mu = 2\pi \Gamma K F_0''$. But in tables, $\Delta f''$ is often zero. Explain.

8.5. (a) From Eq. (8.115), show that for a thick crystal and an unpolarized beam

$$\ln[(1/P_{hkl})/(\mu_0 t/\gamma_0)] = (1-\varepsilon) + \{\ln(2\sin 2\theta/\Gamma F_{hkl})$$
$$+ \tfrac{1}{2}\ln[8(\mu_0 t/\gamma_0)(\varepsilon/\pi)]\}/(\mu_0 t/\gamma_0).$$

(b) Integrated intensity measurements are made of the 220 peaks in symmetric Laue geometry for a number of different thicknesses of highly perfect silicon single crystals with Cu K_α radiation. The data are given below. Calculate ε and compare with theoretical values.

t_0(mm):	0.511	0.830	0.985	1.000	1.400	1.800
$P_{220}(\times 10^7)$:	10.370	5.750	5.450	5.055	3.760	3.016

8.6. (a) Show that for $|\delta| \gg 1$, the intensity I is proportional to $1/4(\delta')^2$ and that for n reflections, $I \propto 1/4(\delta')^{2n}$. Start with Eq. (8.118).
(b) Use the data in Fig. 8.39b to verify these results.

8.7. For a symmetric Bragg reflection from a centrosymmetric crystal, the integrated intensity can be written from Eq. (8.118) as

$$P_{hkl} = \frac{\lambda^2 (e^2/mc^2) N |F'_{hkl}| |T|}{\pi \sin 2\theta} \int \left| \frac{E^e_{hkl}}{E^e_0} \right|^2 d\delta .$$

Here

$$|E^e_{hkl}/E^e_0|^2 = G(\delta) - [G^2(\delta) - 1]^{1/2},$$

and N is the number of atoms per unit volume.

$$G(\delta) = \frac{\delta^2 + x^2 + \{[\delta^2 - (1+y^2-z^2)]^2 + [2x(\delta+z)]^2\}^{1/2}}{1+y^2},$$

$$x = \frac{F_0''}{|T| F'_{hkl}} ; \quad y = \frac{F''_{hkl}}{F'_{hkl}} ; \quad z = \frac{|T| F''_{hkl}}{F_0''}.$$

[See Miller, F., Phys. Rev. 47, 209 (1935).]
(a) Evaluate this expression numerically or graphically for the σ and π states for a silicon 111 reflection for Cu K_{α_1}.
(b) In a rocking curve arrangement, at the first crystal, the incident beam may be assumed to be unpolarized. Hence, show that

$$P_{hkl}^{\sigma+\pi} = \tfrac{1}{2}(P_{hkl}^\sigma + P_{hkl}^\pi).$$

(c) For the second crystal, show that

$$P_{hkl}^{\sigma+\pi} = \frac{P_{hkl}^\sigma + |\cos 2\theta| P_{hkl}^\pi}{1 + |\cos 2\theta|}.$$

(d) Compare the theoretical values from (a) and (c) with the experimental values employing Fig. 8.39a.

$$P_{hkl}^{meas} = \frac{1}{I_0 A_0} \int IA \, d\theta \cong \frac{1}{I_0 A_0} \sum_i (IA)_i \Delta\theta_i,$$

where $I_0 A_0$ is the incident power (10,290 cps) and IA is the measured power. Disregard Compton scattering or TDS. The background intensity was 8 cps.

Integrate Fig. 8.39a from $\delta_{max} = 8.4526$ to $\delta_{min} = -8.4526$. [Keep in mind the scale in Fig. 8.39a is θ, so you must convert to a δ scale with Eq. (8.119)]. Beyond these values of δ assume the results in Problem 8.6 for the variation in intensity with δ in the tails. That is, show that after correcting for background:

$$P_{hkl}^{meas} = \frac{1}{I_0 A_0}\left[\left\{\sum_i (IA)_i \Delta\theta_1\right\}_{\text{meas region}} + \frac{\Gamma F_{hkl}'}{\sin 2\theta}[IA(\text{at } \theta_{min})\delta_{max}' - IA(\text{at } \theta_{max})\delta_{min}']\right].$$

8.8. Equation (8.86) expresses the variation of absorption with incident conditions. Consider a topograph taken in transmission with a thick crystal. To estimate the width of contrast from a defect of length A, take the criterion of the intensity falling to half its value. Derive an equation for this width at the exit surface as a function of A and Z, the depth of the defect in the crystal.

8.9. Repeat the calculations of absorption of the various states presented in Sect. 8.7 with the more precise estimate, $\varepsilon = \exp(-M)$.

Appendix A: Location of Useful Information in International Tables for Crystallography[1]

1. Point groups... Vol. I, pp. 26, 27
2. Other symbols for point groups............................ Vol. I, p. 44
3. Space group symbols.............................. Vol. I, pp. 28, 49, 50
4. Symbols for space groups with different choice of axes........ Vol. I, p. 545
5. Angles between planes in the cubic system.................. Vol. II, p. 120
6. X-ray wavelengths.................................. Vol. IV, pp. 6–43
7. Filters for x-rays................................. Vol. III, pp. 75, 76
8. Rules for extinctions............................... Vol. I, pp. 53, 54
9. Diffraction symbols and their related space groups........... Vol. I, p. 347
10. X-ray atomic scattering factors..................... Vol. IV, pp. 72–146
11. Dispersion corrections for x-ray atomic
    scattering factors............................... Vol. IV, pp. 149, 150
12. Data on incoherent x-ray scattering...................... Vol. III, p. 232
13. Electron atomic scattering factors................. Vol. IV, pp. 155–257
14. Inelastic electron scattering factors................ Vol. IV, pp. 256–268
15. Neutron atomic scattering factors.................. Vol. IV, pp. 270, 271
16. X-ray absorption coefficients........................ Vol. IV, pp. 47–66
17. Neutron absorption coefficients......................... Vol. III, p. 197

---

[1] "International Tables for Crystallography," Vol. I (1952), Vol. II (1959), Vol. III (1962) (K. Lonsdale, ed.), Vol. IV (1974) (J.A. Ibers and W.C. Hamilton, eds.). Kynoch Press, Birmingham, England. There is a newer version of Vol. I that contains other information, particularly on groups but also some other useful items, such as drawings of the cubic space groups (not present in Vol. I): Volume A (ed: T. Hahn) D. Reidel Publishing Company, Boston, MA 1983

Appendix B: Crystallographic Classification of the 230 Space Groups

Appendix B: Crystallographic Classification of the 230 Space Groups

Reproductions of this chart with color coding sized 22 × 34 inches suitable for wall mounting may be obtained from the author, Stanley K. Dickinson, Jr., CRWE, Air Force Cambridge Research Laboratories, L.G. Hanscom Field, Bedford, Massachusetts.

Appendix C: Determination of the Power of the Direct Beam in X-ray Diffraction

Any measurement of the absolute intensity is generally made because considerable precision is required to compare with calculations. Accordingly, comments on general techniques are appropriate. Filtered radiation, measured with a proportional or scintillation counter, even with a pulse-height analyzer, does not provide a narrow enough range of wavelengths for this kind of comparison. (See Sect. 3.11.) The new solid state detectors, however, do have a sufficently narrow range, but for the best precision crystal monochromators should still be employed. In general, these should be bent to focus the beam, thereby reducing the region sampled in reciprocal space and increasing the intensity. Quartz, silicon, or germanium is employed when high resolution is required, for example, for measurements near a Bragg peak. These crystals are in general use at storage ring sources where there is adequate intensity even for the most demanding experiments, and they are bent elastically for focussing as described in the text, Sect. 4.7. When higher intensity is needed for a laboratory source, but not resolution, bent LiF or pyrolitic graphite is preferred. The latter is available commercially and is particularly easy to use because there is uniform mosaic, and it is not necessary to "hunt" for a strong uniform beam from the crystal, as it is with LiF. These crystals are not useful at storage rings, because the divergences due to mosaic spread the beam to dimensions of several cm over the distances employed at such sources. For further details on monochromators, the reader is referred to Sparks and Borie (1966) and Schwartz et al. (1963).

To improve counting stability, a second counter can be employed to measure the fluorescence from a thin foil in the exit beam of an incident beam monochromator (absorbing $\approx 5\%$ of the beam). Counts at each position are made for a given number of these fluorescent counts. This has the advantage of minimizing errors in the data due to aging of the x-ray tube or due to change in barometric pressure and hence, absorption of x-rays in the air path. (This technique or variations of it are in use at storage rings, as the beam decays with time, and a similar technique is employed in neutron diffraction as reactor power can vary considerably during the time of measurement for reasons of control.) Experiments that typically take two days to several weeks or more, even with computer control, can be subject to such errors, which can exceed the statistical counting error with the long wavelengths (Cu, Co, Cr) generally employed for this work; in fact, they can easily reach levels of 2–5% or more without a monitor. Also, these changes are likely to occur systematically during a run; for example, as a weather front passes over the laboratory, air density will change, and hence beam attenuation will vary.

Appendix C: Determination of the Power of the Direct Beam in X-ray Diffraction

All receiving and monochromator slits in a good system should be adjustable both vertically and horizontally and the defining regions should be wires, not edges as recommended by Chipman (1969).

To reduce the general background scattering, a "scatter" slit is recommended on the sample side of the receiving slit. This is set just wider than the receiving slit (generally by closing it and watching until it affects a peak and then opening it slightly). It is a good idea to cover these two slits and the counter with a thin "box" of lead to further reduce background.

The specimen itself offers certain problems. It is best if the crystal is large enough to cut to a specific plane (unless it has natural faces), and thus ease the determination of the precise orientation needed for angle calculations. Surface roughness should be carefully measured; this can easily alter intensities at low angles ($20°2\theta$) 10–15% or more since many materials are fairly highly absorbing. This problem can be examined by measuring the fluorescence from the specimen as a function of 2θ (deWolff 1956; Suortti 1971). In this regard, some care is required to ensure that polishing scratches are not all directionally aligned.

There are three other problems that require investigation with each specimen. A specific example best serves to illustrate the problems. Consider that the diffuse x-ray scattering from a Cu_3Au crystal is to be measured with an incident beam monochromator and Cu K_α radiation. The three problems are: (1) Cu K_α fluorescence from the specimen caused by $\lambda/2$ in the incident beam, and Au L_α fluorescence, (2) fluorescent intensity from impurities in the crystal, and (3) diffuse intensity from anomalous dispersion [Sparks (1974) and Sect. 3.4]. All three problems are best dealt with in this case by using a solid state detector and properly adjusting the window. Air scattering can be determined by trapping the beam in a lead box at the position of the specimen. To keep this term small (0.1–0.2 cps or less including electronic noise), the specimen should be kept in an evacuated chamber; a vacuum of 20 μm suffices. Also, the receiving slit and scatter slit should be separated as much as possible, and be as far from the sample as is practical.

One of the main purposes for measuring the direct beam in an x-ray experiment is to place the intensity on an absolute scale of electron units so as to be able to subtract theoretical values of the Compton scattering. The net values can then be compared to calculations. The measured x-ray intensity is corrected for surface roughness and polarization, air scattering and impurity scattering are subtracted, the power of the direct beam is measured, the data are placed on an absolute scale, and then the calculated Compton scattering can be subtracted. The calculated Compton inelastic scattering for an atom includes all the electrons. It arises from electron recoil (Sect. 3.4). However, the incident beam may not have sufficient energy to move an electron to a higher energy, i.e., if the K electrons cannot be excited to a higher energy state, they cannot be involved in the Compton process. A simple empirical correction to the tabulated values that should be generally adequate is to multiply the calculations by

$$\left(Z_n - \sum_n f_e^2\right)\Big/\left(Z_T - \sum_T f_e^2\right), \tag{C.1}$$

where Z is the atomic number, f_e the electronic scattering factor, n the number of electrons in the atom that can be excited, and T the total number of electrons. This equation is crudely the ratio of the Compton scattering from those electrons that can be excited to the total Compton scattering.

We now turn to a discussion of how to measure the direct beam to place the intensity on an absolute scale. Four methods are discussed and considerable detail is given, so that the reader can proceed with such measurements with a minimum of effort. It is strongly advised that at least two of the methods be used in any one experiment, to be sure of the value and to check that all equipment is properly adjusted.

Method 1: Aluminum Powder

An Al powder (-325 mesh) compacted under 25ksi shows relatively little texture (Batterman et al. 1961). The total energy (E_{meas}) from several peaks can be obtained and corrected for background (which includes, therefore, air scattering, Compton, etc.) and for TDS. This can then be related to the power of the beam through Eq. (4.16). This equation is repeated here with the polarization term for an incident beam monochromator.

$$E'_{\text{meas}} = \frac{1 + P\cos^2 2\theta}{\sin^2\theta \cos\theta (1+k)} \left[\frac{e^4 \eta^2}{m^2 c^4}\right]\left[\frac{wh}{R^2}\right]\left[\frac{\lambda^3}{2\omega}\right]$$
$$\times \frac{16|f|^2}{32\pi} \times \frac{m_{hkl}}{(V_c)^2} \times \left[\frac{1}{(\mu/\rho)\rho}\right] \times I_0 A_0, \tag{C.2}$$

1. $E'_{\text{meas}} = E_{\text{meas}}/(1 + \alpha)$, where α is a TDS correction [see, for example, Walker and Chipman (1973), on how to calculate this term.[1]]
2. The term P is the polarization of the monochromator. For Cu K_α, and a bent pyrolitic graphite monochromator, this has been found to be 0.989. This was determined by carefully measuring the total energy of the 333 reflection from a perfect Si single crystal, first in the normal diffractometer configuration and then with the diffraction plane normal to the diffractometer, in which case the polarization term is $(P + \cos^2 2\theta)/(1 + P)$. From these two measurements k can be obtained; if $2\theta = 90°$, the ratio of the two measurements is simply P. (For a Si 333 reflection of Cu K_α, $2\theta \sim 90°$.) Another procedure is illustrated in Fig. 8.27, Sect. 8.7. For an ideally perfect monochromator and a perfectly collimated beam, $P = |\cos 2\theta|$, whereas $P = \cos^2 2\theta$ for an ideally imperfect monochromator. For graphite monochromators and Cu K_α, P_{perfect} is 0.92 and $P_{\text{imperfect}} = 0.84$. The measured value is larger than either and indicates the need for an actual measurement. See Jennings (1968) for a discussion of P and how it can exceed values beyond P_{perfect} and $P_{\text{imperfect}}$.

---

[1] This calculation requires knowledge of single crystal elastic constants; if these are not available, some of the simpler approaches can be employed. They are referenced in this paper.

Method 1: Aluminium Powder   545

3. $[e^4\eta^2/m^2c^4]$ = the square of the classical electron radius = 7.95×10^{-30} m^2.
4. The terms w, h, and R are width, height, and distance of the receiving slit from the sample, respectively.
5. As usual, λ is the wavelength of radiation used. For K$_\alpha$, average the α_1 and α_2 components, i.e., $\lambda_{K_\alpha} = (2\lambda_{K_{\alpha_1}} + \lambda_{K_{\alpha_2}})/3$. (For Cu K$_\alpha$, $\lambda = 1.54178$ Å.)
6. The term ω is the angular velocity of the *sample*, i.e., $(d\theta/dt)$ in radians per second.
7. $|f|^2 = |(f_0 + \Delta f')^2 + (\Delta f'')^2|\exp(-2M)$.
8. The term m_{hkl} is the multiplicity factor for the hkl reflection being considered.
9. The density is represented by ρ.
10. The term $I_0 A_0$ is the power of the direct beam.

Physical and Experimental Constants

The average of *experimental f* values should be used if these are available. Theoretical values are $\simeq 5\%$ higher at the moment for aluminum and the spread of available experimental results is small.[2] These averages are given here in Table C.1 as the *total absolute scattering factors.* ($\Delta f'' = 0$ as the measurements in the literature were made with Mo K$_\alpha$, but $\Delta f'$ is included.) The values averaged are those of Batterman et al. (1961), De Marco (1967), and Jaarvinen (1969).

For each reflection, $\exp(-2M)$ is also given in Table C.1 for $M = 0.8825 (\sin \theta/\lambda)^2$. It is necessary to remove from f the value of $\Delta f'$ for Mo K$_\alpha$ (taken from the International Tables, Vol. III) and then employ $\Delta f' = 0.1$ and $\Delta f'' = 0.3$ over the entire ($\sin \theta/\lambda$) range for Cu K$_\alpha$. The f values given in column 1 of Table C.1 are *not* corrected this way; they are the averages of the experimental values, but the column "$f^2 \exp(-2M)$" includes this correction.

Table C.1

Indices of peak	$f + \Delta f'$	e^{-2M}	$f^2 e^{-2M}$	m_{hkl}	α in %[a]
111	8.71	0.9226	70.075	8	0.3361
200	8.26	0.8979	61.342	6	0.3508
220	7.34	0.8064	43.59	12	0.6960
311[b]	6.48	0.7441	31.31	24	0.8588

[a] For Al, to calculate α from the equation in the reference mentioned in the text, $c_{11} = 1.08 \times 10^{11}$ Pa, $c_{12} = 0.62 \times 10^{11}$ Pa, $c_{44} = 0.28 \times 10^{11}$ Pa, $T = 298°$K (Pa = Pascals).
[b] For the 311, this is not an average value since only Batterman et al. (1961) measured this peak.

As a typical example, assume $w = 0.483$ mm., $h = 1.397$ mm., $R = 145.542$ mm., $2\omega = 0.1°\ 2\theta$/min. The data in Table C.2 were then obtained in continuous scans.

[2] See Dawson (1975) of Chap. 6 for further discussion of this matter.

No dead time correction was made. (The maximum count rate was 1000 cps, and the correction would have been less than 0.5%.)

Table C.2

Peak	Background low 2θ range	Total intensity	Peak Total 2θ range	Total intensity	Background high 2θ range	Total intensity	2θ Peak
111	35.8–37.5	8170	37.5–39.2	211279	39.2–40.9	7360	38.46
200	42.5–44.0	7051	44.0–45.5	106470	45.5–47.0	5768	44.71
220	62.3–64.0	6346	64.0–65.7	69763	65.7–67.4	6799	65.15
311	76.0–77.5	6954	77.5–79.0	71602	79.0–80.5	6544	78.25

With (μ/ρ) of Al for Cu K$_\alpha$ = 4.91(5) m²/kg, and the lattice parameter calculated from the 2θ positions,

$$P_0 = I_0 A_0 = [(11.44 \times 10^6)/\{m_{hkl} \times |f|^2 \exp(-2M) \times \mathrm{LP}\}][E_{\mathrm{meas}}/(1+\alpha)]. \tag{C.3}$$

Table C.3

Indices of peak	$I_0 A_0$ (cps, $\times 10^8$)
111	5.25
200	5.50
220	5.68
311	5.78
Average	5.55

For the above measurements, the results are given in Table C.3. Care taken in the following procedures is strongly recommended.

(i) Be sure the specimen stage is aligned correctly before you begin, especially that the normal to the specimen surface bisects the incident and diffracted beams, and lies in their plane. Otherwise the absorption is not that assumed in Eq. (C.2) (for a flat specimen plane in Bragg-Brentano geometry).

(ii) Run a chart profile of the four peaks (no attenuation foils are needed, but include any vacuum chamber if you are going to measure in a vacuum). From this you can verify the peak widths and choose the 2θ intervals belonging to each peak and for its background. The low-angle and high-angle background regions should have the same width as the main peak; the average for these two regions can then be subtracted from the integrated peak as the background contribution.

(iii) If you are interested only in a conversion factor to electron units for a scattering experiment, you can obtain $I_0 A_0 wh/R^2$. It is then not necessary to measure w, h, R. That is, in any *measurement*, the conversion factor from measured counts to electron units will include wh/R^2 as well as $I_0 A_0$. There is no need to measure wh/R^2 as it will cancel in expressions for the measurements of the Al powder to obtain $I_0 A_0$ and for the sample you are studying. For example, assume a large flat single crystal of an alloy, with $N_0 \times \rho/\text{at.wt.}$ = number of atoms/cc, a measurement at one point in reciprocal space, with a monochromator for a time t. The result is related to the intensity in electron units per atom, I_{eu}:

$$E_{\text{meas}} = \left(\frac{I_0 A_0}{2(\mu/\rho)\rho}\right)\left(\frac{wh}{R^2}\right)\left(\frac{e^4 \eta^2}{m^2 c^4}\right)\left(\frac{1 + P\cos^2 2\theta}{1 + P}\right)\frac{N_0 \times \rho}{(\text{at.wt.})} I_{\text{eu}} t \,. \tag{C.4}$$

To convert to electron units, the factor $I_0 A_0 wh/R^2$ can be obtained with the Al powder.

Method 2: Polystyrene

An amorphous scatterer such as polystyrene ($C_8 H_8$) in which the atoms scatter independently at high angles can be employed as a second method (Sparks and Borie 1966) both for $I_0 A_0$ or $I_0 A_0 wh/R^2$ (see Problem 7.6). In this case,

$$I_{\text{tot}}^{\text{meas}} = I_{\text{coh}} + I_{\text{inc}} \,. \tag{C.5}$$

Consider a flat piece of as cast (not drawn) polystyrene of thickness x. The beam passes through a length $x/\sin\theta$ upon entering and $x/\sin\theta$ upon leaving, and the *coherent* intensity is attenuated by $\exp(-2\mu x/\sin\theta)$, where μ is the linear absorption coefficient (for the wavelength used). The *incoherent* intensity is similarly reduced by $\exp[-(\mu + \mu')x/\sin\theta]$, where μ' corresponds to the linear absorption coefficient for the modified wavelength. That is, the entering beam has one wavelength, the incoherent beam another. This point needs to be considered (see below) in calculations of the incoherent scattering.

Also, as there is multiple scattering (mainly double diffraction), the measured intensity has to be reduced by a factor that may be calculated from published data (see Warren and Mozzi 1966; Strong and Kaplow 1967). For polystyrene, with Cu K_α at $2\theta = 100°$, the measured intensity (coherent and incoherent) should be reduced by $\approx 5\%$ as a result of this scattering. The expression for the power scattered from a thick flat piece of polystyrene is then

$$P_{\text{meas}} = k^m \left[(I_0 A_0)\frac{e^4 \eta^2}{m^2 c^4 R^2}\right](\text{polarization})\frac{N_0}{\text{MW}}$$

$$\times \left\{\frac{1}{2\mu/\rho}\left(\frac{I_{\text{eu}}}{M}\right)_{\text{coh}} + \frac{1}{(\mu/\rho) + (\mu'/\rho)}\left(\frac{I_{\text{eu}}}{M}\right)_{\text{inc}}\right\} ABS \times w \times h, \tag{C.6}$$

where

1. P_{meas} = counts/time = $I_{\text{meas}} \times w \times h$; w and h are the width and height of the receiving slits.
2. k^m = multiple scattering correction = $(1/0.95)$.

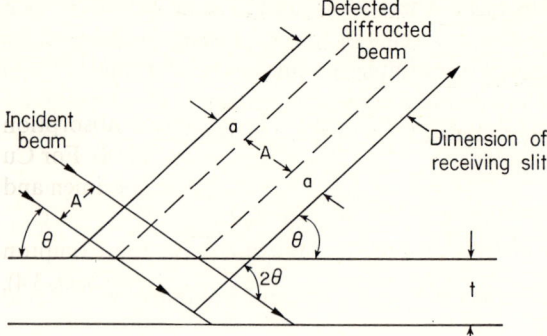

Fig. C.1. Geometry used in determining the absorption correction for a flat faced diffractometer sample

3. The polarization term is $(1 + P\cos^2 2\theta)/(1 + P) = 0.5209$ for a measured $P = 0.989$ and $2\theta = 100° \ 2\theta$.
4. N_0 = Avogadro's number.
5. The molecular weight (MW) of C_8H_8 is 104.152.
6. The terms μ/ρ, μ'/ρ are the mass absorption coefficient of C_8H_8 for the radiation (λ) and the modified radiation, λ'.
7. $(I_{eu}/M)_{coh} = 8(f_c^2 + f_H^2)$ at the determined $\sin\theta/\lambda$. For Cu K_α and $2\theta = 100°$,[3] ($\sin\theta/\lambda = 0.5$), this is $8[(1.685)^2 + (0.071)^2] = 22.75$.
8. $(I_{eu}/M)_{inc} = 8\{[(Z-f)/r]_c + [(Z-f)/r]_H\}$. Here Z is the atomic number and r is the small (Breit-Dirac) correction; see the International Tables for Crystallography, Volume III, p. 245 for precise values and the meaning of r.

Thus,

$$(I_{eu}/M)_{inc} = 8[(4.432/1.037) + (1/1.037)] = 41.91.$$

9. A very thick sample is required of the order of 0.02 m, because of the low absorption of this material. Also, as shown in Fig. C.1, because of absorption there are contributions from a considerable depth in the specimen. The term *ABS* in Eq. (C.6) is the value to correct for this and is the term in brackets in the following expressions:

$0 \leq t \leq \frac{1}{2}a\sec\theta$:

$$I = K[1 - \exp(-2\mu t \csc\theta)]$$

$\frac{1}{2}a\sec\theta \leq t \leq \frac{1}{2}(A + a)\sec\theta$:

$$I = K[(1 - \exp(-2\mu t \csc\theta)) + (2t\cos\theta/A) - (a/A) + (1/\alpha) \\ \times \exp(-2\mu t \csc\theta) - (1/\alpha)\exp(-\beta)]$$

---

[3] The recommended scattering factors for C and H are the theoretical values as there are no measured values. For the value for C see Freeman, A.J., (1959); Hanson, H.P., (1964), for H. For incoherent scattering $(Z - f)$ from C: see Keating, D. and Vineyard, G.H., (1956), and for H: Compton, A.H., and Allison, S.K. (1963).

Method 2: Polystyrene

$t \geq \frac{1}{2}(A + a)\sec\theta$:

$$I = K[1 - (1/\alpha)(1 - \exp(-\alpha))\exp(-\beta)],$$

where $K = I_0/2\mu$, $\alpha = 2\mu A \csc 2\theta$, $\beta = 2\mu a \csc 2\theta$, and μ is the linear absorption coefficient for the sample. These equations were derived by Milberg (1958). For Cu radiation, the third condition generally holds for a thick polystyrene specimen and in this experiment $ABS = 0.98$.

We now turn to the calculation of the changed wavelength for the Compton scattering and its absorption coefficient. For the change in wavelength (see Sect. 3.4),

$$\Delta\lambda = 0.0243(1 - \cos 2\theta),$$

hence $\lambda_2 = 1.57072$ Å for $2\theta = 100°$; $\lambda_1 = 1.54178$ Å [the weighted average $(2K_{\alpha_1} + K_{\alpha_2})/3$)].

The absorption coefficients were then calculated using the relations given in the International Tables (1968), Volume III, p. 161:

$$(\mu/\rho) = [C\lambda^3 - D\lambda^4 + N\sigma_{k-N}(Z/A)]/10.$$

For carbon:

$C = 1.22$, $\quad D = 0.0142$, $\quad N\sigma_{k-N} = 0.3917$, $\quad Z/A = 0.4995$

$$\left(\frac{\mu}{\rho}\right)'_C = 0.4837 \text{ m}^2/\text{kg}.$$

$C = 0.0127$, $\quad D = 0.466 \times 10^{-5}$, $\quad N\sigma_{k-N} = 0.3917$, $\quad (Z/A) = 0.9912$,

$$\left(\frac{\mu}{\rho}\right)'_H = 0.0438 \text{ m}^2/\text{kg}.$$

Then the weighted average is

$$(\mu/\rho)' = 0.4497 \text{ m}^2/\text{kg}.$$

For the unmodified radiation: $\mu/\rho = 0.4027$ m²/kg.

All quantities in Eq. (C.6) have now been discussed. Experimentally, one should follow certain simple procedures.

(i) Measure the polystyrene scattering by chart scanning rapidly at first, to check that the measurement will be in a smooth region of the pattern, indicating independent scattering of the atoms.

(ii) Around the specific $\sin\theta/\lambda$ (in this case = 0.5) count for several hundred seconds at many points to make sure of the lack of detail. That is, the measurements are to be made in a region where the atoms are scattering independently, as that is what is assumed in the formula for $(I_{eu}/M)_{coh}$, the coherent scattering per C_8H_8 molecule. The measurements should be done with *either wide open scatter slits or no scatter slits* at all, as the absorption is low and the beam penetrates deeply. A scatter slit set with a highly absorbing crystal will touch the scattered beam from polystyrene as indicated in Fig. C.1. *Do not* use very broad receiving slits; that is, slits 5 mm. are

Consider the case where $t_{total} = 0.095$ m, $t_{eff} = 0.075$ m, $\mu = 0.57$/m, $\lambda = 1.4986$ Å, Gain $= 10^7$, VF $= 10^5$ and 237897 "counts" are recorded for 200,000 monitor counts at a storage ring. (Both values have been corrected for dead time.) Then

$$\frac{237{,}897}{200{,}000} = \frac{I_0 A_0 \{1 - \exp[-(0.57 \times 0.075)\} \dfrac{12{,}400}{27(1.4986)} 1.6022 \times 10^{-19} \times 10^7 \times 10^5}{\exp\{-[0.57(0.095 - 0.075)/2]\}},$$

hence $I_0 A_0 = 5.82 \times 10^5$ per monitor count.

These techniques should suffice for any x-ray experiment. Some other special techniques are worthy of mention. For small angle scattering while the multiple-foil method is suitable, other methods have been devised (Kratky et al. 1966; Pilz and Kratky 1967; Pilz 1969) and for very small crystals see Coppens (1975).

References

Batterman, B.W., Chipman, D.R., and DeMarco, J.J., Phys. Rev. *122*, 68 (1961)
Chipman, D., Acta Crystallogr. *A25*, 209 (1969)
Compton, A.H., and Allison, S.K., "X-rays in Theory and Experiment", Van Nostrand, Princeton, New Jersey, 1935
Coppens, P., Acta Crystallogr. *A31*, 612 (1975)
DeMarco, J., Phil. Mag., *15*, 483 (1967)
Freeman, A.J., Acta Crystallogr. *12*, 261 (1959)
Hanson, H.P., et al., Acta Crystallogr. *17*, 1040 (1964)
Jaarvine, M., Phys. Rev. *178*, 1108 (1969)
Jennings, L.D., Acta Crystallogr. *A24*, 472 (1968)
Keating, D. and Vineyard, G.H., Acta Crystallogr. *9*, 895 (1956)
Kratky, O., Pilz, I., and Schmitz, P.J., J. Colloid Interface Sci. *21*, 24 (1966)
Milberg, M.E., J. Appl. Phys. *29*, 64 (1958)
Pilz, I., J. Colloid Interface Sci. *30*, 140 (1969)
Pilz, I., and Kratky, O., J. Colloid Interface Sci. *24*, 211 (1967)
Schwartz, L.H., Morrison, L.A., and Cohen, J.B., Adv. X-ray Analysis 7, 218 (1963)
Sparks, C.J., Phys. Rev. Lett. *33*, 262 (1974)
Sparks, C.J., and Borie, B., in "Local Atomic Arrangements Studied by X-ray Diffraction", (J.B. Cohen and J.E. Hilliard, eds). Gordon & Breach, New York, 1966
Strong, R., and Kaplow, R., Acta Crystallogr. *23*, 38 (1967)
Suortti, P., J. Appl. Phys. *42*, 5821 (1971)
Walker, C.B., and Chipman, D., Acta Crystallogr. *A28*, 572 (1973)
Warren, B.E., and Mozzi, R.L., Acta Crystallogr. *21*, 459 (1966)
deWolff, P.M., Acta Crystallogr. *9*, 682 (1956)

Appendix D: Accuracy in Digital Counting

D.1 Some Additional Information on Counting Electronics

When combined with a single channel analyzer (SCA), a proportional counter may be used as an energy selective device, to eliminate higher order components and fluorescence from a monochromatic beam.

The procedure for setting a SCA is as follows:

1. Obtain a nearly monochromatic beam of x-rays by diffraction of characteristic radiation from a crystal of known d-spacing.
2. Establish the counter plateau at one gain level of the amplifier by varying the voltage to the counter. Amplifier gain may shift if the count rate is very high. This occurs because most semiconductor amplification circuits sweep out the current injected in them from the detector, to restore a baseline voltage between the ends of the semiconductor. If the pulses are too closely spaced in time, the baseline is not properly established and the voltage of the pulse is shifted. Modern amplifiers include circuitry for enhancing baseline restoration.
3. Use a narrow window ΔE, and a counter voltage on the plateau. Increase E_1 and record the intensity. A curve like that in Fig. D.1a results.

    Alternately, the window can be made large and E raised, resulting in detection of all pulses with voltage greater than E_1 and the curve of Fig. D.1b. Then E_1 and ΔE are chosen to include some percentage of the desired K_α. If only the white radiation is being excluded, 90% of the K_α intensity is a good figure. If fluorescence from the sample to be used, caused by the white radiation, is the principal problem, a lower value say 80%, can be chosen. However, if the window is too narrow and is set on the steep sides of the pulse distribution, minor voltage drifts can cause large changes in intensity. It should be remembered, however, that the breadth in Fig. D.1a is sufficiently large for a proportional gas counter that with 90% of K_α accepted, only about 90% of radiation of a wavelength $\lambda/2$ or 2λ will be eliminated; it will be very difficult to do much with fluorescence from elements of atomic numbers within two of that of the x-ray tube's target. (The SCA is thus a good device for eliminating the second order $\lambda/2$ radiation from a monochromator.) As the voltage setting E_1 is proportional to the energy of the pulse, it can be written $E_1 = K/\lambda$ and the position of a curve for a different wavelength can be estimated (Fig. D.1a).
4. With these settings, and the x-rays off, the electronic noise and cosmic ray background are checked; they should be close to zero (less than 0.5 cps). If the

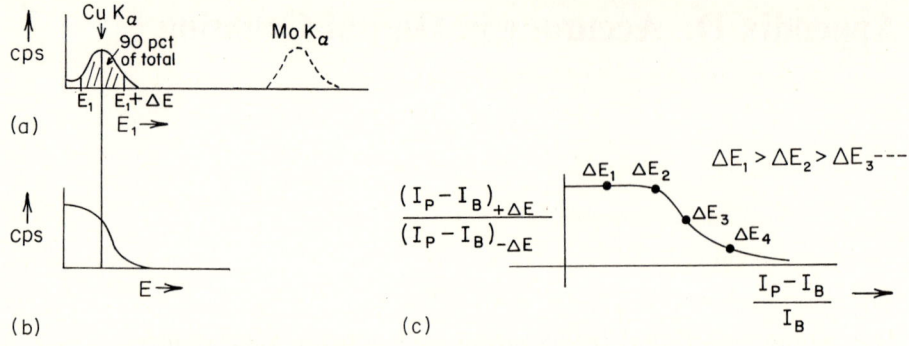

Fig. D.1. a Pulse distribution with narrow ΔE (differential curve). **b** Wide open window (integral curve). **c** Effect of window size on peak intensity I_P (I_B is the background intensity)

noise is high (1–2 cps) the gain (amplification of the counter pulses) is reduced and the procedure started over.

5. Varying the counter angle 2θ from diffraction peak to nearby background, a plot is made of peak intensity (I_P) minus background (I_B) with a window, over the same difference without any window, versus the peak-to-background ratio, for different windows. A typical plot is shown in Fig. D.1c. On this curve we wish to operate on a flat portion, so that minor drifts in the electronics do not affect precision.
6. With an oscilloscope, check that the gain is not too high, resulting in the tops of pulses being cut off; they are then not proportional to energy.
7. For rapid adjustments, choose an arbitrary window, say, $\Delta E = 1$ V in the middle of the SCA range and vary the E setting until the intensity sharply increases, indicating the window is set around the K_α peak. Fine adjustments of E, ΔE, and gain may then be made.

[With the white beam from a synchrotron, a monochromator will diffract the desired wavelength and the harmonics. By setting up several SCA's, one for each wavelength, or by employing a multichannel analyzer, the diffraction from each wavelength can be recorded at the same time, greatly reducing the measuring time. (P. Georgopoulos, private communication.)]

D.2 Measurement of Dead Time

When high count rates and accuracy must be combined, a correction for nonlinearity of the detector is required. This can be done in the following way as suggested by Dr. D. Chipman.

If N_T is the true count, N_O the observed count, and τ the time constant of the system, then it can be shown that for small deviations from linearity,

$$N_T = N_0/(1 - N_0 \tau). \tag{D.1a}$$

D.2 Measurement of Dead Time

Let the subscripts 1 and 2 refer to a foil in and out of the beam. Then,

$$N_{1T}/N_{2T} = (N_{10}/N_{20})[(1 - N_{20}\tau)/(1 - N_{10}\tau)]. \tag{D.1b}$$

If

$$R_T = N_{1T}/N_{2T} \quad \text{and} \quad R_0 = N_{10}/N_{20},$$

$$(1 - N_{10}\tau)R_T = R_0(1 - N_{20}\tau);$$

using the definition of R_0,

$$R_0 = R_T + \tau N_{10}(1 - R_T). \tag{D.1c}$$

As N_{10} goes to zero, R_0 approaches R_T. Plotting R_0 vs. N_{10} should yield a straight line whose intercept is R_T and whose slope is $\tau(1 - R_T)$. To get a range of intensities N_{10}, choose a strong diffraction peak from a single crystal, and vary 2θ around this peak. Count first with the foil, then without, and then readjust the foil to get the first reading (put it back in the same position) before changing 2θ. Be sure the receiving slits are open enough to receive the entire beam. A second receiving slit close to the specimen, but not touching the beam is useful to reduce air scatter. A typical result is given in Fig. D.2 for a modern scintillation counter system. Here, τ was $(6.1 \pm 0.2) \times 10^{-6}$ sec for 90% confidence limits. Losses of 1% occurred at 1800 cps. As we have seen, one might have estimated that with a scintillation counter, 1% losses would not occur until 5000–10,000 cps, so we can see the necessity for actually checking out the linearity of the entire system some way!

With this value of τ, observed readings can be corrected using Eq. (D.1a). As the correction is not generally larger than 10% or so, the typical error in τ contributes an error of 1% or less in the *correction*. Knowing τ, $\ln I$ versus number of foils can then be made accurately enough to get a measure of the direct beam I_0 to a reproducibility of 1–2%, (see Appendix C).

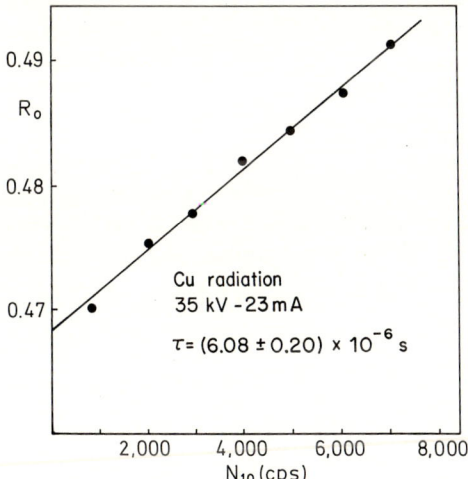

Fig. D.2. Determining the time constant of a counter and its electronics using Eq. (D.1c). (From Gehlen, P.C., Ph.D. Thesis, Northwestern University 1966)

Because long counts may occasionally be necessary (over a long time period), if an accuracy of better than 1% is required, a monitor counter should be used to eliminate the effects of instability in x-ray generation, ±0.6–3% for long times. [By differentiating the equation for absorption, Eq. (3.39c), with respect to ρ, it is easy to show that for Cu K_α and longer wavelengths about half of this error is due to changes in the density of the air path from normal changes in barometric pressure.] A monitor counter can be made by placing a thin foil of an element of Z — 2 (which has an absorption edge at a wavelength slightly longer than the wavelength of x-rays from the target element Z) in the beam from target Z and using the counts from the fluorescence as a timer. That is, those counts are fed to another scaler and counting in the primary detector is stopped for a fixed number of counts to this monitor. The foil should only reduce the beam 10% or so. If the path to the monitor is about the same length as the path to the primary counter, effects of variation of barometric pressure are minimized.

At synchrotron sources, the resolving time or dead time of the counting system may be less than the travel time of an electron bunch past a slit, particularly when only one bunch is in the ring. In this case a different correction for dead time is required. Following Matsubara (Ph.D. Thesis, Northwestern University (1984)), we note that photons arrive in time slices determined by the ring frequency v. One or more counts in each slice are assumed not to be distinguished, that is to be counted as a single pulse. Now let us consider the relation between the true counts (N_T) and the observed counts (N_O) for t seconds. The problem reduces to finding the distribution of N_T balls randomly thrown into vt baskets. Throw the first ball in at random. Since the probability that a particular basket has this ball is:

$$p = 1/(vt),$$

the probability that this basket does not have this ball is:

$$q = 1 - p.$$

Continue with the rest of the balls and we have q^{N_T} as the probability that this basket has none. Because there are vt baskets, the number of baskets which does not have any ball is:

$$vtq^{N_T} = vt(1 - p)^{N_T},$$

and the number of baskets which contain at least one ball is:

$$vt - vtq^{N_T}.$$

Because of the assumption that one or more counts in each slice are still counted as one count, the observed count is equal to the total number of baskets which contains at least one ball, that is:

$$N_0 = vt(1 - vtq^{N_T}). \tag{D.2}$$

As the $\lim(1 - x^{-1})^x = e^{-1}$, for $vt \gg 1$, N_0 becomes:

$$N_0 = vt[1 - \exp(-N_T/vt)] \tag{D.3a}$$

or

D.2 Measurement of Dead Time

$$N_0 = \frac{t}{\tau}[1 - \exp(-\tau N_T/t)],\tag{D.3b}$$

where $\tau = 1/v$. If we use the counting rate instead of total counts (N_0 or N_T), Eq. (D.3b) becomes [B.W. Batterman, CHESS Technical Memorandum No. 6 (1980)]:

$$N_0' = v[1 - \exp(-N_T'/v)]\tag{D.4a}$$

or

$$N_0' = [1 - \exp(-N_T'\tau)]/\tau,\tag{D.4b}$$

where the units of N_0' or N_T' is cps. From Eq. (D.4b), the true count rate is given by:

$$N_T' = -\ln(1 - \tau N_0')/\tau.\tag{D.5}$$

Since the observed counts in Eq. (D.5) include all multiple photon pulses, this equation applies only if a lower level discriminator is set in a pulse height analyzer. The validity of this equation was examined in an experiment at the CHESS synchrotron ($v = 3.91 \times 10^5$ Hz) by measuring the ratio of intensity with a Ni foil to that without it, as a function of various counting rates from 20,000 cps to 90,000 cps. This was accomplished by locating the detector at different points of an aluminum Bragg peak. If the dead time is fully corrected by Eq. (D.5), the ratio of these two intensities must be constant for various beam intensities. These ratios are shown in Table D.1. The ratios of the uncorrected intensities increase with the counting rate because the higher counting rate misses more counts. The ratios of the intensities corrected for the dead time with Eq. (D.5) are fairly constant until the counting rate reaches about 80,000 cps. But at a higher counting rate this correction does not give a satisfactory result. In order to derive Eq. (D.5), it was presumed that the resolving time (or dead time) τ_c of the detector and its associated electronics is less than the time between the bursts of photons τ. Namely, the detector and its associated electronics must completely recover before the next burst of photons gets to the detector. When this condition is not satisfied a further approximate correction may be made (see Matsubara, Ph.D. Thesis, Northwestern University 1984).

Table D.1. Ratios of intensities with a Ni foil to those without it as a function of counting rate

Meas. Int. [cps]		Ratios of these two intensities	
With foil	Without foil	(1)	(2)
48059	87780	0.547	0.516
41953	78666	0.533	0.505
33218	63236	0.525	0.503
27262	53046	0.514	0.496
20338	39713	0.512	0.499
12490	24553	0.509	0.501

(1), Ratios of uncorrected intensities. (2), Ratios of intensities corrected for dead time by $N_T = -(1/\tau) \cdot \ln(1 - \tau N_0)$, where $\tau = 2.54$ μs for CHESS

Answers to Selected Problems

Chapter 1

1.1. The lines furthest apart have the highest density of points.

1.2. Faster growing planes tend to disappear. Combining this with the results of Problem 1.1, the slowest growing planes are those with the highest lattice point density and will predominate in a crystal's habit.

1.4. Ca^{2+} is the smaller ion so it fits into a simple cubic cage of F^-. The F^- are all in tetrahedra on an fcc cell of Ca^{2+}. If R is the radius,

$$a/2 = 2R_{F^-}; \quad \frac{\sqrt{3}a}{2} = 2R_{Ca^{2+}} + 2R_{F^-}; \quad \frac{R_{Ca^{2+}}}{R_{F^-}} = 0.732.$$

1.5. Metals like Cu, Zn, Ni, Au all have twelvefold coordination. It is not possible in an ionic compound AB as each charged particle would be surrounded by 12 of the opposite charge; this structure cannot be repeated in three dimensions.

1.6. $c/a = 1.63$.

1.7.

$$\frac{4.40 \times 10.82 \times 3.85 \, (\text{Å})^3}{\text{unit cell}} \times \frac{10^{-30} \, \text{m}^3}{(\text{Å})^3} \times \frac{5,520 \, \text{kg}}{\text{m}^3}$$

$$\times \frac{\text{mole}}{15156 \, \text{kg}} \times \frac{6.023 \times 10^{23} \, \text{molecules}}{\text{mole}} = \frac{4 \, \text{molecules}}{\text{unit cell}}.$$

1.11. See the International Tables, Volume I, p. 32.

1.12. 54°44′. See also International Tables, Volume II, p. 120.

1.14. Look at Eq. (1.3b). What are the conditions for $u = h, v = k$? These conditions are: all axes equal in length, all angles of 90°. Only in the cubic system does this always occur. Special cases: [110] in tetragonal, hexagonal; [100], [010], [001] in orthorhombic, etc.

1.15. (100).

1.17. $A_1 = -\frac{1}{2}a_1 + \frac{1}{2}a_2; \quad A_2 = +\frac{1}{2}a_2 + \frac{1}{2}a_3; \quad A_3 = -\frac{1}{2}a_1 + \frac{1}{2}a_3.$

1.18. $r_{hkl}^* \cdot r_{hkl}^* = 1/d^2$. Work this out (see Fig. 1.21 for a sketch of the b_i's with the angles between them) and then you will find

$$(1/d^2) = (4/3)[(h^2 + hk + k^2)/a_1^2] + (l^2/a_3^2).$$

1.20. All points on the circle lie at equal angular distances from Q_1. Measurement on the Wulff net shows that R_1 and R_4 are 76° from Q_1. We can locate point S at 76° from Q_1, and it too must lie on the circle. With three points, we can locate the center of the circle C and then draw the arc using a compass.

1.22. (a) 72.5°. (b) 58°S, 70°W.

1.29. (a) Orthorhombic.

1.30. (a) [112], [110], [001], [110], respectively. Plot a (100) projection. The direction which is the intersection of the great circle representing the desired plane with the great circle representing the $(\bar{1}10)$ plane is the trace. [Analytically, solve Eq. (1.8a).]
(b) $[11\bar{2}]$. On a $(\bar{1}10)$ projection draw a line from the center through, say, the [011] (a face diagonal) and where it intersects (110) is the desired direction. Analytically, find some plane $(h'k'l')$ perpendicular to the given (hkl) and which also contains $[uvw]$, the face diagonal. Thus you have two equations to obtain $(h'k'l)$:

$$(h\mathbf{b}_1 + k\mathbf{b}_2 + l\mathbf{b}_3)\cdot(h'\mathbf{b}_1 + k'\mathbf{b}_2 + l'\mathbf{b}_3) = 0$$

and

$$h'u + k'v + l'w = 0.$$

After solving for $(h'k'l')$, the zone axis of this and $(hkl) = (110)$ is the desired direction.

1.31. See Otte, H.M., Acta Crystalogr. *14*, 360 (1961).

1.33. Keep in mind that [112] shear on a close packed plane can only be in one sense, i.e., [112] but not $[\bar{1}\bar{1}2]$. Otherwise you have to "lift" the sheared plane to go over the plane below. The unit cell of Cu_3Au is doubled after a $\frac{1}{6}[112]$ shear to make the structure a mirror image across a {111} type plane. See Fig. A1.33. To draw the stereogram just plot a (111) projection; the twin poles can be placed by a 180° rotation around the [111] direction.

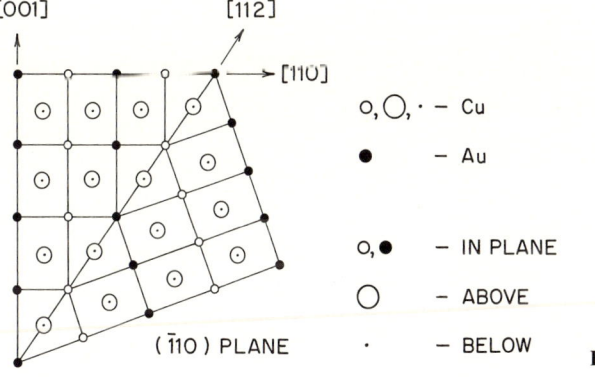

Fig. A1.33

Chapter 2

2.1. We can add the amplitudes as if they are vectors with the phase angle between them. For $n = 1$, the plane at A and the plane at $d/4$ have a path difference of $\lambda/4$, and hence a phase angle difference of $(2\pi/\lambda)(\lambda/4) = 90°$. The sum of two equal unit vectors at $90°$ to each other is $\sqrt{2}$. Other results follow in the same way. Alternately, one can simply add two waves, $\frac{\pi}{2}$ out of phase:

$$\cos\varphi + \cos(\varphi + \tfrac{\pi}{2}).$$

Expanding and simplifying will result in the wave

$$\sqrt{2}\cos(\varphi + \tfrac{\pi}{4}).$$

2.2. Plotting only the numerator the second maximum in intensity occurs for $\beta = 2\pi m + (3\pi/N)$ and the third for $\beta = 2\pi m + (5\pi/N)$. Therefore:

N	β		Relative height	
	2nd max	3rd max	2nd max	3rd max
4	$3\pi/4$	$5\pi/4$	0.073	0.073
8	$3\pi/8$	$5\pi/8$	0.0506	0.0226
16	$3\pi/16$	$5\pi/16$	0.0463	0.0175

Alternately, differentiating $(\sin^2 N\beta/2)/(\sin^2 \beta/2)$ with respect to β and setting the result equal to zero, we obtain

$$\tan N\beta/2 = N \tan \beta/2.$$

This function may be graphed (or the interference function itself graphed) to give the following results:

N	β		Relative height	
	2nd max	3rd max	2nd max	3rd max
4	131.8°	228.7°	0.074	0.074
8	64.72°	111.3°	0.0525	0.022
16	32.7°	55.48°	0.0485	0.0178

2.3. (a) For a maximum $\beta = h2\pi$. Hence

$$(2\pi/\lambda)a \cos v_{max} = h2\pi, \quad \text{or} \quad \cos v_{max} = h\lambda/a.$$

For a minimum,

$$(2\pi/\lambda)a \cos v_{min} = h2\pi + 2\pi/N, \quad \cos v_{min} = (h\lambda/a) + (\lambda/Na),$$

and by difference,

Chapter 2

$\sin v \Delta v \simeq \cos v_{\min} - \cos v_{\max} = (\lambda/Na)$.

As N increases (for fixed v or order), Δv decreases. As order increases (for fixed N) $\sin v$ decreases and Δv increases.
(b) Repeat the steps in (a) for Fig. 2.13 and Eq. (2.7b). Now v is fixed and μ varies.
(c) For planes, the phase angle is $(2\pi/\lambda)(2d \sin \theta)$. Hence $\cos \theta \Delta\theta = \lambda/N2d$. As N increases for fixed order (fixed θ), $\Delta\theta$ decreases. As order increases (for fixed N), $\cos \theta$ decreases (as θ increases) and $\Delta\theta$ increases.

2.4. $\displaystyle\sum_n n \cos nx = \frac{d}{dx}\left[\frac{\sin Nx/2}{\sin x/2}\sin(N-1)x/2\right]$;

$\displaystyle\sum_n n \sin nx = \frac{d}{dx}\left[\frac{\sin Nx/2}{\sin x/2}\cos(N-1)x/2\right]$.

2.5. In general,

$$H = \left[\sum_i f_i^2 + \sum_{i\neq j}\sum f_i f_j \cos 2\pi(x_i - x_j)\right]^{1/2}, \quad 2\pi y = \tan^{-1}\frac{\sum_i f_i \sin 2\pi x_i}{\sum_i f_i \cos 2\pi x_i}.$$

2.6. See International Tables, Volume I, p. 53.

2.7. See International Tables, Volume I, p. 53.

2.8. $F = f\{1 + \exp[2\pi i(h + 2k)/3 + l/2]\}$. Hence reflections occur only when $h + 2k = 3n$ and l is even ($F = 4f^2$); $h + 2k = 3n \pm 1$, l odd ($F = 3f^2$); $h + 2k = 3n \pm 1$, l even ($F = f^2$).

2.9. $F^2 = 64f^2$, for hkl all even ($h + k + l = 4n$). $F^2 = 0$, for hkl all even ($h + k + l = 4n + 2$). $F^2 = 32f^2$, for hkl all odd. $F = f(1 + e^{2\pi i(h+k+l)/4})(1 + e^{\pi i(h+k)} + e^{\pi i(h+l)} + e^{\pi i(k+l)})$.

2.10. $F = 0$, hkl mixed. $F = 4(f_B \pm if_A)$, hkl all odd. $F = 4(f_B - f_A)$, $h + k + l = 2(2n + 1)$. $F = 4(f_B + f_A)$, $h + k + l = 4n$.

2.13. $\displaystyle F_{hkl} = \iiint \rho_1 \exp\left[2\pi i\left(\frac{hx}{a_1} + \frac{ky}{a_2} + \frac{lz}{a_3}\right)\right] dV$.

Now

$dV/dx\,dy\,dz = V/a_1 a_2 a_3 = 1$,

so

$$F_{hkl} = \rho_1 \int_{-a_1/4}^{+a_1/4} \int_{-a_2/4}^{+a_2/4} \int_{-a_3/4}^{+a_3/4} \exp\left[2\pi i\left(\frac{hx}{a_1} + \frac{ky}{a_2} + \frac{lz}{a_3}\right)\right] dx\,dy\,dz .$$

Each integral is of the form

$\displaystyle\frac{a}{2\pi i h}\exp\left(\frac{2\pi i h x}{a}\right)\Big|_{-a/4}^{+a/4}$,

so

$$F_{hkl} = (\rho_1 a_1 a_2 a_3/\pi^3 hkl)(\sin \pi h/2)(\sin \pi k/2)\left(\sin \frac{\pi l}{2}\right)$$

Therefore, hkl must be all odd.

2.14. (a) BCC.

(b) $F_{hkl} = \iiint \rho(xyz) \exp[2\pi i(hx + ky + lz)] \, dx \, dy \, dz$

$= \left(\dfrac{\alpha}{\sqrt{\pi}}\right)^3 \iiint_{-\infty}^{+\infty} \{N_1 \exp[-\alpha^2(x^2 + y^2 + z^2)]$

$+ N_2 \exp(-\alpha^2[(x-\tfrac{1}{2})^2 + (y-\tfrac{1}{2})^2 + (z-\tfrac{1}{2})^2])\}$

$\times [(\cos 2\pi(hx+ky+lz) + i \sin 2\pi(hx+ky+lz)] \, dx \, dy \, dz$

$= \exp[-(\pi^2/\alpha^2)(h^2+k^2+l^2)](N_1 + N_2 \exp[\pi i(h+k+l)])$.

2.15. If the reciprocal lattice of fcc (bcc) is rotated around the [111], the reciprocal lattice of the twin will be produced. There are new spots that are on layers 001/3, 002/3, 004/3, etc.

2.16. $A(s) = \displaystyle\int_{-\infty}^{+\infty} \sum_m \delta(x-mp)(\rho_0 + a\cos qx) e^{-2\pi i s x} \, dx$

$= \rho_0 \displaystyle\sum_0^\infty \exp(-2\pi i s m p) + \dfrac{a}{2}\left\{\displaystyle\sum_0^\infty \exp\left[-2\pi i \left(s + \dfrac{q}{2\pi}\right) mp\right]\right.$

$\left. + \displaystyle\sum_0^\infty \exp\left[-2\pi i \left(s - \dfrac{q}{2\pi}\right) mp\right]\right\}$.

Therefore, there are peaks in s space at n/p, and satellites around each peak at $(n/p) \pm (q/2\pi)$. Their height depends on a.

2.17. Interference function $= \dfrac{\sin^2 \pi s \cdot N_1 a \sin^2 \pi s \cdot N_2 a \sin^2 \pi s \cdot N_3 a}{\sin^2 \pi s \cdot a \sin^2 \pi s \cdot a \sin^2 \pi s \cdot a} = I$.

(a) For $N_1 = N_2 = N_3 = 10^4$, $I = N_1^2 N_2^2 N_3^2$ at reciprocal lattice points. Thus, there are strong diffracted intensities only from reciprocal lattice vectors $\mathbf{r}^*_{hkl}$, where hkl = integers.

(b) For $N_1 = N_2 = 10^4$, $N_3 = 5$, $(\sin^2 \pi s \cdot N_3 a)/(\sin^2 \pi s \cdot a)$ has a maximum of $N_3^2 = 25$, but is spread out in the $\mathbf{b}_3$ direction with half width $\approx 1/(5a) \approx 0.2b$.

Chapter 3

3.1. (a) $P = 122$ cal/s. Enthalpy to melt 0.1 kg of copper is $11.1(10)^3$ cal. Then melting would occur after $11.1(10)^3 \div 122 = 91$ s.
(b) At steady state all the heat goes into the water. $\Delta T = 3.7°C$.

3.7. (a) $1/\lambda_{L_{III}} = 1/\lambda_K - 1/\lambda_{K_{\alpha_1}}$; $\lambda_{L_{III}} = 4.86$ Å, therefore $V_{L_{III}} = 2.54$ keV.
(b) From Moseley's law, $(1/\lambda_{K_{ioniz}})^{1/2} = A(Z - \delta)$. Take $\delta \simeq 1$, $Z = 42$, and $\lambda_{K_{ioniz}}$ from part (a) yields $A = 0.0308$. Then for CuK_{ioniz}, take $Z = 29$ to obtain $\lambda_{K_{ioniz}} = 1.325$ Å, $V_{K_{ioniz}} = 9.20$ keV.

(c) Obtain λ_{K_α} for Cu from Moseley's law, $(1/\lambda_{K_\alpha})^{1/2} = A'(Z - \delta)$. Using λ_{K_α} for Mo and $Z = 42$, $A' = 0.0290$. Then with $Z = 29$, λ_{K_α} for Cu = 1.52 Å. Finally, use $\mu/\rho = CZ^4\lambda^3 = \beta\lambda^3$ since all values of μ/ρ listed are for Al. Hence $(\mu/\rho)_{\text{CuK}_\alpha \text{ through Al}} = [5.30/(0.71)^3](1.52)^3 = 5.2$ cm²/gm. This is in excellent agreement with the tabulated (measured) value of 4.86 cm²/gm!

3.10. (a) $\lambda = 1.08$ Å, $\Delta\lambda = 0.07$ Å. (b) $\lambda = 1.044$ Å, $\Delta\lambda = 0.02$ Å.

3.11. With x-rays, when carbon is in the octahedral positions, the intensities of peaks with hkl all even are raised, those with hkl all odd are reduced. When the carbon is in the tetrahedral positions, there is no effect on the 111, but the 200 is reduced. The structure factor for random occupation may be written as

$$F = F_{\text{fcc}} + \frac{(F_{\text{interstitial sites}})(\text{at fraction C}) \times 4}{\text{number of sites}}.$$

3.13. (a) A monochromator placed in the diffracted beam will not diffract the Compton modified scattering once the change in θ due to the change in wavelength exceeds the divergence of the beam to the crystal and the mosaic spread of the crystal. Thus, above a certain angle we can eliminate this portion from the scattering.
(b) See Warren, B.E. and Mavel, G., Rev. Sci. Instrum. *36*, 196 (1965).

3.14. There is a change in wavelength due to the Compton scattering from the paraffin. At $2\theta = 69°$ the Compton wavelength equals the absorption edge of Ag. The film is darker on the low-angle side.

3.15. (c) The scattered beam can be thought of as the sum of an unmodified beam (I_u) and a Compton modified beam (I_c):

$$I_0 = I_u + I_c.$$

With the given absorption attenuations, one may solve to obtain

$$I_u = I_c = \tfrac{1}{2}I_0.$$

3.18. (a) Mechanical mixture:

$$_A F_{200} = (N_A/4)(4f_A) = N_A f_A = (3/4)Nf_A, \quad _A F_{200}^2 = (9/16)N^2 f_A^2,$$
$$_B F_{200}^2 = (1/16)N^2 f_B^2 = (9/16)N^2 f_A^2, \quad F_{200}^2 = {_A F_{200}^2} + {_B F_{200}^2} = (18N^2/16)f_A^2.$$

(b) Random:

$$F_{200} = (N/4)(3f_A + f_B) = (N/4)(6f_A), \quad F_{200}^2 = (36N^2/16)f_A^2.$$

Laue monotonic $= NX_A X_B(f_A - f_B)^2 = N(3/4) \cdot (1/4)(f_A - 3f_A)^2$

$$= (12/16)Nf_A^2.$$

(c) As in (b) for 200 reflection.

3.20. $(1 + \cos^2 2\theta_2 \cos^2 2\theta_1)/(1 + \cos^2 2\theta_1)$.

3.22. There is no correction for symmetric transmission.

3.23. One form of the correction is

$$(1 - [(1 - n)/\sin^2 \theta])2d \sin \theta = m\lambda,$$

where θ is the measured angle, outside the specimen.

3.27. 18.9% loss.

3.29. Neglecting thermal vibrations, for (a) and (b):

$$F_{nuc}^2 = \begin{cases} 4b^2, & h+k+l \text{ even}, \\ 0, & h+k+l \text{ odd}. \end{cases}$$

For random domain orientations, $\langle q^2 \rangle = \frac{2}{3}$ for all cubic systems.

(a) $F_{mag}^2 = \begin{cases} 4p^2, & h+k+l \text{ even}, \\ 0, & h+k+l \text{ odd}, \end{cases}$ (b) $F_{mag}^2 = \begin{cases} 0, & h+k+l \text{ even}, \\ 4p^2, & h+k+l \text{ odd}. \end{cases}$

3.30. (a) $E(\text{keV}) = 6.2/d \sin \theta$.
(b) Use a high angle for the $\theta - 2\theta$ scan, as 2θ changes more there for a given ΔE, but use a low 2θ for the MCA system, as the energy changes most in that region for a given change in d. [See Cole, H., J. Appl. Crystalogr. 3, 405 (1970).]

Chapter 4

4.1. The marking is a twin.

4.3. (a) $\frac{1}{8}$. (b) 44.

4.7. (a) Tetragonal or orthorhombic.
(b) 3.85 Å.
(c) Orthorhombic.

4.8. The extra spots are from a twin.

4.9. The pattern is of an fcc Cu-Au alloy, with a parameter corresponding to Cu_3Au (3.74 Å).

4.10. (a) 0.8° 2θ for 70° line; 3.26° 2θ for 85° line.
(b) 0.8 mm for 70° line; 3.26 mm for 85° line.
(c) 1.6 mm for 70° line; 6.52 mm for 85° line.
(d) These lines do not appear! ($\sin \theta > 1$.)

4.12. The structure is fcc and the calculated density is 8.8 gm/cm^3. Thus this is not a defect structure.

4.14. The specimen is a Zn powder.

4.15. The unknown is Au.

4.18. To measure along an $h_1 00$ line, set the crystal up with the [100] bisecting incident and diffracted beams and run a $\theta - 2\theta$ scan.

Chapter 5

4.19. The orientation of the foil is $(0\bar{1}3)$. (Consider not only the ratios of the distances to spots, but also the symmetry of the diffraction pattern.) The edges of the dislocations on $(\bar{1}\bar{1}1)$ intersect the foil in a line $[\bar{2}31]$, which is correct for the zone axis of $(1\bar{1}1)$ and $(0\bar{1}3)$. The dislocations end on both faces of the foil, and hence the thickness is 1024 Å.

Chapter 5

5.1. Only a onefold axis. (Note that everyone who parts his hair does not part it in the middle, fingerprints are different on both hands, etc.)

5.2. The points are not lattice points; they do not have identical surroundings.

5.3. See International Tables, Volume I, p. 46.

5.4. See International Tables, Volume I, pp. 58–72.

5.5. See International Tables, Volume I, p. 59.

5.6. The same pattern results as for half a translation but the cell size is reduced.

5.8. The same molecular patterns arise when the glide planes are introduced, since glide planes are already present in these space groups. See International Tables, Volume I.

5.9. See International Tables for Crystallography, Volume I, p. 30.

5.11. The lattice points at the faces have neighbors on other faces at $\sqrt{2}a$ that the corner points do not have. Thus, the former cannot be lattice points.

5.12. 4: $P4, P4_1, P4_2, P4_3, I4, I4_1$ are all possible; $I4 = I4_2$ and $I4_1 = I4_3$ due to the body centers.

$\bar{4}, \bar{4}2m$: Because of the inversion operation, nothing new arises if screws are added.

$$\bar{4}_1 = \bar{4}_2, \quad \bar{4}_2 = \bar{4}_2, \quad \text{etc}.$$

422: As with 4, all possibilities occur; $4_1 22, 4_2 22, 4_3 22$.

4mm: Adding mirrors parallel to a fourfold screw axis leaves it unchanged.

4/m: A mirror perpendicular to a screw axis eliminates the sense of the screw. Hence only $P4/m$ and $P4_2/m$ are possible. Diagonal glides in centered cells are also possible: $I4_1/a$.

(4/m)mm: From the above considerations, it is not possible here.

5.13. $\bar{4}3m$: Notice that the nearest carbon atoms to a carbon atom form a tetrahedron.

5.14. See International Tables, Vol. I, space group No. 62.

5.15. (a) $Pbnm = Pmcn = Pnma$

$a_1 a_2 a_3 \quad a_3 a_1 a_2 \quad a_2 a_3 a_1$.

The space group is not one of those in the Tables but by changing axes it becomes *Pnma*, one of those given in the International Tables, Volume I, p. 548.

(b) Since there are 4 Ge and 4 Se in each unit cell, they must occupy equipoints of rank 4.

4a	0	0	0	0	$\frac{1}{2}$	0	$\frac{1}{2}$	0	$\frac{1}{2}$	$\frac{1}{2}$	$\frac{1}{2}$	$\frac{1}{2}$
4b	0	0	$\frac{1}{2}$	0	$\frac{1}{2}$	$\frac{1}{2}$	$\frac{1}{2}$	0	0	$\frac{1}{2}$	$\frac{1}{2}$	0
4c	x	$\frac{1}{4}$	z	$\bar{x}$	$\frac{3}{4}$	$\bar{z}$	$\frac{1}{2}-x$	$\frac{3}{4}$	$\frac{1}{2}+z$	$\frac{1}{2}+x$	$\frac{1}{4}$	$\frac{1}{2}-z$
	a_2	a_3	a_1	a_2	a_3	a_1	a_2	a_3	a_1	a_2	a_3	a_1

(Note that the axes given in the problem are labeled in terms of *Pbnm*, not *Pnma*, so the coordinates in the Tables, given for *Pnma*, must be relabeled.)

Ge and Se can be distributed about the cell in any of seven different ways:

Ge: 4a   4a   4b   4b   4c   4c   4c

Se: 4b   4c   4a   4c   4a   4b   4c

But some of these ways are redundant. 4a + 4b is equivalent to 4b + 4a (add $00\frac{1}{2}$ to each). 4c + 4a is equivalent to 4c + 4b also (let $z = z' + \frac{1}{2}$ and add $00\frac{1}{2}$). In this way we are reduced to 4 distinct distributions:

Ge: 4a   4a   4c   4c

Se: 4b   4c   4a   4c

After changing the axial system to *Pnma* $[(a_1 a_2 a_3) \to (a_2 a_3 a_1)]$, we have

$$a_1 = 10.82 \text{ Å}, \quad a_2 = 3.85 \text{ Å}, \quad a_3 = 4.40 \text{ Å}.$$

GeSe is covalently bonded, and r_{Ge}(covalent) = 1.23 Å and r_{Se}(covalent) = 1.16 Å. Obviously, Se is too large to fit in the 4a or 4b positions. For example, for the 4a positions, the distance from an atom at 000 to one at $0\frac{1}{2}0$ is 1.92 Å (from $a_2/2$) while the actual atomic radii at 000 and $0\frac{1}{2}0$ would be 1.16 Å Similar considerations rule out Se at 4b and Ge at 4a positions. Hence, Ge and Se both lie on 4c equipoints with different values of coordinates to be determined by analysis of the intensities of a diffraction experiment (see Problem 5.24.)

5.16. For a 4_1 axis along the a_3 axis 00*l* reflections exist only for $l = 4n$, and for a 4_2 axis only for $l = 2n$. A fourfold axis has no effect.

5.17. For 0*kl*, *l* must be 2*n* for a *c*/2 glide on a (100) plane. Only the (0*kl*) plane of reciprocal space has extinctions. A mirror causes no extinctions.

5.18. The general condition (*hkl* exists for $h + k = 2n$ only) occurs due to the fact that the cell is *C*-centered (there are lattice points at the center of the two faces of the unit cell perpendicular to the *c* or a_3 axis.) The special conditions for atoms, say, only in the 8*m* equipoint ($h = 2n$ for all reflections) arise due to the screw axis and inversion center near $\frac{1}{4}, \frac{1}{4}, z$.

5.19. See International Tables for Crystallography, Volume I, p. 370.

5.20. See International Tables for Crystallography, Volume I, p. 380.

5.21. See International Tables for Crystallography, Volume I, p. 519. For the origin of the cell at $\bar{3}m$, and for the 222 peak, $(h + k) = 4n$, $(k + l) = 4n$, $(h + l) = 4n$:

$$B = 0, \qquad A = 192\,(\cos 4\pi x \cos 4\pi y \cos 4\pi z).$$

The occupation of equipoints for a normal spinel AB_2O_4 is as follows, for the origin at $\bar{3}m$: Oxygen in 32e, A in 8b, B in 16c.
Then

$$|F|_{222} = \sum_n f_n A_n = \frac{8}{192} f_A A_A + \frac{16}{192} f_B A_B + \frac{32}{192} f_O A_O,$$

$A_A = 0$ because of the coordinates of the rank 8b equipoint, and $F_{222} = 16 f_B + 32 f_O \cos^3 4\pi x$.

5.22. For *Pmmn* (space group No. 59) looking at Internatonal Tables, Vol. I, p. 406, we find that for h odd the trigonometric terms for each atom will be small for the x coordinate near 0 or $\frac{1}{2}$. This would occur for six of the eight atoms (4Ag + 2Sn). For *Pmn2$_1$* (No. 31, p. 391), x must be near $\frac{1}{4}$ or $\frac{3}{4}$ for *all* atoms to produce a low value of F_{hkl} with h odd. This is not possible for this material as there are two atoms of Ag and two of Sn which go into a rank 2 equipoint with an x coordinate of zero.

5.23. Find the coordinate of the metal atom in the octahedral holes of close-packed oxygen [note that there are six (111) layers of oxygen per unit cell]. This is the critical value. Now expand the temperature factors $\exp(-M_i) \cong (1 - M_i) = (1 - B_i \sin^2 \theta / \lambda^2)$. For each z coordinate (in units of 0.01) of the metal atom near the initial guess, solve for B_0 and B_{metal} from intensity ratios such as I_{002}/I_{004} and I_{002}/I_{006} until you find values of z, B_0 and B_{metal} which fit all reflections well. The expansion of $\exp(-M_i)$ is reasonable in this case since Al_2O_3 has a very high melting point and hence, very small B values. This can be checked by comparing $\exp(-M_i)$ and $1 - M_i$ for the determined values of B_0 and B_{metal}. See Moss, S.C. and Newnham, R.E., Z. Kristallogr. *120*, 359 (1964) for further details and the solution.

5.24. Reflections 200, 400, and 600 can be employed to determine the x coordinate. Make plots of calculated intensity versus x_{Ge} for various fixed x_{Se}, say 0.0, 0.05, 0.1, 0.15, and 0.2. Then narrow the intervals when an approximate solution has been found. This procedure leads to $x_{\text{Se}} = \pm 0.15$, $x_{\text{Ge}} = \pm 0.12$. A similar procedure can be followed for the z coordinates using the 202 and 004 reflections. The choice of signs is made by comparing measured and calculated intensities for the 113 reflection. The results are $x_{\text{Se}} = 0.15$, $z_{\text{Se}} = 0.50$; $x_{\text{Ge}} = -0.12$, $z_{\text{Ge}} = -0.11$.

5.25. (a) 3mm, (b) 2mm, (c) 3mm, (d) 4mm.

5.27. (a) mmm, $P - a$.
These extinction conditions may be established after indexing the film:

hhl, h0l: no conditions
h00: $h = 2n$ (weak 700 and 900 reflections should be checked with a diffractometer, they might be due to diffraction of $\lambda/2$ by the (14,00) and (18,00) planes):
00l: maybe, $l = 2n$.
hk0: $h = 2n$? Weak 510 and 710 reflections should also be checked.
0kl: insufficient information.

Consulting the International Tables, Volume I, p. 350, there are several entries for primitive cells and a glide plane. Examining each of these possible space groups for its extinctions, *Pnma*, and *Pna*2_1 are satisfactory (the latter with a change of axial labeling). The *Pna*2_1 space group is noncentrosymmetric.

Chapter 6

6.1. Substitute the right-hand side of Eq. (6.6) into Eq. (6.7) and integrate subject to the orthogonality conditions of Eqs. (6.30). Then simplify, using Eq. (6.33).

6.2. $B_n = 0$, $A_0 = 0$, $A_n = 0$ for n even, $A_n = 2/n\pi$ for $n = 1, 5, 9, \ldots$, $A_n = -2/n\pi$ for $n = 3, 7, 11, \ldots$.

6.3. $B_n = 0$, $A_0 = X$, $A_n = (a/n\pi)\sin(n\pi X/a)$; therefore,

$$f(x) = \frac{X}{a} + 2\sum_{n=1}^{\infty} \frac{1}{2\pi}\sin\frac{n\pi X}{a}\cos\frac{2\pi n x}{a}$$

6.4. (a) $Tf(cx) = \int_{-\infty}^{\infty} f(cx)e^{+2\pi i s x}\,dx.$

Let $cx = z$,

$$Tf(cx) = T(z) = \int_{-\infty}^{\infty} f(z)\exp\left(\frac{+2\pi i s z}{c}\right)\frac{dZ}{|c|} = \frac{1}{|c|}F\left(\frac{s}{c}\right).$$

(e) $\dfrac{d^n f(x)}{dx^n} = (-2\pi i)^n \int_{-\infty}^{\infty} s^n F(s)\exp(-2\pi i s x)\,ds$,

$T\dfrac{d^n f(x)}{dx^n} = (-2\pi i)^n \int\int s^n F(s)\exp(-2\pi i s x)\exp(+2\pi i s' x)\,ds'\,dx$.

The integration over x has nonzero values only for $s = s'$, so

$$T[d^n f(x)/dx^n] = (-2\pi i)^n s^n F(s).$$

6.5. $F(s) = (A\sqrt{\pi}/B)\exp(-\pi^2 s^2/B^2)$. Thus the Fourier transform of a Gaussian function is a Gaussian function.

6.7. Take the scattering density as

$$\rho(x) = \delta[x + (a/2)] + \delta[x - (a/2)].$$

Then,

$$F(s) = \exp(-\pi i a s) + \exp(\pi i a s) = 2\cos(\pi a s) \quad \text{and} \quad I(s) = 4\cos^2(\pi a s).$$

6.8. Take the scattering density as

$$\rho(x) = f(x) \otimes \{\delta[x + (a/2)] + \delta[x - (a/2)]\},$$

where $f(x)$ has the value unity for $|x| < b/2$ and zero elsewhere.

Then

$$F(s) = T\rho(x) = Tf(x)\, T\{\delta[x + (a/2)] + \delta[x - (a/2)]\}$$
$$= (\sin \pi bs/\pi s)\, 2\cos \pi as$$

and

$$I(s) = 4\frac{\sin^2 \pi bs}{(\pi s)^2}\cos^2 \pi as.$$

The periodicity of the cosine function is $1/2a$, and the intensity is modulated by $\sin^2 \pi as/(\pi s)^2$.

6.9. See Buerger (1960), pp. 422–424.

6.10 $F(s) = J_1(\pi as)/s$, where s is the radial coordinate in reciprocal space, and $J_1(x)$ is the first-order Bessel function of argument x.

6.11. See Buerger (1960), pp. 454–471.

6.12. See Buerger (1960) pp. 89–97.

6.13. (a) All twofold equipoints are equivalent except for change or origin and axes. Thus the positions may be taken as $x, 0, 0$ and $\bar{x}, 0, 0$.
(b) The coordinate x could be obtained from the one-dimensional Patterson section:

$$P(x, 0, 0) = \frac{1}{V_c}\sum_h \left(\sum_{kl} F_{hkl}^2\right) \cos 2\pi hx,$$

in which vectors of the length $2x, 0, 0$ will appear.

6.14. (a) From the International Tables, Volume I, p. 349, the two possible space groups are $P112_1$ and $P112_1/m$.
(b) For both space groups, projections along the 2_1 axis give plane group $P2$. For $P112_1$, projections along the other two axes give Pg. For $P112_1/m$, these projections give Pmg. The Patterson projections $P(x, y)$, $P(x, z)$, and $P(y, z)$ are easily calculated, and may be found on the inside cover of Buerger (1959). A clear distinction between the two structures would occur for Patterson projections along $\mathbf{a}_1$ and $\mathbf{a}_2$, but for projection along $\mathbf{a}_3$ (the 2_1 axis), the Patterson projections are identical.
(c) The Harker section $P(o, o, z)$ will distinguish the two structures, showing only interatomic vectors differing in the z component. Such vectors exist in $P112_1/m$ due to the mirror, but not in $P112_1$.

6.15 (a) The drawing shows the construction when two heavy atoms, M and O are present. Notice that with only the M atom in the structure, two solutions are possible (Fig. A6.15, points a, b).
(b) Two choices for $^R F$ occur, one with positive and one with negative phase. For the case shown, with the observed $|F_{hkl}| < |F_{\bar{h}\bar{k}\bar{l}}|$, the phase must be positive, while for $|F_{hkl}| > |F_{\bar{h}\bar{k}\bar{l}}|$, the negative phase is the correct choice.

6.16 (a) $F_{hkl} = \sum_{j=1}^{N/2} 2f_j \exp\{2\pi i(hx_j + lz_j)\} \cos 2\pi ky_j = A_{hkl} + iB_{hkl}.$

(a)
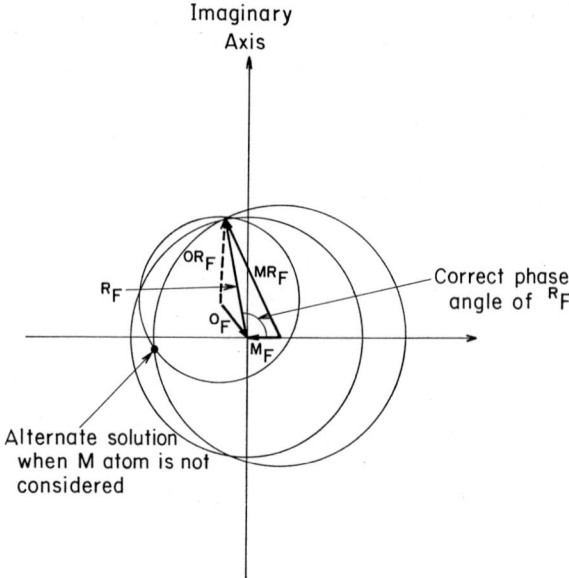

(b)
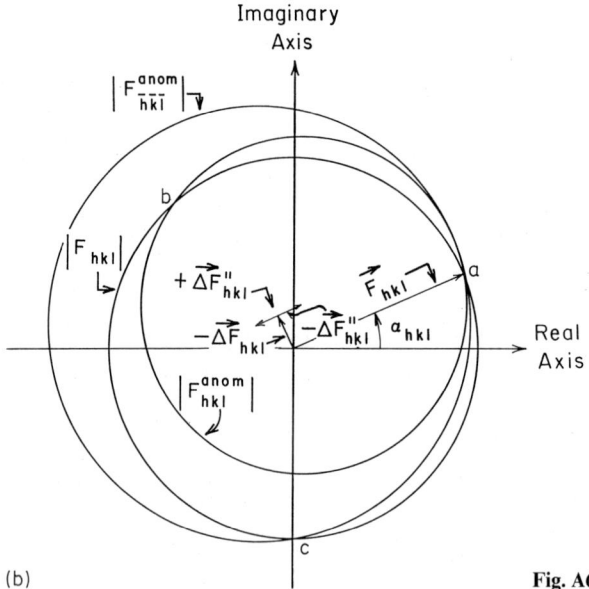

Fig. A6.15a,b

Since $\overline{A}_{hkl} = \overline{B}_{hkl} = 0$, we have $\overline{A^2_{hkl}}$ = variance of A_{hkl}, which by the central limit theorem gives

$$\overline{A^2_{hkl}} = \sum_{j=1}^{N/2} \overline{4f_j^2 \cos^2 2\pi(hx_j + lz_j)} \, \overline{\cos^2 2\pi ky_j} = \sum_{j=1}^{N/2} 4f_j^2 \cdot \tfrac{1}{2} \cdot \tfrac{1}{2} = \tfrac{1}{2}\Sigma.$$

Similarly, $\overline{B^2_{hkl}} = \tfrac{1}{2}\Sigma$, and hence $\overline{|F_{hkl}|^2} = \overline{A^2_{hkl}} + \overline{B^2_{hkl}} = \Sigma$. However, for $k = 0$,

$$\overline{A_{h0l}^2} = \sum_{j=1}^{N/2} 4f_j^2 \overline{\cos^2 2\pi(hx_j + lz_j)} = \Sigma = \overline{B_{h0l}^2}.$$

Therefore,

$$\overline{|F_{h0l}|^2} = \overline{A_{h0l}^2} + \overline{B_{h0l}^2} = 2\Sigma.$$

6.17. Use Eq. (6.74) for the unitary structure factor and
(a) set $a_j = \sqrt{n_j} \cos 2\pi k y_j$, $b_j = \sqrt{n_j} \cos 2\pi(hx_j + lz_j)$;
(b) set $a_j = \sqrt{n_j}$, $b_j = \sqrt{n_j} \cos 2\pi k y_j \cos 2\pi(hx_j + lz_j)$.

6.18. (a) The angle turned in the helix per carbon $= 6(2\pi)/13$ rad $= 166°$. Thus, Teflon is *almost* a planar zigzag arrangement (for which the angle would be 180°). Because there are only single carbon–carbon bonds, we can see how the carbon labeled 3 in Fig. A6.18a can swing around the line joining 1–2 to form the helix. The spacing of carbons (p) along the chain is (1.54) sin 58° $= 1.30_6$ Å. The pitch (P) is $(13/6) \times 1.30_6 = 2.82_9$ Å. The radius of the carbon backbone helix is $\frac{1}{2}(1.54 \cos 58°) = 0.40_8$ Å. [It is also possible to solve for the pitch and the radius without "seeing" the shape. The atomic coordinates can be written as follows: atom 1: $r, 0, 0$; atom 2: $r \cos 12\pi/13$, $r \sin 12\pi/13$, p; atom 3: $r \cos 24\pi/13$, $r \sin 24\pi/13$, $2p$. The distance between atoms 2 and 1 and atoms 3 and 1 (both 1.54 Å) can be written in terms of equations for two unknowns, r and p.]

There are really three concentric helices, two of F, one of carbon slightly displaced along the helix axis with respect to one another. This will affect the pattern; there will be some amplitude cancellation due to this displacement (from which one could work out the positions of these three helices). But this point was avoided by treating the molecule as a single helix. Thus, to the carbon backbone we must add the distance to F^- ions, 1.49 Å. Hence, $r_{\text{helix}} = 0.41 + 1.49 = 1.90$ Å.

(b) The selection rule for an a_b helix with layers l/c apart is: $l = bn + am$, i.e., $l = 6n + 13m$ [n and m (and l) are integers]. (The c axis is $Pb = 16.98$ Å.) For each l, one assumed various m's and solves for the n's. A plot of the selection rule is given in Fig. A6.18b.

It is clear from the selection rule that the pattern repeats itself at the 13th layer. (On a flat film the spot due to $n = 0$ for this layer will be much broader than the one at the zero layer, but the repeat will still be clear.)

(c) As $b_3 = 1/c = 1/16.98$ Å^{-1} and the repeat is at 13 b_3, we should compare this distance to the radius of the Ewald sphere: $13/16.48 > 1/1.542$. Therefore, the repeat will not be seen on the film! The 11th layer will just make it, if the film is *very* big.

The intensity asked for was the maximum and half-width formed by considering the *square* of the lowest order Bessel function for each layer. The order n can be read from the figure in part (b) for each l, and $J_n^2(2\pi Rr)$ is desired. From tables of Bessel functions $J_n(x)$ and knowing r, the values of R for these intensities can be found. There is no problem as to whether these positions will touch the sphere; we will need many such helices to form a pattern so we can assume these take on all orientations around the helix axis. (That is, the Ewald sphere can take on all positions around this axis.)

Now the film to specimen distance is rather large — 0.1 m. As a result of this and the large wavelength, there will be curvature on a flat film; that is, the layers

572 Answers to Selected Problems

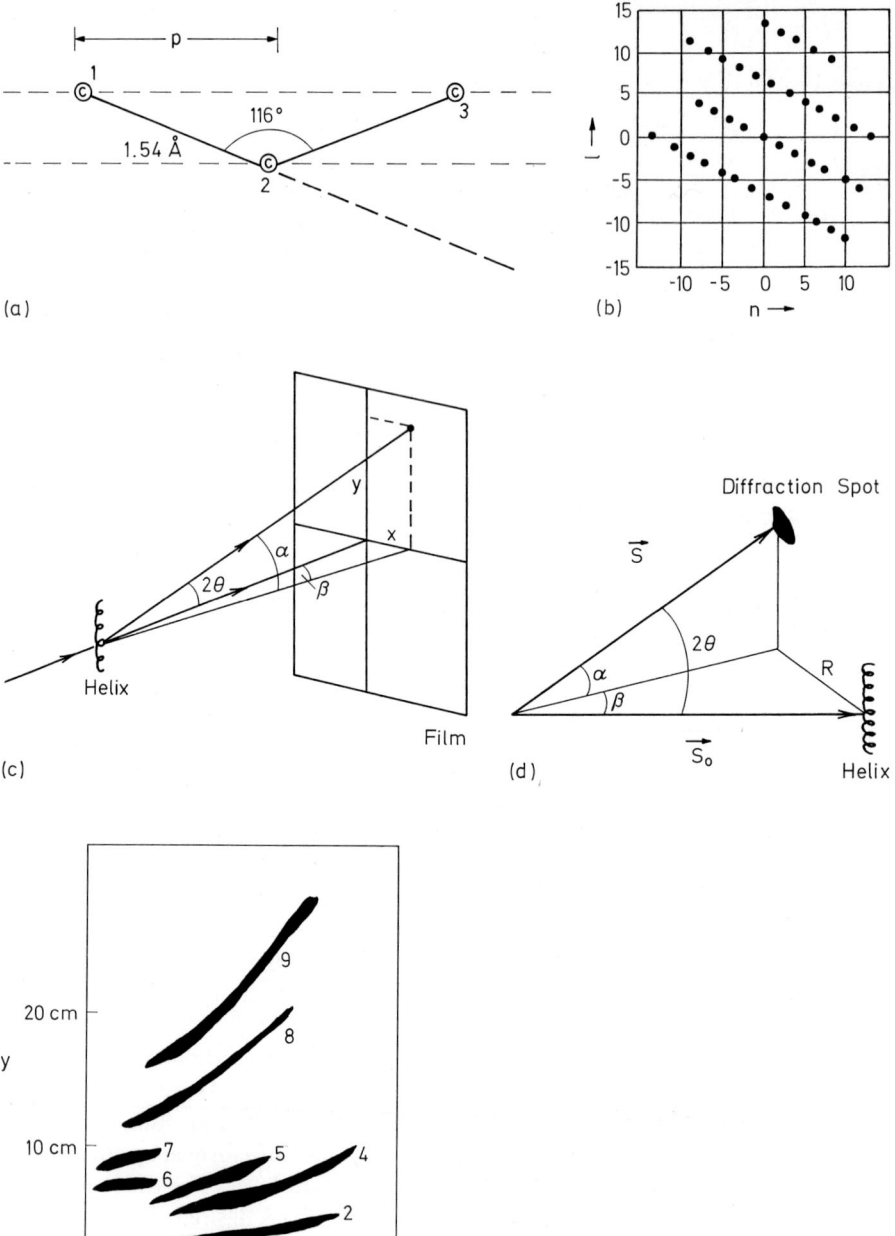

Fig. A6.18a–e

will not be horizontal, except close to the origin, or if the film is cylindrical. In reciprocal space

$$|\mathbf{r}^*| = 2\sin\theta/\lambda = [(l/c)^2 + R^2]^{1/2}.$$

For any l and R this allows us to calculate θ and hence, 2θ. The radial distance from the center of the film to the spot is $0.1\text{ m} \times \tan 2\theta$. The distance from a horizontal line on the film to the spot is (see Figs. A6.18c and d):

$$y = 0.1 \tan\alpha/\cos\beta.$$

Now, $\sin\alpha = l\lambda/c$, so

$$R^2 = \left(\left|\frac{\mathbf{S}}{\lambda}\right|\cos\alpha\right)^2 + \left|\frac{\mathbf{S}_0}{\lambda}\right|^2 - |\mathbf{S}|\cos\alpha\left|\frac{\mathbf{S}}{\lambda}\right|\cos\beta.$$

For known R, $|\mathbf{S}|$, $|\mathbf{S}_0|$, and α, β is obtained.

Therefore, the spot coordinates x, y are fixed for any l, R. The curvature of all layers is upward as you move away from the film's center as in a Laue pattern. A schematic drawing of the result is shown in Fig. A6.18e.

6.19. (b)

$$F^2 = \begin{cases} S^2(f_A - f_B)^2 & \text{for } h+k+l, \text{ odd} \\ (f_A + f_B)^2 & \text{for } h+k+l, \text{ even}. \end{cases}$$

6.20. (a) $A(s) = \int_{-\infty}^{\infty} \sum_m \delta(x - mp)(\rho_0 + a\cos qx)\exp(+2\pi i s x)\,dx$

$$= \rho_0 \sum_0^\infty \exp(+2\pi i s m p) + \frac{a}{2}\left\{\sum_0^\infty \exp\left[-2\pi i\left(s + \frac{q}{2\pi}\right)mp\right]\right.$$

$$\left. + \sum_0^\infty \exp\left[+2\pi i\left(s - \frac{q}{2\pi}\right)mp\right]\right\}.$$

Therefore, there are peaks in s space at n/p, and "satellites" around each peak at $(n/p) \pm (q/2\pi)$. Their height depends on a.
(b) As q decreases, the satellites move closer to the main peaks.
(c) If the waves are finite, say of length L, the integral involving $a\cos qx$ has limits; i.e., the last two sums are not infinite:

$$A(s) = \rho_0 \sum_0^\infty \exp(-2\pi i s m p)$$

$$+ \frac{a}{2}\left\{\frac{\sin\pi(L/p)(s + q/2\pi)p}{\sin\pi(s + q/2\pi)p}\exp\left[-\pi i\left(s + \frac{q}{2\pi}\right)\left(\frac{L}{p} - 1\right)\right]\right.$$

$$\left. + \frac{\sin\pi(L/p)(s - q/2\pi)p}{\sin\pi(s - q/2\pi)p}\exp\left[-\pi i\left(s - \frac{q}{2\pi}\right)\left(\frac{L}{p} - 1\right)\right]\right\}.$$

Then each of the satellites has a width of $2p/L$, because a function

$$\sin\pi sNa/\sin\pi sa$$

has its first zero for $s = \pm 1/N$. Now,

$$q = 2\pi/\lambda \quad \text{or} \quad \lambda/2 = \pi/q .$$

The width of each satellite for $L = \lambda/2$ is $2pq/\pi$. The spacing between them is $2q/2\pi$; if $p = 1$, e.g., they can overlap.

6.21. (a) The $4a$ and $4b$ equipoints are equivalent, differing only by a shift in origin of $0, \frac{1}{2}, 0$, so we need consider only one of them. If iron were on equipoint $4a$, two Fe's giving rise to the most intense peaks in the Patterson would have interatomic vectors $\frac{1}{2}00$ or $\frac{1}{2}\frac{1}{2}0$. No strong peak with these vector coordinates appears in Table P6.21, and we thus disregard equipoints $4a$ and $4b$. Similar reasoning leads to elimination of $4c$ and $4d$, leaving Fe in the $4e$ positions.
(b) For $4e$, the interatomic vectors include:

$$[0, \bar{y}, \tfrac{3}{4}] - [0, y, \tfrac{1}{4}] = 0, 2\bar{y}, \tfrac{1}{2} .$$

Peak No. 1 is of this form with $y = -\tfrac{1}{8}$, and hence the iron positions are:

$$0, -\tfrac{1}{8}, \tfrac{1}{4}; \ 0, \tfrac{1}{8}, \tfrac{3}{4}; \ \tfrac{1}{2}, \tfrac{3}{8}, \tfrac{1}{4}; \ \tfrac{1}{2}, \tfrac{5}{8}, \tfrac{3}{4} .$$

The next strongest peaks after Fe-Fe should be Fe-N or Fe-C. These would be virtually indistinguishable by intensities since $Z_N \approx Z_C$, but we are given the Fe-N distance as 2.1 Å and we see from Fig. P6.21 that the Fe-C distance is much larger. This leads to identification of Peak No. 7 as Fe-N and nitrogen coordinates relative to Fe at 0.0452, 0, 0.1072. The 16 N atoms can be located in any appropriate combinations of rank 8 or rank 4 equipoints, but if any rank 4 equipoints were filled by N, this would give rise to strong Fe-N peaks at $0, 0, \tfrac{1}{2}; \tfrac{1}{2}, \tfrac{1}{2}, 0$; etc. — but these are not found. Thus the N must be in an $8f$ equipoint with coordinates and atom number as shown below and in Fig. A6.21a.

N (atom number)	Coordinates		
4	0.045	−1/8	0.35
2	−0.045	−1/8	0.15
1	0.045	1/8	0.85
3	−0.045	1/8	0.65
8	0.54	3/8	0.35
6	0.46	3/8	0.15
5	0.54	5/8	0.85
7	0.46	5/8	0.65

This procedure locates 2N atoms per Fe, but we must find 2 more. We search for another strong vector of length 2.1 Å for which the x and z coordinates are interchanged (*Note*: this is so because the two sets of two nitrogen atoms are rotated by 90°). Peak #23 satisfies these conditions, yielding coordinates $x, y, z = 0.1232, 0, 0.0556$ relative to Fe. Again using the $8f$ equipoint these 8 nitrogens are at:

0.123	−1/8	0.2;	0.62	3/8	0.2
−0.123	−1/8	0.3;	0.38	3/8	0.3
0.123	1/8	0.7;	0.62	5/8	0.7
0.123	1/8	0.8;	0.38	5/8	0.8

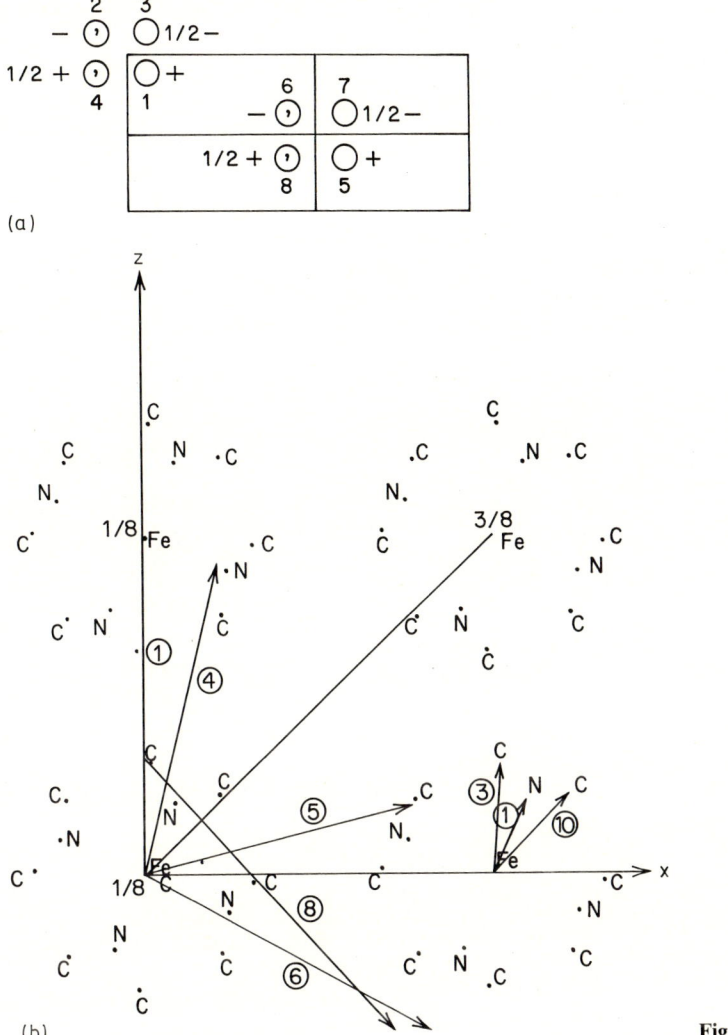

Fig. A6.21a,b

(c) Peaks 1 and 2 are identified as Fe-Fe from their intensity and coordinates; Peak 7 is Fe-N(2), while peak 9 is the origin incompletely removed. The remaining 6 vectors are identified using a drawing of a projection down the y axis as shown in Fig A6.21b. The following points are useful in this analysis:

(i) most of the peaks after 1 and 2 are about the same strength and are therefore Fe-C or Fe-N (differences in strength may be attributed to broadening of the peaks due to intermolecular vibrations).
(ii) Fe-C distances are larger than Fe-N.
(iii) The molecule is relatively flat and perpendicular to the y axis. Thus, intermolecular vectors will have z components $\approx \frac{1}{2}$ in some cases, and $y \approx \frac{1}{4}$ when $z \approx 0$.

(iv) It is most convenient to make the projections with orthogonal coordinates given as $X(\text{Å})$, $Y(\text{Å})$, $Z(\text{Å})$ in Table P6.21.

Using these principles, the final peak identifications are:

1) Fe-Fe   Intermolecular
2) Fe-Fe   Intermolecular
3) Fe-C(9)
4) Fe-N(1)   Intermolecular
5) Fe-C(1)   Intermolecular
6) Fe-C(8)   Intermolecular
7) Fe-N(2)
8) Fe-C(5)   Intermolecular
9) Origin residual
10) Fe-C(1)

Chapter 7

7.1. (d) $D_{\text{eff}} = 230$ Å, $\langle \varepsilon_{50\,\text{Å}}^2 \rangle^{1/2} = 28.7 \times 10^{-4}$.

7.2. The boundaries are on planes (uvw). The reflection being measured is hkl. Then $D_{uvw}/D_{hkl} = \langle \sin \phi \rangle$, where the average implies all (hkl) values in the form. For a cubic material:

$$\sin \phi = \frac{(h\mathbf{b}_1 + k\mathbf{b}_2 + l\mathbf{b}_3) \cdot (u\mathbf{a}_1 + v\mathbf{a}_2 + w\mathbf{a}_3)}{|\mathbf{a}||\mathbf{b}|(h^2 + k^2 + l^2)^{1/2}(u^2 + v^2 + w^2)^{1/2}}$$

$$= \frac{hu + kv + lw}{(h^2 + k^2 + l^2)^{1/2}(u^2 + v^2 + w^2)^{1/2}}.$$

In this example, this expression is now evaluated and averaged for all possible 110's and 100's in their respective forms, with $[uvw] = [111]$.

7.3. $A_L = 1 - L[(1.5\alpha/a) X_{hkl}]$, $\Delta 2\theta° = (90/\pi^2)\sqrt{3\alpha} \tan \theta \, Y_{hkl}$,

$X_{311} = 3/(2\sqrt{11})$, $Y_{311} = -\frac{1}{11}$,

$X_{222} = \sqrt{3/4}$, $Y_{222} = -\frac{1}{8}$,

$X_{400} = 1$, $Y_{400} = +\frac{1}{4}$,

$\Delta 2\theta°_{400} - \Delta 2\theta_{222} = (90/\pi^2)\sqrt{3\alpha}[\frac{1}{4}\tan \theta_{400} + \frac{1}{8}\tan \theta_{111}]$.

Note that the peak *shapes* of the 222 and 400 peaks are the same as for the 111 and 200 peaks (the latter are given in the text). But the peak *shifts* are larger due to the $\tan \theta$ terms and opposite to the 111, 200.

7.4. (a) The complex coefficient $C_n(l)$ can be written as

$$C_n(l) = (N_n/N_3)\{[1 - 4\alpha(1-\alpha)\sin^2 \pi l\varepsilon/a_3]^{1/2} \exp(-i\gamma)\}^{|n|},$$

where $\tan \gamma = \alpha \sin(2\pi l\varepsilon/a_3)/(1 - 2\alpha \sin^2 \pi l\varepsilon/a_3)$.

(b) $P'_{2\theta} = KN_3 \sum_{-\infty}^{+\infty} |C_n(l)| \exp\left(-2\pi i n\left[h_3 + \frac{\gamma}{2\pi}\right]\right)$.

Chapter 7

Let $h_3 + (\gamma/2\pi) = h_3 = l$ for a $00l$ reflection. Then

$$l = h_3 + (\gamma/2\pi) = (2a_3 \sin\theta/\lambda) + (\gamma/2\pi).$$

Hence,

$$\frac{1}{2a_3}\left[l - \frac{1}{2\pi}\tan^{-1}\left(\frac{\alpha\sin 2\pi l\varepsilon/a_3}{1 - 2\alpha\sin^2 \pi l\varepsilon/a_3}\right)\right] = \frac{\sin\theta}{\lambda}.$$

For small $l\varepsilon/a_3$, $\sin\alpha = \alpha = \tan\alpha$ and then,

$$l/2a_3[1 + \alpha(\varepsilon/a_3)] = \sin\theta/\lambda.$$

From Bragg's law: $\sin\theta/\lambda = l/2a_3'$ $(d = a_3'/l)$;

$$a_3' = (1 - \alpha)a_3 + \alpha[(\varepsilon/a_3)a_3 + a_3] = a_3 + \alpha\varepsilon.$$

Hence,

$$\sin\theta/\lambda = l/2a_3' = l/2a_3[1 + (\alpha\varepsilon/a_3)].$$

The two results are the same.
(c) For small α, ε,

$$\beta_{rad} = [4\alpha(1 - \alpha)\pi^2\varepsilon^2/\lambda a_3]\tan\theta\sin\theta,$$

or more precisely,

$$\beta = \left(\frac{d}{l}\right)(1 - \cos 2\pi l\varepsilon/a_3)\tan\theta.$$

Solve for θ with the *exact* solution in part (b) (not the one for small α, ε).
(d) $(a_3')_{001} = 3.58$ Å; $(a_3')_{004} = 3.48$ Å.
(e) The breadth increases and then decreases with 2θ, $2\beta_{001} = 0.044$ rad, $2\beta_{003} = 0.16$ rad, $2\beta_{004} = 0.095$ rad.

7.5. (a) See Fig. A7.5.
(b) There are no peak shifts.

$$A_L = 1 - L\{(1/D_{eff}) + [K(hkl)\eta/a]\},$$

where η is the probability of the boundary in a [100] direction.

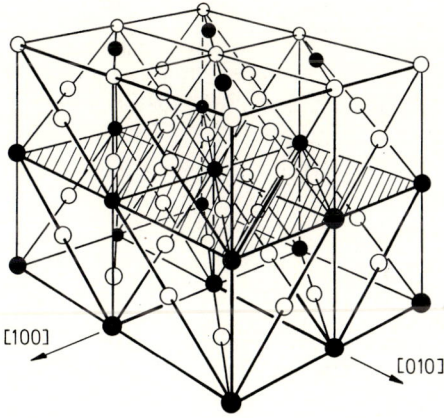

Fig. A7.5

$$K(hkl) = \frac{2}{h_0 p} \sum_{\substack{\text{broadened} \\ \text{components}}} |l| \, .$$

Here, p is the total multiplicity. For $h + k = 2m$ where m is an integer, there is no broadening; this includes all fundamental peaks 111, 200 etc., that would occur even if the structure was random (in which case it would be fcc). There *is* broadening for some of the extra or superlattice reflections. For example,

$$K(110) = 0.943 \, ; \quad K(100) = 0 \, .$$

Thus the presence of such boundaries will cause no broadening of the fundamental reflections and 100, but a large broadening of the 110 reflection. For details of the derivation, see Chap. 9 of Cohen and Hilliard (1966).

7.7. (a) The intensity is proportional to $(N/4)16(\bar{f})^2$; with vacancies $\bar{f} = (0.9 f_R + 0.1 \times 0)$. Thus the intensity is reduced by $\approx 20\%$.
(b) $I_{\text{Laue Monotonic}} = N x_A x_B (f_A - f_B)^2 = N \times 0.9 \times 0.1 f^2$.

$$I_{\text{peak}} = (N/4)(16)(\bar{f})^2 = (N/4)(16)(0.9f)^2 \, .$$

$$I_{\text{LM}}/I_{\text{peak}} = N \times 0.09 f^2 / [(N/4)(16)0.81 f^2] \cong \tfrac{1}{36} \, .$$

7.8. Employ $\alpha_{lmn} = \alpha_{l\bar{m}n} = \alpha_{mln}$, etc.

7.9. The parity condition for the interatomic vectors is:

$$\left. \begin{array}{l} 2(l \pm m) \\ 2(l \pm n) \\ 2(m \pm n) \end{array} \right\} \text{even} \, .$$

Now follow the same considerations in the text concerning this matter for the fcc structure.

7.11. $I_{\text{SRO}} = N x_c^{\text{Mg}} x_c^{\text{Fe}} (f_{\text{Mg}} - f_{\text{Fe}})^2 \sum_{uvw} \alpha_{cc}^{\text{Fe-Mg}}(lmn) \cos 2\pi(hl + km + ln)$.

The subscripts c-c refer to pairs in the cation sublattice. The term N is the number of lattice points, not atoms, and x_c^i is the sublattice fraction of ion i.

$$\alpha_{cc}^{\text{Fe-Mg}}(lmn) = 1 - [P_{cc}^{\text{Fe-Mg}}(lmn)/x^{\text{Mg}}] \, .$$

The term $P_{cc}^{\text{Fe-Mg}}(lmn)$ is the conditional probability of an Mg ion at the end of interatomic vector lmn in the cation sublattice if there is an Fe at the origin of the vector. Only lmn values that span the cation sublattice are involved. The anion sublattice (a) is not involved as it is filled with only oxygen. Anion-cation terms also do not appear because $\alpha_{ca}^{ij}(lmn) = 1 - [P_{ca}^{ij}(lmn)/x_A^j]$ and with j as oxygen, $P_{ca}^{i-0} = 1 = x_A^j$.

For further details for this kind of structure see Hayakawa and Cohen (1975).

7.12. (b) $F^2 = 2f_{\text{Br}}^2 + 2f_{\text{Br}}^2 \sin kr_{\text{Br-Br}}/kr_{\text{Br-Br}}$.
The intensity in electron units (eu) is just N times this expression where N is the number of molecules in the beam.

As the pressure increases, the intensity first increases everywhere due to the increasing number of molecules, then at higher pressures, intermolecular interference becomes important; instead of the intensity increasing steadily as $0°\ 2\theta$ is approached, due to the increase in f^2, at some 2θ associated with this intermolecular interference the intensity begins to decrease. That is, there is a broad maximum at the 2θ position associated with the intermolecular distance.

7.13. $(1/\lambda)x = v$, where x is the fraction of λ_{max}. $C_{44} = 0.55 \times 10^{11}$ Pa.

7.14. Draw the Ewald sphere. Incident and diffracted beams have a range of lengths; hence, one is sampling a line in reciprocal space through 000.

7.15. (a) and (b) See Fig. A7.15.
(e) On the figure for (a), (b), lines 1 and 2 are $(\mathbf{S} - \mathbf{S}_0)/\lambda$ vectors for measuring longitudinal modes, 3 and 4 are for the transverse modes polarized in the [001] direction. In Eq. (7.3), the intensity depends on $(\mathbf{S} \cdot \mathbf{A}_K)^2$, where $\mathbf{A}_K$ is the direction of vibration of the mode. Because the angle for the longitudinal polarization and $(\mathbf{S} - \mathbf{S}_0)/\lambda$ is small at 3 or 4, its contribution is small.
(f) When the neutron gains or loses energy it may not be possible to move the analyzer crystal to the desired angle to satisfy Bragg's law.
(g) See Peckham, G. E., Saunderson, D. H. and Sharp, R. I., Br. J. Appl. Phys. *18*, 473 (1967).

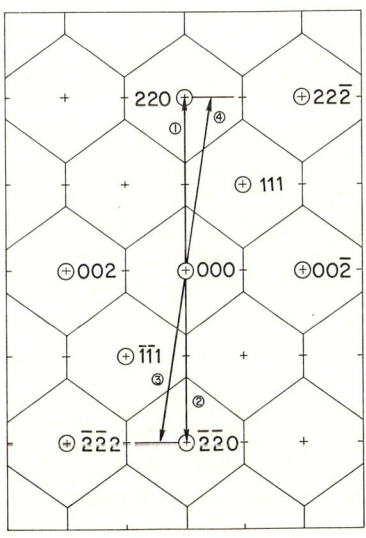

Fig. A7.15

7.16. The radius of gyration is given by

$$R_G^2 = \frac{1}{2n^2} \sum_i \sum_j \langle R_{ij}^2 \rangle.$$

If the units are numbered from 1 to $n/4$ on each arm, there will be 4 terms of the type

$$\sum_{i=1}^{n/4} \sum_{j=1}^{n/4} |i-j|a^2 \qquad (1)$$

when the ith and jth unit are on the same arm. There will be 12 terms of the type

$$\sum_{i=1}^{n/4} \sum_{j=1}^{n/4} |i+j|a^2$$

when the ith and jth units are on separate arms. The sums can be evaluated as follows. Let $m = |i - j|$ and $p = n/4$.

$$\sum_{i=1}^{p} \sum_{j=1}^{p} |i-j| = 2 \sum_{m=1}^{p-1} (p-m)m$$

$$= p^2(p-1) + \frac{p(p-1)(2p-1)}{3} = \frac{p^3 - p}{3}.$$

Now let $m = i + j$:

$$\sum_{i=1}^{p}\sum_{j=1}^{p}(i+j) = \sum_{i=1}^{p}\sum_{m=i+1}^{p+i} m = \sum_{i=1}^{p} \frac{(p+i)(p+i+1) - i(i+1)}{2}$$

$$= \tfrac{1}{2}\sum_{i=1}^{p}(p^2 + p) + p\sum_{i=1}^{p} i = \frac{p^3 + p^2}{2} + \frac{p^2(p+1)}{2}$$

$$= p^3 + p^2.$$

Then

$$\langle R_G^2 \rangle = \frac{a^2}{2n^2}\left[\frac{4p^3 - 4p}{3} + 12p^3 + 12p^2\right]$$

$$= \frac{a^2}{6n^2}[4p^3 + 36p^3 + 36p^2 - 4p]$$

Again, let $p = n/4$; $\langle R_G^2 \rangle = \dfrac{na^2}{6}\left[\dfrac{5}{8} + \dfrac{9}{4n} - \dfrac{1}{n^2}\right]$.

(*Note:* when n is large the radius of gyration of a four arm molecule is $\tfrac{5}{8}$ that of a linear one.)

7.17. (a) Plot $\ln I_{net}$ vs. $\ln q$ (the data as given are adequate for this purpose).

Slope $= -4.1$.

(b) The data needs to be extrapolated to determine the invariant. For large q, this can be done by the q^{-4} dependence, the last data point and $\int_{q(\text{last})}^{\infty} q^2 I(q)\,dq$. For small q, fit the given data, say to a spline function. Result: 4.035×10^{15} m^{-3}. The limit of $q^4 I(q)$ is 1.191×10^{-5} Å^{-4}. Therefore,

$$S/V = \frac{\text{surface area of pores}}{\text{sample volume}} = 0.44534 \text{ m}^{-1}.$$

Assuming the pores are spheres, $R_p = 3(V/S)$, where the term in parentheses has been evaluated. Therefore, $R_p = 340$Å

(d) Assume all the pores are on grain boundaries

$$\text{Surface area} \begin{cases} \text{of one pore:} & 4\pi R_p^2 \\ \text{of } N \text{ pores:} & N 4\pi R_p^2 \end{cases}$$

Therefore: $S/V = N 4\pi R_p^2/V$. From this we can obtain $N/V =$ number of pores/unit volume of sample.

Now as given: Grain boundary area per unit volume of sample (A_{GB}/V) is 1.2 m^{-1}. Hence, the pore number per unit of grain boundary area is $\dfrac{N}{V} \cdot \dfrac{V}{A_{GB}} =$ 2.55 × 10^5 voids per m^2. The square root of this number is an estimate of the spacing.

7.18. The weight pct of Cu is 2.32 wt pct. With this information, $\mu/\mu_k = 6.11$. (μ_k is just the weight fraction of Cu times its absorption, whereas the term μ includes Al.) From Eq. (7.138b) for transmission ($\mu\chi = 2$) $\delta\mu k/\mu_k = 8.3 \, N_0^{-1/2}$.

The solid angle the detector "sees" is the area of the spherical segment with the detector face as base, divided by r^2, (r is the sphere radius). From the information given this angle is 1.84sr. Then, for the fluorescence case:

$$\delta\mu_k/\mu_k = 10.2 \, N_0^{-1/2} .$$

The transmission method gives a smaller error.

Chapter 8

8.3. The experimental width $\Delta\theta = (9.4 \text{ sec}/1.32) \times (2/2.12) = 6.72$ sec compared to 6.13 sec calculated for the σ state and 5.39 sec for the π state.

8.4. Calculated values of $\Delta f''$ generally do not include all the electronic shells.

8.5. See Hountas, A., Fillipakis, S. E., Papathanassopoulos, C., Tsakalakos, Th., J. Phys. Chem. Sol. State 6, 1693 (1973); $\varepsilon = 0.96$.

8.6. (a) Rewriting Eq. (8.118):

$$I_{hkl}/I_0 = E_{hkl}^2/E_0^2 = (\delta')^2 \{1 \pm \text{sign}(\delta')[1 - 1/(\delta')^2]^{1/2}\}^2 ,$$

where sign $(\delta') = 1$ if $\delta' > 0$, or -1 if $\delta' < 0$.
With the expression $(1 - x)^{1/2} \cong 1 - x/2$ for $|x| < 1$,

$$I_{hkl}/I_0 = (\delta')^2 \{1 \pm \text{sign}(\delta')(1 - [\tfrac{1}{2}\delta'^2])\}^2; \quad |\delta'| \gg 1 .$$

For example, for $\delta' > 0$ ($\Delta\theta < 0$), the minus sign applies:

$$I_{hkl}/I_0 = \tfrac{1}{4}\delta'^2 .$$

For a second reflection $I_0/4(\delta')^2$ is the incident intensity and the resultant scattered intensity is $I_0/16(\delta')^4$, etc.

(b) To test the equation, plot $\ln I_{\text{meas}}$ vs. $\ln \delta'$.

8.7. (a) $P^{\sigma}_{111} = 4.15 \times 10^{-5}$; $P^{\pi}_{111} = 3.62 \times 10^{-5}$.
(b) If the beam is unpolarized, $I^{\sigma}_0 = I^{\pi}_0 = \frac{1}{2}I_0$,

$$P^{\sigma+\pi}_{hkl} = \frac{I^{\sigma}_0 P^{\sigma}_{hkl} + I^{\pi}_0 P^{\pi}_{hkl}}{I^{\sigma}_0 + I^{\pi}_0} = \frac{P^{\sigma}_{hkl} + P^{\pi}_{hkl}}{2}.$$

(c) For the second crystal:

$$I^{\sigma}_{0,2} = I_{0,2}; \quad I^{\pi}_{0,2} = |\cos 2\theta| I_{0,2}.$$

Hence, as in (b):

$$P^{\sigma+\pi}_{hkl} = \frac{P^{\sigma}_{hkl} + |\cos 2\theta| P^{\pi}_{hkl}}{1 + |\cos 2\theta|}.$$

(d) To estimate the integrated intensity outside the measured region (after subtracting background) — on the high-angle side, for example,

$$\Delta I_{hkl} = \int_{\delta'_{max}}^{\infty} \frac{I_0}{4(\delta')^2} d\delta' = \frac{I_0}{4\delta'_{max}} = I_{hkl}(\text{at } \delta'_{max}) \cdot (\delta'_{max}).$$

Now, from Eq. (8.119),

$$\delta' = [-\Delta\theta \sin 2\theta + \Gamma F'_0]/\Gamma F''_{hkl},$$

hence,

$$\Delta\theta = \Delta\delta' \Gamma F'_{hkl}/\sin 2\theta.$$

The result is $P^{meas}_{111} = 3.8 \times 10^{-5}$, within 3% of the calculated value.

8.8. The width ω is

$$\omega \cong 2Z \left(\frac{2 \ln 2}{\mu t \varepsilon + 2 \ln 2} \right)^{1/2} \tan \theta + a,$$

where Z is the depth in the crystal and a is the diameter of the defect. See Young, F.W., Jr., Baldwin, T.O., Merlini, A.E., and Sherrill, F.A., Adv X-Ray Analysis 9, 1 (1966). The geometry is shown in Fig. A8.8. α = twice the angle in the x-ray fan for which the intensity is more than half the maximum value.

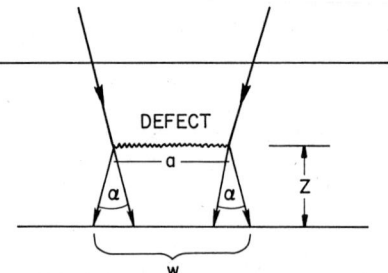

Fig. A8.8

Subject Index

Absorption
  chemical analysis by 136
  correction, flat plate geometry 217–219
  edge, fine structure 447
  of neutrons 131, 138–140
  of x-rays 109–116, 131–138, 474, 476
  of electrons 121, 122
Absorption, anomalous
  in electron diffraction 510
  in neutron diffraction 510
  in x-ray diffraction 114, 115, 501–505
Accuracy and precision in measuring
  d-spacings 199–201
Adsorbed films 524
Ag_3Sn, structure of 271–274
Air scattering 543
Amorphous solids, scattering from 436–446
Angle, Law of Constancy of 3
Anharmonic effects 371
Anomalous dispersion 114, 443
Anomalous transmission 470, 505, 508
Antinodal points 508
Aperiodic crystals 347
Argand diagram 56
Asymmetric crystals 525
Asymmetric unit 248, 249
Atomic displacement 407
Atomic scattering factor 66
Atomic size 16
Auger transitions 132
Axis of symmetry 5

Balanced filters 550
Beam monitors
  neutrons 152
  x-rays 151
Berg-Barrett topography 376
Bernal chart 172–174
Bessel function 336–338
Binary alloys 402
Bonse-Hart monochromator 510, 521
Borrmann effect 740, 495, 505
  physical interpretation 508
Bragg reflection from perfect crystals
  asymmetric geometry 525

  attenuation geometry 523
  intensity of 518
Bragg's law 50, 71
Bravais space lattices 9, 252–254
Bremstrahlung 82
Buerger precession camera 178–182

Center of symmetry 5
Characteristic radiation 80–84
Chemical analysis by absorption 136
Chemical fluctuations 402
Close packing 16
Clustering 402–423
Column approximation in electron
  diffraction 499
Complex notation 56
Compton scattering 118–121
Contrast in topography 532
Convolution 304–306, 382
Counters 143–151
Counting electronics 553
Counting statistics 150
Crystal
  definition of 2
  system 6
Crystal monochromatization 89, 191, 194
Cu_2O, structure of 269–270
Cut and project method 351

Darwin curve 510
Darwin-Prins curve 519
Dead time, detectors 151, 554
Dead time, synchrotron 556
de Broglie's relation 49, 122
Debye-Scherrer camera 189–193
Detection of radiation
  counting detectors 145–152
  counting electronics 143–145
  film techniques 142, 143
Dielectric constant in a crystal 473
Difference Fourier synthesis 327
Difference Patterson synthesis 332
Diffraction
  effect of distortion on 372–382
  Fourier analysis of 290–296

Diffraction
  from a row   54
  from three-dimensional lattice   63–66
  order of   50
Diffraction from perfect crystals
  Bragg geometry   487
  Field amplitude   490–495
  Laue geometry   488–500
Diffractometer   79, 166, 182–187
Diffuse scattering   402
Digital counting   553
Dirac delta function   290
Direct beam power, measurement of   542–552
Directions, indexing of   12–14
Dislocation, effects on diffraction   372–382
Dispersion surfaces   480–482
Distortion, effects of   372–382
Double crystal spectrometer   484
Dynamical theory of diffraction   469–534

Elastic constants   368
Electron density
  Fourier transform of   292
  mapping   293–300
  projections   295–297
  sections   295
Electron diffraction   209–212
Electron microprobe   102, 103
Electron microscopes   102, 103
Electron scattering   101–104
  factor   121
  images and diffraction patterns in electron microscope   102, 103
  polarization of electrons   122
Electron units   66
Electronics   553
Electrons, refraction of   141
Energy flow in a crystal   495
Equipoint, rank of   251, 252
Errors in Fourier coefficients   390
Ewald sphere   60, 69
EXAFS   446–458
Extinction
  primary   522
  secondary   523
Extinction distance   498

Faraday cage   152
Field amplitudes – Laue case   492
Film shrinkage   191
Filters   89, 135
Fluorescent analysis   86, 87
Form   6
Fourier analysis   281–295
  of peak shape   372–402
Fourier integral theorem   284–290
Fourier transform   65, 288
  one-dimensional   286
  three-dimensional   290
Friedel's law   293
Full reflections   509

Glide   249
Glide plane
  two-dimensional   249
  three-dimensional   257
Greninger net   164
Group velocity   113
Guinier's law   424

Habit   7
Half-width, perfect crystal (transmission)   514, 524
Hard-sphere diameter   16
Harker-Kasper inequalities   321
Harker sections   309
Hazards   152, 153
Heavy atom techniques   312–316
Helical structures   335–339
Hexagonal indices   14
Hönl correction   114, 115

Icosahedral symmetry   347–353
Incident intensity   542
Incoherent scattering of x-rays   118–121
Index of refraction
  electrons   141
  neutrons   141
  x-rays   110–113, 141, 474, 481
Inelastic scattering   361–372
  of x-rays   118–121
Insertion devices   96
Integrated intensity
  in Debye-Scherrer method   224–226
  in energy dispersive methods   227
  from flat plate polycrystal   224
  from flat plate single crystal   216–219
  in omega scan   220
  from small crystal   213–216
  in theta – two theta scan   220
  perfect crystal   516, 519, 521, 524
Intensity   55
Intensity inequalities   321
Intensity statistics   317–324
  acentric distribution   319
  centric distribution   318
Interference
  constructive   46
  destructive   46
  function   58
  interplanar   51
  partial   46, 58–60
Interferometer   510

International Tables, useful information in 539
Inversion
  one-dimensional 4, 5, 239
  two-dimensional 5, 241
  three-dimensional 5, 253
Inversion center 5
Interstitial sites 17
Interstitials, detection of 524
Invariant 432
Ionization chamber 147, 551

K absorption edge 85
K shell ionization 85
Kramers-Kronig dispersion relation 137
kx units 90

Lattice 7
Lattice parameters, evaluation of 199–201
Laue conditions 69
Laue monotonic scattering from alloy 126
Laue patterns 160–169
  back-reflection, analysis of 163–166
  transmission, analysis of 165
  uses of 166
Laue reflection from perfect crystal
  exit beams 489, 514
  integrated intensity 516
  position of 489
  width of 484, 574
Least squares refinement 339–341
Leonhardt chart 165
Light scattering 46–48
Limiting sphere 70
Liquids, scattering from 436–446
Local order 402–423
Lorentz factor 216, 217
Lorentz-polarization factor 216

Magnetic interactions 126–130
Mass absorption coefficient 131
Maxwell's equations 475–477
Microstrain 381, 391
Miller-Bravais indices 14
Miller indices 11–14
Minimum volume for diffuse scattering 405, 417
Mirror planes 5, 241
Monatomic liquids (amorphous solids) 437
Monitor counter 151, 152
Monochromators 89
Mosaic size, effects on diffraction 372–382
Moseley's law 83
Multi-bounce crystals 521
Multichannel analyzers 148–150
Multicomponent liquids (amorphous solids) 439

Multiple scattering 470, 480
Multiplicity 189, 224

NaCl, structure of 268
Nelson-Riley extrapolation function 200
Neutron absorption 131, 138–140
Neutron detectors 151, 152
Neutron scattering 122–131
  factor 123–125
  incoherent scattering 124, 125
  magnetic scattering 126–130
  scattering cross sections 126
  spin and isotope scattering 124–125
  spin incoherent scattering cross section 124
Neutrons
  collimation of 131
  detection of 131
  diffraction 130
  monochromatization of 98, 99
  polarized 129, 130
  production of 98–100
Nondispersive diffraction 149
Nuclear reactors 98, 99

Octahedral site or hole
  bcc 18
  fcc 17
Optical transformations 300–302
Omega scan 220
Orienting crystal using diffractometer 185–187
Oscillating crystal method 174

Parametric plane 36
Parseval's theorem 284–288
Particle accelerators 100
Particle size 382, 391
Path difference 48, 50
Patterson synthesis 303, 306
Pendellösung effect 498
Peak breadths due to distortion, mosaic size 377
Peak broadening 400
Peak shift, perfect crystals 489
Peak shifts 396
Peak widths, perfect crystals 484, 488
Permittivity 81
Phase difference 54
Phase transitions 369
Phase velocity 113
Phonon annihilation 367
Phonon condensation 369
Phonon creation 367
Phonons 361
Planes
  indices 11–14

Planes
  parametric 36
  rational 11
Point group
  one-dimensional 239
  two-dimensional 241–244
  three-dimensional 254–257
Point symmetries 254–257
Points, indexing of 13
Polarization of x-rays 104–107
Polarization states 479
Polarizer 506
Poles 26–28
Polymer scattering 429
Porod's law 433
Position sensitive detectors 146
Powder method 188–209
  analytical procedures for indexing
    195–198
  cameras 190–192
  card file 201
  energy analysis 206–209
  resolution of peaks in 192
  sample preparation for 189
  systematic errors in 199
  uses of 198–206
Power in direct beam 542
Poynting's vector and energy flow in crystal
  495–501
Precession method 178–182
  nonzero layers 181
  zero layer 177
Precision lattice constant determination
  199–201
Primitive lattice 8
Proportional counter 145–147
Pulse height analyzer 145–146
Pulsed neutron source 100

Quasicrystals 243

Radial density distributions 437–441
Radius of gyration 431
Ranks of equipoint 252
Reciprocal lattice vectors 22–25
Reciprocal space 61
  one-dimensional 61
  three-dimensional 68, 69
Reflecting sphere 60, 69
Reflection 5, 241
Refraction
  index of 141
  of x-rays 140–142
Reliability index 341
Residual stress 203–205
Resolution in topography 534
Resonant Raman scattering 423
Rietveld method 341, 342

Rocking curve 483–485, 513–521
Rotating crystal method 169–174
Rotation
  one-dimensional 5, 239
  two-dimensional 5, 241–244
  three-dimensional 5, 254–257
Roto-inversion 255
Roto-reflection 255

Safety precautions 152, 153
SANS 426
SAS 423
Sayre's relation 325
Scattered intensity 62, 63
Scattering
  by one electron 105–107
  from crystal 68
  isotopic 124
  magnetic 126–130
  from periodic two-dimensional array 70
  from row of atoms 54
  small-angle 423–436
  spin incoherent 124
  thermal diffuse 361–372
Scattering cross section 120
Scattering factor 114, 115
  electrons 122
  neutrons 124
  x-rays 114, 115
Scattering power 216
Scintillation counters 146
Screw axes, definition of 260, 261
Section topographs 529
Shape transform 423
Short-range order 402–423
Sign triple products 324
Single channel analyzer 145, 146, 553
Size distributions 434
Slit correction 382–387, 435
Small-angle scattering 423–436, 521
Soft modes 369
Solid-state detectors
  neutrons 152
  x-rays 147
Space groups
  one-dimensional 238–240
  two-dimensional 247–252
  three-dimensional 257–264
Space lattice
  one-dimensional 238
  two-dimensional 245–247
  three-dimensional 252–254
Special conditions for diffraction 267
Special positions 252
Spheric waves, dynamical theory 500
Spinels 19
Stacking faults 391–402

disorder 391
   electron microscope image of 392
Stacking pattern 18
Standard projection 28
Standing waves 471, 508, 523
Static atomic displacements 117, 369, 405–423
Stereographic projection 26–38
Stoke's correction 387
Straumanis mounting 192
Structure factor 66, 71–73
Subgrain size, effects on diffraction 372–382
Surface markings, traces 37, 38
Surface roughness 543
Symmetry axis 5
Symmetry elements 4
Symmetry factor 312
Symmetry operations 240
Synchrotron radiation 91–97
Systems
   one-dimensional 6, 239
   two-dimensional 6, 247
   three-dimensional 6, 258

Tagging 431
TDS 361
Temperature effects
   anisotropic factor 341
   depression of peaks 116–118
   dispersion curves 367–369
   elastic constants 368
   inelastic scattering 367, 368
   neutron inelastic scattering 367, 368
   phonons 366–369
   soft modes 368
   TDS 361–366
Termination errors 294, 306
Ternary alloys 422
Tetrahedral site or hole
   bcc 18
   fcc 17
Texture, analysis of 202, 203
Thermal atomic displacements 116–118
Thermal diffuse scattering 361–366
   higher order 371
Thermal vibrations
   effect on elastic scattering 117
   effects of 116–118
Theta-two-theta scan 220
Thickness fringes 501
Thomson formula for one electron 107
Tie point 483
Time-averaged intensity, definition of 55
Time-of-flight powder method 206–209
   resolution of 207–208
Topography 526

Total reflection 517, 528
   Darwin curve 529
   Darwin-Prins curve 529
Transformations
   axes 24
   cell 24, 25
Traveling wave 49
Twin 44
Two dimensional convolutions 386
Two-surface analysis 37, 38

Undulators 96
Unit cell
   primitive 8, 239
   one-dimensional 239
   two-dimensional 247
   three-dimensional 253
Unit translation
   one-dimensional 239
   two-dimensional 245
   three-dimensional 253
Unitary structure factor 320

Vector algebra 20
Vibrations 361

Warren-Averbach analysis of peak shape 378–391
Warren short-range oder parameter 404
Wavelength dispersion 89
Waves 46
Weighting factors 340
Weissenberg method 175–177
White radiation 82, 93
Wigglers 96
Wulff net 29–34

XANES 449
X-ray diffraction, discovery of 48, 49
X-ray fan 501
X-ray interferometer 511
X-ray scattering
   Laue monotonic scattering from alloy 126
   Thomson formula for one electron 107
X-ray topography 526–534
   Berg-Barrett technique 526–529
   Bormann technique 529
   contrast in 532
   Lang technique 529
X-ray tubes 77
   line focus 88
   microfocus 88
   spot focus 88
   take-off angle 88
   target dimensions 88

X-rays
  absorption of 109–116
  absorption coefficient 133
  absorption by multicomponent materials 135
  balanced filters 550
  Bremstrahlung 82
  characteristic radiation of 80–84
  chemical analysis by absorption 136
  Compton scattering 118–121
  discovery of 77
  filters 89, 135
  generation of 78
  Hönl correction 114, 115
  inelastic scattering of 118–121
  minimum wavelength 80
  monochromator 89
  polarization of 104–107
  refraction of 140–142
  scattering factor 107–109
  white radiation 82

Zone 5
Zone axis 5

Materials Research and Engineering

Editors: B. Ilschner, N. J. Grant, H. Riedel

Fracture at High Temperatures

1987. 109 figures. XVII, 418 pages. ISBN 3-540-17271-8

The book starts with an overview of the fracture modes observed in metallic and ceramic materials. It then concentrates on creep rupture by cavity formation on grain boundaries. Physical, mechanical, chemical and metallurgical aspects of this phenomenon are considered in order to provide a broad basis for applications to engineering structures operating at high temperatures. Gaps in the existing literature were bridged whenever possible, and much effort was expended to assess the validity of the models in the light of experiments. The importance of cavitation failure in related subject areas, such as hydrogen attack, helium embrittlement or stress relief cracking, is explored. Part III of the book probably represents the most comprehensive treatise of creep crack growth available.
Finally, many readers will be interested in the chapters on the mechanisms of creep-fatigue failure.

T. Ototani

Calcium Clean Steel

Translated from the Japanese by U. Oelschlägel

1986. 138 figures. XI, 141 pages. ISBN 3-540-16346-8

Contents: Calcium in Steelmaking. - Physical Metallurgy of Calcium and Calcium Alloys. - Addition of Calcium to Steel Melts. - Deoxidation and Desulfurization by Calcium. - Influence of Calcium on Nonmetallic Inclusions in Steels. - Mechanical Properties of Calcium Treated Steels. - Calcium Free Cutting Steels.

E. Dörre, H. Hübner

Alumina

Processing, Properties, and Applications

1984. 178 figures. XIII, 329 pages. ISBN 3-540-13576-6

This is the first comprehensive treatment of alumina as a technical material. Production and processing methods are described in detail. The compilation of physical and mechanical properties serves as a valuable collection of data for researchers and users. The material behavior and the underlying mechanisms are discussed from a materials science perspective. Numerous examples of applications in mechanical and electrical engineering, electronics, and medicine give the engineers suggestions for potential new uses.
The book is primarily directed to materials scientists and engineers. For the researcher, it represents the latest data compilation and summary of current research activities. The practicing engineer involved in resolving material problems and concerned with the search for new materials will find new ideas and technical solutions for the realization of new products.

Springer-Verlag
Berlin Heidelberg
New York London
Paris Tokyo

Materials Research and Engineering

Editors: B. Ilschner, N. J. Grant

Process Modelling of Metal Forming and Thermomechanical Treatment

By C. R. Boër, N. M. R. S. Rebelo, H. A. B. Rydstad, G. Schröder

1986. 195 figures. XV, 410 pages. ISBN 3-540-16401-4

Contents: Preface. – Mathematical Modelling. – Physical Modelling. – Modelling of Forging. – Modelling of Rolling. – Modelling of Drawing. – Modelling of Thermomechanical Treatment. – Outlook.

This book covers the modelling in metal forming processes, such as drawing, rolling and forging, and of thermomechanical treatment like quenching of complex parts and heat treatment of large forgings. An introduction to different modelling techniques is given with a description of the elementary analysis, upper bound analysis and finite element method applied to very large plastic deformations. Several examples from industrial practice are presented, intending to show that Process Modelling for Metal Forming have reached maturity and, in many cases, can be used to good advantage by the designer and engineers. The combination of simulation techniques, mathematical modelling and CAD/CAM technology is shown along with the advantages of their synergistic effect in CAE.

D. G. Altenpohl

Materials in World Perspective

Assessment of Resources, Technologies and Trends for Key Materials Industries

In collaboration with T. S. Daugherty

With contributions by M. B. Bever, J. P. Clark, H. Eldag, G. Friese, P. Kelterborn, I. Reznik, D. Spreng, F. R. Tuler

1980. 33 figures, 40 tables. XV, 220 pages. ISBN 3-540-10037-7

From the contents: Role of Materials in the World Economy. – Present Structure and Future Trends in Key Materials Industries (Introduction, Iron and Steel, Aluminium, Copper, Cement and Concrete, Plastics, Wood and Wood Products, Advanced Materials). – Technology Planning as Part of Industry's Planning Process. – Key Issues for Technology Planning and Assessment. – Research and Development Opportunities. – Outlook. – Index.

Springer-Verlag
Berlin Heidelberg
New York London
Paris Tokyo